DALY AND DOYEN'S

Introduction to Insect Biology and Diversity

Daly and Doyen's
Introduction to
Insect Biology and Diversity

INTERNATIONAL THIRD EDITION

This version of the text has been adapted and customized.
Not for sale in the USA or Canada.

James B. Whitfield
Professor of Entomology
University of Illinois, Urbana-Champaign

Alexander H. Purcell III
Professor Emeritus of Entomology
University of California, Berkeley

Howell V. Daly
Professor Emeritus of Entomology
University of California, Berkeley

John T. Doyen
Professor Emeritus of Entomology
Unviersity of California, Berkeley

New York   Oxford
OXFORD UNIVERSITY PRESS

Oxford University Press is a department of the University of Oxford. It furthers the University's objective of excellence in research, scholarship, and education by publishing worldwide.

Oxford New York
Auckland Cape Town Dar es Salaam Hong Kong Karachi
Kuala Lumpur Madrid Melbourne Mexico City Nairobi
New Delhi Shanghai Taipei Toronto

With offices in
Argentina Austria Brazil Chile Czech Republic France Greece
Guatemala Hungary Italy Japan Poland Portugal Singapore
South Korea Switzerland Thailand Turkey Ukraine Vietnam

1007403761
Published by Oxford University Press
198 Madison Avenue, New York, New York 10016
http://www.oup.com

Oxford is a registered trademark of Oxford University Press

ISBN 978-0-19-987378-4

Printing number: 9 8 7 6 5 4 3 2 1

Printed in the United States of America
on acid-free paper

CONTENTS

INTRODUCTION

This text is an introduction to the study of insects for students who have completed at least a basic course in biology. It is designed for modern courses emphasizing the major features of insects as living systems. Lectures in such courses are commonly devoted to insect structure and function, behavior, ecology, evolution, and some limited aspects of pest management, and the laboratory work is focused on the recognition of different kinds of insects and their biology and morphology. Therefore, our goal has been to provide a resource that includes thorough coverage of basic biological principles as well as a comprehensive set of insect identification keys in one affordable format. Throughout, we have attempted to maintain an emphasis on evolution.

HOW THE TEXT IS STRUCTURED

Parts One and **Two** of the text are appropriate as readings for course lectures. **Part Three** supplies information on the insect groups and keys for their identification, adding biological information on the orders and an appendix on collecting methods. Throughout the text, new references to reviews and the primary literature are included for readers who wish further information. We have tried to limit the terminology and discussions to those topics required for a broad knowledge, without oversimplifying complicated subjects or avoiding controversial topics. The study of the world's most varied animals is complex. With the application of new genetic, molecular and computational methods for the analysis of insect biology, this complexity is increasing. We have introduced a few new text boxes to briefly explain these methods.

In **Part One**, the treatment of insect biology begins with structure and function and progresses through behavior to the social insects. This portion of the book has been heavily rewritten to reflect changes in our understanding of morphological homologies from evo-devo studies, more detailed views of the physics of insect flight, and new molecular insights into insect physiology and behavior.

Part Two describes different relationships that insects have to their environment. This edition adds a new chapter on insect classification. The material previously treated in a separate chapter on Population Biology is now integrated into the other chapters. While this is not a text on applied entomology, **Part Two** relates how basic insect biology affects human welfare or the environment. The principal aims, problems, and methodologies of pest management are summarized in a chapter on pest management.

Our goal was to provide sufficient entomological background to enable students to evaluate new developments in pest control methods and controversies involving the environment and health impacts of pests and their management.

To a large extent, the principles of modern ecology and evolution are based on studies of insects. Mimicry, competition, and coevolution are only three of many important concepts featured in the text that have been developed largely through an understanding of the biology of various groups of insects. We have attempted to integrate this increased understanding into **Part Three**, so that the various taxa may be viewed in a broad biological context. Throughout, we have cross-referenced discussions of biological phenomena to more detailed treatments in other parts of the book. The chapters on the orders of insects and non-insect hexapods have been updated to reflect recent understanding of insect relationships from the study of fossils, comparative morphology of extant insects and, increasingly, molecular phylogeny. For instance, since the second edition, one new order, the Mantophasmatodea, has been described, while a familiar one, Isoptera (the termites) has been subsumed into the Blattodea (roaches). In a few cases where the classification is still being revised, we have taken the approach of describing the new developments in our understanding of phylogeny while retaining a broadly accepted classification; for instance, the Phthiraptera (lice) are now understood to have arisen from within the Psocoptera (bark and book lice), but a new classification of the combined order Psocodea has yet to be fully entrenched so we have still treated the "orders" separately. The **keys** in **Part Three** include nearly all the families of insects known to occur in North America. However, they are not designed to accommodate exceptional species that are normally encountered only by specialists. Likewise, extremely difficult taxa that require special preparation or collection techniques are mentioned, but not keyed. Discussions of the large endopterygote orders are organized by suborder and superfamily. These are important taxonomic categories whose members usually share many biological features.

NEW FEATURES IN THE THIRD EDITION

The third edition focuses more on biological principles, highlights the relevance of the subject to students' everyday lives, introduces the latest scientific research, and includes numerous thoroughly updated insect identification keys. Specific revisions include:

- Completely updated to contain the latest information on the genetics, development, behavior, physiology, evolution, phylogeny, and systematics of insects, including an all-new chapter:

  - "The Study of Classification" (Chapter 13)

- Thoroughly updated insect identification keys, including a completely new chapter on Psocoptera, written and illustrated by experts Ed Mockford and Emilie Bess, with a key to North American families; and an extensively revised chapter on Plecoptera, with expert additions by Ed DeWalt.
- New boxed essays highlight real-world relevance of entomology and examine the experimental basis of how the science behind the presented concepts is done.

- A new thematic discussion of evolution runs throughout the text.
- A new companion website for instructors with all figures and tables in an easy to download electronic format (www.oup.com/us/whitfield-xe).
- New co-author, James B. Whitfield (University of Illinois, Urbana-Champaign), recipient of the 2011 Thomas Say Award from the American Entomological Society, brings expertise as an instructor in the classroom and a researcher who conducts molecular analyses relating to insect development and systematics.

ACKNOWLEDGMENTS

We wish to express our appreciation for the efforts of the dedicated individuals who provided detailed content and accuracy reviews of the manuscript of the third edition:

John Abbott	University of Texas
Julio Bernal	Texas A&M University
Deane Bowers	University of Colorado- Boulder
William Brown	Kutztown University
Stylianos Chatzimanolis	University of Tennessee at Chattanooga
Paul M. Choate	University of Florida
Andrew Deans	North Carolina State University
Jamin Eisenbach	Eastern Michigan University
Michael A. Elnitsky	Mercyhurst College
Patrick Foley	California State University, Sacramento
Ann Fraser	Kalamazoo College
David M. Gordon	Pittsburg State University
Ellen Green	Delta State University
Jeffrey S. Heilveil	SUNY College at Oneonta
Harlan Hendricks	Colorado State University
Alex Huryn	University of Alabama
Sharon Knight Jasper	University of Texas at Austin
William O. Lamp	University of Maryland
Jack Layne	Slippery Rock University
Rodrigo Mercader	Michigan State University
John Meyer	North Carolina State University
Armin P. Moczek	Indiana University
Cesar Nufio	University of Colorado- Boulder
Paul Ode	Colorado State University
Kevin M. O'Neill	Montana State University
Christian Oseto	Purdue University
Christopher Paradise	Davidson College
Megha N. Parajulee	Texas A&M University
Yong-Lak Park	West Virginia University
Clark Pearson	Tulane University
David Rivers	Loyola College
Gregory P. Setliff	Kutztown University

Deborah Smith	University of Kansas
Stephen Taber	Saginaw Valley State University
William Turner	Washington State University
Sean Walker	California State University, Fullerton
Matthew Wallace	East Stroudsburg University
Deborah Waller	Old Dominion University
Robert Wharton	Texas A&M University
Daniel Young	University of Wisconsin

In making our additions and changes to create a new edition, we increasingly realized our indebtedness to the solid framework provided by the first two editions. We wish to acknowledge the generous assistance of many colleagues in preparing the first and second editions of the text. For the first or second editions, or both, the following persons provided us with valuable reviews, key references, or specimens: Paul H. Arnaud, Herbert G. Baker, Cheryl Barr, James Baxter, Leo Caltagirone, John Casida, Byron Chaniotis, Frank R. Cole, Joseph D. Culin, David Durbin, Wyatt Durham, Mark Eberle, Deane P. Furman, David Furth, Kenneth S. Hagen, Michael Haverty, Steve Hendrix, Marjorie A. Hoy, Carlton S. Koehler, Jarmila Kukalová-Peck, Robert Lane, Andrew Liebhold, E. Gorton Linsley, Vernard Lewis, Werner J. Loher, Richard Merritt, P.S. Messenger, Nicholas Mills, Robert Orneduff, William Peters, Gordon Pritchard, George O. Poinar, Jerry A. Powell, Vincent H. Resh, Helmut Riedl, Edward Rogers, Ray F. Smith, John Sorensen, Felix Sperling, Keith Waddington, William E. Waters, Ward B. Watt, Steve Welter, David L. Wood, and Daniel K. Young.

For the third edition, special thanks to Emilie Bess and Ed Mockford for contributing a new version of the Psocoptera chapter (finally with extensive keys to families), and to Ed DeWalt of the Illinois Natural History Survey for heavily revising the Plecoptera chapter. Alan Kaplan was invaluable in helping with proofs. A number of others supplied useful sources or advice during the revision of the text: Rodrigo Almeida, Deane Bowers, Richard Brown, Sydney Cameron, Andy Deans, Jamin Eisenbach, Robert Dudley, Jürgen Gadau, Jeff Heilveil, Ralph Holzenthal, Sharon Jasper, Boris Kondratieff, Sindhu Krishnankutty, Robert Lane, Armin Moczek, Paul Ode, Alison Purcell O'Dowd, Patrick O'Grady, Josephine Rodriguez, Dan Rubinoff, Felix Sperling, Mark Tanouye, Doug Whitman, Alex Wild, Kipling Will, and Daniel Young. Alex Wild and Brian Valentine contributed some superb new photographs showing living insects in their habitats. The third edition continues to feature the excellent photographs of Edward S. Ross, who added useful ideas and comments as well to the text. This edition also continues to benefit from the wonderfully clear line drawings of Barbara Daly. The editorial skills of Jason Noe, senior editor; Melissa Rubes, Katie Naughton, and Caitlin Kleinschmidt, editorial assistants; Lisa Grzan, production team leader; Theresa Stockton, senior production editor; Binbin Li, designer; and Jason Kramer, marketing manager; Frank Mortimer, director of marketing; Patrick Lynch, editorial director; and John Challice, vice president and publisher, were critical not only to producing the book but to keeping us on track.

I, James B. Whitfield would like to personally thank Professors Daly, Doyen, and Purcell for inviting me to head up the team for the third edition—it has been an honor and a pleasure. Completing a revision of a book of this size and complexity requires some serious chunks of time for focusing upon the task. I would like to thank the University of Illinois for granting a sabbatical leave during the fall of 2008, and Doug Emlen, Ken Dial, and Bret Tobalske of the University of Montana for providing a lovely and collegial atmosphere at the Fort Missoula field station within which to work. During the final year of the revision, music with the Stewart Berlocher Experience and the usual traditional Irish session crew at Bentleys provided welcome relief for me.

Finally, we especially wish to express our deep appreciation to our wives, Sydney (Cameron) and Rita, for their patience, wise input, and encouragement during the preparation of the text.

James B. Whitfield
Alexander H. Purcell III

CHAPTER I

Introduction

What are insects? Why are they important? This opening chapter gives a brief overview of insect species diversity and outlines the major reasons for this diversity and why better scientific understanding of insects is important not only to humans but also to our planet.

INSECTS: THE KEY TERRESTRIAL ANIMALS

Estimates of the total numbers of organisms on earth vary widely: from a low estimate of about 1.25 million described species to over 30 million, including mostly undescribed species, but in all of these estimates insects make up the majority of animal diversity. About half of the described species of living things and almost three quarters of all described animal species are insects (Fig. 1.1). In addition to their diversity, insects contribute substantial biomass. Ants and termites alone have been estimated to represent about a third of the animal biomass on earth; certainly ants outweigh manyfold the combined mass of humans. Because they are such a large percentage of the biomass of land animals, insects obviously have important ecological roles as a result of what they eat and what eats them.

In body size, most insects are 1 to 10 mm in length (Fig. 1.2). Some tiny insects are smaller than some Protozoa, and some giant insects are larger than the smallest mammals. The smallest insects are fairy wasps (Fig. 1.2a) (Mymaridae, Hymenoptera). *Dicopomorpha echmepterygis* is a Costa Rican species whose males are 0.14 mm in body length; adult females are smaller than the period at the end of this sentence, smaller than even a one-celled paramecium. Yet, these miniaturized wasps are fully functional insects with digestive, reproductive, and other

complex organs and with complex behaviors like their larger relatives. Other minute insects, measuring less than 0.25 mm long, are wasps of the family Trichogrammatidae (Hymenoptera), which parasitize other insect eggs, and beetles of the family Ptiliidae (Coleoptera), which are found in rotting logs.

The small size of most insects allows them to live in a variety of locations that would be impossible for larger organisms and thus contributes to insect diversification. The smallest insects inhabit (as larvae) the eggs of other tiny insects. In plants, some insects can complete their development entirely within small seeds or tunnel within a leaf blade less than 1 mm thick. Even small leaves can harbor more than ten different species of minute insects, where larger animals might require several large leaves to survive for a week.

At the other extreme of size are giant insects. In terms of wingspread, the all time champions are the extinct dragonflies *Meganeura* and *Meganeuropsis* (Meganeuridae, Protodonata), with wings measuring 700 mm or more from tip to tip. Among living insects, the moth *Erebus agrippina* (Noctuidae, Lepidoptera) from Brazil measures 280 mm, and the Asian moth *Attacus atlas* (Saturniidae, Lepidoptera) spans 240 mm. The tropical American grasshopper, *Tropidacris latreillei* (Acrididae, Orthoptera), has a wingspread of 240 mm. In terms of body length, the long-horned beetle, *Titanus giganteus*

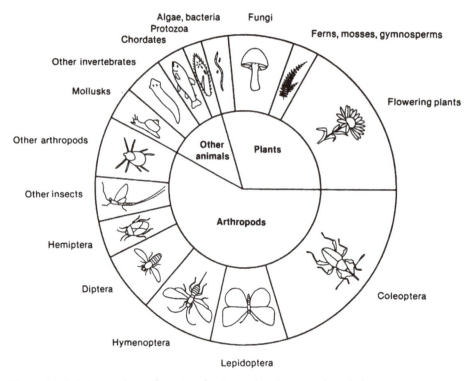

Figure 1.1 Relative numbers of species of arthropods, other animals, and plants.

(Cerambycidae, Coleoptera) reaches 17 cm, exclusive of antennae. The heaviest insects include the scarab beetles, such as the African *Goliathus regius* (up to 110 mm and 100 g) and the South American *Dynastes hercules* (160 mm), and the giant weta *Deinacrida rugosa* (70 g; Fig. 1.2b)

Another huge insect is the Australian phasmid, *Acrophylla titan* (Phasmatidae, Phasmatodea), with a body length of up to 250 mm (a little longer than this page). For comparison, some shrews, mice, and bats measure less than 50 mm.

WHERE INSECTS LIVE

Of all the animal phyla, the arthropods and the chordates have best succeeded in adapting to life on land. Insects now inhabit virtually all land surfaces of the globe except the extreme polar regions, the highest mountains, and oceans, except for the surface and shore. Insects tend to dominate the small terrestrial fauna, being rivaled only by nematodes and another group of arthropods, the mites, in some habitats.

Insects are rare in arctic regions. In Antarctica, for example, insects are represented by two species of flies (Diptera), a species of bird flea (Siphonaptera), and several species of lice that parasitize birds and seals. At 6000 m altitude (19,685 ft) in the Himalaya Mountains, surprising numbers of insects have been found as permanent residents. Insects have also been found living in deep caves, hot springs, salt lakes, and pools of petroleum. About 3 percent of all species of insects live in freshwater, and perhaps 0.1 percent are found in the marine intertidal zone. One genus of water striders, *Halobates* (Gerridae, Hemiptera), lives permanently on the surface of the open ocean, and many other water striders occur on tropical seas closer to land.

The vast majority of insects, therefore, are terrestrial. They occur in an enormous variety of

Figure 1.2 The world's smallest and largest insects: a, the fairy wasp *Dicopomorpha echmepterygis* (Hymenoptera, Mymaridae); the smallest males of this species have been recorded at 0.13 mm in length (the photographed slide-mounted specimen is one of the larger males!); b, one of the world's heaviest insects, the Cook Strait giant weta, *Deinacrida rugosa* (Orthoptera, Anostostomatidae); c, the world's longest recorded insect, *Phobaeticus chani* (Phasmatodea, Phasmatidae).
Source: A, courtesy of John Noyes, The Natural History Museum, London; B, courtesy of Kiwi Mikex; C, courtesy of The Natural History Museum, London; all used with permission.

habitats, including arid deserts. Of course, they must have access to water for drinking or in food. As in other small terrestrial organisms, the surface area of their body is relatively large in proportion to their volume, so the risk of water loss is always great in warm, dry atmospheres. This loss is minimized in insects by the moisture impermeability of the outer integument and by certain water-conserving mechanisms associated with excretion and respiration. Insects avoid unfavorably hot or dry environments by seeking cooler, shady places or moist crevices, and many species satisfy their water requirements by metabolizing carbohydrates. Others avoid water loss by suspending metabolic activity and becoming dehydrated. Insects usually die if their water content falls below about 20 percent. The critical upper limit in temperature for many insects is in the narrow range of 40 to 45°C. The record is held by the ant, *Cataglyphis bicolor* (Formicidae, Hymenoptera), that forages in the sun at temperatures up to 70°C on the surface of the Sahara desert. The lower limits for activity and survival of insects are much broader. Some tropical species do not live for long below 5°C, yet some temperate species are normally active on snow at temperatures barely above 0°C. Insects of the temperate and colder regions survive winter by finding suitable shelter and by physiological adjustments that prevent desiccation and freezing. Some species can tolerate the formation of ice crystals in their tissues and survive −35 to −40°C. These are the most complex animals to tolerate freezing.

HABITATS AND FOOD HABITS

Insects can be found in forests, grasslands, deserts, cultivated lands, urban areas, bodies of fresh and salt water, and flying in the air. In other words, they occupy all major habitats within the limits of geography and physical environment just discussed. Specifically they are found in or on soil, in caves, in surface or underground water, immersed in decaying plant or animal matter, and on or inside the bodies of fungi, living plants, animals, or other insects. Many insects are able to occupy more than one habitat and feed on different foods during different stages of their lives. The greatest diversification of insect species may be due to their associations with plants

In soil, insects are especially rich in variety in the litter of leaves and dead plant matter. Many of the insects are scavengers (Fig. 1.3a) that feed on dead plants or animals or on the living microbes that flourish in decaying matter. Some insects penetrate the upper layers of soil by following natural crevices or entering mammal burrows, while others are adapted to digging burrows themselves. Many types of insects, especially beetles and termites,

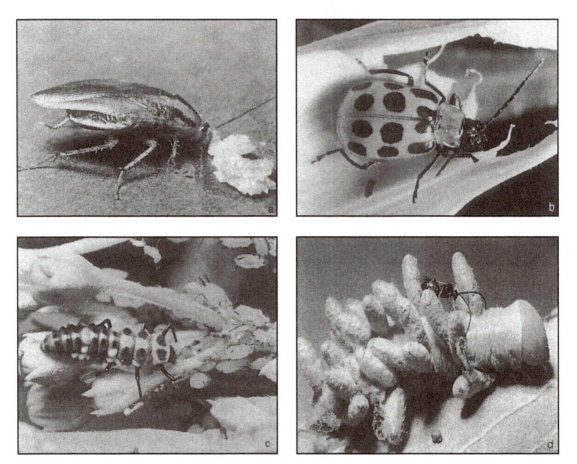

Figure 1.3 Examples of insects with different food habits. a, Scavenger: the German cockroach, *Blattella germanica* (Blattellidae, Blattodea), is a common household pest. b, Phytophagous: the adults and larvae of the cucumber beetle, *Diabrotica soror* (Chrysomelidae, Coleoptera), are phytophagous. Several species of *Diabrotica* are serious pests on cucurbits and corn, e.g., the corn rootworms. c, Predator: the larvae of most species of ladybird beetles (Coccinellidae, Coleoptera), like the adults, are voracious predators. d, Parasitoid: a wasp, *Cotesia* sp. (Braconidae, Hymenoptera), emerging from cocoons on the moribund host, which is a moth caterpillar.

occupy decaying logs. The "subcortical" habitat beneath the bark of dead and, less frequently, living trees is favored by beetles. Animal waste in the form of dung or carcasses on the ground is highly attractive to certain scavenging beetles and flies, and to the insects that prey on the scavengers. The scavenging role of insects is critical in nutrient cycling of terrestrial plant and animal biomass. One illustration of the importance of insect scavengers and their practical benefits to humans is the introduction of new species of insects that were scavengers of domestic animal dung in Australia. Native Australian insects proved to be inefficient in recycling dung. The scavenger species imported from Africa greatly reduced the population of biting flies that bred in the dung and decreased the loss of forage grasses that otherwise would have been covered by undecayed dung.

Insects that feed on green plants are termed phytophagous (Fig. 1.3b). All parts of green plants are attacked: roots, trunks, stems, twigs, leaves, flowers, seeds, fruits, and sap in the vascular system. Insects either feed externally by chewing tissues or by sucking sap or cell contents, or they feed internally by boring into the plant's tissues. The sucking insects, especially Hemiptera, are the only animals that are able to extract sap in quantity from the vascular systems of plants. Likewise, except for some tropical bats, some mammals, and some birds such as hummingbirds, insects are the main consumers (and distributors) of pollen and nectar in flowers.

Insects feed on many other kinds of terrestrial animals. Insects that kill other insects are termed entomophagous. Of these, the predators (Fig. 1.3c) kill their prey more or less immediately, while parasitoids (Fig. 1.3d) feed externally or internally in their host for some period before finally killing it. Insect predators and parasitoids also devour other small invertebrates such as snails, millipedes, spiders, and earthworms. Mammals, birds, reptiles, and amphibians are parasitized by bloodsucking insects and usually not killed. Some parasites live on the host (biting and sucking lice, adult fleas), or burrow in its flesh, or inhabit the alimentary or respiratory tract (fly maggots). Other parasites, e.g., mosquitoes, visit the host only to suck blood.

In the aquatic environment are scavenging, phytophagous, and predatory insects, but few parasitoids and true parasites. Some species are able to skate about supported by the surface film, while others float in the film or swim for brief periods below the surface and return for air. Certain groups are adapted to remain beneath the surface indefinitely and obtain their oxygen from water by means of gills. Of these, some are free-swimming, some crawl about on vegetation and the bottom, and some burrow deeply into the bottom gravels, sand, or silt. Some insects are the only truly aquatic animals capable of prolonged flight—a distinct advantage for inhabitants of temporary bodies of water.

Insect architects are capable of building complex structures out of their secretions and bits of natural objects. Some construct shelters about their bodies of rolled leaves, plant debris, or carefully selected grains of sand. Social insects such as ants, termites, and some bees may construct elaborate nests that permit some degree of control over temperature, humidity, and light, as well as provide a defensible fortress against enemies.

By their feeding activities and being fed upon, insects have evolved complex trophic relationships among themselves and with other animals, plants, and microbes. The origin and diversification of flowering plants, terrestrial vertebrates, and disease agents of plants and animals are due in part to the interactions of these organisms with insects. The blossoms of flowering plants have evolved to attract pollinators, primarily insects. Many of our spices and flavorings for cooking originated in plants as chemical defenses against plant-feeding insects. Lastly, it was not until insects began feeding on the blood of animals or the sap of plants that certain disease agents, especially viruses, could be transferred directly from the vascular system of one host to another.

ELEMENTS OF INSECT ANATOMY: AN OVERVIEW

Small size is both an advantage and a constraint for insects to diversify and proliferate in a variety of environments. As mentioned, small size allows a great variety of insects to share a small physical space. For example, various insects on plants can tunnel within leaves, stems, fruits, or roots, visit or live within flowers, or feed externally on any part of the plant. Muscles can be used more effectively in smaller organisms because their power increases in proportion to their cross-sectional area, whereas body volume (and weight) increases as the cube of body dimensions. Their small size allows insects to utilize different methods of flight and locomotion than larger animals but makes them less able to overcome wind movements in long-distance migrations. For example, insects such as water striders can walk on water due to the reaction of their water-repellent (hydrophobic) legs with the surface tension of water, whereas these water surface forces could not support the weight of heavier animals. On the other hand, minute soil insects can be trapped in the surface film of a drop of water. For minute (less than 1 mm length) flying insects, the viscosity of air is a much more dominant factor for flight than for birds or bats, and insects' wing morphology and movements in flight reflect this scaling effect (Chap. 5).

The system by which insects obtain from surrounding air or water the oxygen needed for cellular respiration and expel carbon dioxide depends greatly on the diffusion of O_2 and CO_2 molecules in air. This process proceeds rapidly enough at a small scale (millimeters), but at a larger scale, diffusion is too slow to satisfy gas exchange requirements. Active larger animals use a fluid circulation system such as the blood circulation systems of vertebrates to accelerate the exchange of metabolic gases with the air or water environment. The tough and rigid external skeleton of insects and other arthropods provides some important protective advantages for small organisms, but exoskeletons are structurally impossible for much larger animals. In general, the proportion of body weight needed for support structures such as bones in vertebrates increases as weight increases: bones are about 8 percent of body weight for a mouse, for example, but twice that percentage for a human. The largest arthropods are marine crustaceans (crabs), whose heavy weights are supported under water but would be too heavy to move about on land.

Like other arthropods, insects have a segmented external skeleton. The insect exoskeleton is composed of chitin (a nitrogenous polysaccharide) and protein and provides not only strong support and protection for the body but also a large internal area for muscle attachments. It is subdivided into plates, or sclerites, that are separated by joints or lines of flexibility and are movable by muscles. In proportion to weight, the tubular construction of the body segments and appendages gives relatively great resistance to bending because of the mechanical strength of a hollow cylinder.

The adult body is divided into three main regions (tagmata; singular tagma): head, thorax, and abdomen (Fig. 1.4a). The head bears a pair of large compound eyes and as many as three simple eyes, or ocelli, a pair of sensory antennae, and the feeding appendages or mouthparts surrounding the mouth. The thorax is composed of three segments, each bearing a pair of legs. The last two segments may also bear a pair of wings. The abdomen is composed of no more than 10 or 11 visible segments and lacks appendages except for a pair of cerci and the reproductive external genitalia that may be present near the tip.

Internally the body cavity, called the hemocoel, is filled with muscles, viscera, and the blood or hemolymph (Fig. 1.4b). The tubular alimentary canal, or gut, extends from the mouth to the anus at the extreme tip of the abdomen. The gut is divided into three regions: foregut, midgut, and hindgut. At the junction of the last two regions are the filamentous Malpighian tubules, which provide for excretion of nitrogen and help to maintain balance in concentrations of hemolymph components. The central nervous system consists of a large brain in the head in front of the mouth and a ventral nerve cord of segmental ganglia extending through the body with nerves to each segment. Large clusters or sheets of fat cells commonly create a fat body near the gut or the integument. Here food reserves are stored and a variety of biosynthetic processes occur. Paired openings, or spiracles, are situated on each side of the last two thoracic segments and the first eight or fewer abdominal segments. These openings lead to a system of tubular tracheae that convey oxygen in air to the internal tissues.

The male and female reproductive organs are located in the abdomen. In each sex the paired gonads usually open to the exterior by a single,

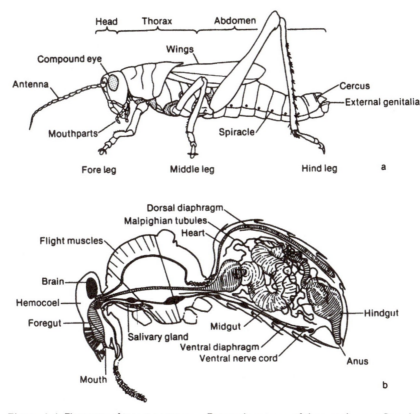

Figure 1.4 Elements of insect anatomy. a, External anatomy of the grasshopper, *Romalea microptera* (Acrididae, Orthoptera). b, Internal organs of the honey bee, *Apis mellifera* (Apidae, Hymenoptera).

Source: a, redrawn from original by M. Lynn Morris, by artist's permission; b, redrawn from The Hive and the Honey Bee, *Dadant and Sons, 1975, by permission of Dadant and Sons, Inc., Hamilton, IL.*

tubular duct in association with the external genitalia. The duct of the male is surrounded by an intromittent organ, or aedeagus, that transfers the sperm directly to the female. The female typically stores the sperm for later fertilization, in exceptional cases over a period of months or even years. Mature females usually have specialized appendages at the end of the abdomen that form an ovipositor for placing the eggs in soil, plant tissues, or other media, including other insects.

ELEMENTS OF INSECT CLASSIFICATION AND EVOLUTION

The hierarchic system for classifying animals includes six primary categories: phylum, class, order, family, genus, and species. You will be mainly concerned with learning the "higher classification" of insects, i.e., the names of the classes, orders, and common families. So far we have referred to insect "species" without defining what we mean by "species," but this will be discussed in a later chapter. There is a wide divergence among biologists in what constitutes a species. To represent accurately the evolutionary lineages of insects, authorities have added intermediate categories, such as superclass or subclass, to the higher classification. Unfortunately, authorities are not in universal agreement on the choice of names for these categories or on the best arrangement. In this text we treat 27 insect orders plus other closely related noninsect orders and generally follow conservative classifications of insects. It will be helpful to remember that the names of the various categories used here have the following endings:

Subclass (-ota)
Infraclass (-ptera)
Division (-ota)
Superorder (-pteroidea)
Order (-ptera, -ura, or -odea, except Collembola, Archeognatha, Odonata, Embiidina)
Superfamily (-oidea)
Family (-idae)
Subfamily (-inae)
Tribe (-ini)

The names of genera and species are italicized, and, in this text, the family is indicated in parentheses. When discussing a family, the name may be abbreviated so that the Acrididae, or grasshoppers, becomes "acridids."

The phylum Arthropoda consists of five subphyla, including the dominant marine arthropod group, the subphylum Crustacea. Members of the subphylum Hexapoda are terrestrial arthropods that have three pairs of walking legs. This distinguishes hexapods from the subphylum Chelicerata, which have chelicerate mouthparts and usually four pairs of walking legs (occasionally three pairs as immatures) and includes the arachnids (class Arachnida: spiders, ticks, mites, and scorpions). Members of the subphylum Myriapoda (millipedes and centipedes) usually have more than four pairs of walking legs. The six-legged condition of hexapods, however, probably evolved more than once from the same ancestral line. Three hexapod classes are recognized. Class Parainsecta includes small, wingless, soil-dwelling arthropods of the orders Protura and Collembola. The latter, also called springtails, are found worldwide and can be extremely abundant. Class Entognatha is another group of small, wingless, soil-dwelling arthropods, of the order Diplura, that have their mouthparts retracted into a cavity in the head (hence the name Entognatha). Class Insecta includes insects in the strict sense. Insects do not have their mouthparts retracted into the head.

The word "insect" is derived from the Latin *insecare*, meaning "to cut into," referring to the bodies of some insects that are almost cut in half by constriction at the neck or waist. Although the word insect is properly applied to the Class Insecta, all hexapods are commonly called insects, and we will do so frequently in this text.

The first members of the Insecta were wingless, and some wingless representatives are common today. These are placed in the subclass Apterygota (meaning "without wings"; Fig. 1.5a). The silverfish is a familiar example.

The winged insects are placed in the subclass Pterygota (meaning "with wings"). The first winged insects were unable to fold their wings flat over their backs. Some survivors from this stage in evolution are also alive today. These are the dragonflies and mayflies, and they are grouped together in the infraclass Paleoptera (meaning "with an ancient type of wing"; Fig. 1.5b). Insects that can fold their wings flat over their backs are classified as the infraclass Neoptera (meaning "with a modern type of wing"). Within this infraclass, two groups can be distinguished on the basis of the manner in which the wings develop. In the division Exopterygota, the wings develop externally and are visible on the young insect as small wing pads—hence the prefix *exo-*. Insects in the division Endopterygota have wing rudiments that develop internally during the early life of the insect (the prefix *endo-* means inside). The life history is divided into stages that are strikingly different in form and habits: larva, resting pupa, and winged adult. This type of life history is called complete or holometabolous metamorphosis.

The orders of the Exopterygota may be divided into two groups: superorder Orthopteroidea (meaning "orders resembling the order Orthoptera"; Fig. 1.5c) and superorder Hemipteroidea ("orders resembling the order Hemiptera"; Fig. 1.5d). The Orthoptera are grasshoppers, crickets, and katydids and rank as the sixth largest order in number of species. The Hemiptera are true bugs, such as stinkbugs and aphids, and rank as the fifth largest order.

The orders of the Endopterygota are divided into three groups: superorder Neuropteroidea

Figure 1.5 Representatives of major kinds of insects. a, Apterygotes: silverfish (Lepismatidae, Thysanura). b, Paleoptera: dragonfly, *Libellula luctuosa* (Libellulidae, Odonata). c, Orthopteroidea: grasshopper, *Trimerotropis* sp. (Acrididae, Orthoptera). d, Hemipteroidea: milkweed bugs, *Oncopeltus fasciatus* (Lygaeidae, Hemiptera). e, Neuropteroidea: lacewing, *Eremochrysa* sp. (Chrysopidae, Neuroptera). f, Mecopteroidea: buckeye butterfly, *Junonia* sp. (Nymphalidae, Lepidoptera), from South America, attracted to sweat on shirt.
Source: a, photo by H.V. Daly.

(Fig. 1.5e), superorder Mecopteroidea (Fig. 1.5f), and superorder Hymenopteroidea (Fig. 1.6). The largest order of the Neuropteroidea is not Neuroptera (lacewings, ant lions, and their allies) but the Coleoptera, or beetles. The superorder is so named because at one time the Neuroptera were considered ancestral to the neuropteroid group of orders. The Coleoptera may be the largest insect order; it includes the largest insect family, the Curculionidae, or weevils, with over 60,000 described species. The Mecopteroidea are similarly named after the Mecoptera (scorpionflies) because

Figure 1.6 Representatives of major kinds of insects. Hymenopteroidea: a worker (sterile female) yellow jacket wasp, *Vespula* sp. (Vespidae), hangs upside down as she adds plant fibers to the wall of her nest.

the order probably resembles the ancestral stock that gave origin to the Lepidoptera (butterflies and moths) and Diptera (true flies). These orders are the second and fourth largest orders, respectively. The last superorder, Hymenopteroidea, includes only the order Hymenoptera (ants, bees, wasps, and their allies). The Hymenoptera comprise the third largest order.

FACTORS IN THE SUCCESS OF INSECTS

The factors that are probably the most significant in the success of terrestrial arthropods are (1) a highly adaptable exoskeleton that provided structural protection and was modified to provide a tracheal or book lung system for gas exchange that minimized moisture loss, (2) colonization of the terrestrial environment before the chordates, (3) small body size, and (4) high birthrate and short generation time. The insects were probably the most successful group of terrestrial arthropods because they evolved (5) highly efficient flight, (6) specializations for eating plants, and (7) complete metamorphosis during development. These factors are interdependent and should be considered in combination.

The arthropods were the first animal phylum to overcome the problems of locomotion, respiration, and water conservation in a terrestrial environment. Their success can be traced to their adaptable exoskeleton. The jointed, paired walking legs were suited for locomotion on land as well as in water. Insects and some arachnids solved the problem of respiration without losing water by evaporation. A wax covering coated with a cement layer was added to the exoskeleton, and the respiratory surfaces were restricted to tubular invaginations of the exoskeleton, the tracheae. The tracheae open to the exterior by spiracles that admit air but reduce the loss of moisture. Thus equipped, insects had ample opportunity to occupy various habitats on land long before any other animal phylum offered serious competition.

In comparison to the calcareous endoskeleton of chordates, the chitin-protein exoskeleton is a distinct asset to small terrestrial animals such as insects. The tubular design and chemical composition combine strength with light weight. Because of their size insects are able to occupy an enormous variety of small places that are not accessible to larger animals.

The appendages of the exoskeleton vary greatly in structure and function in different groups of insects. The mouthparts may be adapted for taking solid or liquid food; the legs may be adapted for walking, jumping, clinging, grasping, swimming, or digging; and the ovipositor may be modified according to the media into which the eggs are deposited. Wings are also parts of the exoskeleton. Insects are the only invertebrates and historically the first of any animal group to possess wings. The power of flight has allowed insects to escape unfavorable habitats and to colonize new habitats, sometimes at great distances. Flying insects are at an advantage in escaping enemies and in finding mates, food, and places to lay eggs. Indeed, except for the Coleoptera, the many species in the large endopterygote orders are strongly dependent on flight for most of the requisites of life. An interesting exception has to be made for the social ants and termites. The reproductive castes fly during mating, dispersal, and establishment of new nests, but worker castes are completely flightless and yet their numbers and ecological impact are enormous.

The majority of the insect species are in the division Endopterygota. In addition to all the features discussed above, the endopterygotes have a complete metamorphosis. The larva is specialized for feeding and in many instances can reach food that is inaccessible to the adult. For example, larvae of certain species are able to bore into stems or leaves or into the bodies of other animals that the winged adult is unable to enter. In contrast, aquatic insect larvae can exploit resources found at lake or stream bottoms that are inaccessible to winged adults, but the adult stages of the same species can use flight to find mates and disperse eggs over great distances. This specialization also reduces the competition between the larva and the adult for the same resources.

INSECTS AND HUMAN WELFARE

Many if not most persons first think of insects as pests and nuisances, but their ecological impacts are by far the most important for human welfare. Simply because of their abundance and diversity, insects have an enormous impact on the ecosystems they inhabit, both as food for higher organisms or as consumers of other organisms or organic matter. Insects are important regulators of plant populations, as proven by the success of controlling exotic weeds by the importation and establishment of insects that specifically feed on targeted weeds. Insects likewise are often the key regulators of populations of insects or other arthropods, as repeatedly demonstrated by studies of how insect natural enemies (parasitoids and predators) regulate pest populations in agriculture or forests. Insects provide crucial interconnections for other organisms as pollinators for plants and as vectors of pathogenic microbes that cause disease in humans, animals, or plants.

Insects' Long Association with Humans

Insect adaptations specifically for human hosts were similar to those between insects and other primates and mammals. Insect parasites fed on humans, annoyed them, and transmitted disease agents among them. In common with other mammals, humans have acquired host-specific parasites such as the head and body louse, *Pediculus humanus*, and the crab louse, *Pthirus pubis* (both in the family Pediculidae, Phthiraptera [or Psocodea according to some authorities]).

Insects that transmit disease agents are called vectors. The most important vectors for human diseases are the mosquitoes (Culicidae, Diptera). Indeed, the mosquitoes are the most important family of all insects in relation to human health worldwide. They transmit the microbes that cause malaria, yellow fever, dengue, encephalitis, and filariasis. These diseases not only kill but also cause sicknesses that result in lost work days. Even though its incidence has declined in the United States, malaria continues to cause death and debilitation elsewhere in the world and is especially severe for children. Some other important vectors and the diseases they spread are the human body louse (epidemic relapsing fever, typhus); several species of fleas (plague, murine typhus); tsetse flies, or *Glossina* spp. (Glossinidae, Diptera), of Africa (trypanosomiasis or "sleeping sickness"); and kissing bugs, or *Triatoma* spp. (Reduviidae, Hemiptera), of the New World (trypanosomiasis or Chagas' disease). These diseases have had lasting economic, population, and cultural effects.

Insect pests directly damage agricultural crops in the field, stored food and shelters, and articles made of wood, plant fibers, and animal hides. In turn, insects are eaten by people, sometimes as a regular part of the diet, and in some cultures today insects continue to provide an important source of nutrition. Honey has long been sought in both the Old and New World. Silk continues to be the basis of fabric industries in Asia, where the silkworm was first domesticated millennia ago.

The development of agriculture and cities brought humans into direct cooperation and conflict with insects. Insects contribute more of direct economic value to the agricultural economy than they cause as damage. Without insect predators and parasites of pest insects, many more crop-feeding insects would reach a damaging level of abundance, increasing the difficulty and expense of pest control or even preventing profitable commercial production of some crops. Without insects that consume and bury cattle dung, the regeneration of grasses on grazing lands would be greatly reduced. Other insects provide useful products or pollinate crops. About two-thirds of all flowering plants depend on pollination by animals. Honey bees are the main pollinators in modern agriculture, but some other kinds of bees and other insects are more effective pollinators of certain crops. Insects are the sole or main food source for many birds, mammals, and fish, including game and sport fish. The "ecological services" of insects in recycling nutrients in streams and soils, in pollinating wild flowers, and in regulating the populations of plants and animals in wild lands are impossible to evaluate, but economic estimates of the value of insects' providing wildlife nutrition, pest control, pollination, and dung burial services exceeded $57 billion in the United States in 2005.

The most destructive agricultural pests are those that feed on stored grains. Damage to grains in storage is especially costly because energy and other resources have already been invested in producing, harvesting, transporting, and storing the crop. Two beetles, the rice weevil, *Sitophilus oryzae*, and the granary weevil, *S. granarius* (Curculionidae, Coleoptera), are the most harmful of the grain insects. They attack rice, wheat, corn, oats, barley, sorghum, and other grains. With over two-thirds of the world's population dependent on rice as a staple food, these weevils must be considered the most destructive insects in the world.

Crop pests are a major concern for food and fiber crops worldwide and a major impetus for entomological research to reduce pest damage with minimal harmful environmental impacts. Insect pest management is outlined in greater detail in Chapter 12. Insects can be pests regardless of what part of a plant they inhabit or consume. Seeds and fruits may seem the most obvious target for crop pests, but some of our most serious pests feed on roots, others on stems or just under bark. Rangelands support periodically large populations of insects. Grasshoppers and locusts (Acrididae, Orthoptera) are of great importance in the more arid regions of the world, with accounts of locust plagues dating to some of the earliest written historical accounts in Egypt and Mesopotamia. They devour nearly all kinds of plants, both cultivated and wild. The migratory locusts, such as *Schistocerca gregaria* and *Locusta migratoria* of Europe, Asia, and Africa, can totally devastate plant landscapes when they land in swarms and feed. In the United States the migratory form of the locust *Melanoplus spretus* was one of the most abundant insects on earth during its major migrations, as well as a major hindrance to the cultivation of the Great Plains of the United States, but conversion of the plains to agriculture drove the locust into extinction by the late 1880s.

In timberlands insects attack trees at every stage of growth, from seeds to mature trees, as well as the freshly cut logs and lumber in storage. Bark beetles (Scolytinae, Coleoptera) are the major pests of coniferous trees because the larval tunnels beneath the bark eventually girdle and kill the host tree. The invasive European elm bark beetle, *Scolytus multistriatus,* has been instrumental in spreading the fungus that causes Dutch elm disease, which has decimated American and European elms across North America. Another insect native to Europe that became established as a destructive invader of North American forests is the gypsy moth, *Lymantria dispar* (Lymantriidae, Lepidoptera). Although the moth was well known as a forest pest in Europe, a naturalist who was experimenting with silkworms brought eggs of the moth from Europe to his home in Massachusetts. Apparently some eggs or young larvae escaped his care in 1868 or 1869. The destructive caterpillars went largely unnoticed, except by his neighbors, until 1889, when the first severe outbreak occurred. Since then this forest pest has invaded many forest regions of North America by humans' inadvertent transport of egg masses (up to about 1000 eggs in each mass) that are laid on outdoor furniture, vehicles, or firewood. Gypsy moths have altered the composition and health of hundreds of thousands of acres of forests. The most important pests of living coniferous trees are bark beetles (Scolytidae, Coleoptera). The beetles attack in large numbers and overcome the resistance of mature trees, often over large areas. The loss of mature trees has both economic and environmental impacts.

In the urban environment insects attack not only ornamental and garden plants but also the framework of houses and their contents. Virtually any article of plant or animal origin can be damaged or destroyed by insects, including wooden structures and furniture, clothing and furs, food, leather goods, books and paper products, and drugs. Furthermore, people are usually annoyed by the sight of live insects in their homes, even when the insects are harmless intruders from outdoors.

Pest Management

Extensive plantings of single crops, i.e., monocultures, favor the rapid population growth of phytophagous insects and mites and of plant diseases. Prior to World War II, agricultural pest control combined several methods because no single approach was usually satisfactory. Cultural methods such as early or late planting avoided the seasonal appearance of certain pests. The role of natural enemies was widely acknowledged. Chemical pesticides were limited in variety and effectiveness, and some were toxic to humans and domestic animals. Yet for annual crops, where prompt relief from pests meant increased production and profits, pesticides were generally expected to be the most economical approach in the future. Problems with using pesticides that at first were

considered as minor have become widely recognized as serious environmental and economic concerns.

A good example is the chlorinated hydrocarbon DDT (dichlorodiphenyltrichloroethane), which was demonstrated to have major medical importance in the military effort of World War II. It was effective against body lice, mosquitoes, and flies and had a low acute toxicity for humans and mammals. Furthermore, the residue had a persistent action against insects. After the war, DDT and other newly developed synthetic organic insecticides were shown to be highly effective on a wide range of agricultural pests. This led to widespread acceptance and use of pesticides that resulted in frequent applications on a fixed schedule regardless of the actual damage caused by pests and neglecting nonchemical methods of pest control. Gradually, however, problems with insecticides began to appear or become recognized. Pests evolved resistance to specific insecticides as a result of natural selection to survive in environments with pesticides. Thus, stronger and more frequent doses or new synthetic compounds were required to control pests, which often only worsened the problems. Finally, evidence began to accumulate that DDT, its breakdown products, and related pesticides were widely distributed in soil and water and had entered the food chains of other animals, including humans. Food chains may act as "biological amplifiers," concentrating a toxic substance in higher amounts at each trophic level. The chlorinated hydrocarbon insecticides were especially likely to be concentrated because of their high solubility in fatty substances and their low solubility in water. In 1962, Rachel Carson argued the case against reliance on pesticides in her book *Silent Spring*. Ten years later, DDT was restricted in the United States to uses requiring special authorization. Insecticide resistance continues to be a major problem that requires constant development and testing of new insecticides. More importantly, alternatives to chemical pesticides continue to be developed through entomological research. The new approaches, usually called integrated pest management (IPM), integrate multiple methods in an ecologically oriented fashion to avoid or minimize adverse environmental effects of control actions. Plant breeding has long been a mainstay of pest management, though its practitioners were not always aware they were using it. The latest addition to plant breeding has been to use molecular methods to introduce into plants new resistance genes from various sources such as bacteria. This rapidly growing endeavor is still controversial because of its novelty but has become rapidly adopted in Asia, Africa, and the Americas.

This diversified approach of IPM requires constant innovations to cope with newly established pest species, new insecticides and crop varieties, changes in agricultural practices, and new information on environmental impacts of pesticides and farming practices. These innovations require basic understanding of all aspects of insect biology and their ecological interactions.

Insects in Research

Insects are important subjects of ecological, behavioral, genetic, physiological, and other biological studies. The vinegar fly *Drosophila melanogaster* was one of the first model systems for the study of genetic mutations and has since remained the most widespread animal model for genetic studies, providing many valuable insights into genetics generally and the genetics of animals specifically. The genome of this fruit fly was the first animal genome to be fully sequenced. Because of their abundance, relative ease of sampling, and lessened ethical concerns about manipulating populations, insects have served as the focus of numerous ecological studies and population modeling efforts. Their small size and ease of rearing has made them the subject of choice for a variety of topics in physiological research, particularly in developmental biology, neurobiology, and toxicology. *Drosophila* species serve as model systems for many genetic and medical research topics.

Because of their rapid generation times, some of the best examples of experimental studies of evolution involve insects. For example, the evolution of resistance to insecticides has provided excellent examples of natural selection that can be studied genetically in the laboratory and whose population genetics can be followed in the field. The extreme diversification of insects at many levels of classification and the wide availability of museum specimens of many thousands of species facilitates the formation and testing of hypotheses on the phylogeny, or ancestral relationships, of groups of insects, from the level of populations within a species to classification levels above the Class Insecta.

CHAPTER 2

The Insect Body

This chapter introduces the general body plan and external anatomy of insects. The description of each part of the anatomy begins with the features seen on common adult insects, such as grasshoppers, and concludes with a brief discussion of some distinctive modifications of other insects. The functional relationships of the parts, evolution of certain structures, and the anatomy of each order of insects are reserved for later chapters.

HOMOLOGY

The application of anatomical terms to different insects requires the determination of homology. This is the structural likeness of anatomical parts due to a common evolutionary origin. In other words, a structure possessed by two or more organisms is homologous because the genes and developmental pathways responsible were inherited from a common ancestor. The same homologous structure may differ in function from one organism to another. For example, the mandibles of all insects are homologous, yet in different species the mandibles may differ in appearance and are used for gnawing, digging, courtship, defense, etc.

Organisms without a common ancestor may independently evolve similar structures by convergent evolution. In this process, structures become similar despite independent origins because they have a similar function. This is called resemblance by analogy. The wings of insects and of birds are analogues.

Homologies have been established traditionally in insects by matching the structures with regard to segmental position, location on the segment, time of embryonic development, size, shape, and the association of other structures whose homologies have been determined. More recently it has become possible to compare the underlying developmental pathways that lead to the morphological traits we observe and to refine our views about homology. Finally, the degree of homology in gene sequences for proteins involved in the functioning or development of some structures provides further confirmation of the evolutionary relationships. The structures named in this chapter are easily recognized, yet some insects possess unique parts whose homology is still unknown.

Another kind of homology, called serial homology, is applied to the same structures on different segments of an individual. For example, each body segment of an arthropod usually has one pair of serially homologous, jointed appendages and one large ganglion of the ventral nervous system. On different segments of insects these appendages have been modified into mouthparts, legs, external organs for reproduction, and cerci. Several ganglia may fuse together to appear as one. Thus, the serial homology of structures on different segments is sometimes not immediately apparent; careful study of embryology, developmental gene expression, and ancient fossil insects is then required.

Terms for Anatomical Position

Before proceeding further, you may wish to review the following terms that denote anatomical position (the suffix -ad means toward) (see also Fig. 2.1):

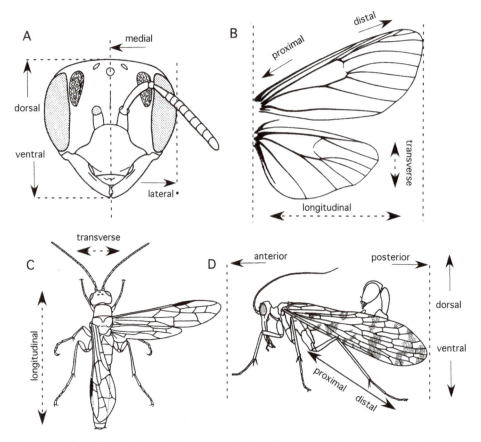

Figure 2.1 Terms for directions of location and orientation on insect bodies, using a bee head a, moth wings b, sawfly body in dorsal view c, and scorpionfly d, in lateral view.

anterior, cephalad, or cephalic: pertaining to the front end or head.

posterior, caudad, or caudal: pertaining to the hind end or tail.

dorsal or dorsad: pertaining to the upper surface or back.

ventral or ventrad: pertaining to the lower surface or belly.

lateral or laterad: pertaining to the side or outer part.

median, medial, mesad, or mesal: pertaining to the middle or inner part.

proximal, basad, basal, or base: pertaining to the part nearest the body or base of a part.

distal, distad, apical, or apex: pertaining to the part farthest from the body or part.

longitudinal: oriented parallel to the length or anteroposterior axis of the body, or in the case of an appendage, the direction parallel to a line extending from base to tip.

transverse: oriented perpendicular to the longitudinal axis of the body or appendage.

PROPERTIES OF THE EXOSKELETON

The body wall of an insect is called the integument. The outer layer is a noncellular cuticle, chemically composed of chitin and proteins that are secreted by the inner layer of epidermal cells (Fig. 2.2). In some areas the cuticle is chemically tanned, or sclerotized, by the epidermis to form rigid, plate-like sclerites. Areas of the cuticle between sclerites remain tough and flexible, creating membranous

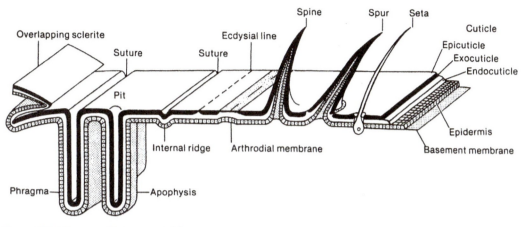

Figure 2.2 Diagram of integumental features.
Source: Modified from Metcalf et al., 1962. Reproduced with permission of The McGraw-Hill Companies.

joints or articulations. The sclerites and membranes together provide an external support, or exoskeleton, for the insect body.

Single cells of the epidermis may be modified into hairs, or setae. The seta-forming cell is the trichogen, and it is accompanied by a socket-forming cell, or tormogen. Setae vary greatly in both form and function. The common types are hairlike or simple setae, featherlike or plumose setae, and platelike scales. Poison setae are hollow and filled with toxic fluids for defense. Setae of diverse shapes are innervated by the peripheral nervous system and function as sensory organs called sensilla (Figs. 3.1a, 6.6a through d).

Many cells are involved in local outgrowths or ingrowths of the integument. Rigid outgrowths are commonly called spines. Some outgrowths, especially on the legs, are movable and are called spurs. Rigid fingerlike ingrowths of the integument associated with muscle attachments are commonly called apodemes or, if larger and armlike, apophyses. In entomology, the latter term is also used for external protuberances. Large internal apophyses in the head and thorax buttress the exoskeleton and provide areas for muscle attachment. These apophyses function as the endoskeleton. Platelike invaginations associated with the dorsal flight muscles are phragmata.

The integument may also be infolded linearly to produce internal ridges for strengthening the body wall or for muscle attachment. The internal ridge is usually traced externally by a line or groove. Virtually any line seen on the surface of an insect is loosely called a suture, regardless of whether it is a line of flexibility (e.g., an articulation), a line marking an internal ridge, or a deep phragma. Sutures are used as convenient boundaries to delimit areas of the exoskeleton. In the strict sense, a suture is a seam along which two plates have fused. The lines between plates of the vertebrate skull are properly called sutures. The cuticle of the insect integument, however, is a continuous layer; adjacent sclerites usually become unified in the course of evolution without leaving a trace. A "suture" transversing a sclerite, therefore, should not be interpreted to mean that the adjacent areas were once separate.

THE BODY REGIONS

The organization of an insect's body is more readily understood if we recall that insects are likely to have evolved from a segmented animal with less specialization into functional body regions. Snodgrass (1935) proposed a sequence of steps leading to the insects that is instructive in this context. Although his sequence was based on then-assumed relationships among animal phyla that are now known to be incorrect, the basic pattern of structural modification is still likely to be relatively accurate. In the earliest stage (Fig. 2.3a),

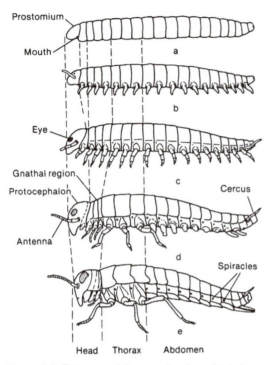

Figure 2.3 Diagrams of the steps in a hypothetical origin of the insect body from a wormlike ancestor. See text for explanation.
Source: Redrawn from R. E. Snodgrass, The Principles of Insect Morphology.

the (possibly) wormlike ancestor has a cylindrical body with the tubular intestine running nearly the full length. Between the mouth and anus the body is divided transversely into segments. In front of the mouth is an unsegmented part called the prostomium. The last part of the body encircles the anus and is the periproct. This is the endpiece and also is not a true segment. In a second hypothetical stage (Fig. 2.3b), the prostomium acquires a pair of light-sensitive receptors, or eyes, and a pair of sensory antennae. Each true body segment develops a pair of movable, ventrolateral appendages. In the third stage (Fig. 2.3c) the ancestor is an arthropod with jointed legs and enlarged sensory apparatus. At some point it becomes terrestrial, because most of the segments have a pair of openings, called spiracles, leading to the internal air tubes, or tracheae, of the respiratory system. The appendages of the last true segment are called cerci.

The features of the insects are foreshadowed in the fourth stage (Fig. 2.3d) as regions of the body become specialized. The modification of segments into functional units is called tagmosis (see also Chap. 15). At this stage, four such units, or tagmata, can be recognized: protocephalon (prostomium plus one to three segments), gnathal region (three segments), thorax (three segments), and abdomen (eleven segments plus periproct).

The exact number of segments that have become fused to form the insect head is a matter of much debate. Snodgrass defended the simple theory that the prostomium housed the primitive brain. This was a mass of nervous tissue in front of the mouth. As the sense organs on the prostomium enlarged, that part of the brain innervating the eyes became the protocerebrum and the part innervating the antennae became the deutocerebrum. The segments behind the mouth each had a pair of primitive ganglia, or masses of nerve cells. Each ganglion innervated, respectively, the appendage on its side of the body. The ganglia communicated transversely within each segment by nervous commissures and longitudinally from segment to segment by nervous connectives (Fig. 2.4a).

Snodgrass speculated that the appendages of the first postoral segment became sensory in function and shifted forward. The ganglia also moved forward to either side of the mouth and became more closely associated with the brain (Fig. 2.4b). In the insects, the appendages atrophied and were lost in the course of evolution, yet the ganglia of the first segment still persist in insects as the third part of the brain, or tritocerebrum. The tritocerebral commissure is evidence of its former segmental position behind the mouth (Fig. 2.4c). In Crustacea the appendages of this segment still persist as the second pair of antennae.

The appendages of the next three segments are specialized for feeding. This is the gnathal region. The three segments following are modified for walking and became the thorax. The last of the four tagmata, the abdomen, contains no more than 11 true segments plus the periproct. Most of the visceral organs are located here. With the walking appendages concentrated in the thorax, the

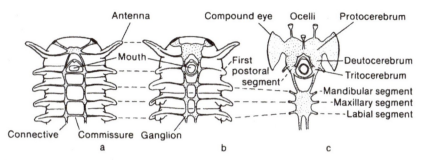

Figure 2.4 Diagrams of the steps in a hypothetical origin of the insect brain and head segmentation. See text for explanation.
Source: Redrawn from R. E. Snodgrass, The Principles of Insect Morphology.

appendages of the abdomen became modified for other functions or atrophied.

In the final stage (Fig. 2.3e), the body regions have been reduced to three by the tagmosis of the protocephalon and gnathal region to form the head tagma. The head combines the sense organs and brain with the gnathal appendages to form a center for sensing the environment and ingesting food. The thorax is the center for locomotion and houses the large muscles that attach to the leg bases, as well as others that act on the thoracic skeleton to move the wings. The abdomen is the center for the reproductive organs and intestines. By pumping movements, the abdomen also aids airflow in the tracheal system.

After study of fossils and living insects, Kukalova-Peck (1991) argued that the insect head has no unsegmented prostomium as described by Snodgrass. She proposed that the head has six segments, each originally with a pair of appendages that have become highly modified. These are named the labral, antennal, postantennal, mandibular, maxillary, and labial segments. In this scheme, the combined clypeus and labrum are derived from the appendages of the labral segment, the antennae are the appendages of the antennal segment, and the appendages of the postantennal segment are probably incorporated in the hypopharynx.

Secondary Segmentation
Among arthropods a distinction is made between primary segments, which correspond to true embryonic segments, and secondary segments, which are functional subdivisions. The trunk segments of annelid worms and soft-bodied insect larvae are clearly defined by transverse constrictions on which the principal longitudinal muscles insert. These are primary segments, and the transverse folds to which the muscles attach are true intersegmental boundaries. By the contraction of the muscles, the soft segmental walls are deformed and wriggling movements are produced (Fig. 2.5a).

The evolution of a rigid exoskeleton introduced a mechanical problem: If the soft body wall becomes entirely sclerotized, then how can the muscles continue to move the body segments? The evolutionary solution was to (1) keep the muscles attached to true intersegmental lines, (2) develop the rigid sclerites to include the true intersegmental lines so that the sclerites can be moved by the muscles, and (3) place lines of flexibility just in advance of the true intersegmental lines. The insect still has a segmented appearance, but the lines of flexibility between the "segments" no longer trace the true intersegmental lines. Accordingly, they are called secondary segments (Fig. 2.5b).

The dorsal plate of a secondary segment is the tergum, and the ventral plate is the sternum. The transverse, muscle-bearing ridge that marks the primary intersegmental fold is the antecosta. Externally, the ridge is traced by the antecostal suture. The margin of the tergum anterior to the antecostal suture is the acrotergite, and the corresponding sternal margin is the acrosternite.

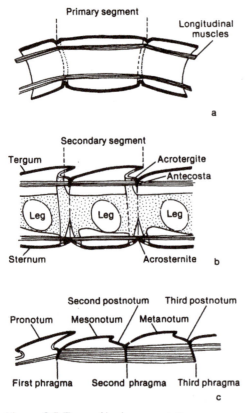

Figure 2.5 Types of body segmentation:
a, primary segmentation; b, secondary
segmentation; c, secondary segmentation and
formation of dorsal phragmata in pterothorax.
Source: Redrawn from R. E. Snodgrass, The Principles of
Insect Morphology.

THE HEAD

The feeding appendages, the most important sense
organs, and the brain are contained in the head. As
the first, or most anterior, body division, the head
leads the way when the insect moves forward. It is
in a position to detect the changing physical and
chemical properties of the environment. Sensations
of images in color, moisture, touch, sounds, odors,
and flavors travel but a brief distance to the brain,
where the insect's behavioral reactions are controlled. The feeding appendages by which the body
is fueled are appropriately located in the midst of
this battery of sensing devices.

The head articulates with the body at the membranous neck, or cervix. Small cervical plates are
sometimes present at each side in the cervical membrane. These function in the extension and retraction of the head. The rear of the head opens into
the cervix and the body beyond by a large hole, the
foramen magnum. Through this opening pass the
ventral nerve cord, salivary ducts, foregut, aorta, tracheae, various neck muscles, and the hemolymph.

Among the various insects, three positions of
the head can be distinguished. The original position
relative to the body is with the mouthparts directed
downwards, immediately ventral to the head capsule. This is the hypognathous position of grasshoppers (Fig. 2.6d), roaches, some Heteroptera
(Hemiptera), and many other insects. In the prognathous position, seen in some beetles (Fig. 9.2b),
the mouthparts are directed forward and project
anterior to the eyes. The last is the opisthognathous position possessed by some Heteroptera and
all Homoptera (Fig. 32.1b). Here the sucking beak
is directed toward the rear, beneath the thorax.

Cranium

The sclerotized head capsule, minus the appendages, is the cranium. In the course of evolution the
distinction between the head segments has been
lost almost entirely. Most of the lines seen now on
the cranium represent secondary modifications
unrelated to segmentation. Some lines trace internal ridges that strengthen the exoskeleton, while
others are actually lines of weakness that allow
the cuticle to split during ecdysis. In most insects
only the postoccipital suture, described below, is
believed to be a segmental boundary. Here a strong
internal ridge has been retained between the maxillary and labial segments, serving as a place for the
attachment of neck muscles that move the head.
Fossils of the order Monura (Fig. 14.4) and the
living fossil *Tricholepidion* (Thysanura; Fig. 17.4)
exhibit intersegmental lines on the cranium.

The compound eyes are easily recognized as the
large paired organs located dorsolaterally on the
cranium (Fig. 2.6a). The surface of the eye is covered with minute facets, each of which represents
the lens of an individual eye unit, or ommatidium.
Situated between the compound eyes on the face

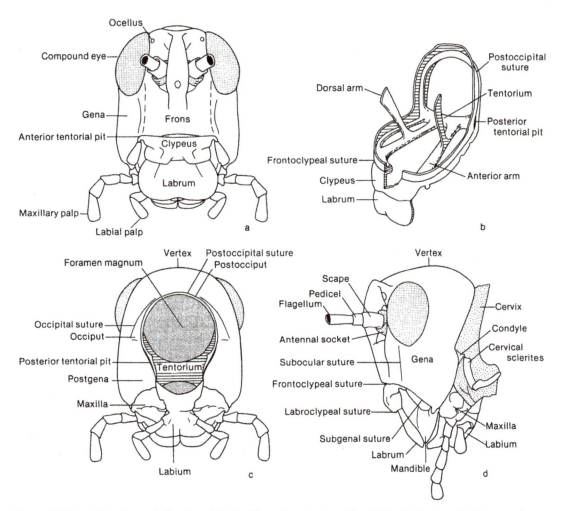

Figure 2.6 Insect head: a, cephalic view of head of *Romalea microptera* (Acrididae, Orthoptera); b, diagram of tentorium of a grasshopper with most of the cranium removed; c, caudal view of head of *Romalea*; d, lateral view of same.
Source: b, redrawn from R. E. Snodgrass, The Principles of Insect Morphology.

are as many as three simple eyes, or ocelli. Each has a single lens. One is usually present on the midline, and the others are placed above and at each side.

On some insects a pale line, shaped like an inverted Y, can be seen medially on the top of the head. This is the ecdysial line, along which the old cuticle first splits when it is shed by immature insects. One or two ventral ecdysial lines occur on the underside of the head of many endopterygote larvae. The lines serve no function in the adult.

The cranium has an internal, sclerotized structure called the tentorium (Fig. 2.6b). This endoskeleton is created by paired invaginations of the integument, at both the front and rear of the head, that fuse together inside the head. Externally the invaginations are indicated by pits or narrow slits. The anterior tentorial pits mark the entry of anterior invaginations that become the internal anterior tentorial arms. Similarly, the posterior tentorial pits mark the invaginations of the posterior tentorial arms. The arms meet medially to form the tentorial bridge. In some insects, each anterior arm has a dorsal arm. The tentorium is variously shaped in different insects. It strengthens the cranium and provides attachment for muscles to the antennae and the gnathal appendages.

Names have been given to different areas of the cranium. In some instances sutures serve as convenient boundaries between areas, but elsewhere the limits are undefined. This should not cause confusion because the names refer merely to topographic areas, not to discrete sclerites.

Beginning at the top of the head and descending in front, the names of the facial areas are as follows: the summit of the head, between and behind the compound eyes, is the vertex. The next area, the frons, is between the antennae and the compound eyes. No definite boundary exists between the frons and the vertex. Below the frons is the clypeus. The clypeus may be defined as the area of the cranium to which the labrum is attached. The labroclypeal suture is the line of articulation between the labrum and clypeus. The frons and clypeus are sometimes separated by a suture extending between the anterior tentorial pits. This is the epistomal or frontoclypeal suture.

Returning to the top of the head and descending on one side (Fig. 2.6d), the gena is the area extending from below the vertex and behind the compound eyes, to the ventral edge of the cranium. The gena is not delimited above from the vertex, but along other boundaries sutures may exist. It may be separated anteriorly from the frons by a frontogenal suture or a subocular suture in *Romalea* (Acrididae, Orthoptera) and from the clypeus by a clypeogenal suture. Ventrally, a subgenal suture may extend between the anterior and posterior tentorial pits, separating off a marginal area, the subgena. Posteriorly, the gena may be delimited by the occipital suture.

Turning now to the back of the head, the areas can be seen best if the head is removed at the neck, or cervix (Fig. 2.6c). At each side near the base of the labium are the slitlike posterior tentorial pits. Extending from the posterior pits and running dorsally around the foramen magnum is the postoccipital suture. The area anterior to the suture is the occiput. The narrow sclerotized rim of the cranium, posterior to the suture, is the postocciput. The labium is suspended at each side from the lower edges of the postocciput, indicating that this is the tergum of the labial segment. The postocciput also bears the occipital condyles on which the cervical plates articulate. Returning to the occiput, an occipital suture, or at least an incomplete ridge, may separate the occiput dorsally from the vertex and laterally from the genae. The lower part of the occiput, near the posterior tentorial pits, is called the postgena.

Modifications of the Cranium

The sutures and areas of the cranium vary considerably in extent and position largely as a result of modifications of the mouthparts and associated musculature. It is entirely possible for distinctive, new areas of the cranium to develop in certain groups of insects. A special terminology is often applied to these areas.

The faces of Hemiptera (suborder Homoptera) and Psocoptera (Fig. 30.2a) are thus altered by the development of a sucking pump. The prominent bulge between the eyes that houses the muscles for this pump is called the postclypeus, and the smaller area below, the anteclypeus. In reality this is a secondary modification of the entire frontoclypeal region where the original frontoclypeal boundary has been obscured.

Insects such as Orthoptera lack sclerotization of the integument beneath the foramen magnum and posterior to the labium. In other words, the throat behind the mouthparts is membranous. However, in hypognathous insects such as some Coleoptera this area may be secondarily sclerotized to form either a hypostomal or a postgenal bridge (called the gula in Coleoptera). Recall that the subgenal suture runs near the edge of the cranium and terminates in the posterior tentorial pit. That part of the suture posterior to the mandible is also known as the hypostomal suture, and the subgenal area thus delimited is the hypostoma. Recall also that the adjacent area of the occiput is the postgena. In caterpillars of Lepidoptera, sclerotic lobes of the hypostoma extend mesally onto the throat but do not meet. In adult Diptera and some Hymenoptera the hypostomal lobes unite to form a hypostomal

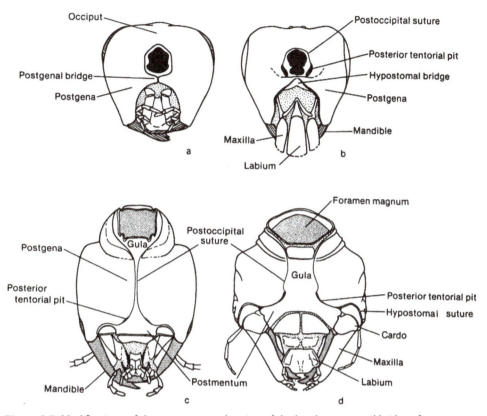

Figure 2.7 Modifications of the posteroventral region of the head: a, postgenal bridge of a wasp, *Vespa* sp. (Vespidae, Hymenoptera); b, hypostomal bridge of the honey bee, *Apis mellifera* (Apidae, Hymenoptera); c, narrow gula of the beetle, *Staphylinus* sp. (Staphylinidae, Coleoptera); d, broad gula of the dobsonfly, *Corydalus* sp. (Corydalidae, Megaloptera).
Source: Redrawn from R. E. Snodgrass, The Principles of Insect Morphology.

bridge beneath the foramen magnum (Fig. 2.7b). In other adult Hymenoptera and Hemiptera (suborder Heteroptera), the postgenae also unite, forming a postgenal bridge (Fig. 2.7a). In the formation of both types of bridges, the posterior tentorial pits remain unaltered in position.

In prognathous insects such as some Coleoptera and Neuroptera, the underside of the head is substantially modified by the forward position of the mouthparts. The postgenae are usually enlarged, and the posterior tentorial pits are shifted forward. The posterior pits, however, still retain their connections at each side with the postoccipital suture. The midventral area of the head between the pits corresponds to the membranous throat of Orthoptera. In Coleoptera and Neuroptera, this area is secondarily sclerotized to form the gula (Fig. 2.7c,d). The gula is continuous with the postocciput and sometimes also the labial base. The parts of the postoccipital suture that delimit the gula laterally are correspondingly renamed the gular sutures. The sutures are external indications of the internal tentorium.

Antennae

The antennae are paired, segmented appendages that articulate with the cranium between or below the compound eyes. Three parts can usually be recognized: the basal, or first, segment is the scape; the second segment is the pedicel; and the remaining segments constitute the flagellum (Fig. 2.8). Each segment of the flagellum is called a flagellomere. The

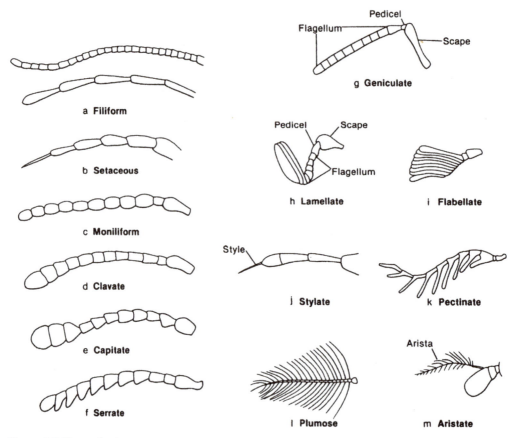

Figure 2.8 Types of antennae.
Source: a through g, j, and l redrawn from Metcalf et al., 1962. Reproduced with permission of The McGraw-Hill Companies.

whole antenna is moved by muscles from the head that insert on the scape. In the antennal socket the antenna may pivot on an articular process, or antennifer. The flagellum is moved by muscles from the scape that insert on the pedicel but not on the segments beyond. In accordance with its sensory function, the surface of the flagellum is supplied with many sensory receptors that are innervated by the deutocerebrum of the brain. In addition, within the pedicel is a mass of sense cells that detects movements of the flagellum. This is called Johnston's organ (Fig. 6.6f). Exceptions to the foregoing account are found in the other six-legged arthropods: in Collembola (class Parainsecta) and Diplura (class Entognatha) the flagellar segments have intrinsic muscles and Johnston's organ is absent. In Protura (class Parainsecta) antennae are missing entirely.

Modifications of Antennae
Special terms for different shapes of antennae are indicated in Fig. 2.8 (see also Chap. 37, Fig. 37.6).

Mouthparts
The mandibulate, or chewing, type of mouthpart is the basic type from which the specialized types have been derived. The mandibulate type consists of an upper lip (or labrum), paired mandibles, paired maxillae, a lower lip (or labium), and a median tonguelike hypopharynx (Fig. 2.9). The mouthparts enclose the preoral cavity, within

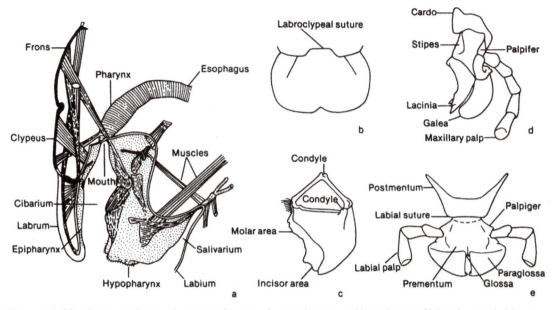

Figure 2.9 Mouthparts and preoral cavity: a, diagram of preoral cavity and lateral view of hypopharynx; b, labrum of *Romalea microptera* (Acrididae); c, mandible of same; d, maxilla of same; e, labium of same.
Source: a, redrawn from R. E. Snodgrass, The Principles of Insect Morphology.

which are the true mouth and the opening of the salivary glands.

The labrum is a flap that closes the preoral cavity in front (Fig. 2.6a, 2.9b). It articulates with the cranium by the membranous labroclypeal suture. The inner surface forms the roof of the preoral cavity and may have a median lobe called the epipharynx (Fig. 2.9a). Muscles from the head capsule insert near the base of the appendage and provide a variety of motions. The labrum is innervated by the tritocerebrum of the brain.

The mandibles are heavily sclerotized jaws that represent limbs that have lost their distal segments (Fig. 2.9c). The inner edge is modified for biting and may be hardened by deposits of zinc, manganese, or iron in some insects. Toward the tip the edge has cutting teeth, or incisors; at the base is a grinding surface, variously called the mola, molar, or molar lobe. Each mandible has two basal points, or condyles, where it articulates with the cranium. The posterior condyle is larger and fits against the gena or postgena. This was the original articulation point of the limb base. The anterior condyle is a secondary modification and fits against the base of

the clypeus, thus forming a hinge line that permits the mandibles to swing laterally. Powerful adductor muscles from the cranium insert at the base of the mandibles on the inner side of the hinge line and by their action bring the opposing mandibles together in a pinching motion. Smaller abductor muscles, attached to the outer margin of the mandibles outside the hinge line, open the mandibles.

The maxillae, a second pair of jaws, are much less massive and retain the segmentation of an appendage (Fig. 2.9d). Each maxilla has a limb base that articulates with the cranium by a single condyle. The base is divided into a proximal sclerite, or cardo, that bears the condyle, and a distal sclerite, the stipes. At the apex of the stipes are two lobes: an inner lacinia, which may be toothed, and an outer galea. Lateral to the galea is the maxillary palp, which represents the distal part of the limb and is sensory in function at the apex. The area on the stipes where the palp articulates may be differentiated into a segmentlike structure called the palpifer. Muscles from the cranium insert on the cardo and stipes, producing various motions of the maxilla about the single articulation. The lacinia, galea, palp, and palpal segments are also individually movable by muscles.

The labium closes the preoral cavity to the rear. The structure of the labium resembles that of the maxillae except that the labial appendages are fused medially with each other and basally with the segmental sternum (Fig. 2.9e). The lobed apex of the labium, called the ligula, consists of an inner pair of lobes, the glossae, and a lateral pair, the paraglossae. The glossae and paraglossae are serial homologs, respectively, of the maxillary lacineae and galeae. Similarly, the labial palp corresponds to the maxillary palp and is likewise sensory in function. The palps of the maxilla and labium function to taste and feel food. The labial palp may be borne on a segmentlike palpiger. The sclerite to which the ligula and palps attach is the prementum. The prementum articulates basally with the postmentum by the labial suture and contains all the muscles of the ligula. The postmentum in turn may be subdivided into a distal sclerite, or mentum, and a proximal sclerite, the submentum. The labium is complexly musculated such that the ligular lobes, the palps and their segments, and the prementum are individually movable.

The hypopharynx is a large median lobe situated between the mouth and the labium (Fig. 2.9a). The cibarium is a special chamber in front of the mouth that is formed by the anterior surface of the hypopharynx. On the posterior side, the salivary duct opens in a cavity, or salivarium, between the hypopharynx and the labium. The hypopharynx is movable by muscles from the tentorium that insert on sclerites in the wall of the hypopharynx. Nerves to the mandibles, maxillae, labium, and hypopharynx are supplied by the subesophageal ganglion.

Modifications of Mouthparts

The structure and function of insect mouthparts vary according to the nature of the food eaten. As a consequence, they not only provide important characters to be used in taxonomic classification but also indicate the feeding relationships of the insect in its ecological community. The latter is of special importance to human welfare. The successful transmission of a disease, for example, may be determined by the minute structure of an insect's

beak. Or an insecticide residue on a leaf surface may or may not be ingested by a pest, depending on the exact manner of feeding.

Some insects do not feed in the adult stage, and the mouthparts are atrophied and functionless. Examples are Ephemeroptera and at least some members of the following orders: Plecoptera, Trichoptera, Megaloptera, Lepidoptera, Strepsiptera, Coleoptera, Diptera, and Hemiptera (male scales).

Various schemes for classifying functional mouthparts have been proposed. Here we have adopted distinctive categories (indicated in bold print) and listed examples and exceptions, with brief descriptions of the mechanisms. Additional information can be found elsewhere in this text under the treatments of the different taxonomic groups.

A. Entognathous mouthparts. The mandibles and maxillae are recessed and largely hidden from view by lateral folds of the head. Examples: Protura (Fig. 15.1c,d), Diplura (Fig. 16.1b), and Collembola.
B. Ectognathous mouthparts. The mandibles and maxillae are visible or secondarily recessed, but lateral folds of the head are absent. Examples: all members of class Insecta.

1. **Mandibulate mouthparts.** The feeding appendages are more or less complete (Figs. 2.9, 2.10), freely movable, and not united in a beak. The mandibles are usually conspicuous and suited for seizing objects or chewing solid food. Those of predators are fitted with sharp apical teeth for catching prey, while the mandibles of plant feeders have more obtusely angled cutting teeth together with basal grinding molars. Mandibles are also used in defense, courtship, and the construction of nests or shelters. Most adult and immature insects have mandibulate mouthparts or some modification of this type. The exceptions are listed under the suctorial and derived types below. Unusual modifications of the mandibulate type deserve mention:

   a. Odonata. The labium of the nymph is greatly enlarged (Fig. 19.5), hinged at the

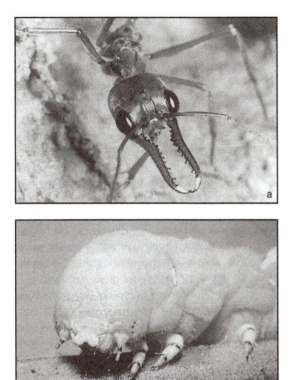

Figure 2.10 Examples of insects with mandibulate mouthparts. a, The bulldog ants, *Myrmecia* sp. (Formicidae, Hymenoptera), from Australia use their large mandibles to gently pick up their eggs and, in combination with powerful stinging, to defend their nest. b, The hornworm caterpillar, *Manduca sexta* (Sphingidae, Lepidoptera), has mandibulate mouthparts suited for eating leaves. The notch on the labrum may serve as a guide while eating along the edge of a leaf. The massive head capsule contains the large mandibular muscles.

labial suture, and held retracted beneath the head. By means of quick forward thrusts, prey are seized in the pincherlike labial palps and brought to the mandibles near the mouth.

b. Hymenoptera. The adult mouthparts are basically mandibulate, but the maxillae and labium are united in a mobile, extendible structure (Fig. 39.3b). The glossae are fused medially, forming a lapping organ, so that liquids can be taken readily in addition to solid food. An elongation of the maxillolabial structure forms the sucking tongue of certain families of bees (see below under 2, b, (2)].

c. Hymenoptera, Lepidoptera, Trichoptera. In the larvae the maxillae, labium, and hypopharynx are united into a lower lip with the opening of the salivary duct at its tip. In these orders the salivary glands produce the silk used in making cocoons or larval shelters. The palps and other distal parts are reduced or wanting in some groups, but the mandibles are usually retained (Fig. 2.10b).

d. Isoptera (Fig. 19.2b), Megaloptera (Corydalidae, Fig. 34.1b), Coleoptera (Lucanidae and several other families), and Hymenoptera (Formicidae, Fig. 2.10a). Adults of certain insects in these taxa have greatly enlarged mandibles that are not used for feeding. The jaws of male Lucanidae actually bite less severely than the small jaws of the female. The mandibles of some male beetles are used in jousting (Eberhard, 1975). The other taxa listed use the mandibles primarily for defense.

2. Suctorial mouthparts. In this type, one or more feeding appendages are modified into a tubular organ for taking liquid food. Insects in different, unrelated orders have independently evolved suctorial, or haustellate, mouthparts, so the exact mechanisms vary. Parts are commonly reduced or missing. The cibarium and pharynx, either alone or in combination, provide a sucking pump in most insects of this type. Note that suctorial mouthparts are present in both the immature and adult stages of exopterygote orders listed below, whereas among the endopterygotes only one stage of the life history has developed suctorial mechanisms.

a. Piercing-sucking mouthparts include one or more appendages that are sharp at the apex and suited for piercing the surface of plant or animal bodies. Saliva is usually injected while feeding. Insects that attack vertebrates often have anticoagulants in the saliva to aid the flow of blood. Examples of insects with piercing-sucking mouthparts are listed below:

(1) Thysanoptera. Sharp stylets are used to rupture leaf cells and the exuding sap is sucked up by a cibarial pump (Fig. 33. 1d). The mouthparts form a short conical beak and are uniquely asymmetrical. Inside the beak the right mandible and inner processes of the two maxillae are slender and sharp, forming piercing stylets. Thrips feed mostly on leaves, seeds, fungi, or pollen, but some are predatory (Chap. 33).

(2) Hemiptera. The piercing-sucking mouthparts must be able to penetrate the plant or animal host and create a tube for fluid to enter the gut and another tube to deliver saliva. Elongated, needle-like mandibles surrounding elongated maxillary stylets form a stylet bundle, which is surrounded by and supported by the labium (Figs. 2.11, 32.2e,g). The labium ensheaths the stylets but does not penetrate the host. The maxillary stylets interlock and create two canals by matching their opposing grooves. The anterior canal leads to the cibarial pump and mouth, and is appropriately named the "food canal." The posterior, or salivary, canal carries saliva under pressure from a syringe-like organ within the hypopharynx. At each side of the united maxillae are the separate mandibles. By alternate thrusts, the mandibular blades pierce the host's tissues and provide an entry for the maxillae.

(3) Phthiraptera (suborder Anoplura). The mouthparts of sucking lice are concealed in a long pouch beneath the foregut in the head (Fig. 31.2a–c). Three stylets, possibly representing the united maxillae, the hypopharynx, and the

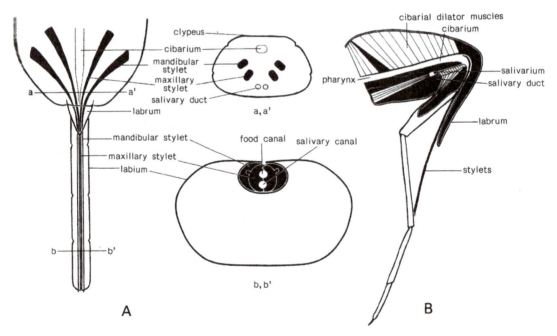

Figure 2.11 Structures contributing to the piercing-sucking beak of Hemiptera.
Source: Modified from CSIRO (1991), The Insects of Australia, 2d ed., Cornell University Press, with permission.

labium, make up the piercing organ. Suction is applied to the feeding wound by a muscular region of the foregut just behind the mouth and including the pharynx.

(4) Siphonaptera. The adult mouthparts of fleas lack mandibles. The piercing organ is formed by the bladelike maxillary laciniae and stylet-shaped epipharynx (Fig. 42.1b). Both the maxillary and the labial palps are present. The latter surround the laciniae but do not penetrate the wound. Once the skin is penetrated, the blood is sucked up by a cibarial and pharyngeal pump.

(5) Neuroptera and Coleoptera (Dytiscidae). The predatory larvae of these insects have each mandible elongate, sickle-shaped, and grooved on the inner surface. In Dytiscidae the mandibles function alone, but in Neuroptera each maxilla is similarly elongated and fits against the mandibular groove to form a closed canal (Figs. 36.1c, 36.3a). The body of the insect victim is pierced by the opposing mandibles, enzymes injected, and the digested body contents sucked as fluid from the victim. The cibarium of Dytiscidae and the pharynx of Neuroptera provide the sucking pumps.

(6) Diptera (Culicidae, Tabanidae, Asilidae). The piercing organ of adult female "biting flies" that take blood meals (Culicidae, Tabanidae) consists of six stylets: labrum-epipharynx (enclosing the food canal), hypopharynx (containing the salivary duct), paired mandibles, and paired maxillae (Fig. 41.3b). The labium surrounds and supports the piercing stylets but does not penetrate the feeding puncture. A cibarial pump provides suction. Mosquitoes (Culicidae; Figs. 2.12a, 2.13a) force the slender piercing organ into the skin, leaving the blunt, flexible labium at the surface. Blood is taken directly into the food canal. Horseflies (Tabanidae) cut the skin surface with their broad, bladelike stylets. The apex of the labium is greatly enlarged in two lobes called the labellum. The undersurfaces of the lobes are traversed by fine grooves called pseudotracheae. These lead mesally to a cleft within which the labral food canal opens. The pseudotracheae direct the flow of blood to the tip of the labrum, and the blood is sucked into the canal. The males of Culicidae and Tabanidae do not suck blood but feed instead on nectar, plant juices, or other liquids

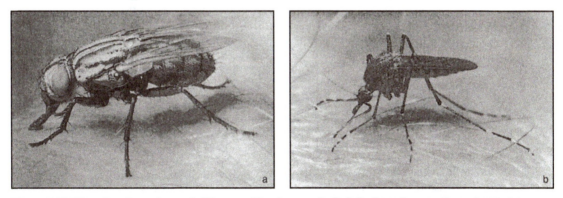

Figure 2.12 Diversity of mouthparts in Diptera. a, Mouthparts of a flesh fly, *Sarcophaga* sp. (Sarcophagidae), is representative of flies that "sponge" food such as sweat. b, The piercing-sucking mouthparts of a female mosquito, *Aedes* sp. (Culicidae), deeply penetrate human skin, as the labium is bent backward.

with their reduced mouthparts. The mouthparts of the predaceous robber flies (Asilidae, Fig. 2.13b) are also of the piercing-sucking type, but differ from the blood-sucking flies. Both sexes of robber flies have the blade-like hypopharynx adapted for stabbing their prey and injecting proteolytic enzymes. Mandibles are absent. The digested contents of the prey are sucked up through a food canal formed by the elongate labrum, maxillae, and hypopharynx, all of which are partly enclosed in a heavily sclerotized labium.

(7) Diptera (*Stomoxys* and *Haematobia* in Muscidae; *Glossina* in Glossinidae). The piercing organ of the stable flies, horn flies, and tsetse flies is a secondary modification of the sponging type described below for adult Diptera. Both sexes of these flies bite mammals. In contrast to the flies described in item (6), the labium of these flies penetrates the host. The labium is slender and stiff, with the labellum reduced in size and provided at the tip with small rasping teeth. The labrum and hypopharynx are partly enclosed by the labium, so that the combined appendages form the piercing organ.

(8) Lepidoptera (certain Noctuidae, Pyralidae, Geometridae). The piercing organ of the fruit-piercing moths of Asia is a secondary modification of the nonpiercing-sucking proboscis of adult Lepidoptera described below. The proboscis is formed by the elongation of the maxillary galeae (as in Fig. 43.3c). These are interlocked to form a tubular food canal. The apex of the piercing proboscis is sharp with hundreds of movable barbs that aid in penetration. By alternate movements of the galeae, the tip is forced into fruit. Some species also visit the eyes of mammals for eye discharges and take blood from

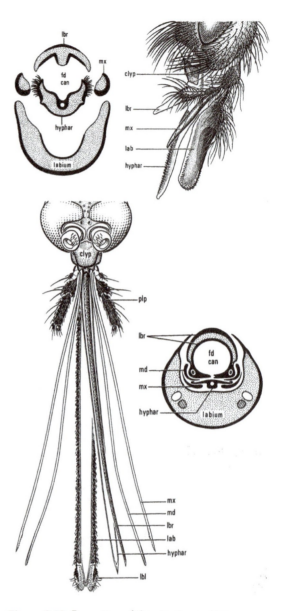

Figure 2.13 Formation of the piercing-sucking mouthparts in Diptera: A, *Laphria thoracica* (Asilidae); B, *Aedes canadensis* (Culicidae). Abbreviations: *clyp*, clypeus; *fd can*, food canal; *hyphar*, hypopharynx; *lab*, labium; *lbl*, labellum; *lbr*, labrum; *md*, mandible; *mx*, maxilla.
Source: Modified from McAlpine, 1981, in Manual of Nearctic Diptera, *Volume 1, Agriculture Canada Research Branch Monograph No. 27, with permission.*

wounds. The unique *Calpe eustrigata* (Noctuidae) pierces the skin of mammals and sucks blood.

b. **Nonpiercing-sucking mouthparts.** The following examples are insects that take water or nutrients dissolved in water, such as the nectar of flowers (Fig. 9.1) or the honeydew (plant sap) excreted by aphids and leafhoppers.

(1) Lepidoptera. Adult butterflies and moths take water, nectar, or honeydew with a long proboscis that is coiled beneath the head when not in use. The proboscis represents the galeae, greatly elongated and tightly interlocked to form a tube (Fig. 43.2c). Suction is created by the pumping action of muscles inserted in front of and behind the mouth. The other mouthparts are reduced or wanting, except the labial palps and the smaller maxillary palps.

(2) Hymenoptera (Megachilidae, Apidae, Anthophoridae). The bees in these families have the maxillolabial structure elongated and are appropriately named the "long-tongued bees" (Fig. 39.19a). The long galeae and labial palps cover the glossal tongue to form a tubular proboscis. The pharynx is modified into a sucking pump. By this means nectar can be withdrawn from deep inside flowers.

(3) Diptera. Most adult flies that do not suck blood have a flexible proboscis consisting of the labrum (enclosing the food canal) and slender hypopharynx (containing the salivary duct) (Figs. 2.12a, 42.3c). Together these fit in an anterior groove on the labium. The labium is enlarged apically into the bilobed labellum, which functions as a sponging organ. The structure of the labellum is described above for the Tabanidae. Liquid food taken up by the labellum is directed into the food canal and sucked up by a cibarial pump. Small particles of food, such as pollen, can also be eaten by some flies. Maxillary palps are present, but mandibles and labial palps are missing.

3. **Mouth hooks of maggots.** Larvae of Diptera (division Cyclorrhapha) have the head highly reduced and invaginated into the thorax, leaving only a neck fold to form a conical snout or functional head (Fig. 41.4c). Within the snout is a preoral cavity, or atrium, from which projects a pair of mouth hooks. The hooks move vertically and arise from the lips of the atrium. No trace of the normal feeding appendages remains. The hooks are secondary modifications of the atrium and are unique to fly maggots. Within the prothorax are sclerotized pharyngeal plates that, together with the hooks, make up the cephalopharyngeal skeleton. Maggots feed on a variety of soft or semiliquid foods, including decaying matter as well as living tissues of plants and animals.

THE THORAX

The second body division, or thorax, is modified for locomotion. The three segments and their ventrolateral appendages are designated from front to rear as follows: prothorax, bearing the forelegs; mesothorax, bearing the middle legs; and metathorax, bearing the hind legs. Internally each segment has a segmental ganglion of the ventral nerve cord. A pair of lateral spiracles is present at the anterior edge of each of the last two segments. In some insects the mesothoracic pair actually opens on the prothorax, but this is a secondary condition.

In the adult pterygote, a pair of wings may be located dorsolaterally on the mesothorax and metathorax. The two wing-bearing segments are called the pterothorax. Wings are absent in the prothorax of insects living today, but certain extinct insects had winglike appendages on the prothorax.

Thoracic Nota

The terga of thoracic segments are called nota to distinguish them from abdominal terga. Accordingly the nota of the respective segments are designated the pronotum, mesonotum, and metanotum.

The pronotum is usually a simple plate, but in some insects like the cockroach it is greatly enlarged to cover the head or, as in the treehoppers, it covers other parts of the body (Fig. 32.13a). In the grasshopper the pronotum extends back over the pterothorax and has several transverse sutures (Fig. 2.14a). The sutures are related to the musculature (which is peculiar to the prothorax) and are not homologous with sutures on the pterothoracic nota.

Each pterothoracic notum is divided into a wing-bearing sclerite, the alinotum, and a phragma-bearing sclerite, the postnotum (Fig. 2.15a). Each alinotum is usually traversed by one or more sutures. These trace the position of internal supporting ridges or lines of flexibility. The extent and position of the sutures vary greatly according to the nature of the flight mechanism. Commonly the alinotum is divided anteriorly by a transverse prescutal suture that delimits an anterior prescutum. Posteriorly a V-shaped scutoscutellar suture separates an anterior scutum and a smaller, posterior scutellum. The lateral edges of the alinotum are modified for the articulation of the wings. At each wing base are anterior and posterior notal wing processes (Figs. 2.15a, 2.17a).

To understand the origin of the postnota and phragmata it is necessary to recall the terminology of secondary segmentation (Fig. 2.5c). The

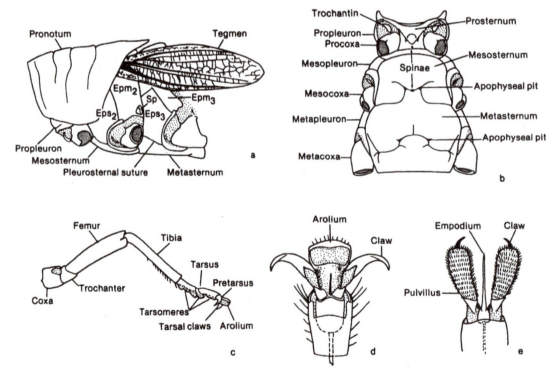

Figure 2.14 Insect thorax and legs: a, lateral view of thorax of *Romalea microptera* (Acrididae); b, ventral view of same; c, foreleg of same; d, pretarsus of a roach, *Periplaneta* sp. (Blattidae, Blattodea); e, pretarsus of an asilid fly (Asilidae, Diptera), ventral view. Epm, epimeron; Eps, episternum; Sp, spiracle.
Source: d and e, redrawn from R. E. Snodgrass, The Principles of Insect Morphology.

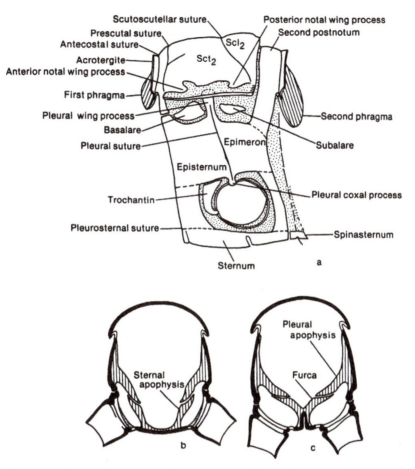

Figure 2.15 Diagrams of the pterothorax: a, lateral view of a mesothoracic segment; b, cross section of a pterothoracic segment, showing separate sternal apophyses; c, same, showing apophyses fused medially to form furca. Scl2, mesoscutellum; Sct2, mesoscutum.

Source: Redrawn from R. E. Snodgrass, The Principles of Insect Morphology.

postnotum corresponds to the acrotergite, and the phragma corresponds to the antecosta. The acrotergite and antecosta anterior to the pronotum were apparently lost during the evolution of the cervix, leaving the pronotum as a simple plate. The narrow acrotergite and the first phragma anterior to the mesonotum are continuous with the alinotum to the rear, just as expected in a secondary segment. In the metathorax, however, the second postnotum and its second phragma are detached from the metathoracic alinotum and associated anteriorly with the mesothorax. Likewise the

third postnotum and third phragma are detached from the first abdominal tergum and associated anteriorly with the metathorax. As a result of this reassociation, the first abdominal tergum is left as a simple plate.

Thoracic Sterna

The ventral plate of a thoracic segment is the eusternum. In the intersegmental regions between the eusterna of the prothorax and mesothorax, and the mesothorax and the metathorax, may be small, separate sclerites. These are named spinasterna because they carry a median, internal apodeme called the spina. One or both of the spinasterna may fuse anteriorly with the eusterna of the preceding segments (Fig. 2.14b). The eusterna plus spinasterna are designated segmentally as the prosternum, mesosternum, and metasternum, respectively. Each euMSternum may have a pair of large invaginations, the sternal apophyses. The apophyses may be well separated and their positions externally marked by a pair of furcal or apophyseal pits (Figs. 2.14b, 2.15b) or fused medially to form a sternal furca (Fig. 2.15c). The longitudinal, midventral line that traces the invagination of the furca is the discrimen. The lateral boundaries of the euster num are usually fused partly or entirely with the pleuron. The exact limits are often indeterminate, but sometimes a pleurosternal suture can be identified.

Thoracic Pleura

The lateral wall of a thoracic segment between the notum and sternum is the pleuron (Fig. 2.15a). Sclerites of the pleura are pleurites. When the pleurites are unified in a solid plate to provide support and points of articulation for the legs and wings, the plate is called the pleuron. The segmental designations are propleuron, mesopleuron, and metapleuron.

The pleuron of a pterothoracic segment has a dorsal pleural wing process, on which the wing articulates, and a ventral pleural coxal process, which provides an articulation for the coxa, or leg base. Extending between the two pleural processes is a strong internal pleural ridge. This provides mechanical support for the pleuron. The ridge may be invaginated at one point to form an internal arm, or pleural apophysis. The arm is connected by muscles or elastic cuticle to the sternal apophysis, and together the apophyses from the pleuron and sternum form a thoracic endoskeleton. Externally the position of the ridge is traced by the pleural suture.

The pleural suture divides the pleuron into two areas, an anterior episternum and posterior epimeron. Small sclerites beneath the wing base are the epipleurites. The epipleurite above the episternum is the basalare, and that

above the epimeron is the subalare. The small crescent-shaped sclerite anterior to the coxa is the trochantin.

The propleuron is smaller and simpler in structure than a pterothoracic pleuron. The wing process and epipleurites, of course, are lacking, but the pleural suture may be present.

Modifications of the Thorax

The pterothoracic segments are similar in structure and nearly equal in size in those insects that provide muscular power about equally to both pairs of wings. Odonata and Orthoptera, for example, fly with the wings unattached to each other. The pterothoracic segments of insects that fly with the wings coupled together tend to have an enlarged mesothorax and a reduced metathorax to accommodate the dominant role assumed by the forewings in flight. In the Diptera, where the hind wings no longer function in flight, the metathorax is almost absent (Fig. 41.2). An opposite trend is seen among insects that fly mainly with hind wings. The metathorax is thus enlarged in Dermaptera, Phasmatodea, Strepsiptera, and Coleoptera (Fig. 37.1b).

Legs

The paired appendages of the thoracic segments are the legs. Each pair has the same general structure, but modified for different functions in different insect groups. Beginning at the base, the six segments of the leg are coxa, trochanter, femur, tibia, tarsus, and pretarsus (Fig. 2.14c). Each segment is individually movable by muscles. The tarsus is divided into no more than five subsegments, or tarsomeres. The tarsomeres are commonly called "segments," but with the exception of the muscles attaching to the basal tarsomere, they do not have individual muscles. Hence the entire tarsus is considered the morphological equivalent of the other leg segments. The pretarsus is the smallest tarsal segment. It is movable by muscles that originate in the tibia and femur and pass to the pretarsus by a slender apodeme. In the grasshopper the pretarsus is represented by a pair of ungues, or tarsal claws, and a median arolium (as in Fig. 2.14d). The claws grip rough surfaces in contrast to the arolium,

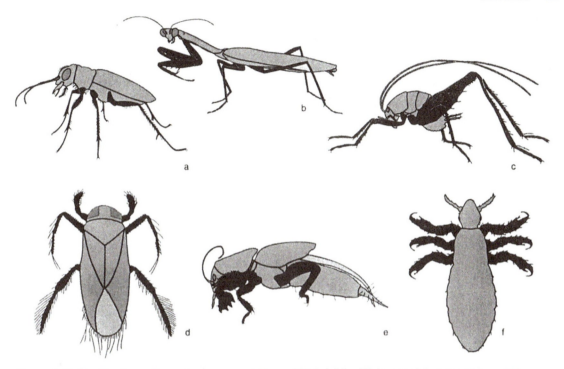

Figure 2.16 Modifications of insect legs: a, cursorial legs of Cicindelidae (Coleoptera); b, raptorial legs of Mantidae (Mantodea); c, saltatorial legs of Gryllacrididae (Orthoptera); d, natatorial legs of Corixidae (Hemiptera); e, fossorial legs of Gryllotalpinae (Gryllidae, Orthoptera); f, prehensile or cheliform legs of Anoplura (Phthiraptera).

which functions as an adhesive pad on smooth surfaces.

Modifications of Legs

The various named modifications of legs for locomotion include the following: ambulatory, or walking, legs (forelegs and middle legs in Fig. 1.4a; Fig. 37.1b); cursorial, or running, legs (Fig. 2.16a); fossorial, or digging, legs (Figs. 2.16e, 9.1); saltatorial, or jumping, legs (hind legs in Fig. 1.4a; Figs. 2.16c, 27.3); and natatorial, or swimming, legs (Figs. 2.16d, 32.10c, 37.5d). Legs modified for grasping are prehensile (Figs. 2.16f, 31.4), and for seizing prey, raptorial (forelegs in Figs. 2.16b, 21.1a). The legs of female bees are used to collect and transport pollen from flowers. Worker honey bees and bumblebees have a pollen basket, or corbiculum, of long setae on the hind legs (Fig. 39.19c). Many other species of bees have a brush, or scopa, of plumose setae on the hind legs

(Fig. 39.19d). Male bees do not collect pollen and are not so modified. Larvae of advanced Diptera (Fig. 41.5c,d) and Hymenoptera (Fig. 39.5b) lack thoracic legs and are called apodous.

The coxa normally articulates dorsally with the pleural coxal process. In many endopterygotes the coxa also has a ventral articulation with the sternum. In Neuroptera, Mecoptera, Trichoptera, Lepidoptera, and Diptera, the coxa has a conspicuous posterior subdivision called the meron. In an extreme development the meron is actually incorporated into the pleural wall of Diptera (Fig. 40.2a).

The trochanter is usually a single segment, but in Odonata it is divided into two subsegments. In some parasitic Hymenoptera, the base of the femur is divided into a short subsegment that resembles a second trochanter (Fig. 39.11b). Wasps with the arrangement are commonly said to have two trochanters. Trochanters are missing entirely on one or more pairs of legs in some insects. In the saltatorial hind leg of grasshoppers, for example, the trochanter is fused with the femur. The femur of insects contains major muscles for locomotion and is usually large and thick. In contrast, the tibia is often slender and armed with large movable tibial spurs near the apex (Fig. 27.3c).

The number of tarsomeres varies from none to five. The basal tarsomere, often longer than the others, is named the basitarsus. The undersurface of the tarsus may be modified to grip smooth surfaces. The tarsi of Orthoptera and some Coleoptera have pads, or tarsal pulvilli. Male beetles of the family Gyrinidae and some Dytiscidae even have suckers on the fore tarsi for clinging to the smooth bodies of their mates. The entire tarsus and pretarsus of larval Lepidoptera and Coleoptera (Polyphaga) is represented by a single claw.

The pretarsus of Diptera has a pair of lateral lobes, or pulvilli, under the claws. The arolium is usually absent, but a spinelike median structure, the empodium, is present (Fig. 2.14e). The tarsal claws may be reduced to a single claw, for example, in Anoplura or nymphal Ephemeroptera. In the mammal-infesting lice, the single-clawed legs are modified for gripping the hairs of their host (Fig. 31.4c,d). Both claws are reduced in Thysanoptera, but the arolium is much enlarged into an adhesive organ.

Wings

The evolutionary origin of wings is not definitely known. Theories concerning their origin are discussed in Chapter 14. During ontogenetic development, the wings grow as saclike expansions of the lateral body walls. The upper and lower layers of the sac partly fuse, leaving a system of narrow, blood-filled channels that are continuous with the body cavity. Tracheae and nerves grow into the channels. At maturity, most of the channels are transformed into wing veins with rigid sclerotized walls that support the fused or membranous portions.

At the base of the wing are small articulatory sclerites (Fig. 2.17a). These include the three axillary sclerites (Ax), the humeral plate (HP), and the median plate (mp). The main, or longitudinal, wing veins are named as follows, beginning at the anterior or leading edge of the wing: costa (C), subcosta (Sc), radius (R), media (M), cubitus (Cu), anal (A), and jugal (J).

The scheme shown in Fig. 2.17b was developed by J. H. Comstock and J. G. Needham and is called the Comstock system (Comstock, 1918). Although there is not universal agreement among entomologists on the homologies of the veins from order to order, nor even agreement on the basic number of veins (see Hamilton, 1972; Kukalová-Peck, 1977), the Comstock system remains the most widely used scheme. We have adopted it for general use in this text.

The longitudinal veins commonly fork, and the branches are indicated by subscripts. The radius often has a main anterior branch, R_1 and a main posterior branch, the radial sector or R_s, that is further subdivided into R_2, R_3, R_4, and R_5. When two or more veins are thought to have fused together to form a single vein, the names of the veins are joined by a plus sign (see, for example, $Sc + R_1 + R_s$ in Fig. 43.12b). Some longitudinal veins are joined by cross-veins that are usually designated according to the longitudinal veins involved (Fig. 2.17b): humeral (h), radial (r), sectorial (s), radiomedial (r-m), medial (m), mediocubital (m-cu), and cubitoanal (cu-a).

The membranous areas enclosed by veins are called cells and are formally named according to the longitudinal vein just anterior to the cell. For example, the veins M_2 and M_3 define a cell, called cell M_2. This may be divided by a median cross-vein to form a proximal cell, first M_2, and a distal cell, second M_2. For taxonomic purposes the cells may be informally referred to as basal, marginal, or submarginal, etc. (Fig. 39.8d), or be given special names such as discal (Fig. 43.10c), depending on the taxonomic group. A closed cell is completely surrounded by veins, whereas an open cell has one side bordering the wing margin.

The area of the wing anterior to the anal vein and its branches is called the remigium (Fig. 2.17b). This is the area, stiffened by veins, that delivers the power of the wing stroke against the air. The anterior edge close to the base of the wing is the humeral region. In the area of the anal veins, the wing is

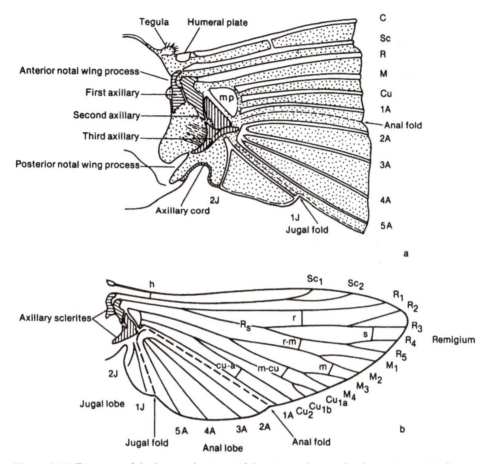

Figure 2.17 Diagrams of the base and regions of the wing and generalized venation: a, wing base, showing articulation and bases of major veins; b, diagram of venation. See text for explanation of abbreviations.

Source: a, modified from R. E. Snodgrass, The Principles of Insect Morphology.

more flexible and may be expanded into a fanlike vannal or anal lobe. An anal fold may separate the anal lobe from the remigium. A smaller lobe at the posterior base of the wing is sometimes present; it is called the jugal lobe and is separated from the anal lobe by the jugal fold.

Modifications of Wings

Wings that are transparent are called membranous (Fig. 2.18a). Wings may be modified to serve protective functions. Thickened, leathery wings such as the forewings of grasshoppers are called tegmina (Figs. 2.14a, 2.18b). The hard forewings of beetles are the elytra (Fig. 2.18d). Hemiptera (suborder Heteroptera) have only the bases of the forewings thickened. These are called hemelytra (Fig. 2.18c).

Wings may be reduced partly or entirely in size or discarded after use. Wings that are functional but are shed after flight are said to be deciduous (compare alates and dealates, Fig. 7.2). Adult insects with short wings are brachypterous (Figs. 1.4a, 20.2), those with vestigial wings are micropterous, and those that never develop wings are apterous (Figs. 23.1, 31.4). The small structures representing

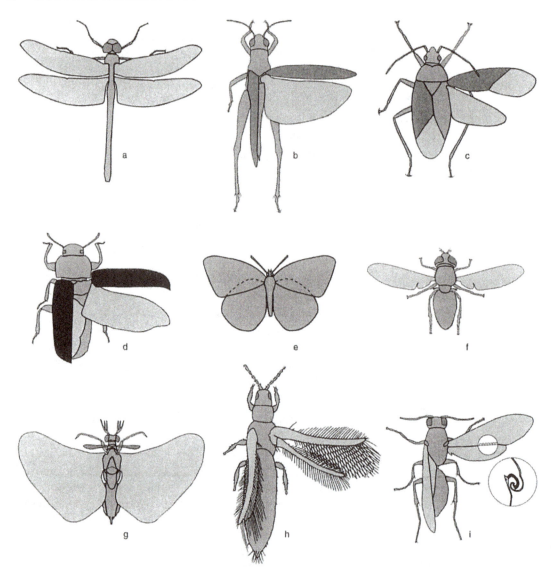

Figure 2.18 Modifications of wings: a, the membranous wings are of equal size in Odonata; b, the forewing is a tegmina and the hind wing is larger and membranous in Acrididae (Orthoptera); c, the forewing is a hemelytron and the hind wing is membranous in Coreidae (Hemiptera); d, the forewing is an elytron and the hind wing is larger and membranous in Tenebrionidae (Coleoptera); e, both wings are membranous, covered with scales, and overlap (amplexiform coupling) in Pieridae (Lepidoptera); f, the forewings are membranous and the hind wings are reduced to halteres (shown in black, between mid and hind legs) in Syrphidae (Diptera); g, the forewings are halteres and the hind wings are large and membranous in Stylopidae (Strepsiptera); h, both wings are fringed with long setae in minute insects such as Thripidae (Thysanoptera); i, both wings are membranous and the larger forewing is coupled to the smaller hind wing by hamuli in Vespidae (Hymenoptera)(hamate coupling; in the magnified detail, black hooks on the leading edge of hind wing engage a fold on the trailing edge of the forewing).

the reduced hind wings of flies (Figs. 2.15f, 41.2a,c) and male coccids are halteres (Fig. 32.16b). The same term is sometimes used for the reduced forewings of Strepsiptera (Figs. 2.18g, 38.1a).

An evolutionary tendency exists for the two pairs of wings to be coupled together and function as a single unit. The simplest coupling is made by the overlap of the adjacent margins. For example, the jugal lobe of the forewing overlies the humeral region of the hind wing in Megaloptera (Corydalidae) and many Neuroptera. In the amplexiform coupling of butterflies the overlap is much broader (Fig. 2.18e). Among some other Lepidoptera, two kinds of modifications are found: in jugate coupling, the jugal lobe is fingerlike in certain moths and clasps the hind wing (Fig. 43.3b). In frenate coupling of other moths, the humeral region of the hind wing has one to several large bristles, called the frenula (Fig. 43.3a). These are engaged by a hook-like flap or a cluster of special bristles called the retinaculum, located near the anterior base of the fore wing. The hind wings of Hymenoptera (Fig. 2.18i) and some Trichoptera have a row of hook-shaped setae, or hamuli, along the leading edge of the hind wing. The hamuli engage the folded posterior edge of the forewing. This is called hamate coupling. Small hooks also are found in Psocoptera (hooks on the forewing engage the hind wing) and some Homoptera, for example, alate aphids (hooks on the hind wing catch the forewing). Other Homoptera have various lobes and folds along the adjacent wing margins to couple the wings. Heteroptera have a groove under the clavus of the forewing that grips the leading edge of the hind wing.

The number of veins and cross-veins in wings tends to become reduced in the more advanced orders and especially among the more derived members. The homologies of the veins with the basic pattern are sometimes controversial. A standardized scheme has usually been established within an order, but differences in terminology exist between orders.

The drag forces caused by the viscosity of air exceed aerodynamic lift forces on wings of minute insects. Such insects have long hairs along the wings that increase drag-induced friction and allow minute flying insects to literally swim through the air. Examples of such wings are found on thrips (Fig. 2.18h), certain parasitic wasps of the superfamilies Chalcidoidea, Proctotrupoidea, and Ceraphronoidea, and beetles of the family Ptiliidae.

THE ABDOMEN

The last body division, or abdomen, contains the viscera (Fig. 2.19b). Included are most of the alimentary canal and dorsal circulatory vessel, as well as the Malpighian tubules, fat body, and reproductive organs. The total number of segments in the abdomen remains controversial: arguments have been made that twelve are theoretically possible, yet no more than eleven segments can be counted in the abdomen of adult insects, and frequently only ten or nine are clearly evident on close examination. The first eight segments may each have a pair of lateral spiracles. Internally, each of the first seven segments may have a ganglion of the ventral nerve cord, but the ganglion of the eighth and the ganglia of the remaining segments are fused together in a single large, compound ganglion (Fig. 6.3).

Paired appendages, homologous to legs, are absent on the first seven, or pregenital, segments of adult pterygotes. On the eighth and ninth segments of the female and the ninth of the male, the appendages are modified as external organs of reproduction, or genitalia. These segments are known as the genital segments. The tenth and remaining segments are the postgenital segments. The eleventh may have a pair of lateral appendages called cerci.

Except where appendages are present, each abdominal segment typically has the tergum and sternum as single plates. The plates overlap posteriorly and often exhibit the features of secondary segments. Externally near the anterior margin is the antecostal suture that delimits the anterior acrotergite and traces the internal ridge or antecosta. Muscles attach to the antecosta and to lateral apodemal lobes of the sterna. On the first abdominal segment of the grasshopper is a pair of lateral acoustical organs, or tympana (Fig. 27.2b).

Ovipositor

The paired appendages of the eighth and ninth segments of females may be modified into an appendicular ovipositor for depositing eggs

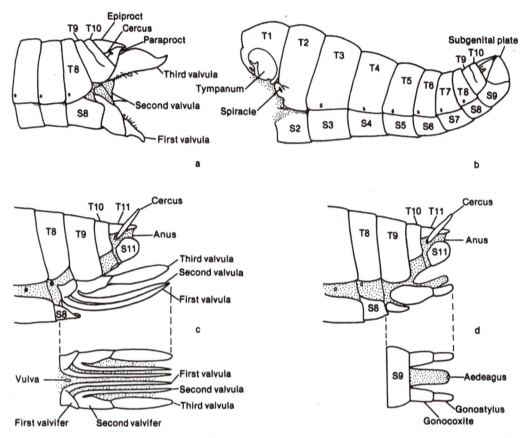

Figure 2.19 Abdomen and external genitalia: a, lateral view of female genitalia of *Romalea microptera* (Acrididae); b, same of male abdomen and genitalia; c, diagram showing general structure of female genitalia; d, same of male genitalia. S, sternum; T, tergum.
Source: c, redrawn from R. E. Snodgrass, The Principles of Insect Morphology.

(Fig. 2.19c). On segment 8 the sternum may be present, but it is absent in segment 9. The appendicular ovipositor consists of basal valvifers, on which muscles insert, and elongate valvulae. The eighth segment bears the first valvifers and first valvulae, and the ninth segment bears the second valvifers and the second and third valvulae. The valvifers correspond to the paired coxae or gonocoxites on each segment. The first and second valvulae correspond to median lobes or gonapophyses of the respective coxae and together may form an interlocking, tubular shaft. The third valvulae correspond to apical lobes of the coxae on the ninth segment and often form a protective sheath for the first and second valvulae. Eggs issue from the genital opening on the eighth segment, pass down a channel in the shaft formed by the first and second valvulae, and are deposited in soil, plant tissues, or, in the case of insect parasitoids, into the bodies of insect hosts. The tips of the valvulae have sensory chemoreceptors to aid the female in placing the eggs. In grasshoppers and their allies the ovipositor is differently constructed for digging in soil (Figs. 2.19a, 2.20d). Here the organ is formed by the first and third valvulae. The first valvulae are ventral, the third valvulae are dorsal, and the second valvulae are small, median in position, and concealed by the other valvulae.

The great majority of insects, however, have the appendicular ovipositor replaced by reduced structures or by special modifications of the tip of the abdomen (Chap. 4).

Male Genitalia

The modified appendages of the ninth segment of males form a copulatory organ (Fig. 2.19d). The basic structure of many insects consists of a median intromittent penis, or aedeagus, and lateral clasping parameres. The ninth sternum may be present as the subgenital plate. The parameres are commonly two-segmented and are thought by most morphologists to be modified legs. The proximal segment is the gonocoxite, and the distal segment, if any, is the gonostylus. These sclerotized structures engage the genital opening of the female and maintain the connection during intromission.

The aedeagus is formed by a median fusion of median coxal lobes and is often partly sclerotized. The ejaculatory duct opens apically. During copulation the membranous portion of the aedeagus is inflated by blood pressure, everts into the vagina of the female, and discharges the sperm and seminal secretions from the ejaculatory duct. Further discussion of reproduction is found in Chapter 4. The specialized male genitalia of grasshoppers, seen externally, consist of a ventral, cup-shaped subgenital plate and a dorsal, membranous pallium (Fig. 2.19b).

Postgenital Segments

The segmentation beyond the genitalia may be difficult to determine. The eleventh segment, when present, surrounds the anus and may bear paired appendages or cerci. Sclerotized plates, if any, on the eleventh segment are divided into a dorsal epiproct and lateral paraprocts below the cerci. In the grasshopper both sexes have the ninth and tenth terga fused and the small cerci situated between the epiproct and paraproct (Fig. 2.19a,b).

Modifications of the Abdomen

The number of visible segments is often reduced in the endopterygotes by the telescoping of the genital and postgenital segments into the apex of the abdomen. In some cases the first segment is reduced or fused with the second; in advanced Hymenoptera, the first segment is incorporated into the posterior wall of the thorax and known as the propodeum (Fig. 39.2). Also in the Hymenoptera, the second and sometimes the third segments are constricted into a petiole, leaving the remaining segments as the gaster (Fig. 39.11c).

Paired appendages are present on the pregenital segments of Parainsecta, Entognatha, Apterygota, and the immature stages of certain pterygotes. The appendages are clearly serially homologous with legs in the orders Protura, Collembola, Diplura, Archeognatha, and Thysanura. Those of Archeognatha, for example, have paired, flattened coxae, each within a stylus that represents the distal part of the limb (Fig. 17.2a). The abdominal appendages of pterygotes are much more highly modified and less readily identified as limbs. Examples include the abdominal tracheal gills of nymphal Ephemeroptera (Fig. 18.4b–d), larval Megaloptera (Fig. 34.2), and larval Sisyridae (Neuroptera), and the fleshy abdominal prolegs of larval Lepidoptera (Fig. 43.5a,b) and Hymenoptera (Fig. 39.5a). The larvae of Diptera, especially Nematocera, may have a variety of locomotory (prolegs, suction discs, spiny ridges called creeping welts), respiratory (anal gills, tracheal gills), or osmoregulatory (anal papillae) structures on the abdomen.

The appendicular ovipositor is well developed in certain Odonata, Orthoptera, Hemiptera (suborder Homoptera), Thysanoptera, and Hymenoptera. The appendicular ovipositor is absent in many other insects, for example, Coleoptera, Diptera, or Lepidoptera, but is replaced in these orders by special modifications of the terminal abdominal segments.

The size and shape of the ovipositor are closely correlated with the nature of the medium into which the eggs are inserted. Insects that deposit their eggs in living plant tissues usually have stout valvulae with sawlike ridges. Parasitic wasps may use long, slender valvulae to insert eggs into insect hosts that are in burrows deep in solid wood (Figs. 2.20c, 39.4a, 39.11a).

The ovipositors of predatory wasps, ants, and bees are modified into an organ for defense, the

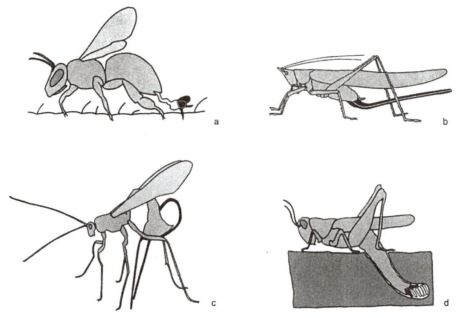

Figure 2.20 Modifications of ovipositors (in black): a, the sting of a female honey bee (Apidae, Hymenoptera) is a modified ovipositor with barbs on the valvulae (the barbs stick in the skin of the victim and as the bee struggles, the sting is pulled out of her body); b, the long ovipositor of Tettigoniidae (Orthoptera) is used to place eggs into a substrate; c, the first and second valvulae of the ovipositor of Ichneumonidae (Hymenoptera) are driven into wood and, if a host larva is detected, the wasp's eggs pass down the hollow shaft to the host (the third valvulae, also long and slender, are pushed to the side and do not enter the wood); d, the ovipositor of Acrididae (Orthoptera) is used to dig in soil where the eggs are laid in rows.

sting (Figs. 2.20a, 39.3d). Thus, only females can sting. The second valvulae are fused to form a stiff piercing organ, with the first valvulae locked beneath in sliding grooves. Glands associated with the ovipositor release toxic venom into the wound created by the valvulae. In the stinging Hymenoptera, the ovipositor no longer functions to deposit eggs; they are merely released from the vaginal opening on the eighth segment.

Worker honey bees have tiny hooks at the tip of the second valvulae (Fig. 2.20a). Once the sting is plunged into the skin of a person, the barbs prevent the sting from being withdrawn. As the bee struggles, the sting plus the poison glands and attached viscera are pulled out of the bee. The bee flies away, only to die shortly thereafter. The muscles of the sting continue to contract and venom continues to be injected. The best first aid is to scrape the sting

off with a fingernail rather than pinch the stinger, which will only force more venom into the wound.

The male genitalia of insects are so greatly varied that they are used by insect taxonomists to classify species. Special terminologies have been applied in some orders where the structures are complex and the homologies uncertain. Male Odonata have the sterna of segments 2 and 3 modified as a secondary genitalia for the transmission of semen to the female (Fig. 19.3b,c). Male Ephemeroptera have paired intromittent organs on the seventh segment. These correspond to the paired openings of the lateral oviducts on the seventh abdominal segment of the female. Some Dermaptera have paired ejaculatory ducts instead of a single duct. The genitalia are especially complicated in Diptera, where in some groups the terminal abdominal segments

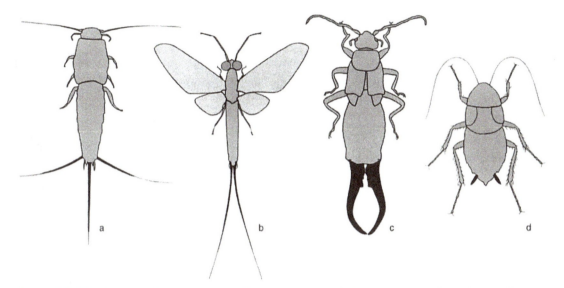

Figure 2.21 Modifications of cerci (in black): **a,** Thysanura have paired, filamentous cerci plus a single median caudal filament; **b,** Ephemeroptera also have filamentous cerci and some species (not shown) have a median caudal filament; **c,** the cerci of Dermaptera are heavy pinchers; **d,** the small cerci of Blattodea have sense organs that are able to detect the movement of air near the insect.

and genitalia actually undergo a clockwise rotation of 90°, 180°, or even 360° during pupal development.

The cerci are appendages of the eleventh abdominal segment. They may be long, slender, and subsegmented as in Thysanura (Fig. 2.21a), Archeognatha, and Ephemeroptera (Fig. 2.21b); shorter and with fewer subsegments as in Blattodea (Fig. 2.21d); or one-segmented as in Orthoptera (Fig. 27.3). The cerci of Dermaptera (Fig. 2.21c) and Diplura (Japygidae) (Fig. 18.5e) are stout and forcepslike. Cerci are absent in the Hemipteroidea. The epiproct is greatly elongated into a cercuslike, median caudal filament in Thysanura (Fig. 2.21a), Archeognatha (Fig. 17.1), and Ephemeroptera (Fig. 18.1a). In nymphal Odonata (suborder Zygoptera) the epiproct and paraprocts are greatly enlarged into tracheal gill plates (Fig. 19.6b).

EVOLUTION OF THE BODY REGIONS

In the last few decades, the Homeobox (Hox) and other genes responsible for regulating the development of animal body forms were discovered and elaborated. It has recently become possible to more directly compare the patterns of gene expression underlying the different patterns of tagmosis found among the major groups of arthropods. In addition, our understanding of arthropod phylogenetic history has improved, so that a presumably more accurate evolutionary perspective on the morphological changes leading to the insect body form is now taking form.

The insect body regions described above are more or less standard for all insects and similar to those of other hexapods (see also Chap. 14). They differ in many details, however, from those found in the other arthropod groups, such as spiders, ticks, crabs, shrimps, centipedes, and millipedes. For instance, the chelicerate groups, including spiders, scorpions, and ticks, have no obvious head or antennae and have different numbers of segments bearing mouthparts and walking legs. The myriapod groups, including centipedes and millipedes, possess a head region and antennae that seem

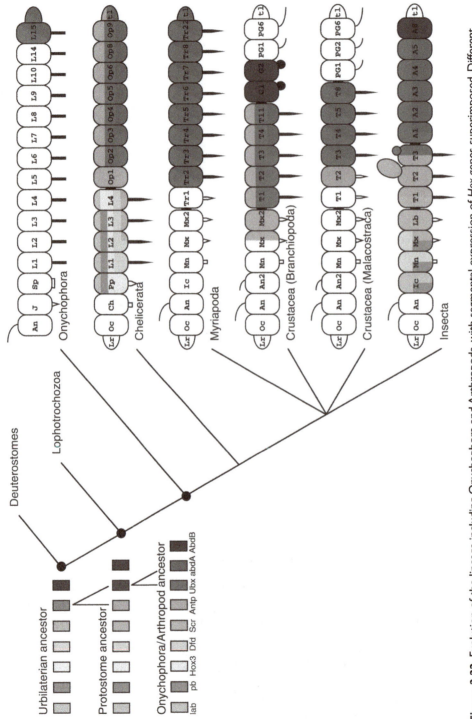

Figure 2.22 Evolution of the lineage including Onychophora and Arthropoda, with segmental expression of *Hox* genes superimposed. Different functions of appendages are indicated by hollow shapes (feeding structures); thickened pegs (walking legs), and thin lines (flattened swimming structures). Segmental expression of *Hox* genes is indicated by shading as indicated at left of figure. Abbreviations (named after mutations) lab: *labial*; pb:; Hox3- *Homeobox* 3; Dfd: *deformed*; Scr: *sex combs reduced*; Atp: *antennapedia*; Ubx: *ultrabithorax*; abdA: *abdominal A*; abdB: *abdominal B*.
Source: Modified from Knoll and Carroll, 1999, with permission.

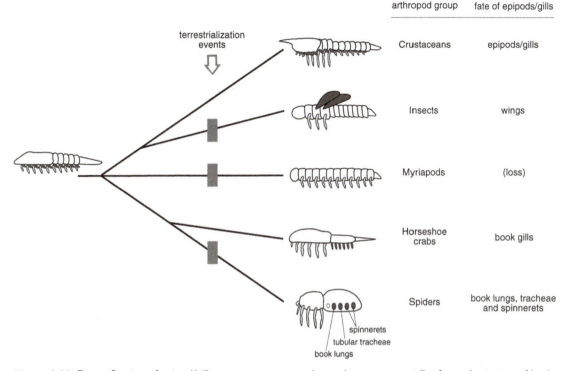

terrestrialization events

arthropod group	fate of epipods/gills
Crustaceans	epipods/gills
Insects	wings
Myriapods	(loss)
Horseshoe crabs	book gills
Spiders	book lungs, tracheae and spinnerets

spinnerets
tubular tracheae
book lungs

Figure 2.23 Diversification of epipod/gill structures among arthropod groups, especially after colonization of land (terrestrial) environments.
Source: Modified from Damen et al., 2002, with permission.

similar to those of insects but show no distinction between the thorax and abdomen regions, instead bearing many segments with walking legs. Crustaceans such as crabs and shrimps show a variety of forms of tagmosis and sometimes have more than one pair of antennae and branched (biramous) appendages. Which segments of an insect are homologous to which segments of these other groups?

Figure 2.22 shows the current understanding of how segments correspond among the major groups of arthropods and the related Onychophora (velvet worms) and how developmental gene expression differs among them. The diagram shows the relationships among Myriapoda, Crustacea, and Insecta as unresolved; traditional views from comparative morphology hold that Myriapoda and Insecta are more closely related and form (along the other hexapods) the Atelocerata based on their shared head capsule, while more recent molecular studies have indicated that Insecta arose from among the crustacean groups, and the Hexapoda and Crustacea are then grouped together as the Pancrustacea.

The same kind of comparative developmental studies have illuminated the diversification and specialization of the segmented appendages (originally legs for walking or swimming) among the different arthropod groups. Shared patterns of gene expression suggest that the epipods or gills of early crustaceanlike arthropods are homologous with the wings of modern insects, the book gills of horseshoe crabs, and the book lungs, tracheae, and spinnerets of spiders (Fig. 2.23)!

More detail on the comparative evolutionary morphology of the various insects, and how this affects the classification of the insects into major groups, is provided in Chapter 14.

CHAPTER 3

The Integument

The integument provides the basis for the rigid exterior skeleton that distinguishes arthropods from other invertebrates. It is physically the largest of all organ systems and exhibits the greatest diversity in structure and in function. The integument has three basic components: a tough, outer, noncellular cuticle; a single layer of cells, or epidermis; and an inner sheet of connective tissue, or basement membrane (Figs. 2.2, 3.1a).

THE EXOSKELETON

The most easily seen portion of the integument is the hardened and jointed cuticle that covers the outside of the body and forms an external skeleton, or exoskeleton, as described in detail in the previous chapter. Our classification of insects is based mainly on the appearance of the exoskeleton. The special arrangements of the hairs and other features provide nearly all the diagnostic characters for the classification of insects down to the level of species. This means that easily a million or more distinctive exoskeleton designs exist. We are indeed fortunate that the exoskeleton is so simply preserved and that so much information concerning the insect's habits can be gained from it.

The exoskeleton is extremely resistant to decomposition. When protected from pests, dried specimens of insects may outlast the pins and cabinets of the museum and usually outlast the collectors and curators. When preserved as fossils, pieces of the exoskeleton may remain intact. The chemical component chitin, for example, has been detected in the fossil of an extinct eurypterid arthropod from the Silurian Period, over 409 million years ago.

In addition to functioning in the support of internal organs and as attachments for muscles, the exoskeleton is involved in such important activities as feeding, locomotion, and reproduction. The intricate shapes of the mouthparts permit the exploitation of a wide variety of foods. Similarly, walking and flying depend in large degree on the strength and flexibility of the exoskeleton. In reproduction, parts of the exoskeleton provide the means by which the male may transmit sperm to the female and the female, in turn, may deposit her eggs in special places.

The exoskeleton also serves many of the same functions as the skin of other animals. It is the main barrier to the loss of water, a problem that is severe in small terrestrial organisms with a relatively large evaporative surface area in proportion to the volume of their bodies. The entrance of disease organisms and the penetration of chemicals such as insecticides may be prevented by the exoskeleton. Assaults by predators are not only resisted by the tough outer covering but also avoided by camouflage or warning coloration that is created by pigments in the integument. The sense organs of the nervous system are intimately connected with specialized areas of the exoskeleton because all the stimuli from the environment must be transmitted through the exoskeleton to the receptors beneath.

The exoskeleton determines the form and maximum size of the insect. At intervals during growth,

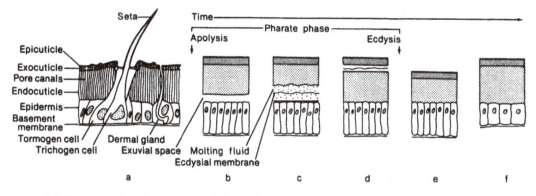

Figure 3.1 Structure of the integument and the molting cycle. See text for explanation.

parts of the cuticle are digested and a new, larger cuticle is deposited. This process is termed the molt, and the actual shedding of the old cuticle is called the ecdysis. At times of molting, many of the chemical components of the old exoskeleton are digested and used again. For this reason, the exoskeleton is actually a kind of food reserve.

The integument includes not only the conspicuous exoskeleton but also thin-walled, tubular structures hidden from view inside the body. These structures have the same basic components: cuticle, epidermis, and basement membrane. The tracheal system, functioning in respiration, is created by invaginations of the integument. The ultimate subdivisions of these tubular ducts number literally in the millions and supply oxygen to virtually every cell of the body. The elimination of carbon dioxide is accomplished through the tracheae and also through the general body surface. In the alimentary canal, the foregut and hind gut are formed by invaginations of the integument during development. The integument also lines the ducts leading from the internal reproductive organs, providing not only passageways and storage vessels for the gametes but also glandular secretions that function in their protection and transmission. Finally, a rich variety of other glands, including the salivary glands, is derived from cells of the epidermis. Their products function in nearly every phase of the insect's life history by supplying enzymes for digestion, materials for shelter and support, and other chemical compounds for attracting the sexes, communication,

defense, and offense. Among these integumental secretions, human beings have found some of great value (e.g., silk, shellac, and beeswax).

The Origin of the Integument

Of the integument's three major components, only the epidermis is cellular. In the embryo, the epidermis is derived from the ectoderm, as described in Chapter 4. The cuticle is periodically secreted by the epidermis, both daily as additional thin layers and at longer intervals when the cuticle is replaced throughout nearly all the body during molting. Certain parts of the cuticle are thought to be secreted by oenocytes (cells of ectodermal origin in the blood). The basement membrane on the inner surface of the epidermis is believed to be secreted by certain hemocytes (blood cells) in the hemolymph and possibly by other cells of the body.

THE CUTICLE

The cuticle— the noncelluar part of the integument—may equal up to half the dry weight of the insect's body and always exists external to the epidermis. When viewed microscopically, it is seen to be made up of layers. Each layer has special chemical and physical properties and is secreted by epidermal cells. Two major divisions of the cuticle are the thin outer epicuticle, no more than about 4 μm thick and composed of proteins and lipids, but with chitin rarely detected, and the inner procuticle (combined exocuticle and endocuticle), which may measure a millimeter or more in thickness and contains protein and chitin. Each division may have additional sublayers, depending on the species of insect and the location of the cuticle on the insect's body.

Epicuticle

The epicuticle is created by secretions from several different sources that are deposited outside the procuticle. Four layers are currently recognized: the outermost cement layer, the wax or lipid layer, the outer layer of epicuticle, and the inner layer of epicuticle. The cement layer is derived from certain modified epidermal cells, called dermal glands, which empty their secretions on the outer surface of the epicuticle by means of gland ducts through the procuticle. The cement protects the critical wax layer from being scratched or lost by absorption by foreign objects. This aspect is well demonstrated by observations that scratches caused by silica dusts to the cuticle of roaches and other insects lead to a loss of internal moisture that eventually kills the treated insects. The copious secretions of the lac insect, *Kerria lacca* (Coccoidea, Hemiptera), yield commercial shellac, a high-quality, clear wood finish. The secretions are probably quite similar in origin to the cement produced normally and in much smaller quantities on the body surfaces of other insects.

The lipids for the wax layer are produced by epidermal cells and must pass through the cuticle to reach the outer surface. The most common route is thought to be by means of the pore canals (see below, "Epidermis") of the procuticle and the wax canals of the epicuticle. Thin wax filaments, about 6 nm in diameter, extend through the canals from the epidermal cells to the surface. The filaments, possibly of tanned protein, are believed to function in the transport of the lipids. The wax secretions vary in chemical composition but are mainly mixtures of hydrocarbons, alcohols, and fatty acids. Depending on the composition and temperature, the physical properties of the wax vary: the greasy surface of the cockroach and the firm honeycombs of beeswax in the beehive are both products of the wax-producing cells of the epidermis.

The wax layer is of crucial importance in preventing the loss of water from an insect in a dry atmosphere. Even thin layers of lipids are sufficient to provide protection. Much interest now centers on the molecular organization of these thin films in relation to their impermeability. Wax may mix with the cement layer or appear on the outside surface as a powdery wax bloom or a greasy film. In such cases a water-repellent or hydrophobic surface is created which has important consequences for the insect in contact with water. For example, a small insect can be supported by the surface tension of water rather than trapped by it. The wax layer also is important in the resistance to invasion by pathogens such as fungi and is probably the main barrier to the penetration of some insecticides. Evidence is now accumulating that the various mixtures of compounds that make up the wax layer can serve yet another function, namely, as pheromones in sexual behavior and as recognition odors among social insects.

The names of the remaining layers of the epicuticle continue to change as our technology allows more detailed studies. The "outer layer" of the epicuticle is beneath the wax layer and always present in insect cuticle. This layer was formerly called the cuticulin layer or protein epicuticle. The outer layer is composed of tanned lipoproteins (lipoproteins stabilized by molecular cross-links). It is the first layer of a new cuticle to be formed by the epidermis during the molting cycle. The layer is only slightly thicker than a cell membrane but may function as a critical barrier between the new and old cuticles. In addition to the wax canals that pass through it, the outer layer has small pores that may filter the digestion products of the old cuticle. Once in place, the outer layer may determine the maximum expansion of the cuticle that is possible before another molt.

The "inner layer" of the epicuticle was formerly included with the outer or cuticulin layer but now is recognized as distinct in some insects. It is much thicker than the outer layer, penetrated by wax canals, and also probably composed of tanned lipoproteins.

Procuticle

The term "procuticle" (meaning early cuticle) specifically refers to the soft, chitin-rich cuticle secreted by the epidermis early in the molting cycle. The term also is used generally to refer to the second major division of cuticle in contrast to the epicuticle. Later in the molting cycle the procuticle becomes differentiated into two layers, an outer hardened and often darker exocuticle and an inner flexible endocuticle, which is usually lighter in color. Both are of much the same composition, namely, long molecules of proteins and chitin.

A variety of proteins exists in the procuticle (the water-soluble proteins are collectively called arthropodin) and may contribute more than half the dry weight. The proteins may differ from instar to instar and from place to place on the insect. For example, one protein may be associated with rigid cuticle and another with flexible cuticle. Among the latter, a rubberlike protein called resilin (from the Latin *resilire*, to jump back) may occur in special places in almost pure form. Resilin confers an exceptional elasticity to certain wing articulations and muscle insertions.

The second major component of the procuticle, chitin, is distributed among diverse organisms such as fungi, diatoms, sponges, hydroids, bryozoans, brachipods, mollusks, annelids, as well as arthropods. Chitin may be up to 60 percent of the total dry weight of an insect's cuticle. Biosynthesis begins with the sugar trehalose and ends with chitin, a nitrogenous polysaccharide of *N*-acetylglucosamine plus some glucosamine. The long molecules of chitin resemble cellulose, a polysaccharide widespread in the supportive structures of plants. Like cellulose, chitin is one of the most abundant organic materials in the world and new commercial uses for it are being developed. Pure chitin has been described as a colorless, amorphous solid that is insoluble in water, dilute acids, dilute alkalis, alcohol, and all organic solvents. It may be dissolved in concentrated mineral acids but is rapidly degraded.

The process of hardening, called sclerotization, involves the tanning or cross-linking of the proteins by the action of sclerotizing agents (one or both of the compounds *N*-acetyldopamine and *N*-β-alanyldopamine) into a stable molecular structure. The sclerotizing agents require the amino acids tyrosine or phenylalanine, which are essential to the diet of insects. Proteins modified by the agents are called sclerotins. The hardened zone

becomes the exocuticle, and the general area where the exocuticle occurs in the integument becomes a discrete, rigid sclerite in the body wall. Endocuticle consists of the unsclerotized inner part of the procuticle that remains flexible. At joints, or arthrodial membranes (Fig. 2.2), the exocuticle is absent or broken into small blocks and the endocuticle predominates. This permits bending between sclerites. Elastic properties may be conferred to hinges by the presence of resilin.

The exocuticle is extremely resistant to digestion, both by the insect itself during molting and by other organisms, like pathogens, equipped with hydrolytic enzymes. The chitin chains are also cross-linked together and bound to the tanned proteins, thus becoming part of the three-dimensional network. However, the tanned proteins rather than the chitin are responsible for the special hardness of the exocuticle. For this reason a hard piece of the exoskeleton is correctly termed "sclerotized" rather than "chitinous" because soft cuticle also has chitin. Hard exocuticles are usually dark in color, partly from sclerotin and partly from pigments like melanin that are deposited in the cuticle. Hard exocuticle is not necessarily melanized: albino mutants of insects may have hard cuticles lacking dark pigments.

Within an insect's cuticle, the chitin molecules are laid down in parallel microfibers, which, in turn, are arranged in thin layers or lamellae. The orientation of the microfibers shifts with time and produces a banded pattern when the cuticle is viewed in cross section. Some lamellae are deposited at times of molting, and additional lamellae may be added each day as a distinct growth layer. The characteristics of the growth layers are influenced by temperature and by the relative lengths of day and night.

Sclerotization occurs elsewhere in insect life histories where strong, lightweight, flexible structures are produced: eggshells, oothecae, puparia, cocoons, and silk. The same biochemical process is widespread among invertebrates.

Although calcium is present in the cuticles of crabs and other crustaceans, it is rarely incorporated in the cuticle of insects. When present in insects, calcification does not modify the hardness of the sclerotized parts. By geological standards, some sclerotized cuticles can scratch the mineral calcite. On a scale of 10 ranked by increasing hardness, these cuticles would be placed third; that is, in hardness they are between a thumbnail and a copper penny. In California, the beetle *Scobicia declivis* (Bostrichidae, Coleoptera) easily chews through the lead sheathing of aerial telephone cables, earning it the name "lead-cable borer." Other insects can penetrate thin sheets of tin, copper, aluminum, zinc, or silver. Sclerotized cuticle is, however, not indestructible, and its surface together with the epicuticle may show extensive abrasion and wear, especially in burrowing insects. Wear is also found on the cutting and grinding surfaces of the mandibles of chewing insects. Renewal of the cuticle at each molt may serve not only to accommodate increases in body size but also to restore the valuable properties of the cuticle.

Because the production and health of the integument is essential for the survival of an insect, opportunities exist for the development of pesticides that disrupt normal processes. Abrasive and absorptive materials may be used to destroy the cement or wax layers. Such desiccant dusts may be essentially inert and nontoxic, yet provide long-term pest control for insects like roaches. The biosynthesis of chitin in insects and fungi is inhibited by benzoylphenyl ureas such as diflubenzuron, known also as the pesticide dimilin. The biosynthetic pathways leading to sclerotization also can be disrupted by metabolic inhibitors, or the process can be abnormally accelerated with a toxic effect.

THE EPIDERMIS

Epidermal cells are organized in a single layer with an apical-basal polarity. Thus the cells have an apical surface, next to the cuticle, bearing microvilli for the secretion of the cuticle. Between the cells are intercellular spaces, and at the base the cells are next to the noncellular basement membrane (also known as the basal lamina), which acts as a barrier and filter between the epidermis and the hemolymph. Once secretion of the new cuticle is underway, the physical routes by which the epidermal cells pass their secretions to the procuticle and the epicuticle are the pore canals (Fig. 3.1). These are minute channels, 1.0 to 0.15 μm in diameter, extending from the surface of the epidermal cell to

the interface between the epicuticle and the procuticle. They are not lined with the plasma membrane of the cell, but wax filaments pass from the cell through the pore canal. The density of canals may be quite high, ranging from 15,000 to 1,200,000 per square millimeter. The canals are not straight but seem to be coiled about an axis perpendicular to the cuticular surface or curve with the microfibers in each cuticular lamella.

THE BASEMENT MEMBRANE

A thin layer of connective tissue covers all the internal organs and the internal surfaces of the integument. So situated, the membrane separates the various organs from the hemolymph and must regulate the flow of molecules to and from the organs. Consequently, the idea that the cells of the insect body are directly bathed by the blood in an "open circulatory system" is incorrect. When beneath the epidermis, the connective tissues are called the basement membrane and consist of an amorphous granular layer up to 1 μm in thickness. In some cases a fibrous structure is apparent in insect connective tissues. In certain chemical and physical properties, these fibers resemble collagen of the vertebrate connective tissues.

THE PROCESS OF MOLTING

The integument of insects undergoes cyclic activities throughout the life of the insect. The events that trigger the molting cycle and the hormonal control of molting are treated in Chapter. 4. The most conspicuous changes are associated with the periodic shedding of the old cuticle. The growing insect is restricted in size by the maximum area to which the cuticle will expand. The new cuticle that is laid down beneath the old is highly wrinkled, thus providing potentially greater area for expansion after the molt. The major growth in the size of the exoskeleton is therefore limited to times of molting and is discontinuous. Changes in the form of the exoskeleton at metamorphosis and in patterns of setae, wound healing, and the regeneration of lost parts are likewise limited to the molt. In addition to these changes in morphological characteristics, the valuable physiological properties of the cuticle that may become weakened by age and wear must be renewed, and the molt serves this purpose as well. Between molts, the epidermis continues to be active in secreting additional layers of endocuticle and in wax production.

Apolysis

The onset of the molting cycle begins with a rapid increase in RNA synthesis followed by exceptional physiological activity at the interface between the epidermis and cuticle. Ecdysteroid hormones initiate molting in the hemolymph., triggering the epidermal cells to thicken and proliferate from a thin layer of polygonal cells to become a thick columnar layer. Apolysis, the detachment of the epidermis from the old cuticle, may be created by a retraction of the cells or by a change in the consistency of the cuticle immediately next to the epidermis. Muscle attachments and connections of the nervous system to sensory receptors in the cuticle remain in place so that their functions continue. The space created between the old cuticle and epidermis is termed the exuvial space. Into this space is deposited the molting fluid from the epidermal cells. The fluid may exist in this space for some time before becoming active (Fig. 3.1c) and later becomes gelatinous.

New Cuticle

The secretion of the new cuticle begins with the outer layer of the epicuticle. This first layer is believed to be critical in determining the surface features and area of the new cuticle. Then the inner layer of epicuticle is deposited, followed by the procuticle. Once the protein epicuticle is in place, the molting fluid becomes active. The main function of the fluid is the digestion of the old endocuticle. Proteases and chitinases digest protein and chitin molecules into their amino acid and N-acetylglucosamine sugars, leaving only the resistant epicuticle and exocuticle. In some insects a resistant part of the old endocuticle survives as an ecdysial membrane. The products of digestion, amino acids and sugars, are probably absorbed through small pores in the outer layer of epicuticle. The products are recycled into the new procuticle by the epidermis. As the old endocuticle is digested, the new procuticle increases in thickness. The proteins destined for the procuticle are

produced by the epidermal cells or are produced elsewhere in the body and transported to the epidermis by the hemolymph. The lamellae, characteristic of the procuticle, probably are the result of cyclic secretory activities of the epidermal cells. The pore canals are initiated with the first layer and become progressively longer with increasing thickness of the procuticle. By means of these canals the cells remain in communication with the epicuticle/procuticle interface.

Sclerotizing of the outer zone of the procuticle to form the exocuticle may begin now in certain areas of the insect's body or be delayed in other areas until after ecdysis. The tanning and hardening of the cuticle take place from the outside inward. To accomplish this, the sclerotizing agents or their precursors are produced in the fat body or hemocytes and transported by the hemolymph to the epidermal cells. The epidermis also is capable of producing the sclerotizing agents from precursors. The sclerotizing agents then diffuse outward from the epidermis through the pore canals to the surface of the cuticle. The sclerotizing agents become activated by enzymes that catalyze oxidation and diffuse inward, transforming the protein and chitin into a rigid polymerized structure and creating the exocuticle. Just prior to the shedding of the old cuticle (now called the exuviae), a layer of wax is deposited on the surface of the new epicuticle. The molecular mechanisms by which a hydrophobic substance such as wax may pass through the hydrophilic procuticle are not known. The pore canals probably provide the exit passageway, but wax is secreted on some cuticles lacking these canals. Finally, a cement layer is deposited over the wax layer by the dermal glands.

Ecdysis

The insect is now ready to shed the exuviae by the process of ecdysis. This behavioral feat requires coordinated movements of the body in a definite sequence. The behavioral acts are programmed into the central nervous system and initiated by the eclosion and Mas-ETH hormones in the hemolymph (Chap. 4). By taking in air or water in the gut, the insect increases its body volume, and this, together with muscular contractions, splits the old cuticle along preformed lines of weakness. Certain areas of the cuticle, even within the boundaries of sclerites, have the exocuticle greatly reduced. Once the endocuticle is dissolved by the molting fluid, only the epicuticle remains along these lines. Such ecdysial lines (Fig. 2.1) provide the initial lines for breakage of the old cuticle. A lubricant may be demonstrated on the inner surface of the exuviae and probably aids in the escape of the emerging insect. It is mucilaginous in nature and may be derived from the molting fluid.

Once freed of the exuviae, the cuticle is expanded to the limits predetermined by the epicuticle. Blood pressure is increased by swallowing air or water and by contractions of various parts of the body. Certain regions of the cuticle are plastic and expand under this pressure. By this means a dramatic change in body size and shape takes place over a relatively brief time. This is especially marked at the last molt, when the insect's wings are inflated to full size. After expansion the cuticle rapidly hardens and darkens. Sclerotization is under control of the hormone bursicon (Chap. 4). After the molting cycle is over, additional layers are added to the endocuticle by the epidermis, and wax may continue to be produced.

THE PRODUCTION OF COLOR

Beautiful colors, or even the general darkening of the cuticle, are created by processes that may be quite apart from the events leading to hardening. In some species of wild bees, for example, certain areas of the cuticle may be perfectly transparent and reveal the pale colors of the tissues beneath. Yet they are as rigid as the adjacent black cuticle. Except for special mutant forms or some subterranean species, all insects exhibit colors. The functions of these colors in survival are treated in Chapter 8 under protective coloration. Here color will be discussed from three points of view: (1) the anatomical location of the color-producing structure, (2) the manner in which the color is generated, and (3) changes in color during the life of the insect. With regard to the first topic, most colors are produced in the cuticle. When the cuticle is translucent or transparent, the underlying epidermal cells, hemolymph, or other internal tissues may produce color. Both the cuticle and the deeper

tissues may contribute to the production of a single color pattern.

White light is composed of various wavelengths of light, commonly identified within the range of human vision as red (0.72 μm wavelength), orange, yellow, green, blue, indigo, and violet (0.41 μm). When an object or insect absorbs all but the red portion of the spectrum, we perceive a red color, and so on. The selective reflection of certain colors may be due to the physical structure of the object, producing structural colors; or it may be the result of selective absorption by certain chemical compounds, producing pigmentary colors; or a color may be produced by a combination of both structure and pigments together. When the cuticle contains the pigment or physical structure for color, the colors often remain lifelike in museum specimens. On the other hand, when the color resides in the cells or hemolymph beneath the cuticle, one may expect fading and discoloration as the internal organs decompose after death.

Structural Colors

The physical shape and thickness of the cuticle can produce colors perceived by observers from light reflected from an insect. Such structural colors are known to be produced in one of three ways: by the interference, the diffraction, or the scattering of light waves. Interference colors are familiar as the iridescence of a thin film of oil on water or a soap bubble. Both the upper and lower surfaces of the film reflect light. The thickness of the film determines the distance separating these reflecting surfaces. Since light reflected from the lower surface must travel a minute distance farther to reach the eye of the observer than the light from the upper surface, the former is said to be retarded. When waves are retarded by an odd number of half-wavelengths, they will be out of phase with the waves of the same length reflected from the upper surface. In this case the two reflections cancel each other and only that part of the spectrum with the waves in phase is visible. Colors vary because of the varying thickness of the film. Interference colors are also produced by crystallike lattices of spheres or regular points. In insects these are represented by spheres of air embedded in the cuticle matrix. Changes in the angle of observation will also change the thickness traversed by the retarded waves and likewise produce iridescence.

Interference colors are widespread among the insects. Least conspicuous, but extremely common, is the play of iridescent colors on the transparent wings of insects. The membranes of the wings have the same effect as thin films. More spectacular are the butterflies and moths whose individual scales are intricately modified to produce striking colors. The blue of the *Morpho* (Nymphalidae, Lepidoptera) arises from a series of vertical vanes on each scale. Each vane has a series of delicate ribs spaced at the correct distance for canceling or transmitting without reflection all wavelengths but the blue that is reflected. A series of superimposed laminae on each scale of the diurnal moth *Urania* (Uraniidae, Geometroidea, Lepidoptera) has a similar effect. Lamellate scales or lamellae within the cuticle itself may produce interference phenomena in certain metallic beetles like the tiger beetles. When underlain by dark pigments, the interference colors are especially sharp, since stray reflections from internal organs do not detract from the purity of the reflected colors.

The helicoidal arrangement of layers of chitin fibrils in insect cuticle produces iridescent metallic coloration in some beetles and flies. As successive layers of chitin fibrils are laid down at a regularly changing angle from the previous layer, an essentially helicoidal pattern of the fibers results. At particular spacings and arrangements of the fibrils, the fibril layers act much like different thin film thicknesses that produce interference colors. In addition, helicoid layers of chitin fibrils can polarize light. While not readily detected by human vision, polarized light may transmit significant information to other insects, many of which perceive light polarization.

Diffraction of light reflected from arrays of minute grooves on the cuticle surface produces colors that vary with the angle from which they are viewed. Light impacting an edge of a groove at an angle will bend or be reflected at an angle that depends on the wavelength of the light. Shorter wavelengths will be reflected at a more acute angle than longer wavelengths, so the perceived colors change as the angular orientation of the insect with respect to the viewer changes. Color produced by diffraction is not common in insects. The beetle *Serica sericea* (Scarabaeidae, Coleoptera) produces color by a parallel series of minute grooves, or striae, that are evenly spaced at about 0.8 μm apart. These produce an iridescence on the elytra when viewed in direct sunlight. An artificial impression of the close-spaced striae, made of plastic and peeled from the elytron, will exhibit the same iridescence.

The last type of structural color, that produced by scattering, is not iridescent and does not change with the angle of observation. Minute particles, less than 0.6 nm in diameter, reflect more of the short waves of light than the long waves and produce a blue color. The sky is blue, for example, because dust particles and atmospheric gases reflect more blue and violet than other colors. Without the impurities in the atmosphere to reflect these colors the sky appears black, as observed in near space by spacecraft. Such a color is called Tyndall blue after an early student of sky color. Among the dragonflies, granules in the epidermal cells produce a blue color. Granules in cells in combination with yellow pigments will give green. Waxy secretions on the surface of the cuticle in other species of dragonflies also give the Tyndall effect. Larger granules reflect all wavelengths of light, producing a structural white. White human hair is white because of minute air bubbles in the hair that scatter light. Depending on the arrangement of the reflecting surfaces, a dull, pearly, or silvery white can be produced in insects.

Pigmentary Colors

The second major category, pigmentary colors, includes a much wider range of colors and a diverse assortment of chemical compounds that serve as pigments. These colors are not metallic or iridescent but otherwise vary enormously. They are often excretory products that are no longer involved in biochemical pathways but that now serve an important function in the ecology of the insect. White and yellow colors of butterflies like the pierids are of this type; the pterine pigments responsible for the colors may be derived from uric acid or guanosine triphosphate. Ommochrome pigments (reddish to yellowish) are synthesized from the amino acid tryptophan, a common breakdown product produced during developmental changes such as metamorphosis. Pigments may also be absorbed from the food of the insect. Carotenoid pigments may be derived from carotenes in the diet. Flavonoid pigments that produce yellowish coloration are derived from plants eaten by insects such as some butterflies.

One of the most common of animal pigments is melanin. The colors produced vary from black and brown to light brown and yellow. The synthesis of each color pigment is incompletely known. Nearly all colors in this range were formerly attributed to melanin. Some are now known to be the result of other chemical compounds, such as sclerotin or ommochromes. Melanogenesis of the black pigment involves the oxidation of tyrosine by the enzymatic action of tyrosinase, plus several additional steps, one of which requires air by way of the tracheal system. Once formed, melanin is insoluble in ordinary solvents, but it can be bleached if a transparent exoskeleton is desired for study.

Tyrosine and tyrosinase are present in insect blood but somehow kept apart. When an insect is injured, the blood that is exposed to air often blackens. Such melanization in experimental insects is usually fatal and may be prevented by excluding oxygen from the surgery or by blocking the action of tyrosinase. The hereditary lack of tyrosinase produces the familiar albino mutant.

Melanogenesis may be partly influenced by the environment. For example, insects reared in the laboratory at lower temperatures often show increased melanization. Day length is another environmental cue that can trigger development of different colors or degrees of melanization. In nature, an insect with two or more generations per year often displays distinctive colors associated with the seasons during which pigments were developed. For example, some leafhoppers have a green summer form and a brown form that overwinters, aiding in camouflage. The presence of darker populations of an insect in a geographic region of lower temperatures cannot be simply explained as entirely an environmental effect. Such individuals may be genetically predisposed to greater melanin deposition, and the darker colors may have survival advantages peculiar to the geographic region, irrespective of the lower temperatures. High humidity and darker colors have also been associated in the study of geographic variation of insects, but oddly there is little evidence for a direct physiological influence. Some environmental factors have indirect effects through the nervous or endocrine systems: certain wavelengths of light received by the eyes of an insect in the pupal stage may promote melanin deposition or, as in migratory-phase locusts, the darker gregarious phase is associated with crowding.

Yellow, orange, and red colors are often produced by carotenoid pigments derived from α- and, more commonly, β-carotene, which occurs in plants. Insects are believed to be unable to synthesize these pigments and therefore must depend on plants for their supply. Some predators acquire their pigment via plant-feeding victims. The pigments are absorbed and deposited in the epidermis or fat cells or in secretions such as silk or wax.

Green colors that closely resemble the chlorophyll of leaves are produced by many kinds of insects. Thus far no experiments have shown that chlorophyll is absorbed and used directly for the color. In several different insects, however, the green is produced by a combination of a yellow carotenoid and a blue or blue-green pigment belonging to a group known as a bilin or bile pigment. The origin of the latter is not clear, but it is probably synthesized by the insect. Curiously, a diet of fresh green leaves seems to be important for the synthesis, but it is thought that some component other than chlorophyll is the needed constituent.

The color patterns of insects are usually fixed for the life of the instar. An exception is the walking stick, *Carausius morosus* (Phasmatodea), in which the color changes daily under neuroendocrine control: dark at night and pale in day. This physiological color change is the result of dark granules in the epidermis. When the granules are clustered together at the base of the cell, the insect is pale; when the granules migrate apically to form a sheet, the insect is dark. Darkening can also be stimulated by low temperature or high humidity.

Finally, pigments produced by some insects are of commercial value. The red pigment carmine or cochineal, used in lipstick and rouge, is extracted from female *Dactylopius coccus* (Coccoidea, Hemiptera), a scale insect that feeds on cactus. Kermes, a crimson color and also from a scale insect, *Kermes* sp. (Coccoidea), was used as a dye in ancient Greece and Rome.

BOX 3.1   Formation of Eyespots in Butterflies

As any child has observed, some species of butterflies and moths have concentric-ring spots on their wings that appear to resemble eyes. Some of the larger ones, such as on the hind wings of saturniid moths, appear to join other color patterns on the wings and body to form an overall pattern that mimics the face of an owl and thus might frighten a small bird or mammal predator. Very much smaller ones resemble the eyes of jumping spiders and might serve a similar role in startling potential small insect predators.

How are these patterns produced on the wings of lepidopterans? In Chapter 2 we briefly examined how gene expression alters the formation of segments and appendages during development. But how is a circular pattern formed on a wing that must unfold after it is formed?

In the early stages of wing formation, small groups of epithelial cells called foci become evident at the center of where the eyespots will eventually form. These can be detected by visualizing the locations on the wing where the homeodomain transcription factor gene *Distal-less* (*Dll*) is being expressed (Box Fig. 3.1). *Dll* specifies where on the wing the spots will form (please note that this gene has other roles in limb formation in arthopods as well).

"Eyespots" probably evolved from original spots of uniform color caused by the induction of a particular pigment around the focus after expression of the *Dll* and *engrailed* (*En*) genes (among others). The evolution of signaling ability from the cells in the focus then allowed the recruitment of new target genes over different distances from the focus. This allows formation of the concentric ring pattern of different pigment colors, each influenced by different developmental gene regulation depending on distance from the center of the focus (Box Fig. 3.2).

A great variety of eyespot patterns and colors is now known, only some of which have been studied in detail.

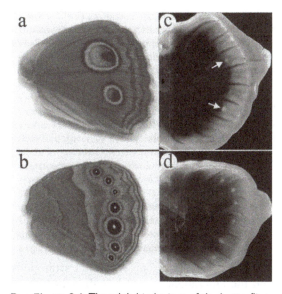

Box Figure 3.1 The adult hind wings of the butterflies *Junonia coenia* (a) and *Bicyclus anynana* (b) and corresponding developing wings with *Distal-less (Dll)* expression visualized (c and d). Arrows in (c) indicate the foci where *Dll* is expressed corresponding to the two eyespots in *Junonia*; the seven foci in *Bicyclus* are also clearly visible in (d).
Source: From Brakefield et al., 1996; Reprinted by permission from Macmillan Publishers Ltd: NATURE.

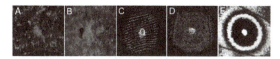

Box Figure 3.2 Genes expressed in the eyespot region during pupal development in the butterfly *Bicyclus anynana*. a, *pSmad* (a signaling gene); b, *Wingless*; c, *Engrailed* (center bright dot) and *distal-less* (darker band surrounding dot); d, *Spalt* (darker area around center) and *Engrailed* (brighter halo around outside); e, the resulting eyespot.
Source: From Monteiro, 2008, reprinted by permission from Monteiro, Antonia: Alternative models for the evolution of eyespots and of serial homology on lepidopteran wings. Bioessays 30:358:366. John Wiley and Sons, 2008.

CHAPTER 4

Continuity of the Generations

Development and Reproduction

This chapter traces the major events in the life history of insects, beginning with the egg and ending with a sexually mature adult (Fig. 4.1).

Then we return to the reproduction system, describe the structure and function of the male and female organs, discuss different types of reproduction, and conclude with examples of life histories of insects with polymorphic development. We will outline the general features of insect development, but insects vary from group to group and even among closely related species in having exceptions to general trends of embryo development (embryogenesis) and larval or pupal development. Few things in biology are more fascinating than watching the often rapid developmental changes inside an egg or from embryo to larva through different stages to the adult form. The rapid and multifaceted development strategies of insects may have provided significant competitive advantages over other invertebrates in leading to the greater diversity and ecological dominance of insects.

EGGS AND EMBRYONIC DEVELOPMENT

The majority of insects are diploid animals that reproduce sexually and are oviparous, that is, they lay eggs. Haploid eggs and sperm are produced by meiosis. The eggs of the class Insecta and the order Diplura are rich in yolk, or centrolecithal. The early cleavage, or division, of the nuclei in the fertilized egg does not involve division of the large yolk mass into separate cells. This proliferation of nuclei without cell division is known as syncytial cleavage or karyokinetic (also called meroblastic) cleavage. The developing early insect embryo may contain thousands of syncytial nuclei before forming the first discrete cells that contain a single nucleus. The eggs of noninsect hexapods such as Collembola and myriapods have less yolk (microlecithal), and the early cleavage is cytokenetic (or holoblastic or total), that is, cells within the developing egg divide completely after each mitotic division. Syncytial cleavage occurs in Diplura and occurs in some chelicerate arthropods and the Onychophora, which is thought to be a sister group to Arthropoda. The speed of syncytial cleavage is an advantage in being able to complete embryogenesis quickly.

Egg Structure

The eggs of insects are usually elongate and ovoid, with rounded ends (Fig. 4.2a), but other shapes also occur. Egg sizes range from 0.2 to 20 mm in length. The eggs of exopterygotes tend to be slightly larger than those of endopterygotes. From the outset eggs are bilaterally symmetrical, with a dorsal and a

Figure 4.1 The male (above) and female (below) of mantids, *Stagmomantis californica* (Mantidae), during copulation. Direct transfer and storage of sperm are important adaptations that permit insects to reproduce in terrestrial environments, including deserts.

ventral surface and differentiated anterior and posterior ends.

The follicle cells of the ovary secrete an outer shell, or chorion, which provides protection from physical injury and water loss. Eggs have passages for the sperm to enter, for gas exchange, and, in some, for water absorption. At the anterior end is the micropyle of one or more holes through the chorion that admit sperm. Aeropyles are pores of one to a few micrometers that permit the entry of oxygen and the exit of carbon dioxide while minimizing water loss. In eggs that are laid in water or moist places, such as decomposing organic matter or in areas that are temporarily submerged, the chorion is often modified into a plastron (Chap. 5). This is a spongy meshwork that retains a film of gas and resists wetting. The gas film operates as a physical gill, permitting oxygen dissolved in water to pass into the gas and then into the embryo (Fig. 5.5a,b). Special areas called hydropyles are sometimes present on terrestrial eggs to absorb water.

Beneath the chorion is the vitelline membrane, which is secreted by the oocyte, or egg cell, in most insects. Most of the substance of the egg is yolk, consisting of lipid, protein, and carbohydrate. A yolk-free island of cytoplasm surrounds the female nucleus, which is usually located peripherally in an anterodorsal position. The mother insect provides her eggs with mRNA, messenger proteins, and mitochondria (with their own DNA), and may also facilitate the introduction of symbiotic microorganisms. A thin yolk-free periplasm often surrounds the egg beneath the vitelline membrane.

Fertilization

Hexapod females store sperm after mating and release sperm as needed to fertilize the eggs as they pass out the genital tract. This method of fertilization is highly efficient compared to the sperm wastage of vertebrates and other animals and allows fertilization to continue, as in honey bees and some ants, even years after mating. Female hymenoptera (bees, ants, and wasps) can control their fertilization of eggs as they pass by the sperm duct from the sperm storage organ (spermatheca); unfertilized eggs become males; fertilized eggs become females. Eggs laid by the female are usually in the metaphase of the first maturation division of meiosis. Females release sperm near the micropyles as the egg passes through the female oviduct so that one to several sperm enter each egg (Fig. 4.2b).

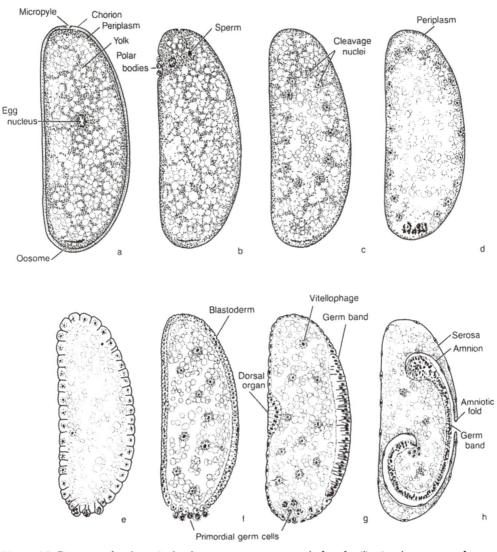

Figure 4.2 Diagrams of embryonic development: a, egg structure before fertilization; b, entrance of sperm; c, cleavage of zygote; d, migration of cleavage nuclei; e, beginning of blastoderm formation; f, blastoderm complete; g, formation of germ band; h, elongation of germ band.
Source: Redrawn from Johannsen and Butt, 1941.

The penetration of more than one sperm into an egg, known as polyspermy, is unusual among animals but common among insects. The fruit fly *Drosophila* (Drosophilidae, Diptera) is an exception; after a sperm enters the egg, additional sperm not do enter.

After penetration of the egg by the sperm, the egg nucleus completes meiosis, resulting in two or three polar bodies and the haploid female pronucleus. The latter migrates toward the interior of the egg, and the polar bodies remain grouped at the surface of the oocyte. The female pronucleus unites with one sperm pronucleus to produce the diploid zygote. The other sperm degenerate.

Formation of the Blastoderm

The early syncytial cleavages, or mitotic divisions, of the zygote result in increasingly numerous nuclei, each surrounded by an island of cytoplasm but without a cell membrane (Fig. 4.2c). The cleavage nuclei and their associated cytoplasm, known as energids, migrate through the yolk to finally cover the yolk surface (Fig. 4.2d) as a syncytial blastoderm. In *Drosophila* at this stage, the nuclei are totipotent (each capable of developing into a whole organism). Development up to this point has been under the control of gene products and paragenetic elements such as microbial symbionts that originated in the female parent, that is, were maternally encoded and stored in the egg. The basic spatial organization of the embryo (anterior-posterior and dorsal-ventral axes), DNA replication, nuclear assembly, and nuclear migration up to this point have all been supported by maternal messenger RNA (mRNA) and proteins that form a gradient that defines the anterior-posterior orientation (polarity) of the egg.

After cell membranes surround each nucleus, the resulting cells form the cellular blastoderm, one cell thick (Fig. 4.2e,f). In *Drosophila*, once the cells are formed, zygotic genes begin transcription and dominate control of development. Depending on their position in the embryo, embryonic cells now begin to express different sets of genes to develop into different tissues. The effects of maternal products can still be detected, for example, in the development of the nervous system. Some cells remain in the yolk or migrate back into it as vitellophages that digest yolk for the embryo.

Formation of the Germ Band and Germ Layers

Along the midventral line, the blastoderm thickens into a single layer of columnar cells called the germ anlage, which is destined to become the embryo; the remainder of the blastoderm cells form a thin, extraembryonic membrane (serosa) enclosing the yolk (Fig. 4.2g). A second layer of cells originating from cells next to the anlage forms another protective sheet, the amnion. Also at this time, a cluster of cells migrates through a special region at the posterior pole of the egg, called the oosome, that becomes the sex or primordial germ cells that will later form eggs and sperm in the adult insects.

The serosa and amnion cover the yolk and fold over to enclose the developing germ band (Fig. 4.2h). The germ anlage next folds inward along a longitudinal groove, corresponding roughly to gastrulation in other animals, and becomes the germ band. The outer layer of this band becomes the ectoderm layer, and the infolded cell layer becomes the inner mesoderm layer.

At this point it is useful to distinguish three major categories of insect embryos based on the relative length and development of the germ band. Short germ band insects have a band a third or less the length of the egg. At first the band consists of only the protocephalon (preantennal and antennal segments) plus a small region that will gradually elongate by cell proliferation into the rest of the body. Long germ band insects have a band that extends over the length of the egg. All the segments of the body appear almost at once as the germ band develops by subdivision of the anlage. Short germ band insects tend to develop more slowly and are usually Exopterygota with panoistic ovaries (see "Types of Ovary," below). Such ovaries lack nurse cells so that the oocytes most likely must synthesize essential components. Long germ band insects tend to develop more quickly and are usually Endopterygota with meroistic ovaries. In this case nurse cells supply the oocyte with large amounts of maternal products that may allow rapid development. *Drosophila* is a long germ band insect. Most insects, however, belong to the third or intermediate germ band category in which the band is from a third to half the length of the egg. Otherwise, this category is difficult to characterize and includes a mixture of species in the Pterygota.

Segmentation

During gastrulation, transverse divisions produce segments. In the segmentation of *Drosophila* fruit flies, experimental manipulations of many developmental mutant genes have clarified the genetic control of segmentation. The general features of control of segmentation and the body plan appear to have much in common with patterning in vertebrates as well as invertebrates.

The primary segments of insects that have been identified by comparative morphology (see Figs. 2.2, 2.4) are not the visible "segments" of the embryonic epidermis of *Drosophila* flies. One would have expected that the primary segments could be traced back to the embryo. However, in the fly's embryo, the developmental unit defined by transverse grooves is called a parasegment and consists of the posterior part of one morphological segment and the anterior part of the following segment. The embryo initially has fourteen parasegments, corresponding approximately to three for the mouthparts, three for the thorax, and eight for the abdomen. For example, parasegment six will develop into the posterior part of the metathorax and the anterior part of the first abdominal segment.

During embryonic development, the ectoderm differentiates into the epidermis of the integument, tracheal system, endocrine and exocrine glands, sense organs, and the nervous system. Paired appendage rudiments become visible on the lateral epidermis of the anterior segments for those insects with gradual changes in larval (or nymphal) forms to develop into an adult. In *Drosophila*, clusters of cells that will later become eyes, wings, or legs invaginate below the ectoderm and separate to become imaginal discs. Many, if not most, other holometabolous insects produce imaginal discs as invaginations of the larval epidermis rather than that of the embryo. As the appendages lengthen, the ectoderm grows dorsally, finally meeting along the mid-dorsal line and closing the embryo with the yolk now inside the body. Each segment (parasegment of *Drosophila*) initially has a bilateral cavity in the mesoderm that corresponds to the coelom of other animals. The coelom later breaks down as the mesoderm differentiates into muscles, heart and dorsal diaphragm, gonads (including the primordial germ cells), fat body, and blood cells. Invaginations appear at the anterior and posterior ends of the band of ectoderm. These blind sacs elongate to become the foregut and hindgut, respectively. Near each invagination is a mass of cells that probably corresponds

to the endoderm of other animals. The cells will grow inward from each of the gut invaginations, forming a tubular sheet around the yolk. This tube becomes the midgut. The Malpighian tubules develop from the blind end of the invaginated hindgut. In this position it would appear that the tubules arise from ectoderm, but the true embryonic origin has not been determined. Near the time of hatching from the egg, the blind ends of the foregut and hindgut break down to fuse with the tubular midgut to complete the alimentary canal. Blood cells (hemocytes) originate early in embryogenesis in the cephalic (head) mesoderm. Later an organ, the lymph gland, forms along the dorsal vessel, or insect heart.

POSTEMBRYONIC DEVELOPMENT

Embryonic development ends and postembryonic development begins when the insect hatches from the egg shell. All hexapods except Protura emerge from the egg with a body that has the adult number of segments (metameres). This is called epimorphic development, in contrast to the anamorphic development of Protura, in which metameres are added after hatching from the egg. Although Collembola are epimorphic, they have a reduced number of body segments in comparison with other hexapods. As explained in Chapter 3, the integument undergoes repeated molting cycles during the life of an insect. The number of molts may be as few as four or five or may exceed thirty. Only at times of molting can the exoskeleton be increased in size or be altered structurally. Other organ systems are less clearly synchronized with molting, but changes in their size and organization are under the same hormonal controls. Therefore, the postembryonic development of an insect is conveniently marked by the ecdysis at the end of each molt. The form of the insect between ecdyses is called an instar and is designated by numbers. For example, the first instar is the insect after hatching from the egg and before the first ecdysis.

Arthropods, amphibians, and certain other animals may have a distinctive, actively feeding, sexually immature stage that is generally called a larva. The change in appearance from the larva to the sexually mature adult, or imago, is broadly termed metamorphosis. When the insect reaches sexual maturity, molting ceases in the Pterygota but continues in certain of the Apterygota, Parainsecta, and Entognatha.

Some developmental terminology may cause confusion. The term eclosion has been used for the act of hatching from the egg, or the emergence of the adult from the pupal cuticle at ecdysis, or the emergence of any instar from the old cuticle at ecdysis. Among entomologists the designation "immature insects" or "immatures" refers to the larval instars of insects, not to sexually immature adult insects. Further, some entomologists used the word "larva" for all immature instars that are active, while others restrict the term to the immature instars of holometabolous insects. Nymphs are larvae of terrestrial, hemimetabolous exopterygotes. Finally, the distinctive aquatic larvae of Odonata, Ephemeroptera, and Plecoptera are often called naiads.

Types of Metamorphosis

Three general categories of metamorphic development describe the degree to which the larva differs from the adult. In the most extreme metamorphosis, programmed cell death (apoptosis) destroys the tissues of virtually the entire larva, and imaginal tissues (imaginal discs or histoblasts) derived from the larval epidermis grow and differentiate to create new structures.

Ametabolous development literally means "without change." The life history typically has the egg, several larval instars, and several imaginal instars (Fig. 4.3a–d). The living hexapods with this type of growth are all primitively apterous: Collembola, Diplura, Thysanura, and Archeognatha. The larvae change little except in size and proportions as they grow from instar to instar and mature sexually. Members of these orders continue to molt an indeterminate number of times even after reaching adulthood.

Hemimetabolous, or incomplete, metamorphosis involves a partial change in appearance from the larva to imago. The life history typically has the egg, several larval instars, and the final winged and sexually mature imago (Fig. 4.3e–i). Hemimetabolous metamorphosis is characteristic of Paleoptera and nearly all Exopterygota (including female Coccidae).

In hemimetabolous insects, the most conspicuous change in the imago is the development of functional wings and external genitalia. The larvae usually possess well-developed exoskeletons with legs, mouthparts, antennae, compound eyes, and, except in Hemipteroidea, ocelli and cerci. These structures are carried over without much change into the imago. The transition in form is gradual up to the last molt; each successive larval instar increasingly resembles the form of the imago. Wings develop as external pads on the pterothoracic nota and then expand enormously at the last molt (Fig. 4.3e–i).

Most Exopterygota have larvae and adults that live in the same places and eat the same food. Except for the wings and genitalia, the alterations at maturity are usually slight. The larvae of Odonata (Fig. 19.6), Ephemeroptera (Fig. 18.4), and Plecoptera (Fig. 25.1b), however, are aquatic, and the adults are aerial. Consequently the change is marked, involving the loss of gills and modification of the mouthparts as well as the acquisition of wings and new body proportions.

All living pterygotes differ from the living ametabolous orders in that once sexual maturity is reached, molting ceases. Furthermore, functional wings are restricted to the imago of all living pterygotes except the Ephemeroptera. In this order the first winged instar, or subimago, is sexually immature except in a few species, where it is sexually mature. After a brief aerial existence, the subimago molts and the second winged instar is the sexually active imago.

Holometabolous, or complete, metamorphosis is characterized by larvae that undergo histological reorganization before becoming sexually mature adults. The life history typically has an egg, several larval instars, one pupal instar, and the final winged and sexually mature imago (Fig. 4.4). The pupa is a usually immobile, nonfeeding instar during which metamorphosis transforms the larva into the adult form. In contrast to the hemimetabolous exopterygotes, the holometabolous endopterygotes delay the development of wings until the last one to three larval instars (Fig. 33.2). In most holometabolous

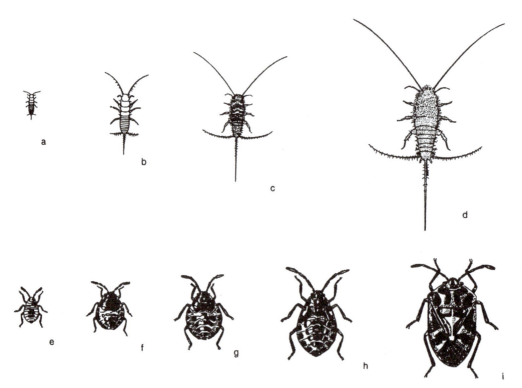

Figure 4.3 Types of life histories: a–d, ametabolous development, illustrated by the firebrat, *Lepismodes inquilinus* (Lepismatidae) (a–c, nymphs; d, imago); e–i, hemimetabolous development, illustrated by the harlequin bug, *Murgantia histrionica* (Pentatomidae) (e–h, nymphs; i, imago).
Source: a–d, redrawn from Metcalf et al., 1962. Reproduced with permission of The McGraw-Hill Companies. e–i, redrawn from U.S.D.A., Farmer's Bul. 1061.

insects, the wings do not become visible until the pupa. All Endopterygota and certain exceptional Exopterygota (Thysanoptera, Fig. 33.2; Aleyrodidae, Fig. 32.14c–e; and male Coccidae, Fig. 32.16b) have holometabolous development.

The last larval instar of a holometabolous insect usually seeks a pupation site and often secretes a silken cocoon or constructs a special cell of earth or wood particles. The functions of the cocoon or cell are probably to reduce loss of moisture, provide protection from ice crystals in winter, and avoid predators. Certain specialized endopterygotes do not construct pupal shelters. For example, cyclorrhaphous Diptera pupate within the sclerotized last larval cuticle, called a puparium.

The molt into the pupal instar is called pupation, and during this time, after apolysis and before ecdysis, the insect is called a pharate pupa. Metamorphosis begins and the larval tissues are partly or entirely destroyed by histolysis and replaced by growth of imaginal tissues. The latter are derived from clusters of cells, the imaginal discs (or anlagen) that are distributed in the larval body near where the adult organs develop. The imaginal discs form during embryogenesis but remain inactive during the larval instars. In larval *Drosophila*, the cells of the imaginal discs remain diploid, whereas larval cells become polyploid and generally do not divide although the larva grows larger. The fly larva has pairs of imaginal discs that will develop into adult mouthparts, antennae and compound eyes, thorax, legs, wings, and halteres. A single disc develops into the genitalia. Other clusters of imaginal cells are scattered in the brain, along the gut, salivary glands,

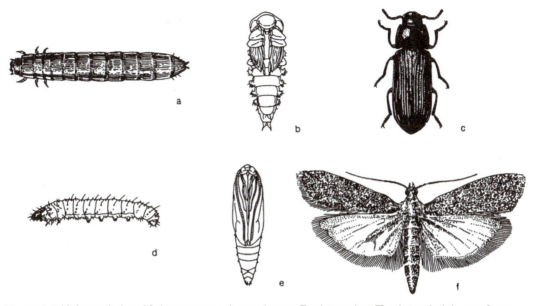

Figure 4.4 Holometabolous life histories: a–c, the mealworm, *Tenebrio molitor* (Tenebrionidae) (a, eruciform larva; b, adecticous exarate pupa; c, imago); d–f, the Mediterranean flour moth, *Ephestia kuhniella* (Pyralidae) (d, eruciform larva; e, obtect pupa; f, imago).
Source: Redrawn from Calif. Agr. Exp. Sta. Bul. 676.

and in the abdominal segments. At metamorphosis the discs and gonads are activated and begin to divide and differentiate into adult organs.

After ecdysis, the pupa emerges, and adult structures such as long antennae, compound eyes, ocelli, mouthparts, legs, and miniature wings replace larval features. Adults lack pregenital abdominal appendages of the larva, if they were present. The pupa has a special pupal cuticle that differs from that of the larva and the adult. With apolysis and the onset of the final molt, the insect is called a pharate adult. The wings grow enormously in size within the wing sacs of the detached pupal cuticle. Internally the alimentary canal and nervous system are reorganized, the growth of the endoskeleton apophyses is completed, and the musculature assumes its final distribution and attachment. Setae, wing scales, and pigmentation characteristic of the imago are completed. Following the final ecdysis, the body usually contracts in volume as the wings are inflated by hemolymph and the cuticle is sclerotized. At this time a fluid substance,

called the meconium, is sometimes excreted from the alimentary canal.

HORMONAL CONTROL OF MOLTING, ECDYSIS, METAMORPHOSIS, AND REPRODUCTION

During its development the immature insect converts energy from food into often rapid growth to a mature (adult) stage that specializes in reproduction and dispersal. Most insect species have a complete change or metamorphosis of the larval stage into a very different adult (imago) stage. Circumstances such as the season or food availability can affect when and how the next developmental step proceeds. This plasticity in development allows an insect to respond to environmental signals processed by the central nervous system to initiate or influence subsequent developmental changes. Hormones play a primary role in regulating postembryonic development, sexual maturation, and a variety of aspects of behavior and physiology. Hormones are chemicals secreted

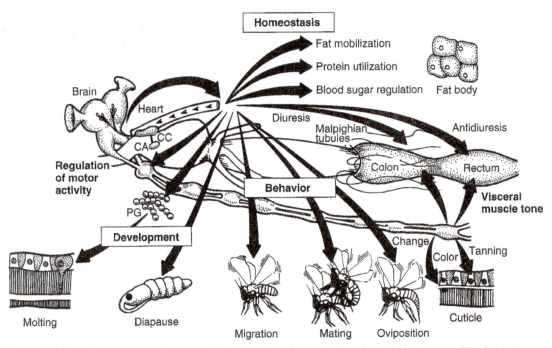

Figure 4.5 Insect hormones mediate many developmental, physiological, and metabolic activities. The following are major sources of hormones: in the brain are the corpora cardiaca (CC) and corpora allata (CA); the prothoracic gland (PG) is in the thorax. Some hormones are produced by secretion from specialized nerves.

by one part of the body (the source) that have a profound effect on another part of the body (the target). Endocrine cells or organs (Fig. 4.5) produce hormones in response to environmental or developmental signals and store or release them in specialized neurohemal structures. Cells in the brain and other parts of the nervous system secrete neuropeptide hormones. The circulation of body fluids, or hemolymph, distributes hormones from their source tissues. Hormones bind to transmembrane or surface proteins on the outer cell surface to trigger a variety of cellular responses by repressing or inducing gene expression. Enzymes or general peptidases or proteases in the hemolymph break down hormones to prevent their persistence or accumulation.

Biochemical, immunocytochemical, and molecular techniques constantly provide new findings in insect endocrinology. Isolation of hormones and identification of their amino acid sequences allows the estimation of their structural gene sequences.

With this information, RNA transcription of hormone genes can be quantified from various source tissues or in hemolymph to follow the timing of hormone production and transport. Earlier models of how hormones act involved single hormones from single glands and with single targets. We now recognize that groups of related hormones, some from more than one source, may have different targets depending on the stage of the insect's life history. Moreover, a single hormone can affect multiple target tissues or have different functions in different developmental stages. Hormones specifically involved in development are listed below (marked in boldface), together with mention of other hormones from the same organs (Fig. 6.2). Abbreviations are provided for organs and hormones commonly mentioned in the scientific literature.

1. Several paired clusters of neurosecretory cells (NSC) in the brain produce at least two

groups of neuropeptides. (1) Prothoracicotropic hormones (PTTH) of different molecular weights are produced in different clusters of NSC. PTTH was known first simply as the "brain hormone" that stimulated the prothoracic glands to synthesize ecdysteroids. PTTH probably has other functions in the brain and even in embryonic development; in adult females it affects egg maturation. In Lepidoptera, nerves pass from the NSC to the retrocerebral complex consisting of the corpora cardiaca (singular corpus cardiacum) and corpora allata (corpus allatum). Secretory cells in the corpora allata release PTTH into the hemolymph from neurohemal endings. In some other insects, the corpora cardiaca are reported to release the hormone. (2) Eclosion hormones (EH) are neuropeptides with different molecular forms produced by NSC in the ventral median region of the brain and the proctodeal region of the ventral nervous system. EH is released just before ecdysis by neurohemal sites of the proctodeal nerves. At this time ecdysteroids are at a low level. EH is one in a series of hormones that ultimately trigger the central nervous system to begin the programmed sequence of muscular movements that comprise ecdysis behavior.

2. Neurosecretory cells in the brain and ventral ganglia secrete the neuropeptide bursicon, which initiates sclerotization following ecdysis. NSC in the brain or ventral ganglia also secrete peptide diuretic and antidiuretic hormones that control water balance.

3. Paired corpora cardiaca (CC) are located behind the brain and are connected to NSC in the brain. The CC are closely associated with the corpora allata in the retrocerebral complex. As neurohemal organs, the CC store and release PTTH to the hemolymph in some insects. Hormones are also produced that regulate hemolymph sugar (trehalose) concentration, lipid mobilization in blood (adipokinetic hormone), heartbeat, and Malpighian tubule movement.

4. Paired prothoracic glands (PG), or ecdysial glands, are located in the thorax or head. Upon stimulation by PTTH in the hemolymph, the PG secrete the hormone ecdysone (E), which is synthesized from cholesterol or a related steroid obtained in the diet. In moths, E is converted to the more active hormone 20-hydroxyecdysone (ecdysterone or 20-E) by the fat body, gut, and Malpighian tubules. Collectively, these hormones, plus other forms that have been identified, are called ecdysteroids. The ecdysteroids initiate molting in the epidermis and stimulate differentiation of imaginal tissues.

5. Paired corpora allata (CA) are located behind the brain in the retrocerebral complex. CA are innervated by nerves from the brain that pass through the CC and also by nerves from the subesophageal ganglion. The CA secrete sesquiterpenoid compounds called juvenile hormones (JH). Several structures are known and designated JH 0 to JH III, plus others. Hemiptera have JH I, Diptera have a *bis*-oxide of JH III, and Lepidoptera have up to five structural forms in a single species; the other orders appear to have only JH III. Rather than operating directly on a target protein, JH interacts with ecdysones to preserve larval characteristics during the molting cycle and inhibit metamorphosis. Adults secrete JH from reproductive organs to stimulate vitellogenesis in the female and accessory glands in the male. JH production also acts to stimulate changes in the appearance and behavior of immatures or adults provoked by changes in season, food availability, or crowding.

6. Epitracheal glands are associated with each spiracle of *Manduca sexta*. The gland is a cluster of cells, including a large cell that produces a neuropeptide hormone ("Manduca sexta ecdysis-triggering hormone," or Mas-ETH). When released in the hemolymph, Mas-ETH directly acts on the ventral nervous system to initiate the programmed ecdysial behavior of coordinated contractions.

Hormonal Control of Molting

The endocrine organs are ultimately under the control of the central nervous system. Depending on the kind of insect involved, certain events serve to stimulate molting. External stimuli, such as a pattern of changing temperature or light, or internal cues, such as the stretching of the abdomen after feeding, are received by the brain, which then causes PTTH to be released. PTTH stimulates secretory activity in the prothoracic glands, and ecdysteroids increase in the body. This in turn stimulates the epidermis to begin a molting cycle, starting with apolysis. Ecdysteroids target genes in epidermal cells. The injection of ecdysteroids into the midge, *Chironomus* (Chironomidae, Diptera), quickly induces RNA transcription, causing chromosomes in the salivary glands to exhibit puffiness.

Hormonal Control of Ecdysis

The final step of the molting cycle, ecdysis, is accompanied by a stereotyped pattern of muscular movements programmed into the ventral nerve cord. The behavior is thought to be controlled by a sequence of hormonal interactions. In *Manduca sexta*, the eclosion hormone is released into the hemolymph when 20-hydroxyecdysone declines. In turn, the eclosion hormone stimulates the epitracheal glands to release the peptide Mas-ETH. The last hormone acts directly on the nervous system to initiate the ecdysial behavior. Sclerotization or tanning of the new cuticle is stimulated by release of bursicon from neurohemal organs.

Hormonal Control of Imaginal Differentiation

Molting and differentiation of imaginal tissues are stimulated by ecdysteroids. Juvenile hormone released from the corpora allata inhibits differentiation as long as the hormone is present in the hemolymph of the larva. The exact means of control of the corpora allata is not understood. Nervous connections or neurosecretions called allatotropins probably serve to regulate the secretion of JH. When molting is stimulated by ecdysteroids in a young larva, sufficient JH is normally present to cause the cells to retain larval characteristics. The larva increases in body size in the next instar without metamorphosis.

In hemimetabolous insects, JH disappears in the last larval instar, and at the next or final molt the insect differentiates into the adult. Holometabolous insects are similar, but the process of metamorphosis is more complex. In the last larval instar, JH drops to an undetectable level followed by secretion of ecdysteroid, which commits the insect to metamorphosis into the pupal instar.

Metamorphosis of Holometabolous Insects

The steps in the hormonal control of holometabolous metamorphosis have been determined in detail in *Manduca sexta* (Fig. 4.6). After ecdysis to the fifth larval instar, JH drops to undetectable levels. Two small pulses of ecdysteroids, called the "commitment pulses," are then released. This initiates metamorphosis and triggers stereotyped behavior. The larva stops feeding, wanders about, burrows in soil, and constructs a chamber in which to pupate. The commitment pulse is followed by a much larger pulse of ecdysteroid, called the "prepupal peak." The second release of ecdysteroids initiates apolysis and the molt. A pulse of JH is also released at the same time and presumably regulates the differentiation of imaginal discs. Metamorphosis in the pharate pupa is now underway while still hidden in the last larval cuticle. The epidermis secretes the distinctive pupal cuticle. Ecdysis occurs after the ecdysteroids and JHs have again declined. The dark brown pupa exhibits some of the external features of the adult, although the appendages are glued to the body (see discussion of obtect type of pupa in "Characteristics and Types of Endopterygote Pupa," below). Ecdysteroids again are secreted, now accumulating to high titers in the hemolymph as the differentiation of imaginal tissues continues because JH is absent. The final apolysis detaches the pupal cuticle, and the epidermis begins to secrete the adult cuticle. The pharate adult, concealed in the dark pupal cuticle, completes development as the ecdysteroids decrease to low levels. The final ecdysis, like the previous ecdyses, is controlled by the eclosion hormone followed by the Mas-ETH hormone. After burrowing out of the soil, the moth perches and begins to pump hemolymph into the wings. The hormone bursicon is now released from neurohemal organs in the abdomen, causing the cuticle of the wings to become plastic during their expansion and later to harden by sclerotization.

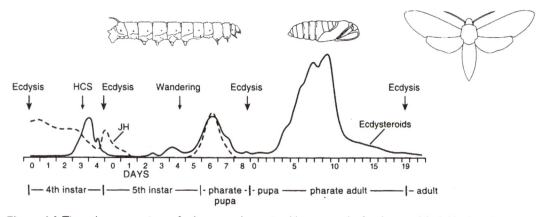

Figure 4.6 Titers (concentrations of substances determined by titration) of ecdysteroid (solid line) and juvenile hormone (broken line) during late larval life and metamorphosis of *Manduca sexta* (Sphingidae). Above the graph are a fifth-instar larva, pupa, and adult. Head capsule slippage (HCS) marks apolysis in the fourth-instar larva. After apolysis in the fifth-instar larva, the pharate pupa develops inside the larval cuticle. Likewise, after the next apolysis, the pharate adult (= developing adult in diagram) develops inside the pupal cuticle. The times of ecdysis are marked with arrows.
Source: Redrawn from Goldsmith and Wilkins, 1996. Reprinted with the permission of Cambridge University Press.

Hormones and Reproduction

Ecdysteroids in male pupae stimulate the development of the accessory glands and testes and the differentiation of the germ cells. For example, in certain male Lepidoptera spermatogenesis begins in the pupa. Here ecdysteroids stimulate the testes, control spermatogenesis, and inhibit the release of sperm.

In adult pterygote insects, the prothoracic glands degenerate, and, although ecdysteroids are produced elsewhere, the integumental epidermis no longer responds by molting. After regulating the growth and molting of the immature instars, JH and ecdysteroids assume new roles in the maturation and function of the adult reproductive organs. The corpora allata renew secretion of JH, and ovarian tissues secrete 20-hydroxyecdysone. These hormones now stimulate organs involved in reproduction, in other words, they become gonadotropic.

In female insects, both hormones have been associated with the synthesis of yolk proteins in the fat body and deposition of yolk in developing eggs. Both hormones may act in vitellogenesis in the same species of insect, or only one of the hormones may be present. Juvenile hormone, alone or in combination, is the most common and is also responsible for development of the ovarian follicles and preparation of the oocyte to receive yolk. Ecdysteroids are present in embryos.

In males, the production of fluids by the genital accessory glands is dependent on stimulation by JH. Silk moth males have an extraordinary amount of JH stored in the abdomen. This is a feature that was exploited to purify large amounts of JH for research on the chemical nature of JH, but the function remains uncertain. The spermatophores from male moths may provide carry-over amounts of JH to the developing embryo following fertilization.

In the ametabolous Thysanura, sexual maturation occurs gradually. Once adulthood is reached, the neurosecretory cells of the brain, ecdysial glands, and corpora allata remain active. Molting cycles, initiated by the brain, occur regularly. If females are mated within a few days after ecdysis, the corpora allata become active and yolk is deposited in the eggs. If the females are not mated, the eggs are resorbed at the onset of the next molting cycle. Thus molting and reproduction alternate in their demands on the body reserves.

CHARACTERISTICS
OF ENDOPTERYGOTE LARVAE

In addition to having concealed wing buds, endopterygote larvae differ in other respects from hemimetabolous larvae. For example, fly larvae (Diptera) are headless, legless maggots, totally unlike adult flies (Fig. 41.5). In general, larvae differ from imagoes in being soft-bodied and in having simplified sensory, feeding, and locomotory structures. The antennae are usually reduced or absent. Compound eyes and ocelli are usually absent; the most common light receptors are clusters of single-faceted stemmata (Fig. 15.1b). The mouthparts and legs are often reduced in structural complexity. Legs are absent in larval Diptera (Fig. 41.5c) and Hymenoptera (suborder Apocrita; Fig. 39.4b). Pregenital appendages are present on the abdomen as gills in some aquatic larvae (Fig. 34.2) or as fleshy prolegs in caterpillars of Lepidoptera (Fig. 43.5a,b) and Hymenoptera (sawflies; Fig. 39.1b).

Types of Endopterygote Larva

Some larval types are recognized. For example, campodeiform larvae have prognathous mouthparts, elongate flattened bodies, long legs, and usually some kind of caudal appendages. Larvae with these characteristics resemble the Campodeidae (Fig. 15.1b) of Diplura. They are usually active predators. Specializations of this basic form include carabiform beetle larvae that have shorter appendages (Carabidae) and scarabaeiform larvae of the scarab beetles (Scarabaeidae).

Eruciform larvae have hypognathous mouthparts, cylindrical bodies, short thoracic legs and abdominal prolegs, and reduced or no caudal appendages (Fig. 4.4a,d). The most familiar examples are caterpillars of Lepidoptera (Fig. 43.5a,b) and larvae of Mecoptera (Fig. 40.1c) and Hymenoptera (suborder Symphyta, Fig. 39.5a). Most eruciform larvae feed on plant materials.

Vermiform larvae are wormlike and cylindrical and lack locomotory appendages, using a wavelike sequence of extending and contracting body segments to move along a surface. Most larvae or maggots of Diptera (Fig. 41.5) are of this type, as are the larvae of fleas (Siphonaptera; Fig. 42.1d), certain wood-boring beetles (Fig. 37.11c), and advanced Hymenoptera (bees, wasps, and ants; Figs. 4.7, 39.5b). Other types of larvae are described under the various orders in Part Three.

Another classification system for larval body types primarily considers the appendages. Polypod larvae have unsegmented locomotory appendages, or prolegs, in addition to segmented thoracic legs.

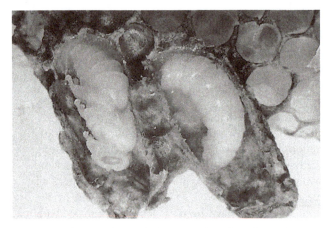

Figure 4.7 Queen cells at the bottom edge of comb in a honey bee hive (Apidae). The pupa (left) and fully grown larva (right) are both females that have been fed a diet of royal jelly. Both will develop into queens. The pupa is of the exarate type, and the larva is vermiform, without appendages.

Examples are eruciform larvae such as Lepidopteran or Hymenopteran caterpillars. Oligopod larvae have only thoracic legs, as in campodeiform larvae. Apod (or apodous) larvae lack any legs, as in all vermiform larvae and some larvae of various beetle families or ants, bees, or wasps.

Hypermetamorphosis

In some insects the larvae pass through two or more instars that differ markedly in appearance. The first instar is often campodeiform and actively seeks food or a host. The subsequent instars are almost always relatively inactive and grublike. Examples are commonly parasites or parasitoids and are found in Neuroptera (Mantispidae), Coleoptera (certain species of Carabidae and Staphylinidae, Meloidae and Rhipiphoridae), Strepsiptera, Diptera (Bombyliidae, Acroceridae, Nemestrinidae, and some Tachinidae and Calliphoridae), Hymenoptera (certain Chalcidoidea), and certain ectoparasitic Lepidoptera (Cyclotornidae and Epipyropidae). When the first instar has legs, as in Coleoptera, Mantispidae, or Strepsiptera, it is called a triungulin (Fig. 38.1b,c). The legless planidium larvae of Diptera and Hymenoptera use long "walking" setae or jumping movements for locomotion (Fig. 39.6a).

CHARACTERISTICS AND TYPES OF ENDOPTERYGOTE PUPA

Pupae with movable mandibles are called decticous. Such pupae are found in Megaloptera, Neuroptera, Mecoptera, Raphidioptera, Trichoptera, and certain Lepidoptera (suborders Zeugloptera and Dacnonypha). The pupal mandibles are used by the pharate adult to cut through the cocoon or gnaw out of the pupal cell. The other appendages are free from the body in a condition called exarate (Figs. 4.4b, 4.7). In some of the above taxa the legs are used by the pharate adult to walk or swim before the pupal cuticle is shed.

Pupae without movable mandibles are adecticous. The pupae of this type may be exarate, or have the appendages cemented to the body in a condition known as obtect (Fig. 4.4e). Exarate adecticous pupae are found in Hymenoptera and Coleoptera. To escape the cocoon or cell, the insect first sheds the pupal cuticle and then uses the adult mandibles and legs to exit. In Strepsiptera and certain Diptera (Cyclorrhapha) the pupa is exarate, but it is enclosed in a puparium, or sclerotized cuticle of the last larval instar. Obtect pupae are found in certain Lepidoptera (suborders Ditrysia and Monotrysia, Fig. 43.5e) and Diptera (Nematocera, Orthorrhapha). Backwardly directed spines on the cuticle of some obtect pupae help to force the emerging insect out of the cocoon or cell.

REPRODUCTIVE ORGANS AND THEIR FUNCTION

Male Organs

The functions of the male organs of insects are to produce sperm and to transfer the sperm to the female of the same species (Fig. 4.8a,b). The mesodermal gonads are the testes, which are paired but may be secondarily united in a single body. Spermatogenesis, or the production of sperm, takes place in the sperm tubes. In insects with a short adult life, spermatogenesis may be completed as early as the larval or pupal instars, while in long-lived adults sperm may be produced in the adult. The sperm tubes may be single as in some Apterygota, Coleoptera, and Diptera, or multiple as in other insects. The sperm tubes of each testis are usually held together by a peritoneal sheath.

Mature sperm pass from the sperm tube through a short vas efferens to the vas deferens. The vas deferens is usually mesodermal in origin, but in Diptera it is ectodermal. It may be dilated into a seminal vesicle, where sperm are stored prior to their exit via the ectodermal ejaculatory duct. Accessory glands of mesodermal or ectodermal origin may be associated with the vas deferens or ejaculatory duct. These are especially well developed in insects that produce a spermatophore, or jellylike capsule in which the sperm are transferred to the female.

Sperm Tube Structure and Spermatogenesis

The sperm tube has a cellular epithelial sheath of one or two layers. Inside the tube, the germ cells, or spermatogonia, are located at the apex in the germarium. Here also is a large cell, the apical cell. The spermatogonia multiply mitotically. The cells from consecutive divisions remain clustered and

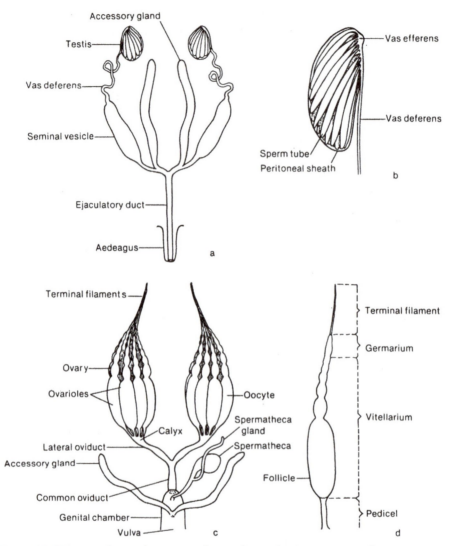

Figure 4.8 Diagrams of general structure of internal reproductive organs: a, male system; b, testis; c, female system; d, ovariole.
Source: Redrawn from R. E. Snodgrass, The Principles of Insect Morphology.

enclosed in a cellular capsule or sperm cyst. Within the cyst the spermatogonia produce spermatocytes, which undergo meiotic division to form the haploid spermatids. The spermatids then differentiate into flagellated spermatozoa or sperm. Insect sperm are quite slender and long. Their narrow diameter is probably correlated with the diameter of micropyles in the egg chorion, an opening that must be kept small to reduce water loss. The several steps in spermatogenesis can be seen in successive zones arranged lengthwise in the sperm tube. Near the base where the vas efferens is located, the cyst disintegrates and the sperm are released, still grouped in bundles. The sperm pass into the vas deferens and are stored in the seminal vesicles until ejaculation.

Transfer of Sperm

Among the Pterygota, mates find each other by sight, sounds, or chemical sex pheromones. Following courtship, sperm are transferred by the male to the female directly and internally by a copulatory organ (Fig. 4.1). Odonata are unique in having the male transfer sperm to a secondary organ on his third sternum (genital fossa, Fig. 19.3), from which the female receives the sperm.

The Apterygota and Entognatha have an indirect or external transfer of sperm, not involving copulation. The sperm are contained in a spermatophore and deposited on the substrate or on threads to be picked up by the female. This method of transfer requires humid conditions to prevent the drying of the spermatophore. In Collembola and Diplura the sperm are contained in a stalked spermatophore that is deposited by the male on the substrate. In some Collembola the female picks up the spermatophore with her vulva in the absence of the male; in other species the spermatophore is picked up after courtship and in the male's presence. Diplura males deposit the spermatophore only in the presence of the female. In Thysanura and Archeognatha, the male spins silk threads during courtship (Fig. 17.2). In the former the spermatophores are placed on the substrate under the threads and the female is guided under. In the Archeognatha the spermatophores are deposited on a thread held by the male, and the female is guided to pick up the spermatophores.

In the Pterygota, sperm are transferred directly to the female's genital tract during copulation. Internal fertilization results in less wastage of sperm and is much less dependent on humid conditions. The spermatophore is retained in many orders: Orthoptera, Blattodea, Mantodea, Dermaptera, Psocoptera, Neuroptera, Lepidoptera, and some Diptera (suborder Nematocera). Mecoptera and most Diptera transfer sperm freely without a spermatophore to the female. Depending on the species, spermatophores may be present or absent in Hemiptera, Trichoptera, Coleoptera, Hymenoptera, and Neuroptera.

Once the spermatophore is deposited in the genital opening, the sperm are freed by the partial rupture or digestion of the spermatophore. The sperm then migrate or are moved to the spermatheca. The empty spermatophore is dissolved in the vagina or eaten by the female.

Traumatic or hemocoelic insemination occurs in the Strepsiptera; in the bedbugs (Cimicidae) and some of their allies, the Anthocoridae and Polyctenidae; and in some Nabidae (Hemiptera). The male genitalia are used to pierce the integument of the female, and the sperm are injected into the hemocoel. In some bedbugs a spermalege, or specialized organ of the integument, is developed where the puncture normally is made.

Reproductive females of social insects (Chap. 7) may mate multiple times during a single mating bout (honeybees) or repeatedly over a lifetime (termites). Female insects of most species, however, become less receptive or unreceptive to further mating and increase production of their eggs after they are inseminated. Changes in receptivity can be traced to substances contained in the seminal fluids or spermatophores. The spermatophore of male crickets, *Teleogryllus commodus* (Gryllidae, Orthoptera), contains a prostaglandin-synthesizing complex. When transferred to the spermatheca of the female, this enzyme complex synthesizes prostaglandin from the precursor arachidonic acid. As a result, the female initiates massive oviposition. In another example, the experimental injection of seminal fluid into the hemocoel of female leafhoppers makes recipient females reject mating advances by males.

Female Organs

The functions of the female organs of insects are to produce eggs and to ensure their fertilization and placement in the environment (Fig. 4.8c,d). The mesodermal gonads are the paired ovaries, each of which may be single but usually are divided into multiple ovarioles. Oogenesis, or the production of eggs, takes place in the ovarioles, beginning in some insects in the pupa or last nymphal instar. The number of ovarioles is commonly four to eight on a side, depending on the species, and is correlated with the number of eggs produced. In Lepidoptera the number is usually four on each side. The highly fertile queen ants or termites, for example, may have more than 1000 on each side.

At ovulation the eggs pass from the ovariole and follow ducts toward the exterior. The paired lateral oviducts, each with a calyx, lead to the single common oviduct, which empties into the genital chamber. Attached to the genital chamber are accessory glands and the spermatheca with its spermathecal gland. Fertilization takes place in the genital chamber before the eggs are oviposited. The external opening of the reproductive tract is the vulva. Except for the lateral oviducts, which may be partly mesodermal, the other ducts and glands are ectodermal in origin and are lined with cuticle. The arrangement of the female tracts varies in the Lepidoptera (see Fig. 43.4) in which separate openings for copulation and oviposition may be present.

Ovariole Structure and Oogenesis

The ovariole is essentially a tube, attached anteriorly in the body by a terminal filament. The combined filaments of the ovarioles of each ovary may form a suspensory ligament. The wall of the tube is the tunica propria, which is lined internally by a follicular epithelium. The ovariole is divided along its length into the germarium, vitellarium, and pedicel. In the germarium are the oogonia, which are derived from primordial germ cells that can be traced back to the embryo of the female insect. By mitotic divisions of the oogonia, the oocytes (literally: egg cells) are produced, and these pass into the vitellarium. Each oocyte becomes surrounded by follicular cells, forming a cystlike follicle. Within the follicle, yolk is deposited in the oocyte during vitellogenesis and the chorion is added. As the follicles swell, arranged one behind the other, the vitellarium elongates and becomes beaded in appearance. The first or most advanced follicle is separated from the opening of the pedicel by a follicular plug.

Studies on *Drosophila* indicate that many genes are expressed during oogenesis. The oocytes rapidly increase in size as protein, carbohydrate, and lipid yolk bodies are formed by vitellogenesis in the female's fat body. Recall that this process is triggered by JH and ecdysteroid hormones. The process of vitellogenesis is complex. The nutrients that make up the yolk are derived from food eaten by the female or, if the adult does not feed, from food eaten and stored in the fat body during the immature instars. The female insect deposits not only nutrients in the egg for the future embryo but also products of maternal genes (proteins, ribosomal RNA, messenger RNA) on which embryonic development will depend up to the stage of the cellular blastoderm. Yolk components enter the oocyte from the hemolymph by way of the follicle cells. Intercellular spaces develop between the follicle cells to allow movement of proteins from the hemolymph to the surface of the growing oocyte; the spaces close after vitellogenesis is completed. Protein synthesis inside the oocyte may involve additional nucleic acids contributed by the trophocytes or follicle cells. Lipids and the glycogen found in the yolk of many insects are transported to the oocyte by the trophocytes and especially by follicle cells. When yolk deposition is complete, the chorion is secreted around the oocyte by the follicle cells. The chorion is layered like a cuticle with respect to the follicle cells: next to the follicle cells is protein, and the layer next to the vitelline membrane of the oocyte is lipoprotein. The sculpturing and various passages through the chorion are created by the follicle cells. The vitelline membrane is usually considered to be the cell membrane of the oocyte, but in some Diptera it also is secreted by the follicle cells.

Types of Ovary

Two major structural types of ovarioles are recognized: panoistic and meroistic. The former is the simple ovariole just described and found in Entognatha, Apterygota, Paleoptera, most Orthopteroidea, Thysanoptera, and in one endopterygote order, Siphonaptera. Holometabolous and some hemimetabolous insects have meroistic-type ovaries, characterized by nurse cells or trophocytes, associated with each oocyte. These are usually derived from the oogonia and transport cytoplasm into the oocytes via connecting strands large enough to transport cell organelles such as ribosomes and mitochondria.

Meroistic ovarioles are further classified as either polytrophic or telotrophic types. In the former the trophocytes accompany each oocyte into the vitellarium and are included in the follicle. Polytrophic ovarioles are found in the exopterygotes Phthiraptera, Dermaptera, and Psocoptera, and in most endopterygotes except Siphonaptera and Coleoptera (Adephaga). In the telotrophic ovarioles, the trophocytes remain in the germarium and are connected with the oocytes by nutritive cords. Ovarioles of this type are found in Hemiptera and Coleoptera (suborder Polyphaga).

Ovulation and Fertilization

At ovulation the follicular covering breaks down and the egg pushes past the follicular plug into the pedicel and to the calyx beyond. The degenerate follicular covering forms a reddish or yellowish mass, which persists for some time in the ovariole. The presence of the colored mass can be used to discriminate between parous females that have ovulated and nulliparous females that have not. In those insects that lay their eggs in groups, such as many Orthopteroidea, the eggs may be accumulated in the pedicels or oviducts before fertilization and oviposition.

As the eggs pass into the genital chamber from the common oviduct, they are usually oriented in a manner to receive sperm directly from the opening of the spermathecal duct. The mechanism of sperm release involves the special muscles associated with the spermatheca. In most insects the release is probably stimulated by ovulation or the passage of eggs in the oviduct. In Hymenoptera, however, the female can control the release so that some eggs are fertilized and others are not. The latter become males.

Oviposition

The segmental position of the genital opening varies among the orders. All insects have a common oviduct except the Ephemeroptera, in which the lateral oviducts open separately just behind the seventh sternum. In Dermaptera the short common oviduct opens behind the seventh sternum. Those insects with an ovipositor derived from the appendages of the eighth and ninth segments usually have the genital opening on the eighth segment. The great majority of insects, however, have the appendixlike ovipositor reduced or replaced by special modifications of the abdominal apex: most Odonata, Plecoptera, Phthiraptera, Thysanoptera (suborder Tubulifera), Coleoptera, most Neuroptera, Mecoptera, Trichoptera, Lepidoptera, and Diptera. In these insects the opening is in the eighth or ninth segment. Many have the caudal abdominal segments tubular and telescopic, with the genital opening at the apex. This permits the eggs to be placed in crevices or cemented to leaves or other surfaces. Before the eggs are released, they may receive adhesive substances or a protective coating from the accessory glands. Groups of eggs of Mantodea (Fig. 21.1b,c) and Blattodea (Fig. 20.1) are enclosed in a protective case, or ootheca, derived from the accessory (also called colleterial) glands.

Oviposition behavior may be quite simple, as in Phasmatodea, which merely drop the eggs at random. More complex behavior involves the selection of appropriate sites through the use of various sense organs, including those on the ovipositor or abdominal tip. The sites are usually near the food required by the newly hatched offspring or in locations where the food is likely to be in the future. For example, some mosquito species lay their eggs in locations such as open cavities in trees at the start of a dry season, and the eggs will later be stimulated to hatch by water that fills the cavity in the wet season. The aquatic mosquito larvae then feed on microorganisms and decaying plant matter in the water. Depending on the food habits of the species, the favored site may be in or on water, soil, decaying organic matter, certain species of live plants or animals, or other insects. Oviposition may occur only at certain times during a 24-hour cycle. All the eggs may be laid at once, as in some Ephemeroptera, or they may be laid singly or in small groups over a considerable period of time. Social relationships, involving the special care of eggs and offspring, are discussed in Chapter 7.

SEX DETERMINATION

Sex in insects, as in other organisms, is commonly determined by heterogamy, or the production of gametes of two types, but the main signaling methods that determine sex are diverse among insect groups. Signals can be genetic, epigenetic (such as environmental), cytoplasmic, or the result of bacterial infections. Sex chromosomes that occur in pairs in one sex are designated X; those that occur as single chromosomes are Y; and the missing chromosome of a pair is O. Males are heterozygous XY and XO and females homozygous XX in most Diptera, Neuroptera, Mecoptera, Hemiptera, Odonata, Coleoptera, and Orthoptera. The reverse can occur within the same order. Females are heterozygous XY or XO and males homozygous XX in Lepidoptera and Trichoptera. In contrast to mammals, Y chromosomes in insects rarely carry sex-determination genes, facilitating the evolutionary loss of Y chromosomes so that one sex has the XO condition. For some insects the chromosomes bearing sex-determining genes are not readily identifiable.

In Hymenoptera, fertilized eggs develop into females and unfertilized eggs into males. The females are therefore diploid and the males haploid. The sex ratio may deviate considerably from 1:1. This mechanism of sex determination, called haplodiploidy, is also known in some Thysanoptera and *Micromalthus* (Micromalthidae, Coleoptera). Male Coccidae (Hemiptera, suborder Homoptera) are also haploid, but apparently this is a secondary condition because the male zygote is initially diploid.

Certain bacteria that infect insect reproductive tissues can determine male:female ratios of the host insect's offspring. Bacteria in the genus *Wolbachia* are widespread in insects as intracellular parasites. Some of these infections cause skewed sex ratios in the host insect's offspring by killing early-developing embryos. Other *Wolbachia* and other microbes (Chap. 11) induce the production of only female offspring or cause recombinational incompatibility between the eggs of uninfected females and the sperm of infected males.

TYPES OF REPRODUCTION

Thus far we have discussed the most common type of reproduction, oviparity, in which eggs are deposited by the adult female shortly after fertilization. A variety of other reproductive methods, however, exists among the insects; these are listed below.

Ovoviparity

In this type of reproduction the eggs are retained in the female's genital tract until embryonic development is complete. The female then deposits a larva or nymph instead of an egg. The embryo is nourished solely by the yolk initially deposited in the egg. Examples are found in some Thysanoptera, Blattodea, Coleoptera, Diptera (some Tachinidae, Calliphoridae, and Muscidae), and other taxa.

Viviparity

In contrast to the above, the embryo of viviparous insects is fed by the female after development has begun. Three types of viviparity have been recognized:

1. **Adenotrophic viviparity** is found in the tsetse fly, *Glossina* (Glossinidae, Diptera), and several other ectoparasitic flies—Hippoboscidae, Nycteribiidae, and Streblidae. One embryo at a time is carried by the female and deposited as a larva, which soon pupates. After the egg yolk is consumed, the larva ecloses from the thin chorion and is fed orally by special glands of the genital chambers.

2. **Pseudoplacental viviparity** involves eggs that have little yolk in which the embryo is nourished by the wall of the genital tract, but not orally. A placentalike organ for the transfer of nutrients may be developed by maternal or embryonic tissues or both. Examples are found in Hemiptera (Polyctenidae, Aphidoidea), Dermaptera (suborders Arixeniina and Hemimerina), *Diploptera* (Blaberidae, Blattodea), and *Archipsocus* (Archipsocidae, Psocoptera).

3. **Hemocoelous viviparity** occurs in all Strepsiptera and in the paedogenetic larvae of certain Diptera (Cecidomyiidae), discussed below. In these insects the ovaries disintegrate and the embryos in their egg membranes take their nourishment directly

from the maternal tissues. The female is consumed as a result.

Paedogenesis

Reproduction by larval insects occurs in the beetle *Micromalthus* (Micromalthidae) and the flies *Miastor*, *Mycophila*, and certain other Cecidomyiidae (Diptera). In the flies, the life history may be of normal, sexual adults when the food supply is average. When the food supply changes, larvae may give birth to more larvae for several generations without the appearance of the adult instar. The ovaries within the larva become functional, the eggs develop parthenogenetically, and the young larvae devour their mother (hemocoelous viviparity). The life cycles of these species are quite complex.

Parthenogenesis

Development without fertilization occurs in at least some species in all orders that have been investigated except the Odonata and Hemiptera (suborder Heteroptera). The types of parthenogenesis are classified by the sex produced, by the cytological mechanism, and by the frequency of occurrence in the species. According to the sex produced, three types are recognized. Arrhenotoky is the parthenogenetic production of males. This is the sex-determining mechanism in all Hymenoptera, some thrips, aphids, and *Micromalthus* (Coleoptera). Thelytoky, the parthenogenetic production of females only, is the most common type. Deuterotoky (or amphitoky), the production of both sexes, is known in certain aphids and cynipid wasps.

The mechanisms of asexual reproduction fall into two main types. In apomictic parthenogenesis the egg fails wholly or in part to undergo meiosis, resulting in no reduction in chromosome number and no opportunity for new gene combinations. Except for mutations, the offspring retain the genes of the mother. This is the most common type of mechanism in insects. In automictic parthenogenesis, the egg undergoes meiosis and the diploid condition is restored in a variety of ways, that is, by fusion of two cleavage nuclei, two polar bodies, or a polar body with the egg pronucleus. The last is probably the most common method. A given

lineage will become increasingly heterozygous or homozygous depending on which nuclei fuse.

Parthenogenesis may be facultative, that is, the eggs, if fertilized, develop normally and, if not fertilized, develop parthenogenetically. Thus the eggs of Hymenoptera are facultatively arrhenotokous. Obligatory parthenogenesis involves only thelytoky; males are rare or altogether lacking in the population. Such populations are often morphologically very similar to a normal bisexual population and apparently arose as a parthenogenetic race. Without males or the need for mating, a thelytokous race can rapidly reproduce but at the expense of genetic variability. Examples are found in some Phasmatodea, Lepidoptera (Psychidae), and Coleoptera (Curculionidae). A number of species of minute wasps of the genus *Trichogramma* (Trichogrammatidae, Chalcidoidea, Hymenoptera) have both sexual and thelytokous populations. Thelytoky in these wasps may arise from either hybridization between species or, more commonly, microbial infection. In the latter, wasps become thelytokous when infected by *Wolbachia* bacteria, but treatment by temperature or antibiotics can render the wasps permanently sexual.

Polyembryony

In certain Strepsiptera and Hymenoptera (Chalcidoidea, Braconidae, Dryinidae), a single egg results in two or more individuals. This unusual phenomenon probably is made possible by the totipotent property of the early cleavage nuclei in the insect embryo. The insects are endoparasites of other insects. Polyembryony permits a large number of offspring, sometimes several thousand, to emerge from oviposition on one host (Fig. 10.2b). The female of *Copidosoma floridanum* (Encyrtidae, Hymenoptera) lays one or two eggs in the egg of a host caterpillar. If one egg is laid, one sex emerges; if two eggs are laid, both males and females emerge. Eggs of both sexes develop into numerous larvae, but of two types. First to differentiate are the "precocious" larvae or defender morphs that eliminate the larvae of other species of competing parasitoids inside the host. The precocious larvae fail to pupate and die, thus making an altruistic sacrifice of their own chance to reproduce (see discussion of altruism in Chap. 7). The second type or "reproductive" larvae complete development, emerge from the host body, mate, and disperse. Evidence exists that female precocious larvae may kill male reproductive larvae, thus altering the sex ratio of emerging wasps. In *Copidosomopsis* (Encyrtidae), the defender morphs have heavily sclerotized mouthparts and have been observed to attack other parasitoids (Cruz, 1981).

Functional Hermaphroditism

Individual insects are sometimes found or experimentally produced that have both male and female characteristics. Gynandromorphs are genetic mosaics of male and female tissues derived from a zygote plus other, genetically different, nuclei. All tissues of intersexes have the same genetic composition, but unstable development results in the differentiation of male and female features. Gynandromorphs and intersexes are not normal in any insect species. However, in the cottony-cushion scale *Icerya purchasi* (Coccoidea, Hemiptera) and related species, a functional hermaphroditism does occur normally. Both male and female gonads develop in the female scale, and the eggs are self-fertilized. Haploid male scales are rarely produced, and no pure females are known.

GENERATION TIME AND LONGEVITY OF INSECTS

The numbers of generations per year in insects is called voltinism. Univoltine insects have one generation per year; this is probably the most common type of life history. Bivoltine insects have two generations per year, and multivoltine insects have more than two per year. Some insects require one or more years to complete their life cycle. For example, Thysanura and Archeognatha live for 2 to 3 years. The long-lived 13- and 17-year periodical cicadas (Cicadidae) spend most of their life in soil, sucking the juice of roots.

The number of generations per unit time tends to be inversely related to body size. In other words, smaller insects have more generations than larger insects during the same period. This is only approximate because other factors such as moisture and nutrition influence the rate of growth. The number

of generations per year for some species depends on suitable temperatures during development. This may vary markedly with location and to a lesser extent from year to year. The European corn borer, *Ostrinia nubialis* (Pyralidae, Lepidoptera), for example, has three or more generations in southern North America but only one at the northern extent of its range. Some insects that live at high altitudes or inhabit cold waters require multiple years to become adults.

Insects that live in dry environments and eat nutritionally deficient food require a longer time to mature. Wood-boring beetles (Cerambycidae, Bostrichidae, Buprestidae), wasps (Siricidae), and moths (Cossidae) require one to several years. Records of 23 years for a cerambycid and 22 years for a bostrichid were obtained when the beetles emerged from wood that had been dried and made into furniture.

The adult life of insects is devoted to mating and reproduction. Adults may survive only a few hours, as in male Strepsiptera (Chap. 38), some species of Ephemeroptera, or some species of the *Clunio* midges (Chironomidae, Diptera). On the other hand, univoltine insects may have adults that live for almost a year. Adult *Cryptocercus* (Cryptocercidae, Blattodea) roaches live 3 to 6 years.

Among the longest-lived adults are the reproductive castes of social insects. Queen honeyees normally live 2 to 3 years, and lifetimes of up to 5 years or more have been recorded. The maximum age of an ant queen is 18 years and that of a termite queen is 12 years.

Several families of beetles are long-lived as adults: Elmidae (up to 9 years), Scarabaeidae (up to 2 years), and Dytiscidae (up to 2.5 years). Marked individuals of tenebrionid beetles have been recovered in the field after 3 years, and it is possible they live for up to 10 years. The tenebrionid illustrated in Fig. 5.11, after serving as the artist's model for the first edition of this book, lived for 15.5 years on a diet of wheat bran, lettuce, and green peas.

INSECT SEASONALITY AND ARRESTED DEVELOPMENT

The life cycles of insects are synchronized on a long-term or seasonal basis. They survive annual periods of winter cold, drought, summer heat, or food shortage by undergoing a state of dormancy. When activity is suspended at the onset of the unfavorable period and resumed immediately afterward, the temporary arrest is called quiescence. During quiescence the hormonal stimulation of growth or reproduction remains unaffected. A few chilly days in spring, for example, may lower metabolism, but growth continues with the return of sunny weather.

Development or reproduction may be arrested on a long-term basis in winter or summer. A prolonged dormancy regulated by hormones is called diapause. Depending on the species of the insect, diapause may occur in any stage of the life. In eggs, larvae, and pupae the diapause is an interruption of growth and development; in adults the diapause is a period of no reproduction, called reproductive diapause. The stage at which diapause occurs is genetically determined, but whether it will occur or not may be influenced by environmental conditions. In moth eggs, diapause is controlled by a maternal diapause hormone stored in the eggs. In moth larvae, juvenile hormone is the main regulatory agent for the beginning and duration of diapause. In moth pupae, PTTH is not released by neurohemal organs in the retrocerebral complex. Consequently, ecdysteroids are not produced and molting and the associated developmental changes do not occur. In diapausing adults the corpora allata are inactive and reproduction is not stimulated by JH. The regulation of gene expression to induce, maintain, and terminate diapause is an area of active investigation. It clearly involves both upregulation of some genes and reduced expression of others. A conspicuous set of genes upregulated are those that regulate the heat shock proteins. Most of these proteins increase markedly with diapause onset and rapidly decrease when diapause terminates.

An obligate diapause takes place regardless of the environment. Univoltine insects commonly have an obligate diapause in an immature stage, as, for example, in many Lepidoptera of the Temperate Zone. The onset, or induction, of a facultative diapause is dependent on certain environmental cues. Day length is the most reliable predictor of season and most often the prime

cue to induce diapause. Winter diapause can be triggered by the shorter days and longer nights of autumn. These cues usually induce diapause well in advance of the diapausing stage but at a time when environmental conditions are still favorable for continued growth. The stage of the insect that experiences the critical day length is not necessarily the stage that will diapause. For example, whether or not eggs will diapause is determined by the female parent. Adult diapause is usually induced in an immature stage.

The photoperiodic cues trigger changes in the brain's endocrine secretions to induce and maintain diapause. The compound eyes and ocelli do not seem to serve as photoreceptors to induce diapause because surgically removing the eyes or covering them to prevent light detection does not stop diapause induction. As in other animals, the responses to photoperiod are mainly to blue light.

Diapause is recognized by a variety of symptoms. Active insects become relatively inactive but not necessarily immobile or unresponsive to prodding. Oxygen consumption falls as the rate of metabolism is lowered and is no longer correlated with temperature. Water loss decreases. Thus, in diapause a resistant state is attained before the onset of unfavorable weather or food shortage.

Diapause is induced in most insects by a specific proportion of day length to night length, called the critical day length. This is, strictly speaking, insect photoperiodism, in contrast to the brief cues that entrain circadian (daily) rhythms (Chap. 6). Some insects respond to long days/short nights, others to short days/long nights, and some require more complex schedules in which long days are followed by short days, or vice versa. Near the equator only slight differences in day length may be sufficient, but often temperature changes are the key cues to trigger diapause in the equatorial tropics. The critical day length for a widely distributed species is usually adapted to the local climate because day length changes more rapidly over the year toward the poles and less toward the equator. The adaptation of invasive or introduced species such as the European corn borer, *Ostrinia nubialis* (Crambidae, Lepidoptera), which was first seen in North America in 1917, to different critical photoperiods at different latitudes shows how quickly evolution can operate to adapt organisms to local conditions.

The intensity of diapause and its probable duration may be judged by the extent to which metabolism is reduced. The length of time that an insect remains in diapause is the result of a complex interaction involving (1) seasonal changes in photoperiod and temperature, (2) the insect's changing responses to these physical factors, and (3) the rate of diapause development. The term "diapause development" is intended to indicate that a physiological process, the nature of which is unknown, must be completed before diapause is terminated. Ordinarily diapause development takes place only at temperatures below the minimum temperature for growth. A period of chilling, therefore, is often required for diapause development. Once diapause is broken, the brain resumes neurosecretory activity and development resumes or, in adults, the reproductive functions proceed.

Diapause in the egg stage may take place late enough to be controlled by the young larva's neuroendocrine system. In the embryo, however, the nervous system is not developed. In the silkworm, *Bombyx mori* (Bombycidae, Lepidoptera), a diapause hormone is secreted by the subesophageal ganglion.

In insects with more than one generation each year, the insects that develop in different seasons may exhibit different characteristics. Seasonal forms are known in Lepidoptera, Gryllidae, and Cicadellidae. Early taxonomists frequently described the different forms as separate species. For example, the European butterfly, *Araschnia levana* (Nymphalidae, Lepidoptera), has two forms. Caterpillars reared under short-day photoperiods diapause and emerge as the spring, or levana, form (Fig. 4.9a). When reared with 16 or more hours of light, the caterpillars do not diapause, and the summer form develops (Fig. 4.9b).

POLYMORPHISM

The term polymorphism in the broad sense means the presence of two or more forms of an organism. The differences are not restricted to physical appearance but may be physiological, behavioral, or genetic. The forms may be different individuals

Figure 4.9 Seasonal forms of the European nymphalid butterfly *Araschnia levana*: a, spring, or levana, form; b, summer, or prorsa, form.

in a population or different stages in the life history of a single individual. For example, the holometabolous life history is polymorphic. Here we are concerned with different forms among adults of the same species other than the usual two sexes (sexual dimorphism). Adult forms may be genetically different, as in mimetic butterflies, or individuals may be genetically similar and develop differently according to external influences. Aphids and cynipid wasps are just two of many types of insects that display polyphenism, that is, distinctive forms that appear at certain times of the year. As mentioned previously in the discussion of seasonality, the spring and summer generations of bivoltine butterflies may differ in appearance. The quality and quantity of food or pheromones emitted by other members of the colony influence the development of morphologically and behaviorally different forms, called castes, in social insects (Chap. 7). In most of these examples, developmental hormones play a central role in directing the development of different forms of the same species.

Cyclical parthenogenesis is a kind of polymorphism involving the alternation of parthenogenetic and bisexual generations. Examples are found in the Hymenoptera (some Cynipidae), Hemiptera (Aphidoidea), and Diptera (paedogenetic Cecidomyiidae, mentioned above). Here the advantages of both sexual and asexual reproduction are combined in an annual cycle, and the life histories may be complex.

Seasonal Forms of Cynipid Wasps

Female cynipids place their eggs in the meristematic tissues of plants, mainly in the oak and rose families. Larval feeding induces the plant host to develop a characteristic gall. The adults later escape from the gall by chewing an exit hole. The life histories of cynipids vary greatly: some species are bisexual and univoltine, others are mainly thelytokous and without males, and still others are cyclic, alternating between a sexual generation and a parthenogenetic generation (called the agamic generation) each year. The females of each generation in the cyclic type and their galls on plants may differ in appearance. In some instances, the females were originally assigned to different species. The annual cycle is as follows. In the spring, males and females of the sexual generation develop quickly and emerge from their galls and mate. The sexual females lay fertilized eggs that become agamic females. These develop slowly in their galls, overwinter, and emerge the following spring to produce unfertilized eggs that become either males or females of the sexual generation. In some species of aphids and cynipids, one type of female produces both males and females by deuterotoky, while in other species the females are of two types: male producers and female producers.

Seasonal Forms of Aphids

In aphids usually one or more parthenogenetic generations take place in spring, followed by a bisexual generation that produces overwintering eggs. Most species of aphids live throughout the year on one species of plant. The typical cycle begins in the spring when the eggs hatch to produce the first generation, or fundatrices (singular, fundatrix). These become alate (winged) females that reproduce by parthenogenetic viviparity. Several successive generations are produced by this means until autumn. As colonies become crowded or plants become senescent or dry, individuals may fly to other plants of the same species and establish new colonies. In autumn, alate males and apterous (wingless) oviparous females are produced

parthenogenetically. These mate, and the females lay eggs that overwinter. Some species of aphids, however, live on woody perennial plants (primary hosts) in the autumn, winter, and spring and fly to herbaceous annual plants (secondary hosts) in the summer.

For example, the annual cycle of the apple-grain aphid, *Rhopalosiphum padi* (Aphididae, Hemiptera), is shown in Fig. 4.10. The all female fundatrices (a) are found in the spring on various trees, including apple, pear, and plum. The fundatrices and most of the second generation (b) are apterous, but by the third generation, alates are produced (c). These migrate to various grasses such as the grains, wheat, oats, and corn, or weedy grasses. New colonies of apterous aphids (d) are established by parthenogenetic viviparity. When the colonies become crowded, alates (e) are again produced, and they fly to new grasses. In autumn, males (g) appear for the first time, and females (f) are produced; both sexes fly to the primary host. The females produce apterous females (h) that mate with the alate males. The apterous females are oviparous and lay eggs (i) that overwinter in diapause. In some aphid species, the overwintering egg is too large for the small bodies of the female aphids, so that the dead body of the mother serves as a protective covering for the egg over winter.

Aphids maximize their reproductive rate during favorable circumstances by channeling morphological and metabolic resources needed for flight (flight muscles and wings) into producing offspring. Alates produce fewer total offspring than do wingless adults. Yet aphids retain the critical ability to produce alates in response to environmental signals (e.g., crowding, short day lengths, and declining plant condition) that predict deteriorating resources. Parthenogenesis further maximizes production of progeny by eliminating the time, effort, and males needed for mating. The annual sexual cycle preserves sexual recombination. Viviparity minimizes the time for development to sexual maturity, but the egg stage is retained to survive the winter's cold and absence of suitable plants for food. Evolution has rapidly adapted some invasive species such as the pea aphid to local conditions. For example, in North America the pea aphid, *Acyrthosiphon*

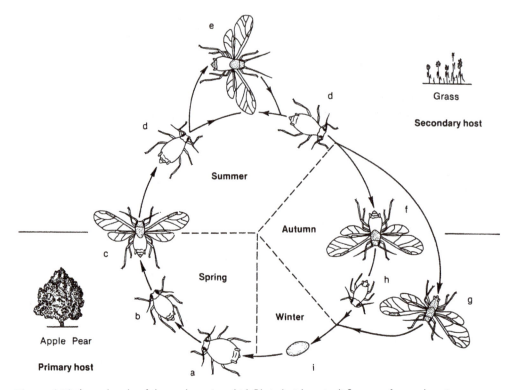

Figure 4.10 Annual cycle of the apple-grain aphid, *Rhopalosiphum padi*. See text for explanation.
Source: Redrawn from Dixon, 1973, by permission of Edward Arnold/Hodder & Stoughton Educational Publishers, Ltd., London.

pisum, has retained the overwintering egg stage in severe winter climates, but this form is not found in milder climates such as California.

Phase Polymorphism

We conclude this chapter with a discussion of phase polymorphism in those species of Acrididae (Orthoptera) known as locusts: *Anacridium aegyptium*, *Dociostaurus maroccanus* (Moroccan locust), *Locusta migratoria* (African migratory locust), *Nomadacris septemfasciata* (red locust), and *Schistocerca gregaria* (desert locust). Theirs is one of the most extraordinary transformations in insects. Under certain ecological circumstances, a harmless, even rare, insect becomes a threat to human survival equal to a military invasion.

In locusts, individuals of each species may develop into one of a spectrum of adult forms or phases, depending on the circumstance under which the young nymph or hopper grows to

maturity. The extreme forms are called the solitary phase and the gregarious phase (Fig. 4.11). The visible differences are in coloration, proportions of body parts, and behavior, but these are only reflections of underlying physiological differences.

Hoppers of the solitary phase of *Schistocerca*, for example, are uniformly colored green, whereas the gregarious hoppers have a black pattern on a yellow or orange background. In the adults, the coloration of the phases is more or less reversed: the solitary phase has a more pronounced dark pattern and the gregarious phase a less pronounced one. Furthermore, male adults of the gregarious phase continue to change color, but the solitary adults do not. Structurally, the head of the solitary locust is narrow, the pronotum is convex or arched in profile, and the hind femur is relatively long. In the gregarious phase the head is wider, the pronotum is depressed in profile, and the hind femur is relatively shorter.

Figure 4.11 An outbreak of plague locusts (Acrididae) in Argentina. The large nymphs have developed into the migratory form and after their final molt will form a migratory swarm in the air.

The solitary hoppers spend relatively less time walking and walk at a slower rate, whereas the gregarious hoppers form groups that march together for long periods and at a faster rate. The most dramatic difference is in the adult behavior. The solitary adults remain in the vicinity of their birthplace, but the gregarious adults aggregate in immense swarms and take flight. Like great clouds of smoke that darken the sun, swarms of *Schistocerca* may have up to 10,000 million or more individuals. Over a season, a swarm may migrate for distances totaling up to 2000 miles (3226 km). In October 1988, desert locusts from western Africa were found in the West Indies of North America after an exceptional flight over the Atlantic Ocean estimated to be at least 3100 miles (5000 km) and of 6 days duration. The locusts probably were able to glide some of this distance by taking advantage of up currents in the wind. Normally, the direction of flight over land is with the prevailing winds of weather fronts in the intertropical convergence zone, increasing the chance

that they will arrive in areas of recent rainfall and new plant growth. To sustain this flight, locusts may consume their body weight in green plants in a day. When a large swarm (up to 100,000 tons) settles on the ground, they completely defoliate vegetation and crops.

Even today, locust plagues are among the most feared of natural disasters. Locusts are figured in Egyptian tombs as early as the sixth dynasty (2354–2181 BCE), and the Old Testament contains vivid references to them (e.g., Book of Joel). Famine and disease epidemics with high human mortality often followed the devastation of crops.

Until 1921, entomologists considered the migratory locusts to be several species that were distinct from solitary acridids. About this time and almost simultaneously, researchers in South Africa, Central Asia, and Russia came to the same independent conclusion: each species of locust may exist in two phases, of which one is the destructive migratory form and the other is a relatively harmless, nonmigratory or solitary form. Largely through the research of Uvarov and his associates at the Anti-Locust Centre in London, it is known that locusts directly influence the development and sexual maturation other locusts (see Uvarov, 1966). Hoppers reared in isolation develop into the solitary phase, but as few as two hoppers reared together shift development toward the gregarious phase. Up to a certain density, increases in the number of hoppers reared together increase the shift toward characteristics of the gregarious phase. Beyond a certain density, no further change is noticed. As a result, a continuous series of adult forms can be reared experimentally.

A pheromone, called the gregarization pheromone, stimulates the young instars to congregate. Placing locusts together also changes the adult stage. Caging sexually immature gregarious females and males together with sexually mature males increases their rate of maturation and changes their color from pink to yellow. Loher (1960) found that the epidermal cells of mature males secrete a volatile pheromone that is stimulating to other locusts, even at a distance. The pheromone, now called the maturation pheromone, stimulates the yellowing of males and the maturation of both sexes. Crowding of hoppers and adults, therefore, increases the sensory stimuli (probably including tactile, visual, and auditory stimuli as well as olfactory stimuli) that, acting through the corpora allata of the endocrine system, promote the development of the gregarious phase. Serotonin produced in the central nervous system has been shown to be necessary and sufficient for this behavioral change.

Certain geographic areas, known as "outbreak areas," have been identified from which migratory swarms regularly emerge. These are usually mixtures of habitats favorable to oviposition and offering food and shelter. In such areas, the locusts are sparsely distributed in the solitary phase and cause no economic damage. At intervals several years apart, however, conditions such as rainfall become more favorable for increased breeding. As a result of population buildup, the immature locusts become more concentrated and more of the resulting adults are of the gregarious phase.

Similar phase changes in response to crowding are known in about two dozen other species of acridids, but the gregarious phase is less regularly produced. Up until the late nineteenth century, *Melanoplus sanguinipes* (migratory grasshopper) and *Melanoplus spretus* developed migratory forms and moved in vast swarms from areas east of the Rocky Mountains to the Mississippi Valley and Texas. For hundreds of years before European colonization, some of these migratory grasshoppers landed on glaciers high in the mountains of Wyoming, and their frozen bodies have been found entombed. Since the natural environment was altered by agriculture, conditions have not again favored the formation of the migratory phase. *Melanoplus spretus* is now considered extinct, but *Melanoplus sanguinipes* continues to be an agricultural pest over wide areas of the semiarid West.

CHAPTER 5

Maintenance and Movement

This chapter is devoted to the nutrition of insects and the various organs involved in the general flow of nutrients and oxygen into the insect's body, the utilization of energy to produce movements, and the elimination of waste products (Fig. 5.1). Other aspects of feeding behavior are treated in Chapters 6, 10, and 11.

NUTRITION

Insects have an impressive diversity of diet. Examples of insect foods include virtually all types of plant tissues—including wood. Some insects eat only blood; other insects ingest fur or dung. Yet all insects' basic nutritional requirements are remarkably similar to those of other types of animals, perhaps because of deep evolutionary similarities in cellular biology and biosynthetic systems (see Table 5.1 for a general list of essential nutrients). Essential nutrients are those whose omission from the diet prevents further growth, development, or reproduction. Yet the specific dietary requirements for a given species in terms of origin, molecular structure of nutrients, and relative amounts remain difficult to define because of the dynamic nature of internal metabolism and the interactions of nutrients within the total diet.

Nutritional Ecology

Some species of insects have a close association with symbiotic (meaning "living together") microorganisms that provide essential nutrients. The symbionts are passed to succeeding generations by special devices, especially when the insect feeds on nutritionally inadequate diets. The diet may lack essential nutrients or the nutrients are not available because they cannot be digested by the insect. Some symbiotic relationships are external to the insect, as in the fungus gardens cultivated by ambrosia beetles, some gall midges, fungus termites, wood wasps, and fungus ants. In other insects, the microorganisms are present in the intestinal tracts of insects, in their hemolymph, or even in special organs called mycetomes or bacteriomes. Dependence on symbionts is associated most often with feeding on wood, plant sap, vertebrate blood, or other food sources that may be deficient in essential amino acids, sterols, or vitamins. Blood is deficient in several B vitamins. Microbes are believed to supply these and other missing nutrients. In contrast, fleas and mosquitoes feed on blood only during the adult stage and lack obligate symbionts. Special experimental cultures of insects devoid of other organisms may be prepared for nutritional studies. Such insects, deprived of their normal microbes, fail to grow or may die but will complete development if the diet is supplemented with vitamins or other nutrients.

Plant-feeding Hemiptera that suck plant vascular fluids exist on a diet deficient in nitrogen or particular forms of nitrogen and vitamins. Microorganisms can supply vitamins and aid in recycling nitrogen from waste products inside the insect. Some wood-feeding termites depend on flagellate protozoa in the hindgut to digest cellulose

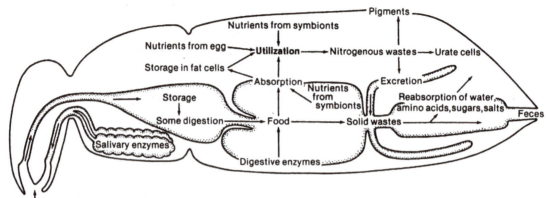

Figure 5.1 Diagram of routes of nutrients and wastes in an insect.

Table 5.1 Nutritional requirements of insects

Water	**Water soluble growth factors:**
Energy sources	**B Vitamins:**
	Thiamin (B1)
Amino acids (L-isomers):	Riboflavin (B2)
Arginine	Pyridoxine (B6)
Lysine	Nicotinic acid (Niacin)
Leucine	Pantothenic acid
Isoleucine	Biotin
Tryptophan	Folic acid
Histidine	
Phenylalanine	**Ascorbic acid (C)**
Methionine	**Lipogenic growth factors:**
Valine	Choline and myo-inositol
Threonine	**Ribonucleic acid (Diptera)**
	Minerals:
Lipid growth factors:	Potassium
Sterols:	Magnesium
Cholesterol, 7-dehydro	Phosphorous
cholesterol,	**and trace amounts of:**
beta-sitosterol, or ergosterol	Sodium
	Calcium
Polyunsaturated	Zinc
fatty acids:	Copper
Linoleic or linolenic acids	Sulfur
Palmitoleic acid (some Diptera)	Aluminum
	Manganese
Fat soluble vitamins:	Iron
Vitamin A (Retinol)	Cobalt
Vitamin E (alpha-tocopherol)	Chlorine

and make oxidizable energy sources available, commonly in the form of acetic acid. Most termites depend on bacteria as symbionts, although many termite species produce their own cellulases.

Insects may store nutrients in their fat body to be utilized in a subsequent stage of the life history. Female mosquitoes, for example, suck blood only as adults and use essential nutrients obtained while they were aquatic larvae. Adults of many Lepidoptera, Diptera, and Hymenoptera also depend to a large extent on reserves built up by the larvae but need additional nutrients for egg production and longevity of females, as well as carbohydrates for energy.

In other insects, eggs taken from well-fed parents may contain enough reserves of certain essentials to permit the resulting offspring to survive on an experimentally deficient diet. Effects of the maternal nutrition on offspring are documented in ants, locusts, and cockroaches. Some nutrients can replace others, such as many sterols for cholesterol, where the requirement is actually for only a part of the molecule. Substitutes or combinations may also replace certain functions of essential nutrients or have a "sparing action" when a structural similarity exists. Utilization of a diet depends not only on the chemical constituents, but also on the quantity of each with respect to others. Excesses of a single constituent may inhibit growth; deficiencies in one can decrease the use of others.

Dietary requirements can evolve over time. For example, in order to thrive and reproduce on a specific host plant, a phytophagous insect (plus its symbionts) must be able to adequately digest and utilize the unique combination of nutrients supplied by the plant. When the success of the insect becomes detrimental to the survival of the plant population, the plant host may evolve defensive responses such as protective structures, lowered dietary value to the insects, antifeeding compounds, or toxins. This in turn may provoke corresponding changes in the insect that allow survival on the altered plant host (see Chap. 10).

Essential Nutrients

Water
Water is sometimes overlooked as a nutrient because of its universal requirement by all living things. Depending on an insect's behavior and physiology, water can be obtained directly from drinking, moisture in food, oxidative metabolism, or absorption of water vapor. The water content of food varies greatly. For phytophagous insects, water is available in plant leaves, sap, and nectar, but seeds, nuts, grains, and pollen have low water content. When large amounts of water are taken in by feeding directly on the plant's vascular system, the excretory system must regulate the excess and conserve needed substances in the hemolymph.

The sugar-rich honeydew excreted by sap-feeding insects such as aphids makes water and unused nutrients available to other insects. Predators and parasitoids obtain water from fresh insect prey, while insect parasites of mammals and birds suck blood. For insects otherwise associated with mammals, wounds and secretions like sweat or mucus around the eyes, mouth, or nose are attractive sources of moisture. Scavengers on dry animal matter or wood-feeders must deal with food low in water content.

Oxidative metabolism of organic compounds is an important source of water that can be conserved by the excretory system. Insects may lose substantial water during vigorous flight while the spiracles remain open to supply oxygen to the flight muscles, yet metabolic water from the same muscular effort helps to offset the loss. By conserving metabolically produced water, insect pests in dry, stored grains have higher water content than the food they eat.

In a moist atmosphere, some insects can absorb water passively, except as limited by the waxy epicuticle. On the other hand, special structures capable of active absorption by expending energy allow some insects to take water from unsaturated atmospheres. Larvae of *Tenebrio molitor* (Tenebrionidae, Coleoptera), the mealworm, are able to absorb water vapor in the rectum as a function of their cryptonephric excretory system. Bladders in the mouth of the desert cockroach, *Arenivaga investigata* (Polyphagidae, Blattodea), absorb water vapor. Likewise, specialized mouthparts are involved in water vapor absorption by Psocoptera and Phthiraptera.

Energy Sources
Carbohydrates, fats, organic acids, and suitable amino acids can supply needs for energy from oxidative metabolism. Carbohydrates are readily available in a variety of sources, including plant tissue, nectar from flowers, and honeydew (plant sap) excreted by aphids. Glucose, fructose, and sucrose are commonly utilized monosaccharides and disaccharides, but some sugars, such as pentose sugars, are not used and may be toxic. Energy from sugars is often stored in the fat body as glycogen. Insect flight muscles convert more energy per unit weight

than any other animal tissue. For this reason many beetles, flies, wasps, and bees frequent flowers for sucrose-rich nectar. Long-distance fliers (many Lepidoptera, locusts, Coleoptera, and the bug *Lethocerus* [Belostomatidae, Hemiptera]) usually consume their stored sugars and then utilize fat. The latter is suited as a reserve energy source because it yields twice as many calories per gram as do carbohydrates. Fat is frequently stored in the fat body of larval insects for later use during pupation, energy for nonfeeding adults, or development of eggs in females. Trehalose, the major sugar in insect blood, also is used in the synthesis of chitin in the cuticle.

Amino Acids

Enzymes and structural components of cells are proteins, which are synthesized from amino acids. Hence, dietary proteins or amino acids are required for growth and maintenance. Insects require at least the same essential ten amino acids required by many animals such as the rat (Table 5.1). The total nitrogen in food, measured as a percentage of dry weight, is a useful guide to the amount of nitrogen-containing compounds in the food. Predatory, parasitoid, and scavenger insects that feed on the bodies of other insects or animals have rich sources of nitrogen in an easily used form such as protein or amino acids. Phytophagous insects that feed on leaves, seeds, pollen, or fungi are supplied with less nitrogen, much of which is not readily available for digestion and assimilation. Nitrogen content in leaves generally declines with age, making young leaves the most nutritious. Insects that feed on decay-free dead wood (i.e., without fungi as an additional source of food) or leaf litter receive still less nitrogen. Nitrogen and other compounds in phloem fluid, and especially in xylem fluid, are so dilute that a sap-feeding insect must take in large amounts of liquid to extract adequate nutrients.

Methionine in the diet is important because insects cannot synthesize sulfur-containing amino acids. Phenylalanine or its derivative tyrosine is required because tyrosine is a precursor of acetyl-dopamine, the tanning agent in sclerotization of the cuticle. Some species have been found also to require asparagine, aspartic acid or glutamic acid, and proline. Bacterial symbionts of aphids synthesize both methionine and tyrosine.

Lipids

Lipid requirements include fatty acids and sterols. Reserve energy sources in the form of fat, mentioned above, are derived from either sugars or dietary saturated and monosaturated fatty acids. Fatty acids for energy are not deemed essential because alternate sources of energy are available, but some butterflies and moths require polyunsaturated fatty acids for normal development. Diptera have a unique lipid metabolism not dependent on polyunsaturates. Their major fatty acid is palmitoleic, and their phospholipids are of the ethanolamine phosphoglyceride type.

In contrast to mammals, insects are unable to synthesize sterols. In addition to being essential components of cell membranes, a sterol is required for the production of ecdysone and other growth hormones. Foods vary in the predominant kind of sterol supplied to insects: animal foods have cholesterol, plants have β-sitosterol, and fungi have ergosterol. Regardless of the origin, the presumed biogenetic pathway to ecdysone involves converting cholesterol to 7-dehydrocholesterol. Cholesterol is ordinarily adequate in an artificial diet, but some insects with highly specific diets have apparently lost the ability to convert cholesterol: the fly *Drosophila pachea* (Drosophilidae, Diptera) requires the cactus sterol schottenol from its host *Lophocereus schotti*.

Vitamins and Growth Factors

Vitamins are essential for living processes, but they are not used for energy or basic structures. Insects at one time were thought to need only water-soluble vitamins. Among these, seven B vitamins are required for most insects (Table 5.1), and some species also require cyanocobalamin, B_{12}, and carnitine, B_T. Plants are usually a rich source of ascorbic acid (vitamin C), and most plant-feeding insects require vitamin C in their diet in concentrations higher than those required by vertebrates, apparently having lost the ability to synthesize it.

Among the fat-soluble vitamins are D (calciferol), involved in calcium metabolism; E (α-tocopherol), an antisterility factor for mammals; A (retinol) or precursors, involved in carotenoid visual pigments; and K (phylloquinone, etc.), essential for clotting mammalian blood. Vitamin D is not required because insects do not have calcareous skeletons and are apparently unable to use it even as a sterol. The clotting of insect blood differs from that of mammals and does not involve vitamin K, but some experimental evidence exists that vitamin K stimulates growth in insects. Vitamin E deficiencies in insects have been difficult to demonstrate consistently, but both egg and sperm production, as well as larval growth, may be affected. A need for vitamin A has long been suspected because the visual pigments of insects are rhodopsins. Morphological and visual defects have been found in several species of insects reared on vitamin A–deficient diets.

The growth factors choline and *myo*-inositol are required as constituents of phospholipid molecules. Ribonucleic acid has been found to be required by some Diptera.

Minerals

Insects require approximately the same minerals as do vertebrates because minerals, of course, cannot be synthesized. In comparison to mammals, insects require much less calcium (not needed for the skeleton), less iron (few insects have hemoglobin, but otherwise iron is essential for many cellular processes), and much less sodium and chloride (many insects do not use these in homeostatic functions of blood). Although required in only trace amounts, zinc is apparently so critical in the moth *Heliothis virescens* (Noctuidae, Lepidoptera) that it is conserved by transferring it in the male semen to the female during copulation.

ANATOMY AND FUNCTION OF THE ALIMENTARY CANAL

The alimentary canal of an insect is a tube passing through the body that provides the special internal environment for food to be mechanically and chemically disintegrated and brought to the vicinity of absorbent cells. Despite the diversity in physical consistency and composition of food, the gut exhibits certain general anatomical features throughout the insects (Figs. 5.2, 5.3).

Anatomy of the Alimentary Canal

The gut is attached at each end to the body wall but otherwise is held in place in the hemocoel largely by the pressure of adjacent organs and flexible tracheae. The three major divisions are: (1) the stomodaeum or foregut (Fig. 5.2a), derived from embryonic ectoderm; (2) the mesenteron, or midgut, derived from endoderm, and (3) the proctodaeum, or hindgut, derived from ectoderm. Attached at the juncture of the midgut and hindgut are the filamentous excretory organs called Malpighian tubules. In the embryo, the Malpighian tubules arise from the anterior end of the developing hindgut.

The wall of the alimentary canal is composed of a single layer of cells separated from the hemolymph by a basement membrane. The foregut and hindgut are both lined with a thin cuticle, called the intima, that is continuous with the integumental cuticle and is usually impervious to water. These portions of the gut are properly classed as organs of the integument, and the cell layer is homologous with the epidermis. The intermediate section, the midgut, is differentiated by its endodermal origin and lack of an intima. The cells of the midgut are called epithelial cells or the epithelium. In place of an intima, however, is secreted in most insects a loose film, the peritrophic membrane, that, in addition to other functions, protects the delicate cells from abrasion by food particles and also compartmentalizes digestive processes.

Muscular connections between the gut and exoskeleton are limited to near the mouth and pharynx to provide a pumping action. Layers of circular and longitudinal muscles in the gut wall push the food along by peristaltic contractions. These are controlled by sensory and motor innervations from the stomodaeal nervous system (Chap. 6) to the foregut and, in some insects, the midgut. The hindgut receives nerves from the posterior abdominal ganglia of the ventral nerve cord. Stretch receptors along the gut provide inputs to the nervous system that can affect development and reproduction and numerous other physiological processes. The scarcity of sensory nerves probably means that insects rarely suffer stomachaches. Tracheae from visceral trunks supply oxygen to the alimentary canal according to the metabolic needs of the different regions.

The true mouth is hidden from view and opens into the preoral cavity above the hypopharynx and below the epipharynx (Fig. 5.2b). The preoral cavity

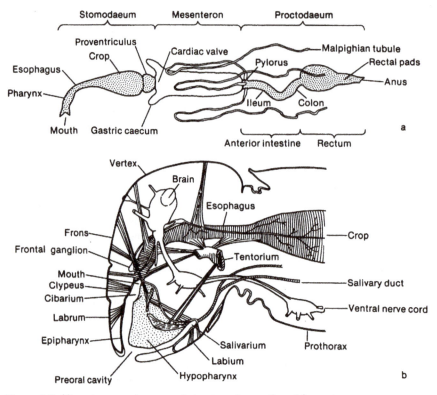

Figure 5.2 Alimentary canal: a, general structure; b, mouth and foregut.
Source: Redrawn from R. E. Snodgrass, The Principles of Insect Morphology.

directly in front of the mouth is the cibarium, where food is passed to the mouth. The preoral cavity also contains another chamber beneath the hypopharynx and above the labium. This is the salivarium, where the saliva is emptied into the cavity from the labial or salivary glands.

The first distinctive region of the foregut is the pharynx. This heavily muscled organ is situated behind the mouth. The esophagus is merely a tubular portion of the foregut leading to the saclike crop (Fig. 5.3a). The release of food from the crop is controlled by an intricate organ, the proventriculus. Like the pharynx, the proventriculus may have a heavy muscular coat. Internally the cuticle is often shaped into heavy teeth, plates, or spines that can act as a filter that regulates food passage or grinds the ingested food into smaller particles. In fluid-feeding insects and many larvae, it is a simple sphincter. The foregut may be expanded into a large sac or have an attached sac that is an outpocketing

from the foregut for use as a storage organ from which fluids may be regurgitated (as in the case of honey bees) or slowly released into the midgut. The most posterior part of the foregut is invaginated in the midgut and called the cardiac valve, or proventricular valve (Fig. 5.3b). It may either be heavily muscled to open and shut the connection between the foregut and midgut, or it may lack muscles but open passively in response to the positive pressure of incoming food from the foregut and close when the pressure difference reverses to prevent food movement back into the foregut. Special secretory cells between the valve and the midgut wall are partly responsible for the formation of the peritrophic membrane in larval and adult flies (except blood-feeders) and some Lepidoptera.

The midgut may be a simple tube or variously modified. The epithelium may be differentiated into distinctive histological regions. The columnar cells are especially active, since the secretory

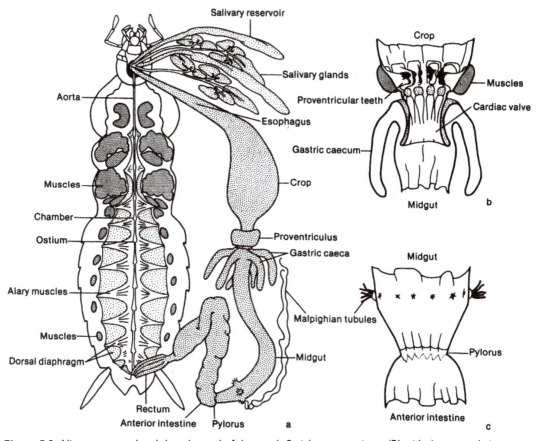

Figure 5.3 Alimentary canal and dorsal vessel of the roach *Periplaneta americana* (Blattidae): a, ventral view of dissected roach with fat body, reproductive organs, and most of Malpighian tubules removed; b, interior of proventriculus and cardiac valve; c, interior of pylorus.

processes that yield the digestive enzymes often involve the eventual disintegration and rejuvenation of the cellular lining. Next to the lumen, the cells characteristically are densely covered with short, hairlike filaments, or microvilli. Secretory cells lost during enzyme production or by molting are replaced by regenerative cells. Substances may be accumulated in epithelial cells for storage.

In insects such as grasshoppers, true bugs, and larval scarab beetles, hollow, fingerlike pouches may be present at the anterior or posterior ends of the midgut. These are the gastric caeca, which shelter symbiotic bacteria in some species such as stinkbugs (Hemiptera, Pentatomidoidea). At the junction of the midgut and hindgut are the Malpighian tubules. These serve as a convenient

external marker to distinguish the posterior end of the midgut. Also in this area is the pyloric valve (Fig. 5.3c). The valvelike function is best developed in caterpillars and beetles.

The hindgut may have several recognizable sections: (1) an anterior intestine, composed of the ileum, which is larger in diameter, and a narrow posterior portion, the colon; and (2) the posterior intestine, or rectum, which is the final section before the anus. The hindgut is lined with cuticle, which is usually impermeable. The rectal pads (see Fig. 5.8b, below), however, have a cuticular lining that is permeable to water, salts, sugars, and amino acid molecules. Together with the Malpighian tubules, the ileum and rectal pads form parts of the excretory system. In the rectum, the waste is dehydrated and often compressed by muscles into a pellet that bears the shape of the rectal walls.

Modifications of the Alimentary Canal

The alimentary canals of primitive insects such as Collembola as well as many larvae of Endopterygota are simple, direct tubes with but slight modifications even in the regions of the cardiac and pyloric valves. More advanced chewing insects, including Orthoptera, Odonata, Hymenoptera, and many Coleoptera, have an enlarged crop and proventricular specializations to handle particulate meals, plus frequent caecal pouches and enlarged rectal pads. The ingestion and passage of a fluid diet poses other engineering problems. Some fluid-feeders, such as some Diptera and Lepidoptera, possess lateral, bladderlike diverticula protruding from the crop that receive liquids. The midgut is enlarged in Siphonaptera and many Hemiptera (suborder Heteroptera). The most elaborate arrangement is the filter chamber of many Hemiptera (suborders Auchenorrhyncha and Sternorrhyncha), which is discussed later.

During the larval instars of Neuroptera and some Hymenoptera, the midgut is a blind sac, not connected to the hindgut until pupation. As a result, waste products are retained in the larval gut and excreted on emergence as adults. For insects such as larval bees that develop in a confined space, this has an obvious sanitary function.

The ileum is modified to contain symbiotic microbes in termites and larval scarab beetles. In the aquatic, immature stages of dragonflies, tracheated gills project into the rectum and serve to acquire oxygen from water circulated in the enlarged rectum. Forceful contractions of the rectum also eject the water and propel the nymphs forward.

Salivary Glands

The enlarged integumental glands of the head function in a great variety of life processes, but they are discussed here because they contribute generally to ingestion, movement of food, and digestion. The largest and most common glands that secrete saliva are the paired labial glands (Fig. 5.3a). The single, median salivary duct opens in the salivarium between the labium and hypopharynx.

The glandular cells extend back into the thorax, or even far into the abdomen in some caterpillars, like clusters of grapes and may be differentiated into regions. In addition to saliva, the modified labial glands are responsible for the silk secreted by larvae of Lepidoptera, Trichoptera, and Hymenoptera. These fibrous proteins are used in the construction of cocoons, webs, and nets. Other glands of the head found in various insects are the mandibular glands, which are the main salivary glands of larval Lepidoptera, also known to secrete the pheromone called queen substance in the queen honey bee (Chap. 7); the maxillary glands, which are small and probably provide lubricants for the mouthparts; and the pharyngeal glands, which are best developed in worker honey bees for production of the food royal jelly, fed to larvae (Chap. 7).

Functionally the saliva of the labial glands moistens and lubricates the mouthparts, dissolves food, transports flavors to gustatory (taste) sensilla, contains enzymes that act on the food both before and after ingestion (Fig. 5.1), and also serves as an excretory organ. Aphids, mirids, a leafhopper, and a lygaeid bug are aided in the penetration of their piercing/sucking mouthparts into plants by a pectinase in the saliva. This enzyme hydrolyzes the pectin in the cell walls of plant tissues. Many sucking insects in the order Hemiptera secrete a salivary sheath around their feeding punctures. The sheath presumably functions to retain the turgor pressure of the plant's vascular system, thus enhancing the insect's sucking mechanism.

Bloodsucking insects, such as the biting flies and parasitic Hemiptera, facilitate ingestion by the inclusion of anticoagulins and antiagglutinins in the saliva. Blood clots would otherwise plug the fine canals of the insect's mouthparts. Digestive enzymes are notably absent in the saliva of these insects, and some have a local anesthetic, perhaps to avoid irritating the host while they feed. Enzymes and other proteins in their saliva, however, act as antigens that invoke an immune response that causes the irritation and swelling that follows the bites of mosquitoes and other blood-feeding insects.

The injection of saliva by insects into plant or animal hosts provides a natural entry point for pathogens. Complex cycles of transmission have been evolved by viruses and microbes that take advantage of the insect vector's feeding behavior (Chap. 11, Insects as Vectors of Microbes).

Peritrophic Membrane

The peritrophic membrane (meaning membrane around the food) is a thin (about 1 μm thick), tubular film surrounding the food and forming a loose lining inside the midgut and hindgut. The membrane is often multilayered or laminar. Structurally it is a fibrous network, or an amorphous sheet of polysaccharide, including chitin, glycoprotein, and proteins (peritrophins). In most insects that have been examined, the membrane is produced from microvillate columnar epithelial cells distributed locally or generally in the midgut epithelium. In many fewer insects, secretory cells located in a ring or rings between the invaginated cardiac valve and the anterior end of the midgut form a "cardiac press." In certain Diptera, for example, the secretory cells in the cardiac press extrude a tubular membrane of several layers around the incoming food bolus as it passes through the cardiac valve. In Dermaptera, both methods have been reported, and in mosquitoes, the membrane is produced by the cardiac press in larvae and by the epithelium in adults. Regardless of the method, the transparent membrane continues to move posteriorly along with the food and is ultimately discarded, often still in place around the fecal pellets. The activation of the secretory cells is related to the hormonal controls of metamorphosis and in response to a feeding stimulus or gut expansion.

The fluid-feeding Hemiptera lack a peritrophic membrane, yet they have a unique multilayered matrix on the apical surfaces of the midgut epithelial cells that may function like the peritrophic membrane of other insects. The membrane presumably acts to protect the exposed midgut epithelium from abrasion by food particles and to lubricate and facilitate movement of food. In this, the membrane functions like mucus, a substance not found in insects. The membrane proteins (peritrophins) also resemble mucus proteins (mucins) in chemical composition and in the kind of secretory cells that produce it, suggesting that mucus and the peritrophic membrane may have evolved from a common ancestor.

The peritrophic membrane covers the food and separates it from direct contact with the epithelium. It has minute pores that appear to be selectively permeable. The pores may act as a filter controlling the movement of enzymes into the ingested food and the movement of the products of digestion into the midgut cells.

Disease organisms that are ingested commonly invade the bodies of insects by penetrating the midgut and entering the hemocoel. These include useful microbial agents for biological control of pests as well as agents of harmful human diseases carried by insect vectors. The peritrophic membrane is a critical barrier to invasion and may effectively exclude certain microbes. The pore sizes are usually smaller than bacteria, yet some bacteria, viruses, trypanosomes, and filarial worms are able to penetrate the membrane either mechanically or by action of their enzymes.

Digestion

Digestion of food may begin outside the insect, or extraorally, but most digestion occurs in the midgut. The salivary labial glands and midgut epithelium are the primary sources of digestive juices. For solid food, mechanical breakdown begins with the mouthparts and may be continued with the cuticular proventriculus. The reduced size of food particles exposes larger surface areas to the action of digestive enzymes. The enzymes are hydrolases, commonly including proteases, lipases, carbohydrases, and nucleases. Thus, complex organic constituents in the food are reduced stepwise to a molecular structure that can be absorbed by the epithelial cells of the midgut.

As a tubular membrane enclosing the food within the tubular midgut, the peritrophic membrane is believed to have important functions in digestion by compartmentalizing digestive enzymes and food particles and channeling the flux of food flow. Enzymes initially break down the polymers in the food while it is covered by the membrane. Intermediate oligomers pass through the membrane to be finally digested into monomers next to the apical surfaces of the epithelial cells by cell membrane–bound enzymes and immediately absorbed. In general, digestive enzymes are secreted by epithelial cells, pass through the peritrophic membrane, and begin hydrolyzing the food. The enzymes move posteriorly with the food and, on reaching the posterior midgut, pass back through the membrane. Fluids secreted by the epithelium in the posterior midgut move anteriorly between the membrane and the epithelium. This countercurrent flow of fluids outside the membrane carries forward the enzymes and products of digestion. Once in the anterior midgut, the enzymes pass again through the membrane to the food and repeat the cycle, reducing the need for enzyme production.

Extraoral digestion is widespread among the animal phyla and is common among predatory insects. Enzymes, especially proteases, in saliva or regurgitated from the midgut are injected into prey. The body of the victim becomes a container in which the tissues are liquefied and sucked out. The enzymes also are recovered, thus conserving protein, and continue their action after ingestion. Paralysis and extraoral digestion of prey are accelerated when assassin bugs (Reduviidae) inject hyaluronidase into prey. This spreading agent probably dissociates the cells of the prey and hastens the action of digestive enzymes, which are concurrently injected. It is because of the enzymes that the bites of predatory Hemiptera are painful to people. The venomous saliva can also be defensively ejected at vertebrate enemies. The housefly ejects saliva on solid food, and the partly digested liquid is eaten. Enzymes such as amylase, which converts starch to maltose, or invertase, which converts sucrose to glucose and fructose, are commonly present in saliva.

Digestion in Orthoptera, Blattodea, and some Coleoptera occurs mainly in the crop with enzymes produced by the salivary glands or regurgitated from the midgut. In Diptera, Lepidoptera, and Trichoptera, digestion apparently is limited to the midgut.

The honey bee adds salivary invertase to sucrose-rich nectar that is carried back to the hive in the crop. The digested mixture of glucose and fructose is regurgitated in the hive, dehydrated, and stored in the comb as honey.

Digestion of Unusual Foods

Insects are exceptional in being able to digest some abundant, yet resistant, compounds. Structural proteins in vertebrate tissues, such as collagen of connective tissues or keratin in wool, hair, and feathers, are highly resistant to degradation by common proteases. This is evident when a dead bird or mammal decays and only the resistant tissues remain. The feces of blowfly larvae (Calliphoridae) feeding on such corpses contain a collagenase that attacks the collagen and elastin of muscle tissues, resulting in extraoral digestion. Bird lice (Mallophaga) and the larvae of clothes moths (Tineidae), house moths (Oecophoridae), and dermestid beetles possess a keratinolytic enzyme that attacks keratin. The moths and beetles are household pests of woolen clothing and carpets. A strong reducing agent is able to break the sulfur bonds of keratin because poor tracheation in the insect's midgut provides a low oxidation-reduction potential. Once the peptide chains are free, the keratinase is effective.

Beeswax is ordinarily resistant to digestion, but the larval wax moth is able to utilize some wax components. At an optimal pH of 9.3 to 9.6 in the gut, the wax is hydrolyzed by enzymes originating from the larva's cells, from intestinal bacteria, or both.

Woody plant tissue is regularly exploited as a food source by larvae of some wood-boring beetles (some Anobiidae, some Scarabaeoidea, Buprestidae, Cerambycidae, Lyctidae, Bostrichidae, some Tenebrionidae, and others); termites; wood-feeding roaches; and silverfish. Cellulase, hemicellulase, lignocellulase, and lichenase have been demonstrated in insects, some or all of which have been identified in members of the above list. The most primitive families of termites house symbiotic flagellate protozoa and bacteria in the gut that are critical in digesting cellulose. Bacteria in the midgut caeca of Scarabaeoidea beetles are thought to digest cellulose. Some of these enzymes, however, are directly secreted by insects, for example, more recently evolved termites and the larvae of some cerambycid beetles. Among wood-boring insects, the ability to digest cellulose to varying degrees permits partitioning of the food resource: some thrive only on the more easily digested, but scarce, starch and sugar in the wood (Lyctidae, Bostrichidae); others also use the hemicelluloses (Scolytidae); and a third group consumes the whole cell wall, minus the lignin (Anobiidae, Cerambycidae).

Absorption

The midgut epithelium and the hindgut are the principal absorptive regions. Although digestion may take place in the crop, absorption is largely prevented by the cuticular intima. In the midgut, monosaccharide sugars, commonly glucose, rapidly pass through the permeable epithelial membrane without an active transport mechanism as long as the concentration inside the cells is lower than outside. A steep osmotic gradient is maintained by converting the simple sugar to the disaccharide trehalose. In the migratory locust this conversion takes place in the fat body, which is near the midgut. In the silkworm, oligosaccharides may be absorbed and further hydrolyzed in the epithelium.

Amino acids are similarly absorbed by passive diffusion processes related to the uptake of water into the hemolymph. In *Rhodnius* (Reduviidae, Hemiptera) even proteins that are not fully degraded to amino acids may be absorbed and then digested further by the midgut cells. The phloem and xylem sap in plants contains amino acids in monomeric form (in concentrations of 3–50 mM) that need no digestion and are efficiently absorbed by these insects: up to 55 percent efficiency in aphids (Aphididae) and at least 99 percent efficiency in xylem sap–feeding leafhoppers (Cicadellidae).

The absorption of lipids is more difficult to analyze because it is possible that these molecules could pass through the cuticular intima of the crop or between epithelial cells of the midgut, in addition to the usual route via the epithelial cells. Lipids have been seen as droplets in cells of both the midgut and crop. Although cholesterol has been recently demonstrated to be absorbed by the crop of a cockroach, the midgut is known to be the major absorptive region. The fats need not be fully degraded before absorption. No evidence exists that an active transport mechanism is involved.

TRACHEAL SYSTEM AND GAS EXCHANGE

Oxygen for cellular respiration in insects is physically transported by an internal system of air-filled tubes, the tracheae, to within a few cell diameters of each cell (Fig. 5.4a). Respiratory pigments and blood are not involved in gas exchange in most insects. This is in contrast to oxygen transport in vertebrates, which depends on the pigment hemoglobin as an oxygen carrier and the rapid circulation of blood throughout the body. The respiratory system of terrestrial vertebrates has clearly evolved from that of aquatic ancestors, but the aerial systems of insects and certain other arthropods are thought to be peculiarly terrestrial adaptations that may have independently evolved in several arthropod stocks. Tracheae are highly branched and interconnected tubular invaginations of the integument that arise from paired, lateral openings, the spiracles, and ultimately terminate in fine fluid-filled tubules, the tracheoles (Fig. 5.4f).

Small insects, such as Protura and most Collembola that live in damp places or some aquatic larvae of Chironomidae (Diptera), lack a developed tracheal system and exchange gases through the cuticle. In larger, free-living, terrestrial insects, the demand for oxygen cannot be met by the body surface, partly because the area is inadequate for sufficient transfer and the thicker cuticle permits only a negligible diffusion of oxygen. More importantly, a body wall permeable to oxygen diffusion would also be permeable to water loss; a molecule of water is smaller than a molecule of oxygen gas. The tracheal system solves this problem by creating an enormous internal surface of permeable cuticle where loss of water can be controlled. The fifth-instar larva of the silkworm has an estimated 1.5 million tracheoles.

Diffusion, the random movement of molecules from a higher to a lower concentration, is a major process by which oxygen moves into and carbon dioxide moves out of respiring tissues. Diffusion is a millionfold more rapid in the gas phase than in tissue. Calculations of oxygen consumption, diffusion rates, and measurements of tracheae indicate that diffusion alone could supply small insects (<3 mm long) with active flight muscle. Ventilation, or moving drafts of air through the system, can greatly reduce the distance that oxygen must diffuse over a gradient from atmospheric concentration to respiring tissue. However, the tracheal system imposes maximum limits on the size that insects can attain. Even forced movements of air cannot overcome the size limitations of diffusion as the primary means of gas exchange. To illustrate by analogy of diffusion with sound communication,

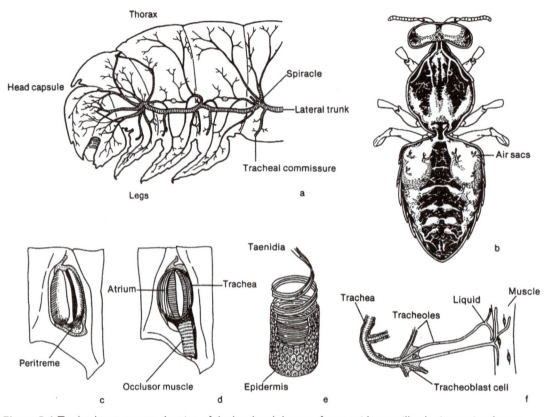

Figure 5.4 Tracheal system: a, tracheation of the head and thorax of a noctuid caterpillar; b, air sacs in a honey bee; c, external view of closing mechanism of a thoracic spiracle; d, internal view of same; e, structure of a trachea; f, structure of a tracheole.
Source: a, c–f, redrawn from R. E. Snodgrass, The Principles of Insect Morphology; b, redrawn from Snodgrass, 1956.

sound (330 meters per second) is as effective as electronic communication (300,000 kilometers per second) within a room, but it is not adequate for distances greater than tens or hundreds of meters. Similarly, diffusion aided by ventilation is adequate over distances measured in millimeters but not more than a few centimeters.

The passage of air into the tracheal system is regulated by the spiracles. In some apterygotes, such as the springtail *Sminthurus* (Sminthuridae, Collembola), the opening is simply a hole in the body wall. The rate of gas diffusion through this pore, as through the stomata of plant leaves, is a function of the perimeter rather than the cross-sectional area. Elaborate modifications of the spiracle, therefore, are possible without necessarily reducing

the intake of air. In most insects the tracheal orifice is recessed in the integument, creating a small chamber, or atrium, which communicates with the exterior. The atrium may be situated on a separate sclerite, the peritreme, and provided internally with filtering hairs or projecting, valvelike lips, closable by muscles (Fig. 5.4c). The latter are characteristic of thoracic spiracles and are not found on the abdomen. The atria of abdominal spiracles remain open, with a closing device located inside at the tracheal orifice. The opening is pinched shut by the action of an occlusor muscle on one or two sclerotic bars. When the muscle relaxes, the elastic bars spring open or they are pulled apart by a second muscle, the dilator. Spiracular muscles are innervated from their segmental ganglion or the ganglion immediately anterior. Tracheae leading from the spiracles may be connected longitudinally by lateral trunks and transversely by tracheal commissures.

The histological parts of tracheae are homologous to corresponding parts of the integument: an inner, cuticular intima secreted by a surrounding layer of epidermal cells that are bounded outside by a basement membrane. The intima is thin, permeable to gases, and marvelously flexible because of fine, spiral thickenings called taenidia (Fig. 5.4e). The epicuticle contains at least an outer layer and probably a hydrofugic wax layer next to the lumen. Corresponding to the external procuticle is a chitin and protein layer, which is largely (if not entirely in smaller tracheae) involved in constructing the taenidia. At each molt, portions of the intima are routinely shed along with the exuviae. During metamorphosis of flying insects the entire system is reorganized to accommodate the increased demand for oxygen by the flight muscles. New tracheal branches or fusions are developed, and old tracheal pathways are not replaced, yet the design of major trunks and anastomoses is consistent in taxonomically related forms and may be traced from the immature stages to the adult.

The tubular network of tracheae may be interrupted at intervals by enlarged dilatations, or air sacs (Fig. 5.4b). Taenidia may be discernible on the walls of some sacs as if the trachea is merely inflated, but other cavities are reinforced by reticulate thickenings. The sacs are collapsible, a feature of importance in ventilation and one that also permits internal organs to expand in the body cavity, as when the ovaries are enlarged by eggs or the gut is engorged with food. Associated with the tracheae in Diptera, Lepidoptera, and Hymenoptera are peritracheal glands, which consist of several cells each and are distributed pairwise in thoracic and abdominal segments. The cells are thought to secrete exuvial fluids during molting and possibly hydrofugic (water-repellent) lipids that coat the intima.

Tracheoles

As they branch, tracheae become narrower in diameter and terminate in tiny subdivisions, the tracheoles (Fig. 5.4f). These tubules are up to 350 μm in length and 1 μm in diameter, tapering to a diameter as small as 0.1 μm. Whereas the tracheal tube is enclosed in a multicellular coat, the epidermis, the tracheole is intracellular in a tracheoblast cell and may be exempt from renewal at each molt. When tissues are experimentally deprived of tracheae or organs are implanted in a new host, tracheoles migrate to the affected tissues and establish an oxygen supply. The same processes are presumably involved in normal development and injury repair. Although the general surface of the tracheal intima probably is diffusible by gases, the tracheole is believed to be the major site of oxygen transfer. Organs of high metabolic activity are richly tracheated. For example, the flight muscles of fast-flying insects are enveloped in tracheae and air sacs. The profuse tracheoles indent the plasma membranes of muscles to bring oxygen to the mitochondria deep in the fibers.

Tracheoles are often filled with a lymphlike liquid. During the depletion of oxygen in surrounding cells, this liquid is withdrawn against a capillary force by the tracheoblast cell, bringing gas to the tracheolar tip. The physical mechanism of withdrawal is not clear but is correlated with changes in osmotic pressure of surrounding tissues. A porous structure observed in the tracheolar outer layer may facilitate this process. Oxygen moves the final distance from the tracheoles to respiring cells by diffusing through tissue. The carbon dioxide produced by respiration exits by the reverse course, hastened by a greater solubility in water and a permeability through tissue thirty-six times faster than that of oxygen but diffusing slightly slower than oxygen in gaseous phase. The greater mobility of carbon dioxide in tissue favors some loss of this gas directly through the cuticle, especially at less sclerotized intersegmental membranes.

Ventilation

The distance gases must diffuse is shortened by ventilation of the tracheal system. Motions of the body or churning of internal organs of active insects incidentally move air through the interconnected major trunks. Air movement through the system is greatly accelerated by pumping movements of the telescoping abdominal segments, creating inspiratory and expiratory flows. When the spiracles remain open, air may enter and exit tidally through the same orifice. The closing valves attached to spiracles in most insects permit coordinated opening and closing of the tracheal orifices, resulting in unidirectional flow. Ordinarily air is taken in through the anterior spiracles and expelled via the posterior spiracles. Air sacs increase the volume of air pumped and may have other mechanical functions in ventilation, since they are partly collapsible. During flight, abdominal pumping is inadequate to supply the flight muscles. Movements of the thoracic segments, in conjunction with open spiracles, ventilate the air sacs and tracheae surrounding the muscles. Large, quiescent insects like diapausing moth larvae and pupae as well as temporarily inactive insects may have discontinuous ventilation that may span up to several hours. This cyclic process begins by closing the spiracles while a slight vacuum develops as oxygen is consumed, then partly opening and closing the spiracles to allow air to be sucked into the system, and finally fully opening the spiracles only briefly to release a burst of carbon dioxide.

X-ray cinematography of insects revealed that the major tracheal branches of a diversity of insect subjects rhythmically pulsate in successive contractions, which presumably increase air movement in the tracheal system. This finding challenges our current understanding of the importance of diffusion versus forceful ventilation in the gas exchange physiology of insects.

Ventilatory movement of air in and out of the insect's body increases water loss. Air in the tracheal system is near saturation while outside the humidity may be quite low. The structure of the atrium and closing mechanisms reduce or prevent the diffusion of water vapor from the system. It is therefore advantageous for the spiracles to remain closed, with the occlusor muscle contracted, until breathing is demanded. Spiracles may be controlled separately or in concert by the central nervous system. Cellular respiration leads to lowered levels of oxygen and higher levels of carbon dioxide. Changes in both gases will reduce impulses from the central nervous system to the occlusor muscle (thus relaxing the muscle and opening the spiracle by its elastic hinge) or, where

dilator muscles are involved, increase impulses to the dilator to stimulate opening. Occlusor muscles are also influenced directly by carbon dioxide and will relax when the local concentration is high. As a consequence of this reaction, care must be exercised when carbon dioxide is used as an anesthesia to immobilize insects: the open spiracles permit water loss that may cause death.

Hemoglobin

Hemoglobin functions in the gas exchange of only three insects studied thus far: the bloodworm, *Chironomus* (Chironomidae, Diptera); the backswimmer, *Anisops pellucens* (Notonectidae, Hemiptera); and the horse-botfly larva, *Gasterophilus intestinalis* (Gasterophilinae, Oestridae, Diptera). The hemoglobin of insects contains two heme groups, half as many as the vertebrate pigment. In each case, hemoglobin provides a temporary respite from oxygen deprivation. The bloodworm lives in stagnant or polluted water deficient in oxygen. Movements of the larva in its burrow refresh the cutaneous supply and charge the hemoglobin with oxygen, which is not released unless the tension in tissues is low. The bloodworm's hemoglobin has a high affinity for oxygen, and the uptake of oxygen is more rapid than diffusion in the hemolymph; hence, recovery from hypoxia is aided and aerobic respiration is sustained under poorly oxygenated conditions. The horse-botfly maggot is similarly situated in a liquid, oxygen-poor medium: a horse's stomach. Hemoglobin is in richly tracheated fat cells, permitting rapid uptake and storage of oxygen derived from passing gas bubbles. The bug *Anisops* utilizes hemoglobin not only to extend its period under water but also to sustain a favorable specific gravity that allows submerged floating.

Types of Tracheal Systems

The structure, number, and position of spiracles vary taxonomically: the apterygotes exhibit the most diverse arrangements and provide some insight into the evolution of the system. As already mentioned, most Collembola and Protura lack tracheae altogether. Some springtails, for example, *Sminthurus*, have a single pair of simple openings on the cervix, between the head and thorax, leading to branched tracheae at each side. The proturan *Eosentomon* (Eosentomidae) has two thoracic pairs, likewise leading to independent tracheal tubes. *Campodea* (Campodeidae) of the Diplura is similar, with three thoracic pairs of spiracles, but the system of *Japyx* (Japygidae) has four thoracic (two pairs for both mesothorax and metathorax) and seven abdominal pairs, making a total of eleven—the maximum for any insect. Moreover, the spiracles are connected at each side by a longitudinal trunk, permitting movement of air lengthwise but not transversely.

The Machilidae are relatively less specialized, with two thoracic and seven abdominal spiracles individually leading to mostly independent tracheae. The Lepismatidae, on the other hand, exhibit features that also characterize many of the winged insects: a pair of spiracles on each of the mesothorax, metathorax, and first eight abdominal segments plus longitudinal trunks and transverse segmental commissures. Ventilating drafts of air through this interconnected system are forced by movements of the body and internal organs.

Three major types of respiratory systems are recognized. The holopneustic respiratory system (ten pairs of functional spiracles) described for the Lepismatidae is found in the immature stages and adults of many terrestrial insects, including Orthopteroidea, Hemipteroidea, and some Hymenoptera. The hemipneustic system is derived by the loss of one or more functional spiracles and is characteristic of larval Neuropteroidea. The hemipneustic system is subdivided into the peripneustic arrangement, prevalent among terrestrial larvae (metathoracic spiracles nonfunctional; as in larvae of Cecidomyiidae), and the oligopneustic type of many Diptera (only one or two pairs of functional spiracles) that is specialized for life in water or liquid media. The oligopneustic type is further subdivided into the amphipneustic type (mesothoracic pair plus a pair of posterior abdominal spiracles functional, as in larvae of Psychodidae) and the metapneustic type (only the last pair of abdominal spiracles functional, as in larvae of Culicidae). The apneustic system of submerged aquatic insects and most endoparasitic larvae has no functional spiracles but does have a closed tracheal system. Oxygen

enters the body by cutaneous diffusion, either over the general body surface or in special, integumental gills. Tracheae within gills consist of very fine tracheae at the gill extremity that coalesce into larger tracheae that lead to the main tracheal branches.

Gas Exchange in Aquatic Insects

Nearly all aquatic insects (see Table 9.1) have a gas-filled tracheal system. The few exceptions are the young fly larvae of *Chironomus* (Chironomidae, Diptera) and *Simulium* (Simuliidae, Diptera) and some young aquatic caterpillars (Pyralidae, Lepidoptera). In some groups the system is open and air can be taken inside the body, but in other groups the system is closed (apneustic) to the entry of air and oxygen must be absorbed from water. In either case, many aquatic insects have the body surface well supplied with tracheae and are able to obtain at least part of their oxygen by cutaneous gas exchange.

In contrast with air, the supply of oxygen from water is sharply limited by a series of physical constraints. Oxygen dissolves in water in an amount inversely proportional to temperature (warm water has less oxygen than cold water). The concentration, even when saturated, is many times less than in air, and movement by diffusion in solution is exceedingly slow. Immediately next to a submerged insect's body is a thin layer of still water called the boundary layer in which oxygen is renewed only by the slow process of diffusion. Finally, the diffusion of oxygen through cuticle is many times slower than in water.

Some insects, such as many larval Diptera, depend mainly on cutaneous gas exchange, but others have special tracheal gills that augment gas exchange. These are thin-walled outgrowths of the integument that are richly tracheated. Oxygen in water diffuses through the cuticle and into the gas-filled tracheoles. Gills may also be important in the elimination of carbon dioxide. Platelike gills are found in immature mayflies (Ephemeroptera) (Fig. 8.6a,b) and damselflies (suborder Zygoptera, Odonata Fig. 19.6b). Filamentous gills are also found in mayflies (Fig. 18.4b) and are characteristic of groups of, for example, aquatic flies, beetles, and caddis flies. The rectum of Odonata is thin-walled and tracheated as a rectal gill. In general, gills occur most commonly on the sides of the abdomen, less commonly on the thorax, and rarely on the head.

Adult and immature insects with an open tracheal system obtain their oxygen by frequent trips to the surface for air, by tapping air spaces in submerged plants, or by carrying a bubble or film of gas with them while they are submerged. The oligopneustic system of larval Diptera is suited to obtaining air by returning to the surface (Fig. 5.5b). The spiracular opening is usually surrounded by hydrofuge areas or hairs that prevent entry of water into the spiracle.

Air spaces in plants are punctured by the specialized respiratory tubes of various unrelated insects: certain larval beetles (Donaciinae, Chrysomelidae; *Noterus*, Noteridae; *Lissorhoptrus*, Curculionidae) and flies (*Chrysogaster*, Syrphidae; *Notiphila*, Ephydridae; *Mansonia* and *Taeniorhynchus*, Culicidae).

Bubbles of air are carried beneath the surface by many aquatic bugs and beetles (Fig. 5.5a). Such temporary air stores are held in place by special hairs or are carried in cavities beneath the wings or elytra. The spiracles are situated to open into the air bubble. All aquatic Hemiptera in North America that swim beneath the water carry air stores. The manner of obtaining air varies, for example, through a caudal siphon (Nepidae, Fig. 32.10b), caudal flaps (Belostomatidae), abdominal tip (Naucoridae, Notonectidae), or the pronotum (Corixidae). Aquatic beetles of the suborder Adephaga carry air under the elytra and renew the store by breaking the surface film with the tips of their elytra and abdomen. Aquatic members of the suborder Polyphaga have elytral air stores and usually also a coating of hydrofuge hairs on the ventral surface that holds a film of gas. Hydrophilid beetles break the surface with an antenna.

The bubble that is taken below serves not only as a store of atmospheric oxygen and as a hydrostatic organ but also as a physical gill. The latter functions as follows: The gas in the bubble initially has the composition of air: about 20 percent oxygen, 79 percent nitrogen, and less than 1 percent carbon dioxide. The carbon dioxide produced by the insect rapidly diffuses into the water and is of negligible importance as a gas in the bubble. As oxygen is used by the insect, the content in the bubble falls below that in the surrounding water. Oxygen will then diffuse into the bubble from the water. Nitrogen makes up almost 80 percent of the bubble volume and thus plays an important role in maintaining the bubble's volume. Because nitrogen is inert, the bubble maintains most of its volume even at low oxygen concentrations. A remarkable modification that utilizes the principle of the physical gill is the plastron. In this device, a thin film of gas is held on the body surface indefinitely by dense hairs or a fine cuticular meshwork (Fig. 5.6). The European bug *Aphelocheirus* (Naucoridae, Hemiptera) has fine, short hairs over much of the body surface. The density of water-repellent hairs is 200 to 250 million per square centimeter. The hairs are bent at the tips in an arrangement that supports the gas/water interface and resists compression of the gas film. The bug and similarly equipped beetles of the families Elmidae and Dryopidae can remain submerged for months.

The aquatic pupae of certain beetles (*Psephenoides*, Psephenidae) and flies (some Tipulidae, Blephariceridae, Deuterophlebiidae, Simuliidae, Empididae) have long cuticular processes developed from the peritreme and atrial region of one or more spiracles. These processes are called spiracular gills. The gill is associated with a plastron that is supported by a cuticular meshwork rather than by dense hairs. Likewise, the chorion of the eggs of some insects has a plastron that allows the embryo to respire when the egg is flooded or covered by a film of moisture.

The hair or cuticular plastron functions equally well when the insect or egg is out of the water because the spiracles or embryonic respiratory organs are then in direct contact with air. In water deficient in oxygen, however, the physical gill fails because oxygen no longer diffuses into the bubble. For this reason, insects with plastron gas exchange are typically slow moving or found in swiftly flowing streams or other well-oxygenated waters.

Various adaptations for gas exchange similar to the foregoing can be observed among insects in semiliquid media: scavengers in decaying plant or animal bodies or animal waste; phytophagous insects that burrow in juicy fruits or other plant tissues; and endophagous parasites and parasitoids immersed in hosts' blood. For example, small Hymenoptera

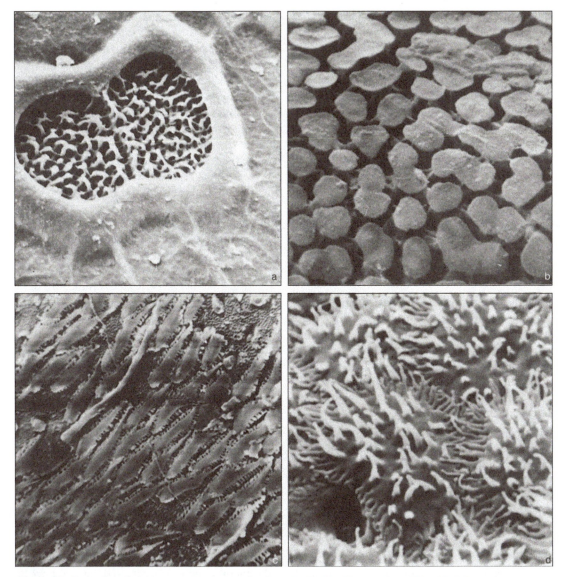

Figure 5.5 Examples of plastrons: a, plastron crater of eggshell of fly, *Musca sorbens* (Muscidae); b, respiratory horn of egg of fruit fly, *Drosophila melanogaster* (Drosophilidae); c, metasternum of aquatic beetle, *Tyletelmis mila* (Elmidae); d, elytron of aquatic beetle, *Portelmis nevermanni* (Elmidae).
Source: Scanning electron microscope photos courtesy of and with permission of Howard E. Hinton.

larvae living as internal parasitoids of other insects may depend on cutaneous gas exchange but still retain tracheae inside their bodies.

ANATOMY AND FUNCTION OF THE CIRCULATORY SYSTEM

The circulatory system has multiple functions, including transportation of nutrients, waste products, and hormones; storage of nutrients and waste; hydrostatic pressure for molting and loco-motion; and resistance to disease, wounds, and freezing.

The fluid, or hemolymph, that circulates in the bodies of insects differs in many respects from that of vertebrates. It is a single fluid with cells, not divisible into blood and lymph systems, that moves

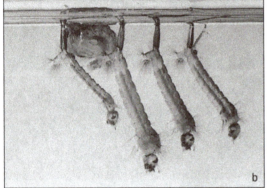

Figure 5.6 Examples of aquatic respiration. a, Aquatic beetle, *Tropisternus* sp. (Hydrophilidae), submerged with an air film that functions as a physical gill. b, Larvae and a pupa of a mosquito, *Culiseta* sp. (Culicidae), "hang" from the surface tension of water and obtain atmospheric air. The larvae have the metapneustic type of respiratory system with only the terminal spiracles open. The pupa has respiratory horns on the thorax.

freely around the internal organs without a pressurized, closed network of blood vessels. The cells, or hemocytes, are highly variable in size, shape, number, and function. The plasma is biochemically unique in the regulation of osmotic pressure by amino acids and organic acids instead of inorganic ions. Neither hemocytes nor plasma is significantly involved with the transport of oxygen and carbon dioxide in most species. Consequently, insect blood is rarely red (see discussion of hemoglobin above) but may be colored yellow, blue, or green by various pigments, including α-carotene,

riboflavin, and chromoproteins, but probably not chlorophyll.

The blood is limited to the major cavity of the body, the hemocoel. Insect hemolymph, however, does not directly bathe the cells of the body, as is often claimed. The inner surface of the integument and all the organs are covered in a basement membrane of connective tissue. This membrane probably is interposed between the hemolymph and all other tissues. Thus situated, a "blood-cell barrier" is created that regulates the exchange of materials between the body cells and the hemolymph.

Dorsal Vessel

The circulation of the hemolymph is not haphazard but follows a pattern of flow. Motion is imparted to the fluid by movements of the body segments and of the gut and by the peristaltic contractions of the dorsal vessel (Fig. 5.7a). This is a tube developed middorsally from the embryonic mesoderm and extending almost the full length of the body. In the abdominal region the vessel is closed at the posterior end but perforated by paired segmental openings, or ostia. Insects may have as many as twelve pairs. These have valvelike flaps projecting into the lumen of the vessel. The portion of the dorsal vessel with the ostia is called the heart. It is slightly inflated, especially so between the pairs of ostia. These dilations create the chambers of the heart. The heart is suspended middorsally in the abdomen by a transverse membrane of connective tissue and thin muscles. This dorsal diaphragm, with its alary muscles (meaning "winglike" in shape), is not complete but is interrupted segmentally by openings. By this means blood can pass from the vicinity of the gut or perivisceral cavity to the space above the heart, the pericardial cavity. Also in the abdomen is a ventral diaphragm, formed by an incomplete sheet of connective tissue. This membrane extends transversely, usually above the ventral nerve cord, and partly separates the perivisceral cavity from the ventral perineural cavity.

Movement of Hemolymph

Muscles embedded in the fibrous tissues of the heart wall provide the sequential contractions that impel the blood forward. As the wall of a chamber begins to contract (the systolic phase), the ostia close to block the rear chamber. Thus, blood moves forward to the next chamber. When the wall relaxes (the diastolic phase), the ostia open and the flaps admit blood from the outside and from the chamber to the rear. The flow of blood is normally toward the head, but reversals can be observed. The rate of heartbeat varies widely with age, species, the state of nervous excitement, and environmental factors such as temperature. The initiation of contraction may come from an independent but diffuse "pacemaker" in the heart or may be partly or entirely controlled by the central nervous system.

The aorta, a continuation of the dorsal vessel, conveys blood to the head. This part of the vessel lacks ostia and is generally smaller in diameter. It is not supported by the dorsal diaphragm, but it may be suspended by a vertical sheet, or dorsal septum, in the thorax. The aorta opens beneath the brain and above the pharynx. Blood released here moves posteroventrad, entering the appendages of the head and the thorax. Flow into and out of these cylindrical structures is made possible by a septum that divides the appendage lengthwise into two channels. Blood also enters the wing veins and flows in an orderly, but variable, pattern throughout the wing and returns. Accessory pulsating membranes aid local movement of the blood. After passing around the leg muscles and flight muscles in the thorax, the blood returns to the abdomen. The movement here is upward from the perineural and perivisceral cavities through the dorsal diaphragm and back to the heart. Although there is almost no blood pressure, complete mixing time in an insect has been estimated at from 5 to 30 minutes. In humans the time is 2 to 4 minutes.

The circulation of hemolymph can also provide thermoregulation. Bumble bees (Apidae, Hymenoptera) are among the first foraging bees to appear in cool climates. In spring overwintering

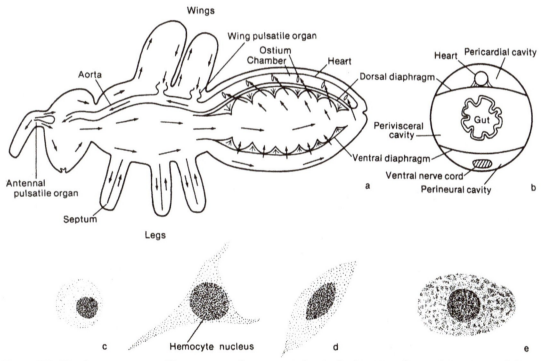

Figure 5.7 Circulatory system and hemocytes: a, diagrammatic longitudinal section of general structure and direction of circulation; b, cross section of abdomen; c, prohemocyte; d, plasmatocyte; e, granulocyte.
Source: a, modified from Weber, 1933.

adult female bumble bees warm up their flight muscles of the thorax with rapid contractions without wing movements. When the temperature of the flight muscles reaches a temperature at which flight can be sustained, the bee can fly in cool weather. If not dissipated, the heated thorax can rapidly reach lethal temperatures, but hemolymph moving through the dorsal vessel as it passes along a contorted route through the thorax absorbs excess heat from the surrounding muscle mass. The hemolymph empties from the dorsal vessel into the head, from which it then is shunted along the ventral thorax to the ventral abdomen. During flight the head and mouthparts and the ventral abdomen are cooled by air flowing over the insect. In the nest, the female can generate heat from the flight muscles and transfer the heat through hemolymph movements to her ventral abdomen. During the cool days and nights of the early spring months, she can use this method of heat transfer to speed the incubation her first offspring by pressing her ventral abdomen against the brood cell of her first developing larva. This is far from the conventional view that insects are strictly "cold-blooded" organisms. The bumble bee displays remarkable control of not only its body temperature but the part of its body that is warmed.

Hemocytes

Unlike the blood cells of humans, the cells or hemocytes of insects vary greatly in size, cytological details, and the numbers that one can see in a sample of hemolymph. The naming of cell types is somewhat disputed, but common names are prohemocytes (small, spherical precursors of various hemocytes), plasmatocytes (variable shapes from round to spindle-shaped), and granulocytes (containing conspicuous granules that are discharged in defensive reactions) (Fig. 5.7c–e). Hemocytes function in resisting disease and injury, in storage and distribution of nutrients, and probably in many other aspects of internal physiology. Most circulate with hemolymph movement, but some remain attached to specific tissues.

The cuticle of the body, gut, and tracheal system is the main barrier to the entry of disease organisms. After entry, insects are able to resist the progress of disease by humeral defenses (due to substances within the hemolymph) and cellular defenses of phagocytosis (engulfing of foreign objects) and encapsulation (surrounding objects with hemocytes). Plasmatocytes and granulocytes protect against infections by phagocytizing viruses, bacteria, fungi, and protozoa. Immune reactions of the hemolymph also appear to regulate numbers of symbiotic microorganisms within insects. For example, symbionts of the human louse, *Pediculus humanus* (Pediculidae, Phthiraptera), periodically move in mass from protected intracellular locations within a specialized organ (bacteriome) to quickly invade the base of the ovarioles and ultimately the developing eggs. Symbionts that do not succeed in penetrating the ovarioles are soon engulfed by hemocytes.

Foreign objects, microbes, and endoparasites too large to be engulfed by a single cell are immobilized or killed by the same kind of hemocyte in a process called encapsulation. A capsule of connective tissue is secreted as the inner layer of cells flattens over the object's surface and new cells are added outside. The capsule may remain clear or become darkened by melanization.

To act on foreign antigens, insects do not have the equivalent of mammalian lymphocytes or their serum proteins, the immunoglobins. However, insects do have induced humeral responses to parasite invasions. Lectins have been identified in insect hemolymph, are inducible by injury, and are known to agglutinate microbes. Also inducible by bacterial infection are defensive proteins such as the peptides called cecropins that disrupt the cell membranes of bacteria and other pathogens. Constitutive (not inducible) humeral defenses include the release of phenoloxidases and lysozymes (enzymes that weaken bacterial cell walls) into the hemolymph as a result of a cascade of enzymatic reactions. The clearance of bacteria from an insect's hemolymph can be quite rapid; most are cleared within an hour after experimental introduction into the hemocoel. This implies that the induced defensive reactions serve to continue to destroy remaining pathogens with another defensive method to counter any rapid evolution of resistance in the population of invading pathogens.

The risk of bleeding is greatly reduced among insects because of the exoskeleton and low blood pressure. However, most insects have hemolymph that form clots and repair wounds. Prohemocytes gather at a wound, causing local coagulation of the hemolymph. Later epidermal cells and possibly hemocytes seal the wound and repair the cuticle.

Composition of the Plasma

The cell-free fluid, or plasma, of the hemolymph contains water, inorganic ions, and organic molecules, including metabolic wastes. The volume of hemolymph and water content varies widely— up to 94 percent of the body weight. In vertebrate blood inorganic ions determine osmotic pressure, but as the evolution of insects is traced, the more advanced orders have inorganic ions replaced by amino acids and other small organic molecules. Thus, in Lepidoptera, Hymenoptera, and many Coleoptera amino acids are the primary osmotic effectors. The high concentration of amino acids is a distinctive feature of insect hemolymph.

Organic molecules in the hemolymph include carbohydrates, amino acids, peptides, proteins, lipids, organic acids, organic phosphates, and glycerol. The main carbohydrate is the trisaccharide trehalose, although glucose and some other sugars have been found in quantity in the honey bee and certain flies. Glycogen and lipids are largely stored in the fat body. Depending on the stage of the life history, the large amounts of amino acids in the hemolymph support growth and metamorphosis. Insect hemolymph has a remarkable variety of peptides, polypeptides, and proteins. These function in a wide array of physiological processes, including immunity, regulation of growth, development, reproduction, metabolism, and homeostasis. The most abundant proteins, especially in larvae, are storage proteins (heximerins) synthesized by the fat body. Transport proteins are necessary for the movements of hydrophobic compounds. Lipophorins are the most abundant transport proteins and bind with lipids to shield them from water. Insects that use lipids as fuel for flight release diglycerol from the fat body under the influence of adipokinetic hormone. Juvenile hormone (JH) is a lipid transported through the hemolymph by specific carrier proteins. Organic acids, especially those associated with the tricarboxylic acid cycle, are most prevalent in larval endopterygotes. Acid-soluble organic

phosphates are synthesized and found in high concentrations in certain moths. Levels of waste end products of protein metabolism in plasma— including uric acid, allantoin, allantoic acid, urea, and, in some immature aquatic insects, ammonia— are sometimes high. Glycerol is accumulated in the plasma of certain insects as an antifreezing agent.

ANATOMY AND FUNCTION OF THE EXCRETORY SYSTEM

Excretory processes expel or render harmless metabolic wastes and other substances that are harmful, and at the same time they maintain an appropriate balance of water, ions, and other components of the hemocoel (Fig.5.8a). Excretion thereby plays the key role in physiological homeostasis (i.e., the maintenance of a constant internal environment) by regulating body chemistry. In contrast to defecation (i.e., physical expulsion of undigested food and excreted waste), excretion involves the movement of waste molecules across one or more plasma membranes. The most critical excreted products are nitrogenous compounds resulting from the breakdown of proteins and nucleic acids. The processes of excretion occur mainly in the alimentary tract. Most of the excretory products in insects are regulated by the Malpighian tubules and the rectum acting in concert.

Nitrogenous Waste

Nitrogenous end products may be discussed in three groups: (1) ammonia, (2) urea, and (3) uric acid, allantoin, and allantoic acid. Ammonia is a direct product of the oxidative deamination of amino acids. Containing 82 percent nitrogen, it is an efficient waste product, but it is also highly soluble in water and highly toxic. Terrestrial animals must conserve water; hence ammonia is usually transformed to the less toxic urea (e.g., in mammals, amphibians, some reptiles) or uric acid (e.g., in birds). Among insects, ammonia is the primary waste of species that ingest large quantities of water such as the xylem sap feeders and the aquatic immature stages of some dragonflies (Aeshnidae), alderflies (Sialidae), and blowflies (Calliphoridae). The rapid flow of water through the gut in such insects rapidly dilutes and eliminates excreted ammonia to avert its toxicity without the energy requirement for synthesizing less toxic forms of nitrogen excretion products.

Urea, containing 46 percent nitrogen, is the primary waste of carpet beetles (Dermestidae). Most insects excrete uric acid, one of its breakdown products (allantoin and allantoic acid), or a combination of these substances. These are the least soluble, as well as the least toxic, end products, and they contain less nitrogen (32–35 percent). Some uric acid is derived from the breakdown of nucleic acids, but the bulk of these wastes must be synthesized by energy-consuming pathways.

Waste Storage and Reuse

In certain insects some of the nitrogenous waste is not expelled but is stored in a harmless state or utilized elsewhere in the body. Uric acid is deposited in special urate cells or urocytes of the fat body in the roach *Periplaneta americana* (Blattidae, Blattodea), adult *Culex* (Culicidae, Diptera) mosquitoes, the silkworm *Bombyx mori* (Bombycidae, Lepidoptera), and the wasp *Habrobracon juglandis* (Braconidae, Hymenoptera). The rate of accumulation of uric acid thus stored provides an index of aging but can be heavily influenced by the amount of food consumed. Adult males of *Blattella germanica* (Blattellidae, Blattodea) store uric acid in accessory sex glands in amounts up to 5 percent of the 'insect's weight. This is discharged as a white covering on the spermatophore during mating. The white color of crystalline uric acid is used as a pigment in the epidermis of some insects or conspicuously in the wing scales of butterflies in the family Pieridae.

Waste Elimination

Most nitrogenous waste is eliminated from the 'insect's body as part of regulatory processes involving other substances of small molecular size. The organs responsible are the Malpighian tubules (named for the seventeenth-century insect anatomist Marcello Malpighi) and the hindgut, especially the rectum. The filamentous, blind tubules insert and empty at the anterior end of the hindgut.

The walls of the Malpighian tubules and rectum are both one cell in thickness. Thus an internal and an external plasma membrane are situated between the hemolymph and ducts leading to the

exterior. The cells of the Malpighian tubules are richly tracheolated and bounded by a basement membrane. The plasma membrane next to the hemolymph is deeply and complexly infolded; the luminal membrane is microvillate and associated with long mitochondria. The tubules are usually free in the hemocoel and may move by means of slender muscles placed lengthwise in a spiral fashion. Malpighian tubules are lacking in Collembola, *Japyx* (Japygidae, Diplura), thrips, and most aphids; are rudimentary in certain Protura, Diplura, and Strepsiptera; and number only two in Coccoidea and some larval parasitoid Hymenoptera. Other insects usually have multiples of two, commonly having a total of six. Fifty or more tubules are formed in some Ephemeroptera, Plecoptera, Odonata, Orthoptera, and aculeate Hymenoptera. Grasshoppers (Acrididae, Orthoptera) may have hundreds.

The composition of the hemolymph is regulated in the following manner (Fig. 5.8a). Water, salts, sugars, amino acids, and nitrogenous wastes pass into the lumen of the Malpighian tubules, beginning at the tip or distal end. The filtrate then moves through the tubule to the hindgut and rectum. Along this route useful components are selectively reabsorbed back into the hemolymph while the waste and surplus are voided. Potassium or sodium ions or both are transported across a steep electrochemical gradient from hemolymph to lumen. The movement of these ions by a cation pump and the transport of chloride ions is linked to the secretion of urine in the lumen. Water, sugars, amino acids (except proline, which is transported), and nitrogenous wastes are thought to be passively carried along with this flow while some other substances are actively transported. The rates of passage of solutes in the hemolymph into the tubules are typically at least tenfold less than the uptake of solutes in the vertebrate kidney glomerulus. This is because of the relatively limited area for permeation in the insect's tubules. This slow permeation of molecules into the tubule means that less energy is required to reabsorb needed solutes back into the hemolymph. Glucose and trehalose may be returned to the hemolymph at once from the tubule. The fluid is nearly iso-osmotic with the hemolymph when it is initially secreted.

Selective reabsorption may begin in the tubule and the anterior parts of the hindgut, but the major site of activity is the rectal pads in most insects. Active transport by transmembrane transporter proteins moves potassium, sodium, and chloride ions against an osmotic gradient. Here the molecules must move from the lumen of the rectum, through the rectal cuticle, through or between the rectal cells, and finally to the hemolymph. The recovery of useful amounts of water,

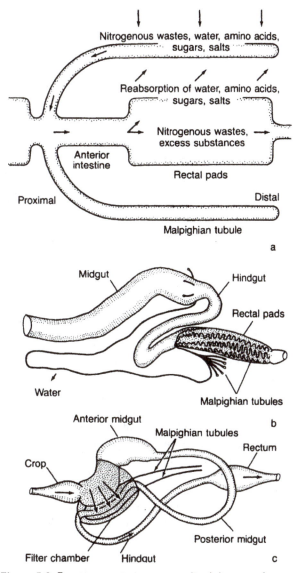

Figure 5.8 Excretory systems: a, generalized diagram of flow of nitrogenous wastes in the insect excretory system; b, cryptonephric system of *Tenebrio molitor* larva (Tenebrionidae) with five of the six tubules partly removed; c, diagram of filter chamber of Homoptera.
Source: c, modified from Weber, 1933.

amino acids, salts, and sugars is linked to this process, but the molecular pathways remain unclear. Undoubtedly, aquaporins transport water into and out of the tubules.

In larvae and adults of Coleoptera and larval Lepidoptera, the distal ends of the Malpighian tubules may be held in contact with the rectum. This cryptonephric arrangement of the tubules has been carefully studied in the mealworm, *Tenebrio molitor* (Tenebrionidae, Coleoptera). The adults and larvae live in dry, stored grains where conservation of water is required. The rectal complex is specially developed to dehydrate feces before elimination. Water is even extracted from air around the feces if the humidity is greater than 88 percent. The six tubules are held in contact with the rectum by a multilayered perinephric membrane and an outer membrane (Fig. 5.8b). This arrangement allows the potassium pumps in the tubules to generate an osmotic gradient strong enough to draw water from the rectum across the rectal cells and into the tubule. The fluid in the tubule is moved proximally into the free portion of the tubule, where it is passed to the hemolymph. The waste is eliminated as a dry powder.

Insects that feed directly on the sap in the vascular tissues of plants have the opposite problem: excessive fluid and dilute nutrients. In the Cicadoidea (Hemiptera, suborder Auchenorrhyncha), the basal portions of the Malpighian tubules, the posterior end of the midgut, and the beginning of the hindgut are folded into and surrounded by the anterior midgut (Fig. 5.8c). The area of contact is enclosed in a membrane. This is called the filter chamber. This structure concentrates ingested plant sap by shunting water from the anterior midgut to the hindgut. As a result, the density of nutrients in the absorptive regions of the midgut is increased. Cations from the hindgut are actively transported from the hindgut contents into the distal ends of the Malpighian tubules. As pressure builds up within the tubule, the tubule fluid moves toward the hindgut. Water in ingested sap moves osmotically from the proventriculus into proximal parts of the Malpighian tubules because the proximal parts contain increased salt concentrations. Following the discovery of water-transporter membrane proteins, the aquaporins, we presume that these too are

involved in water transport by the filter chamber. From the lumen of the tubules, the fluid is expelled into the hindgut. The cuticle-lined hindgut usually is impervious to water movement except for the rectum, where cations in the urine again may be recycled. Thus, the bulk of the liquid is excreted, bypassing the rest of the midgut where nutrients are concentrated, digested, and absorbed.

The Malpighian tubules have an unusual function in spittlebugs (Cercopidae, Hemiptera), where they add a substance to prolong the durability of bubbles ("spittle") that spittlebug nymphs expel from their anus to surround their bodies. This presumably protects them from some natural enemies. All leafhoppers (Cicadellidae, Hemiptera) and some treehoppers (Membracidae, Hemiptera) expel unusually shaped particles called brochosomes. These peculiar particles are produced in the Malpighian tubules and fall in two general functional categories. Leafhoppers spread or "anoint" excrement containing spherical brochosomes over their bodies, perhaps as a water repellent. The adult females of a few species of leafhoppers use their hind legs to capture their excrement containing high concentrations of elongated, rod-like brochosomes in a large drop and manipulate it until it evaporates to a paste, which they transfer to a spot on the outer wings. The brochosomes appear in the excrement just before egg-laying, and the females spread the stored brochosomes as a powder over plant surfaces containing their freshly inserted eggs. The brochosomes appear to irritate and distract tiny wasp parasites that attack the leafhopper eggs by sticking to the parasite's antenna and body hairs.

MUSCLES AND LOCOMOTION

Insects move from place to place in virtually every natural medium on the earth's surface except solid rock, ice, or, rarely, salt water. On the surface of the ground they wiggle, hop, walk, or run, often with surprising agility. By their feeding and burrowing activities, they move through the flesh of animal hosts, the tissues of plants, including solid wood, and the leaf litter, soil, or subsoil of the earth. In the aquatic environment some skate on the surface, propelled by their legs or by chemicals that alter the surface tension. Under the surface they swim with the legs or wings, with body movements, or by jets of water expelled from the anus. The best speeds, the longest distances, and probably the safest trips are achieved in the air. Yet, compared to locomotion on the ground or in water, flight consumes much more energy and requires a lightweight body equipped with wings and muscles for flight together with keen senses and a nervous system for coordination.

Insects were the first animals to evolve flight and are still the only invertebrates that fly actively and can control the duration and direction of their flights. The only other invertebrates to fly are the ballooning spiders and some mites, which, however, are passive drifters. Flight has clearly had a major influence on the diversity and abundance of insects. As shown by the fossil record, insects had wings and presumably were flying in the Pennsylvanian Period (323–290 million years ago; see Table 15.3), over 150 million years before the first flying reptiles (Pterosauria) appeared in the Jurassic (208–146 million years ago). Much later the birds and mammals (bats) each independently evolved flying mechanisms.

Insect Muscle

The power for locomotory and other movements is supplied by the muscular system. Internal organs, such as the gut, heart, or Malpighian tubules, also have sets of muscles. All muscles of insects are striated, in other words, each fiber has a cross-banded appearance when viewed under the microscope (Fig. 5.9a). In contrast, vertebrates and many other invertebrates have both the striated type and a second type, smooth muscle, which is composed of single spindle-shaped cells.

The muscular system of insects differs in several important respects from that of vertebrates. The most conspicuous difference is the relation to the skeleton: the muscles of insects are inside the skeleton, while those of the vertebrates are outside. As a consequence of the tubular construction, which resists bending, and the general absence of calcium salts, the insect skeleton compared to the vertebrate skeleton is relatively lighter and stronger and provides greater space and attachment areas for the muscles. The attachment of muscles to the skeleton is not by connective tissues, as in the vertebrates,

but by tonofibrillae. Tonofibrillae are special secretions of the epidermal cells whose attachments may extend through the cells into the cuticle as far as the epicuticle. If so, the muscles remain attached to the old cuticle even while molting is in progress, thus allowing movements during this critical period. At the point of attachment the cuticle may be invaginated into a tendonlike apodeme. When composed of normal cuticle such apodemes lack elasticity, but when the special cuticular protein resilin is present, a highly elastic tendon is formed. Although insects might be assumed to have fewer muscles than larger animals such as vertebrates, some insects have more than two or three times as many separate muscles than do humans, largely owing to the more numerous body segments and appendages.

Insects may appear to perform feats of unusual strength, but the power per gram of muscle is about the same as that of vertebrates. A muscle's power is proportional to its cross-sectional area. The explanation for insects' seemingly greater strength lies in the effective use of lever systems and the small body size. As the body of an organism decreases in size, the volume or mass decreases as with the cube root of length, whereas the cross-sectional area

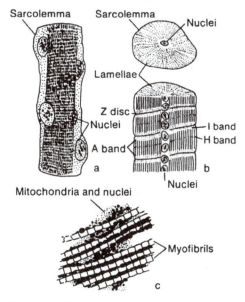

Figure 5.9 Types of muscles: a, larval, b, tubular, c, fibrillar.
Source: Redrawn from Snodgrass, 1956.

decreases as the square root. Thus muscle weight (a function of muscle volume) decreases faster with miniaturization than does muscle power (proportional to cross-sectional area). The amount of work per gram achieved by insect muscles can greatly exceed that of vertebrates because some insect muscles can contract much faster, leading to relatively more energy consumed and work accomplished on a per weight basis. Reductions in the size of insect muscles involve mainly reductions in the number of fibers rather than reductions in the size of the fibers. A species of small insect may have only a single fiber functioning in the position occupied by hundreds of fibers in a species of large insect.

Structure of Muscles

Muscle cells have multiple nuclei (syncytium). Further, each muscle is the product of many cells that share a common plasma membrane and an outer sheath, the sarcolemma. This outer covering may be invaginated into the muscle in certain places, forming a system of intracytoplasmic intermediary tubules (called the IT or T, for transverse, system). Tracheoles may reach deep into the interior of the muscle without penetrating the sarcolemma by entering these tubules. This is one feature that enables insect flight muscles to have the highest rates of oxygen consumption per gram for any animal locomotory muscle. When the muscle is stimulated by an excitatory nerve, the T system allows the depolarization of the cell membrane to pass from the surface to the deeper regions of the muscle. Within the muscle are the contractile myofibrils traversing the length of the muscle. These are striated and may exist as flat sheets (lamellae) or as threadlike fibers. Under great magnification with an electron microscope, each fibril is seen to be composed of protein filaments of two sizes. The thick filaments are myosin, and the thin filaments are actin. In cross section each actin filament of an insect muscle is usually situated between two myosin filaments. In vertebrate muscle the actin filament is between three myosin filaments. When the muscle contracts, the two sets of filaments slide past each other. The distribution of the two kinds of filaments also provides an explanation for part of the banded, or striated, appearance. The I bands represent only

actin filaments, the H band only myosin filaments, and the A band represents both together. At regular intervals, both thin and thick filaments attach to a disklike structure called the Z disk. The interval between disks is called a sarcomere. Dispersed among the myofibrils are numerous nuclei and mitochondria and various amounts of a sarcoplasmic (endoplasmic) reticulum. Mitochondria and tracheolar endings are more numerous in muscles that have a high degree of activity, such as flight muscles. Individual muscles are grouped together into a "muscle unit," which has a separate tracheal supply. A single muscle unit or several units may be separately innervated by branches of one, two, or three motor axons (Chap. 6), making a motor unit.

Finally, a single motor unit or several motor units may be grouped closely together into a functional "muscle," which is anatomically discrete from other such units, has definite areas of attachment, and exerts a pulling force on the attachments in a definite direction. Unlike vertebrate muscles, insect muscles lack connective tissues among the fibers. Ordinarily a skeletal muscle has one of its attachments, or its origin, on an immovable part of the skeleton and its insertion on a movable part such as a segment of an appendage or an articulating sclerite. Some are bifunctional—under certain circumstances the structures at either end may be moved. For example, some thoracic muscles function during walking by moving the leg bases and also function in flying by moving sclerites at the opposite ends of the muscles. The large indirect flight muscles move large portions of the thoracic skeleton simultaneously. Visceral muscles, which move various organs such as ovaries or alimentary organs, usually attach to other visceral muscles.

Innervation and Contraction of Muscles

Although all insect muscle is striated, evolutionary changes in structure and function have led to several distinctive kinds of muscle. Perhaps the most primitive locomotory muscles are tubular, characterized by the tubelike form of myofibrils being arranged in sheets around a central canal that is filled with nuclei (Fig 5.9b). Larval muscles, known from immature bees and flies, are distinguished by minute myofibrils with the nuclei located externally beneath the sarcolemma (Fig. 5.9a). Visceral muscles are composed of parallel or anastomosing fibers similar to larval muscles in construction, but they have larger myofibrils and are associated with the internal organs of adult insects. Microfibrillar muscles have small myofibrils measuring 1 to 1.5 μm in diameter that are found in the flight musculature of some orders. Fibrillar muscles, so called because their coarse fibers are easily sheared and separated when dissected, have large myofibrils measuring 1.5 to 5.4 μm in diameter (Fig. 5.9c). They make up the flight muscles of the orders Hymenoptera (most species), Diptera, Coleoptera, and certain Hemipteroidea (some Psocoptera, Thysanoptera, Heteroptera, and some other Hemiptera).

Insect muscle cells or units may be innervated (connected to nervous system) by more than one neuron, or a single nerve output process (axon) may branch out to attach to the same muscle fiber in more than one place. This contrasts with vertebrate muscles, which have a single end-plate connection. Insects usually have two motor axons, rather than the one typical of vertebrates, attached to each muscle: one "fast" and another "slow" in its rate of contraction. Some motor nerves inhibit rather than stimulate contraction. The rapidity of nerve transmissions and the number and type of motor nerves used result in differences in insect muscle response ranging from slow and precise to rapid and powerful. Vertebrates achieve such muscle contraction differences by varying which muscles or units are stimulated.

Normally skeletal muscle contractions are synchronous with each nerve impulse. Visceral muscles may be innervated by typical neurons or by special neurosecretory cells that secrete neurotransmitter substances not closely restricted to nerve junctions. The major flight muscles of several groups of insects contract several times for each nerve impulse. Fibrillar muscle, called asynchronous (or resonating) muscle, is pivotal to the flight abilities of many "advanced" insect groups such as the flies, beetles, and wasps and many true bugs (Heteroptera). The asynchronous type has evolved independently from the synchronous type in seven to ten different lineages of insects. Asynchronous muscles allow greater speed of contraction, enabling higher wing beat frequencies. They also make possible greater mechanical power output by having a higher ratio of contractile fibers and energy-providing mitochondria. The wing muscle of active fliers can contain 30 to 40 percent mitochondria by volume. After contraction, the shortened asynchronous muscle is deactivated and can then be restretched to its original length with less force than produced by the contraction. Stretching reactivates the muscle contraction. The positive net difference in work output between the contraction and stretching phases is the net work used for powering movements. Asynchronous muscle is more efficient and powerful than synchronous muscle at the high contraction frequencies of advanced fliers. Its greater efficiency is due to less energy being used for nervous control of muscle activation, largely due to the amounts of calcium recycled in muscle activation. It is more powerful partly because it contains a higher proportion of power-producing fibrils within the sarcoplasmic reticulum.

Walking, Jumping, and Swimming

Hexapods are distinct among the arthropods for having three pairs of locomotory appendages. In terrestrial locomotion, each leg must function in two ways: (1) to support the 'body's weight above the substrate and (2) to thrust against the substrate for propulsion. The best explanation for the evolutionary choice of six legs lies in the fact that a minimum of three legs can provide support in a stable tripod arrangement while the other three legs are stepping forward (Fig. 5.11). The insect walks by shifting from alternate tripods of leg contact with the substrate. The sequence or rhythm of leg movements changes as the insect increases speed. Depending on the gait, more than three or less than three legs may be on the substrate at once. At high speeds, cockroaches run by alternate steps of only the hind legs. The rhythmic movements of insect legs originate in the central nervous system and are modified by feedback from sensory receptors in the legs. The segmental ganglia control their respective legs but are coordinated by intersegmental nervous coupling.

The forelegs, in front of the center of gravity, exert their force in a pulling direction, while the middle and hind legs, which are near or behind the

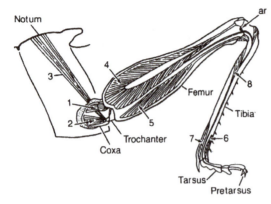
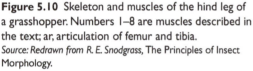

Figure 5.10 Skeleton and muscles of the hind leg of a grasshopper. Numbers 1–8 are muscles described in the text; ar, articulation of femur and tibia.
Source: Redrawn from R. E. Snodgrass, The Principles of Insect Morphology.

center of gravity, exert a pushing thrust. The propulsive thrust of the hind legs is greatly developed as a jumping device in Orthoptera (Figs. 27.1, 27.3), many Auchenorrhyncha (Fig. 32.13a, b), fleas (Fig. 41.1a), and flea beetles (Alticinae, Chrysomelidae). Jumping is also accomplished in other ways: click beetles snap their bodies at the joint between the prothorax and mesothorax, and springtails utilize an abdominal appendage (Figs. 16.2, 16.3a).

During walking, a stepping motion involves movements of the whole leg as well as movements of individual leg segments. The whole leg is moved by a complex musculature that originates inside the thorax and exerts pulling forces on the basal rim of the coxa. By the positions and insertions of the muscles, the whole leg can be moved forward, backward, away from the body, or toward the body. Movements of the individual segments of the leg are made by muscles operating in pairs that arise in one segment and exert a pulling force on a more distal segment. One member of the pair operates as a levator, raising the distal segment, and the other member operates as a depressor, lowering the distal segment. As shown by the skeletomuscular mechanism of a grasshopper's hind leg (Fig. 5.11), the femur is moved by a levator muscle from the coxa (1) and two depressor muscles, one from the coxa (2) and the other from the thorax (3). The tibia is

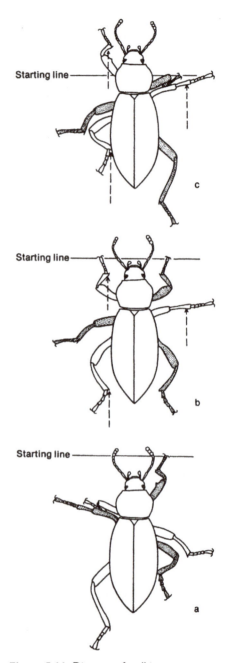

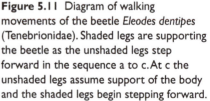

Figure 5.11 Diagram of walking movements of the beetle *Eleodes dentipes* (Tenebrionidae). Shaded legs are supporting the beetle as the unshaded legs step forward in the sequence a to c. At c the unshaded legs assume support of the body and the shaded legs begin stepping forward.

moved by a levator muscle (4) and depressor muscle (5), both originating in the femur. The tarsus is moved by a levator muscle (6) and a depressor muscle (7) that originate in the tibia and insert on the base of the tarsus. The individual tarsomeres do not have muscles. The pretarsus has only a depressor muscle (8) that originates in the femur and tibia and inserts by means of a long tendon on the base of the pretarsus.

Modifications of hind legs enable impressive jumping ability in many insects. Enlarged hind femora of grasshoppers (Fig. 5.11; Figs. in Chapt. 25), for example, house large muscles whose slow contractions compress the elastic cuticle in the femurtibial joint much like compressing a spring or drawing a bow. The sudden release of energy propels the insect for jumping. The jumping mechanism of a flea is illustrated in Fig. 42.2. Vermiform (Figs. 40.5c) or eruciform (Fig. 4.4d, 44.5c-d) larvae can crawl through coordinated wavelike successive contractions of the body egments. External spines along the body may help provide traction. Lepidopteran and Hymenopteran (suborder Symphyta) larvae have abdominal prolegs (Fig.39.5a,-b), which are unsegmented but leglike modifications of the body wall equipped with hooked spines called crochets (Fig.15.2).

Insect tarsi adhere to surfaces by the action of tarsal claws or adhesive pads (Fig. 2.11c,d). Adhesion is facilitated in some cases by molecular forces (Van der Waals forces) between the substrate and numerous and very fine hairs on the feet or by lubricating fluids applied from the tips of fine hairs. Aquatic insects that live in flowing water face the problem of adhering to exposed substrates. Larvae of black flies (Simuliidae, Diptera) anchor their abdomens to rocks or logs in fast-flowing portions of streams with anal suckers or hooks at the end of the abdomen as they extend their fan-like mouthparts to capture small particles of food suspended in the current. Some aquatic larvae such as stoneflies (Plecoptera) and mayflies (Ephemeroptera) that crawl among rocks in swift currents have highly flattened bodies, both to be able to crawl into tight spaces and to minimize drag from the moving water.

Insect tarsi and legs can be modified to a paddle shape for swimming (Figs. 2.13d, 32.8b, 37.7a).

The paddles are usually formed by hairs projecting from the tibia or tarsi. Because of the viscosity of water relative to the size of the legs, the hairs develop adequate traction for swimming. An advantage of hairs in swimming (and for flying in minute insects, as described later) is that the hairs can be extended to increase the effective area for propulsion, then be streamlined along the leg so that the net drag caused by bringing the leg forward (protraction) is less than the opposite force for the propulsion stroke (extension).

Flight

The historical origin and evolution of wings are discussed in Chapter 13. Here we are concerned with the skeletomuscular and aerodynamic mechanisms of flight. Insect wings are not homologous with appendages, as they are in birds and bats. There are no muscles in insect wings, although the wings have trachea and circulating hemolymph. Wings are powered entirely by muscles within the thorax that either deform the meso- and metathoracic segments of the thorax to move the wings (indirect flight muscles) or by directly deforming of the basalare and subalare sclerites near the wing base that act as a lever to move the wings (direct flight muscles).

Lengthwise in the thorax, the dorsal longitudinal indirect flight muscles extend between skeletal phragmata. Their contraction raises the center of the flexible notum. The dorsoventral indirect flight muscles operate as antagonists and depress the notum by their contraction. Mechanically, the wing base rests on the pleural wing process in the same manner as a lever on a fulcrum (Fig. 5.12a). Thus, contraction of the longitudinal indirect muscles causes an upward movement of the thoracic notum, which lifts the wing base and forces the rest of the wing downward (Fig. 5.12c). Likewise, contraction of the dorsoventral indirect muscles depresses the notum, which lowers the wing base and forces the wing upward (Fig. 5.12b).

A simple up-and-down movement of flat wings would not, however, generate the forward thrust needed for flight. For propulsion, the wings must twist at the base so that the surface of the wing is at an angle to the body. On the downstroke, the leading (anterior) edge of the wing twists ventrally (pronates). On the upstroke, the leading edge twists in the opposite direction (supinates). At the anterior edge of the wing base is located a small sclerite, the basalare, and at the posterior base of the wing is the subalare (Fig. 2.13a). Each sclerite has a direct muscle from the pleuron. Contraction of the direct basalare muscle depresses the leading edge of the wing to produce pronation (Fig. 5.12a). Similarly, contraction of the direct subalare muscle depresses the trailing edge, producing supination.

On meeting the resistance of the air and depending on the structure of the wing veins and flexibility, the wings may bend and fold in a regular pattern of deformation during each stroke. In the grasshopper, the hind wing traverses a slightly greater amplitude and leads the forewing, thus avoiding its turbulence. The stiff forewings of beetles (elytra) are held open without moving during flight. The elastic properties of the thorax and antagonistic muscles are greatly enhanced by special deposits of the elastic protein resilin on the pleural wing process.

The general design of the flight mechanism of the grasshopper is found throughout the Orthopteroidea and some Neuropteroidea. Exceptions are the Odonata, Blattodea, and Mantodea, where the downstroke is powered by direct muscles rather than the longitudinal indirect muscles.

In summary, with the exceptions just noted, the orders discussed above have two sets of synchronous muscles, the indirect and direct muscles, which provide the power and the twisting of the wings. Both the mesothoracic and metathoracic segments are usually so equipped. The wingbeat frequency and, consequently, the speed of flight in all these orders are limited to the rate at which the nervous system can send impulses to each of the participating muscles in a coordinated fashion—about 50 per second.

A second major type of flight mechanism is found in the Diptera, Hymenoptera, Coleoptera, and some Hemipteroidea. These orders have a high wingbeat frequency that depends on flight muscles contracting asynchronously with many contractions per nerve impulse. The wingbeat frequencies of bees and flies is on the order of 100 to 300 beats per second. With wings shortened experimentally

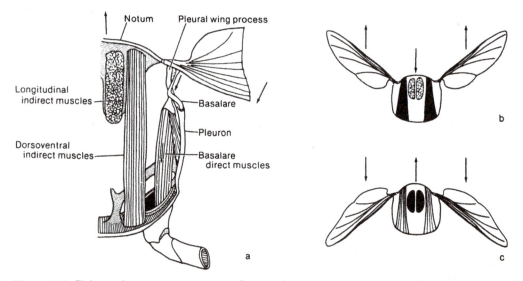

Figure 5.12 Flight mechanism: a, cross section of a mesothoracic segment during the downstroke and pronation of the wing (view is of left side of insect as seen from in front); b, wing upstroke and supination (contraction of dorsoventral indirect muscles indicated in black); c, wing downstroke and pronation (contraction of longitudinal indirect muscles indicated in black).
Source: Redrawn from R. E. Snodgrass, The Principles of Insect Morphology.

and at a high temperature, the midge *Forcipomyia* (Ceratopogonidae, Diptera) was rated at a frequency of 2218 strokes per second.

In these orders, the main source of power for the wings resides in one thoracic segment and is often limited to one set of longitudinal and one set of dorsoventral muscles operating antagonistically, such that one set relaxes (stretches) when the other set contracts. Flies have elastic parts of the thoracic box that are stretched during one part of the wing stroke, storing energy, then quickly released by a "click" mechanism to assist the next movement. The complex path of the wing is now determined mechanically by the shapes of the interconnected sclerites at the base of the wing and the adjoining thoracic processes. You can demonstrate this for yourself in a freshly killed bee or fly. Depression of the mesonotum will automatically supinate the wings as they are forced upward or raise the mesonotum with a pin and watch the wings pronate as they are forced downward. The direct muscles of these insects serve to modify the wing stroke over many beats. This allows bees and flies to hover and maneuver in flight.

The insect nervous system has also evolved some impressive adaptations for flight. Day-flying insects have much larger compound eyes and the associated connections with the brain to transport, process and respond to visual information useful for flight. The flicker fusion rate (speed with which the brain renews an image) of a housefly is almost twenty times faster than that of humans. The hind wings of Diptera have evolved to become tiny stubs (halteres) that vibrate rapidly to provide information about changes in orientation to the horizon, much as applying force to a gyroscope causes a predictable rotational reaction force. Complex pattern-generating neural circuits set up a pattern of nerve inhibition and excitation to rapidly coordinate the activation of upstroke and downstroke muscles.

Additional adaptations for flight include providing robust ventilatory air movements in the tracheal system to provide oxygenation and to expel carbon dioxide as well as providing energy for the oxidative energy producing metabolism of the flight muscles. Hemolymph circulation provides ways to dissipate heat from the relatively massive and highly energetic flight muscles, as discussed below.

THERMOREGULATION

Insects are active within wide ranges of temperature. The snow scorpionflies *Boreus* (Boreidae, Mecoptera; Fig 8.5), snowfly *Chionea* (Tipulidae, Diptera), snow "fleas" *Achorutes* (Poduridae, Collembola), and winter stoneflies *Allocapnia* (Nemouridae, Plecoptera) are active at low temperatures on snow. *Boreus* dies of heat stress if kept for long in the human hand. At the other extreme, the firebrat *Lepismodes inquilinus* (= *Thermobia domestica*; Lepismatidae, Thysanura) prefers 12 to 50°C and dies at 51.3°C. At unfavorably high ambient temperatures, an insect is severely limited in the extent to which it can cool itself. Cooling by the evaporation of water quickly leads to desiccation unless water is freely accessible. Nectar-gathering bees regurgitate fluid from their crop ("honey stomach") and hold in within their mouth while flying so that the evaporation of the fluid cools their heads, absorbing heat from the adjacent thorax. The tsetse fly *Glossina morsitans* (Glossinidae, Diptera) feeds while exposed to sun on the hot skin of African mammals. The opening of the spiracles at temperatures above 39°C permits evaporative cooling within the tracheal system to lower the body temperature about 2°C. This is possible because the meal contains ample liquids to restore lost water. Honey bees transport water to their hives, deposit it in the combs, and vigorously fan with their wings so that evaporative cooling takes place. Most insects, however, escape heat by seeking shade or cooler places.

Ectothermy

Although insects usually cannot lower their temperature below that of ambient, many can actively regulate their temperature above that of the environment. Insects are able to increase their body temperature well above ambient by absorbing heat from the sun. Organisms that derive heat almost entirely from their environment are called ectothermic. Butterflies assume distinctive postures while basking in the sun. "Dorsal baskers" spread their

BOX 5.1   How to Study Insect Flight

How can bumble bees fly? Legend has it that engineers in the 1930s calculated that, according to conventional aeronautical theory, bumble bees should not be able to fly. The point of this story was that insects must have used some aeronautical principles unknown at the time, but we have made major advances to date in understanding how muscular, aerodynamic, and neural processes operate to achieve insect flight.

Most lift on the wings of aircraft is attributed to the Bernoulli principle of lowered static pressure created by the higher velocity of air over a wing's upper surface creating lift for flight. You can easily demonstrate this lifting force by blowing across the top of a strip of paper held horizontally, which causes the paper to rise. This example, and the studies of airfoils in wind tunnels or on cruising aircraft, describe steady-state conditions. Insect flight, however, is the epitome of constant changes that enable flight. Many insects, especially those with large wings, use their momentum to glide between wing strokes, and the Bernoulli principle can account for some of the lifting forces in this flight with large wings and slow wingbeats. But due to their small size and type of wing movements, insects utilize other aspects of other aerodynamic forces for flight (Box Fig. 5.1).

To begin with, at the size scale of most flying insects—a few millimeters—the viscosity of air is a far more influential force than it is for larger objects. Minute insects, like Thysanoptera and some Coleoptera (Ptiliidae) and Hymenoptera (Mymaridae, Trichogrammatidae), about a millimeter or so in length, are so small that the viscosity effects of air predominate in flight. Their tiny paddle- or rod-shaped wings are fringed with fine bristles that utilize drag forces on the wings rather than lift, so that their flight is analogous in part to swimming through the air.

Engineers use a scaling factor called the Reynolds number (Re) to calculate how fluid viscosity relates to the size and speed of an object moving in either a gas or liquid fluid. Re is a dimensionless number that is the ratio of inertial forces (velocity × size) to viscous forces (force/area × time). Examples of Re range from over 10^8 for a whale to 10^5 for a typical baseball pitch and 10^{-4} for human sperm. Insects have an Re of about 10 for the smallest and 10^4 for the largest. Engineers can adjust the dimensions of aircraft models to have the same Re, and thus similar flight dynamics, as the full-size aircraft as a way to predict and test aerodynamic characteristics by working with a model. Similarly, in experiments to study insect flight at a scale and speed that was easier to observe, researchers used oil instead of air as the model fluid, thus enabling the much larger 25-cm wingspan of model insects submerged in oil to have the same Re as the real insect in air. Other experiments utilize smoke plumes in generated airflows coupled with high-speed photography from different perspectives of a flying insect. These experiments showed that vortices and downdrafts coming off of the moving wing produced forces significant for providing lift (Box Fig. 5.1). You can visualize a vortex as the visibly swirling water coming off the blade of a paddle or oar during a stroke.

In part because of their small size, insect wings can develop significant lifting forces at very high angles of attack (a), the angle between a wing's horizontal axis and the oncoming air (relative wind). In excess of an angle of attack of about 15 degrees, a conventional fixed aircraft wing typically loses smooth flow over the wing, resulting in turbulent airflow and a sudden loss of lift, called a stall. Many insect wings develop lift at an angle of attack two to three times higher (a of 30–45°) by using the upwelling vortex of air that separates from the leading edge of the wing at these high angles of attack to provide lift. This type of lift, also called "delayed stall" (Box Fig. 5.1C-1) is rapidly transient in fixed-wing aircraft, but insect wingbeats are so rapid that after shedding the leading edge vortex at the end of power stroke, the wing rotates and generates another vortex. The rotation of the wing produces another lifting force, rotational lift (Box Fig. 5.1C-2 and -3), much like a tennis ball hit with backspin rises because of the faster airflow across its top side.

BOX 5.1   *(Continued)*

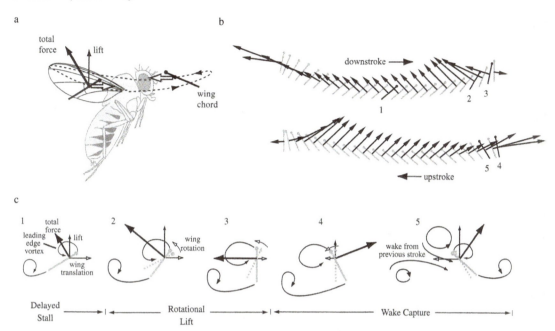

Box Figure 5.1. Aerodynamic mechanisms of insect flight. a, In hovering flight, the fruit fly wing tip's path (dotted line) is formed by forward and rearward movements (white arrows), with a high angle of attack (angle between wing chord and flight direction). The wing rapidly rotates between strokes to reverse the wing surface for downstroke (dorsal surface up) and upstroke. The total wing force (thick black arrow) is shown with the lift component (part of wing force that is opposite gravity) as a vertical arrow. b, The magnitude and direction of the wing's force (black arrows) and the orientation of the wing to flight direction change continually throughout flight (wing chord, or wing axis, is shown by gray lines with circles at leading edge). c, Curved lines show air flow generated by wing movements (as in b). The vortex generated at the wing's leading edge creates delayed stall (1). Rapid wing rotation at the end of the wing stroke provides rotational lift (2–3). In wake capture, the swirling wake generated by a previous wing stroke (4) is captured by the wing's next stroke (5) to aid lift.
Source: Copyright Elsevier. Used with permission.

In addition, in some conditions insects can use a mechanism called "wake capture" (Box Fig. 5.1C-4 and -5) to conserve much of the energy that would otherwise be dissipated in vortices by positioning its wings to capture some of the lifting energy of its wake.

A method to study insect free flight is to make high-speed motion pictures of insects flying through smoke plumes to record air movements about the wings. One such study of a butterfly (Srygley and Thomas, 2002) showed the versatility of this insect to use all of the above-mentioned methods of generating flight forces in addition to the "clap-and-peel" method of slamming the wings together on the forward stroke and retrieving the wings by peeling the stiff leading edge of the wing apart. The more flexible parts of the wings open in a billowing manner to generate lift. The method of producing lift or thrust often varied from one wing stroke cycle to the next. These and other studies show that insects use no single method to generate flight forces. Some methods act simultaneously, while others alternate depending on the type of insect or phase of flight.

wings and point their heads away from the sun. The skippers (Hesperiidae) commonly spread only the hind wings, leaving the fore wings vertical. "Lateral baskers" keep both wings vertical over the body and orient to present one side to the sun's rays.

Endothermy

Insects that elevate their temperature through metabolic activity are endothermic. Large flyers such as bumble bees and sphinx moths (Sphingidae, Lepidoptera) heat their thorax to temperatures optimal for flight muscle performance. With their wings folded back across the body to disengage the flight muscle contractions from moving the wings, these insects contract both the dorsoventral and the horizontal indirect flight muscles at the same time, rapidly building up heat in the thorax to enabling them to fly even in very cool weather. If not dissipated, the heated thorax can rapidly reach lethal temperatures, but hemolymph moving through the dorsal vessel as it passes along a contorted route through the thorax absorbs excess heat from the surrounding muscle mass. The hemolymph empties from the dorsal vessel into the head, from which it then is shunted along the venter of the thorax to the ventral abdomen. During flight the head and mouthparts and the ventral abdomen are cooled by air flowing over the insect. In the nest, the female can generate heat from the flight muscles and transfer the heat through hemolymph movements to her ventral abdomen. During the cool days and nights of early spring months, she can use this method of heat transfer to speed the incubation her first offspring by pressing her ventral abdomen against the brood cell of her first developing larva. The dorsal abdomen is insulated by a covering of hairs (setae), but the ventral abdomen is bare, facilitating heat transfer. Without this added heat for incubation to develop the first brood, there would not be enough time to rear subsequent broods within the brief warm season of the subpolar climates these bumble bees inhabit.

The complexity and efficiency of the bumble bee's thermoregulation belies the simplistic view that insects are strictly "cold-blooded" organisms. The bumble bee displays remarkable control of not only its body temperature but what part of its body is warmed. Vertebrates shift some blood flow to the peripheral circulation for cooling during strenuous activity at the cost of diverting oxygen availability to the muscles. In contrast, the heat-transfer mechanisms of insects are separate from gas exchange mechanisms, allowing high flight muscle activity, for example, to be unrestricted by oxygen demand.

Most insects are small and have a high surface:volume ratio, so that radiation and convection rapidly dissipate heat produced by the flight muscles. Very small flying insects thus have flight muscles that operate even at cool temperatures.

Many endothermic insects that operate in cool climates have a covering of insulating scales (moths) or hairs (bumble bees). Modification of tracheal trunks into air sacs can also function as insulators between the thorax and abdomen to retard the loss of heat from the thorax.

PROTECTION FROM COLD

Insects must adapt to the temperature regimes of their location and habitat. They may become immobilized at temperatures above freezing and, if kept in this state, will die from desiccation because they are unable to restore lost water by drinking. Insects living in temperate or colder regions, such as the Arctic or high mountains, risk freezing when air temperatures drop below 0°C. Yet some insects do survive subzero weather, and others live normally after being frozen. Freezing is not the only cold temperature stress for insects. Tropical insects can often be easily killed by brief exposures to temperatures of 4°C.

Some insects avoid low air temperatures by virtue of their habitat or behavior. Aquatic insects and those in deep, moist soil escape because freezing usually occurs only at the water or soil surface. In winter, honey bees cluster together in a mass in the combs of their hive. Even when air temperatures outside are well below 0°C, they maintain 20 to 30°C inside the cluster. Heat is derived from the metabolism of stored honey. Winged adult insects that migrate, such as the monarch butterfly, fly south in autumn to warmer regions and return northward in spring.

Immature stages, however, are often on the soil surface, on plants, or on other objects and must endure exposure to cold weather. Diapause is a physiological state that is associated with overwintering (Chap. 4). In the next section we will describe other processes that are usually combined with diapause to ensure survival even in the Arctic.

Susceptibility and Tolerance to Freezing

Freezing injury to insects is most obvious when it results from ice formation. Ice crystals may disrupt cell membranes and alter the concentrations of intracellular fluids by initially concentrating water into the growing ice crystals. These shifts in water concentration in turn set up osmotic imbalances.

The supercooling point (temperature of crystallization) of a liquid is the temperature at which it turns to ice. Although 0°C is ordinarily considered the freezing point of water, insect fluids and cell contents may not freeze down to -20°C or lower because of cryoprotectants that reduce the supercooling temperature. Glucose, trehalose, low-molecular-weight lipids, and sorbitol in the hemolymph act as cryoprotectants. No other natural chemical, however, equals glycerol to prevent freezing. Glycerol has been found in most overwintering, freezing-tolerant insects, especially larvae and pupae. Glycerol extends the temperature range of supercooling without freezing to retard the rate of freezing and to reduce the size of crystals. Both actions reduce freezing injury to tissues. When the insect is indeed frozen, glycerol presumably reduces the deleterious osmotic effects and prevents the intracellular freezing that is fatal.

A second strategy that insects use to avoid freezing injury is to avoid having ice nucleation centers (nucleators) that accelerate ice formation. Mineral particles, and especially bacteria, act as catalysts for ice formation. The presence of nucleators in the gut or hemolymph of an insect raises the supercooling point. On the other hand, the addition of dissolved substances in water lowers the points of freezing and supercooling. Even without special antifreeze substances, the body fluids of overwintering insects can often be supercooled to −20°C. Thus many insects survive subzero weather by evacuating their guts in autumn and supercooling without harm, to temperatures above the supercooling

point. When the supercooling point of the insect is reached, an ice crystal will form internally around a nucleator.

Studies of overwintering adults of the Arctic ground beetle *Pterostichus brevicornis* (Carabidae, Coleoptera) provide an example of the ecology, behavior, and physiology of an insect in a cold environment. The beetle is a scavenger that lives in forest litter in circumpolar regions. Near Fairbanks, Alaska, summer is spent in the upper 2 to 3 cm of surface litter that is warmed to 10 to 14°C by air and intermittent sunshine. Only a few centimeters below the surface, the temperature of litter drops to 0 to 1°C because of the underlying permafrost. After the first frost in autumn and independent of photoperiod, the adult beetles migrate by the thousands into the old galleries of wood-boring beetles in decaying tree stumps. The beetles continue to move toward the interior of the stump, favoring regions of −1 to 2°C, until immobilized and finally frozen by the approaching winter.

In winter the stumps may be exposed to −40°C for several weeks and may reach −60°C for shorter periods. Temperatures in the interior of the stumps lag behind the outside temperature. Stumps also are often covered by an insulating blanket of snow. Thus the hibernating beetles are buffered against rapid temperature changes. When beetles are kept at 5°C, no glycerol is detectable in the hemolymph. Without glycerol, the freezing point of hemolymph is −2°C and the supercooling point is −7.3°C. Synthesis of glycerol is stimulated by exposure to 0°C for more than 24 hours. Glycerol then continues to accumulate at temperatures below freezing (Fig. 5.13). A sudden drop in temperature to −15°C or more is fatal. Beetles inside stumps, however, are protected against such rapid changes. If given time to synthesize glycerol, the beetles become freezing-tolerant.

Other adult insects are generally not known to be freezing-tolerant. In this respect the carabid is unique, but the annual cycle of glycerol production and loss resembles that known in less detail for overwintering larvae and pupae of other insects. It is also interesting to note that the carabid is able to digest food, develop eggs, and exhibit other signs of physiological activity while in a cold-hardened and

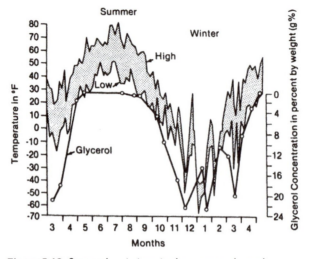

Figure 5.13 Seasonal variations in the average glycerol concentration in hemolymph of the Arctic carabid, *Pterostichus brevicomis*, and high–low ambient temperatures near Fairbanks, Alaska. Note that the glycerol scale is inverted.
Source: Data from Baust and Miller, 1970.

frozen state. Although only the adults have thus far been analyzed, the beetles also overwinter as larvae or pupae. Some individuals overwinter twice and live up to 36 months.

Beetles start hibernation with a water content at about 62 percent of the body weight, abundant fat, and the guts filled with the summer diet of small insects and other arthropods. During the first few months of hibernation, digestion and excretion continue until the gut is emptied. Body weight declines as the gut is cleared, fat is consumed, and water content drops about 10 percent. Then in January and February whenever temperatures exceed –3 or –4°C, the beetles begin to crawl about and feed. The diet is now rotten wood containing bacteria and fungi. The ovaries become active and eggs start to develop. Fat droplets in fat cells fluctuate in abundance as a result of the renewed intake of food (increasing fat) and the growth of eggs (decreasing fat). The water content of the beetle's body begins to rise as water in the stump thaws and becomes available. In the spring, the glycerol concentration declines steadily. After one day at 21°C, the beetles

have completely lost their tolerance to freezing; all die if returned to below-freezing temperatures.

Cold Hardiness and Shock Proteins

A change in tolerance to cold as a result of prior exposures to cold is called cold hardening. This process may take months to complete in some species and only minutes in others. The season or generation of the insect may determine how quickly and to what extent it may be conditioned to withstand cold temperatures. This adaptation appears to be widespread and important for survival in insects of temperate to arctic climates. Molecular methods have revealed proteins associated with cold hardening. Not only cold but, interestingly, heat and other stresses such as poisons and desiccation can induce the expression of shock proteins, often called "heat shock proteins." Studies to date, mostly in bacteria and vertebrates, indicate that these proteins (chaperonins) act to preserve protein assembly and folding to prevent their denaturation by temperature changes. This should be a fertile field for further investigations.

CHAPTER 6

Reception of Stimuli and Integration of Activities

How does the nervous system enable insects to perceive and relate to their environment? After a brief review of the structure and function of nerve cells, we describe the major sense organs and address how insects respond to internal and external stimuli. After an introduction to the kinds of behavior exhibited by insects, we conclude the chapter with a discussion of communication and circadian rhythms.

ORGANIZATION OF THE NERVOUS SYSTEM

Neurons and Neuronal Connections

Nerve cells are either (1) neurons that are excitable by chemical or electric signals and rapidly transmit electric impulses or (2) glial cells that support, insulate, and nourish the neurons. The nucleated cell body of a neuron is the soma (or perikaryon) (Fig. 6.1a). Axons are the slender fibers that extend from the soma to connect with other neurons or with muscles or glands.

The synapse, or synaptic cleft, is the space between connecting axons or between axons and muscles at the tips of the axons. Most tips of the axons in insects are finely divided into numerous branches and separated by a gap. Nerve transmission crosses the synapse from the presynaptic axon to the postsynaptic axon by diffusion of a chemical substance called a neurotransmitter. The postsynaptic or receptive axon is also called the dendrite. Various types of glial cells wrap tightly around nerve cells.

Monopolar neurons have only one axon issuing from the soma. The single axon is at least forked so that one branch serves as a dendrite. Some monopolar neurons, however, are highly branched and complexly interconnected with other neurons. Bipolar neurons have two separate axons extending from the soma. Again, one serves as the dendrite and the other leads to the next neuron. Multipolar neurons with more than two axons are rare in insects (stretch receptors; see "Mechanoreceptors," below).

A nerve cell is basically a tiny battery whose charge is created by membrane-bound ion pumps and ion-specific channels in the nerve cell membrane. A nerve cell at rest has a deficiency of sodium ions (Na^+), which results in a resting potential (voltage difference) between the interior and exterior of the cell membrane of about -70 millivolts. Sodium-potassium pumps that span the nerve cell membrane selectively expel Na^+ and transport potassium (K^+) into the cell to create a deficiency of Na^+ within the cell. Ion channels are pores in the nerve membrane that selectively allow passage of certain ions, mainly Na^+, K^+, Cl^-, and Ca^{2+}. Pores may be open or "gated," that is, closed under some conditions and open under other conditions. Open potassium channels create

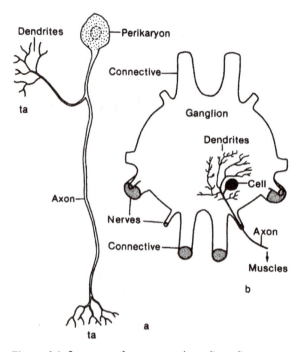

Figure 6.1 Structure of neurons and ganglia: a, diagram of a monopolar neuron; b, diagram of a monopolar motor neuron inside a thoracic ganglion. ta, terminal arborizations.
Source: a, modified from R. E. Snodgrass, The Principles of Insect Morphology; b, simplified and modified from Altman and Tyrer, 1974.

the resting potential by allowing K^+ to diffuse out of the cell to create an imbalance of Cl^- ions. Voltage-gated sodium channels close in the resting state but open in response to voltage changes. A nerve impulse triggered (fired) by the reception of a neurotransmitter substance at the postsynaptic membrane moves outward as a localized current flow, or graded response, to portions of the nerve cell, the axon hillock, to induce the opening of gated sodium channels to allow a surge of Na^+ ions into the cell. This resulting voltage spike, the action potential, reverses the voltage difference from about –70 mV to +50 mV. The change in voltage (depolarization) triggers adjacent gated Na^+ channels to open, propagating a wave of depolarization along the axon. Some insecticides, such as plant-derived pyrethrins, are toxic because they cause gated Na^+ channels to remain open

and to continue to fire. As the nerve impulse moves along the axon, gated Na^+ channels close and gated K^+ channels open to restore the resting potential. Numerous gated ion channel genes from *Drosophila* have been identified and cloned for use in experiments to help decipher how this system operates in humans and other animals. The most studied of these is the *Shaker* gene (*Sh*), which codes for a transmembrane protein whose conformational changes allow or halt passage of K^+.

Nerve transmissions across a synapse are slower than the nerve impulses along axons because neurotransmitter substances must be released from vesicles, diffuse across the synapse, and bind to receptor sites. As a consequence, longer axons and nerve cords with fewer interconnecting neurons have evolved to maximize the speed of transmission for rapid reactions to stimuli. Giant axons in the cockroach *Periplaneta americana* (Blattidae, Blattodea) bypass synapses in the abdomen to go directly to ganglia via giant interneurons in the thorax to activate legs for escape maneuvers without intervention of the insect brain. In this reflex arc, air movements stimulate sensory hairs on the cerci to trigger nerve impulses that travel directly to the thorax and then on to leg muscles and nerve pattern generators to provoke a movement away from the stimulus without involving the brain. In this case the increased diameter of the axons also increases the speed of depolarization along the giant axon.

Nerve impulses arriving at an axon tip trigger the opening of Ca^{2+} channels. The resulting higher concentrations of Ca^{2+} within the nerve tip cause vesicles containing neurotransmitters such as acetylcholine to fuse with the presynaptic membrane and release acetylcholine into the synapse. Neurotransmitters that diffuse across the synaptic cleft bind to receptor proteins on postsynaptic membranes of the adjacent dendrite to trigger the opening of a gated sodium channel, which then initiates a nerve impulse. In the case of acetylcholine, enzymes (acetylcholinesterases) then break down this neurotransmitter. Some insecticides such as malathion inhibit cholinesterases by binding strongly to them. As a consequence, acetylcholine is not broken down and the affected synapses continue to elicit repeated firing of the postsynaptic neuron. Acetylcholine serves as a neurotransmitter at excitatory synapses. The branching of dendrites from a single neuron means that some dendrites may lie opposite axon tips that release inhibiting neurotransmitters such as γ-aminobutyric acid (GABA) that cause hyperpolarization (the opposite of depolarization), and other connecting axons may release activating neurotransmitters. This introduces flexibility and complexity into neural networks. Discovering previously unrecognized neurotransmitters and determining their roles are popular current topics for active research in neurobiology. Numerous neuropeptides have been identified as neurotransmitters and often have dual roles as hormones (Chap. 5). L-Glutamic acid and L-aspartic acid serve as neurotransmitters for neuron-muscle junctions in insects.

Sensory neurons within sense organs send nerve impulses to the nervous system in response to physical or chemical stimuli. Sensory stimuli to specific parts of the sensory cell may directly alter the permeability of ion channels to trigger a nerve impulse (ionotropic sensory detection), or the stimuli may induce a nerve membrane protein to activate a cascade of molecular intracellular signals that open or close ion channels (metabotropic detection). For example, a sex-attractant molecule (a pheromone, discussed later) that binds to a receptor protein on a sensory neuron of an insect's scent organ triggers the enzymatic activation of a cascade of chemical messengers within the cell that controls ion channels to fire an action potential. Mechanical stimuli such as stretching and pressure may act directly on ion channels to fire neurons in proprioreceptors. Chemosensory receptors detect specific chemicals, photoreceptors detect light, and so on. The sensory soma is usually located just beneath the sense organ and sends nerve impulses along a long fiber to a segmental ganglion. Within the ganglion the sensory axon transmits impulses either directly by one synapse to a monopolar motor neuron or indirectly by two or more synapses involving other monopolar neurons, the interneurons. The nervous stimulation of muscle fibers in insects differs from that of mammals. In the latter the axon terminates in a single motor end plate, and the transmitter substance to the muscle is acetylcholine. Motor axons of insects are branched, providing many endings on a muscle fiber and the neurotransmitter is not acetylcholine. A muscle may receive stimulation from two or three different kinds of motor axons. Stimulation by a fast axon produces a "fast" response in the form of a brief powerful contraction. Each stimulation by a slow axon results in a small contraction, but when the stimulation is increased in frequency, a "low" contraction of increasing force is produced. Slow axons thus give greater control over movements. The transmitter substance for these neuromuscular junctions is glutamate. Motor programs are neural mechanisms that coordinate repetitive activities such as ecdysis, wing movements, rhythmic breathing, and walking. Interconnected neurons within the ganglia generate rhythmic bursts of excitatory or inhibitory outputs. These combine with neural feedback from sensory organs to modify the motor activity. Motor programs are complex, and only a few, such as walking or Drosophila flight, have been described in detail for insects.

Nerves and Ganglia

On a larger scale, the bundles of sensory and/or motor axons that issue from the segmental ganglia are called nerves. A ganglion is a mass of nerve tissue (Fig. 6.1b) often including thousands of neurons. The word is used loosely to apply not only to a single large mass but also to smaller masses that may be present inside. Thus each segmental ganglion is formed by the median fusion of a pair of embryonic ganglia. The central portion is the neuropile, consisting entirely of axons and processes of their associated glial cells. Included are sensory, motor, and connecting axons and their various synaptic junctions. The cell bodies of the neurons and glial cells are located only around the outside of the neuropile. The paired longitudinal connectives between ganglia are bundles of axons plus sensory and motor axons that communicate with other segments.

The ganglia, nerves, and connectives are surrounded by a supportive sheath composed of an outer, noncellular neural lamella of connective tissue and an inner perineurium of glial cells. The sheath is a "blood-brain barrier," isolating the neurons from the chemical components of the hemolymph and maintaining a chemical environment around the axons which is essential for impulse transmission. The glial cells also pass nutrients to the neurons and probably prevent contact between neurons except at synapses.

The nervous system is divided into four subsystems: (1) sense organs, including the light, chemical, temperature, and mechanical receptors; (2) the peripheral nervous system, including the sensory and motor neurons that communicate with the integument and muscles (this system is not discussed further); (3) the central nervous system, including the brain and the segmental ventral ganglionic chain that generally regulate the bodily activity; and (4) the stomatogastric nervous system, including several small ganglia that regulate the foregut, the midgut, and several endocrine glands.

CENTRAL NERVOUS SYSTEM

The central nervous system of insects includes the brain, situated above the mouth, and the ventral nerve cord, beginning behind the mouth. The evolution of the central nervous system was discussed earlier (Fig. 2.4). Recall that the basic structure of the system consists of paired embryonic ganglia connected transversely by commissures and longitudinally by connectives.

Brain

The brain has three ganglionic masses (Fig. 6.2). (1) The protocerebrum is the largest and most complex region. Laterally it receives the nerves from the large optic lobes of the compound eyes; dorsally, it receives nerves from the ocelli. Internally, the protocerebrum has a pair of mushroom-shaped neuropiles, the corpora pedunculata, which are important centers for the integration of behavior. In most insects the main sensory input is from the antennal lobes. In social Hymenoptera the corpora pedunculata are especially large, suggesting that they are important in the complex behavior of these insects. Medially is the pars intercerebralis, a complex cellular mass. Clusters of neurosecretory cells here and elsewhere in the brain communicate by long nerves with the neuroendocrine corpora cardiaca and corpora alata. Three other neuropiles are present in the protocerebrum: the median protocerebral bridge and central body, and the ventral, paired accessory lobes that are connected by a commissure. (2) The deutocerebrum has motor and sensory axons to the antennae. The paired neuropiles, or antennal lobes, are connected by a commissure. Recall that both the protocerebrum and deutocerebrum were primitively in front of the mouth. (3) The tritocerebrum has sensory and motor axons to the labrum and to the frontal ganglion of the stomatogastric system. The paired neuropiles, or tritocerebral centers, are connected by the tritocerebral commissure, which is visible behind the esophagus. The tritocerebrum was primitively the first ventral ganglion behind the mouth and secondarily became incorporated into the brain.

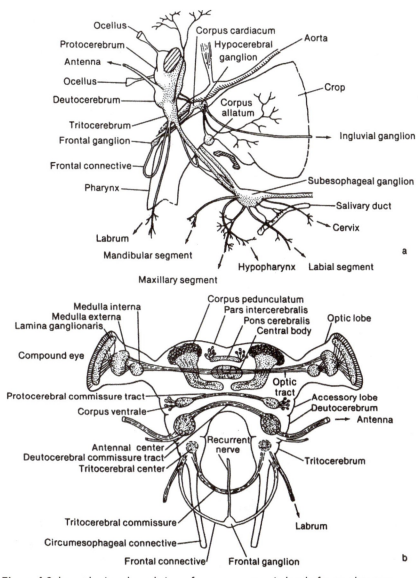

Figure 6.2 Insect brain: a, lateral view of nervous system in head of a grasshopper; b, diagrammatic cross section of brain showing ganglia and commissures.
Source: Redrawn from R. E. Snodgrass, The Principles of Insect Morphology.

Estimates of the number of neurons in the brain are about 850,000 for a worker honey bee, 1.2 million for a drone honey bee, 330,000 for a house fly, and 360,000 for a locust.

Ventral Nerve Cord

The ventral nerve cord is a chain of segmental ganglia connected anteriorly with the tritocerebrum by the paired circumesophageal connectives. The ganglia of the three gnathal segments are fused into a compound mass named the subesophageal ganglion. Motor and sensory nerves extend to the mandibles, maxillae, labium, salivary glands, and cervical muscles. The three ganglia of the thoracic segments serve the legs and flight mechanism and are usually the largest ganglia of the ventral nerve cord. Sensory and motor nerves extend to the various sense organs and muscles. In the abdomen are no more than eight ganglia. The last is a large, compound ganglion serving the eighth and following segments, including the genitalia and cerci.

There is an evolutionary tendency for segmental ganglia to fuse anteriorly into compound ganglia (Fig. 6.3), effectively increasing the speed of nerve impulse transmission to the thorax and head by reducing the number of ganglia or eliminating interneurons and synapses in the abdomen. In the grasshopper, for example, the first three abdominal ganglia are fused with the metathoracic ganglion. Coleoptera and Diptera have variable amounts of fusion among the ganglia. Nearly all the Hemiptera have all of the ganglia of the thorax and abdomen fused into a single compound ganglion, with nerves extending to the respective segments. During the life history of an insect, a developmental tendency also exists for some separate ganglia in the larva to be fused in the adult. This arrangement presumably permits more rapid communication among the ganglia by reducing the lengths of interconnecting axons.

STOMATOGASTRIC NERVOUS SYSTEM

The stomatogastric includes two median ganglia connected to the brain and a single ganglion or paired ganglia on the surface of the foregut. (1) The frontal ganglion is situated medially in front of the pharynx, usually just behind the frontoclypeal suture. It receives the paired frontal connectives from the tritocerebrum and the single recurrent nerve from the hypocerebral ganglion. Some small nerves from the frontal ganglion innervate the foregut and mouth cavity. The frontal ganglion controls the emptying of the crop. If the ganglion is removed experimentally, undigested food accumulates in the crop. The ganglion probably also contributes neurosecretions to the endocrine system. (2) The hypocerebral ganglion is situated behind the brain, between the aorta and foregut. It usually has three major connections: the single recurrent nerve running from the frontal ganglion, one or two esophageal nerves posteriorly to the ingluvial ganglia on the surface of the foregut, and dorsally to the brain by way of the neuroendocrine corpora cardiaca. The hypocerebral ganglion also sends

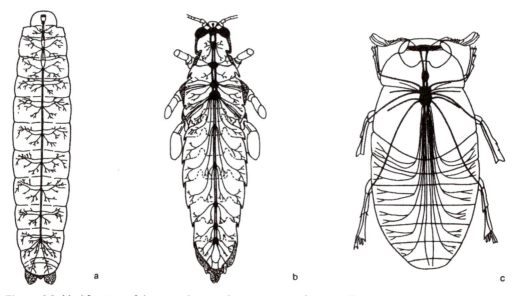

Figure 6.3 Modifications of the ventral nerve chain: a, system of a caterpillar, *Malacosoma americana* (Lasiocampidae); b, system of a grasshopper, *Dissosteira carolina* (Acrididae); c, system of a beetle, *Lachnosterna fusca* (Scarabaeidae).
Source: a and b, redrawn from R. E. Snodgrass, The Principles of Insect Morphology; *c, redrawn from Packard, 1898.*

nerves directly to the wall of the esophagus. (3) The single or paired ingluvial ganglia are located posteriorly on the foregut. They control gut movements through nerves to the foregut wall, proventriculus, and midgut. The neuroendocrine corpora cardiaca and the corpora allata make up the retrocerebral complex and are closely associated with the stomatogastric system (Chap. 4).

Often discussed along with the stomatogastric system are two groups of small nerves closely associated with the ventral nerve cord and called the visceral nervous system: (1) the unpaired ventral visceral nerves that originate from each ventral ganglion and run caudally to innervate the tracheal system, spiracles, and some somatic muscles and (2) the caudal visceral nerves that originate from the last abdominal ganglion and run to the reproductive organs and hind gut.

SENSE ORGANS

Insect Eyes

Insect eyes are of two types: compound and simple. Compound eyes (Fig. 6.4 a–d) consist of from 8 (Collembola) to over 28,000 (Odonata, Fig. 6.5) closely packed photoreceptive units, the ommatidia (Fig. 6.4b). Each ommatidium has its own corneal lens, which usually forms a hexagonal facet on the eye surface. Compound eyes connect directly to the optic lobes of the brain, minimizing the time needed for nerve impulse transmission from eye to brain. (2) Simple eyes each have a single round corneal lens with several to many light-sensitive cells beneath. Simple eyes are of two types: (a) Dorsal ocelli (Fig. 6.4e) are located on the top of the head in a pair or triangle. They are innervated by the median part of the protocerebrum and commonly occur together with compound eyes in both adult and larval stages (ocelli absent endopterygote larvae). (b) Larval eyes, called stemmata or lateral ocelli, are located laterally on the head, where usually less than a dozen at each side occur in loose clusters. They are innervated by the optic lobes and commonly occur in endopterygote larvae in the place of compound eyes (Fig. 6.4f). Insects do not have devices like eyelids to shield their light-sensitive receptors from direct sunlight.

Protura (Fig. 15.1c) and Diplura (Fig. 16.1) have no discrete eyes. Compound eyes are found on at least some adults of all other orders except Siphonaptera and Strepsiptera. The lateral eyes of Siphonaptera, if present, are simple (Fig. 42.1c), and the berry-like eyes of adult Strepsiptera are unusually large stemmata. Compound eyes are also found on immature stages of Exopterygota, but they are missing in the larval Endopterygota. The lateral eyes of endopterygote larvae are usually stemmata (Fig. 15.1b,d). Larval Mecoptera are the exception: they may have stemmata or unique compound eyes. The evolutionary loss of compound eyes and ocelli has occurred in various groups of insects, especially in apterous insects such as the parasitic Phthiraptera (lice), female scale insects, cave insects (Fig. 8.3b), or worker castes of some termites and ants.

Compound Eyes and Vision

The general structure of an ommatidium (Fig. 6.4b) includes two lenses, the corneal lens and crystalline cone, and a set of eight (rarely six, seven, or ten) sensory, or retinula, cells. Surrounding the crystalline cone are primary pigment cells, and around the retinula are secondary pigment cells. Two modified epidermal cells, the corneagenous cells, secrete the biconvex cornea. The cone is composed jointly of four cells, and by their internal structure light from the cornea is focused on the retinula. The density, and thus the refractive index, of the cone may increase along its longitudinal axis to bend light rays more strongly in focusing. The elongate retinula cells are usually packed together, and each contributes to a central rod-shaped structure called a rhabdom. Each cell contribution is a photoreceptive rhabdomere of tightly packed microtubules oriented at right angles to the central axis. Microtubules within a rhabdomere are parallel to one another and usually also parallel with those on the opposite side of the axis. In the honey bee, for example, eight retinula cells of each ommatidium are fused pairwise into four compound units. The opposite units have parallel microtubules such that adjacent units have microtubules at right angles (Fig. 6.4f).

Rhodopsins, the visual pigments, are located on the membranes of long narrow microvilli in the

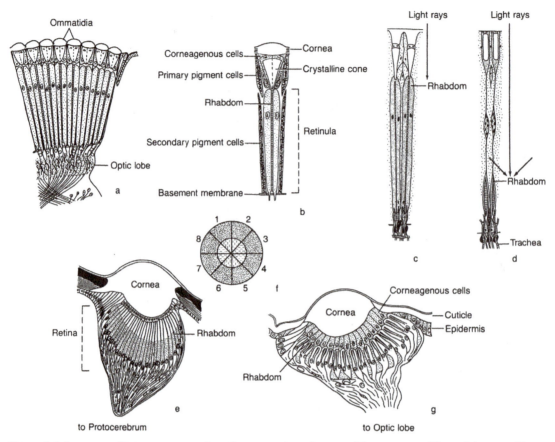

Figure 6.4 Structure of insect eyes: a, section of compound eye; b, ommatidium; c, ommatidium of the apposition type; d, dark-adapted ommatidium of the superposition type; e, dorsal ocellus; f, stemma or lateral ocellus. In c light rays that pass through the cornea of an ommatidium strike only the rhabdom beneath, but in d the light rays pass into adjacent ommatidia and strike other rhabdoms.

Source: a and b, redrawn from R. E. Snodgrass, The Principles of Insect Morphology; c and d, redrawn after Nowikoff, 1931; e, redrawn after Caesar, 1913; f, redrawn after Corneli, 1924.

rhabdomeres. These opsins (G-protein coupled receptors, or GPCRs) react to light waves between 300 and 700 nm to initiate a nerve impulse. In a single insect, different opsins may respond to different portions of the spectrum, thus providing a mechanism for color vision. Behavioral and electrophysiological experiments have demonstrated that many insects are sensitive to three or four spectral classes: ultraviolet (340–360 nm), violet to blue (420–460 nm), blue-green (490–550 nm), and, in addition in butterflies, red (585–610 nm). Although the color vision of insects can be described as a trichromatic system like our own,

their visual world is different because the spectrum is shifted toward the shorter wavelengths to include ultraviolet. Except for butterflies, most insects lose the ability to distinguish differences in wavelengths between 550 to 650 nm even though their vision extends into this region.

Some insects are able to discriminate the plane of polarized light coming from the sky and to use this for navigation during walking or flight. Polarized light is light vibrating in a single plane, whereas direct sunlight has planes of vibration in all directions around the line of travel. Depending on atmospheric conditions the proportion of light

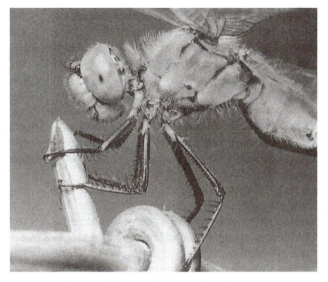

Figure 6.5 Dragonfly head with ommatidia of compound eyes visible. Such aerial predators have a large field of vision for spotting prey.

that is polarized increases in certain regions of the sky up to about 70 percent. The regular orientation of the microtubules plus the configuration of the rhodopsin molecule provides an inherent mechanism for sensitivity to polarized light. Certain ommatidia along the dorsal margin of the eye appear to be specialized for this function in bees, desert ants, and some other insects.

Image Formation in Compound Eyes

What do insects see? Light entering an ommatidia travels down the rhabdom and is reflected within the rhabdom so that all spatial information from light entering a rhabdom is lost. The resulting nervous signal is effectively delivered to the optic nerve as a single "pixel" rather than a crude 6- to 10-pixel image from light that enters the ommatidium. Each separate rhabdomere may have distinctive differences in spectral sensitivity, however, so the resulting information transmitted from the ommatidium includes average intensity and duration of the wavelengths received. An exception occurs with flies, in which the tips of seven rhabdomeres within an ommatidium are separated.

The field of view of a single rhabdom can vary from 1 or 2° to over 20°. This field of view will overlap those of neighboring ommatidia, but the orientations of adjacent ommatidia shift with the curvature of the surface of compound eye.

The resulting image from a compound eye is thus not a mosaic of tiny, identical images in each ommatidium that are so often depicted in science fiction films. Instead the optic nerve transmits an apposition image of apposed (literally "side by side") dots of varying intensities and colors. Thus the resulting image must be very coarse. Dragonfly eyes have the greatest known resolving power for insects. They can distinguish angular differences between adjacent points in an image of only 0.25°. The best resolution for a human eye is about thirty-five times better (Table 6.1). The human eye achieves its higher resolving power by having a larger lens, longer focal distance, and a larger retina (more photoreceptor cells). For each facet of an ommatidium to have similar resolution, the compound eye of an insect would be from a little over a meter to numerous meters in width. At the other size extreme for insect eyes, diffraction of light waves blurs the image in very small lens-bearing eyes, limiting how small a functional eye with a lens can be. There are two types of compound eyes: apposition

Table 6.1 Resolution of Eyes of Selected Insects Compared to Humans and a Spider with Excellent Single Lens ("Camera") Vision

Organism	Interreceptor angle
Human	0.0007°
Jumping spider (*Portia*)	0.04°
Dragonfly (*Aeshna*)	0.25°
Honey bee (*Apis*)	0.95°
Vinegar fly (*Drosophila*)	5°

The smaller the interreceptor angle, the better the ability to distinguish small details (higher resolution).

Data from Land and Nilsson, 2002.

and superposition. In the apposition type of eye described above, the rhabdom is placed immediately below the crystalline cone. Structurally, a jacket of pigment cells shields each ommatidium from light outside the ommatidium's field of view or from adjacent ommatidia, so that only light from the lens reaches the rhabdom immediately beneath (Fig. 6.4c). An apposition image is most useful for insects that are active in daylight. Day-active predators such as dragonflies and wasps have clusters of anterior ommatidia that subtend smaller angles and have more overlap with adjacent ommatidia, plus have larger corneal facets, all of which increase resolution in the forward field of vision.

Although the visual acuity of insects would seem adequate for seeing objects in some detail, behavioral experiments with honey bees demonstrate that their discrimination of form is poorly developed. The compound eye, however, is well adapted for the detection of movement because successive ommatidia are stimulated as the object sweeps across the visual field. Depth perception is possible when the visual fields of the paired eyes overlap. The position of the object seen by both eyes is thus triangulated. The wide separation of the eyes in predators such as mantids and dragonfly naiads probably improves distance perception.

The superposition type of compound eye does not restrict light reception to individual ommatidia. Instead, light from numerous corneal lenses stimulate a rhabdom layer that forms a retinalike surface beneath a clear zone across which light rays

are focused (Fig. 6.4d). Parallel light rays arriving at the surface of the compound eye are refracted (bent) by the lenses so as to emerge as parallel rays towards the central axis of the eye. In the apposition eye, the lenses bend the rays to converge on the photoreceptor cells. The superposition eye is 50- to 100-fold more optically sensitive (detects low light) than apposition eyes of the same size. The diameter of the eye facets determines the size limits determined by diffraction, just as it does for apposition eyes. Superposition eyes are most useful for nocturnal insects. In the Lepidoptera, butterflies have apposition eyes, and moths and skippers have superposition eyes, as do many beetle groups that are night-active. Butterflies and moths are clearly closely related, but the problem of how these great differences in eye structure between apposition and superposition eyes evolved remains unsolved.

The migration of the secondary pigment cells in some insects permits the eye to form images over a wide range of light intensities. During the day the pigment and to some extent the individual pigment cells move proximally into the clear zone and limit the passage of light from the lenses to single rhabdoms. This is the day-adapted eye, and the image is formed by apposition. At dusk, the pigments and secondary pigment cells contract to permit light from adjacent ommatidia to stimulate the rhabdoms. This is the night-adapted eye, and the image is formed by superposition. In some moths, a tapetum of tracheal tubes between and parallel to the ommatidia aid in reflecting light onto the rhabdoms. Most insects of this type are active at twilight or night. Shining a light directly at the eyes of these insects causes the entire eye to reflect light from the tapetum, giving the appearance of brightly glowing eyes. Superposition eyes have had the reputation of poor resolving power or even of lacking the ability to resolve images, but day-flying moths (e.g., skippers, Lepidoptera, Hesperiidae) have optics that produce observable images of excellent quality.

Ocelli and Stemmata

Simple eyes have a single, biconvex, cuticular lens that transmits light on the rhabdoms of sense cells beneath (Fig. 6.4e,f). This is in principle the camera type of eye found in vertebrates. Insect ocelli, however, focus optical images below the rhabdom layer. The prevailing hypothesis is that this lack of focus helps to determine the position of the horizon because small objects such as leaves and twigs are out of focus and so do not interfere with determining the contrast of light above and below the horizon. Monitoring the horizon provides essential guidance information to stabilize flight.

Mechanoreceptors

A variety of simple sense organs, or sensilla (singular sensillum), can detect mechanical distortion of the body. In the usual arrangement, the dendrite of a single bipolar neuron attaches to a movable part of the body, often by a minute cuticular sheath called a scolopale. Movements of the body part in response to mechanical stress, touch, wind, or vibrations in air, water, or solid substrates initiate nerve impulses via mechanically activated ion channels. Some sensilla respond to disturbances from the environment as exteroceptors, while others respond to the position of one part of the body relative to another, thus functioning as proprioceptors. These are important in maintaining posture and in orienting to gravity.

Two general types of mechanoreceptors are based on morphological structure. Type I receptors are cuticular and ciliated. Type II receptors are multipolar and nonciliated. There are three major groups of Type I mechanoreceptors (Fig. 6.6). (1) Trichoid sensilla are hairlike setae with a single dendrite attached at the base. Some project well above the cuticle surface and serve as tactile, or touch, organs (Fig. 6.6a). (2) Campaniform sensilla are oval, domelike areas of cuticle that rise or lower as the adjacent exoskeleton is mechanically strained (Fig. 6.6b). The dendrites detect movements of the dome. (3) Chordotonal organs (Fig. 6.6e) are internal and formed by scolopidia consisting of three cells: a bipolar neuron, a scolopale cell, and an attachment cell. The dendrite sometimes ends inside the scolopale in a process that resembles a cilium. One or more scolopidia cluster into a chordotonal organ and are variously attached to the epidermis. Some are suspended between two points in membranous areas or between leg joints, thus permitting detection of movements between the points of attachment. Subgenual organs of the tibia, however, have a single attachment and do not span a joint. They are especially sensitive to vibrations in the substratum. Johnston's organ is a chordotonal organ in the antennal pedicel (Fig. 6.6f). By its attachment to the base of the flagellum, movements of the flagellum are detected. Tympanal organs are specially designed for the reception of sound (Fig. 6.6e). A chordotonal organ is attached to a thin cuticular membrane or tympanum that freely vibrates when struck by sound waves. Type II mechanoreceptors are stretch receptors that are attached to connective tissue or muscles, registering the tension of such soft tissues, especially in the abdomen. All four types of sensilla (Types Ia, Ib, Ic, and II) may function as proprioceptors. Groups of trichoid sensilla located at each side behind the head on the cervical sclerites register the position of the head with respect to the body. Similar groups at leg joints register the positions of the limbs. Campaniform sensilla are widely scattered over the body surface at points of stress and bending. The halteres of Diptera and insect wings in general are well supplied. Strategically placed chordotonal organs and stretch receptors permit detection of movements of the limbs, segments, and viscera. The positions of the antennae are detected by Johnston's organ.

Vibrations are detected mainly by trichoid sensilla and chordotonal organs. Hairs in the cerci of cockroaches, for example, respond to sound waves of low frequency as well as to wind currents. In male Culicidae and Chironomidae (Diptera), the antennae

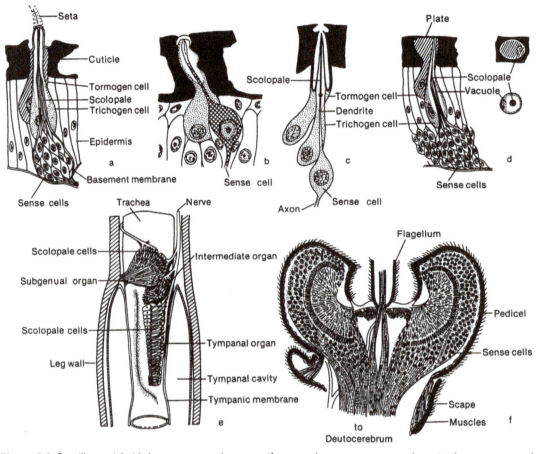

Figure 6.6 Sensilla: a, trichoid chemoreceptor; b, campaniform mechanoreceptor; c, coeloconic chemoreceptor; d, plate organ; e, chordotonal acoustical organ in tibia of right leg of a tettigoniid; f, Johnston's organ in antenna of the fly *Chaoborus*.

Source; a, b, d, and e, redrawn from R. E. Snodgrass, The Principles of Insect Morphology; c, redrawn from Snodgrass, 1925; f, redrawn after Child, 1894.

have fine, long hairs that vibrate in response to the flight tone of the females. This stimulates a mating response. The hair vibrations move the flagellum, which in turn stimulates Johnston's organ in the pedicel. The same response can be evoked with a tuning fork that vibrates at 100 to 800 Hz.

The special acoustical receptors of insects are the tympanal organs found in various unrelated taxa and in different places on the body. The organs are on the abdomen of grasshoppers (Orthoptera, Fig. 27.2b), cicadas (Hemiptera, Fig. 6.8a–c), and moths (Lepidoptera, Fig. 43.6f); on the tibia of the forelegs in crickets and katydids (Orthoptera,

Fig. 27.2a); and on the mesothorax of the bugs *Corixa* (Corixidae, Hemiptera) and *Plea* (Pleidae, Hemiptera) and the metathorax of some moths (Noctuoidea, Lepidoptera, Fig. 43.6e).

The noctuid moths mentioned above are night-fliers. Their tympanal organs detect approaching bats, the sounds of which can be sensed at about 30 m. The tympanal organs of Orthoptera and Hemiptera are used in courtship behavior. Sounds produced usually by the male, or by both sexes, are detected and serve to bring mates together. In *Plea* there is only one scolopidium per organ, but in others the organ may be extremely complex,

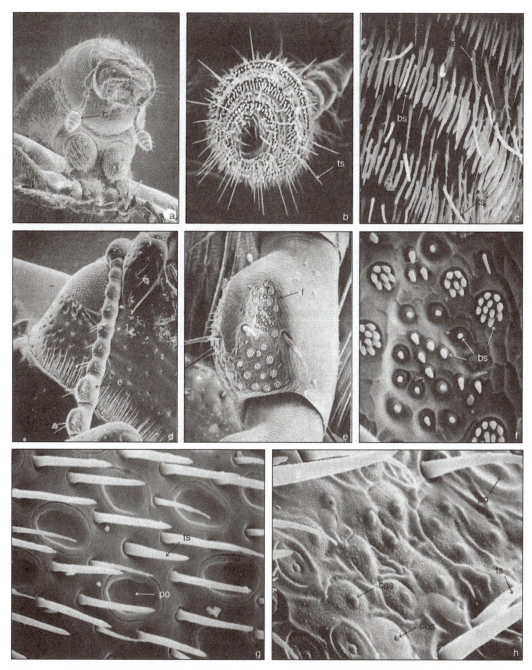

Figure 6.7 Antennal sensilla: a, head of the bark beetle *Pseudohylesinus* sp. (Scolytidae), showing antennae (arrow marks position of photo b; 37×); b, antennal club of same showing sensory bands (arrow marks position of photo c; 304×); c, sensilla of club of same, blunt basiconic sensilla form dense bands (743×); d, head of a male beetle *Temnochila virescens* (Ostomidae), showing antenna (arrow marks position of photo e; 32×); e, ninth antennal segment enlarged (arrow marks position of photo f; 169×); f, basiconic sensilla of several types whose functions are unknown (878×); g, eleventh antennal segment of the worker honey bee, *Apis mellifera* (Apoidea; 1439×); h, apical (twelfth) antennal segment of same (2483×). bs, basiconic sensilla; cas, campaniform sensilla; cos, coeloconic sensilla; po, plate organ; ts, trichoid sensilla.
Source: Photos a–f courtesy of and with permission of Clyde Willson; photos g and h courtesy of and with permission of Kim Hoelmer.

involving up to 1500 scolopidia in cicadas. These organs respond to a wide range of sound frequencies, generally less than 20 kilocycles per second. The average upper limit for humans is about 14 kilocycles per second. Insects discriminate poorly, if at all, among the different frequencies, but respond mainly to the intensity and duration of sound and the intervals between sounds. Insects can detect an interval as brief as 0.01 second, whereas the human ear fails to detect intervals of less than 0.1 second.

Chemoreceptors

Insects are famously sensitive to chemical signals for finding food and mates through gustation (taste) and olfaction (smell). Both taste and smell chemosensors require a pathway for the stimulant chemicals to move from the outside environment to contact the membrane or sensory cells beneath a cuticular surface or organ so that receptor proteins (such as GPCRs) embedded in the dendrites' membranes can react with specific molecules in the surrounding air or fluid to stimulate a nerve impulse. A water-protein pathway protects the sensory membrane from drying. Accessory cells supply the concentrations of cations critical to the chemical reactions needed for the stimulus to trigger a signal to the nervous system. Taste (detection of chemicals from a solid or liquid substrate) receptors usually have a single pore filled with liquid. Olfactory receptors have the sensory membranes contained beneath or within a cuticular covering in such a way that gas molecules must pass through narrow channels or pores to reach the dendritic surface. This tends to restrict chemoreception in olfactory organs to gas molecules. Four common types of setae are commonly associated with chemoreceptor neurons. (1) Trichoid sensilla may serve both as mechanoreceptors and chemoreceptors (Figs. 6.6a, 6.7). Such hairs on the labella and legs of blowflies (Calliphoridae) have a mechanoreceptor attached to the base and the dendrites of a few chemoreceptor neurons situated at a single pore at the tip. Trichoid sensilla are commonly taste receptors. (2) Basiconic sensilla are peglike and thin-walled, have many pores, and project

above the cuticle. They are the most common type of olfactory receptors, typically found on the antennae as well as on the mouthparts and other places on the body. (3) Coeloconic sensilla have the sensory peg sunk in a cuticular pit (Figs. 6.6c, 6.7). These are found on the antennae and mouthparts of some insects. (4) Plate organs are found on the antennae of aphids and honey bees (Figs. 6.6d, 6.7). Each is a round, flat cuticular plate perforated by pores. Sensory pits contain many sensilla in a subcuticular pit. Such pits have been identified on the third antennal segment of Diptera (suborder Cyclorrhapha) and labial palpi of Lepidoptera and Neuroptera. Sensilla that sense odors usually have multiple pores, often opening into a chamber with a network of channels that lead to multibranched dendrites. This arrangement acts to filter larger molecules and particles but allow smaller odor molecules. Taste sensilla most often have a single pore that leads to an interior space where a viscous liquid surrounds the dendrite. Chemoreceptors are found on the antennae, mouthparts, legs, and ovipositor. They play crucial roles in many activities. The detection of odors, or olfaction, is involved in locating food and sites for oviposition and in finding mates through chemical sex attractants. Social insects recognize colony members by odor. The ability of insects to detect minimal concentrations of chemicals varies with the physiological state of the insect and, of course, with the specific chemical compound or mixture of compounds. In tests with honey bees involving many odorous compounds, the general conclusion was that the threshold concentration detected is on average about equal to that of humans. Bees do sense certain substances important to them at lower concentrations, such as the odors of beeswax and certain glandular secretions.

Certain key compounds that are crucial in the life of an insect may be sensed at unbelievably low concentrations. Calculations for the male silk moth, *Bombyx mori* (Bombycidae, Lepidoptera), indicate that single molecules of the sex attractant bombykol produce impulses in the appropriate antennal chemoreceptors (also GPCRs). The moth responds behaviorally to as few as

200 bombykol-induced impulses per second. This is barely above the theoretical minimum needed to transmit information in a system that already has 1600 impulses per second of spontaneous nerve activity or "noise."

The taste or contact chemoreception of substances in liquid uses the same physiological mechanisms as for olfaction. Behavioral and electrophysiological analyses have shown that insects discriminate among sweet, salt, acid, and bitter compounds in solution and exhibit a sense for water. The minimum taste thresholds are generally lower in insects than in humans. A convenient behavioral response has aided in surveying chemicals in various concentrations. When sugar solutions are applied to the tarsi of flies, butterflies, and honey bees, they respond by extending the proboscis. The proboscis is not extended for water or solutions too weak to detect. As a consequence, responses to sugars have been extensively studied.

Honey bees respond positively to only seven, and marginally to two, of thirty substances that taste sweet to humans. Starved bees can detect 0.06 to 0.12 M sucrose in solutions from water with their mouthparts. Their antennae are much more sensitive. With behavioral training, bees respond to sugar solutions applied to their antennae in concentrations as low as 0.0001 M.

The trichoid sensilla on the tarsi and labella of blowflies (Calliphoridae) can be stimulated individually with solutions. They accept a much greater variety of carbohydrates than honey bees. When it is applied to the tarsi, the acceptance threshold for sucrose in solution is 0.01 M in *Phormia regina* but as low as 0.0006 M in *Calliphora erythrocephala* (Calliphoridae, Diptera). *Pyrameis* butterflies (Nymphalidae, Lepidoptera) respond by tarsal stimulation to 0.00008 M sucrose when starved. The threshold for humans at 0.02 M is 250 times poorer by comparison.

BEHAVIOR

What an insect does in response to external and internal stimuli depends not only on the nature of the stimulus but also on the insect's physiological state. The response to food may depend on the taste of food, hunger, and hormonal inputs. At the onset of molting, for example, feeding may cease altogether. Thus both nervous and endocrine systems control an insect's behavior. Both function to sequence behavior and prevent conflicting behavior to favor the survival of the individual and the propagation of the species. Both communicate directly to various tissues. The nervous system communicates by physical contact and the endocrine system by hormones in the hemolymph. The systems differ in that the nervous system functions instantaneously to integrate specific sensory and motor impulses, while the endocrine system reacts more slowly, with long-term effects in a variety of tissues, including the central nervous system.

The behavior of insects under natural conditions is a complex mixture both of stereotyped responses to stimuli that are controlled and coordinated by the central nervous system and of learning. Segmental ganglia may be largely independent in controlling the responses to certain stimuli. Stepping movements or reflexes of the legs, for example, are controlled directly by the respective thoracic ganglia but are coordinated by the inhibition and stimulation of the brain and other ganglia by neural motor programs. Genetics can program stereotyped responses into the nervous system but with considerable flexibility in the timing, coordination, and sequencing of acts, depending on sensory input.

Orientation and Directed Movements

A complex but stereotyped movement that orients the whole body to an environmental stimulus is called a taxis (plural taxes). Commonly this is an orientation to gravity (geotaxis), to light (phototaxis), to wind currents (anemotaxis), or to sound (phonotaxis). An orientation may be either positive (toward the stimulus) or negative (away from the stimulus). A positive phototaxis, for example, involves orienting the body so that both eyes are equally stimulated in front. The orientation to light may also involve positioning the body so that the light is dorsal or ventral. The dorsal light reaction is common among aquatic insects, providing a means to orient to the earth's surface under circumstances where the effect of gravity on proprioceptors is reduced by the buoyancy of water. The backswimmer

BOX 6.1   How Do We Know This? Electroantennograms

As already noted, insects can be incredibly sensitive in detecting odors from great distances. The attraction of male moths from long distances by caged or tethered female moths was long believed to be because of attractant odors (sex pheromones) emitted by the female moths. How does one identify the attractant chemicals? Or prove that an insect responds to a particular chemical?

The electroantennogram (EAG) technique has been used to address these questions by taking advantage of the concentration of olfactory sensory cells in insect antennae and the ability of electrophysiological methods to record nerve impulses as they occur. As a simple example, a freshly detached male moth antenna (a live insect can also be used) can be mounted so as to keep the nerve tissues fresh and biologically active. Fine electrical probes connected either to individual sensory cell output axons or more often to larger nerves consisting of many axons are integrated with equipment to amplify and record the electrical signals generated by nerve impulses from the sensory system. Air containing either known volatile chemicals or unknown volatiles from various extracts (such as from female pheromone-emitting glands) is blown across the antenna and any resulting nerve impulses can be recorded. A sharp increase in nerve transmission from olfactory sensory cells indicates that the sensory system is reacting to specific chemicals.

To identify which chemicals are pheromones that attract mates, chemicals are extracted from biologically attractive sources such as the glands used during "calling" postures taken by female moths. The chemicals in these extracts can be partitioned using a gas chromatograph, a device used to discover the chemical identity of complex mixtures of chemicals or to confirm the chemical composition or purity of chemical preparations. The chromatograph releases volatile chemicals from a stationary liquid phase in a sequence dependent upon the mass and chemical structure of the chemicals, each chemical having a characteristic "retention time" in the stationary phase. The output of volatiles from the chromatograph is recorded graphically (Fig. 6.1) as the chemicals are eluted from the device and also blown across the antenna preparation. The detection of a particular chemical by the chromatograph along with a simultaneous sharp increase in antennal nerve transmissions indicates that the chemical is stimulating sensory cells. The chromatograph can also provide quantitative information as to how much of each chemical is eluted during the process. The EAG provides a simple and elegant method for identifying pheromones.

The process is not usually as simple as outlined above, however, because pheromones often occur as mixtures of two or more chemicals. Indeed, different insect species can have nearly identical pheromones but in different proportions. Behavioral studies with artificial mixtures of compounds tentatively identified as pheromones help to confirm that the molecules identified by EAG are in fact the operative chemicals responsible for driving the behavior of the insects perceiving these compounds.

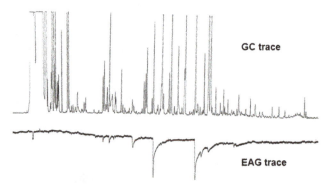

GC trace

EAG trace

Box Figure 6.1 Diagram of electroantennogram set up in conjunction with a gas chromatograph.

Notonecta (Notonectidae, Hemiptera), so named because it swims upside down (Fig. 8.6d), appropriately has a ventral light reaction. Orientation at a fixed angle to sunlight is called the light compass reaction. When it is used as a means for navigation to and from a distant point, compensation must be made for the sun's movement. Honey bees, some ants, and a few other insects have such a compensatory mechanism, probably involving their sense of time or circadian rhythm (discussed later in this chapter). Orientation to a celestial body combined with compensation for its movement is called an astrotaxis.

Although a taxis is a stereotyped behavior, a given taxis is not an invariant response to a given stimulus. Taxes are under the control of the nervous system and can be modified by learning. Depending on the circumstances, a taxis can be quickly switched on or off; the orientation switched from positive to negative or vice versa; and receptivity to one stimulus replaced by another. Positive geotaxis and negative phototaxis, and each with opposite signs, are widely coupled among insects. In honey bees, gravity and the sun are interchanged as stimuli for orientation.

Using this terminology, an exploratory foraging trip by a worker ant might be described as follows. The ant is negatively geotactic and positively phototactic as she makes her way to the surface from her underground colony. As she walks over the ground she is oriented at an angle to the sun by astrotaxis. On her return, the astrotaxis is switched 180°, permitting her to navigate back to the nest, and on entering the nest she becomes positively geotactic and negatively phototactic.

Instincts

Many of the activities of insects are a series of stereotyped behaviors of greater or lesser complexity. Such sequences are loosely called instincts or fixed action patterns. Examples include feeding, migration, mating, oviposition, nest building, cocoon spinning, and movements during ecdysis. Here the physiological and developmental state of the insect, especially the hormonal milieu, is important in determining whether a stimulus will trigger, or release, a given program of behavior. The program may be complex and flexible, including alternate sequences of behavior depending on the response of the mate, suitability of food, or other circumstances.

Learning

A simple definition of learning is a change in behavior as a result of experience. Learning can occur even in the segmental ganglia of insects. Headless cockroaches and grasshoppers can be trained to lift a leg to avoid an electric shock. Such elementary forms of learning may prove to be widespread. Nonassociative learning includes habituation, the waning of a response due to repeated stimulation, and sensitization, where repeated exposure to a stimulus enhances a response. More complex associative learning associates a stimulus with another stimulus or a motor pattern. The most important centers for insect learning are the brain's mushroom bodies, or corpora pedunculata. In insects that exhibit complex behavior and learning, the bodies are larger in volume and have more cells and more synapses than in insects with simpler and clearly stereotyped behavior. Manipulative experiments have demonstrated spatial learning. For example, moving a rock or twig near a wasp nest in the ground after the mother wasp leaves the nest may make the nest more difficult for the wasp to locate. Spatial learning is critical for foraging insects such as bees and ants or insects that repeatedly return to a nest or food source.

Communication

The behavior interactions of insects among themselves and with other organisms usually involve some sort of communication. Communication occurs when an organism provides a visual, tactile, auditory, and/or chemical signal that influences the physiology or behavior of another organism. The distance over which the communication takes place may be such that only one sense will effectively receive the signal. This is long-range communication, in contrast to short-range communication, in which more than one sense may be involved.

Instances of communication may be broadly divided into those that occur between members of the same species and those that occur between members of different species. The former, or intraspecific, cases of communication are for the purpose of attracting or recognizing mates, aggression or defending territories, assembling or dispersing aggregations, or, in social species, for various cooperative activities and to control caste development and behavior.

Communication between different species, or interspecific, may be divided further into (1) those that benefit the sender, (2) those that benefit the receiver, and (3) those that result in a mutual benefit. In the first relationship the sender is usually under threat of attack by a predator and produces one or a combination of warning signals to discourage attack. Other instances in which the sender benefits involve signals that confuse the communication of a predator or competitor. High-frequency clicks produced by the tymbals of night-flying Arctiidae moths (see below) probably confuse the echolocation system of bats. In the second relationship, the sender may become a victim when some aspect of its activities signals its presence to a predator or parasitoid, or a potential victim may escape after sensing the enemy's presence. The detection of the sounds of bats by the tympanal organs of moths is an example. In either case the receiver of the signal is the beneficiary. For the third type of relationship, the insect pollination of flowers supplies an example. The flower provides visual, tactile, and chemical signals that attract insects. The insects obtain food, and the flower achieves cross-pollination (see Chap. 10).

As we shall see below, communication may become quite complex: a given signal may be sensed both by members of the same species and by members of different species; different species of organisms may produce an identical signal; and the same signal may be used for different purposes by the sender and receiver.

Communication by Sight

Insects that are active by day commonly depend on visual signals. For example, the color patterns, size, and motion of female butterflies are known to attract males of the same species. Male butterflies also respond to visual cues to identify and pursue other males in defending territories during courtship. Visual signals also present the risk of attracting predators and parasitoids.

Lightning bugs or fireflies (Lampyridae, Coleoptera) of both sexes use flashes of light to find each other at night. Males emit flashes while flying that have a species-specific duration and frequency. The females, which are less active or even wingless and larviform, respond by flashes. Depending on the species, the female's flashes may be emitted after a brief pause of precise duration after the male signal. The luminescent organs consist of large, well-tracheated cells in the abdomen. Light is generated by oxidation of luciferin in the presence of the enzyme luciferase. The reaction is highly efficient; 98 percent of the energy is released as light. A product of the reaction inhibits luciferase and stops further reaction. Nerve impulses stimulate light production, probably by removing the inhibition of luciferase.

The flash signals of fireflies also may function interspecifically in what is called aggressive mimicry. The females of *Photuris versicolor* and related species are known to flash-respond correctly to mating signals of males of other species and to eat the males that are attracted. *P. versicolor* preys on eleven other species of fireflies.

Communication by Sound

Airborne vibrations so familiar to humans as sound can also be important means of insect communication. Crickets (Orthoptera, Gryllidae) and cicadas (Hemiptera, Cicadidae) are widely recognized examples. Because of the small sizes of most insects, however, vibrations transmitted along liquid or solid substrates are more common than airborne sounds for communication. This is because vibrations of a given energy level travel much farther (and faster) in solid compared to gaseous media. The very low level of vibrational energy expended by small insects requires more sensitive and sophisticated equipment to detect and record. Because of these difficulties a large realm of insect sounds goes unnoticed, even by professional entomologists.

Insects can create vibrations as a by-product of an activity such as flying to generate communication signals for other purposes. We have already mentioned how the flight tone of female mosquitoes attracts males, thus serving an intraspecific function. Loud buzzing by bees and wasps serves interspecifically as an effective warning signal that is even imitated by other insects. Striking the body against the substratum creates vibrations in wood or soil. The death-watch beetle *Xestobium* (Anobiidae, Coleoptera) bores in furniture and wooden houses. At sexual maturity it strikes its head against the burrow walls, producing a rapping sound that at night influences the behavior of humans as well. Many species of termites drum their heads in response to disturbances of their colonies.

Special devices for creating sound are prevalent in numerous orders but are especially frequent in species of Orthoptera and Hemiptera (Fig. 6.8). Vibrations are generated mostly by one of two mechanisms: movement of roughened surfaces together or vibration of a membrane. In the first type of mechanism, one surface may be modified into a scraper and the other into a file (Fig. 6.8d–f). The movement of the scraper on the file is called stridulation. Some adult as well as immature insects stridulate to produce arrhythmic, rasping

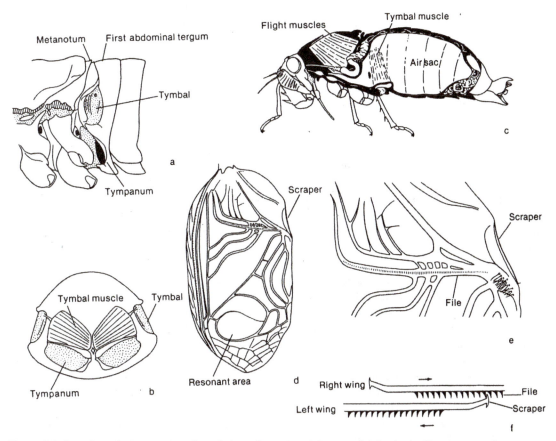

Figure 6.8 Sound-producing organs: a, lateral view of anterior abdomen of male cicada, *Okanagana vanduzeei*, showing sound-producing organ (tymbal) and sound receptor (tympanum); b, cross section of same; c, longitudinal section of *Magicicada septendecim* (Cicadidae) showing large resonant air sac; d, ventral view of right wing of cricket *Acheta assimilis* (Gryllidae); e, enlargement of stridulatory organ of same; f, diagram of relationship of right wing and left wing during stridulation.
Source: c, redrawn from R. E. Snodgrass, The Principles of Insect Morphology.

warning sounds. These are heard for only a short distance or sensed by contact. The devices are simple and may be formed wherever two parts of the body normally rub together.

In Orthoptera stridulation by adults can be heard at long range and serves to bring the sexes of the same species together for mating. The devices for sound production and reception are complex, and the sounds of each species have a characteristic frequency, rhythmic spacing, and duration. In crickets (Gryllidae) and katydids (Tettigoniidae), one or both forewings have a resonant area of cuticle that vibrates as the scraper of one wing strikes the teeth of the file on the opposite wing. Tympanal organs of hearing are located in the fore tibiae (Fig. 27.2a). Receptive females orient to the calling songs of the males and are attracted to them. Grasshoppers (Acrididae) rub a row of pegs on the inner side of the hind femur against raised veins of the forewings, causing the wings to vibrate. The tympanal organs are on the first abdominal segment (Fig. 27.2b). Both males and females produce long-range calling songs and approach each other.

Sound-producing membranes called tymbals are situated in the metathorax of Arctiidae (Lepidoptera) and in the abdomen of both sexes of cicadas (Cicadidae) and leafhoppers (Cicadellidae) and some shield bugs (Pentatomidae). Tymbals are best developed in male cicadas (Fig. 6.8a–c), in which they produce long-range, airborne calls. In these insects the tymbal is a thin resilient area of cuticle situated dorsolaterally on the first abdominal segment. Contraction of an asynchronous fibrillar muscle attached to the center causes the tymbal to click inward. When the muscle relaxes, the elastic cuticle then clicks outward. Large air sacs in the abdomen resonate with the tymbal vibrations to increase sound intensity. The synchronous singing of many individuals forms a "chorus."

The tymbals of the much smaller leafhoppers (Cicadellidae) produce weaker vibrations that are transmitted for longer distances along plants than through the air as sound waves. This communication requires that the insects be on the same or touching plants. Planthoppers (Hemiptera, Fulgoroidea) and lacewings (Neuroptera, Chrysopidae) create

vibrations by rapidly drumming their abdomens against a plant surface.

Communication by Chemicals

Chemical messengers are used for communication throughout the animal kingdom but are especially important for insects. The term semiochemical is applied to all chemicals produced by one organism that incite responses in another organism (Matthews and Matthews, 1978). Most semiochemicals of insects are produced by exocrine glands of the integument. The chemicals, either single compounds or mixtures, are usually active in minute quantities and are usually species-specific. Nonvolatile chemicals are limited to contact chemoreception. Volatile chemicals function in either short- or long-range communication, but the direction and distance may be limited by wind.

Semiochemicals that function intraspecifically are called pheromones. These are often blends of several compounds and, at least in social insects, may have more than one effect on the behavior and physiology of the recipient. Pheromones are the dominant mode of communication for insects. Pheromones may be grouped broadly into primer pheromones, which have a long-term effect on the physiology and development of the receiver that is mediated by hormones but may not have an immediate effect on behavior, and releaser pheromones, which have immediate, reversible effects on the receiver's behavior. Among insects, primer pheromones are well known in social species, where they are involved in caste determination and the inhibition of reproduction by workers (see Chap. 7). In other insects, especially those that aggregate, primer pheromones function to bring adults into reproductive synchrony. For example, the presence of mature males of the locust *Schistocerca gregaria* (Acrididae, Orthoptera) increases the rate of sexual maturation of immature males. Experimental evidence indicates that the pheromone responsible for promoting maturation is transmitted by air and by contact to the young adults and affects their corpora allata (Chap. 4).

Releaser pheromones are greatly varied in function and widespread taxonomically. The best known are the sex attractants, or sex pheromones, which bring the sexes together. In the most common situation, odors are emitted at a certain time of day or night by the female as a long-range signal to advertise her receptivity to mating. The odors are carried away from her by wind currents, forming an odor trail that becomes wider and less concentrated with increasing distance. A male downwind that perceives the odor with chemoreceptors on his antennae becomes active and flies against the direction of the wind, oriented by a positive anemotaxis. The flight is a zigzag pattern toward the female as the male turns back and forth to remain in the odor trail. In the vicinity of the female where the odor concentration is high, the male arrests his forward flight and searches for the female by other senses, especially sight. Courtship and copulation are then stimulated by close-range communication involving visual, tactile, or auditory signals, alone or together with pheromones. The long-range sex attractant released by the female may continue to excite the male, and the male, on close contact, may produce his own releaser pheromone or aphrodisiac that makes the female more receptive.

The foregoing sequence of sex attraction is not universal. In some insects males produce long-range sex attractants, in some the females produce aphrodisiacs, and in some one sex produces pheromones or aggregating scents that attract both sexes to a common place (see behavior of bark beetles, below).

The active compounds in the sex pheromones of a number of insects have been isolated, identified, and synthesized. Sex pheromones are extremely useful in pest management (Chap. 12).

Intraspecific pheromones also have nonsexual functions. In some insects, the aggregations of immatures or of females at oviposition sites are probably assembled and maintained by chemical releasers. Rapid dispersal from an assemblage may likewise be signaled by chemicals, in this case by alarm pheromones. When disturbed, the aphid *Myzus persicae* (Aphididae, Hemiptera) discharges a secretion that disperses nearby aphids.

Social insects, especially ants, employ an array of chemical signals for aggregation, alarm, and other communications among colony members. Odor trails that recruit and guide foragers to food are known in certain ants, termites, and stingless bees (*Trigona*, Apidae, Hymenoptera). On returning to the nest after discovering a food source, a forager ant lays down a minute trace of a volatile, species-specific chemical, or trail pheromone. Other ants are stimulated to follow the odor trail from the nest to the food and on their return add their own trail pheromone. A rich food source is soon visited by an increasing number of foragers. When the food becomes covered by ants or diminishes in size or quality, some ants return without contributing to the trail. Unless renewed by successful foragers, the trail declines in attractiveness because the odor is rapidly dissipated to a level below the ant's threshold of perception. Thus the foragers readily shift in appropriate numbers to exploit new food sources.

Territorial marking with scents is widely recognized for mammals, and insects have some similar behaviors with respect to food sources. For example, fruit flies (Tephritidae, Diptera) that lay their eggs in fruits drag the ovipositor across the fruit surface to mark the fruit with a pheromone that inhibits other females from laying eggs in the same fruit. Wasps that parasitize other insects by inserting eggs into the host insect they attack avoid attacking hosts that have already been parasitized by the same species.

Semiochemicals involved in interspecific communication are called allelochemicals and are divided according to the effect on the sender or receiver as follows: (1) allomones are substances released by the sender that evoke a response in the receiver which is beneficial to the sender, (2) kairomones are substances that, when transmitted, benefit the receiver, and (3) synomones are substances that are mutually beneficial to the sender and receiver. Theoretically the same allelochemic compound can be an allomone, kairomone, or synomone depending on the context in which it is transmitted.

Allomones include the venoms injected by the sting of female aculeate Hymenoptera; defensive secretions of nasute termites and various beetles and true bugs; and toxic chemicals in the body, such as cardiac glycosides in the monarch butterfly, that discourage ingestion by vertebrate predators. Plants also produce allomones, such as the secondary metabolites, juvenile hormone mimics, and phytoecdysteroids that protect against plant-feeding insects.

Kairomones include odors or tastes that attract and stimulate attack, be it an insect victim or a plant host. Sex pheromones become kairomones when predators or parasitoids are attracted to feed on the mating insects. Secondary plant compounds can serve as kairomones for insects that tolerate the toxic compounds and attack the plant.

Odors of flowers that attract pollinating insects are synomones; both benefit because the plant is cross-pollinated and the insect obtains food (Chap. 10). Here again, the same odors may be kairomones by attracting flower-feeding insects that destroy the flowers and seeds.

Aggregation Behavior of Bark Beetles

A complex system of semiochemicals coordinate the mass attacks on living trees by the bark beetles *Ips*, *Dendroctonus*, and *Scolytus* (Curculionidae, Coleoptera) (Raffa et al., 1993; Wood, 1982). Coniferous trees resist attack by expelling or drowning boring insects in sap (Fig. 6.9). This resistance is overcome by bark beetles when they attack simultaneously in large numbers and introduce a fungus that interrupts the water-conducting system of the tree. In the usual sequence of events, beetles of both sexes emerge from the old host or

overwintering site and fly for some period before orienting to a new host. Visual and possibly odor cues guide the beetles toward trees. "Pioneer" beetles (males of *Ips*; females of *Dendroctonus*) initiate attacks to feed. Galleries are excavated in the bark of acceptable hosts, probably as a direct result of chemical feeding stimuli in the bark and/ or phloem. During this initial boring activity, an aggregation pheromone or a species-specific mixture of such pheromones is produced both by *de novo* biosynthesis (Seybold et al., 1995) and by conversion of host terpenes (Hendry et al., 1980) at some unknown site inside the insect and liberated with the feces. The compounds ipsenol, ipsdienol, *cis*-verbenol, and *trans*-verbenol, alone or in combination, are pheromones of various species of *Ips*; *exo*-brevicomin, *trans*-verbenol, verbenone, frontalin, and others are found in *Dendroctonus* species; multistriatin has been identified in *Scolytus multistriatus*. These compounds may be derived from bacterial, fungal, and/or insect biosynthetic pathways. Mycangial fungi are closely associated with the beetles as symbionts in the gut or gallery (Chap. 11). The waste, or frass, from the borings is made attractive at a long range to other beetles. Thus the pioneers attract more beetles of both sexes. By their boring activity the aggregation pheromone concentration increases. In some cases

Figure 6.9 A bark beetle, *Dendroctonus brevicomis* (Scolytidae) trapped in a pitch tube on a pine tree.
Source: Photo courtesy of and with permisson of Clyde Willson.

certain volatile compounds from the host tree itself also contribute to the overall attractiveness of the tree, for example, myrcene for *D. brevicomis* and α-cubebene for *S. multistriatus*. The total aggregation pheromone now consists of products from the beetle's gut and from the tree.

Mates are located and recognized by both sound and odor. *Ips* females are attracted to frass marked with a male pheromone, and they stridulate before being permitted entry into the gallery occupied by a male. Males of *Dendroctonus pseudotsugae* are similarly attracted to frass marked with female-produced pheromones, and they also stridulate at the entrance to her gallery before entering. The factors that terminate a mass attack are not well understood. Possibly the aggregation pheromone declines in production or one or more compounds from either or both sexes increase in concentration and interrupt the chemoreception of the pheromone.

Birch and Wood (1975) discovered that *Ips pini* and *I. paraconfusus* both attack *Pinus jeffreyi* and *P. ponderosa* at the same season in the Sierra Nevada of California but not the same individual trees. The volatile compounds produced by male *I. paraconfusus* inhibit the response of *I. pini* to its own pheromone and vice versa. Thus the first males to arrive at a tree effectively prevent colonization by the other species. The pheromone components simultaneously act as intraspecific attractants and as an allomone against a competitor.

The aggregations of bark beetles attract a large number of other insects that feed on the beetles, or on other insects, or on the weakened host tree. Certain of the predators and parasitoids of bark beetles are specifically attracted by one or more of the constituents of the aggregation pheromones. In this situation, the pheromone components function simultaneously as kairomones (Borden, 1982; Raffa et al., 1993; Wood, 1982). Studies indicate that fungi inoculated into trees during host colonization may attract parasitoids, which generally arrive on the tree when the new bark beetle broods have developed to late-stage larvae (Stephen et al., 1993).

Learning and Communication in Honey Bees

The ancient Greek philosopher Aristotle noticed that a bee forages on one species of flower on each foraging trip and is attended by other bees on her return to the hive. These simple observations describe aspects of what has proved to be a remarkable ability for learning and communication in the honey bee, *Apis mellifera* (Apidae, Hymenoptera). Much of our knowledge has resulted from the investigations of the late Professor Karl von Frisch in Germany. For his efforts, he shared a 1973 Nobel Prize in Physiology (see von Frisch, 1974).

After about 3 weeks spent as an adult in the hive, a worker honey bee begins the last phase of its life foraging for pollen and nectar in flowers. In an extensive series of experiments that began prior to 1914, von Frisch demonstrated that bees could be trained to seek a sugar solution and to associate specific stimuli with this food source. The following experiment for color discrimination is typical. A table was placed near a hive. A blue square of cardboard was placed flat on the table and surrounded by gray cards of different shades. On each square was placed a glass dish. The dish on the blue card was then filled with sugar solution. After a few hours during which the bees foraged on the food, the cards and dishes were removed and replaced by clean cards and empty dishes arranged in a different pattern. The bees returned and landed on the blue card, demonstrating that they discriminated the blue color from all shades of gray. This indicated that they have true color vision and that they had learned to associate the color with food.

In similar experiments, von Frisch (1971) trained bees to one or a combination of odors, flavors, geometric figures, and colors within their range of discrimination. He also trained bees to visit a feeding station at a specific time. Some bees were even trained to seek food at three to five separate times of day, thus demonstrating a time sense in bees.

The most famous of von Frisch's discoveries was the complex signaling system that communicates the distance and direction to food sources up to about 14 km from the hive. On returning to the hive from a rich food source 100 m or more away, the worker inside the dark hive moves excitedly in a waggle dance on the vertical comb (Fig. 6.10c). After performing the dance she may move to a different area on the comb and repeat the dance. Although it is dark, some bees follow the dance with their antennae, and they detect the flower

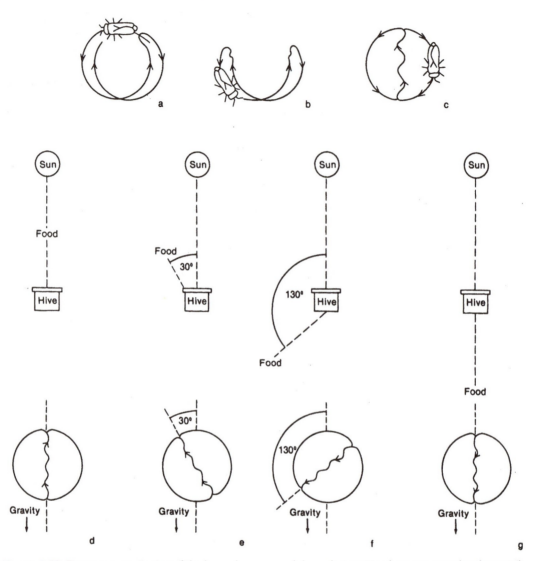

Figure 6.10 Dance communication of the honey bee: a, round dance; b, transition between round and wag-tail dance; c, wag-tail dance; d-g, diagrams of wag-tail run (on vertical surface of comb in dark hive) in relation to the direction of food from the hive and the compass direction of the sun.

odor on the waxy surface of the forager's body. The forager may also offer some of the nectar she has collected and carries in her crop. A rich concentration of sugar is highly attractive and stimulating to other bees. In experiments with feeding stations located at various distances and directions from the hive, von Frisch demonstrated that bees recruited by the dancer search in a certain direction and at a certain distance for food of the same odor and

nectar flavor as that collected by the dancer. After successful foraging trips, these newly recruited bees begin to dance too. More bees are recruited, and the number of foragers rapidly increases. When the food supply diminishes, the frequency of dancing declines.

During part of the waggle dance the bee makes a straight run, vigorously wiggling her abdomen and emitting sound pulses at a rate of about 32 per second. This is the called the waggle run of the waggle dance. Then she circles back to the starting point and repeats a waggle run of the same duration. The time spent during each run and the total number of pulses is directly proportional to the distance traveled to the food source. For a flight of 100 m, the time of a single waggle run is about 0.25 second, and for a 1200-m flight, 2 seconds. Thus, with increasing distance the waggle runs take longer, and fewer runs are completed within a given period.

The direction the bee moves during the waggle run on the vertical comb has a consistent orientation with respect to gravity (Fig. 6.10d–g). This is related to the direction from the hive to the food with respect to the compass direction of the sun when it is more than about 2° from the zenith position. Recall that positive phototaxis and negative geotaxis are commonly linked. For example, if the forager flew toward the compass direction of the sun en route to the food, her waggle run is oriented vertically up the comb, and her head points up during the run. If she flew at an angle to the right of the sun, the run is oriented at the same angle to the right of vertical.

As the food site is moved closer to the hive at distances of about 100 m or less, the dance shifts from the waggle dance to a transition form (Fig. 6.10b) and then to the round dance (Fig. 6.10a). The waggle runs become less conspicuous, but as the bee moves on the comb in a direction appropriate for the direction of the food she still emits sounds (Kirchner et al., 1988). It is likely that bees following a round dance obtain information about the direction of food quite close to the hive. However, the duration of the sound is too variable to indicate different distances to food less than 100 m from the hive.

Wenner and associates (1971, 1990) criticized the design of previous experiments and argued that their own experiments indicated that odor (specific food odor, Nassanoff gland scent, and overall odor of the food area) was the main stimulus both for recruitment and for the finding of food at some distance from the hive. These criticisms stimulated others to look more closely at the behavior of the dancers and recruits.

By ingenious experiments, Gould (1975) demonstrated that distance and directional information is indeed passed from the dancer to other bees. He was able to have a forager send recruits to a food site that the forager had never visited and consequently could not have the odor of the food site on her body. This is how he did it: When bees dance on the comb in a light beam of sufficient brightness, they will orient to the light instead of gravity. The bees that follow the dancer also orient to the light and correctly interpret the direction to the food source. If, however, the ocelli of the dancer are covered by paint, she becomes much less sensitive to light, and a much brighter light is required to stimulate her to reorient to it. In Gould's experiments, the light was adjusted to not affect the ocelli-covered dancer but to cause reorientation of the recruits to the light. Suppose the wag-tail run of her dance is vertically upward, correctly indicating a food station directly toward the sun. If the light beam is directed from the left at 90° from the vertical, or horizontally, then the recruits with normal ocelli will orient to the light and interpret the direction of the food station to be 90° to the right of the sun. If they rely only on her odor, the recruits should appear at the dancer's station, but if they rely on their interpretation of her dance, they should be "misdirected" to other stations situated at 90° to the right of the sun. Gould found that normal recruits were "misdirected" according to the dance information of an ocelli-covered dancer. The light could be placed at various angles, and the recruits appeared at the predicted stations with an error of about 11.9°, or 31 m at 150 m, and 4.2°, or 29 m at 400 m. The reasons why the more distant stations are indicated with less error are unknown.

Kirchner, Michelsen, and their associates (Kirchner and Towne, 1994) carried the analysis further with the construction of a computer-controlled, mechanical model bee that danced with waggle runs, vibrated, and dispensed sucrose to recruits. The researchers were able to show that both the waggle run and the pulsed vibrations of the dancer's wings were required for successfully directing a recruit to the food site. Honey bees were thought to be deaf to airborne sounds, but they demonstrated that Johnston's organ in the antennal pedicel was responsive to the movement of air particles in the vicinity of the dancer's wings.

The dance is important in initially recruiting new foragers to a new, distant food source. By following the dance, the recruit learns the distance, direction, flower odor on the dancer's body, and nectar flavor from samples given by the dancer but not the color or shape of the flower or the landmarks and obstacles en route. Once they have made the trip, foragers are able to return to the food by their memory of landmarks, obstacles, and various odors encountered en route; distance and the compass direction with reference to the sun's position; and attern and shape of the flower. For orientation to the sun, they utilize an astrotaxis and compensate for the sun's movement. Cells connected to the nervous system in the abdomen contain magnetite and are believed to function in magnetic orientation.

On returning to the hive, foragers unload the liquid carried in their crop to nest bees. The rate at which the nest bees accept the liquid regulates how quickly the forager can depart for another trip. When the hive is overheated and water is used for evaporative cooling, the nest bees readily take water or dilute nectar rather than concentrated nectar from the returning foragers. This stimulates the foragers to collect water or dilute nectar on their next trip rather than concentrated nectar. When the hive temperature is normal, nest bees favor foragers with concentrated nectar over those with dilute nectar. At times, the amount of incoming nectar may exceed the capacity of the nest bees. When delayed in unloading concentrated nectar, a forager performs a tremble dance, a distinctive shaking motion, plus a vibration against the comb called a stop signal. This signals to nest bees to increase the reception of nectar and inhibits other foragers from dancing (Nieh, 1993; Seeley, 1992).

Rhythmic Activity and Sunlight

Sunlight plays a key role in the orientation of insects in their environment and in the timing of events in their life cycles. Momentary responses to light, perception of images, color vision, and behavioral reaction to light and polarized light were discussed earlier in this chapter. Of interest here is the influence of the natural alternation of light and darkness in the rhythmic activity of insects.

As a consequence of the earth's inclined rotation, physical factors important to the life of terrestrial organisms fluctuate daily and annually on a regular basis. The earth's axis is tilted 23.5° from a plane vertical to the plane of the earth's annual orbit around the sun. Any given area of the earth's surface receives sunlight for only a part of the 24-hour rotation. With the appearance of the sun's heat each day, temperature usually rises and relative humidity falls. The tilted axis produces an annual variation in the quantity of heat reaching the nonequatorial regions, resulting in the seasons. Furthermore, the earth's rotation around the tilted axis precisely changes the proportion of light and darkness each 24 hours. This results in long summer days and short winter days in the Temperate and Polar Zones.

The daily activities of insects usually take place during specific periods of the 24-hour cycle. Such synchronization, for example, permits the sexes to find each other, limits the search for food to periods of peak abundance, and allows insects to avoid enemies or to take advantage of a favorably high humidity for ecdysis or oviposition. Insects usually spend a part of each day inactive. For example, at night some bees and wasps (Fig. 6.11) clasp leaves or the stems of plants with their mandibles or crawl into flowers and seem to "sleep." During their regular rest period at night, honey bees exhibit behavioral and physiological changes comparable to sleep in birds and mammals.

An activity during the day is termed diurnal; at dawn or dusk, crepuscular; at night, nocturnal; and at dawn only, matinal. These terms should be used in reference to specific activities or events rather than loosely to overall activity: the emergence of adult *Drosophila pseudoobscura* (Drosophilidae, Diptera) from pupae is matinal; attraction to food baits, mating, and oviposition are crepuscular; and pupation is mainly crepuscular or nocturnal.

One might interpret a daily rhythm in activity as simply a response to a certain level of light or other physical factor that occurs regularly each day. In other words, the insect's activity is a response to exogenous factors. This is true for oviposition

Figure 6.11 Gripping with their mandibles, these male eucerine bees (Anthophoridae) in Peru cling to the underside of a leaf. They may remain together during midday heat or all night. Such sleeping aggregations of male wasps and bees often assemble at dusk.

behavior in the walking stick *Carausius morosus* (Phasmatidae, Phasmatodea). Most eggs are laid at or near dusk, or at the time the lights are switched off in laboratory experiments. After *C. morosus* has been kept only a day in constant light or darkness, the laying of eggs no longer occurs at 24-hour intervals but is distributed over the 24-hour period. A direct response to an environmental signal may also be delayed a specific interval as if timed by a "sand hourglass."

Circadian Rhythms

In most insects, activity rhythms have a physiological, or endogenous, basis that is partly independent of environmental signals. In a light/dark cycle of 12:12 hours, mature male crickets, *Teleogryllus commodus* (Gryllidae, Orthoptera), begin to stridulate their mating song 1 to 2 hours before the onset of darkness. Singing continues during darkness and is terminated 2 to 3 hours before the lights switch on. When these crickets are placed in constant darkness, singing begins at intervals of about 23.5 hours and lasts the normal period. In constant light, singing occurs at intervals of about 25.3 hours (Fig. 6.12, days 1–12). A return to the 12:12- hour cycle of light/dark

results in a return to the 24-hour rhythm, beginning 2 hours before darkness (Fig. 6.12, days 13–31).

When a given activity or event can be demonstrated experimentally to recur at about 24-hour intervals after all environmental signals have been excluded, it is said to occur in a circadian rhythm (from the Latin *circa diem* for "approximately daily"). Under constant conditions the time interval is nearly always slightly more or less than 24 hours, resulting in a progressive drift each day in the time of initiation (Fig. 6.12, days 1–12). The presence of the drift is evidence that other possible signals with 24-hour periodicities are not influencing the insect. Cosmic radiation and geomagnetic forces, for example, fluctuate on an exact 24-hour cycle with the earth's rotation. Individual cells exhibit rhythmic changes in nuclear volume and protein synthesis. Specific organs function cyclically; for example, endocrine glands synthesize hormones cyclically. Susceptibility to insecticide poisoning is rhythmic, as is general oxygen consumption. Clearly, rhythms are evident in a variety of physiological processes.

The physiological basis for circadian rhythms has been a subject of intense and prolonged study. Tissue and organ transplant experiments showed that the

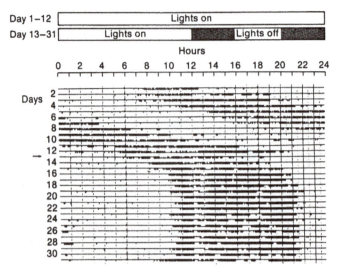

Figure 6.12 Circadian rhythm of stridulation in a male cricket, *Teleogryllus commodus* (Gryllidae). Marks in rows are tic marks from automatic recorder that is actuated by sound of stridulation. Each row records stridulation during 1 day. When the record began, the cricket had been in constant light and the stridulation rhythm was free-running. On day 13 the lights were switched off at 12:00 and remained off for 12 hours. Thereafter, the cricket was exposed to 12 hours of light and 12 hours of dark. Note that stridulation is temporarily interrupted when the lights are switched off at 12:00 each day. *Source: Data courtesy of Werner Loher.*

cerebral lobes of the brain regulated the rhythms of adult emergence (eclosion) from the pupal stage of moths and all other insects examined in this way. The fruit fly *Drosophila melanogaster* has been an important model system for circadian rhythm studies. Mutants of *D. melanogaster* that have drastically altered circadian rhythms led to discovering identification of numerous genes (*per, tim, cyc, clk*) that influence rhythmicity. The rhythm itself, however, is not determined by the metabolic rate of the process. Unlike nearly all physiological processes, circadian rhythms are largely independent of temperature within normal limits of tolerance.

The underlying mechanism is conceptualized as a "biological clock" or an "oscillator." The clocks of circadian rhythms are reset daily or entrained by brief environmental signals. The most common cues are dawn or dusk, when light intensity varies rapidly. The location of the photoreceptors varies

with the species. The homology of proteins identified as critical molecules for circadian rhythms in *Drosophila* (such as "period" and "timeless") with proteins found in mammals suggests that the underlying mechanisms evolved early and should be widely applicable to many animals.

An entrained circadian rhythm is called a diel rhythm. It is possible that more than one clock controls the different rhythms of cells, development, general physiology, and behavior. The clock or clocks serve to coordinate life processes within the insect while at the same time scheduling the insect's activity in a changing environment. Locomotor activity, feeding, mating, and the insect's varying responsiveness to stimuli are programmed to take place at opportune times in the daily cycle.

A different type of diel rhythm involves populations and "once-in-a-lifetime" events. If *Drosophila melanogaster* (Drosophilidae, Diptera) is reared in constant darkness, the adults emerge from pupae at any time. When the pupae are given as little as 1 minute of exposure to light, emergence becomes synchronized to occur during one period each 24 hours. The period of emergence has been compared to a developmental "gate" that opens or closes. If the open gate is missed by an individual, then its emergence is delayed 24 hours. Gated phenomena are known for egg hatching in the pink bollworm *Pectinophora gossypiella* (Gelechiidae, Lepidoptera), pupation of the mosquito *Aedes taeniorhynchus* (Culicidae, Diptera),

and emergence of the intertidal midge *Clunio marinus* (Chironomidae, Diptera) in Europe.

The genetic basis of circadian rhythms is illustrated by the mating behavior of moths. Most moths have specific periods for mating activity. Certain closely related species have identical sex pheromones. Consequently, the males can be attracted to females of different species. Some of these species are allopatric (do not occur in the same place) or mate at different seasons. In the sympatric species (occur in the same place), mating is at different times of day under circadian control, thus preventing cross-attraction.

CHAPTER 7

Social Relationships

In this chapter we will be concerned with those cooperative relationships among members of the same species that are called social and are beyond relationships directly involved in sexual behavior. Social interactions form the basis for important evolutionary strategies in diverse animals, both invertebrate and vertebrate, including humans. Only about 2 percent of described insect species are social: termites (2200 species); ants (8800 species); vespid hornets, yellow jackets, and paper wasps (860 species); sphecid wasps (at least one species, the neotropical *Microstigmus comes*); and the social bees (1000 species, of which the European honey bee is only one). Yet social insects are disproportionally more abundant in numbers and biomass than nonsocial insects in many habitats such as grasslands and rain forests.

STEPS FROM SOLITARY TO EUSOCIAL BEHAVIOR

The evolution of social behavior has attracted the attention of many scientists, including Charles Darwin, who sought to explain how workers that did not reproduce could be favored by natural selection. Others have asked: Did the behavior evolve from the relationships of parents to their offspring, or as interactions among adults of the same generation, or some combination? In the process of development, or sociogenesis, of an individual colony, how do members of a colony change in behavior, caste, or physical location over time? To address these questions we first need to define an array of behaviors that lead from solitary to true social behavior. These behaviors can be considered from two points of view: (1) phylogenetically—for a given lineage, what are the steps in the evolution of social behavior over geologic time—and (2) sociogenetically—for a given group of individuals, what are the steps in the development of social behavior over the life of a colony?

Solitary Behavior

The great majority of insects are solitary, meaning that interactions among adults are largely limited to sexual behavior and competition, and adult-offspring contacts are limited to activities associated with oviposition. For example, the sexes of Phasmatodea meet during copulation and remain joined for long periods, sometimes for more than a day. Later the female, aloft in vegetation, drops her eggs to the soil litter without further care. Selection of oviposition sites is important for many species to ensure a favorable chance for the survival of their offspring. As one example, the cabbage butterfly (*Pieris rapae*, Pieridae) lays eggs on plants of the cabbage family (Cruciferae), providing immediate access to a preferred host plant of this caterpillar. Insects such as the monarch butterflies, which have conspicuous warning coloration to advertise their distastefulness to predators (Chap. 8), often gather in large aggregations during seasons when they are inactive. These butterflies are distasteful to insectivorous birds, and birds that eat them quickly learn to avoid their conspicuous black-orange-white coloration. Adult monarchs aggregating during the winter months in highly concentrated roosts in trees covering a relatively small area need only sacrifice a few individuals to educate local birds of their distastefulness.

The more advanced solitary species of insects exhibit some maternal care—for example, preparation for the safety of offspring—even though they do not directly interact with the immatures. Most species of bees and wasps, for example, are solitary. They construct individual chambers or "cells" in earth or plant materials, which are then provisioned with food for their larvae. This is called mass provisioning, because the adult does not return to add food once the egg has hatched. The exacting preferences of some ground nesting bees (Hymenoptera, Halictidae) for specific soil types for nest construction can lead to aggregated nests averaging a separation of only 3 cm.

Subsocial Behavior

Adults of subsocial insects protect and/or feed their own offspring for some period of time after hatching, but the parent leaves or dies before the offspring become adults. Some provide food to the offspring in a behavior called progressive provisioning. Careful observation has revealed a surprising number of species that exhibit various degrees of maternal and/or paternal care. Among the subsocial insects are some species of the following: cockroaches, crickets, earwigs (Fig. 7.1), mantids, jumping plant lice, web spinners, thrips, thirteen families of true bugs, treehoppers, nine families of beetles, and certain bees and wasps, plus some other arthropods such as spiders. Protection of young during critical early stages has obvious advantages to insects as well as to some amphibians, reptiles, and fish and to all birds and mammals. The subsocial parent-offspring relationship is one possible phylogenetic or sociogenetic route to social behavior.

Parasocial Behavior

A different possible route from solitary to social behavior, at least in the Hymenoptera, is via interactions of adults of the same generation. This behavior is broadly termed parasocial and includes communal, quasisocial, and semisocial behaviors. A simple type of parasocial relationship is the communal colony or aggregation of a group of female bees or wasps. The adult females inhabit the same nest burrow, but each constructs and provisions cells for her own eggs. The frequent presence of at least one of the females in the communal burrow provides mutualistic defense for the colony against natural enemies. This relationship can be

Figure 7.1 Female earwig, *Forficula auricularis* (Forficulidae), exhibits subsocial behavior by protecting her eggs.

carried another step: a female in the communal burrow may cooperatively make and provision brood cells with other females while still laying eggs herself. Such cooperation, called quasisocial behavior, probably occurs early in nests that later exhibit a reproductive division of labor among the females.

The most advanced stage of parasocial behavior is called semisocial. Some females are mated and lay eggs, thus serving as queens. Others of the same age and size may mate or not, but they fail to lay eggs and become workerlike in behavior. Thus a reproductive division of labor is established among females of the same generation.

Eusocial Behavior

Widely accepted categories of sociality codified by E. O. Wilson define true social or eusocial behavior by three traits in assemblages of a species:

- Cooperative care of the young.
- Overlapping of at least two adult generations in the same colony so that the offspring aid the parents in the colony's work.
- Division of reproductive labor between reproductive castes (queens and males or, in termites, kings) and nonreproductive castes (workers).

Eusociality evolved once in orthopteroids to produce the termites (Blattodea, often separated as order Isoptera), all species of which are eusocial. Eusociality also evolved once in aphids (Hemiptera, Aphididae) and in thrips (Thysanoptera). In the Hymenoptera, eusociality evolved independently at least twelve times.

Pheromones play a major role in the interactions of individuals in a colony. For example, colony odors allow members to distinguish between nestmates and individuals from other colonies. In kin recognition individual pheromones, which are probably blends of hydrocarbons in the epicuticular wax, allow members to recognize close relatives within a colony. Many, but not all, social insects exchange liquids from the mouth or anus among members of a colony. This is called trophallaxis and is important in the distribution of food and social pheromones in the colony.

Primitively Eusocial Insects

The sweat bees (Halictidae), allodapine bees (Anthophoridae), bumble bees, and some social wasps are called primitively eusocial. These have queens and workers that are similar in appearance (queens may average larger in size) and differ mainly in behavior and the development of the reproductive organs. Colonies have fewer individuals and often last only a year. Queens often spend the winter apart from the colony and establish new colonies by themselves. Among these bees and wasps and their close relatives can be found a full range of behavior from solitary to eusocial. Certain species even exhibit these behaviors from one nest to another, revealing that eusocial behavior does not necessarily require a long progression through evolutionary steps. Primitively eusocial bees and wasps differ markedly in their caste distinctness and divisions of labor and may eventually need to be categorized differently.

Highly Eusocial Insects

Termites, ants, some social wasps, honey bees, and stingless bees are considered highly eusocial. They have large colonies that last for more than a year and often last many years. The distinct castes are easily distinguished on sight and display different behaviors or caste polyethism. Individual members do not live for long as independent insects. The highly eusocial insects have convergently evolved extraordinary abilities in a number of features, of which many species exhibit the following:

- Coordinated defense of the colony
- Use of pheromones in communication, recognition, and behavior
- Use of pheromones in control of post-embryonic development and physiology of adults
- Regulation of the numbers of individuals in different castes and sexes
- Discovery and efficient exploitation of resources: food, water, nest sites, and materials to construct nests
- Maintenance of a sanitary, stable environment inside their nests (nest homeostasis)
- Transfer of the nest and accumulated resources to the next generation

In many respects the colony is like an organism with individuals rather than cells specialized for reproduction and other functions. Arguments have been made that highly eusocial colonies represent a superorganism, or level of biological organization beyond the individual organism. As a result of their superior competitive traits, highly eusocial insects and especially ants and termites dominate small organisms in most terrestrial habitats and make up a major portion of the total animal biomass (Wilson, 1990).

GEOLOGICAL HISTORY
OF SOCIAL BEHAVIOR

The evolution of social behavior began in the Mesozoic Era. Fossils of social termites, ants, wasps, and bees are all found first in the Cretaceous Period. The oldest are termites: two fossil species of the genus *Meiatermes* of the Hodotermitidae from Brazil and Spain date from 110 million years ago in the Lower Cretaceous. The ants are the second oldest: workers and males of *Sphecomyrma freyi* and males of the genus *Baikuris* have been found preserved in amber in New Jersey dating from about 92 million years ago, leading to an estimate of about 130 million years ago for the origin of the ants. A fossil nest of a social wasp has been found in Upper Cretaceous deposits in Utah; the family Vespidae as a whole has been recorded back to 118 million years ago. Finally, what is clearly the worker of a stingless bee, *Trigona* (now *Cretotrigona*) *prisca* (Meliponini, Apidae), was found in amber in New Jersey, originally dated at 96 to 74 million years ago but recently reestimated as about 65 million years old. The oldest fossil interpreted as a bee dates to 100 million years ago, but it was not apparently social and possesses a mosaic of traits of wasps and bees as would be expected of a relatively early bee lineage.

The fossil record indicates that distinct castes and presumably highly eusocial behavior were established early in these insect lineages and, once fixed, have remained largely unchanged up to modern times. Among living species of bees and wasps of the primitively eusocial type, social evolution continues in a dynamic state and behavior is remarkably flexible.

EVOLUTION OF EUSOCIAL BEHAVIOR

The critical transition from subsocial or parasocial behavior to eusocial behavior involves the production of a nonreproductive caste. The sterile individuals are said to be altruistic because they sacrifice their reproductive ability to benefit the survival of the colony. In order for altruism to evolve, a shift must occur from individual organisms as the units of natural selection to individual colonies as the units of selection. This was the solution proposed by Darwin. In modern terms, selection is said to act on inclusive fitness, that is, the sum of an individual's fitness plus the effect of the individual on the fitness of relatives other than direct descendants. Altruistic behavior is not limited to eusocial insects: the "defender morphs" of the polyembryonic *Copidosoma* (Encyrtidae, Hymenoptera; see Chaps. 4 and 10) and the "soldiers" of certain aphids (see Chap. 32) are also examples of altruistic behavior.

Mechanisms that would facilitate the evolution of eusocial behavior are mutualism, parental manipulation, and kin selection. These are not mutually exclusive and probably act jointly in some groups. Mutualism has been mentioned above as responsible for the higher survival of individuals living in groups with common defense versus living alone. In parental manipulation, the queen suppresses reproduction by her daughter workers or other female nestmates through behavioral dominance, sterilizing pheromones, or partial starvation. Queens usually have better developed ovaries and a more active corpora allata than the subordinate workers. Some workers may still lay a few eggs, which may be eaten by the queen. As a genetic trait, parental manipulation could be favored by selection provided that the workers do not leave the nest to reproduce elsewhere. For the workers to stay, the risk of failure in founding a new nest must be greater than the security of staying in the colony as a worker, laying fewer eggs, or waiting for the queen to die. The evolution of eusocial behavior by parental manipulation would be enhanced if the worker helped to raise closely related sisters (see below).

The third mechanism, kin selection, acts on the inclusive fitness of a colony. This is best explained by considering the parent-offspring relationships in a family of Hymenoptera. Recall that reproduction in this order is by haplodiploidy. Unfertilized eggs develop into males; fertilized eggs yield females. The haploid male offspring each receive one of each of the paired chromosomes from the mother (resulting, incidentally, in a situation where a son has a grandfather, but no father). The mother therefore shares half of her genes with each son. The diploid daughters have half their chromosomes from the mother and half from the father. Accordingly, the mother shares half of her genes with each daughter. Of special importance is the genetic relationship among sisters sired by a single father. They all share the same paternal chromosomes (one-half of the chromosomes among sisters are the same). Among the maternal chromosomes, because the chromosomes are segregated randomly during meiosis, the sisters are likely on the average to share half of her contribution (one-fourth of the chromosomes among sisters are likely to be the same). On average, therefore, sisters share three-fourths of their genes. The average fraction of genes shared is called the coefficient of relationship.

In order for genes for altruistic traits to increase under kin selection, the altruist must actually promote the survival of genes like its own by increasing the numbers of closely related individuals who will reproduce. Altruistic sisters (related at three-fourths) who live with their mother and who increase the survival of their reproductive sisters by caring for them, defending the colony against enemies, and foraging for food for the colony are increasing altruistic genes more effectively than if they cared equally for their own daughters (related at one-half).

On the other hand, the genetic relationship is not favorable in semisocial groups, even when sisters are involved, because an altruistic sister would be helping the production of nieces (related at three-eighths) rather than her own offspring (related at one-half). Evidence exists that unrelated females may sometimes establish semisocial groups, further lowering the coefficient of relationship. Similarly, if the female parent of a eusocial group is inseminated by several males or if several females contribute female offspring,

the average genetic relationship among the group of female offspring is rapidly lowered. It remains to be shown that local inbreeding might raise the general level of genetic relationship to a point where the coefficients are again favorable. For insects in which both sexes are diploid, as are the termites, inbreeding might also facilitate the evolution of altruism.

For about three decades, studies of insect sociality were dominated by investigations of the relationships of genetic relatedness to altruistic behaviors. In the last decade or so, the realization that other factors affecting the costs and benefits of altruism can often be more important than relatedness has led to a shift back to a more ecologically integrative approach.

In the first decade of the twenty-first century, a genomic approach to identifying the genetic changes responsible for the behaviors associated with sociality was pursued vigorously, especially in bees, where there have been convergences in social evolution. The results indicate that many genes can show differences between social and nonsocial taxa, but it may be some time before we have a comprehensive picture.

BIOLOGY OF TERMITES, THE EUSOCIAL COCKROACHES

All Isoptera are eusocial. Termites differ from the social Hymenoptera discussed later in three important features: (1) termites live in colonies with castes of both sexes; (2) both sexes are diploid; and (3) they have a hemimetabolous life history in which the young instars may participate in the work of the colony. Almost certainly termites evolved along the subsocial route from cockroaches, with which they share many anatomical and physiological similarities. The step from subsocial to eusocial may have been aided by high degrees of genetic relatedness within a colony as a result of inbreeding.

The present-day cockroach, *Cryptocercus punctulatus* (Cryptocercidae, Blattodea), may give us some clues to the behavior of the termites' ancestors. The roaches exist as subsocial groups in cavities in decaying wood. They derive nourishment from cellulose by means of flagellate protozoa that live symbiotically in the roaches' intestine. Flagellates are transmitted among roaches and to the next generation by anal trophallaxis. This bond between generations requires a subsocial life but not necessarily a eusocial life. Although the circumstances under which true sociality first arose in termites are not clear, it seems probable that termites evolved from a common ancestor near *Cryptocercus* within roaches during the early Mesozoic Era.

Castes of Termites

The reproductive castes of termites (Figs. 7.2, 20.6) are as follows:

1. Primary reproductives are the only individuals that are fully winged after the last molt. They correspond to the alate, imaginal instar of other insects. Correlated with flight are a sclerotized and often pigmented body, compound eyes, and frequently

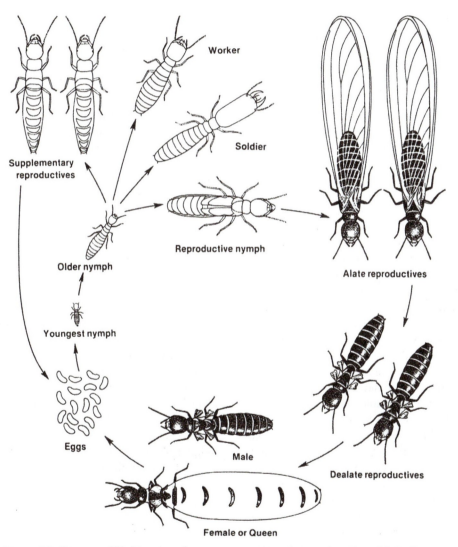

Figure 7.2 Diagram of life histories of castes in a termite colony, such as that of *Reticulitermes hesperus.*
Source: Redrawn from Kofoid (ed.) 1934; copyright 1934 by the Regents of the University of California; reprinted by permission of the University of California Press.

a pair of ocelli. After the nuptial flight the wings break at preformed lines of weakness near the base, leaving characteristic wing stumps. After insemination, the female's body undergoes extensive changes leading to sustained egg production. The swollen abdomen of queen Termitidae, for example, may reach 9 cm in length. The male remains small. Usually only one functional primary reproductive of each sex is in a colony, and both are relatively long-lived in comparison to the sterile castes.

2. Supplementary reproductives are imaginal individuals capable of producing eggs and sperm, but whose development has been partly inhibited. In comparison to primary forms, the pale body is less sclerotized and the compound eyes are reduced. Functional wings are not developed at the last molt, but those individuals with wing pads are termed secondary reproductives, and those without wing pads are called tertiary reproductives. One or more supplementary reproductives of the appropriate sex replace a dead or declining primary reproductive.

3. Workers are the most numerous members of a termite colony. They are usually of either sex and are wingless and sterile (Figs. 7.2, 20.4). They perform all the duties of foraging, nest construction, and tending the young. The mandibles are not exceptional but are suitable for gnawing and are strongly muscled. Must workers are pale and sightless, but tropical Hodotermitidae, which forage during the day for grasses, are pigmented and possess compound eyes. Some scientists limit the worker caste to a more or less distinctive group of older instars that have lost the capacity to develop further into soldiers or reproductives. Such worker castes are found almost entirely in the advanced family Termitidae. The workers are already different after the first molt and may be divided into subcastes, with one sex predominating in each. The "workers" of more primitive families are properly termed pseudergates, or "false workers." These older nymphs without wing pads may later differentiate into soldiers or reproductives or may remain "arrested" and workerlike throughout the rest of their instars. Pseudergates may also be derived from other partly differentiated castes by regressive molts resulting in loss

of wing pads. Soldiers and sexually functioning reproductives are final instars and do not regress to pseudergates.

4. Soldiers are workers with special development of the head for defense. One type found in some species has enlarged mandibles. A second type, known as the nasute, has a nozzlelike frontal projection from which is secreted a defensive fluid. In some species the soldiers have both large mandibles and the nasute head. Soldiers cluster about openings when a nest is damaged and engage intruding ants or other enemies. Soldiers may be of both sexes, but in advanced termites, soldiers may be all females or all males, depending on the taxonomic group.

Caste Determination in Termites

How is caste determined? Early investigations sought to distinguish between intrinsic or genetic mechanisms versus extrinsic or environmental influences. The research of Luscher (1961) on a relatively primitive termite, *Kalotermes flavicollis* (Kalotermitidae), points conclusively to the latter explanation: eggs are equipotent, capable of developing into any caste depending on the nest environment. Yet in the Termitidae, early caste differentiation and sex-specific castes suggest an additional genetic basis.

Determination of the caste of an individual *K. flavicollis* involves the familiar hormones of insect development plus pheromones from caste-determined members of the colony. Not unexpectedly, ecdysone promotes differentiation from pseudergate to the reproductive caste. On the other hand, low doses of juvenile hormone in a pseudergate result in no further differentiation, but high doses oddly promote differentiation into soldiers. Judging by the volume of the corpora allata just following a molt, the pseudergate at this time has the lowest level of juvenile hormone and is most likely to differentiate toward a reproductive if ecdysone is present. Proportions of the hormones are presumably regulated by the brain, which in turn is responsive to pheromones ingested by anal feeding from other termites. The pheromones have not been chemically identified, but experimental evidence points to (1) a substance produced by the queen that inhibits female pseudergates from becoming functional queens, possibly by suppressing their secretion of ecdysone after molts; (2) two substances produced by the male reproductive or king, namely, a corresponding male inhibitory substance and another substance that stimulates female pseudergates to become functional queens; (3) a substance produced by soldiers that inhibits pseudergates from becoming soldiers; and (4) a recognition pheromone such that when more than one queen is present in the colony, queens recognize each other and fight; similarly kings recognize each other by their own recognition pheromone and fight. Male or female supplementary reproductives become active when the inhibitory influence of the reproductive male or female, respectively, is removed. In other species of termites, both reproductives are known to stimulate soldier production. Thus, recruitment into castes may be both stimulated and inhibited by the presence of individuals already caste-determined. Excess numbers are destroyed by cannibalism.

Life History of Subterranean Termites

To illustrate the life cycle of a colony of termites, let us examine the rhinotermitid, *Reticulitermes hesperus*, a common and especially destructive species in the western United States (Fig. 7.2; Kofoid et al., 1934). Other species of *Reticulitermes* are found in almost every state and are equally important as structural pests. These are known as subterranean termites because colonies are nearly always in contact with damp earth by means of closed tunnels. When decaying logs are rolled, pieces of lumber turned over, or, worst of all, when part of a house collapses, the presence of *Reticulitermes* is revealed by clusters of hundreds of white insects the size of rice grains. New colonies are established by a male and female primary reproductive. After the autumn swarming flight, the reproductives flutter to the ground and by flexing the abdomen break off their wings at the lines of weakness near the wing base. Females raise the abdomen and release a sex attractant. Males running on the ground orient to the odor and locate females. The pair then moves off in tandem with the female in the lead. On finding a crevice under or in a piece of wood, the two enlarge

the cavity and plaster the walls with feces and wood bits. Copulation follows the first efforts at nest construction and is repeated at intervals for the rest of the lives of the reproductives. Initially fewer than ten eggs are laid and carefully cleaned to prevent mold. The young nymphs are fed from the mouth or anus of the parents. By the second instar they have acquired intestinal protozoa. Growth of the population is slow, and members are long-lived. Workers are estimated to live 3 to 5 years, and reproductives probably live longer. The first clearly differentiated castes are apterous, nonreproductive workers and soldiers. After no less than 3 or 4 years under favorable conditions, the colony produces its first large numbers of new primary reproductives. Fully winged in the seventh instar, they remain in the nest until early autumn rains soak the soil. Then on a bright warm day, they emerge by the thousands. During a weak flight of 200 m or so, they mingle with reproductives from other colonies in the vicinity.

Large colonies may also reproduce by a budding process whereby a part of the constituency becomes isolated and develops its own supplementary reproductives. Outlying food sources, such as houses, may become rapidly infested in this way. Supplementary reproductives of the appropriate sex also replace primary reproductives, which die or lose physiological control of the colony.

BIOLOGY OF SOCIAL HYMENOPTERA

Eusociality has emerged repeatedly in Hymenoptera, probably once leading to Formicidae, or ants; twice in Vespidae or hornets and paper wasps (independently in the lineages leading to the Old World tropical Stenogastrinae and to Polistinae and Vespinae); once in Sphecidae (the neotropical *Microstigmus comes*); and possibly up to eight or nine times in Apoidea, or bees, of which honey bees, stingless bees, bumble bees, some allodapine bees, and sweat bees are eusocial. Just among the four tribes of corbiculate bees, eusocial behavior may have arisen independently in honey bees and in the stingless bees/bumble bee clade of Apidae; among the Halictidae or sweat bees, there may be one or more separate origins.

Social behavior in this order is an activity among adult females. As holometabolous insects, the larvae are helpless and rarely contribute to colony welfare. Males are usually winged and function solely in reproduction.

Why are aculeate (stinging) Hymenoptera predisposed to evolve social behavior? The search for possible factors among the presocial Hymenoptera has not yielded a simple explanation, because other insects have some of the same features. However, four features in combination may have uniquely set the stage for recurring sociality.

1. **Morphology**: Adult aculeates are strong fliers with versatile mandibulate mouthparts useful in nest construction, food manipulation, and brood care. Female aculeates have the ovipositor modified into an effective weapon of defense.
2. **Behavior**: Nest building and various levels of presocial behavior exist, including parental manipulation and mutualism among long-lived females.
3. **Genetics**: The haplodiploid reproductive system creates genetic relationships in a family that favor evolution of altruistic behavior among sisters living with their mother.
4. **Geological age and statistical opportunity**: Aculeates date from the late Jurassic Period and now number about 50,000 species, providing ample chance for eusociality to arise.

BIOLOGY OF ANTS

The life histories of different species of ants are extraordinarily varied (Fig. 7.3). As an extreme example, species are known in which reproduction is by thelytokous parthenogenesis, queens are absent, males very rare, and unfertilized workers reproduce. In general, however, most colonies have reproductive queens and males and nonreproductive workers. Virgin queens usually mate with males from other colonies in the area. Depending on the taxonomic group, essentially two kinds of mating behavior exist. In the female-calling syndrome, virgin females on ground or vegetation release a sex pheromone from the pygidial gland between the sixth and seventh abdominal tergites. Males are attracted and copulation ensues. In the male-aggregation syndrome, flying males

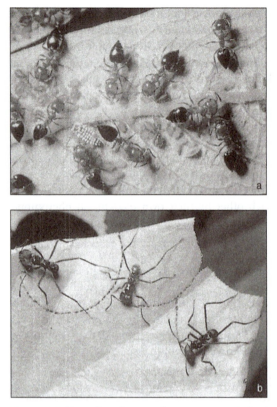

Figure 7.3 Examples of ants. a, *Crematogaster* ants feed on honeydew from aphids on the underside a willow leaf. The ants may protect the aphids from enemies, but in this case two kinds of predators have escaped attention: a white-spotted coccinellid beetle larva and, at lower right, a syrphid fly larva. b, Leaf-cutter ants, *Atta* sp., cut pieces of leaves and carry them to their underground nest. The leaves are masticated to serve as a cultural medium for the fungus that the ants eat.

form swarms in the air usually above conspicuous landmarks like trees. Females fly into swarms and copulation ensues.

Colony growth can be divided into three stages. The founding stage begins with the end of the nuptial flight. The virgin queen mates with one or more males, which die without returning to the colony. The queen breaks off her wings, finds a suitable nest site in soil or plant material, and constructs the first cell. The first workers are reared by the queen and as adults are quite small in size. Depending on the species, queens may forage for some food outside the nest (known as partially claustral founding type) or remain entirely in the nest (fully claustral founding type), drawing on her own tissues (flight muscles and fat body) to produce eggs and feed the larvae. Feeding is by special salivary secretions or by trophic eggs (eggs fed to larvae). In some species young queens are accepted back into the parental nest.

In the second, or ergonomic stage, of colony growth, the first brood of workers reaches adulthood. The queen begins to devote herself to egg-laying, leaving foraging for food, enlargement of the nest, and brood care to the workers. The population of workers grows in number and the average size of individuals increases. New castes may appear among the workers. At this stage all activities in the nest are devoted to colony growth.

After a period of one season or up to 5 or more years, depending on the species, the colony enters the final, or reproductive stage. Some of the colony's resources are now invested in new queens and males. These leave the colony, and the colony activities revert to the ergonomic stage until the next reproductive stage is reached.

Colonies of ants are perennial, but nest sites may be changed from time to time. Some species, such as the army ants and driver ants, are more or less constantly on the move. The number of eggs per queen per year varies from 400 in one species to 50 million in the African driver ants. The time from egg to adult varies with the species, on the order of 1 to 2 months. Adult queens are long-lived; some have been estimated to live as much as 18 to 29 years. Adult workers may live 1 to 2 years, although most probably have much shorter lives. Males have a much shorter life span than workers or queens.

Worker ants lavish care on all the immature stages, from egg to larva to pupa. Workers lick, clean, carry the immatures about the nest, and remove their waste products. They feed the larvae by regurgitation, glandular secretions, and by placing the larvae near other kinds of food that have been prepared for them by cutting up or chewing. Some workers are capable of laying trophic eggs that are fed to larvae.

Larvae of some species are not entirely helpless but contribute to the colony's welfare. In some species, larvae produce their own secretions, which are consumed by the workers. It has been speculated that the larvae are able to digest certain foods and share the digested food with the workers. Larvae of weaver ants help in the construction of arboreal nests. Worker ants pull leaves together, then hold the larvae at the leaf edges, where they secrete silk to bind the leaves together.

Males and Castes of Ants

1. Males are generally considered to be a sexual form but not a caste because they are all alike and function only in reproduction. They are winged with a fully developed thorax and with fully developed compound eyes and ocelli. In some species they undergo some degeneration of internal structures (fat body depleted, corpora allata and midgut degenerated) prior to the nuptial flight and become "sexual missiles." In most colonies they are fed by workers and do no work. In some species, however, males are long-lived and exchange food with workers so that they have some, but not all, of the behaviors of a caste.

2. Queens are the female reproductive caste in the colony. They have a fully developed thorax with flight muscles and wings, compound eyes, and ocelli. The wings are broken off after the nuptial flight, and the flight muscles degenerate. Queens have a fully developed reproductive system with a spermatheca capable of storing sperm for the duration of the queen's life. Depending on the species, a colony may have a single queen (monogyny) or multiple queens (polygyny).

3. Workers are sterile females with reduced or no ovarioles, no wings or flight muscles, and an otherwise simplified thorax. Workers are divided into subcastes according to size: minor, media, and major. When majors function mainly in defense of the colony they are called soldiers. Workers of different sizes also differ in the proportions of some body parts such as the head and mandibles. These grow relatively larger in relation to the rest of the body in large ants, resulting in ants suited for defense or grinding of seeds with mandibles. Large workers in the nests of the honeypot ants, *Myrmecocystus*, and some other species become so engorged with honey or water in their crop that they are almost immobile. Called "repletes," they are confined to the nest and serve as storage vessels.

Caste Determination

Caste determination is virtually always environmental rather than genetic. In rare instances, single alleles have been shown to determine some special reproductive types in one species. The determination of caste can be viewed in the following way. In the absence of modifying influences, a diploid female egg will develop into a queen. The *anlagen*, or imaginal disks, of the larva are divided into two sets: (1) a dorsal group, including wing buds, incipient gonads, and ocellar buds, and (2) a ventral group, including leg buds, mouthparts, and central nervous system. In the morphogenesis of a queen, both sets develop with equal vigor. However, an interwoven set of environmental and/or maternal factors may inhibit or fail to promote (presumably via the endocrine system) the complete development of the dorsal set. Such an individual becomes a worker.

Certain experiments indicate that pheromones secreted by a functional queen inhibit development of more queens among her brood. Nutrition of larvae also probably varies in both quantity and quality as the colony grows. A large foraging force may provide optimum food, or the queen's inhibitory pheromones may be diluted, or both. In any event, new queens are produced from large colonies. Other environmental influences are chilling temperatures. Brood produced in the fall at temperate latitudes become dormant and later chilled by winter's cold. Dormancy is later broken by the warmth of spring, and the surviving larvae are able to grow rapidly, metamorphosing into queens. Maternal effects include the size of eggs laid by the queen. During periods of high egg production, eggs average smaller and produce mostly workers. Younger queens also tend to lay eggs that result in workers.

Life History of the Argentine Ant

The Argentine ant, *Linepithema humile* (Fig. 12.3; formerly known as *Iridomyrmex humilis*), is a familiar species that differs in several respects from the normal life history of ants. Commercial traffic from South America distributed the ants to countries in the Northern and Southern Hemispheres within latitudes of 30° to 36°. In North America the first colonies were seen in 1891 in New Orleans, Louisiana. They are now widely distributed in the southern United States and in California, where they first appeared in 1905. Native ants are regularly displaced by Argentine ants, often by physical combat.

Nectar and honeydew from plant-sucking Homoptera are more than 99 percent of the diet of these ants. This search for sweet fluids has earned the Argentine ant a place among the important household and agricultural pests. Massive numbers may invade well-kept kitchens or restaurants for food, often in late summer and after heavy rains. Beehives are entered and honey is taken from the combs. Mealybugs, scale insects, and other honeydew-producing Homoptera are tended and protected by the ants. The homopterous pests of citrus and other crops are protected by ants from attack by predators and parasitoids, thus greatly reducing the effectiveness of biological control by natural enemies.

Argentine ants occupy a wide range of habitats provided sufficient moisture is available in soil, logs, or under cover objects. The seminomadic colonies do not have clearly defined limits or individuality. In a given area many nests are excavated in soil at favorable sites and near food sources, but these are parts of an extended colony. Workers move freely from one nest to another. These nests are abandoned when the food is depleted or when flooding or drying renders a site unsuitable. When the general area is unfavorable, occasionally a single large nest is made or the colony moves 100 m or more to a new area. Argentine ants exploit food resources in their territory by a rapid, highly organized exploration by workers. A column of ants extends from the nest, then spreads out into a fan-shaped front in which the individuals search randomly. Once food is discovered, a trail pheromone (Z-9-hexadecenal) from a sternal gland leads new recruits to the food until it is consumed.

New colonies are established by colony fission or "budding" rather than the usual aerial dispersal of young queens to new sites. Reproductive males and new queens are produced in the spring, but these usually mate entirely within the nest. The inseminated queens shed their wings and begin laying eggs. During the growing season the number of queens in a colony may comprise as much as 3.5 percent of the biomass. In autumn, the worker population decreases. Workers then attack queens and dismember them, reducing the number of queens in the colony before the next year's cycle.

BIOLOGY OF SOCIAL VESPIDAE

The family Vespidae includes six subfamilies, of which most eusocial species are in two subfamilies: Polistinae (800 species in 29 genera; cosmopolitan) and Vespinae (60 species in 4 genera; Holarctic and Oriental tropics). Examples of social genera are *Vespa*, *Vespula*, and the largest genus, *Polistes*. The Stenogastrinae (50+ species in 6 genera; Oriental tropics) also has at least some independently derived eusocial species, but these have not been as comprehensively studied.

The wasps are predatory on small invertebrates, scavengers on carrion, and also feed eagerly on sugary solutions such as nectar, honeydew, and fruit juices. Both males and females are winged and strong fliers. In some species, the queens and workers are similar in size and appearance, while in others the queens are larger than the workers, especially early in the season. Nests are constructed in cavities or exposed and with paper made of plant or wood fibers. Brood is reared in combs of hexagonal cells that open downward.

The typical life history in the temperate zone begins with inseminated females (queens) overwintering in protected places and founding new nests in spring either alone or in groups. The queen builds the first nest and feeds the larvae. The first brood usually consists of all females that remain on the nest as workers. The queen then devotes herself to laying eggs and the workers build the nest and feed the larvae of subsequent broods. The last brood of the season includes males and females. On maturity, these leave the nest and mate. On the death of the queen, brood production ceases and the nest population declines as the workers and males die. The newly inseminated females overwinter, living on fat reserves and in a reproductive diapause. Nests may have a single queen (haplometrosis) or several queens (pleometrosis) that either initially found the colony together (often sisters) or join an existing colony. Nest sharing by closely related females is common. In the tropics, nests may be perennial and new nests may be formed by swarming of wasps—either a single queen and workers or multiple queens and workers.

BIOLOGY OF SOCIAL BEES

The apoid bees include about 25,000 to 30,000 species in over 4000 genera. Most are solitary in behavior, but perhaps 1000 species are eusocial. The most highly eusocial bees are the honey bees (*Apis* in Apidae) and tropical stingless bees (*Trigona* and *Melipona* in Apidae) that regularly form perennial colonies. These establish new nests by large flying swarms of workers plus the queen because the queen cannot live apart from the colony. The "primitively" eusocial bees include bumble bees (*Bombus* in Apidae), the social sweat bees (*Halictus* and *Lasioglossum* in Halictidae, some of which are facultatively social depending on conditions), and some allodapine bees of Africa, Asia, and Australia (*Allodape* and *Exoneura* in Apidae) that have smaller colonies usually lasting for one season and females that spend part of the year not in colonies.

The life history of bumble bees is similar in some respects to that of social wasps. The inseminated queen overwinters in a protected place. In the spring she makes a nest in a preexisting cavity, constructs rounded wax cells for brood and food storage, and forages outside the nest for pollen and nectar. The larvae are fed progressively. The first brood is of small females that function as workers. They take over the duties of nest building, foraging, and defense, leaving the queen to lay eggs. Early in the season there may be additional broods of entirely female workers, but toward the end of the season both males and new, larger and morphologically distinct queens are produced. These mate and the males die. After the founding queen and the workers die, the species survives the winter as the young inseminated queens.

Sweat bees are similar in life history with inseminated females overwintering, often in the old nest, and founding new nests in the spring. Nests are burrows in soil with individual cells mass provisioned for brood. Mostly females are produced early in the season and both sexes toward the end of the season. These mate, the males and old females die, and the young inseminated females overwinter. Depending on the species, the history of the individual nest, and the length of the growing season, the behavioral interactions among the females are enormously varied. Within a single species, the populations in a region of short growing season (higher latitudes or altitudes) may live as solitary bees while those in a region with a longer growing season have several generations and eusocial colonies. Furthermore, examples are known within a single species at one location where nests have solitary females and other nests have groups of females displaying various degrees of social behavior, including semisocial and eusocial. Females have the capacity to function as solitary bees, as workers, or as morphologically identical queens and to change behavior during their adult life. A worker can replace the queen if she dies. The flexible behavior of sweat bees contrasts sharply with the rigid castes of highly social insects like the honey bee.

Allodapine bees have a different annual cycle: males and females overwinter together, and mating is at the beginning of the growing season. One or more females excavate a burrow in the soft pith of a dead stem. No cells are constructed; the larvae are dumped together at the bottom of the burrow and all fed progressively by the female or females. A reproductive division of labor may occur even when only two adult females occupy a nest. Like the sweat bees, a great range of behavior has been observed within a single species: solitary (subsocial), semisocial, and eusocial.

The stingless bees (Apidae, Meliponinae), found in tropical regions of the world, are so named because the ovipositor is atrophied and does not function as a sting. The bees nevertheless defend themselves with mass attacks on intruders, biting, and in some cases secreting toxic chemicals. Colonies are highly eusocial with nests built of wax, resins, and other materials. Each colony has a single queen, many workers, and males. Depending on the species, nest populations range in number from 300 to 180,000. Nests of some species somewhat resemble the clusters of oval brood cells of bumble bees, while in others with more populous nests, brood is reared in horizontal combs of resin and wax in which the cells open upward. Food is stored in wax pots. The brood cells are mass provisioned, so workers do not directly contact the developing larvae. Worker bees of *Trigona*, when returning to the nest from a food source in the forest, deposit droplets of odorous secretion from the mandibular glands on vegetation at 2- to 3-m intervals. This "odor trail" guides new foragers to the site. Odor trails are especially efficient in marking surface routes over broken terrain or in the three dimensions of a tropical forest. Female caste determination in *Trigona* is apparently trophic: males and workers are reared in smaller cells and receive less food; the larger queens are reared in larger cells and receive more food. In *Melipona*, males, workers, and queens are reared in the same sized cells. It is likely that a certain proportion of females are genetically determined to develop into queens provided that food is adequate; otherwise they develop into workers as do all other females.

Life History of the Honey Bee

The best known and most useful social insect is the honey bee, *Apis mellifera*. Possibly it has been studied more than any other insect. The native home of the honey bee is in the Old World, where it has been exploited for thousands of years. Colonies in natural cavities, such as a hollow tree, or in hives made by human beings were killed to extract stored honey and wax. Only relatively recently, in the 1500s, did European beekeepers devise a hive with movable wooden frames so that individual combs could be removed without destruction of the entire colony. Apiculture is now the most widely practiced agricultural activity in the world. The role of the honey bee as a pollinator of crops has become increasingly important in North America. During the earlier stages in the development of agricultural lands, adjacent wild areas were usually inhabited by native pollinators. Under modern intensive cultivation, these pollinators have been eliminated, and honey bees now are regularly transported from crop to crop for this function. As repeated recent threats (*Varroa* mites, colony collapse disorder, etc.) to managed honey bee populations have highlighted our dependence on honey bees for food production, there has been an effort to enhance native pollinator diversity and abundance through less destructive cultivation methods and encouragement of native flowering plants as pollinator food sources (see also Chap. 12). The common honey bee is one of four distinct species or species complexes in the genus *Apis*. All are properly called honey bees because they are eusocial and construct wax combs for rearing brood and storing honey. All except *A. mellifera* are confined to the eastern Palearctic and Oriental regions. *Apis mellifera* originally inhabited western Asia, Europe, and Africa south of the Sahara Desert.

Races of European Honey Bees

Several distinctive races of honey bees evolved in Europe during the Pleistocene, when advancing glaciers fragmented their distribution. The bees survived in separate refugia in what are now Spain, Italy, Austria, and the Caucasus. The respective races are *A. m. mellifera*, the German, Spanish, or English honey bee; *A. m. ligustica*, the Italian honey bee; *A. m. carnica*, the Carniolan honey bee; and *A. m. caucasica*, the Caucasian honey bee. Beekeepers in these temperate regions have selected for productivity and ease of management, so that European bees are preferred in agriculture.

When colonies of European settlers were established abroad, honey bees were taken from the home countries. In the New World, *A. m. mellifera* was the race initially introduced. As communication and transportation increased, beekeepers exchanged queens and colonies around the world in an effort to improve productivity in the new climates. The Italian honey bee became especially favored and now is the predominant commercial honey bee worldwide.

African Honey Bees

In Africa, *A. m. scutellata* evolved under quite different conditions. Here it was beset by a variety of natural enemies, including human beings, and existed in a tropical climate with dry seasons. The African honey bee is consequently highly aggressive in the defense of its nests and able to undertake migrations to more favorable places. The colonies also reproduce more frequently than European bees. The high productivity in the tropics has attracted bee breeders on several occasions to make hybrids with European stocks. These experiments had gone largely unnoticed until a recent episode had calamitous results.

European bees in tropical South America are unsatisfactory honey producers. In 1956 African honey bees were taken to southern Brazil for the breeding of a strain combining their higher productivity in the tropics with desirable features of European bees. In 1957 queens with workers escaped. Hybrids were formed between the African bees and the resident European bees. In the Old World, however, *A. m. scutellata* and the European races had evolved apart for at least 2000 years, separated by the Sahara Desert. As a result, free cross-mating between the races was reduced, owing to the apparent preference of reproductives for their own race. Such hybrids as were formed perpetuated mostly African characteristics, hence the name "Africanized" for the bees in the New World.

The Africanized bees directly compete for food and nest sites, even invading the hives of European bees and displacing them. Large numbers of wild colonies were also established in small cavities not suitable for European bees. As the Africanized bees migrated annually into new areas of Brazil, it became evident that the European bees were being completely replaced by a much higher population of Africanized colonies. Beekeeping became difficult or impossible. Reports appeared in the world press of deaths and injury to humans and domestic animals. Although some minor differences in the chemistry of bee venom may exist, the danger is created by the large numbers of bees that quickly sting and pursue a victim. Since 1957 Africanized bees have spread throughout most of tropical South America, up through Central America to Mexico, reaching the United States at Brownsville, Texas, in October 1990. The bees have now been identified in the southern parts of the United States from Texas to Florida. Southward in Argentina, the Africanized bees appear to be unable to survive the cold winters of the temperate climate. A similar temperature barrier may prevent northward invasion in the United States beyond the southern and southwestern states.

In southern Brazil, large numbers of Italian queens were distributed to beekeepers in a program to reduce the aggressive behavior of Africanized bees by further hybridization and by killing aggressive colonies. This has been apparently successful in producing bees that are manageable and still productive. The beekeeping industry has recovered in this area. The genetic approach offers the best means of controlling the undesirable traits of the bees, but over vast regions of tropical America this is impractical. Moreover, in South America the Africanized bees have outperformed the European strains.

Males and Female Castes of Honey Bees

Both sexes of honey bees are winged with well-developed compound eyes and ocelli.

1. Males, or **drones**, are produced during the spring and summer seasons. They are distinguished by larger eyes and their larger body size. Drones are fed throughout their 4- to 5-week lives by the worker bees and are unable to take nectar from flowers. Aerial excursions are taken from the hive in search of queens, with which they mate in the air. The male genitalia are uniquely constructed to evert violently and be torn off during copulation. Successful drones die after mating. At summer's end, unsuccessful drones are expelled from the hive, also to die.

2. The **queen** is the functional female reproductive in the colony. A queen is slightly larger than a worker and has a longer abdomen that extends well beyond the folded wing tips. From their mandibular glands is secreted the queen pheromone, which is a mixture of 9-keto-(E)-2-decenoic acid, 9-hydroxy-(E)-2-decenoic acid, and lesser amounts of other components. Possibly another pheromone is produced from tergal glands, but its chemical identity is unknown. The pheromone or pheromones have multiple functions: inhibition of the development of the ovaries of workers; inhibition of the construction special wax cells for rearing new queens; attraction of workers to the queen in the hive and during swarming; and as a sex attractant during aerial mating. The absence of the queen in a colony or swarm is quickly detected by the absence of her pheromones. The queen never forages for food but does leave the hive for one or more mating flights early in her life and during periodic swarming. Ordinarily, only one queen is present at a time in a colony. She produces all the eggs that are fertilized and most of the unfertilized eggs. These may total 600,000 over the 2 to 3 years of her normal life span.

3. Workers are females whose reproductive organs are atrophied to the extent that copulation is prevented but which can occasionally produce eggs. Some of the unfertilized, or drone, eggs in a colony come from this source. An exceptional race of honey bees restricted to the Cape region of South Africa has laying workers whose eggs develop by thelytokous parthenogenesis into females. The sterile workers perform all the foraging, defense, wax secretion and comb construction, brood care and feeding, and regulation of the physical environment and also locate new sites during swarming. Colonies normally contain 60,000 to 100,000 workers. During the active foraging seasons, a worker may live 5 to 6 weeks. Workers reared in the fall and overwintering may live 6 months or more.

Caste Determination in Honey Bees

The vertical comb is made of wax secreted by glands on the abdominal sterna of workers. Using their mandibles, the bees shape the scalelike pieces of wax into cells that are hexagonal in cross section. Each comb is composed of two layers of cells that open horizontally on each side. Human engineers have imitated this design for creating strong, lightweight structures. The bees' cells are used as a nursery or for the storage of honey or packed lumps of pollen. Most of the cells are closely similar in size. In the brood area the queen lays a single, fertilized egg in each cell to produce workers. When food is plentiful, larger cells near the margin of the comb receive unfertilized eggs to produce drones. New queens are reared in much larger cells, which are round in cross section and are specially constructed to hang from the surface or lower edge of the comb. The brood cells remain open while the larvae are fed progressively by the workers.

The determination of caste among the females is directly related to differences in food.

For the first 2 days, drone and worker larvae are fed a mixture of glandular secretions and crop fluids from workers functioning as nurse bees. This "bee milk" includes (1) clear hypopharyngeal gland secretions plus clear crop fluids and (2) a lesser amount of white mandibular gland secretions and some hypopharyngeal secretions. The white secretions diminish and disappear in the food on the third day, and afterwards the larvae are fed "bee bread," consisting of the clear fluids variously mixed with pollen and honey. A female larva destined to become a queen is fed throughout her life on "royal jelly," which is composed of equal parts of the clear and white foods. These secretions include basic nutrients such as sugars, amino acids, proteins, nucleic acids, vitamins, and cholesterol plus a fatty acid, (E)-10-hydroxy-2-decenoic acid. The quantity, quality, and timing of the food, not a queen-determining pheromone, divide worker-destined and queen-destined larvae at an early age. With high-quality food, the neurosecretory system of the queen larva becomes active earlier than in the worker larva, leading to higher titers of juvenile hormone in the queen. For the queen, this difference in hormone results in a higher respiratory rate as a larva, protection of her ovaries from degradation, and initiation of vitellogenin synthesis in the pupal instar. Intermediates between queens and workers are rarely found in nature but can be experimentally produced.

Once the larva has reached full size, it defecates for the first time. A cocoon is spun that separates the larva from its feces. The workers cap the cell with wax. The larva now transforms to a pupa and then to an adult inside the closed cell. After the final molt, the new bee chews its way out of the cell to join the colony. Three weeks are required for growth from egg to adult.

Life of the Worker Honey Bee

A worker bee spends about 3 more weeks inside the hive before becoming a forager, or field bee. The bee's behavior and the tasks performed in the hive change as the bee ages. This is called age polyethism and is regulated by juvenile hormone acting as a behavioral primer. Much time is spent resting or patrolling the comb. Young worker bees first clean the hive and feed larvae with secretions from their hypopharyngeal and mandibular glands, which are well developed at this time. With age these glands normally decline in activity and change secretions. The hypopharyngeal gland later secretes invertase, which converts the sucrose in nectar to the glucose and fructose of honey, and the mandibular gland secretes an alarm pheromone, 2-heptanone, when the worker later serves as a guard or forager. As the nursing function decreases, the sternal wax glands may become active if comb building is prevalent in the hive. She also begins short flights from the hive. After the wax glands decline, activity shifts to the entrance, where the worker may guard the hive before finally becoming a forager. For the remaining 1 to 3 weeks of her life, she collects nectar and pollen from flowers (see "Learning and Communication in Honey Bees," Chap. 6). Plant resins are also collected and used to seal openings in the hive. Beekeepers call the resins propolis, or bee glue.

Annual Cycle of a Colony

The annual cycle begins when the spring flowers provide abundant food. The queen begins laying fertilized eggs that greatly augment the worker force. Normally the increased food supply, stored provisions, and high population inside the crowded hive lead to swarming. The inhibitory effect of queen pheromone declines, and workers construct the special cells in which new queens are reared. Egg production by the old queen declines, as does foraging by the workers. Before the new queens emerge, the old queen, part of the worker force, and some drones leave the hive. The mass flight, or swarm, is spectacular. The bees may periodically settle in a cluster on a tree limb or other object. Scout worker bees fly from the swarm in search of suitable new quarters. The scouts return and dance on the cluster, indicating the distance and direction to the favored site. Once the swarm has occupied the new hive, wax combs are constructed, food is collected, and the old queen begins to restore the worker population. In the original hive, the first new queens to emerge fight physically for the role as queen. Pupal queens are destroyed by the workers. The survivor flies from the hive several times over several days and may be inseminated by a drone each time. The sperm are stored in her spermatheca for use during the remainder of her life. Her new colony has the benefit of existing combs and stored food. The colony may produce swarms more than once during favorable seasons. In the fall when flowers no longer supply enough food, brood rearing declines and drones are expelled. More propolis is added to reduce cold drafts. The colony clusters together on combs filled with stored honey during cold weather. The bees utilize honey as heat-giving energy, remaining alert through winter.

CHAPTER 8

Diversity and Adaptations of Insects in Selected Habitats

Ecology is the study of the interactions between organisms and the physical, chemical, and biotic factors of their environment. Chapters 10 through 13 deal with the biotic environment, namely, interactions of insects with plants, other insects, vertebrates, and microbes. In this chapter we focus on certain habitats or the places where insects live: soil, caves, deserts, high altitudes, and freshwater and salt water. We will discuss some of the physical and chemical features of these environments, the kinds of insects present, and their special adaptations for survival. Insects in these habitats are frequently scavengers, and at the end of the chapter we will list the variety of foods eaten.

Soil and freshwater are home to many kinds of insects and also are vital natural resources for humans. Insect scavengers play an important role in the health of these resources by acting as decomposers and recycling nutrients. Unfortunately, soil and freshwater are easily contaminated by pesticides or pollutants, with the result that these beneficial insects are destroyed. Because they are so sensitive, certain species of insects by their presence, abundance, or absence can serve as biological monitors of environmental quality.

THE SOIL HABITAT

Soil is the medium that connects all habitats of the land, from the intertidal to the alpine. It is also a distinct realm and one on which green plants and ultimately all forms of terrestrial life depend. Thousands of kinds of soils have been identified, each the product of different circumstances during formation. In general, the process of soil formation leads to horizontal layers that are readily visible in a cross section such as a road cut. On the soil surface is usually a duff, or litter of whole or partly decayed leaves and other plant parts. The nature of the litter varies with moisture and the covering vegetation, such as grassland, coniferous forest, deciduous forest, or lichens and mosses. Beneath the loose litter is the soil's uppermost layer, or topsoil, which

is darkened by the accumulation of a complex, organic substance called humus. Topsoil is the place where most plant roots and soil organisms occur. Below the humus-rich zones, the subsoil is paler in color, being only slightly enriched from above and with a higher content of clay. The topsoil and subsoil make up the true soil. Beneath the topsoil and subsoil may be a layer of varying thickness called the soil parent material, which grades below into the unaltered rock of the earth's crust.

To sustain a community of living plants and animals, the nutrients locked in dead plants and animals and animal waste must be released by decomposition. This recycling of nutrients is accomplished primarily by bacteria and fungi. Arthropods and other animals indirectly influence the process by feeding on the microbes and thereby regulating their populations. Arthropods also contribute directly by feeding on plant and animal waste at the soil surface. This is mechanically and chemically broken down, often with the aid of microorganisms in the gut, and the feces are deposited in their burrows deeper in the soil. Their excavations also mix the surface layer between litter and soil. Microbes in the moister regions below then act again on the partly decomposed fragments, now with greater surface area, to produce humus. Arthropods are especially important in the recycling of nutrients in grasslands and deserts.

Cracks, old root channels, and tiny cavities up to 1 mm in size between soil particles create pore spaces in soil. The total pore space of topsoil is usually about a third of the soil volume, but in some places it may exceed half the volume. At greater depths, the soil is more compact and has smaller pore spaces. Depending on the nature of the soil particles, rainfall, groundwater, and drainage, the pore spaces are filled partly by air and partly by water held by capillary forces. During periods of flooding, insects in the soil obtain oxygen from air trapped in their burrows or pore spaces. Soil air is usually saturated with water vapor except near the soil surface. In well-drained topsoil, oxygen in soil air is replenished by gaseous diffusion from the atmosphere. Soil water derives its oxygen from soil air. Oxygen, however, diffuses much more slowly in water and can be depleted by bacterial decomposition of organic matter. When this happens, anaerobic conditions are created in as short a distance as 1 mm from a water/air interface. Deeper in the soil, gaseous diffusion from the atmosphere decreases. Here the respiration of organisms reduces oxygen and increases the carbon dioxide content of soil air.

If the vegetation overhead is not too dense, sunlight penetrates the litter, open burrows, and superficial soil crevices. Below the surface, at depths of only a few centimeters at most, is permanent darkness. In deserts, grasslands, alpine regions, and deciduous forests after the leaves have fallen, the surface is largely exposed to the sun's radiation. Thus heated by the sun, the surface temperature of soil rapidly climbs well above that of either the air or the soil at slight depths beneath. Surface temperatures above 50°C are not infrequent even in the Temperate Zone. Owing to the low thermal conductivity of soil, the heating and cooling of deeper layers lags behind the fluctuations in temperature of the air, litter, and the soil surface. Heat acquired during day penetrates into the soil quite slowly. A high moisture content in soil will reduce the amplitude of these daily fluctuations in temperature. "Cover objects," such as large rocks or logs resting on the surface, shelter the soil beneath from extreme temperatures and maintain a high humidity by preventing evaporation. The spaces beneath cover objects provide insects a place to escape from predators and environmental extremes, especially in grasslands. When looking for insects under rocks and logs, it is a good practice to return such cover objects to their former position so that other insects can find shelter.

Insects and Other Small Organisms of the Litter and Soil

The litter of plant debris, soil surface, and the soil itself are inhabited by at least some members of all insect orders except the Ephemeroptera and all orders of hexapodous arthropods. Also found here are many other small organisms: annelids, bacteria, fungi, mollusks, nematodes, protozoa, and terrestrial arthropods (arachnids, centipedes, isopods, millipedes, pauropods, symphylans). Indeed, in a Brazilian rain forest, about three-fourths of the total animal biomass was found to be small, soil-dwelling animals. In this and most other terrestrial habitats, social ants and termites dominate the animal community in terms of number of individuals and biomass (Wilson, 1990).

After termites and ants, the next in abundance in soil are Acarina, Collembola, all stages of Coleoptera, immature Diptera, and immature Lepidoptera, especially pupae. The following insects and small arthropods, although generally less numerous, spend their entire lives in litter, soil, or in crevices under cover objects: Archeognatha, some species of Blattodea, Dermaptera, Embiidina, Grylloblattodea, Orthoptera, Thysanura, and Zoraptera; and all species of the arthropod orders Diplura, Pauropoda, Protura, and Symphyla. Some other orders are represented in soil, at least by the immature stages, that feed as scavengers or predators or that feed on algae, fungi, or mosses: Coleoptera, Mantodea, Mecoptera, Neuroptera, Psocoptera, Raphidioptera, Siphonaptera, and Thysanoptera. Parasitoids and parasites, including Strepsiptera, attack other litter insects. Phthiraptera are found as parasites on ground-dwelling birds and mammals.

Some species of insects that are ordinarily considered aquatic can be found in terrestrial habitats such as pockets of decayed leaves, moist litter, or soil away from the water. The exceptions are immatures of some species of Odonata, Plecoptera, and Trichoptera. The larvae of the damselfly *Megalagrion peles* (Odonata) of Hawaii regularly live in damp leaf mold caught in leaf axils. Several species of stoneflies in the Southern Hemisphere live far from water on cold, wet mountains. The larvae of the caddisfly *Enoicyla pusilla* (Trichoptera) inhabit forest litter in Europe.

Immobile stages of insects are commonly spent on the ground. Eggs are often deposited here: Phasmatodea drop eggs in litter, and Acrididae lay eggs in soil. Lepidoptera pupae and the resting "pupae" of Thysanoptera are often in litter. Periods of quiescence or diapause also may be spent among dead leaves. Aggregations of diapausing adults of the lady beetle, *Hippodamia convergens* (Coccinellidae), are found on the ground in Sierra Nevada of California and higher parts of the Coast Ranges from June to February.

Adaptations of Insects in Litter and Soil

Insects that move about on the surface of the litter or soil are exposed to predation by visual hunters such as birds. Such insects may have well-developed compound eyes; circadian rhythms in behavior, often with nocturnal activity; and bodies pigmented with concealing patterns. Insects at the surface also are generally resistant to desiccation. Respiration is by tracheae, and the spiracles are usually fitted with closing devices that conserve water.

The active stages of insects and hexapodous arthropods in litter and crevices in soil often are somewhat dorsoventrally compressed and parallel-sided or anteriorly narrow, a form that facilitates movement among varied obstacles. Most are agile and possess well-developed means for locomotion

on ground. Escape by jumping is often a characteristic (Archeognatha, Collembola, Orthoptera, and Schizopteridae in Hemiptera). The antennae and tactile sense organs are usually well developed.

Large, membranous wings are apparently an encumbrance in narrow passageways. Except for mating and dispersal, flight is an unnecessary investment of energy and structure. The Embiidina spin silken runways to move out from cover objects into the litter. The wings of male web spinners, if present, are flexible and bend freely inside the tunnels. The Apterygota, Collembola, Diplura, and Protura are, of course, wingless, and the immature pterygotes have small wing pads at most. Among adult pterygotes, some have protective forewings (Coleoptera, Dermaptera, and Orthoptera,); others fail to develop wings (Dermaptera, Coleoptera, Grylloblattodea, Orthoptera, some Psocoptera, and the sterile castes of Isoptera and Formicidae) or discard them after mating and dispersal (Zoraptera, reproductive castes of Isoptera and Formicidae).

Females of most soil-dwelling arthropods other than insects acquire sperm indirectly from the male without copulation. The male produces a spermatophore, which is placed on the ground or a special web and later picked up by the female. Thanks to the generally moist environment, the sperm are not in danger of desiccation for at least a short period. Among the insects and related arthropods, this behavior is known in the primitive orders Archeognatha, Collembola, Diplura, and Thysanura. Ancient arthropods were aquatic, and sperm was transferred in water without copulation. During the early evolution of terrestrial arthropods from aquatic forms, litter and soil evidently provided a moist place where sperm could continue to be transferred indirectly. Other insect orders (Pterygota) transfer sperm directly from the male genitalia to the female reproductive tract through copulation, thus protecting the sperm and allowing a fully terrestrial existence. Odonata are an exception; males place the spermatophore in a secondary organ on their bodies (Fig. 19.3).

Subterranean Insects and Other Organisms

The high humidity of soil air and films of soil water and the moderate temperature of soil at greater depths create conditions approaching an aquatic environment. Many invertebrate phyla are represented in soil by minute aquatic forms found elsewhere in freshwater. Protozoa, for example, are the most numerous microscopic animals in moist soil. With increasing depth below the soil surface, the variety of arthropod taxa abruptly declines. In the soil air of pore spaces live species of the minute arthropods Acarina, Collembola, Diplura, Pauropoda, Protura, and Symphyla. The underground nests of ants and termites are found here, each populated with large numbers of individuals. Other insects deep in soil include some species of Coleoptera, Diptera, Hemiptera, Hymenoptera (certain bees and wasps), and Orthoptera. Of these, the Coleoptera are the most numerous in species. All stages of the life cycle of Coleoptera can be found in soil (Fig. 8.1c,d), but generally only the immature stages of Diptera are present. Many species of bees and predatory wasps excavate simple burrows in earth that are stocked with provisions for their young. Roots are the food of nymphal Cicadidae (Fig. 8.1a) and certain other subterranean

Hemiptera. Orthoptera are represented by the omnivorous Stenopelmatinae (Gryllacrididae) and by mole crickets (Gryllotalpinae, Gryllidae; Fig. 8.1b) that feed on roots but also take insect larvae.

Soil dwelling termites and ants make deep nest galleries, often into the subsoil and below. Their extensive excavations and the organic matter brought into the nest contribute substantially to soil development and drainage. In the tropics and some arid regions, they are of primary importance as soil builders.

Large earthen mounds built by termites in Africa, South America, and Australia are conspicuous and long-lasting features of the landscape (Fig. 8.2).

In southern Africa, mounds still occupied by termites have been carbon dated at 4000 years old. In warm grasslands termites are among the primary decomposers of plant litter and animal dung. In some areas termites are so numerous that they influence the regional types of soil and vegetation. The mounds are rich in organic matter from the termite's feces and in minerals from soil

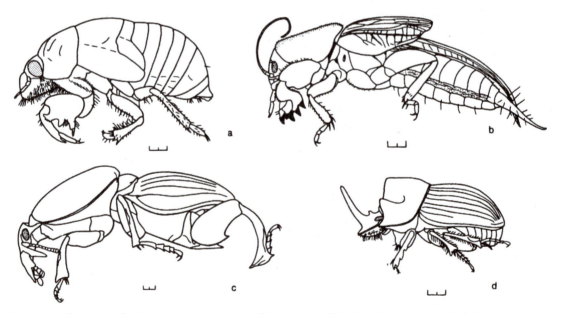

Figure 8.1 Examples of soil insects: a, cicada nymph, *Okanagana* sp. (Cicadidae); b, mole cricket, *Gryllotalpa* sp. (Gryllotalpinae, Gryllidae); c, unusual soil-dwelling cerambycid beetle from Brazil, *Hypocephalus armatus*; d, dung beetle, *Copris lugubris* (Scarabaeidae). Scales equal 2 mm.

Figure 8.2 The large size of this mound made by fungus-growing termites, *Macrotermes* sp., in Africa exemplifies the diligence, long life, and importance of termite colonies in the African savanna.

excavated below. Abandoned nests resist decay so that nutrients are locked up for many years. Moreover, in living colonies carbon is lost to the atmosphere as carbon dioxide and as methane gas produced by intestinal symbionts.

Subterranean nests of *Atta* ants in the American tropics may be enormous and reach 3 to 6 m in depth. Earth brought to the surface by ants in one nest weighed 44 tons. In prairies of North America, excavations by ants are less impressive but significant in nutrient dynamics, soil drainage, and vegetation.

Adaptations of Subterranean Insects

An important factor limiting insects deeper in soil is the obvious difficulty in moving about. Unlike small arthropods that live in pore spaces, most soil insects are not small enough to freely negotiate interconnecting pore spaces for any distance. They must either tunnel by pushing aside particles and

squeezing through or excavate the soil in front and deposit it behind. The physical resistance of the soil, therefore, raises the energetic cost of locomotion and reduces the distance to which an individual can hunt for mates or food. Sight and windborne odors are also eliminated as aids in the search. For this reason, mating of pterygote insects is usually accomplished above ground. Except for the roots of green plants, insects must find their food in the form of other soil organisms, dead or alive. Active excavators such as ants, termites, silphid beetles, larval cicindelids, wasps, and bees obtain their food from energy-rich sources above ground.

Reduction in wings is a characteristic both of insects that live deeper in soil as well as of those that live in litter. Burrowing insects are generally round in cross section but are greatly varied in shape otherwise. "Tunnelers" have reduced or no legs and are of two kinds. Larval Elateridae (Coleoptera) and Therevidae (Diptera) are smooth, stiff-bodied, and slender. By sinuous movements of the trunk, they force their hard, tapered heads forward through the substrate. In contrast, larval Tipulidae and Bibionidae (Diptera) are soft-bodied and use peristaltic movements to penetrate soil. "Excavators" may be fairly thick-bodied, armored, and often exhibit conspicuous molelike modifications of the head and forelegs for digging. Such fossorial adaptations are seen in Gryllotalpinae (Orthoptera), Cydnidae and nymphal Cicadidae (Hemiptera), and Scarabaeidae (*Copris, Geotrupes;* Coleoptera), to list only a few (Fig. 8.1). Worker ants and termites are not so conspicuously equipped for digging, but by their great numbers and coordinated activity, they use their strong mandibles to remove and carry away immense quantities of particles.

In the absence of light, insects and the hexapodous arthropods that live deeper in the soil tend to have reduced or no compound eyes; antennae variably developed, even absent (Protura); well-developed tactile sense organs; and pale pigmentation of the integument without colored patterns. They are usually repelled by light, higher temperatures, and humidities of less than 90 percent. The upper limits of tolerance to temperature are relatively low, but some soil insects can remain active a few degrees above 0°C. Rhythmic behavior has been poorly studied. Vertical migration in response to daily temperature fluctuations may occur if the soil is exposed to solar radiation. Resistance to desiccation is low, probably because the epicuticle becomes abraded by soil particles or is naturally permeable to water. Some soil insects and arthropods respire cutaneously. Those that normally inhabit the deeper subsoil are tolerant of higher levels of carbon dioxide, at least for several days (some Collembola and larvae of the beetles Elateridae and Scarabaeidae).

THE CAVE HABITAT

In addition to tiny pore spaces in soil, larger cavities exist much deeper below the soil layers that are inhabited by insects and other organisms. Such cavities range from networks of air-filled spaces in rock, measuring a few millimeters in width, to giant caverns. Both terrestrial and aquatic habitats exist here because cavities below the water table are filled with still or running water. The full extent of the cavernicolous, or cave-dwelling, biota remains unknown because only a few cavities have openings to the surface and still fewer are large enough for a person enter. Where access is limited, samples of the underground aquatic fauna can be taken from wells and springs.

Underground cavities are commonly found in three geologic situations. (1) Classic caverns exist where limestone is dissolved by carbonic acid produced when water is charged with carbon dioxide. An extensive area with underground streams, sinkholes, and caves is called a karst topography. These systems may be millions of years old. (2) Volcanic deposits have cavities formed by gas bubbles or by gaps in the lava flow, or by lava flowing in covered channels that may leave large empty tubes many kilometers in length. These are generally a few thousand years old. (3) Cavities or fissures also are found in broken rock strata or talus and are of variable ages.

Insects of the Cave Entrance

The entrance to a cave is a transition zone between the outer world and the cave beyond. The humidity is often higher and the temperature less varied, and partial darkness prevails. Animals here may be only temporary residents or have various degrees of adaptation to life in caves. They are often the same species that normally are found in moist litter, dark crevices, or the soil outside. Representatives from many orders and families can be found here. Flying insects, especially flies (Culicidae, Mycetophilidae), seek shelter at entrances during the day. The famous New Zealand glowworm is the cavernicolous larva of the fungus gnat, *Arachnocampa luminosa* (Mycetophilidae). These luminous predators catch small flies by secreting sticky threads that hang from the ceiling of the cave. Bodies of water within the entrance to a cave may have the usual aquatic insects. However, some species of Trichoptera and Ephemeroptera are rare outside of caves and may be less pigmented than their normal relatives.

Insects of Deep Caves

Beyond the entrance, food must come ultimately from the surface. The absence of green plants except for their roots largely limits cave-dwelling insects to scavenging, parasitism, or predation. Phytophagous insects do exist where roots are available. Otherwise, new organic matter must be washed in or be deposited as waste by organisms that enter and leave. Bats attract several taxa of parasite (Polyctenidae, Streblidae, and Nycteribiidae; some Siphonaptera and Cimicidae), and the accumulated feces of bats (called guano) attracts scavengers (Arixeniina, Dermaptera; Blattodea; Coleoptera; Diptera, etc.).

The environment deep in a cave is stable and usually has the following conditions: darkness is total, temperature is constant and near the average annual surface temperature, the air is saturated with water vapor, the walls are covered with water, and oxygen is lower and carbon dioxide higher than in the atmosphere above ground. Even though they may be terrestrial animals, deep cave inhabitants live in an essentially aquatic environment with the saturated atmosphere and wet walls. Animals adapted to continuous life in caves and that reproduce there are called troglobites.

In continental caves, terrestrial hexapodous troglobites are frequently Collembola (Entomobryidae), Diplura (Campodeidae), Orthoptera (cave crickets, Rhaphidophoridae, Fig. 8.3a), and

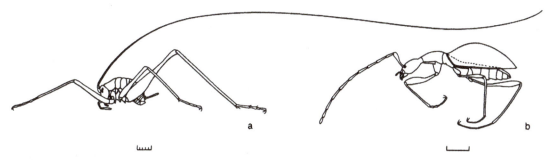

Figure 8.3 Examples of cave insects: a, cave cricket, *Tropidischia xanthostoma* (Gryllacrididae; scale equals 5 mm); b, blind cave beetle, *Glacicavicola bathyscioides* (Leiodidae; scale equals 1 mm).

especially Coleoptera (pselaphine Staphylinidae; trechine Carabidae; and Leiodidae). Icy lava tubes of Idaho have leiodid beetles appropriately named *Glaciacavicola bathyscioides* (Leiodidae, Fig. 8.3b).

Terrestrial troglobite arthropods exhibit special modifications. On average, individuals of cave species tend to be larger than their close relatives outside. Cave crickets and trechine Carabidae have more slender bodies and longer appendages. Antennae are generally longer, sometimes exceeding the length of the body in Collembola. Beetles may have erect, long hairs scattered over the body that are probably tactile. Most troglobites are repelled by light. The tracheal system and spiracles of trechine carabids are rudimentary; respiration probably takes place through the membranous abdominal terga. Resistance to desiccation and higher temperature is low. Long periods without food are tolerated, and growth is slow.

Activity and rest in bathysciine beetles do not follow a circadian rhythm but are associated with temperature fluctuations. Some bathysciine beetles that live far from the entrance, where food is rare, have larvae that do not feed. The female lays one egg at a time; each is large and full of yolk. Shortly after hatching the larva builds a clay capsule and remains inside 5 to 6 months, then pupates. The adult searches widely for food in a manner not possible for a young larva. Adults live for about 3 years.

In the Hawaiian Islands, an amazing fauna has been found in lava tubes, including lycosid spiders, amphipods, centipedes, millipedes, predatory earwigs (*Anisolabris,* Carcinophoridae, Dermaptera), omnivorous crickets (*Thaumatogryllus*

and *Caconemobius*, Gryllidae, Orthoptera), predatory bugs (*Nesidiolestes*, Reduviidae, Hemiptera), and scavenger bugs (*Speleovelia*, Mesoveliidae, Hemiptera). Phytophagous planthoppers (*Oliarus*, Cixiidae, Hemiptera) and caterpillars (*Schrankia*, Noctuidae, Lepidoptera) feed on roots hanging from the roof. The cixiid bugs have special tarsal claws and empodia for crawling on wet walls. Many of these insects are flightless and have reduced or no eyes and the pale bodies of true cave animals, yet some species may have evolved from surface-dwelling ancestors in less than 100,000 years.

Aquatic troglobite insects have been discovered in underground waters throughout the world, including species of aquatic Coleoptera (Dryopidae, Dytiscidae, Elmidae, Hydrophilidae, Noteridae) and Hemiptera (Nepidae). The beetles are distinguished by reduced or no eyes; reduced pigmentation; a thin, soft exoskeleton; and reduced or no hind wings. The water scorpion (Nepidae, Hemiptera) from Romania is also blind.

Evolution of Cave Faunas

The restriction of troglobite insects to caves has stimulated research into their evolutionary origin and geographic distribution. No single explanation of their evolution is adequate for all species, but a common element in their history is that initially they were adapted to life in cool, moist, dark places like leaf litter or crevices in soil. This preadaptation permitted them to enter caves where they exploited new food resources and had less threat from natural enemies and competition. In Hawaii, for example, evolution of cave dwelling is believed to be a continual process as species take advantage of new food resources in caves. The closest relatives of Hawaiian troglobites are species living nearby on the surface. In contrast, most troglobites in continental North America inhabit caves where colder climates and glaciation occurred in the Pleistocene Epoch (0.01–1.64 million years ago). Under these circumstances, insects became adapted to cold, humid, forest litter near the edges of glaciers. At that time glaciers extended to lower latitudes and altitudes. When the glaciers receded, some cool-adapted insects found refuge in caves, while the remainder of the fauna moved to higher latitudes and altitudes or became extinct. Therefore, the closest relatives of the troglobites are now far away or extinct. The scarcity of food and lower temperatures of caves in the Temperate Zone places strong selection pressure on efficient use of food at a low metabolic rate. In separate, isolated cave systems, insects independently evolved economizing adaptations: reductions of eyes, wings, flight muscles, pigmentation, and, in some beetles, number of offspring.

THE DESERT HABITAT

In contrast to life in well-watered regions of the earth, deserts have low, often unpredictable rainfall and long droughts; dry winds and abrasive sandstorms; sparse vegetation and barren sandy or rocky soil; and periodic deficiencies of nutrients and are often extremely hot with intense solar radiation. Despite this harsh environment, insects have become among the most successful of deserticolous animals. This is largely due to water-conserving features such as their waxy epicuticle, closable spiracles, and the production of uric acid by their excretory system. In addition, they avoid daily and seasonal environmental extremes by their behavior and life history. Larval growth and activity of adult insects are often correlated with the rainy season in deserts.

Certain insects are unusually tolerant of heat, perhaps because they have enzymes that are not denatured at higher temperatures or epicuticular waxes that do not melt. However, they can use evaporative cooling for only short periods because the loss of water would be too great for organisms of such small body size. The most extreme case is the ant, *Ocymyrmex barbiger*, which forages in the sun at temperatures up to 67°C on the surface of the Sahara Desert. Single workers move quickly in search of heat-killed arthropods before returning to the safety of their subterranean nest.

Many insects depend on the sandy substrate for protection against lethal temperatures and desiccation. During the hot season many forage openly only at night, spending the day in burrows or simply immersed in sand. Surface temperatures as high as 72°C and humidity as low as 10 percent are avoided by the desert cockroach, *Arenivaga investigata* (Polyphagidae, Blattodea). At depths of 30 to 45 cm in sand, the roach inhabits a zone where daily temperatures are well below the upper lethal limit (about 45.5–48.5°C) and humidity may reach above 82 percent, a level at which the roach can actually absorb water vapor. Like some tenebrionid beetles, the roach literally swims in sand. The body is biconvex in cross section, oval, and smooth and has sharp lateral margins, resembling that of an aquatic beetle. Insects that run on the surface during the day may have long legs that lift the body well above the hot surface (Mutillidae, Formicidae, Tenebrionidae).

Tenebrionid beetles thrive in deserts as scavengers (Fig. 8.4). Here they are important decomposers of plant litter because bacteria and fungi are limited by aridity. At night, adults of *Onymacris unguicularis* stand on sand dunes in the fog, head down, and drink the water that condenses on their bodies and runs to their mouth. Although most desert tenebrionids are flightless, they retain the elytra, which are fused together to enclose an air space over the abdomen. Air is released by the abdominal spiracles into the space and escapes only at the tip of the abdomen, thus reducing the loss of water vapor.

THE HIGH-ALTITUDE HABITAT

Terrestrial animals and plants are generally confined to the dense, oxygen-rich atmosphere of low elevations. By comparison, the highland, or high-altitude environment of mountains is extremely inhospitable to life. It is the highest habitable region on earth and borders on outer space. Nevertheless,

Figure 8.4 Tenebrionid beetles of the Namib Desert in Africa. a, Species active by day run fast on the hot surface of the desert. b, These dune beetles are active at night and escape the heat of day by burrowing in sand. The beetles are disc-shaped and covered with fine hairs.

decrease in atmospheric pressure are decreased air temperature and a decreased content in air of oxygen, carbon dioxide, and water vapor. With fair weather, the thin transparent atmosphere transmits intense solar radiation (especially ultraviolet) in daytime and allows rapid loss of radiation in the shade and at night. Over a 24-hour period the temperature at the soil surface fluctuates over a wide range. The air can be exceedingly dry so that evaporation is high. The growing season is short in summer, and frost is common. At still higher altitudes, the average air temperature never rises above freezing and snow remains on the ground throughout the year.

Among the permanent animal residents, or hypsobionts, of the highlands are both terrestrial and aquatic insects. In the Himalayas near the highest elevations inhabited by arthropods, Collembola are dominant. At altitudes up to 6300 m, they occur in enormous numbers on snow and ice, in glacial waters, on rocks, and on plants that form flat cushions. Feeding on windborne pollen and spores, the Collembola are at the base of a food pyramid that supports insect predators and scavengers, especially on snow. Collembola have been seen crawling about at –10°C.

In cold glacial torrents and glacial lakes at lower elevations are found the immature instars of Ephemeroptera, Plecoptera, aquatic Coleoptera (Dytiscidae, Hydrophilidae), and aquatic Diptera (Blephariceridae, Chironomidae, Deuterophlebiidae, and Simuliidae). The adults that emerge and mate later become food for scavenging insects and birds. Terrestrial insects include Orthoptera (Acrididae), Dermaptera, Lepidoptera (Papilionoidea, Geometroidea, Noctuoidea), and many Diptera (Syrphidae, Stratiomyidae, Empididae, Rhagionidae, Asilidae, Dolichopodidae, Calliphoridae, Anthomyiidae, Muscidae, and Tachinidae). Active on snow are the wingless snow tipulid, *Chionea* (Tipulidae, Diptera), and snow "flea," *Boreus* (Boreidae, Mecoptera; Fig. 8.5). The small order Grylloblattodea is found primarily in highland environments, usually under rocks near snow. Archeognatha have been found basking on rocks at altitudes up to 5800 m in the Himalayas. The most important in terms of the numbers of

relative to other animals, arthropods have been among the most successful in living permanently at high altitudes. The highest elevation at which arthropods have been found is 6700 m, and the prize goes to the spiders. Vertebrates are poorly represented at high altitudes, and perhaps relief from their predation has permitted the arthropods to flourish.

The highland is generally the treeless area above the timberline at altitudes above about 2000 m. An important difference between the lowland and the highland is that the latter has a progressively thinner atmosphere. Correlated with the

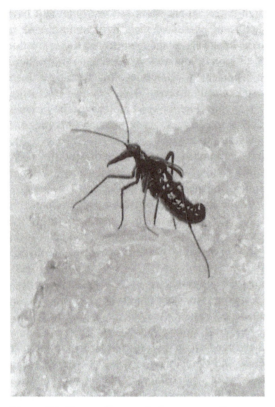

Figure 8.5 A snow flea or wingless scorpionfly, *Boreus* sp. (Boreidae), is active on snow in winter at higher elevations.
Source: Photo by H.V. Daly.

different genera and species is the Coleoptera, and especially the Carabidae. Other common beetle families are Staphylinidae, Histeridae, Tenebrionidae, Chrysomelidae, and Curculionidae.

Unlike birds and mammals, which must have adequate oxygen to generate body warmth, the most specialized hypsobiont insects have evolved a metabolism that functions normally at low temperatures in an oxygen-deficient atmosphere. They are closely associated with snow, which enables them to avoid the drying effect of higher temperatures. Active only in a narrow range of temperatures near 0°C, they die quickly if exposed to temperatures slightly above normal. For example, aquatic instars are active in water between –1.5 and 5°C and die if held in a human hand.

In winter, inactive hypsobiont insects are able to survive prolonged exposure to low temperature. Some are physiologically cold-hardy (see Chap. 5), but even without this mechanism, insects sheltered under insulating snow or protected in crevices can survive a few degrees below freezing without harm. As the short summer season progresses and patches of snow melt, terrestrial insects exhibit tolerance to higher temperatures and move away 20 to 30 m from the edge of the snow onto warmer soil. Here their development accelerates and the life history is completed.

In comparison with related lowland species, hypsobiont insects are usually more darkly pigmented with black, brown, or reddish tones. Dense melanic pigments presumably prevent injury of soft tissues by the strong ultraviolet radiation. The melanism also aids in absorbing warmth from the sun. Other common features of hypsobiont insects in comparison to lowland forms are univoltine life history; slower growth and smaller body sizes of adults; mainly ground-dwelling habits with a reduction or loss of wings; and increased density of body hairs and scales.

Most terrestrial insects in the highland are predators or scavengers. Much of their food is transported from lower elevations by wind. Many kinds of insects and hexapodous arthropods plus pollen, spores, and bits of vegetable matter are deposited in this manner on the soil, rocks, or snow at high altitudes. Newly arrived living insects are numbed by the cold and become prey to the resident insects and birds. Plant life is generally sparse, but flowering plants are visited by flies, moths, and butterflies. Bees and bee-pollinated plants are usually absent.

THE AQUATIC HABITAT

In this section we discuss insects that spend at least the immature stages of their lives floating or submerged in water. All aquatic insects evolved from terrestrial ancestors, and nearly all still have significant phases of their lives in terrestrial or aerial habitats. Aquatic species are a minor fraction of all insects, probably numbering no more than 3 to 5 percent of all species. Yet these insects are taxonomically diverse and fascinating in structure (Fig. 8.6) and biology, and some of them, such as

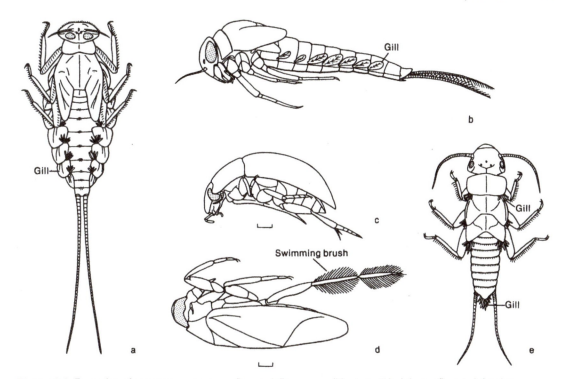

Figure 8.6 Examples of aquatic insects: a, mayfly naiad, *Epeorus* sp. (Heptageniidae); b, mayfly naiad, *Ameletus* sp. (Siphlonuridae); c, water beetle, *Tropisternus ellipticus* (Hydrophilidae); d, back swimmer, *Notonecta undulata* (Notonectidae); e, stonefly naiad, *Acroneuria pacifica* (Perlidae). Scales equal 1 mm.

mosquitoes, are of extreme importance in public health.

The limited number is probably the result of the limited amount of freshwater habitat in comparison with the land surface. On a global scale the total extent of freshwater is remarkably small. A few insects are truly marine. Some others live in the intertidal zone, in brackish coastal waters or saline lakes or hot springs (ephydrid flies in Iceland at 47°C). Most aquatic insects, however, are restricted to freshwater, where they are among the most important organisms in the aquatic ecosystem. As herbivores, predators, scavengers, and even parasitoids, insects are consumers in the food chain and, in turn, are food for fish, birds, and amphibians. Insects are important as decomposers of tree leaves and other plant material that fall into running water.

The most important limitation for submerged insects is a supply of oxygen. Oxygen readily dissolves in water, but at concentrations far less than that of air. Once dissolved, its movement in still water by diffusion alone is exceedingly slow, hence currents and turbulence in water are critical to the distribution of oxygen throughout the habitat. Respiration of aquatic insects is discussed in Chapter 5.

Aquatic insects also must regulate the balance of water and salt in their body fluids. The waxy epicuticle, evolved to prevent water loss in air, now operates in reverse to prevent an influx of water. Excretory organs are able to produce urine more dilute than hemolymph, thus preserving valuable salts. Ions may be absorbed from water by active transport through special areas of the cuticle, for example, the rectal gills of mosquito larvae. With abundant water for dilution, some species excrete nitrogenous waste in the form of ammonia.

Marine Insects

Whereas insects and other hexapodous arthropods constitute nearly three-quarters of the earth's animal species, they are greatly reduced in number of species in the earth's largest habitat, the ocean. By no means are insects scarce as individuals on the coasts or even far out at sea. The beach or littoral zone has abundant insect life, especially Coleoptera and Diptera. Each marine species has overcome certain physical barriers such as:

- **Tidal submergence.** Examples: the collembolan *Anurida maritima*; insects in several orders, including the bug *Aepophilus bonairei* (Saldidae, Hemiptera) of England and Europe.
- **Wave action.** Examples: especially midges of the genus *Clunio* sp. (Chironomidae, Diptera) and the barnacle-eating larvae of the fly *Oedoparena glauca* (Dryomyzidae, Diptera).
- **Salinity.** Examples: especially salt-marsh mosquitoes (Culicidae, Diptera), shore flies (Ephydridae, Diptera), water boatman *Trichocorixa* sp. (Corixidae, Hemiptera), and, in Australia, the intertidal rock-pool caddisfly larva of *Philanisus* sp. (Trichoptera).
- **Depth**. Examples: the midge, *Chironomus oceanicus* (Chironomidae, Diptera), dredged from 36 m and another midge, *Pontomyia* sp., from 30 m.
- **Life offshore.** Examples: five species of the pelagic water striders, *Halobates* sp. (Gerridae, Hemiptera), which occur hundreds of miles from land.

Of these barriers, perhaps the most limiting is the inability to respire for long deep beneath the surface where oxygen may be deficient. No marine insect is known to spend its entire life under water. Insects in general also have not been successful in colonizing deep freshwater lakes. Chironomid fly larvae have been found below 1300 m in Lake Baikal of Russia, but the life history is not completed there. Only the peculiar stonefly, *Utacapnia* sp. (Capniidae, Plecoptera), in Lake Tahoe, California, can exist indefinitely in deep, still water.

Added to the physical barriers is the intense biological competition from marine arthropods and other organisms. During the Paleozoic Era (570–245 million years ago), while insects flourished on land, the marine fauna was diversifying and occupying the habitats of the sea. The oceans now seem effectively closed to insects.

Insects of Lotic Water

Freshwater habitats can be divided into running, or lotic water, and still, or lentic water. Although the total volume of lotic water in the world is less than that of lentic water, the great river drainages exist for a million years or more. This provides a continuously habitable environment for aquatic organisms over a wide geographic area. Because major drainages are more or less permanent and are otherwise favorable, insects have evolved the greatest variety of taxa in lotic waters. Indeed, most kinds of aquatic insects are represented, and some occur only here. This is especially true in the rapidly flowing, shallow, rocky streams of the headwaters, where erosional processes predominate, oxygen concentration is high, and the mean monthly temperature seldom exceeds 20°C.

Water seeping into a stream from the ground or issuing from springs is commonly cooler than air in summer and warmer than air in winter. This difference may be sustained for some distance if the stream is shaded in summer or insulated by a snow cover in winter. Although subject to daily and seasonal fluctuations, the range of variation is usually less than that of the shallow regions of lakes. Furthermore, cooler water absorbs much more oxygen than warmer water (e.g., water saturated at 4°C holds 1.5 times the amount held at 22.5°C). Turbulence over the rocky bed maintains dissolved oxygen at the point of saturation and distributes it to the depth of the stream. In summer, algae and other aquatic plants thrive in clear, shallow water. In autumn, deciduous trees along streams drop leaves that provide nutrients into the water. Consequently, with food, oxygen, and moderate temperatures, insects are able to feed and grow throughout the year, even in winter in temperate regions.

The immatures of Ephemeroptera, Plecoptera, Trichoptera, and Megaloptera are much more numerous in streams than in lakes. Among rocks in swift, cool waters are also larvae of the fly families Blephariceridae, Simuliidae, and Deuterophlebiidae and the beetles Dryopidae, Elmidae, Ptilodactylidae, and Psephenidae. Larvae of the odonates Agrionidae, Cordulegastridae, and Gomphidae also favor running water. Among aquatic holometabolous insects, Trichoptera and some Diptera remain in water as pupae, whereas Megaloptera, aquatic Coleoptera, and some other Diptera leave the water to pupate in soil or litter.

Adaptations of Insects in Lotic Water

Organisms associated with the bottom of a body of water are often referred to as the benthic community or benthos. In strongly flowing streams, an ever-present hazard to benthic insects is to be swept off the bottom and drift downstream exposed to predators. The number of insects drifting usually rises at night, because many are nocturnally active. Drifting may involve only short distances and not be entirely deleterious. Competition for space and food is relieved, and the population is more evenly distributed downstream. Evidence is inconclusive that adults tend to fly upstream to lay their eggs and recolonize the uppermost waters. The eggs of mayflies (Ephemeroptera), stoneflies (Plecoptera), and some dragonflies (Odonata) are equipped to sink rapidly, and firmly attach to the bottom, thus remaining near the ovipositional site.

Insects that inhabit swift water are generally prevented from visiting the surface for air. Their respiratory structures are tracheal gills or plastrons (Chap. 5) that function while bathed in oxygen-rich water. Larvae of some mayflies and stoneflies that normally reside in still pools create their own currents. The abdominal gills of mayflies have an undulating beat that circulates the surrounding water. Similarly, stoneflies pump up and down on their legs at a rate proportional to the oxygen need. Insects living in the swiftest waters do not exhibit such respiratory movements and depend entirely on the current to bring oxygen. They die quickly even when placed in still water of the same temperature and oxygen saturation because the oxygen supply becomes depleted in the proximity of the gills and is not renewed by currents.

A number of similar morphological adaptations evolved independently in various insects in torrential waters. The body may be dorsoventrally compressed so that the insect can creep beneath stones or in crevices out of the current. Larvae of the mayflies Heptageniidae (Figs. 8.6a, 18.4c) and larval beetles of Psephenidae are thus flattened (Fig. 37.10c). Small size also permits insects to enter small spaces protected from the current. Nearly all adult beetles in swift water are small.

Various attachment devices allow insects to cling to the substrate. The beetles Elmidae and Dryopidae have the last tarsal segment and claws enlarged for anchoring themselves. In addition, the larvae of Elmidae have prehensile hooks near the anus. Stout tarsal claws and anal hooks also are found on the free-living caddis larvae of Rhyacophilidae (Trichoptera, Fig. 44.2a) and the hellgrammites or larvae of Corydalidae (Megaloptera). Lacking true legs, the aquatic larvae of the fly families Simuliidae, Deuterophlebiidae, Blephariceridae, Chironomidae, Rhagionidae, Empididae, Syrphidae, Ephydridae, and *Limnophora* (Anthomyiidae) have developed leglike prolegs for attachment. The prolegs are usually fitted with fine hooks. Suckers are developed near the anus of larval simuliids and on the ventral surface of *Maruina* (Psychodidae) and Blephariceridae. Frictional resistance to detachment is developed by ventral gills or hairy pads on the mayfly naiads *Rhithrogena* (Heptageniidae), *Ephemerella doddsi*, and *E. pelosa* (Ephemerellidae), the African *Dicercomyzan* (Tricorythidae), and the larval beetles of Psephenidae (Fig. 37.10c). Silk and sticky secretions are used for attachment by some Chironomidae and the caterpillars of Pyralidae (Lepidoptera). Larval simuliids move about with their prolegs on silken mats. *Hydropsyche* (Trichoptera; Fig. 44.3e) spin nets in the current and harvest waterborne food. Most Trichoptera attach their pupal cases to rocks with silk. The stone cases built by Trichoptera in streams are usually composed of heavier grains and a few heavy pebbles (Fig. 44.4). This provides ballast so that the caddis larva quickly sinks into still water if dislodged.

Few insects actively swim against the current. The streamlined, minnowlike mayflies *Ameletus* (Fig. 8.6b), *Isonychia*, *Centroptilus*, and some *Baetis* move in and out of the current with powerful, up-and-down movements of their caudal filaments. On the surface of riffles are the bugs *Rhagovelia*. Leaves packed against obstructions and wood debris have a special fauna of mayflies, caddis flies, and stoneflies. The stoutlimbed *Ephemerella* (Ephemerellidae, Fig. 18.2a) is commonly found here.

Insects in crevices can utilize the rich supply of oxygen and food of fast-running streams without the immediate danger of the current. Flattened mayflies are also found in the quiet water beneath rocks. Deeper still, in the sand or gravel soils under the stream that are saturated with water is hyporheic zone. Here are found tiny young instars of insects, the narrow-bodied stone flies and mayflies, and crane fly larvae, *Antocha* and *Hexatoma* (Tipulidae, Diptera). When the stream is dry, some immature aquatic insects can survive buried in the stream bed.

The lower region of a drainage, or where it flows through relatively flat land, is characterized by slower flow, greater depth, occasional low oxygen concentrations, and monthly mean temperatures close to that of air and rising above 20°C. Depositional processes now predominate so that the bed of the river is formed by fine sediments of sand and mud.

Fewer kinds of insects are specialized for life here, but some species are extremely abundant. Larvae of the burrowing mayflies Ephemeridae (Fig. 18.4b) excavate U-shaped burrows in silt. Undulating movements of their feathery gills circulate water in the burrow, bringing in organic matter, which is eaten, and oxygen. Other mayflies, such as Caenidae, Tricorythidae, and some Ephemerellidae, inhabit silty bottoms of streams and rivers. These have the functional gills kept clean of particles by the thick uppermost gill. The body surfaces of some bottom-dwelling mayflies, stoneflies, and dragonflies are covered by a protective layer of fine hairs. Larvae of chironomid midges are often abundant in soft, mucky bottoms where they build tubes.

Insects of Lentic Water

Lakes are relatively short-lived in comparison to rivers because they normally fill with sediments during prolonged wet periods or evaporate during prolonged dry periods. Most lakes existing today originated in the Pleistocene Epoch (0.01–1.64 million years ago) and are less than 25,000 years old. Only a few large, deep lakes are much older. Lake Baikal of Russia and Lake Tanganyika in Africa probably existed in the Tertiary Period (1.64–65 million years ago).

In contrast to lotic water, the insect life of lentic water is influenced more by temperature and oxygen supply than by water current. Temperatures in shallow, marginal waters of lakes or small bodies of water such as ponds tend to fluctuate with air temperature. Large bodies of still water will warm slowly in summer, becoming stratified into an upper, warmer layer and deeper, colder water that may be 4°C near the bottom. In winter, lakes cool slowly. Insects in lakes are protected against freezing because water reaches maximum density at 4°C and is less dense at lower or higher temperatures. Surface water cooled to 4°C sinks. Ice forms at the surface after deeper water is uniformly 4°C and the colder, lighter water is at the top.

Biological communities of lakes can be divided into three zones: the littoral zone or nearshore, shallow waters where light reaches the bottom; the limnetic zone or offshore, open water to the depth of light penetration; and the profundal zone or deep, offshore water below the penetration of light.

Oxygen is absorbed at the water's surface and is produced by photosynthetic plants in the littoral and limnetic zones. In summer the oxygen supply of the littoral zone may exceed saturation, but the concentration is inversely proportional to temperature. The supply varies daily because at night oxygen is consumed by both plants and animals and not replaced. In the deep profundal zone, oxygen is often depleted by bacterial respiration.

Lakes that are poor in nutrients are called oligotrophic. Young, deep lakes are characteristically of this type. Precipitation and runoff from the watershed carry dissolved substances into the lake, promoting the growth of free-floating algae in sunny, open water. The gradual enrichment of water with nutrients, especially nitrogen and phosphorus, is called eutrophication. Continued addition leads to the production of organic matter in excess of decomposition. The decomposing organisms in deep water deplete the water of its oxygen, leaving the organic matter to accumulate with other sediments. Older lakes, rich in nutrients and shallow, are called eutrophic. When lakes receive large amounts of acidic humic substances from the watershed, the water becomes stained brown and poor in variety of aquatic life. Organic matter accumulates from plants in the shallow marginal waters. Such lakes or bogs are called dystrophic. Ultimately both eutrophic and dystrophic lakes fill completely with organic matter, are overgrown by vegetation, and disappear.

Slender fly larvae dominate the insect life of the deeper, open limnetic and profundal zones and the bottom beneath. The thin sediments of oligotrophic lakes are characterized by colorless midge larvae of *Tanytarsus* (Chironomidae). On the surface of the bottom ooze of eutrophic lakes, where oxygen is often depleted, the predominant or only insects are red, hemoglobin-containing larvae of *Chironomus*. These are able to survive the irregular oxygen supply and continue to function as important decomposers. The open waters of eutrophic lakes are inhabited by the phantom midge larvae, *Chaoborus* (Culicidae). As one of the few free-floating insects, the predatory *Chaoborus* spend the day near the bottom and swim toward the surface at night. The phantom larva is nearly transparent—hence the name. Gas sacs are visible at each end and act as organs of equilibrium. The volume of gas is controlled by the nervous system. When the sacs are compressed, the larva sinks; when they are expanded, it rises.

Adult midges emerge synchronously in enormous numbers from productive lakes throughout the world. *Chaoborus edulis* forms spectacular clouds seen at great distances over Lake Malawi in Africa. The emergence of *C. astictopus*, the Clear Lake gnat of Clear Lake, California, creates an annual nuisance to resort owners. When carried over land by winds, the masses of chironomids and *Chaoborus* remove significant quantities of organic matter from the lake ecosystem.

The littoral zone of lakes and ponds contains more different kinds of insects than deeper, open waters. Certain areas approximate the physical and chemical conditions of running water. On the wave-washed, rocky shores of oligotrophic lakes we find some of the same mayflies, stoneflies, and caddisflies that are in streams. Stoneflies are sensitive to low oxygen and are rare in lakes except in such places. The numbers and variety of insects in shallow water increase with eutrophication. Mayflies, caddisflies, dragonflies, and damselflies live among submerged plants or on the bottom. With their gills these insects depend on high levels of dissolved oxygen for respiration and do not rise to the surface for air.

The quiet, warm waters of shallow bays have a greater daily variation in dissolved oxygen. Here are found the many Hemiptera, Coleoptera, and Diptera that breath air, either directly or in air bubbles. On the surface are water striders (Gerridae and Veliidae, Hemiptera), whirligig beetles (Gyrinidae), Collembola, and marsh treaders (Hydrometridae, Hemiptera).

Shallow bodies of water may be created each year by rain and snow melt. These last but a few weeks or months before the water is lost through evaporation or absorption into the soil. Such bodies of water formed in the spring are called vernal lakes or ponds. Insect life ordinarily teems in these temporary habitats because predatory fish are absent. Even the small volumes held in tree holes, cavities at the bases of leaves of plants such as the tropical bromeliads, and hoofprints are sufficient for some insects to complete their life. With warm temperature, growth is usually quite rapid. For example, less than 2 weeks are needed for the yellow fever mosquito, *Aedes aegypti*, to complete development.

INSECT SCAVENGERS

Many, but by no means all, insects that live in soil and water are scavengers. Perhaps the word "scavenger" is poorly chosen. Does it indicate an animal that eats whatever comes along, one that eats mostly dead things, or one whose diet consists of items so small and miscellaneous that an accounting is difficult? Here we intend all these meanings when we call certain insects scavengers. The term is broad enough to convey the diversity of food that some insects are known to eat while also concealing our ignorance of the exact diets of others.

A complex terminology has developed to describe the diets of scavengers. Some may be omnivorous, eating plants or animals whether they are dead or alive. Saprophagous scavengers feed on dead organisms and have been subdivided into xylophagous forms, boring into and feeding on sound or decaying wood; phytosaprophagous forms, feeding on decaying vegetable matter; scatophagous or coprophagous forms, feeding on feces or dung; and zoosaprophagous or necrophagous forms, feeding on dead animals. Detritivores feed on particles of plant or animal matter plus the associated microbiota of fungi, bacteria, protozoa, and other tiny invertebrates. The food of saprophagous insects, therefore, is not all dead matter, because microbes flourish in such places and are nutritious.

Small living food is taken by microphytic insects that selectively eat bacteria, yeasts, fungi, algae, diatoms, lichens, spores, and loose pollen. Animal foods include Protozoa and small invertebrates and their eggs. Fungivorous insects are adapted to feeding exclusively on living and dead fungi. Aquatic filter feeders strain microbes and organic matter from water with specialized brushes on their mouthparts or webs. Thus among the insects loosely called scavengers are those with specific choices of food.

Saprophagous insects have unity in an ecological sense because they function in the complex world of the decomposers (Fig. 8.7). These should be counted among our most beneficial insects. Intact bodies of dead vascular plants and vertebrates, smaller dead organisms, and excrement are progressively disintegrated. The organic nutrients in streams and soils are derived from plant material that is decomposed in part by insects and other hexapodous arthropods. Scavenging species are usually specific to either plant or animal matter. Tissue is ingested and partially or completely digested, often with the aid of symbionts or unusual enzymes such as cellulase, keratinase, or collagenase. The feces of scavengers are further decomposed by other scavengers or microbes, thus aiding in nutrient recycling. Associated with this community are insect predators and parasitoids that prey on the scavengers.

In terrestrial environments the scavenger species appear at a dead animal or plant in a predictable sequence as characteristic chemicals are emitted from the decaying organic material at each stage of disintegration. The kind of scavengers associated with a human corpse can provide estimates of the time of death in criminal investigations. This and other uses of evidence from insects in legal issues is included in the field of forensic entomology.

INSECTS AND VERTEBRATES

Insects have coexisted with freshwater and terrestrial vertebrates throughout most of their evolutionary history. In the course of this long

Figure 8.7 Insects as decomposers in water and on soil. a, Dead leaves are recycled in streams by aquatic insects and microbes. This mayfly larva (Heptageniidae) is resting on a leaf that is in the process of disintegration; b, Many kinds of insects not normally considered as scavengers visit excrement for moisture and other substances. Included here, on hyena dung, are butterflies, flies, bees, and wasps as well as scavenging scarab beetles.
Source: Photo by H.V. Daly.

association, feeding relationships have evolved such that certain insects are parasitic on vertebrates and certain vertebrates are predatory on insects. The orders Phthiraptera and Siphonaptera and various families and species of Dermaptera, Hemiptera (suborder Heteroptera), Lepidoptera, Coleoptera, and Diptera are dependent on the tissues or blood of vertebrates as food during part or all of their lives. This parasitic behavior also has promoted the evolution of insect transmission among vertebrate hosts of certain microbes and helminths that cause some of the most important diseases of humans.

Insects played a key role as food for vertebrates in the early evolution of reptiles, birds, and mammals. Fossils suggest that the early forms of each group appear to have been mainly insectivores. Today insects continue to be an important source of nutritious food for vertebrates. Many are specialists on insects.

Many familiar birds in urban and rural areas feed on insects and in some instances significantly reduce pest populations. The annual spring migration to northern forests and grasslands is timed to feast on the emergence of insects. Some freshwater fish like trout include insects as a regular part of their diet. The mosquito fish, *Gambusia affinis,* is a valuable agent for the biological control of mosquitoes in ponds and irrigated rice fields. By at least 200 CE, human anglers had learned to imitate insects by making "flies" of feathers or hair. Thus the popular recreation of fly-fishing takes advantage of the predatory behavior of fish on insects.

INSECT PARASITES OF VERTEBRATES

The larger size of vertebrates does not prevent parasitism by insects. All terrestrial vertebrates are subject to attack. Only vertebrates that remain partly or completely submerged in water throughout their lives—fish, whales, and sea cows—are usually exempt, but mosquitoes have been seen biting fish at low tide. Other marine vertebrates, such as seals and marine birds, including penguins, periodically leave the water and are hosts to lice, fleas, and occasionally other kinds of parasites. Parasites consume hair, feathers, and skin (chewing lice), blood (numerous flies, fleas, lice), or flesh (flies). In some instances, the eating of foreign matter, fungi, or microbes on the host's skin may be beneficial, but most parasites are detrimental to their hosts to some degree. Nevertheless, the health and survival of the host are essential to the parasite.

Carbon dioxide is a common attractant for insect parasites to locate vertebrate hosts. This behavior can be used to capture parasites in traps baited with dry ice. Host odors also may be important attractant stimuli. Research on tsetse flies led to the identification of chemicals in cow breath as major attractants. The results led to the development of cheap and effective traps for tsetse flies that are used for monitoring tsetse populations or as part of control programs. Tsetses and other day-active flying blood feeders also use visual cues to locate large animal hosts. Vibrations trigger pupal fleas to adult molt into the adult stage. This behavior has obvious benefits in timing adult emergence to the availability of hosts for feeding.

As a rule, the host's reaction to a new parasite is severe, while long-established parasites have evolved in such a way as to minimize irritation. On its normal host the bite of a bloodsucking insect is often apparently painless, but on other hosts the reaction may be swift. As a rule, ectoparasites or external blood bood feeders that feed for prolonged periods, such as lice and ticks, have relatively painless bites. The disadvantages of causing the host pain during feeding provide obvious selective pressure for not alarming the host while feeding. For painful biters such as horseflies or stable flies, whose swordlike mouthparts rapidly stab into the host to create an external bleeding wound from which the insect can lap up a bloodmeal, there is much less to be gained from avoiding a painful reaction from the host.

TYPES OF PARASITIC INSECTS

Some terms will be needed for our discussion of insect parasitism. Parasites located on the outside of the host's body are ectoparasites; those inside the body are endoparasites. Continuous parasites remain on the host's body throughout the parasite's life. Transitory parasites spend a part of their life on the host and part elsewhere as a free-living insect. Intermittent, or temporary, parasites visit the host only to feed. Facultative parasites are insects that normally complete their life without parasitism but are able to survive as parasites under certain circumstances. Obligatory parasites require a host to complete their life.

Parasites have evolved repeatedly from different lineages during the long association of insects and vertebrates as the codominant terrestrial animals. For our purposes, we divide parasites into three broad categories based on their general adaptation to parasitism:

- **Crawling ectoparasites** are apterous or sometimes winged and cling to their host or crawl on their hosts at least during feeding.
- **Flying ectoparasites** are winged, active fliers and alight at or near the site of feeding.
- **Myiasis-producing Diptera** are tissue- or blood-feeding larvae.

Crawling Ectoparasites

Crawling ectoparasites are the most taxonomically diverse. They exhibit a number of convergent adaptations associated with life in the hair or feathers of their hosts. In most species both sexes are parasitic. Adults are more or less flattened with a tough, leathery cuticle. Siphonaptera are laterally compressed for movement through hair; the others are dorsoventrally flattened and manage equally well in hair or feathers (Figs. 8.9d, 42.1a). The adults of continuous parasites and Siphonaptera generally show signs of reduced sensory and locomotory apparatus, such as: (1) antennae are short; (2) compound eyes are reduced (absent in Hemimerina, Fig. 8.9b; Polyctenidae, Fig. 8.8c; some Anoplura; and Siphonaptera); (3) ocelli are absent in endopterygote parasites (present in some Hippoboscidae); (4) wings are absent (except in some Hippoboscidae and Streblidae, Fig.8.9c, in which the wings may be well developed, vestigial, or absent); (5) legs are usually short, more or less stout, with well-developed tarsal claws, which may be single and chelate on mammalian hosts (Nycteribiidae have long legs, Fig. 8.9e); (6) stiff, posteriorly directed hairs or scales or a comblike ctenidium (ctenidium present in Polyctenidae, Fig. 8.8c; Siphonaptera, Fig. 42.1a; Nycteribiidae; and

Platypsylla, Fig. 8.9a) are present; and (7) where the life history is known, mating and oviposition usually takes place on or near the host. The temporary parasites Cimicidae (Fig. 8.8b) and Triatominae (Reduviidae, Fig. 8.8d) move to and from their hosts. Consequently, they retain long antennae, compound eyes, and walking legs. The Cimicidae are otherwise apterous, flattened, and have small eyes and no ocelli. Nymphs of Triatominae are crawling ectoparasites, but the adults are strong fliers and have well-developed eyes and ocelli.

Flying Ectoparasites

All flying ectoparasites are intermittent parasites in the adult stage. Most species are Diptera, but also included here are some odd Lepidoptera. Four groups are:

1. **Mandibulate biting flies.** The parasites are females with bloodsucking mouthparts, which include functional mandibles (Figs. 2.10b, 40.3b). They attack a wide range of vertebrate hosts, both warm- and cold-blooded. Species in all families attack humans. Blood meals are usually required for egg maturation and production. Species that can produce the first clutch of eggs without a blood meal are termed autogenous. The immature stages are free-living and usually aquatic or associated with damp soil. Larvae of tabanids and snipe flies are also found on dry litter or soil. Families of Diptera included are Psychodidae, or sand flies; Ceratopogonidae, or biting midges; Simuliidae, or blackflies; Culicidae, or mosquitoes (Fig. 2.12b); Tabanidae, or horseflies and deerflies; and Rhagionidae, or snipe flies.

2. **Bloodsucking muscoid flies.** Both sexes suck blood. Most feed on mammals, but *Glossina palpalis* prefers crocodiles and monitor lizards. Humans are bitten by various species of *Glossina* and *Stomoxys*. Larvae of *Stomoxys* and *Haematobia* live in manure, but *Glossina* is viviparous, producing a mature larva, which soon pupates. Families of Diptera included are Muscidae (*Stomoxys calcitrans,* or stable flies and their allies; *Haematobia irritans,* or horn fly; *H. exigua,* the buffalo fly

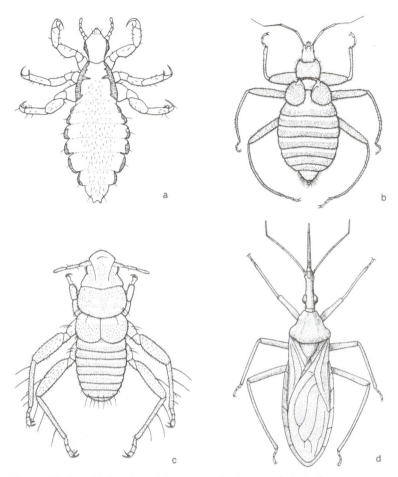

Figure 8.8 Parasitic hemipteroids: a, human body louse, *Pediculus humanus* (Pediculidae); b, primitive cimicid bat bug from Texas cave, *Primicimex cavernis*; c, polyctenid bat bug, *Hesperoctenes fumarius*; d, blood-sucking reduviid, *Rhodnius prolixus*.

of Australia; and *H. stimulans*, the cattle fly of Europe) and Glossinidae (about twenty species of *Glossina*, or tsetse flies, in Africa).

3. **Eye gnats and flies**. These insects are attracted to the eyes of humans and other mammals, where they feed on lachrymal and sebaceous secretions, pus, and sometimes blood. Families of Diptera included are Chloropidae (*Hippelates* sp.) and Muscidae (*Musca* sp.).

4. **Eye moths** (fruit-puncturing moths). The moths have a sharp galea with which they puncture fruit and suck the juice. They are also attracted to fluids in the eyes of mammals. The noctuid moth, *Calpe eustrigata*, is able to puncture mammalian skin and take blood. Eye moths include about twenty-three species from Africa and Southeast Asia and in the following families of Lepidoptera: Noctuidae, Pyralidae, and Geometridae.

Myiasis-Producing Diptera

Larvae of certain flies infest the bodies of animals, including humans, and produce a disease called myiasis. Pupation is usually on the soil, and adults of

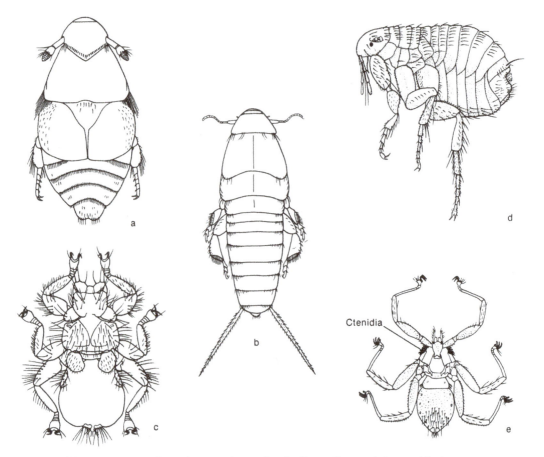

Figure 8.9 Various parasites of vertebrates: a, beaver beetle, *Platypsylla castoris*; b, parasitic dermapteran, *Hemimerus talpoides*; c, streblid bat fly; d, oriental rat flea, *Xenopsylla cheopis*, a vector of plague; e, nycteribiid bat fly.

the flies are free-living. Four groups of myiasis flies have been recognized (James and Harwood, 1969):

1. **Enteric myiasis–** and **pseudomyiasis–** producing larvae of many species have been reported in alimentary canals of vertebrates. Pseudomyiasis involves fly larvae that are ingested accidentally or enter through the anus and are able to survive for varying periods in the intestine. The larvae may eat the ingested food of the host or, less frequently, tissues of the intestine. True enteric myiasis is created by the larvae of horse bot flies, *Gasterophilus* (Oestridae, Gasterophilinae), which live attached to the stomach wall while

feeding on the wall and absorbing the host's food. Included are the families Muscidae, Calliphoridae, Sarcophagidae, and Oestridae.

2. **Facultative myiasis–producing larvae** of many species normally live as scavengers in dead animals. Adult females, attracted by odors of decomposition, lay eggs on the carcass. Wounds or deep, soiled wool of living sheep also attract ovipositing females. The resulting larvae become parasites in such wounds on sheep. Included are the families Muscidae, Calliphoridae, and Sarcophagidae.

3. **Obligatory myiasis–producing larvae** regularly infest living hosts. The primary screw worm, *Cochliomyia hominivorax*

(Calliphoridae), of the New World; Old World screwworm, *Chrysomya bezziana* (Calliphoridae); and *Wohlfahrtia magnifica* (Sarcophagidae), also of the Old World, enter the body at wounds or through mucous membranes and create local infestations. Larvae of the cattle grubs, *Hypoderma lineatum* and *H. bovis* (Oestridae), penetrate unbroken skin, migrate extensively in the host's body, and return to complete their development in tumors beneath the skin before dropping from the host to pupate. The female of the human bot, *Dermatobia hominis* (Oestridae, Cuterebrinae), attaches her eggs to another fly, such as a blowfly, tabanid, or mosquito. While the carrier insect is feeding, the bot larva hatches and drops to the host's skin, burrows into the subcutaneous tissues, and remains in the same place until fully fed. Mature larvae drop from the host to pupate. These flies occur in Latin America and attack other animals as well as humans. Included are the families Calliphoridae, Sarcophagidae, and Oestridae.

4. **Bloodsucking larvae** that are temporary feeders include the Congo floor maggot, *Auchmeromyia luteola*, which lives in huts and feeds on sleeping persons, and various species of *Protocalliphora* in the Old and New Worlds, which feed on nestling birds. All species are in the family Calliphoridae.

EVOLUTIONARY EFFECTS OF PREDATION BY INSECTIVOROUS VERTEBRATES

Insects are a regular part of the diet of many species of vertebrates: freshwater fish, salamanders, frogs, toads, lizards, turtles, snakes, birds, and mammals, including bats. Some human cultures also regularly include insects in their diet. The effectiveness of these predators in catching insects is aided by some of the following: (1) physical agility; (2) keen senses of hearing, smell, taste, and/or vision (including color vision in some fish, some reptiles, most birds, and primates); (3) hunting strategies; and (4) the capacity to learn from experience. Insects are subject to predation in aquatic, terrestrial, and aerial habitats and during day and night. To survive, exposed insects have evolved special patterns of coloration, behavior, and defensive properties to take advantage of what the predator fails to see or hear, cannot capture, or will not eat. Similar protective strategies can be found repeatedly among different insect taxa. This is because the insects are subjected to selection by predators with similar behavior and senses. The convergent evolution of protective strategies by insects provides some of the best opportunities to study natural selection in action. Much speculation has occurred, but too few instances of the protective strategies have been analyzed with critical experiments.

For our discussion, we first divide insect prey into those that are palatable (edible or acceptable) to a given kind of predator and those that are unpalatable (inedible or rejected). This distinction must be made with qualifications. Some individual insects will be relatively more or less palatable than others, and individual predators will vary in accepting prey depending on hunger and previous experience. Hard or spiny bodies, sticky secretions, stings, stinging hairs, and distasteful or toxic chemicals are defenses that tend to render an insect unpalatable. Even a low frequency of unpalatable individuals, however, will be an advantage favored by selection in the prey population. Unfortunately, we do not know the palatability of most insects relative to their main predators, but observations indicate that most species of insects are edible by a wide variety of insectivores.

Palatable Insects

It is assumed that relatively palatable species evolve either an effective means of escape, or a disguise, or both. Another option that will be discussed further on is Batesian mimicry, in which a palatable species avoids attack by resembling an unpalatable species.

The most effective means of survival would be to avoid detection altogether. This can be accomplished by a combination of cryptic behavior and cryptic coloration (also called camouflage or concealing coloration). Insects with cryptic coloration are overlooked by predators unless the predator learns to seek them. A low density of the species of insect or of other species with similar protective strategies therefore prevents accidental detection and reinforcement of the predator's search image. The coloration may be one or a combination of the following: (1) background resemblance such that the insect either closely matches the background against which it is seen or special resemblance of an inanimate object of no interest to the predator such as a flower, twig, stone, or bird dropping (Fig. 8.10); (2) obliterative shading, in which the rounded shape of the insect, such as a caterpillar,

is obscured by having the illuminated surface of the body more darkly pigmented and the shaded surface lighter in color; (3) shadow elimination by checkered borders or marginal fringes of hairs, which reduces the contrast along shaded edges next to the substrate; and (4) disruptive coloration of contrasting patterns and colors to alter the perception of the true body outline or contour. These patterns are usually accompanied by one or a combination of special behavioral traits in which the insect (1) selects an appropriate background and orients its body in a definite position relative to the background, (2) adopts a special prostrate posture to eliminate shadows or more closely resemble the object imitated, (3) rests motionless even when closely approached, and (4), when attack is imminent, escapes quickly, silently, and erratically.

Cryptic coloration and behavior are the rule among palatable insects exposed to diurnal, visually hunting predators. Aquatic nymphs and larvae may be complexly patterned to resemble aquatic vegetation, detritus, or the variegated gravels or other substrates of the bottom of the body of water. Immature and adult insects that inhabit plants, whether as predators or herbivores, are often green (commonly created by the pigment insectoverdin) or colored or sculptured to resemble leaves, stems, or bark. The larger insects of litter, soil, sand, or gravel may closely match their surroundings. The color patterns are usually characteristic of the instar and may change from instar to instar, but usually do not change rapidly by physiological processes in the manner of chameleon lizards. The walkingstick, *Carausius* (Phasmatidae), however, does become darker or paler depending on various stimuli.

Once discovered, jumping, dropping from vegetation, or flying are the most common means of escape. When approached, leafhoppers on a stem will move behind the stem out of sight. Some species do the opposite—they become motionless to avoid detection or feign death. Large mantids and Orthoptera may adopt a threatening posture that deters attack. These behaviors are often aided by coloration that diverts the predator's strike or attention. Eyespots may be of two main types: (1) small, dark spots, with or without a light mark simulating

Figure 8.10 Examples of concealing coloration: a, a katydid (Orthoptera, Tettigoniidae) that resembles a leaf. The katydid is active at night, remaining still amongst green leaves in the daytime. b, a deadleaf butterfly, *Anaea* sp. (Lepidoptera, Nymphalidae) from Peru that has a false stem and midrib line.

a reflection from the "eye," and (2) large spots with concentric rings depicting a colored iris and dark pupil marked with a reflected highlight. The first are thought to imitate an insect's eye, and the second are unmistakable representations of a terrestrial vertebrate's eye (Fig. 8.11; Box 3.1). The small eyespots may be situated at the posterior end of the insect next to antennalike processes to imitate a head. Examples are hairstreak butterflies (Lycaenidae), which have tails that resemble antennae on the caudal part of their wings, and, at the bases of the tails, spots that resemble eyes. When perched, the butterfly moves its wings to wiggle the false antennae. This presumably directs a predator's attention to the wrong end of the body and permits the insect to escape in the opposite direction with minimal damage. Small spots near the outer wing margin of other butterflies, such as the satyrs (Satyrinae, Fig. 43.11i), presumably attract the bird's peck away from the vital body and give the insect a chance to free itself. Evidence that such butterflies do escape is provided by collected specimens that have notched margins or V-shaped bill marks on their wings.

Larger eyespots have been shown to frighten vertebrate predators. Cryptically colored insects may hide their eyespots and suddenly display them when contacted by a predator. The nymphalid butterfly, *Caligo* (Fig. 8.11), and the Io

moth, *Automeris* (Saturniidae), have conspicuous eyespots.

Unexpected displays of bright flash colors are similarly designed to startle enemies at the moment of escape. Underwing moths, *Catocala* (Noctuidae, Fig. 45.13b), have conspicuous red or yellow bands on the hind wings. The colors are suddenly revealed and visible during the moth's erratic flight, but abruptly disappear when the moth alights and resumes a cryptic posture. Similarly, the upper surfaces of the wings of *Morpho* butterflies are brilliant blue, while the lower surfaces are cryptically colored. In flight the butterfly is conspicuous, but at rest with its wings held together only the cryptic pattern is visible.

Industrial Melanism

The best documented study of the evolution of cryptic coloration concerns industrial melanism, or the evolution of dark or melanic populations of insects in industrially polluted areas. Prior to the Industrial Revolution in England, trunks of trees such as oak were naturally light in color and often encrusted with light-colored lichens. Resting against this background, the pepper moth, *Biston betularia* (Geometridae), is virtually invisible. With the development of smoke-producing industries in the latter half of the eighteenth century, soot

Figure 8.11 Example of special resemblance: the Neotropical nymphalid butterfly, *Caligo* sp., has large eye spots on the hind wings. Seen from the side, the pattern resembles the eye of an *Anolis* lizard. False eyes startle and frighten would-be predators.

accumulated on vegetation in the areas surrounding and downwind from the factories. Lichens were killed and the bark of trees was stained black. The light, or *typica*, form of the pepper moth was conspicuous against this darkened background. In 1848 a dark, or melanic, form of the moth, called *carbonaria*, was first recorded at Manchester in the polluted district. At that time the melanic form probably was not more than 1 percent of the population, but by 1898 it was estimated to be 95 percent.

Melanic forms of over 100 of the 780 species of larger moths in England have also increased in frequency during the last 130 years. Such dark forms of moths are becoming evident elsewhere in industrial areas throughout the world's Temperate Zone, including North America. Industrial melanics are not known in tropical regions, possibly because the heat-absorbing dark color is a disadvantage in warmer regions. The melanics are always members of species that are concealingly colored. They rest during the day on such objects as lichen-encrusted bark, rocks, or fallen logs. In the majority of species, the switch from predominantly light forms to dark forms does not involve intermediate shades and is usually controlled genetically by a single dominant gene.

Kettlewell (1973) carefully investigated various explanations for the rapid spread of the *carbonaria* form of the pepper moth. Mutagenic effects of the pollutants were discounted by experiments; the melanic forms must arise in populations by normal, low rates of mutation. Except where the melanics are favored by selection, the frequency remains quite low. However, migratory melanic males may fly into unpolluted environments. In industrially blackened areas, the resting melanic forms are less visible to insectivorous birds. Observation in the field and experiments verified that birds of several species selectively eat the conspicuous form (*typica* against dark trunks in polluted districts and *carbonaria* against light trunks in unpolluted districts) and overlook the concealed forms.

Thus in a relatively brief period, the cryptic patterns of many species of moths over wide areas shifted to a dark, patternless design as a result of bird predation in polluted areas. This dynamic relationship is a sensitive monitor of pollution. With improved pollution controls now in England and northern Europe, nonmelanic moths have reappeared in districts where only melanic forms previously occurred. In West Kirby, England, melanic forms were 93 percent of the population in 1959 and 29.6 percent in 1989; in Holland, the change was from 60 to 70 percent in 1969 to less than 10 percent in 1988.

Studies of other melanic insects have shown that different insects are affected differently by pollution. Melanic forms of the ladybird beetle, *Adalia bipunctata* (Coccinellidae), also increased and then decreased in Britain in correlation with the level of smoke pollution. These beetles are believed to be unpalatable so the changes in the frequency of the melanic forms cannot be explained by selective predation. It is speculated that the melanic beetles are more efficient in heat absorption than nonmelanics when the sun's radiation is partly blocked by smoke.

Unpalatable Insects

The coloration and behavior of unpalatable species often contrast strikingly with those of cryptic palatable species. The strategy here is twofold: (1) to create an unpleasant, *but not lethal*, ordeal for the vertebrate predator by means of poisonous or nauseous secretions, repulsive odors, effective stings or bites, or spiny surfaces, and (2) to be easily recognizable so that an experienced predator probably will not attack again. Vertebrate predators are relatively long-lived and are capable of learning so once they have experienced an unpleasant insect, they avoid future encounters. As a result, the population of the unpalatable insect suffers less mortality. Unpalatable species usually have warning or aposematic coloration and aposematic behavior, including one or more of the following: (a) bold, simple patterns of contrasting black and red, orange or yellow, which are highly visible to vertebrates with color vision and which also contrast against the predominantly green color of the landscape (green is the complementary color of red); (b) gregarious behavior that leads to local, dense aggregations; (c) free exposure when walking or resting and slow, conspicuous flight in full view of predators; (d) characteristic movements, buzzing sounds, or smells that warn against attack; and (e) when attacked, only sluggish efforts to escape, repulsive taste, and a durable body often permitting survival without fatal injury. For example, beetles of the family Lycidae have black and orange colors, aggregate, are sluggish, and have leathery, resistant bodies. The colorful caterpillars of Papilionidae, when disturbed, release defensive odors from brightly colored osmeteria. Many beetles with defensive secretions are dark or black in color. Against pale soil or sand, the black color is visible during both day and night.

A naive predator has ample opportunity to sample an aposematic species and, even without color vision, has a number of cues by which to remember the experience. Bird predators are also known to learn vicariously by witnessing another predator's response. Learning can be reinforced by merely seeing an aposematic insect, again without touching it. Not all the cues may be necessary to elicit an avoidance reaction. It is not known to what extent a predator's response includes some innate avoidance of aposematic colors or behavior. Aposematic coloration is oddly missing among most insects that live in water, even though some aquatic insects are unpalatable and some fish are known to have color vision and learning ability.

In a few cases, insects have been reported with venoms or toxic chemicals in their bodies that do kill vertebrates. These would seem to be exceptions to the rule that vertebrate predators survive their experience with unpalatable insects. In the first case, stings of honey bees cause some human fatalities each year. The venom is a complex mixture of enzymes, proteins, amines, and amino acids. Some persons are hypersensitive to bee stings and may succumb after a single sting. These rare events are an unfortunate by-product of human physiology. On the other hand, a mass attack by Africanized bees that results in large numbers of stings and invasion of the mouth and nose by bees can kill humans and domestic animals. The attacks by Africanized bees evolved in response to the persistence of their natural enemies in Africa such as the honey badger, *Mellivora capensis* (Mustelidae, Carnivora). African bees do not distinguish between tenacious enemies and defenseless animals. In a second case,

meloid beetles contain the chemical cantharidin and, when eaten by horses in their pasture or in dried hay, may cause horses' death. The beetles are normal aposematic insects. However, horses appear to be unusually susceptible to cantharidin, are not a natural enemy of the beetles, and must be considered accidental victims.

Mimicry

The sensory and learning abilities of relatively long-lived vertebrate predators have allowed aposematic species to survive. Experiments have supported the conclusion that predators can generalize their learning to avoid insects with a similar appearance after an unpleasant experience with only one and that additional unpleasant encounters strengthen the avoidance behavior. The tendency of aposematic species such as adult monarch butterflies, *Danaus plexippus* (Nymphalidae) (Fig. 8.12a, or some ladybird beetles (Coleoptera, Coccinelidae) to form groups or clusters while overwintering presumably reduces predation by local birds and rodents that learn to avoid eating these insects after a single or few encounters. This also gives some advantage to insects that are in the process of evolving aposematic coloration and behavior but have not acquired the full set of advertising characteristics.

The predator's ability to generalize (or failure to discriminate) is also the basis for the evolution of two types of mimicry: (1) Müllerian mimicry, named for Fritz Müller (1879), involving two or more unpalatable and aposematic species; and (2) the previously mentioned Batesian mimicry, named for Henry W. Bates (1862), involving one or more unpalatable species that serve as models and one or more palatable species that mimic the model's coloration and behavior (compare). Models and mimics with a common pattern of coloration are often not closely related taxonomically and have clearly acquired their similarity through convergent evolution.

The mutual resemblance of Müllerian mimics confers protection on all members of the association in a given area. Because all are unpalatable and predators can generalize, the resemblance among the species need not be exact, although it is often

Figure 8.12 Multiple mimicry: a, monarch butterfly, *Danaus plexippus*; b, queen butterfly, *Danaus gilippus*; c, viceroy butterfly, *Limenitis archippus*. See text for explanation.
Source: Photos by University of California Scientific Photography Laboratory.

astonishingly precise. Naive predators presumably attempt to eat mimics in proportion to the relative abundance of each species, and each species suffers fewer losses in the process of educating predators when several mimics are together. Examples are found among the Neotropical butterflies, especially the genus *Heliconius* (Nymphalidae) and the nymphalid subfamily Ithomiinae; the monarch, viceroy, and queen butterflies (Fig. 8.12); netwinged beetles (Lycidae) and certain other insects; and the convergent similarity of many ants, wasps, and bees, the females of which are equipped with stings.

Batesian mimics are avoided by predators familiar with the unpalatable model. An argument can be made that the populations of mimics must remain smaller than the model's population. Otherwise a naive predator will be more likely to eat palatable mimics. Generally, Batesian mimics are rarer than their models (in some, only the female is the mimic), but recent studies show that strongly unpalatable models may confer protection even when outnumbered. Apparently tasting a model is truly a memorable experience for the predator! The resemblance of Batesian mimics to their models is often quite close. Examples include neotropical moths and katydids that resemble wasps; the syrphid flies and other insects that resemble wasps, bees, and ants; and long-horned beetles

(Cerambycidae), leaf beetles (Chrysomelidae), and moths that resemble lycid beetles.

On careful analysis of mimetic associations, the distinction between Müllerian and Batesian becomes more difficult to make. Lincoln P. Brower and his associates (1964, 1968) have studied the mimetic relationships of the monarch butterfly, *Danaus plexippus*, and queen butterfly, *Danaus gilippus* (Figs. 8.12a,b). The larval food plants of both species are milkweeds, mostly in the genus *Asclepias*. Two species of milkweeds in the southeastern United States, *A. curassavica* and *A. humistrata*, contain cardiac glycosides (cardenolides). These chemical compounds cause vomiting or, in larger amounts, death in birds and cattle. In Costa Rica, cattle learn to avoid eating *A. curassavica*; hence the plant is protected. Monarch larvae feed on these milkweeds and acquire various amounts of cardiac glycosides, which are retained in the body of the adult butterfly. Insectivorous birds vomit after eating a sufficiently toxic adult monarch and will avoid pursuing other monarchs when sighted. Thus, not only are emetic monarchs (those that cause vomiting) effectively unpalatable and aposematic, but the aposematic larvae on the poisonous milkweed are also undisturbed by grazing cattle and birds.

Not all milkweeds, however, contain enough cardiac glycosides to cause emesis, and some lack the drugs altogether. Monarchs reared on these plants are palatable in laboratory tests with naive birds but are rejected by birds previously fed emetic monarchs. The natural population of monarchs in an area may include both palatable and unpalatable individuals, yet the predators learn to avoid most of both types. Even as low as 25 percent of unpalatable monarchs theoretically will gain protection for 75 percent of the population. Brower has called this kind of intraspecific relationship automimicry. A further complexity was recently revealed: as unpalatable adult monarchs age, the concentration of cardiac glycosides decreases; young adult monarchs are Müllerian mimics, old monarchs are automimics.

The queen butterfly also feeds on toxic milkweeds and resembles the monarch. Where they occur together, they are usually considered Müllerian mimics. In Trinidad, however, Brower found both monarch and queen larvae feeding on both poisonous and nonpoisonous species of milkweed. Adult monarchs were 65 percent emetic, but only 15 percent of queen adults were emetic. The mimetic relationships here include automimicry between palatable and unpalatable members of each species, Batesian mimicry between palatable members of one species and unpalatable members of the other, and Müllerian mimicry between unpalatable members of both species.

For many years the viceroy butterfly, *Limenitis archippus* (Fig. 8.12c), was considered a palatable Batesian mimic of the monarch butterfly (Fig. 8.12a). The larvae feed on nontoxic plants such as willows (*Salix*), populars (*Populus*), and some kinds of fruit trees. Nevertheless, a high percentage of viceroys have been found to be unpalatable to birds. The viceroy now joins the monarch and queen butterflies as members of a Müllerian and Batesian mimicry complex.

Batesian mimicry may be thought of as involving coevolutionary interactions similar to those occurring in host-parasite complexes. The mimic, of course, plays the role of the parasite. Its strategy is to take advantage of the model without destroying it. The model gains nothing and faces the danger of a "credibility gap" developing in its potential predators. At some point, if the mimics become too common, most predators will associate only pleasant experiences with the aposematic pattern

of the model. Such a development, of course, ruins the game for both model and mimic, since a conspicuous pattern now means "tastes good" to the local insectivores. One would expect that the model would evolve a pattern different from that of the mimic at a maximum rate, everything else being equal. It is to the mimic's advantage to maintain a maximum of resemblance to the model until that critical point mentioned above is reached. Then the advantage becomes a disadvantage—the mimic is conspicuously patterned, but predators now associate that pattern with tastefulness. As a result, selection would tend to move the mimic away from the model into a more cryptic pattern.

In all cases it obviously is of advantage to the Batesian mimic to become distasteful if it can do so without sacrificing too much physiologically. In butterflies, at least, it appears that the usual source of noxious compounds is plant biochemicals, so that food-plant relationships must play a large role in the evolutionary dynamics of any given situation. A butterfly has several different routes to evolve distastefulness open to it. If it eats a food plant that does not produce an appropriate compound, it may switch food plants. If it is feeding on a plant with an appropriate compound or its precursors, the butterfly may evolve the ability to use the compound or synthesize a noxious compound from precursors. Finally, the food plant of the butterfly may evolve an appropriate compound, which then may be picked up by the butterfly. In the latter case the mimetic butterfly would be involved in a complex of "selection races," involving the model, the food plant, and predators. As the food plant becomes more and more toxic, the butterfly must find ways of "breaking even" by avoiding poisoning or "winning" by turning the poison to its own advantage. Predators may simultaneously be undergoing selection for ability to discriminate between model and mimic and for "resistance" to the noxious properties of the model. Of course the presence or strength of such selection will depend on many variables. For instance, in some cases butterflies in a single population may make up such a small proportion of the targets of a single predator that selective influence on the predator will be negligible.

CHAPTER 9

Insects and Vascular Plants

Plants are important sources of food and shelter for insects. Approximately half of the described species of insects are phytophagous, that is, they feed on green plants. Insects in the six largest orders (Fig. 1.1) derive much or most of their food from plants. The largest insect family, weevils (Curculionidae, Coleoptera), is entirely phytophagous. Because of their abundance, ubiquity, and biomass, insects are the most important primary consumers of land plants, usually exceeding vertebrate herbivores and competing directly with humans. Most plant parts are generally low in dietary nitrogen and have numerous physical features and defensive chemicals evolved for protection from herbivores, but most insect herbivores have become specialized to deal with the challenges presented by various plants. The impact of insect feeding is checked by natural enemies and ameliorated by varied plant defenses. Not all relationships with insects, however, are harmful to plants. The dominant plants today, the angiosperms or flowering plants, owe their origin to the feeding behavior of Mesozoic insects that functioned as pollinators. The subsequent coevolution of this mutualistic association of plants providing rewards—mainly pollen and nectar as food—and insects providing pollination has resulted in a huge diversity of plants and their pollinators.

Phytophagy has evolved repeatedly from other food habits, usually scavenging. In the following sections we will examine the relationships of insects with plant hosts. These relationships, both mutualistic and detrimental to plants, are envisioned as coevolutionary, that is, they involve the interaction of two or more kinds of organisms in which evolutionary changes in one affect the evolution of the others.

FLOWER-VISITING INSECTS

Evolutionary Origin of Insects as Pollinators

Sexual reproduction in plants, as in most organisms, is widespread and presumed to be important in maintaining variation within populations. The majority of flowering plants are monoecious: they have male and female organs within the same flower or plant. Less common are plant species with separate female and male plants (dioecious).

Although self-fertilization occurs in some plants, the majority have cellular mechanisms that prevent self-fertilization and are said to be obligately outcrossing. Wind pollination (anemophily) provided most cross-pollination in ferns and gymnosperms. Insects may well have transported spores among early plants during the Paleozoic Era (570 to 245 million years ago; see Chap. 14). The relationships between insects and the angiosperms probably began in the Mesozoic Era (245 to 65 million years ago). The earliest fossil evidence of flowering plants is leaves and pollen found in strata about 124 million years old from the Cretaceous Period. All existing orders and the major insect families found today had already evolved before the appearance of most angiosperms. The evolution of angiosperms was so rapid that an astonishing sixty-seven families are represented by the end of the Cretaceous Period (65 million years ago).

Early phytophagous insects doubtless fed on the reproductive organs, pollen, and seeds, as well as the vegetative parts of plants. The first step in the mutualistic association of insects and plants probably took place when insects accidentally carried pollen from male to female organs of the same species of plant. This transfer, called pollination, had previously required large quantities of windborne pollen. In exchange for food, insects became the carriers, or pollen vectors. Insect pollination (entomophily) had the advantages that smaller quantities of pollen had to be produced, pollination did not depend on wind, and widely separated plants could be pollinated. About 10 percent of plants are pollinated by wind, including gymnosperms, most grasses, and many deciduous trees.

The first pollen vectors may have included terrestrial or flying reptiles or early birds, but the most important were probably the flying insects. Among these, the Coleoptera are thought to have been the most significant. They were well diversified in the Mesozoic Era. Numerous primitive angiosperms today are beetle-pollinated. Typical beetle-pollinated plants are open to allow easy access to beetles. However, beetles may consume the ovules of the plants that they pollinate. For this reason, certain flower structures, such as carpels, can be explained as defensive measures that initially evolved against the powerful chewing jaws of these insects.

The next steps in the evolution of mutualism included features of plants that attracted insects by taking advantage of their mosaic vision, color perception, and the important roles of chemicals in their behavior. A broken outline and colored petals aided in the visual recognition of flowers. Odors attracted insects at a distance. The first floral odors may have imitated odors of fruits or decay that

appealed to scavenging beetles. Primitive flowers visited by beetles generally lack floral nectar. Flower architecture had to not only attract pollinators but also provide for precise placement or collection of pollen by the pollinator and then its deposition on the female part (stigma) of the flower that receives the pollen.

The evolution of nectaries allowed the plant to locate an attractant near the reproductive structures and to regulate its quantity and content. Nectar is an aqueous fluid rich in sugars (commonly sucrose, glucose, and fructose) plus nutrients attractive to pollinators: amino acids, proteins, and lipids. Thus, both water and food are provided. Other substances include ascorbic acid, possibly serving as an antioxidant, and alkaloids, which might be toxic to certain unwanted flower visitors. The secretion of nectar inside the flower was an incentive to actively flying insects in need of carbohydrate fuel such as Lepidoptera, Diptera, and Hymenoptera. Indeed, the abundant supply of nectar as food must have contributed to the success of these orders. Floral nectar differs in composition from extrafloral nectar (from nectaries elsewhere on plants) in ways that suggest it is secreted expressly for the food needs of favored pollinators. Advanced families of plants have higher concentrations of amino acids than primitive families, and butterfly-pollinated flowers have higher concentrations than bee-pollinated flowers. This is associated with the inability of most butterflies to ingest protein-rich pollen, whereas bees obtain their amino acids from pollen. Flies that frequent protein-rich dung are attracted to the nectars of fly-pollinated flowers that are high in amino acids but are attracted to the fetid odors of the flowers that they visit.

The most common pollinators are the adults of the endopterygote orders Coleoptera, Lepidoptera, Diptera, and Hymenoptera. By far the most important pollinators are bees (Apoidea). Their search for mates, oviposition sites, and plant or animal food is aided by a strong flight apparatus, highly developed senses, and, in some groups, learning ability. These same attributes aid pollinators as they search for and remember flowers.

Behavior of Flower-Visiting Insects

Individual insects that visit flowers of the same plant species during a single flight or longer period are said to be flower-constant. When all individuals of an insect species are restricted to visiting a single species of plant for pollen, the insect is said to be monolectic. If several, possibly related, plant species are visited by individuals of the insect species, the insect is said to be oligolectic. If many plant species are visited, the term polylectic applies. An individual bee may be flower-constant during successive visits but also be a member of a polylectic species. Flower constancy benefits both plants and insects. Cross-pollination requires that an insect that acquires pollen from one flower visit another flower of the same plant species to deliver the pollen. Moreover, the pollen from the insect must be placed on the receptive stigma of the plant for pollen to fertilize the flower. It is advantageous to insects to become temporary or permanent specialists because they can take advantage of learning and remembering what types of flowers have adequate food rewards and how to operate its floral mechanism to gather pollen or nectar.

Mutual Adaptations of Flowers and Their Insect Visitors

The flowers of plants that require cross-pollination may attract a variety of insects, but not all flower-visiting insects can serve as pollen vectors. The hairy bodies of bees are especially suited to carry pollen, but most bees have special structures on the legs (e.g., honey bees) or abdomen (e.g., bumblebees) to store pollen grains that the foraging bees collect. Pollen deposition on the stigma in such cases is a by-product of flower visitation. As we will discuss, in the most extreme examples of coevolution, the pollinating insect deposits pollen as a specialized behavior rather than simply a by-product of foraging.

The most obvious attraction that plants provide to attract pollinators is food: nectar and pollen. However, lipids, attractive scents, and food bodies are used as rewards by some plants. In a general way, the flower size, shape, position of reproductive parts, color patterns, odor, nectar composition, and time of flowering can match the sizes,

anatomies, diets, sensory physiology, rhythmic activity, and foraging behaviors of its pollinators. Combinations of such traits are called pollination syndromes. For example, a typical butterfly flower would be open during the day, with red or orange petals, a long, tubular flower tube (corolla) accessible to the long proboscis of butterflies, nectar rich in amino acids as well as sugars but with little odor, and a landing platform that could support the large size and weight of butterflies. Hummingbird moth (Sphingidae) flowers, in contrast, would not require a landing platform because these moths feed while hovering, but they ideally would have strong scents that attract the moths at night, have a pale color, and be open at night. Even related species of plants might depend on quite different kinds of pollinators. This specificity attracts effective pollinators and tends to reduce losses of pollen and nectar to nonpollinating visitors. The reward given to pollinators is thereby more closely regulated to promote cross-pollination. Pollination syndromes were early guides used to develop our ideas about coevolution of plants and insects for pollination. They are not absolute guides to pollinator specificity, however, only a starting point for investigations, as there are many exceptions to these general syndromes.

Nectar is commonly situated deep within a floral tube so that casual visitors are unable to reach it. Elongation of the mouthparts into a sucking tube is a frequent adaptation among specialized flower-visiting insects. The elongation is achieved in many ways and often independently (Fig. 9.1), although the Hemiptera, with their long sucking beaks, never became regular flower visitors for nectar. A famous example of this type of adaptation is the orchid *Angraecum sesquipedale* of Madagascar, which Charles Darwin predicted would have a moth pollinator with a proboscis 30 cm long in order to reach the bottom of the orchid's floral tube. By experiment, only a thin wire this long could reach the nectar at the bottom of the flower's long spur and detach the pollen bundle. Contemporaries of Darwin ridiculed this idea, and one of them argued that this orchid represented an example of "divine design" counter to Darwin's theory of evolution. Over 40

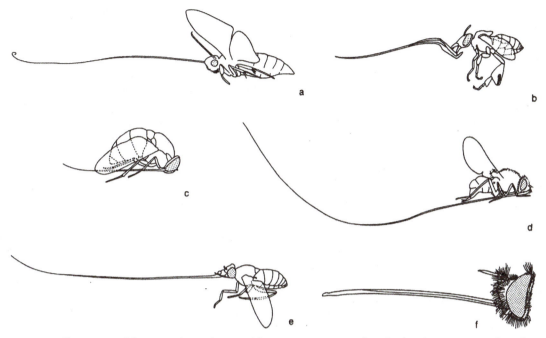

Figure 9.1 Extreme modifications of mouthparts of flower-visiting insects (length of proboscis in parentheses): a, sphingid moth, *Manduca quinquemaculata*—proboscis (12 cm) is coiled under head when not in use; b, male apid bee, *Euglossa asarophora*— proboscis (3 cm) is folded between legs when not in use; c, acrocerid fly, *Lasia kletti*—proboscis (2.4 cm) is swung forward in use; d, nemestrinid fly, *Megistorrhynchus longirostris*— proboscis (8.4 cm) is swung forward in use; e, tabanid fly, *Philoliche longirostris*—proboscis (5.8 cm) is carried in flight as shown; f, head of bombyliid fly, *Bombylius lancifer*—proboscis (9.7 mm) is carried as shown.

years later, the usual pollinator of this orchid was discovered—with a proboscis just the length that Darwin had predicted.

Apoidea as Flower Visitors

The largest group of efficient pollinators consists of bees, superfamily Apoidea. About 30,000 species are known in the world. Most species are solitary in behavior; females make and provision their own nests. Bees evolved from visually hunting, predaceous wasps that frequently visit flowers for nectar. Both sexes of bees take nectar as flight fuel. Females eat pollen as a source of the protein used in producing their eggs. Most females, except the cleptoparasites and social queens, also collect pollen and nectar for their larvae. These are stored in the nest as provisions, or, in certain social species, the pollen is fed directly to the larvae. Some bee-pollinated plants, especially in the neotropics, offer energy-rich fatty oils in addition to, or instead of, pollen and nectar.

The adaptations of bees that are associated with flower visitation are plumose or featherlike hairs; special pollen-transporting devices; modifications of the tongue, or glossa, for extracting nectar; and a diet of pollen and nectar. Correlated with their social behavior, honey bees have a highly developed system for foraging on flowers that involves communication (Chap. 6). The species of bees that visit oil flowers have special brushes to collect and transport lipids. Some flowers that require intense vibrations provoked by buzzing flight to release their pollen are only pollinated by bees.

A few species of bees are monolectic, but most bees are oligolectic or polylectic. Among the latter are honey bees, which visit an enormous variety of flowers, including those designed to attract insects other than bees. During times of food scarcity, honey bees collect honeydew or fruit juices as nectar substitutes. So compelling is their foraging behavior that when pollen is not available, honey bees have been observed to collect flour or even inert dust.

Certain primitive bees, such as *Hylaeus* (Colletidae), are relatively hairless, like wasps. They eat pollen and later regurgitate it with nectar while preparing the nest provisions. But most bees have abundant, plumose hairs that retain pollen grains brushed on the body during flower visits. Female bees methodically groom themselves and pack the pollen into special devices for transport to the nest. These are of two kinds: (1) pollen brushes, or scopae, of long, dense hairs on the hind legs of most bees or on the underside of the abdomen in Megachilidae, and (2) pollen baskets, or corbiculae, which are created by a circle of stiff hairs on the outer surface of the hind tibiae of honey bees, bumblebees, and their relatives in the family Apidae. Andrenidae, Colletidae, and Halictidae have short tongues suited to take nectar from exposed nectaries or flowers in which the bee can bodily enter. These are considered less specialized than the long-tongued Megachilidae, Anthophoridae, and Apidae, which can reach nectar hidden in the inner recesses of specialized flowers. The expandable crop is used to carry nectar back to the nest and also functions on outbound flights as a fuel tank.

"Bee flowers" characteristically open at certain times during the day, emitting sweet or aromatic odors and presenting their pollen and nectar. Petals of bright blue, purple, yellow, and other colors within the bee's range of color vision are common. Recall that red is invisible to many bees; it is an uncommon color for bee flowers. In contrast, bees can see ultraviolet markings on flower petals that serve as "nectar guides" and a familiar and distinctive flower image to both attract bees and facilitate their finding the floral tube. Separate petals that create a broken outline are suited to detection as a mosaic image by the bee's compound eyes. The two-lipped form of certain bee flowers, such as those of legumes or mints, provides a landing platform. This places the bee in a position favoring access to the food and pollination.

Diptera and Lepidoptera as Flower Visitors

Two other large groups of pollinating insects show conspicuous modification for flower visiting: certain Diptera and most Lepidoptera. In both orders, the special modifications are mainly elongation of the mouthparts to reach hidden nectar. Although the mouthparts of flies are variously modified for bloodsucking or sponging, the general ability seems to be retained to ingest nectar and small particles such as pollen grains. Certain species in each of several families of flies have developed exceptionally long mouthparts for probing deep flowers. Some species with long mouthparts are found in Bombyliidae, Apioceridae, Nemestrinidae, Acroceridae, and Tabanidae (Fig. 9.1e). Shorter, but distinctly specialized, mouthparts are also seen on species of *Rhingia* (Syrphidae), Conopidae, and Tachinidae. The long-tongued flies visit the same kinds of flowers that bees visit.

The so-called "fly flowers" are mainly adapted to less specialized, short-tongued insects, which normally feed on fluids from dead animals, feces, or plant juices. The flowers depend more on odor than on appearance to attract these insects. The shallow flowers are often white or dull-colored, or they may be reddish-brown like rotting meat; the nectar is exposed; and the smell is often musty or rank. The flies attracted to carrion odors and colors quickly determine that the flower is not a dead animal, so some plants trap them to delay their departure so as to be able to acquire and later to deliver pollen. The fungal odors of the California pipevine (*Aristolochia californica*) attract fungus gnats (Mycetophilidae, Diptera) to enter its S-shaped tubular flowers, but long, inward-pointing hairs prevent the flies' exit until the hairs wilt the following day.

The mouthparts of butterflies and moths are suited only to sucking fluids. Most species take nectar, but some moths do not feed at all as adults. They do not ingest pollen, perhaps explaining their preference for higher than average concentrations of amino acids instead of protein-rich pollen.

Coleoptera as Flower Visitors

The Coleoptera that visit flowers are active fliers that frequent open, sunny places, in contrast

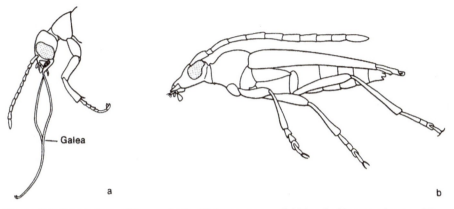

— Galea

a

b

Figure 9.2 Adaptations of flower-visiting Coleoptera: a, meloid beetle, *Nemognatha* sp., with elongate galeae 7.5 mm long; b, cerambycid beetle, *Cyphonotida laevicollis*, with prognathous elongate head.

to their terrestrial, cryptic relatives. Beetles tend to linger in flowers, feeding on pollen and flower parts with their powerful mouthparts and sometimes taking nectar. Elongation of the mouthparts has occurred in only a few cases. The heads of the cerambycids *Strangalia* sp. and *Cyphonotida laevicollis* (Fig. 9.2b) are prognathous, with the anterior portions somewhat elongated. The most spectacular adaptations of this sort in the Coleoptera are the greatly elongated galeae of *Nemognatha* (Meloidae), which form a sucking tube (Fig. 9.2a).

Orchids and Hymenoptera

The final examples of insect pollinators to be discussed are those in which pollen and nectar are not the prime incentives for the insects. Pseudocopulation involves certain species of orchids, such as the Mediterranean *Ophrys* sp., that have flowers that attract male bees. Depending on the species of orchid, the plant emits a chemical mixture that closely mimics the sex pheromone produced by females of a certain species of bees. Some orchid species also have markings resembling a female wasp on the "lip" (labial petal) of the orchid. Male bees of the same species emerge earlier than females and are attracted by the odor and attempt to copulate with the flower. In the course of this behavior the males pick up the adhesive pollinium (a packet of pollen) from the orchid. As they move from flower to flower they pollinate

the orchid. Similarly, in Australia, *Cryptostylis* orchids attract male ichneumonid wasps. Once female wasps emerge, pseudocopulation lures fail to attract male wasps to flowers.

Another improbable relationship is that of the male euglossine bees of the American tropics, which visit orchids. Males of *Euglossa* (Fig. 9.1b) and *Eulaema* are spectacular bees with bright metallic green or blue colors. They are extremely fast flyers and travel long distances in the forest. Nectar is taken from various flowers as flight fuel, but orchids are visited to obtain odors. The floral odors of each orchid species are created by blends of volatile compounds such as cineole, benzyl acetate, eugenol, methyl salicylate, methyl cinnamate, skatole, and vanillin. Males of each species of bee are attracted to certain odor blends. On arrival at an orchid, the males secret lipids from labial glands onto the odor-producing surfaces and pick up the mixture with brushes on their forelegs. The lipids and fragrances are then transferred to special cavities in their large hind tibiae. During their contact with the orchid, the bees fall into a trap device in the flower. In the course of their escape, a pollinium is attached to a specific place on their body. At their next visit to an orchid of the same species, the pollinium is removed by the orchid, thus achieving pollination. Males are also attracted to chemicals in rotting wood. The bees' use of the odors is not clear. Investigators have speculated that the chemicals are metabolized by the males or converted to attractants for either males or females or both sexes.

Figs and Fig Wasps

The last two examples of plants and their pollinators are of obligate mutualism in which host specificity is compounded by the specificity of the larvae that feed on the plant tissues: the fig wasps and yucca moths lay their eggs in the ovaries of their hosts. Initially they probably were only accidental pollinators. Seed production, however, is advantageous to both the plant and the seed-infesting insects. The plants and their pollinators now are completely interdependent but, as we will see, open to exploitation by closely related insects.

The largest group of highly specialized pollinators are the tiny wasps of the family Agaonidae, which pollinate figs. The genus *Ficus* (plant family Moraceae), found in tropical regions, includes about 900 species. Virtually every species of wasp is confined to a single species of fig. The fig "fruit," or synconium, that we eat is actually an inflorescence composed of an enlarged, cup-shaped receptacle that encloses many small, unisexual florets. The small opening of the synconium is called the ostiole and is guarded by scales to prevent entry of nonpollinating insects.

Pollination of the wild caprifig, *Ficus carica*, by the wasp *Blastophaga psenes* takes place as follows (Fig. 9.3). The plant has florets of two kinds: pollen-producing male florets (staminate flowers) located near the ostiole and seed-producing female florets with short styles (pistillate flowers) located at the base of the synconium. Each wasp larva feeds inside a pistillate floret, causing a tiny gall and destroying the seed. At maturity, the adult males are flightless and have reduced legs, eyes, and antennae (Fig. 9.3d). They chew their way out of their galls first (Fig. 9.3b) and seek the galls containing adult female wasps. A hole is bored in the gall by the male, and he copulates with the female by means of an extensible abdomen. The male wasp then dies. The inseminated female emerges from her gall (Fig. 9.3c,e) and makes her way toward the ostiole. She passes through the staminate florets and becomes covered with pollen. Some of the pollen enters folds

of the pleural intersegmental membranes between the terga and sterna of the abdomen. The female's body shrinks slightly, probably by loss of water, and pollen in the folds is trapped. She leaves the fig, flies to another fig in the proper state of development, and attempts to enter it. The process of squeezing through the ostiole pulls off her wings and antennal flagella (Fig. 9.3a). Once inside, she penetrates deeply in the fig to reach the female florets. There she inserts her ovipositor through the short style to the ovarian region, where she lays an egg. In the moist fig, her body is rehydrated and swells, releasing the trapped pollen from the membranous folds of her abdomen. As she moves about, pollen is dusted on the female stigmas. After laying her eggs, she dies, still in the fig.

The fruit of the caprifig is generally not edible, but agricultural varieties have been selected that are edible. The Smyrna variety has only female florets with long styles. The wasps and suitable pollen must be obtained from the caprifig, in which the normal cycle can be completed. Fruits of the caprifig, containing pollen-laden females of *Blastophaga*, are hung in perforated bags among the limbs of the Smyrna fig trees. The *Blastophaga* emerge and enter the Smyrna figs. They are unable to lay eggs because their ovipositor is too short for the long-styled florets. The fig is nevertheless pollinated and normally ripens with seeds into an edible fruit. Some varieties of figs do not require pollination but, if pollinated, yield larger fruits with better flavor.

The behavior of *Blastophaga* inside the caprifig is considered primitive in comparison to most agaonid wasps. In other species the females have special containers in which pollen is packed and later actively placed on the stigmas. Some species of agaonid wasps have evolved a "cuckoo" or "cheater" relationship, ovipositing in the flowers without pollinating the stigmas.

Yucca and Yucca Moths

Another group of host-specific pollinators are the yucca moths, *Tegeticula* and *Parategeticula* sp. (Incurvariidae). All species of *Yucca* (plant family Liliaceae) are American in origin, but they have been introduced elsewhere in the world. More than

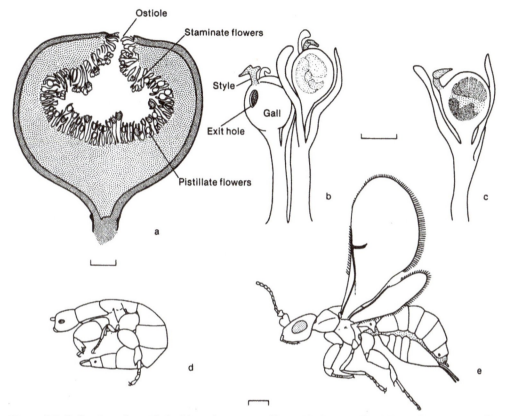

Figure 9.3 Pollination of caprifig by *Blastophaga* wasps (Agaonidae): a, caprifig at time of emergence of wasps (scale equals 5 mm); b, gall flower with male wasp inside and another with exit hole; c, gall flower with female wasp inside (scale for b and c equals 1 mm); d, male of *Blastophaga*; e, female of *Blastophaga* (scale for d and e equals 0.25 mm).

two dozen species of *Yucca* east of the Rockies and the Mojave Desert are pollinated by what was considered one moth species, *T. yuccasella* (McKelvey, 1947; Riley, 1892). In the West, *Yucca brevifolia* is pollinated by *T. synthetica*; *Y. whipplei* by *T. maculata* (Powell and Mackie, 1966); and *Y. schottii* and *Y. elephantipes* by *Parategeticula pollenifera*, as well as *Tegeticula yuccasella* (Davis, 1967; Powell, 1984).

The eastern moths are active at night, the time at which flower scent is also strongest. The female *T. yuccasella* enters the white flower, climbs up the stamens to the anthers, and gathers the pollen in a ball (Fig. 9.4a). The pollen mass from one to four anthers is carried under her head, clasped by a prehensile elongation of the maxillary palpi and the bases of the forelegs. She then flies to another

flower in the proper state of ovarian development. After inspecting the ovary, she bores in with her sclerotized, elongate ovipositor and lays an egg (Fig. 9.4b). Climbing the style, she packs some pollen on the stigma. This behavior is often repeated after each egg is laid. On average, one egg is inserted into each of three compartments of the ovary. A few of the many seeds in pollinated flowers serve as food for the moth larvae. Unpollinated flowers do not develop seeds. When fully fed, the larvae leave the seed pod (Fig. 9.4c) and pupate in the ground. Emergence of the adults is timed to coincide with the flowering season.

Only the genera *Tegeticula* and *Parategeticula* include species with the specialized maxillary palps associated with pollination. Related moths of the

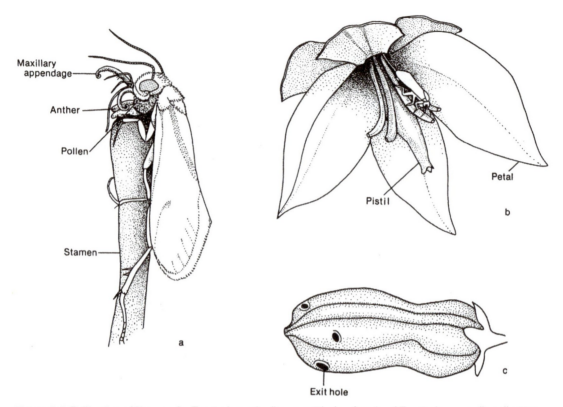

Figure 9.4 Pollination of *Yucca* sp. by *Tegeticula* moths (Incurvariidae): a, female of *Tegeticula yuccasella* collecting pollen from anther; b, position of female moth during oviposition in ovary of *Yucca*; c, mature pod of *Yucca* showing emergence holes.
Source: Redrawn from Riley, 1892.

genus *Prodoxus* are cheaters that do not pollinate their host. Females lay eggs in the inflorescence, and the larvae feed on sterile tissues of the fruit, scape, or leaves. Recently, certain populations of *Tegeticula yuccasella* were discovered to be distinct species that lacked functional palps for pollination and were cheaters on yuccas pollinated by their relatives. In this case, the cheaters feed on seeds, causing a heavy cost to the host plant and destabilizing the mutualistic relationship.

PHYTOPHAGOUS INSECTS AND THEIR MODES OF FEEDING

External and Internal Feeding

An adequate classification of phytophagous feeding habits would be overly complex for our purposes.

Here we shall divide insects by their taxonomic groupings and by their general mode of feeding.

Phytophagous insects can be broadly divided into external feeders, or those whose body is outside the plant (Fig. 9.5), and internal feeders, or those whose body is physically inside the plant (Fig. 9.6b). The first major category, the external feeders, includes a wide variety of insects, both exopterygotes and endopterygotes. During the life history of some species, especially small moths, the larvae may feed first inside then outside.

External feeders may be freely visible to predators and parasitoids. Such exposed feeders are usually protectively colored and patterned if they are large enough to be edible by vertebrates. The manner of feeding is characteristic of the species. Many caterpillars and sawfly larvae feed along the

Figure 9.5 Phytophagous exopterygotes: a, cicadellid leafhopper, *Graphocephala atropunctata*; b, pseudococcid mealybug, *Pseudococcus fragilis*.
Source: Photos by H.V. Daly.

leaf edge. Surface feeders with chewing mouthparts may ingest whole pieces of leaf, leaving holes or notches. Others remove most of the photosynthetic tissues, leaving a delicate skeletonized vascular network. The cellulose-rich feces are called frass. Exposed feeders simply drop frass from their feeding stations.

The feeding injuries of leaf-feeding Hemiptera are often recognized as small, discolored spots where the inner palisade and spongy mesophyll cells are broken down and emptied. Within the leafhopper family Cicadellidae, some subfamilies are primarily phloem feeders, the subfamily Cicadellinae and a few others are mainly xylem sap feeders, and subfamily Typhlocybinae feed on cell contents of mesophyll cells. The composition of the excrement reflects the tissues from which the insects predominantly ingest the most: viscous, sugary excrement for phloem feeders, mostly water for xylem feeders, and solid pellets for cell content feeders. The insect's saliva sometimes has toxic effects on the leaf (phytotoxemia). Thysanoptera also remove the contents of the inner leaf cells, resulting in silvery air spaces inside the leaf. Some external feeders gain protection by feeding on roots or within enfolding leaves. Those that are able to spin silk may bind together leaves in a sheltering cluster. The leaf rollers are caterpillars and sawfly larvae that roll or fold leaves to create a tube. The larvae hide in the tube, feeding at its edge or at night on nearby leaves. Some moth species pupate in the roll. Adult attelabid weevils cut the leaf bearing their egg and the leaf forms a roll naturally.

The weevil larva feeds inside. Casebearers attach bits of plants or frass in a tough silk bag and hide inside for protection.

Members of the second major category, internal feeders, are completely surrounded by living tissues. They are exclusively endopterygotes and usually the larval stages of the life history. Leaf miners are small larvae that eat some or all of the mesophyll tissues between the outer layers of leaf blades or needles. Female insects select the host and lay their eggs on the leaf surface, inserted in the mesophyll, or nearby on the plant. Usually mature leaves are mined. The shape and size of the excavated tissue is characteristic of the species. Larvae of Diptera are cylindrical but softbodied and adaptable to the restricted mine. The other larval miners are highly flattened, usually colorless, legless, and with a flattened, wedgelike head that slopes forward. No adult insects are leaf miners. Linear mines begin as a tiny channel and progressively widen as the larva grows in size and appetite (Fig. 9.6a). The direction of the mine may be altered when vascular bundles are encountered. Blotch mines are created when the larva feeds in various directions, eating both vascular and mesophyll tissues. The frass is often deposited in the mine or ejected through an opening. Lepidoptera may lay frass inside in a continuous line or pack it at one end or plaster it randomly on the mine floor, whereas Diptera deposit two rows. Hymenoptera scatter the frass about the cavity or in piles. Some miners pupate in the mine, while other species leave to form a cocoon elsewhere.

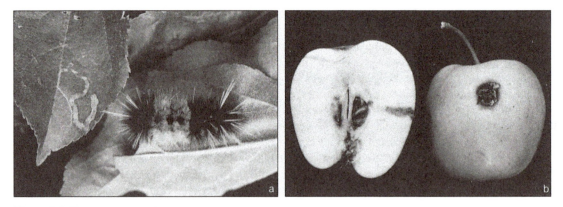

Figure 9.6 Phytophagous endopterygotes on apple: a, leafmine of *Nepticula* sp. (Nepticulidae) on leaf at left, caterpillar of *Halisidota maculata* (Arctiidae) at right; b, gallery of codling moth larva, *Cydia pomonella* (Tortricidae). *Source: Photos by H.V. Daly.*

Plant borers are insects that burrow in living or dead plant tissues other than leaves. The xylophagous (wood-eating) scavengers such as long-horned beetles (Cerambycidae) bore into dead limbs or plants. Their life span is usually long because of the lower nutrient value of the food, variable and sometimes low moisture content, and the stable environment. Most plant-boring larvae tunnel into living buds, stems, roots, fruits (Fig. 9.6b), seeds, nuts, or grains. Life spans of these borers may be short, not only because the food value and moisture are higher, but also because the living plant tissues change rapidly over time.

Although this is a heterogeneous group, the larvae exhibit several similarities arising from their sheltered mode of life. The bodies are often cylindrical, pale, legless, and with bumps or rough areas of skin that provide traction against the burrow walls. The mouthparts, mounted on a retractable head capsule, excavate tissue bit by bit. Antennae are short and retractable into grooves. These specializations give borers access to food rich in nutrients: newly developing leaves, cambium tissue, the flesh of fruits, and endosperm of seeds. Frass is often packed behind the larva as it advances.

Galls

Certain secretions of insects and mites stimulate abnormal development of growing plant tissue, resulting in misshapen leaves or in swellings called galls. These are usually initiated in the spring and early summer when the meristematic tissue is active. The gall results from both cell enlargement (hypertrophy) and cell proliferation (hyperplasy). Aphids feeding on the underside of a new leaf can cause the leaf to curl around them and enclose a favorably high humidity. Hemiptera (chiefly aphids, scales, phylloxera, and psyllids), Thysanoptera, and mites create galls by their feeding secretions. The gall is typically a hollow cavity that may remain open to the exterior and consequently is called an open gall. A group of insects or mites or several generations may live inside the gall, feeding and taking shelter.

The galls of endopterygote insects are usually, but not always, occupied by a single larva. Either the larva or both the ovipositing female and larva seem to be responsible for the biochemical stimulants. The gall stops growing if the insect dies or pupates. The galls do not have a permanent opening and are called closed galls. Closed galls are made by species in the orders Coleoptera, Lepidoptera, and especially Diptera and Hymenoptera (Fig. 9.7). The host plant responds with a growth of specific color and design such that the insect can be identified by the gall it makes. The gall surrounds the larva with moisture and food and protects it from nonspecific natural enemies. In addition to structure and nutrient content, the developing gall also has lower concentrations of plant-defensive chemicals.

Figure 9.7 Cynipid gall. a, female gall wasp, *Andricus* sp. (Cynipidae), selects live oak for oviposition; b, appearance of a mature gall of the summer generation of wasps; c, wasp larvae develop inside the gall.

Parasitoid wasps, however, can lay their eggs in or near the gall-making larvae so that several kinds of insects may emerge from a single gall.

HOST SPECIFICITY OF PHYTOPHAGOUS INSECTS

The great majority of phytophagous insects are fairly host specific. Highly specific insects are usually small. Monophagous insects feed on one or more species in a single genus of plants. Insects with the more restricted range of hosts are usually the leaf miners, borers, and gall makers, which are surrounded by plant tissues, or insects adapted to toxic plants. Oligophagous insects feed on a limited number of plant species, often related to one another as genera within a family. Insects that feed on many species of plants from different families are called polyphagous. They are most often some of our larger insects and are external feeders. The larvae of the gypsy moth, *Lymantria dispar* (Lymantriidae), are known to feed on 458 species of plants in the United States (Leonard, 1974). Migratory locusts (Chap. 4) are polyphagous, opportunistic feeders, but even so they exhibit marked preferences among available plants. The early plant-feeding Neoptera are assumed to have been polyphagous, exposed feeders that now are exemplified by Orthoptera. The oligophagous and monophagous taxa evolved as specialists, with occasional reversions to polyphagous habits.

RESISTANCE OF PLANTS TO INSECT ATTACK

What factors favor the evolution of host specificity? Part of the explanation lies, on the one hand, in the defense strategies of plants against attack. We have seen that all parts of a plant may be eaten: roots, stems, leaves, sap, flower parts, fruits, and seeds. Plant growth is affected by loss of plant tissues. Damage to roots and xylem deprives the plant of water and minerals. Chemicals in an insect's saliva or secreted by the ovipositor may have general or specific effects on the host's physiology. Injured cambium or meristematic tissue results in abnormal growth. These influences lead to stunting, lowered production of seeds, or death. Damage to flowers, fruits, and seeds directly affects the population growth of plants. Phytophagous insects also transmit viruses and microorganisms that cause plant diseases. Yet, despite all the hungry insects, plants still seem to flourish. A major question posed in ecology is: Why is the world so green? In other words, why aren't plants more damaged when there are so many herbivores, parasites, and pathogens?

As the most abundant and important herbivores of plants, insects are thought to have a substantial role in regulating plant abundance and distribution. Compared to insect-pollinator coevolution, it has proven surprisingly difficult to find detailed or direct evidence that plants have adapted defenses to insects countering with adaptations to overcome or bypass new plant defensives. It is possible that plants adapt defensive measures against feeding guilds: groups of insects with similar feeding methods and food types. Because plants have so many different attackers, it is difficult to definitively ascribe a defensive chemical, structural trait, or other trait to avoiding or resisting a particular fungus, bacteria, or insect. In addition, phytophagous insects and plants interact not only with each other but also with climate, natural enemies, and other organisms. Plant diseases, for example, can greatly affect the level of plant resistance or nutritional value for a plant feeder. The evidence for important ecological interactions between insects and plants is substantial. Moreover, the specializations of phytophagous insects for herbivory to overcome particular physical or chemical defenses strongly indicates that plants and insects are continually engaged in evolutionary interactions. Clearly insects have adapted to specific plant defenses, explaining in large part why phytophagous insects are so specialized.

Plants that are less damaged by insects or other organisms because of heritable traits are said to be resistant. Three basic kinds of resistance have been classified by Painter (1958):

1. Resistant plants may not be chosen by an insect for oviposition, shelter, or food; in other words, either some chemical or physical feature of the plant is lacking and it fails to attract insects or some feature is present that is repellent. This is called antixenosis (formerly called nonpreference).

2. Resistant plants may adversely affect the biology of the insect. Physical or chemical properties of the plant may result in the insect's early death, abnormal development, decreased fecundity, or other deleterious conditions. For example, toxins, repellents, copious sap or pitch, or tissue that is nutritionally inadequate for insects could reduce or eliminate attack. This kind of resistance is called antibiosis.

3. Tolerant plants survive even when infested by insects at levels that kill or injure susceptible plants. Rapid compensation for or replacement of lost parts, excess production of seeds, rapid wound healing, and detoxification of or insensitivity to insect salivary toxins are some of the ways in which a plant might tolerate damage.

Antixenosis and antibiosis properties prevent or reduce attack. These properties vary (as do other genetic traits) from plant species to species, geographically with the ranges of species, locally depending on ecology, and from individual to individual. For example, the oleoresins or pitch of conifers traps, expels, or is toxic to boring insects in needles or bark. The variation in physical and chemical composition of oleoresins gives almost every tree a unique chemical individuality (Hanover, 1975). For example, individual trees have been shown to vary greatly in susceptibility to scale insects. Even within a single tree, different branches have been shown to vary in their resistance to insects.

Physical traits can provide significant protection from insect feeding. Tough leaves, bark, and epidermis or plant tissues fortified with lignins, cellulose, or suberin can make chewing or removing plant tissues difficult. Surface hairs (trichomes), waxes and wax crystals, and sticky exudates can impede the movements and feeding of small insects on plant surfaces or facilitate or impede natural enemies in finding and attacking plant feeders. Experimental manipulations of silicon fertilization of plants showed that increased silicon content in plants reduced the feeding of stem-boring caterpillars in rice, fly larvae in rye grass, and other insects in various cereal crops.

Domatia (singular domatium) are plant structures that provide shelters for predaceous arthropods. They are found on many woody plants. Small tufts of hairs at the junction of leaf veins commonly shelter predaceous mites that feed on phytophagous mites and some small insects. Experimental management of domatia showed a major effect on prey populations. In hundreds of tropical plant species, domatia consist of hollow stems or thorns that provide refuges for ants that protect the plants from herbivorous insects. Interestingly, plants defended by ants generally had few protective chemical defenses, suggesting that the ants alone provided adequate protection from herbivores.

Angiosperms have so-called secondary compounds or metabolic products, such as essential oils, amines, alkaloids, terpenoids, phenolics, and glycosides, whose primary function seems to be in chemical defense, although some may be otherwise integrated into the plant's metabolism (Ehrlich and Raven, 1967; Schoonhoven, 2005; Whittaker and Feeny, 1971). These chemicals, incidentally, also give us spices and flavorings. Secondary metabolites are another example of allelochemicals (Chap. 6). The toxic, biochemical shield partially protects flowering plants against herbivores, both insect and vertebrate. The chemicals may be localized in the plant to deter attack on vulnerable tissues. For example, the pyrethrum plant, *Chrysanthemum cinerariaefolium* (plant family Asteraceae or Compositae), has insecticidal pyrethrins concentrated in the seeds of the flowers.

Insects have evolved a variety of physiological countermeasures to detoxify plant secondary compounds. For example, microsomal oxidases, esterases, glutathione transferases, epoxide hydrolases, and other catabolic enzymes break down toxins into harmless products. One hypothesis is that the high rate at which herbivorous insect species develop resistance to man-made insecticides is because herbivores already have adaptations such as catabolic enzymes to overcome plant defensive chemicals (Gordon, 1961). Changes in some of the same enzyme systems that insects use to detoxify plant chemicals also are responsible for evolved resistance to insecticides. As evidence of yet another step in coevolution, the pyrethrum plant

mentioned above also has the compound sesamin, which is an inhibitor of insect oxidases.

Chemical substances in plants that reduce or prevent feeding are called antifeeding compounds. These substances vary greatly in chemistry, including many secondary metabolites. They operate as allomones—they are beneficial to the plant or emitter. Injury to plant cells caused by insect feeding is believed to promptly activate production of certain antifeeding compounds at the site or even some distance from the site. Of special interest are the protease inhibitors that are found in leaves and especially in seeds and tubers, like the potato. Some polypeptides or phenolics in plants inhibit common protease enzymes of insects. These reduce digestion, keeping the insect's gut full, and the nervous system's feedback from the distended gut stops further feeding. Protease inhibitors may function in concert with other allelochemicals to resist insect attack.

Alkaloids in some grasses that deter feeding in insects come from an unexpected source: endophytic fungi that live within the plant. For example, the endophytic fungus *Acremonium lolii* grows intercellularly in rye (*Lolium* sp.) and produces insect repellent alkaloids. The toxins cause poisoning in sheep and cattle but also prevent damage to the grasses by the Argentine stem weevil, *Listronotus bonariensis*, and other insects. By cultivation and selection of endophyte strains, plants can be reinfected with fungal strains that repel serious insect pests but are not toxic to cattle. The endophyte toxins are concentrated in the basal leaf sheath where insect feeding would be most damaging. The endophyte-plant-insect relationship involves a three-way coevolution.

Plant defensive compounds also include insect growth regulators. A wide variety of organisms, including plants, have been found to contain compounds with the physiological activity of some of the developmental hormones of insects (Chap. 4). Extracts with juvenile hormone activity have been isolated from vertebrates, bacteria, yeast, protozoa, and plants, especially the balsam fir. These are called juvenile hormone mimics, or juvenoids. Both ecdysone and 20-hydroxyecdysone, as well as over 65 compounds with similar structure, have been isolated from 111 plant families. Ferns and yews are the major sources of the active substances, called phytoecdysteroids.

Growth regulators of a different kind, called juvenile hormone antagonists, have been extracted from *Ageratum houstonianum* (plant family Asteraceae or Compositae). Named precocene I and II, the compounds cause atrophy of the corpora allata and precocious metamorphosis in hemimetabolous insects. Our last example is the chitin inhibitor plumbagin, which has been found in *Plumbago* (plant family Plumbaginaceae). Insects die at ecdysis because the cuticle lacks mechanical strength. Investigation of natural insect growth regulators from plants has led to the development of several novel pesticides.

Plant-feeding insects must compete with other herbivores for food. The resistance factors evolved by plants further limit the number of kinds of plants available to eat. Insects that are able to survive the antibiotic properties and develop a preference for a resistant plant will not only acquire food but also achieve some relief from interspecific competition. Those able to bore into plant tissues must tolerate immersion in the host's chemical environment and physical constraints, but here also are food and relief from competition, plus escape from certain enemies and protection against desiccation, freezing, etc. Some insects store toxic plant substances and thereby acquire a defense for themselves against vertebrate predators (Rothschild, 1973). Thus a number of advantages accrue to the monophagous or oligophagous insect that is adapted to the special conditions of life associated with one or a few kinds of plant hosts.

In specializing on plants toxic to other competing species, insects resistant to the plant's chemicals use the antibiotic chemicals as cues to locate and identify suitable hosts. For example, plants of the cucumber family (Cucurbitaceae) possess a variety of terpenoid compounds called cucurbitacins. Various species of chrysomelid beetles have specialized on certain cucurbit species and are not harmed by curcurbitacins. The chemicals even serve as cues to find the hosts and stimulate feeding.

The distribution of plants in space and their phenological changes with the seasons increase the demands on insects to locate their specific host plants. Genetic and environmentally induced variability among populations of a plant species further complicate this task for insects. This aspect of plant exposure is called apparency. The highest degree of apparency is in plants that grow close to each other and that have uniform stages of growth and genomes. Such plants present no difficulty for plant-specific herbivores to locate. This describes the most common situation in production agriculture. Even if successful in locating widely separated plants, specialist herbivores that only live for days to weeks can spend a significant fraction of their limited time and energy in searching for hosts, thus reducing the number of eggs and offspring they can produce. Recall that obligately outcrossed plants that occur as widely separated individuals invest considerable energy to provide specific visual and chemical signals and rewards to attract pollinators. A general trend is that highly apparent plants invest heavily in quantitative defenses such as physical toughness, defensive structures, and high concentrations of defensive chemicals that decrease herbivores' digestive abilities and metabolic utilization of nutrients. Plants that are less apparent tend to rely more on toxic and repellent defensive chemicals, but specialist herbivores use these chemicals as cues or signals to locate their hosts.

The defensive mechanisms of plants and host specificity of insects have evolved in response to essentially two periods of contact (Beck, 1965): (1) the period of oviposition, during which the female seeks suitable host plants on which to lay her eggs, and (2) the period of feeding, during which the immature or adult insects eat the plant. Oviposition behavior involves detection and orientation to a host plant at a distance, the search for specific sites in the plant, and, finally, the deposition of eggs, often followed by dispersal. The behavior is a series of complex events that involves the insect's sensory receptors and integration of physiological responses. Any heritable physical or chemical feature of the plant that reduces the numbers of eggs, and thus ultimately the number of feeding insects, will be selectively favored. This is resistance of the antixenosis type and is achieved by either deterring or failing to attract ovipositing females.

INDUCED RESISTANCE

Plant defenses can be costly in the energy required to synthesize defensive chemicals and produce physical defenses. Some defenses are constitutive, that is, they are present at all times, regardless of the intensity of herbivory. More recent is the discovery that some defenses are induced or activated by herbivory, which conserves the deployment of defensive measures for times when that is most needed. Both induced and constitutive defenses can act directly against plant feeders or indirectly by aiding natural enemies of plant feeders. The use of microarrays of plant DNA or RNA have shown that plants respond very quickly (within minutes to hours) to feeding damage by altering the transcription of RNA, thus changing gene expression. Simple mechanical damage such as leaf abrasion or tissue removal can increase the expression levels of defensive chemicals, but in other cases the expression is greater or only occurs when a particular phytophagous insect causes the damage. Feeding by tobacco hornworms (*Manduca sexta*) caused altered transcription levels in over 500 RNA in tobacco plants. Not all of these genes, of course, are involved in plant defense, but clearly many do. Induced effects may be localized near the feeding injury or may be systemic throughout the plant. Some induced changes are long term in nature. For example, the concentration of phenolic compounds in birch trees in Finland was a function of the intensity of feeding by caterpillars. The phenolic content, in turn, directly lowered the nutritional value of birch foliage for the caterpillars. A short-term increase and decrease in phenolics occurred over a matter of days, but a longer-term effect lasted for months to as long as 2 years.

Induced indirect resistance increases the effectiveness of natural enemies to reduce herbivory. For example, oviposition by the green stinkbug, *Nezara viridula* (Pentatomidae, Hemiptera), induced the production of volatiles in bean plants that attracted an egg parasitoid. Maize plants fed on by caterpillars of the fall army worm *Spodoptera exigua* emitted terpenoids that attracted the parasitoid *Cotesia maginiventris*. Mechanical wounding of plants did not cause the emission of volatile chemicals, but wounding combined with applying regurgitated fluid from the caterpillars caused an identical response in the parasitoid.

Some induced resistance responses are signaled not only systemically through plants but also from plant to plant. Lab experiments confirm field findings that volatile emissions induced by insect feeding damage can also induce resistance responses in nearby plants and induce the attraction of natural enemies of the feeding insects.

FEEDING BEHAVIOR OF PHYTOPHAGOUS INSECTS

Insects' location and acceptance of host plants follow stereotyped sequences of behaviors called reaction chains. There are two main phases: (1) searching and (2) contact-testing. Visual, tactile, and chemical cues provide the stimuli and information for finding and accepting the plant. Different insect species vary considerably in the nature and relative importance of different kinds of cues. A typical sequence for aphids locating a plant illustrates this type of analysis. Experimental flight chamber studies show that winged aphids initially fly toward the sky (ultraviolet light). For landing, aphids appear to be trying to move opposite to the sky (away from ultraviolet light). At this stage reflective plastic sheeting repels aphids from landing, presumably because it reflects ultraviolet light, a fact that is useful in preventing aphids from transmitting viruses to some crop plants. Landing aphids respond both to distinctive plant odors and to certain colors. Upon landing the aphid walks over the leaf surface and makes frequent, brief probes of the plant's epidermal cells with its needlelike mouthparts. Typically the aphid makes a series of short flights to nearby plants and repeats its probing. Sensory cells at the tip of the rostrum (labium) perceive surface physical and chemical features. Repellent cues cause the aphid to fly. As it contacts the phloem elements of the plant, it sustains ingestion of phloem sap and either accepts the plant by remaining on it to feed using signals from taste sensory cells in its cibarium or it leaves the plant to continue searching. The behavioral response at each step depends on releasing stimuli provided by the plant and on the insect's response thresholds, which vary with its physiological state.

Resistance of the antixenosis type would be given a plant that lacked the appropriate releasing stimuli or that discouraged feeding at some step.

Physical and chemical stimuli have been classified according to the response they elicit from insects (Beck, 1965; Dethier et al., 1960). During orientation to a host plant at a distance, certain stimuli, like volatile chemicals, may act positively as attractants or negatively as repellents. When in close contact with the plant, positive stimuli may stop further locomotion (i.e., be arrestants) or act as repellents to hasten the insect's departure. At the initiation of feeding, positive stimuli are called incitants or stimulants and negative stimuli are suppressants or deterrents. Feeding is maintained by phagostimulants and may be terminated by high concentrations of antifeeding compounds, also known as feeding deterrents. For growing insect immatures that feed on low-nutrient foods, stimulants may be critical nutrients that determine whether or not feeding is sustained. Some caterpillars consume several times their weight daily; xylem-feeding leafhoppers can consume up to a thousand times their weight per day, so sustaining food intake can be critical where massive intakes of low-nutrient diets is required.

The sensory apparatus and orientation behavior of insects are finely tuned to the characteristics of the desired host plants. The sensory receptors on the antennae and maxillae of caterpillars are chiefly involved in discrimination. When such receptors are experimentally removed, oligophagous caterpillars often accept as food a broader range of plant species. Although the female usually selects the host plant when she lays her eggs, the caterpillar must select the parts of the plant to eat, avoiding concentrations of toxins and finding the richest food.

Secondary plant substances that, as allomones, are repellent to most insects are, in fact, often the feeding stimulants for the appropriate monophagous insects. The same compound can be an attractant or feeding incitant, thus serving as a kairomone and benefiting the perceiver. The attraction of certain chrysomelid beetles to cucurbitacins in cucurbits was mentioned above. The terpenoid gossypol, produced by *Gossypium* or cotton plants, serves as a repellant to the noctuid moths, *Helicoverpa zea* and *Heliothis virescens*, and the meloid beetle, *Epicauta*, but is an incitant for the boll weevil, *Anthonomus grandis*. In deciding to feed on a plant, the absence of antifeedants is apparently more important than the presence of feeding incitants. Young needles of pine with antifeedants are not consumed by sawfly larvae, but older needles, with reduced concentrations of antifeedants, are consumed. Nutrients, including sugars, amino acids, phospholipids, and ascorbic acid, can also be stimulants to certain insects.

CHAPTER 10

Entomophagous Insects

In this chapter we discuss insects that kill or injure one or more other invertebrates before completing their life cycle. Most of these carnivorous insects feed on other insects and are said to be entomophagous (Fig. 1.3c,d). Snails, earthworms, millipedes (Fig. 10.1a), spiders (Fig. 10.1b), mites, and other terrestrial and freshwater invertebrates are also eaten by insects. Such animal bodies, with their high nitrogen and water content, are among the most nutritious of foods eaten by insects. By feeding on other insects and invertebrates in their communities, entomophagous insects contribute to the control or regulation of the density of individuals in the prey populations. Applied biological control makes use of the regulatory action of entomophagous insects to lower the density of pest organisms below a level that causes economic damage (Chapter 12). In this chapter we discuss the life histories, behavior, and evolution of entomophagous insects.

TYPES OF ENTOMOPHAGOUS INSECTS

For simplicity in discussing entomophagous insects, we provide several definitions: Victims are called prey if killed directly, hosts if fed upon while still living. According to their mode of feeding:

- Predators kill and consume more than one prey organism to reach maturity.
- Parasitoids require only one host to reach maturity, but ultimately kill the host.
- True parasites feed on one or more hosts, but do not normally kill the host.

Although these definitions will apply to most insects without confusion, some problems should be noted. First, the word "parasite" is widely used in ecology for an organism that obtains nutrients from another organism, causing harm but not necessarily death. Unfortunately, in entomological literature a parasitoid is often called a parasite and may be said to parasitize an insect host, even though the host dies. Furthermore, terms like "hyperparasite" or "superparasite" are used in reference to parasitoids. We prefer to distinguish between parasitoids and true parasites, and, further, to define the terms hyperparasitoid and superparasitoid. Second, the behavior of sphecid wasps does not easily fit our definitions. The female wasps prepare nests that they provision with paralyzed (i.e., living but immobile) insects as food for their larvae. Should the victims be called hosts or prey? Some species of wasps place only one victim in a nest, while others place two or more victims. Should the former be called parasitoids and the latter predators? To avoid confusion within this single family of related wasps, we simply call all these wasps predators and their victims prey. Biologically the distinction between these wasps and ectoparasitoid relatives is small.

Parasitoid insects are intermediate between predators and parasites in the sense that they live at first parasitically at the host's expense but ultimately kill the host. From an ecological viewpoint,

Figure 10.1 Insect predators of other arthropods. a, With its sting this female *Pepsis* sp. (Pompilidae) wasp will paralyze the tarantula. The spider will serve as immobilized food for the wasp's larvae in an underground burrow. If the spider were killed, it would soon spoil and be unsuitable. b, An adult, but larviform, female of *Zarhipis integripennis* (Phengodidae) is eating a millipede. The adult male has wings and bipectinate antennae.

predators and parasitoids both act to eliminate individuals in the prey or host population, whereas true parasites permit the host to continue functioning in the community, though often at a suboptimal level.

Cleptoparasites and Other Kinds of Parasitic Behavior

Before further discussion of predators, parasitoids, and parasites, some additional kinds of behavior should be mentioned in which an insect consumes food collected at the energetic expense of the host or otherwise causes harm. Although the insect may not be directly entomophagous, the net result is a reduction in the number of host offspring and an increase in the offspring of the insect that benefits. Cleptoparasites, or "cuckoo" parasites, lay their eggs in the nests of other species, in the manner of cuckoo birds. The cleptoparasitic parent or larva may kill the egg or larva of the host immediately, or the host larva may die of starvation after the cleptoparasite larva eats the nest provisions. In the former instance, the young cleptoparasites often possess enlarged mandibles suited for attack. For example, among the megachilid bees, the genus *Coelioxys* is cleptoparasitic on the related genus *Megachile*, which provisions a nest with pollen and nectar. A social parasite is a female that enters the nest of a social host and takes over the role of the queen. The social parasite's offspring are fed by host workers at the expense of host offspring. Females of the bumblebee subgenus *Psithyrus* are social parasites of their close relatives in other subgenera of the genus *Bombus*. Parasites that spend much of their life in their host's nests and consume food are known as inquilines.

Slavery is practiced by certain ants. Worker pupae of another species are taken by slave-making workers. The resulting adult slaves then become members of the colony and do most of the work. Social species may also rob one another. The robbing species enters the nest of another colony and removes the stored food.

Phoresy is the transport of an insect on or physically inside the body of another insect. The insect that provides the transportation is usually not harmed. When the passenger is an adult female parasitoid or predator, however, the usual result is that the passenger remains aboard until the host lays its eggs. The passenger then oviposits on the host's eggs. This habit occurs frequently in the wasp family Scelionidae. The first instar larvae of meloid and rhipiphorid beetles, stylopid parasites (Strepsiptera), and eucharitid wasps are commonly phoretic on adults of their favored hosts. By this means the larvae gain entrance to the nests of the hosts.

RESISTANCE OF HOSTS OR PREY TO ATTACK

Prey insects may avoid or prevent attack by entomophagous insects by actively biting, stinging, or kicking. Enemies may be excluded by tough cocoons, hard puparia, nests of mud or leaves, or by the inaccessible position of the insect in burrows in soil or plant tissues. Some still succumb to parasitoids and predators that are suitably equipped with strong mandibles or long, powerful ovipositors.

Exposed insects may violently resist attack or quickly escape. When approached by an enemy, an aphid may drop from the plant, walk away, or kick and secrete oily droplets from its siphunculi. The odor of the siphuncular secretion alarms other aphids, and they drop off or walk away. Larvae and pupae of many Lepidoptera or Coleoptera thrash about when contacted. Larvae may bite or regurgitate or secrete defensive fluids. One of the advantages of social behavior is mutual defense against insect enemies. In a remarkable behavior, Asian honey bees defend their colonies against vespid wasps by covering each invader for up to 20 minutes with a mass of worker bees whose metabolic heat raises the wasp's temperature to a lethal 46 to 48°C.

It seems possible that some of the defensive strategies to avoid predation by vertebrates are also effective against insect enemies. Certain aphids are distasteful and are avoided by insect predators after an initial contact. The gustatory senses and learning ability of the predator are therefore important to the success of this defense. The role of protective coloration and behavior, however, is generally less effective because many insect predators, unlike vertebrate predators, do not depend on vision to find prey.

Fourteen orders of insects are known to encapsulate (see Chap. 5) foreign bodies, endoparasitic worms, and parasitoids. Among the parasitoids, larvae of certain species definitely stimulate encapsulation in hosts that are not the normal hosts: maggots of Tachinidae and some, but not all, wasp larvae of Ichneumonidae, Braconidae, Chalcidoidea (Encyrtidae, Eulophidae), Proctotrupoidea, and Cynipoidea. Yet in their normal hosts, these parasitoids usually escape encapsulation. How do they succeed?

Among wasps, some attack only eggs. By rapidly completing their growth before the host's hemocytes develop, they avoid encapsulation altogether. Rapid destruction of a larval host also avoids encapsulation by simply killing the host quickly. If the parasitoid lingers for a while and allows the host larva to grow, a larger food supply becomes available, but at a greater risk of encapsulation. Certain parasitoids insert their eggs precisely into an organ of the host, where the endoparasite will be separated from the host's hemolymph by a layer of connective tissue. This prevents encapsulation. Others first invade the alimentary canal and are sheltered from the host's hemocytes. When the host is suitably larger, the parasitoid feeds voraciously and so debilitates the host that it is unable to react in time. Some parasitoids are initially invested in cellular membranes, the cells of which become dissociated, enlarged, and circulate in the host's hemolymph. Apparently the cells absorb nutrients and similarly act to reduce the host's ability to muster an encapsulation.

Parasitoids in the Ichneumonoidea have viral DNA (called polydnaviruses) incorporated into their own chromosomes. The gene products of the virus disarm the ability of the host caterpillar to encapsulate the egg and the wasp larva (Box 10.1).

PREDACEOUS INSECTS

Predators are usually larger in body size than other entomophagous insects. The need for more than one prey organism makes predators dependent on prey populations of higher density than other entomophagous insects. Prey are usually smaller in body size than the predator but proportional to the predator's size. In other words, larger predators take larger prey and smaller predators take smaller prey. These generalizations about relative body size have some exceptions. Dragonflies are relatively large predators, yet they mainly feed on small prey. Wasps with stings (Fig. 10.1b) and ants in groups can kill prey larger than themselves. Predators usually eat their prey immediately unless, as in some wasps, the prey are stored for later consumption by their larvae.

The same terminology can be used here as in describing the choice of foods by phytophagous insects. Monophagous predators feed on one species of prey, oligophagous predators take several species of prey, and polyphagous predators take many species of prey. The last type are the "generalists." They take individuals of prey species largely in proportion to the relative abundance of the different prey species. Thus polyphagous predators, by continually shifting to the most abundant species of prey, tend to regulate populations of prey species in the community. Furthermore, if mobile, polyphagous predators are able to thrive in disturbed communities where prey species vary in abundance as ecological succession proceeds. At the other extreme, monophagous predators are

density-dependent on one species of prey and may regulate the prey at lower levels than polyphagous predators. Monophagous predators tend to be associated with undisturbed communities where their host maintains continuous, stable populations.

A majority of predaceous taxa are carnivorous in all feeding stages. This does not mean they are always exclusively carnivorous; exact diets vary species by species. Adults of some of these visit

---

BOX 10.1   How Do We Know This? Unraveling the Strange World of Parasitoid Viruses

Life on earth has evolved, interacted, and diversified to contain some surprising and bizarre partnerships. One of the most ecologically ubiquitous, yet initially unexpected, of these is that between viruses and parasitoid wasps. Mutualisms between viruses and eukaryotes are rare to begin with, but the virus/wasp partnership adds to the intrigue paradoxically by being evolutionarily driven also by the opposite trend—parasitism, in this case parasitism of caterpillars by the wasps. Literally thousands of species of caterpillars, many of them of considerable agricultural and ecological importance, are affected by parasitoids that compromise their immune systems using these viruses, called polydnaviruses due to their "*poly*disperse DNA" genomes (multiple circlets of double-stranded DNA). The wasps that carry these viruses diversified during the Cretaceous and Tertiary into huge lineages of tens of thousands of species, distributed throughout the world.

The dependence of the wasps on their associated viruses for their larval survival inside host caterpillars is complete. A number of studies have demonstrated that not only is the virus involved in interference with normal function of the immune system of host caterpillars, but they intricately manipulate the host endocrine system in ways that favor the developing wasp larvae. This obligate relationship is also so genetically intimate that the two biological taxa can no longer be considered as separate entities. Indeed, the viral DNA is integrated into the chromosomes of their host wasps and is only replicated, packaged, and injected as needed

into caterpillars as a sort of molecular "venom" against these wasps' hosts.

How did such a unique and complex interaction arise? What kind of virus was the progenitor of this viral symbiont? These questions have only recently been begun to be answered. The symbiotic viruses associated with the parasitoids of the huge braconid wasp microgastroid lineage (Box Fig. 10.1) share a number of core housekeeping genes with the recently characterized nudiviruses, which are also associated with insects and other arthropods. This historical relationship with nudiviruses was initially difficult to detect due to the high level of genetic divergence between the two viral groups, but the number of genes with apparent closely shared history is now impressive.

The clarification of the nearest living relatives of the parasitic wasp–associated bracoviruses explains the evolutionary significance of these viruses. Despite many years of research on parasitoid wasps and their relationships with host insects, it was not until 1967 that viruslike particles were discovered to be coating the exterior of wasp eggs injected into caterpillars by the wasps. Just by unfortunate chance, the first investigated system involved a case where the "viral" particles no longer contained any DNA, so it was not until the mid- to late 1970s that the apparently viral nature of these particles became evident. Even then, the fact that these "viruses" do not replicate in the host caterpillars, but are inherited vertically within the wasp genomes, was taken by some as a sign that they might not be viruses at all (either currently or in origin), but instead might represent sophisticated

(Continued)

BOX 10.1   *(Continued)*

Box Figure 10.1 *Cotesia congregata* oviposits into its host, *Manduca sexta*. As the *Cotesia* female introduces its eggs into the lumen of the *Manduca* hemocoel, it also introduces capsids of a symbiotic polydnavirus. Expression of the viral genes affects both the immune and endocrine systems of the caterpillar in ways that benefit the wasp's larval offspring.
Source: Photo by Alex Wild, used with permission.

"genetic secretions" developed by the wasps to deliver virulence genes into host insects.

Another feature of the wasp-virus association that was noted very early on was the restriction of the association to only certain taxonomic groups of the wasps. With the advent of PCR and easy DNA sequencing of gene homologs, it was later possible to show that the wasps carrying the viruses formed a monophyletic lineage that shared a common ancestor some 74 to 100 million years ago. This finding suggested that the viral association (or alternatively the wasps' "genetic secretion" invention) might have originated only once within braconid parasitoid wasps. There was thus either a very old, and subsequently diversified, symbiosis between wasp and viruses, or alternatively an anciently shared (and remarkably complex and viruslike) wasp innovation that enhanced the parasitism of caterpillars.

Following this, the genomic age brought the reporting of the first complete DNA sequence of a polydnavirus and then the first comparative genomics of these mysterious entities. The genomic findings were initially bewildering: not only were polydnaviruses (including bracoviruses) incredibly complex, with large gene families resulting from repeated gene duplications and subsequent divergences, but almost none of the obviously functional genes appeared to be shared with other viruses! While this result would first appear to suggest that the polydnaviruses may not after all have been derived from other viruses, it was also realized that virtually none of the genes packaged inside the "viral" particles were the usual housekeeping and packaging genes necessary to manufacture the virions in the first place. Those genes must then be found elsewhere, perhaps nearby, in the wasp genome. When found, would they be unique, as would be consistent with a wasp invention, or would they be found to share sequence motifs of genes of similar functions from other viruses?

We now know the answer to be the latter, and the question of where bracoviruses came from seems to be settled. Many fascinating questions still remain: Why do bracoviruses package essentially none of their original genes into virions? Where did all the other virulence genes used against host caterpillars come from? Does expression of those genes confer the exceptional host specificity we see in these parasitic wasps? Bracoviruses and other parasitoid viruses provide a wealth of other research surprises for the foreseeable future.

flowers for energy-rich nectar, as do certain adults of taxa that are predaceous only as immatures. Scavenging, honeydew, symbionts, and plant foods also may supplement a predator's diet.

Some of the complexities of diet are revealed in the following example. Hagen and his associates (1970) have carefully examined the diets of lacewings of the genus *Chrysopa* (Chrysopidae). Adults of about half of the species have larger mandibles and are predaceous like the larvae. Adults of other species have shorter mandibles and feed only on honeydew and pollen. The foreguts of the second group contain yeast symbionts and are supplied with larger tracheal trunks for increased respiration. By this means the essential amino acids are acquired even though the adult diet lacks animal prey. Predaceous adults of the first group lack yeast and tracheal modifications. Females of both groups of *Chrysopa* require substantial amounts of food before eggs are produced. As a consequence, females lay eggs only near abundant food sources that will later supply the lacewing larvae. An advantage to the honeydew-feeding species is that prey of the larvae may include not only aphids but a variety of other insects attracted to honeydew.

In contrast to other entomophagous insects, predators of both sexes must repeatedly find and subdue prey. Predators may be active during day and night. The eggs of predaceous insects are usually deposited by the females in close proximity to or at least in the vicinity of suitable prey. In this way, the adult's well-developed compound eyes, chemoreceptive organs, and ability to fly are used to search for prey-rich habitats within which the less well-equipped immatures can forage. Plants of a certain height containing odors of honeydew, odors of decay, and even the pheromones of prey are attractive to searching predators. The adult may eat prey or other attractive foods when they are found or may only lay eggs near the prey.

Many predatory insects utilize extraoral digestion. This allows insects with piercing-sucking mouthparts like the Heteroptera to inject powerful protease enzymes into the prey and later suck out the liquefied tissues. The predaceous larvae of Neuroptera and some Coleoptera have mouthparts

for piercing-sucking and injecting enzymes that have been modified from the mandibulate type. Some larval and adult Diptera with their specialized mouthparts also feed in this manner. Extraoral digestion offers several advantages to predators: relatively larger prey can be subdued by smaller predators; concentrated nutrients can be extracted more quickly and completely from the prey; and the ingested food contains no bulky indigestible pieces to be passed through the gut.

Strategies of Insect Predators

Several kinds of strategies are used by insect predators in finding and capturing prey: (1) random searching, (2) hunting, (3) ambush or "sit and wait," and (4) trapping.

Insects of the first type roam in the appropriate microhabitat and seize prey after physical contact. Their orientation to objects in the microhabitat and their movements may increase the probability of encountering prey. For example, the predator may patrol leaf edges, veins, or stems and eat insect eggs or sedentary Homoptera. Random searchers may have either monophagous or polyphagous habits. The prey is accepted if proper incitants are detected by receptors on the forelegs, mouthparts, or antennae. The victim is then devoured using mouthparts that are usually specially adapted for predation. After an initial contact or meal, further searching often involves more frequent turns so that the predator stays in an area of previous success. While seemingly inefficient, the random searchers are probably the most common type of insect predator and include species highly effective in regulating prey populations.

Most predatory Coleoptera are random searchers. Predatory carabids are distinguished by long, sharply hooked mandibles that contrast with the broad, blunt jaws of plant-feeding relatives. Coccinellid beetles, syrphid larvae, and neuropteran predators that feed on aphids also search at random. The mandibles of predaceous coccinellids may be incisors with one or two apical teeth and a basal tooth, or they may be small with ducts for sucking prey juices. Neuropteran larvae have sickle-shaped jaws, each of which is formed by the mandible and maxilla locked together to create a tube through which body fluids of prey are sucked (Figs. 36.1b, 36.2, 36.3a).

The Lepidoptera are almost exclusively phytophagous. Caterpillars of certain moths retain their close association with plants but attack other phytophagous insects such as coccids and leafhoppers as well as mites. In Hawaii, eighteen species of geometrid caterpillars in the genus *Eupithecia* have evolved into ambush predators of small insects. Although they appear like ordinary inchworms perched on a leaf, the caterpillars wait for long periods until an unsuspecting victim approaches and is quickly seized. Caterpillars of the lycaenid butterflies may be phytophagous or partly or wholly predaceous on ant larvae and pupae or aphids. Under crowded conditions, larvae of noctuid moths become cannibals.

The relatively small size and high nutritive value of insect eggs make them vulnerable to predation by insects of varied food habits. Scavenging, phytophagous, predatory, or parasitoid insects may consume eggs. Collectively they are called egg predators. These predators are of special ecological importance because the prey never function in the community. The large clustered eggs of Orthoptera are a special case: They are subject to frequent attack by many predatory larvae: clerids, meloids, bombyliids, rhagionids, anthomyiids, calliphorids, otitids, phorids, sarcophagids, and eurytomids. The adult female of the predatory species finds the eggs and lays her own eggs, and the larvae feed essentially as random searchers among the mass of eggs of the host.

The hunting insects differ from random searchers by using sight or other stimuli to orient to prey at a distance. Visual hunters have enlarged compound eyes with overlapping fields of vision that permit distance perception. Mandibles may be sharply toothed, and the legs may be strong and spiny for seizing elusive prey. Strong fliers often carry their prey to perching places or nests. Conspicuous examples of such aerial hunters are the adults of dragonflies, asilid flies, and the aculeate or stinging wasps. Stinging wasps are able to subdue prey such as katydids or spiders that are physically larger than themselves by injection of paralyzing chemicals with their sting. Dragonflies

such as *Anax* (Aeshnidae) fly almost continuously, "hawking" for prey, while *Libellula* (Libellulidae) remain perched until prey approaches, then quickly dash in pursuit. Asilids commonly perch and await flying prey, then return to the perch after the victim is caught (Fig. 2.10c). Adult cicindelid beetles pursue their prey by running fast on open ground while the larvae ambush prey from burrows.

Stimuli other than visual may be used by hunting predators. Kairomones may attract predators to prey (Chap. 6). Although blind, the famous army ants (genus *Eciton*) of the New World tropics are enormously successful predators. Odor and movement of prey are detected by chemical and tactile receptors while the colony is on one of its periodic raids. Prey many times larger than the individual worker ants are attacked by massive swarms. Cooperative foraging by "packs" of other kinds of ants also subdues large prey.

The backswimmers (Notonectidae) are voracious aquatic predators. They flush and stalk prey by sight and are able to detect the vibrations of prey movements with receptors situated along the forelegs. The water striders (Veliidae) also perceive prey by sight and vibrations.

Insects that ambush prey conserve energy by simply waiting for prey to approach within striking distance. Reduviid and phymatid bugs often remain motionless on flowers awaiting visitors. Such predators are exposed to predation by vertebrates and have either concealing or warning coloration. Concealing coloration may also prevent detection by alert prey. Raptorial or clasping forelegs are frequently characteristic of insects that ambush, especially among the Hemiptera (Fig. 32.2a–c). Some reduviids aid prey capture by smearing sticky secretions or plant resins on their legs. The beaks of predatory bugs are stout and inject saliva that contains proteolytic enzymes. This is why their bites are painful to humans in contrast to the mild bites of bugs that are parasites of humans. Some reduviids also have a potent toxin that quickly renders prey helpless. Once the tissues of the victims are digested, the resultant fluid is sucked by the bug.

The praying mantis is a familiar predator that strikes prey from ambush. The coordination of the depth-perceiving vision, mobile head and prothorax, and toothed, raptorial forelegs has been carefully analyzed. The complete strike, timed by high-speed photography, takes 50 to 70 milliseconds. A fly or cockroach would require about 45 to 65 milliseconds to respond if startled by the first movement of the mantis—alas, too slow to escape the strike already in motion. The dragonfly naiad similarly grasps prey with quick strikes of its extensible labium. Although tiger beetle larvae (Cicindelidae) have stemmata as their primary visual organs, the larvae apparently have keen vision because they are able to seize prey that pass near the entrance to their burrows in soil.

Only a few kinds of insects trap prey. The "ant lion" larvae of *Myrmeleon* (Myrmeleontidae, Neuroptera) excavate conical pits in fine, loose sand (Fig. 36.3b). The larva waits motionless and buried at the bottom until a small insect ventures into the pit. The antlion flicks sand toward the victim to cause the unstable sand to slide down. Once the prey is seized, the antlion extracts the body fluids with its sickle-shaped jaws. Larvae of the fly *Vermileo* (Rhagionidae) also construct pits and trap prey. Wheeler (1930) aptly describes these insects as "demons of the dust."

Larvae of the New Zealand glowworm *Arachnocampa luminosa* (Mycetophilidae) live in moist caves. They spin slimy webs that dangle from the ceiling. The larvae glow in the darkness, attracting small flies that become entangled in the sticky webs. Larval mycetophilids of the genus *Platyura* also secrete webs that entangle prey and contain a toxic fluid. Even though silk is secreted by various insects, webs are surprisingly rarely used to trap prey in the manner of spiders. The aquatic caddis fly larvae of the Hydropsychidae spin webs that filter small prey and other food from passing currents. Though they do not spin their own webs, one group of slender reduviid bugs, the emesines, wait in ambush at spider webs to attack trapped prey.

PARASITOID INSECTS

Only the larvae of holometabolous insects, especially Diptera and Hymenoptera, are parasitoids. Their unique attributes combine certain features of parasites and predators. In addition to their requirement for only one host, parasitoids differ from predators in the following features: (1) the host is larger than the parasitoid; (2) parasitoids are frequently host-specific, attacking one or several related host species; (3) a lower density of host population will sustain a parasitoid population; and (4) the victim is usually searched for and selected by the diurnally active, adult female. A species of parasitoid may be specific not only in the choice of host species, but also in the choice of the life stages of the host to be attacked. Eggs or young or both are most frequently attacked, but pupae and sometimes adult hosts are also eaten.

Types of Parasitoid Insect

Parasitoids may feed externally as ectoparasitoids or internally as endoparasitoids. Hosts openly exposed to other predators are usually attacked by endoparasitoids that conceal their presence by being inside the host. Hosts in protected situations such as leaf mines, galls, or nests are attacked by either endoparasitoids or ectoparasitoids. A further distinction is also recognized—that between idiobionts, which rapidly consume their (often paralyzed) host, and koinobionts, which interact with and feed upon the host for an extended period of time, often allowing the host to develop beyond the stage originally attacked. There is a general tendency for ectoparasitoids to be idiobionts and endoparasitoids to be koinobionts, but there are exceptions.

When the larva of a species characteristically develops in the ratio of one to a host, the species is termed a solitary parasitoid. When several larvae of the same species normally develop in a single host, the species is called a gregarious parasitoid. If the host is a phytophagous insect, the parasitoid is called a primary parasitoid. Parasitoids that attack other parasitoids are called hyperparasitoids or secondary parasitoids. Multiple parasitoidism occurs when two or more species of primary parasitoids attack one host individual. Superparasitoidism occurs when a host is attacked by more larvae of the same species than can reach maturity in the one host.

Adult parasitoids are free-living, and most of them are winged except for female velvet ants (Mutillidae) and certain other solitary aculeate wasps. Nectar and honeydew are frequently consumed for energy to fly. Some parasitoid wasps feed on the body fluids that are released from the host's body when it is punctured by the parasitoid's ovipositor. This is called host feeding. In some instances, the host is inside a cocoon or for some other reason can be reached only by the parasitoid's ovipositor. A tube of coagulated host's hemolymph, or possibly of a secretion produced by the parasitoid, forms around the ovipositor. When the ovipositor is withdrawn, the parasitoid is able to suck the host's hemolymph as it wells up in the tube.

How Parasitoids Find and Attack Hosts

The adult parasitoid searches first for the habitat in which the appropriate hosts exist. Primary parasitoids are often attracted mainly by the plants that shelter favored insects hosts. In this behavior the parasitoids respond like phytophagous insects to shapes, colors, and odors of plants, yet they do not feed on green plant tissues. Their search movements may be essentially random or somewhat systematic in response to stimuli detected at a distance. Parasitoids are also known to be attracted by the sex pheromones or defensive secretions of their preferred hosts. Once near a potential host, female Hymenoptera examine suitable individuals on which to oviposit, using the tactile and olfactory senses of their antennae. Female *Trichogramma* (Trichogrammatidae, Chalcidoidea) walk back and forth on a host's egg while drumming with their antennae. By this means they determine the acceptability of the host, the number of eggs to lay, and often the sex of the host egg. The wasp then inserts her ovipositor through the egg chorion. Olfactory receptors at the tip of the ovipositor respond to the egg contents. At this point, the female may reject the host and withdraw the ovipositor without oviposition. Otherwise, she will lay an appropriate number of eggs.

The ovipositor of Hymenoptera plays an important role in locating and selecting the host, sometimes in paralyzing it, and in delivering the egg to the host (Figs. 39.3, 39.10, 39.12d). Chemoreceptors on the ovipositor tip respond to a variety of chlorides, aliphatic alcohols, and hydrochloric acid in much the same way as taste receptors on the anterior appendages. These receptors function in the final discrimination of hosts before the egg is laid. Hosts inside tough cocoons or sclerotized puparia, in leaf mines, or even deep in burrows in solid wood can be reached by parasitoids equipped with ovipositors of suitable length and strength. By alternate movements of the sharply ridged valvulae, parasitoids can drill through resistant substrates. Receptors at the tip sense the presence and suitability of the host. The highly elastic eggs then pass down the minute channel inside the ovipositor by becoming greatly elongated. After deposition in the host, eggs of some species increase in size by as much as 1000 times. Some wasps temporarily or permanently paralyze or kill the host before ovipositing. For example, the venom of *Bracon hebetor* (Braconidae) causes permanent paralysis of host caterpillars when diluted up to 1 in 200 million parts of host hemolymph.

Female Diptera lack the needlelike ovipositor of wasps for precisely inserting eggs in hosts. Consequently, some flies stick their eggs on the bodies of their hosts while others depend on the first instar larvae to actually locate hosts after the eggs are laid in the proper habitat.

Insects that have two or more successive larval instars specialized for different modes of life are hypermetamorphic (Fig. 39.5). Among the hypermetamorphic parasitoids, the first instar larva is specialized for active host finding and the subsequent instars are specialized for feeding.

Superparasitoidism and multiple parasitoidism may be advantageous to the parasitoids in that the host's encapsulation reaction is dissipated. The resulting competition among larvae also may be disadvantageous. First instar larvae of certain wasps are specialized to compete successfully in such situations. Some have large sickle-shaped mandibles and kill competitors; others release inhibitory toxins or, if larger, deprive smaller competitors of oxygen. Competition is also commonly reduced by the behavior of the female wasp. For example, *Trichogramma* (Trichogrammatidae, Chalcidoidea) reject hosts on which other females have walked. Some wasps seem to be able, by means of sensilla on their ovipositors, to select host individuals that have not been previously attacked.

Endoparasitoids are immersed in the body tissues and fluids of their hosts. Many synchronize their development with that of the host by responding to the host's hormones. Some obtain food in the normal manner via the mouth, but others, especially those in eggs, are greatly simplified and absorb food through the outer integument. Endoparasitoid wasp larvae commonly exchange gases through their outer integument, with or without a tracheal system and rarely with open spiracles. Respiration may be aided by caudal filaments, anal vesicles, or an everted hindgut. Endoparasitoid fly larvae usually have at least a metapneustic tracheal system (Chap. 5). Some perforate the host's trachea or integument and obtain atmospheric air. Others take advantage of the host's encapsulation reaction and mold a respiratory "funnel" or tube of host tissue connected to a trachea or the integument.

Reproductive Biology of Parasitoid Hymenoptera

Among the parasitoids, Hymenoptera have several features of reproductive biology that are noteworthy. Recall that sex is determined in this order by a haplodiploid mechanism. The ratio of sexes is often unbalanced in favor of females. A peculiarity of Aphelinidae (Chalcidoidea) is that the female may select the host according to the sex of the egg to be laid. Furthermore, the sex of the larva may influence its feeding behavior; males may be regular hyperparasitoids. Some wasps are exclusively parthenogenetic, in contrast to other entomophagous insects that are rarely so.

The unusual phenomenon of polyembryony (Chap. 4) is found only among parasitoids (Fig. 10.2b) and parasites; it occurs in some endoparasitoid Hymenoptera and a few Strepsiptera. As many as 1500 embryos in a single caterpillar have been counted after oviposition by one female of *Litomastix* (Encyrtidae). The most remarkable case of self-defense among parasitoids is the polyembryonic wasp *Copidosomopsis* (Encyrtidae; Cruz, 1981), in which two kinds of larvae develop from one egg. The first larvae to appear are "defender morphs" and attack any other parasitoid in the host. They fail to pupate and thus sacrifice

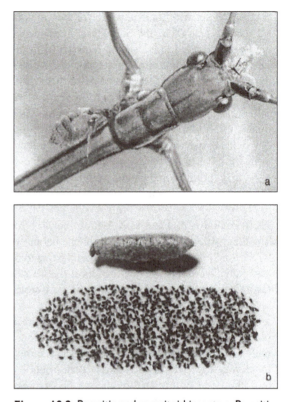

Figure 10.2 Parasitic and parasitoid insects. a, Parasitic fly, *Forcipomyia* sp. (Ceratopogonidae), sucking hemolymph of a stick insect (Phasmatodea) in Peru. An engorged fly firmly attaches to the host like a tick. b, Emergence of most of 789 brood produced by the polyembryonic wasp parasitoid, *Pentalitomastix plethroicus* (Encyrtidae), from caterpillar of navel orangeworm, *Paramyelois transitella* (Pyralidae). *Source: Photo courtesy of and with permission of Frank E. Skinner.*

themselves in favor of the normal feeding larvae that emerge later and complete development to adult wasps.

PARASITIC INSECTS

True parasitic relationships between one insect and another are surprisingly rare. As we have seen in the previous sections, the usual result of attack is the death of the victim. Entomophagous parasites may be divided into ectoparasites and endoparasites.

Adult biting midges of the genus *Forcipomyia* (Ceratopogonidae) and several other genera are ectoparasites that take blood from both vertebrates and large insects such as walkingsticks. When feeding on the latter, the flies puncture the wing veins or intersegmental membranes (Fig. 10.2a). The flies visit the host only during feeding.

The small wasps of the family Scelionidae lay their eggs among the freshly laid eggs of host insects. Females of several genera have been observed attached phoretically to the bodies of female hosts. After the host's eggs are laid, the wasp immediately oviposits. Young winged females of the European *Mantibaria manticida* (Scelionidae) search for and attach themselves to mantids of both sexes. Female hosts are more frequently selected. Once attached by their mandibles, the wasps shed their wings and await the host's oviposition, which may not take place for several months. In the interim, the wasp feeds as an ectoparasite on the mantid's hemolymph. When the mantid's ootheca has been deposited and is still soft, the wasp leaves the mantid to lay its own eggs. The wasp is said then to attempt to return to its host's body.

An unusual instance of ectoparasitic behavior is provided by the aquatic larvae of the midge *Symbiocladius* (Chironomidae). The larvae attach themselves behind the wing buds of mayfly naiads and feed.

Life History and Hosts of Strepsiptera
Strepsiptera are the only entomophagous insects that are true endoparasites. Numbering about 500 described species, they probably evolved in the Mesozoic as the sister group to Coleoptera. Fossils have been found in Cretaceous amber. Hosts include insects in the orders Thysanura, Blattodea, Mantodea, Orthoptera, Hemiptera, and Diptera, but most of them are aculeate Hymenoptera. The infested hosts are said to be "stylopized," because the name of a common genus is *Stylops* (Fig. 38.2d).

The life history of *Stylops pacifica* (Stylopidae) was studied by Linsley and MacSwain (1957) near Berkeley, California. The hosts are two species of bees in the genus *Andrena* (Andrenidae). On bright warm days in February and early March, adult bees emerge from their burrows in soil and visit flowers of the buttercup, *Ranunculus* (Fig. 10.3a,c). As many as 16 percent are parasitized. The visible

evidence consists of the puparia that protrude from the body, commonly between the fourth and fifth terga of the abdomen (Fig. 10.3a). The male *Stylops* is winged and less than 3 mm long (Fig. 10.3b). On emergence from the puparium, it immediately begins a rapid flight in search of a female. The latter is reduced to virtually a sac of reproductive organs inside the puparium. Females release a sex attractant while the host flies from flower to flower. Males fly upwind, tracking the odor, until the female is located on the dorsum of a feeding bee. The male *Stylops* lands and inserts the aedeagus by puncturing the female's puparium (Fig. 10.3c). Once inseminated, the female no longer releases the attractant. Males probably die the same day as they emerge and mate but once. Except during copulation, their intense flight activity never ceases and they rarely feed. The tarsi even lack claws for clinging to objects. Linsley and MacSwain were able to capture the rare males by putting bees with virgin female parasites in cages and placing the cages among flowers.

The first instar larvae are triungulins, and they develop from the fertilized eggs while still inside the female's body. During the next 30 to 40 days, the female dies and the larvae move into a median brood passage in preparation for their exit. A large female *Stylops* probably produces 9000 to 10,000 triungulins. On warm days while the host is visiting flowers, the active larvae emerge through the ruptured puparium (Fig. 10.3d). Parasitized hosts move more rapidly among flowers than normal hosts. The triungulins are brushed singly or several at a time onto the flowers.

In contrast to the triungulins of Meloidae and Rhipiphoridae (Coleoptera) and other Strepsiptera, those of *Stylops pacifica* do not readily attach themselves to new hosts but are ingested with nectar by bees (Fig. 10.3e). The bee returns to its nest and prepares a ball of pollen mixed with regurgitated nectar. In this way the triungulin is deposited in a nest cell with the bee's egg (Fig. 10.3f). The triungulin penetrates the egg, transforming as it does into the second instar, and becomes an endoparasite (Fig. 10.3g).

As the host feeds and grows, the *Stylops* larva feeds on the host's nonvital tissues and passes through an

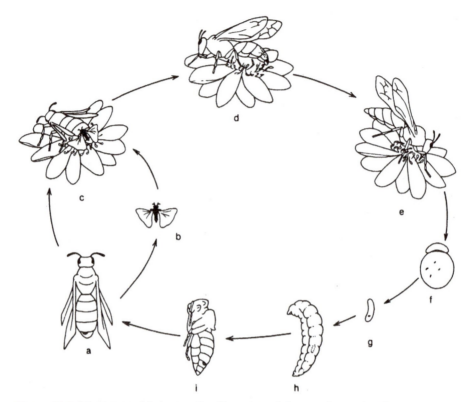

Figure 10.3 Life history of *Stylops pacifica* (Strepsiptera). See text for explanation.
Source: Redrawn from Linsley and MacSwain, 1957; published 1957 by the Regents of the University of California; reprinted by permission of the University of California Press.

unknown number of instars (Fig. 10.3h). When the bee pupates, the *Stylops* protrudes the anterior portion of its body through the intertergal membrane and also pupates (Fig. 10.3i). The last larval skin is not shed but forms a tanned puparium.

Strepsiptera have no known natural enemies. Major causes of mortality are the initial losses of triungulins that fail to find a female host of the proper species and superparasitism. Linsley and MacSwain noted that usually only the larger female bees with a solitary female parasite lived

long enough for the triungulins to escape. Smaller male hosts, those with several *Stylops*, or those with *Stylops* plus nematode parasites usually died before the cycle was completed.

Bees that survive parasitism are variously affected. The genitalia of both sexes may be reduced. Secondary sexual characters, especially in the female, may be shifted toward the opposite sex. Stylopized females may have more malelike yellow color and reduced pollen-collecting hairs on the legs.

CHAPTER 11

Insects and Microbes

A diversity of microorganisms populated the earth long before higher animals appeared. These microbes altered the environmental conditions to make possible the oxygen-based respiration of plants and animals. The antiquity, variety, and abundance of microbes have led to the evolution of many unique insect-microbe relationships. The term microorganisms, as used here, simply refers to taxonomically diverse organisms that are difficult or impossible to see without magnification. Microbes may be unicellular organisms such as bacteria, yeast, and protozoa, multicellular organisms such as fungi and nematodes, or viruses. Pathogens are microbes that harm their insect hosts. Although pathogenic relationships are the exception rather than the rule, pathogens are often important in regulating insect populations. At the other extreme, some bacteria, fungi, and viruses have evolved with their insect hosts to create mutually beneficial relationships to such an extent that neither partner can survive without the other. Selected examples listed in Table 11.1 illustrate the wide spectrum of insect-microbe relationships.

MICROBES AS FOOD

Insects that feed on living or dead organic matter consume decay microorganisms. Aquatic filter feeders such as the larvae of mosquitoes and black flies consume planktonic microorganisms along with organic debris. Among the particles ingested, microbes often contain the greatest concentrations of protein, carbohydrates, and other nutrients. Habitat and environmental conditions determine the types and concentrations of microbes present, but microorganisms typically account for no more than about 4 to 25 percent of the nutritive value of decaying plant matter. Associated microbes improve the food value of detritus by serving as sources of concentrated digestible nutrients and by transforming complex organic substrates into metabolites that are more readily digested or assimilated. Some microbes may serve as either food or infectious pathogens. For example, the protozoan

Lambornella clarki is free-living in water in tree hole cavities, where it feeds on organic detritus (Fig. 11.1). The free-living stages (trophonts) are food for mosquito larvae (Fig. 11.1a). Under certain conditions, this protozoan parasitizes mosquitoes. It first forms an invasion cyst (Fig. 11.1b) on the host mosquito larva (*Aedes sierrensis*) that penetrates the mosquito's cuticle and then multiplies within and eventually kills the mosquito larva. Adult female mosquitoes that survive infection are castrated but introduce *L. clarki* into tree holes while attempting to oviposit (Washburn et al., 1988).

Termites in the family Termitidae (Chap. 22) and leaf-cutter ants (Chap. 44) may cultivate fungi in "gardens" within the nest. Leaf-cutter ant workers cut leaves or petals from a variety of plants, return them to their underground nest, and arrange the harvested and processed plant materials on a nest to fuel a fungal garden. Cultivation of fungi by ants

Table 11.1 Types of Insect-Microbe Relationships

Effect or role of microbe	Examples
Food	Fungus-feeding ants and termites, aquatic filter feeders, detrivores, many others
Aids digestion of food	Termites, beetles, wasps, others
Facultative symbiosis effects range from pathogenicity to protection from parasitism, heat shock, etc. (see text)	Fruit flies (Tephritidae) and bacteria; bacteria in aphids
Obligate symbiosis provides nutritional benefits; aids parasite in overcoming host defense	Beetles and mycangial fungi; aphids and symbiotic procaryotes, polydnaviruses in parasitoid wasps
Affects sex ratio of insect offspring (includes feminization and male killing)	Bacteria in parasitoid wasps
Limits reproductive compatibility of insect	Bacteria and spiroplasmas in vinegar flies, parasitic wasps
Insect pathogens	Nematodes, fungi, protozoa, bacteria, viruses
Insect-vectored pathogens of plants or animals	Nematodes, fungi, protozoa, bacteria, viruses

allows the ants to exploit a wide variety of plant species that might otherwise be toxic or indigestible. Some ants and fungi evolved from a common ancestral relationship, while other associations arose independently (Chapela et al., 1994).

Ambrosia beetles, the common name of beetles in several families (Scolytidae, Platypodidae, Lymexylidae), feed on fungi that grow along the galleries that the beetles excavate in wood. Adult females carry fungal spores in specialized cavities called mycangia (fungi-transporting structures) behind the head, along the thorax, or at the bases of the coxae. Some beetles lack mycangia and transport spores that adhere to sticky secretions on their bodies. Female wood wasps in the genus *Sirex* (Hymenoptera: Siricidae) harbor spores of fungi in the genus *Amylosternum* in mucus-filled pouches along the female genital tract. Female wasps introduce the fungal spores during egg-laying. The mucus secreted by the wasp with its eggs aids in the fungus's survival and reproduction. The fungus develops in the wood surrounding the egg and provides nourishment for the wasp larvae.

SYMBIOSES

The original and most general definition of symbiosis is a close association or living together of two dissimilar organisms. More current usage of the term tends to describe relationships that are mutually beneficial. More specifically, such a relationship should be called a mutualistic symbiosis. Examples are the protozoan and bacterial symbionts (or symbiotes) of some termite species that digest cellulose for their termite hosts. Killing the termites' intestinal symbionts slowly starves the termites unless they can replenish their intestinal microflora. Obligate symbiosis (or primary symbiosis) describes a relationship where both symbiont and host are required for the growth or reproduction of either host or symbiont. Where the host can grow or reproduce without the symbiont, the relationship is a facultative symbiosis (or secondary symbiosis). Symbionts that cause no harm or benefit to their hosts are called commensals. Some symbiotic relationships are either harmful or beneficial depending on external circumstances. For example, pea aphids (*Acyrthosiphon pisum*, Aphididae, Hemiptera) may harbor a bacterium (*Serratia* sp.) that reduces the survival of aphid that it inhabits on some plants (but not on others). On the other hand, this bacterium enables pea aphids that are subjected to pulses of heat as young nymphs to reproduce after they become adults, whereas aphids without the bacterium will not produce offspring following the same heat exposures.

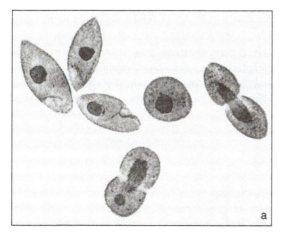

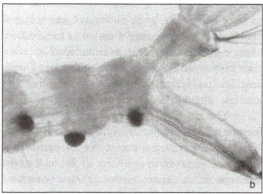

Figure 11.1 The protozoan *Lambornella clarki*: a, oval and round theront stages that may be eaten and digested by mosquitoes; b, three cuticular cysts of *L. clarki* on the abdominal segments of a mosquito larva (*Aedes sierrensis*).
Source: Photo courtesy of and with permission of J. O. Washburn.

The role of microorganisms in aiding food digestion has been closely studied in termites, but we can expect that many insects in diverse orders use microbes for this purpose (Martin, 1987). The role of microbes in aiding insect digestion varies widely among insect groups. The consistent occurrence of bacteria, even in structures apparently specialized for containing bacteria, cannot be interpreted as proof of a beneficial role. The heads of fruit flies (Diptera: Tephritidae) contain a bulbous diverticulum of the esophagus that houses bacteria. Experimental investigations of the role of gut-associated bacteria in fruit flies failed to

find any differences in growth and survival of fruit flies on a sterile diet compared to flies reared on a diet that included bacteria commonly found in the flies' guts. Although a variety of bacterial species can colonize the esophageal bulb, individual flies typically harbor only one type of bacteria. Many heteropteran species have ceacal pouches of the midgut that contain distinctive bacteria. Newly emerged nymphs immediately feed on the surface of the eggshell to acquire the proper gut microflora. If the egg surface is sterilized before hatching, the nymphs that emerge do not grow or develop normally.

Symbionts that live within the host insect's cells are called endosymbiotes; those that live only outside cells are ectosymbiotes. Aphids (see Box 11.1), leafhoppers (Cicadellidae), scales (Coccoidea), and other hemipterans house obligate endosymbionts in specialized cells—bacteriocytes for cells that house bacteria and mycetocytes for those that house fungi. The mycetocytes may be loosely aggregated or may form a specialized organ, the bacteriome or mycetome. In some aphids, facultative symbionts inhabit the sheath cells that surround the mycetome as well as the hemolymph. The term facultative describes the nonobligate nature of the association, meaning that not all aphids within a species have them. The intracellular habitat of the symbionts probably protects the symbionts from destruction by insect immune responses such as phagocytosis and encapsulation. Symbionts are transmitted from mother to offspring by infecting the oocytes (or embryos in parthenogenic reproduction). The symbionts invade the ovariole or lateral oviduct to colonize eggs. Symbiotes of the human louse periodically migrate within the hemocoel from the mycetome to the lateral oviduct (Fig. 11.2). They penetrate the oviduct to invade the egg before the chorion is deposited. Circulating phagocytes consume any remaining symbiotes that fail to penetrate the oviduct. The developing embryo develops mycetomes, which the symbionts invade intracellularly. Many leafhoppers, planthoppers, and other hemipterans have two or more different types of symbiotes that travel to the ovary within cells that detach from the mycetome, perhaps to evade attack by the insect immune system during transit to the ovaries. Some hemipteran species harbor as many as five morphologically distinct types of symbiotes (Buchner, 1965). Fungi and yeast are symbiotes found within caecal pouches in wood-feeding beetles such as anobiids and cerambycids. Endosymbiotic yeast inhabit the hemocoels of planthoppers and subsocial aphids.

SEX RATIO AND REPRODUCTIVE INCOMPATIBILITY ORGANISMS

Procaryotes that alter the ratio of male to female offspring are sex ratio organisms. A bacterial (spiroplasma) parasite of some species of *Drosophila* fruit flies causes the offspring of

---

BOX 11.1   Aphids as Ecosystems

Aphids house obligate bacterial symbionts (*Buchnera aphidicola*) within specialized cells (bacteriocytes) within an organ (bacteriome) devoted to maintaining the bacteria. By means of transovarial passage, the symbionts invade the aphid's eggs or embryos early in their development. Experimental treatments with heat or antibiotics can eliminate the symbionts from their aphid hosts. These treatments stop the aphids' development and reproduction, showing that the symbionts are obligate. What do the symbionts provide? The *Buchnera* genome contains the genes necessary for the synthetic pathways of all essential amino acids for the aphid. The symbionts thus provide an abundance of nutrients that may be in short supply in the aphid's diet of plant sap. Providing these supplements enables aphids' trademark fast rates of development and reproduction.

The phylogenies of *Buchnera* symbionts (based on gene sequences) and of various aphid taxa (based on a diversity of characters) are perfectly congruent, with the exception of a clade of Asian aphid species that occur on bamboo and have specialized soldier and reproductive forms (castes). These social aphids instead have fungal symbionts. These matching *Buchnera*-aphid phylogenies strongly suggest that this insect-bacterium symbiosis evolved from an initial association that may date to 200 million years ago. Typical of many parasites with a long evolutionary association with a specific host, the symbionts appear to have lost many functions now provided by their host. For example, the symbiont genome lacks almost all genes for synthesizing nonessential amino acids. Presumably, the aphid host provides the nonessential amino acids and possibly even some structural proteins for its symbionts.

Buchnera have a genome that is only about 15 percent of the size of their nearest bacterial relatives (*Escherichia coli* and closely related enteric bacteria), having lost genes whose functions the host aphid now provides. Yet despite this extensive loss of genome size, *Buchnera* retains the genes (often having multiple copies of key genes) for the synthetic pathways of essential amino acids that it provides its aphid host. The aphid-*Buchnera* symbiosis is an outstanding example of close coevolution of different organisms.

In addition to *Buchnera*, several other kinds of bacteria may be present within the aphid's hemocoel. These bacteria, called secondary or facultative symbionts because they do not occur in all aphids of the same species, also are transovarially transmitted from mother to offspring. Under some conditions, secondary bacteria may harm the host aphid by reducing its survival and reproduction, but they may benefit the aphid under other conditions. For example, in the pea aphid (*Acyrthosiphon pisum*) one type of secondary bacterium kills larvae of parasitic wasps within the aphid. Another bacterium protects the aphid from heat damage. A third bacterium protects the aphid from fungal pathogens. Additional types of secondary bacteria continue to be discovered. In addition, viruses (bacteriophage) that attack the bacteria within the aphid have been detected. Thus a single aphid can consist of a complex of microorganisms at different trophic levels. This means that a single aphid can house several different genomes within its body cavity. In addition to an aphid's chromosomal genes and mitochondrial genes, the aphid can house a variety of microbes with their own chromosomal and plasmid genomes. A single aphid thus constitutes a mini-ecosystem and a mosaic of paragenetic bacterial elements.

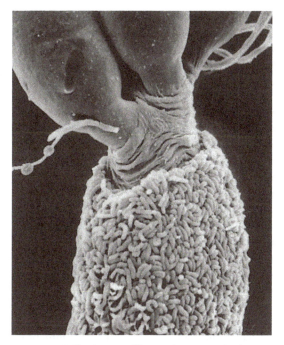

Figure 11.2 Symbiotes of human louse attached to lateral oviduct of female louse prior to invading eggs. The louse's phagocytes remove the symbiotes that fail to bore into the ovary.
Source: Photos courtesy of and with permission of Mark Eberle.

Bacteria can cause reproductive incompatibility between members of the same or closely related insect species. Eggs fertilized by males from mothers that do not harbor the same microorganism prevent maturation of embryos. Eliminating the bacteria from insects with heat or antibiotics restores reproductive compatibility. Typically, the incompatibility is unidirectional but can be bidirectional. Incompatible crosses produce sterile males or no progeny, but the reciprocal crosses produce normal offspring. The associated bacteria invade the egg during oogenesis. Incompatibility bacteria interfere with the incorporation of the paternal chromosome in fertilized eggs. For example, transovarially transmitted bacteria confer reproductive incompatibility between the alfalfa weevils *Hypera postica* and *H. brunneipennis* (Curculionidae). Similar relationships have been demonstrated for *Wolbachia* and other bacteria found in *Culex pipiens* mosquitoes, *Drosophila* fruit flies, planthoppers, and numerous other insects. Reproductive incompatibility may favor speciation of the host insects by isolating otherwise reproductively compatible populations of insects. *Wolbachia* infections are very widespread and common in insects and parasitize many other types of animals as well. A key reason for their abundance is probably the fact that they can manipulate the sex and reproductive success of their insect hosts. Insects appear to be capable, however, of countering this *Wolbachia* by evolving resistance to their effects. This was observed to occur within ten generations of a butterfly in Samoa.

infected female flies to be all or mostly females. Male embryos die early during embryogenesis in infected mothers. A bacterium called "son killer" has the same effect on the parasitic wasp *Nasonia vitripennis*. Parasitoid wasps (Chalcidoidea) that harbor bacteria in the genus *Wolbachia* produce only females (thelotoky). Antibiotic or heat treatments that eliminate the bacteria cause the insects to produce both sexes. The parthenogenesis bacteria prevent chromosomes in unfertilized eggs from segregating. In all of these cases, parents transmit sex ratio microorganisms to their offspring. Because the transmission usually is only from mother to offspring, the production of only female offspring is thought to maximize the production of the sex ratio organisms as long as the continued existence of the host insect is not threatened by the excessive elimination of males.

INSECT PATHOGENS

Viruses, bacteria, fungi, protozoa, and nematodes are the principal pathogens of insects. Some of these are facultative pathogens that cause no harm unless they enter an insect's hemocoel or shift to a pathogenic form. For example, the bacterium *Serratia marcescens* can persist in soil without insects but can invade the hemocoel of insects through wounds. Once in the hemocoel, it is usually fatal to the infected insect. Most facultative pathogens enter the insect hemocoel after being ingested. Obligate parasites of insects multiply only within an insect host.

Viral Pathogens

Numerous types of viruses infect insects (Table 11.2). The most important viruses for biological control are the baculoviruses (Baculoviridae), entomopoxviurses (Poxviridae), and reoviruses (Reoviridae). Baculoviruses include nuclear polyhedrosis viruses (NPVs) and granulosis viruses (GVs). Virus particles (virions) of both types of baculovirus have double-stranded DNA and are embedded in a protective matrix of virally encoded protein, called an occlusion body. Some nonoccluded viruses occur in arrays of virus called inclusion bodies that are similar to crystalline structures but that do not contain protein. In the NPVs the protein matrix contains large numbers of embedded virions. The occlusion bodies of GVs contain a single virion but may be almost as large as those of NPVs. The baculoviruses have been found exclusively in arthropods and are considered not to infect vertebrates. They are found mostly in endopterygotes, especially the Lepidoptera. Viral occlusion bodies protect baculoviruses from deteriorating in the environment, but they are sensitive to ultraviolet radiation and thus persist for longer periods in soil and environments shielded from light.

Symptoms of viral diseases vary with the causal virus, the developmental stage of the host insect that is infected, the dosage of virus, and environmental factors, especially temperature. External symptoms of baculovirus diseases generally do not appear for the first 2 to 5 days after ingestion of virus by caterpillars. The color and pallor of infected insects typically begins to change several days after infection. Diseased larvae are less active and feed less. Some viral infections may prolong larval life by interfering with ecdysone metabolism. This has the advantage for the virus of prolonging viral reproduction. Very young caterpillars may die within a few days of infection, but later instars of some species may live for weeks. Just prior to dying, the integument of

Table 11.2 Selected Pathogenic Viruses of Insects

Representative Insect Viruses	Nucleic Acid, Shape of Virus	Taxa of Principal Insect Hosts
Baculoviruses	Double-stranded (ds) DNA	
Nuclear polyhedrosis viruses (NPV)	Large polyhedra with numerous virions (virus particles)	Numerous orders; most in Lepidoptera
Autographa californica NPV		Noctuidae
Nonoccluded baculoviruses	No occlusion bodies (masses of virus/protein)	Numerous orders
Granulosis viruses (GV)	Single virion in a granular occlusion	Lepidoptera, Hymenoptera
Sawfly GV		Tenthridinidae
Reoviruses	ds RNA, spherical particles	Lepidoptera, Diptera, Coleoptera, Neuroptera, Hymenoptera, Hemiptera
Rice dwarf virus	Transovarial transmission	Leafhopper vectors, rice
Silkworm cytoplasmic polyhedrosis virus (CPV)	Inclusion bodies in cytoplasm	Lepidoptera
Entomopoxviruses	ds DNA; spheroid or spindle inclusion body	Lepidoptera, Diptera, Coleoptera, Orthoptera, Hymenoptera
Polydnaviruses	ds DNA	Ichneumonidae, Braconidae
Rhabdoviruses	ss RNA, rod-shaped capsid	
Potato yellow dwarf virus		Leafhopper vectors, plants
Picornaviruses	ss RNA, no envelope	
Cricket paralysis virus		Orthoptera, Lepidoptera, others

ds, double-stranded; ss, single-stranded.

baculovirus-infected caterpillars becomes fragile and easily torn. At death, the body contents are largely liquefied. The spread of virus to other caterpillars may also be aided by the disease-induced behavior displayed by some insect species whose larvae climb to the tops of plants just before dying. The virus-laden contents of the insects killed by the virus disease can then spread more widely to other insects feeding below. Granulosis viruses of sawflies (Tenthredinidae) that appear to infect only the midgut cells trigger viral epidemics that are sufficient to keep populations of forest pests below damaging levels for many years. To improve the usefulness of baculoviruses as biological pesticides, genetic engineering is being applied to NPVs to increase the speed at which they kill insects after infection and to increase the viruses' host ranges.

The reoviruses have double-stranded RNA genomes enclosed in protein coats. One type of reovirus, called cytoplasmic polyhedrosis virus (CPV), has been reported from over 250 species of insects, mostly Lepidoptera and Diptera. The polyhedra of CPVs, like the occlusion bodies of baculoviruses, protect the virions from environmental degradation. Some reoviruses replicate in both plants and insects and may be pathogens of both kinds of hosts.

The entomopoxviruses are a subfamily within the large virus family Poxviridae. They are structurally similar to poxviruses of vertebrates. These include the causal viruses of familiar human diseases such as smallpox and fowlpox (chicken pox), but the entomopoxviruses have restricted host ranges and do not infect vertebrates. Like the baculoviruses, they contain a double-stranded DNA genome.

Bacterial Pathogens

Numerous bacterial species are facultative pathogens capable of rapidly invading wounded or freshly killed insects. The bacterial pathogens of insects most commonly used for insect control so far are species of the genus *Bacillus*. The spore-forming stage of *Bacillus thuringiensis* (*Bt*) contains crystalline proteins that are toxic to certain insect species. For example, *B. thuringiensis* subspecies *kurstaki* kills numerous species of Lepidoptera. Other strains of *Bt* are used to control larval mosquitoes and black flies in aquatic environments (*Bt* subspecies *israeliensis*) or chrysomelid beetles (*Bt* subspecies *san diego*). More than forty types of toxin proteins have been identified by their chemical properties and insect host range. After ingestion by the host insect, the toxin crystals dissociate in the proper chemical environment (pH may be especially important), bind to the midgut epithelium, and disrupt the midgut cells.

Bacillus popilliae is more host-specific than *Bt* and can play an important role in the natural regulation of its host insects, scarab beetles (Coleoptera: Scarabaeidae). Ingestion by root-feeding beetle grubs activates the spores of *B. popilliae* to multiply within the beetle's gut and invade the hemocoel, causing a condition called "milky disease" because infected grubs have a pale-colored abdomen. The large numbers of *B. popilliae* spores produced in an infected grub can survive in soil for years. Unlike *Bt*, however, *B. popilliae* cannot be fermented on an artificial medium for commercial production of biological insecticides; it multiplies only within host insects. In contrast to *Bt*, *B. popilliae* can set in motion an epidemic of milky disease that can persist in soil for years. The bacterium *Bacillus sphaericus* also produces crystalline proteins toxic to larvae of mosquitoes and other flies. It is less host-specific than *B. popilliae* but also cycles within natural populations, whereas *Bt* subspecies *israeliensis* does not multiply or persist in significant numbers in natural environments after application for insect control. Another host-specific bacillus, *B. larvae*, causes American foulbrood disease of honey bees. The environmentally resistant spores survive for long periods in abandoned nests or hives of honey bee colonies destroyed by the disease.

Rickettsiae are bacteria that are intracellular parasites of animals, including insects. Many rickettsiae in insects cause disease. For example, *Rickettsiella melolonthae* causes the lethargy disease of scarab beetles (Scarabaeidae). Infected grubs turn an abnormally blue color and tend to rise to the soil surface rather than burrowing deeper, as healthy grubs would normally do. This behavior may aid in the dispersal of the pathogen to younger larvae, which are more numerous in the upper layers of soil.

Mollicutes are bacteria that lack any traces of a cell wall. As a result, they usually have a spherical or amorphous shape but may be pleomorphic (taking on many shapes). Commonly known as "mycoplasmas," the mollicutes are the smallest free-living organisms. They are common and widespread in insects, but only a few are known to be pathogens. This may be partly because of the difficulty of detecting them microscopically or of culturing them. Some species of mollicutes in the genus *Spiroplasma* have filamentous cells coiled in a helix. Some *Spiroplasma* species are plant pathogens transmitted by leafhoppers (Cicadellidae) (Table 11.3) and are also pathogenic to some of their insect vectors. *Spiroplasma apis* causes May disease (maladie de mai) of honey bees and may occur widely in flower nectaries during spring months. Most species of spiroplasmas and other mollicutes, perhaps numbering into the millions of species (Whitcomb, 1981), are associated with insects. They are seldom noticed, however, because they are barely discernible by ordinary light microscopy.

Protozoan Pathogens

Protozoa are a taxonomically diverse group of eucaryotic, single-cell microorganisms classified as a subkingdom in the kingdom Protista. They have subcellular organelles, a membrane-bound nucleus, and other features common to multicellular organisms. Protozoa commonly inhabit insect guts.

Only a tiny fraction of protozoan species are insect or animal pathogens, but protozoa transmitted by insect vectors cause important human and animal diseases such as malaria, African sleeping sickness, and Chagas' disease. One of the earliest studies of a protozoan disease of insects was Louis Pasteur's investigation of pébrine disease of the silkworm. Although he never established the identity of the causal pathogen (*Nosema bombycis*), Pasteur's systematic studies showed that pébrine was an

Table 11.3 Selected Pathogens Transmitted by Insect Vectors

Disease	Pathogen	Vector	Transmission
		Viruses	
Animal			
Yellow fever	Togavirus	Mosquito (*Aëdes aegypti*)	Circulative, propagative
Bluetongue of cattle	Reovirus	Biting midges (Ceratopogonidae)	Circulative, propagative
Myxomatosis of rabbits	Myxomavirus	Mosquitoes (Culicidae), fleas	Nonpersistent, nonpropagative
Plant			
Barley yellow dwarf	Luteovirus	Aphids	Circulative, nonpropagative
Watermelon mosaic	Potyvirus	Aphids	Nonpersistent, nonpropagative
Squash mosaic	Comovirus	Leaf beetles (Chrysomelidae, Chrysomelinae)	Noncirculative, semipersistent (hours to days), nonpropagative
Lettuce infectious yellows	Geminivirus (ds DNA)	Whiteflies (Aleyrodidae. *Bemesia* spp.)	Noncirculative, semipersistent
		Bacteria	
Animal			
Plague	*Yersinia pestis*	Fleas	Noncirculative, persistent, propagative in gut
Epidemic typhus	*Rickettsia prowazekii*	Human body louse (*Pediculus humanus*)	Noncirculative, propagative in gut cells only
Relapsing fever	*Borrelia recurrentis* (spirochaete)	Human body louse (*Pediculus humanus*)	Noncirculative; propagative, but dies in hindgut
Tularemia	*Franciscella tularensis*	Horse flies	Nonpersistent on mouth-parts

(Continued)

Table 11.3 *(Continued)*

Disease	Pathogen	Vector	Transmission
Plant			
Fire blight of pome fruit trees	*Erwinia amylovora*	Flying, flower-visiting insects	External contaminant on vectors
Stewart's wilt of corn (maize)	*Pantoea stewartii*	Flea beetles	Noncirculative, persistent
Pierce's disease of grapevines, citrus variegated chlorosis, others	*Xylella fastidiosa*	Xylem-feeding Hemiptera (Cicadellidae: Cicadellinae, Cercopidae)	Noncirculative, persistent, propagative in foregut
Pear decline (PD)	PD phytoplasma	Psyllid (*Cacopsylla pyricola*)	Circulative, propagative
Elm yellows (EY)	EY phytoplasma	Leafhoppers (*Deltocephalinae*)	Circulative, propagative
Corn (maize) stunt	*Spiroplasma kunkelii*	Leafhoppers	Circulative, propagative

Protozoa

Disease	Pathogen	Vector	Transmission
Animal			
Chagas' disease	*Trypanosoma cruzi*	Triatomine reduviids	Propagative in gut, feces infect wound
African sleeping sickness	*Trypanosoma* spp.	Tsetse flies (*Glossina* spp.)	Circulative, propagative
Malarias	*Plasmodium* spp.	*Anopheles* mosquitoes	Cyclodevelopmental
Leishmaniasis	*Leishmania* spp.	Sand flies (Psychodidae: *Phlebotomus* spp.)	Cyclodevelopmental in gut and foregut
Plants			
Oil palm hartrot	*Phytomonas staheli*	Pentatomidae	Cyclopropagative
Milkweed trypanosomes	*Phytomonas emassiana*	Milkweed bugs (Lygaeidae)	Cyclopropagative

Nematodes

Disease	Pathogen	Vector	Transmission
Animals			
Filariasis (Brugian and Bancroftian)	*Wuchereria bancrofti, Brugia malayi*	Mosquitoes (*Culex* spp.)	Cyclodevelopmental
Onchocerciasis (river blindness)	*Onchocerca volvulus*	Black flies (Simuliidae)	Cyclodevelopmental
Canine heartworm	*Dirofilaria immitis*	Mosquitoes	Cyclodevelopmental
Plant			
Pine wilt	*Bursaphelenchus xylophilus*	Pine sawyers (Cerambycidae)	Carried in beetle tracheae

Fungi

Disease	Pathogen	Vector	Transmission
Plant			
Dutch elm disease	*Ophiostoma ulmi*	Elm bark beetles (Scolytidae)	Sticky spores carried externally, invade feeding wounds
Oak wilt	*Ceratocystis fagacearum*	Sap beetles (Nitidulidae)	Vectors carry spores to fresh wounds
Endosepsis of fig	*Fusarium moniliformae*	Fig wasps (Agaonidae)	Carried externally during pollination (Chap. 10)

infectious disease. Sanitary methods he developed for disease prevention continue to be used to this day. His early investigations of pébrine and other silkworm diseases led to his famous later studies of important human diseases such as rabies and cholera.

Most protozoan pathogens of insects are not highly virulent, instead causing chronic infections that do not kill the host insect. Some of the most common protozoan inhabitants of lepidopteran caterpillars are assumed to nonpathogenic or only mild pathogens because they appear to universally infect some species that are very prolific pests. The infective stages of protozoa are usually spores or cysts. For example, the ciliates *Lambornella* spp. penetrate the integument through invasion cysts attached to the cuticle (Fig. 11.1). Vertical transmission (from parent to offspring) of protozoan parasites is common, either through invading the ovary (transovarial) or by contaminating the egg surface. All members of phylum Microspora, commonly called microsporidia, are obligate pathogens that multiply only in living cells. They are the most important group of insect pathogenic protozoans. *Nosema apis* causes serious, chronic disease in honey bees. Infected worker bees age more quickly and are sluggish. The hypopharyngeal glands of diseased bees atrophy, so these workers do not attend the queen bee, perhaps explaining the rarity of queens having the nosema disease.

Fungal Pathogens

Fungi are perhaps the most common insect pathogens. Agostino Bassi's experiments with a fungal disease of silkworms first demonstrated the role of a microorganism as a cause of an animal disease. The evidence produced by Bassi and others disputed the then prevalent theory of spontaneous generation of insects and microbes. Spores, the infective stage of fungal pathogens, germinate on the insect cuticle. The developing fungal extensions, called hyphae, penetrate the cuticle to invade the hemocoel. Death ensues from fungal toxins or the conversion of insect tissues into fungal growth (Fig. 11.3b). Some fungal diseases kill the host insect while it is attached to plants or other external substrates. Grasshoppers, flies, and other insects characteristically climb to the tops of plant stems or other elevated locations to die after infection by some fungi. This facilitates the dispersal of the spores released from the cadaver.

The usefulness of fungal pathogens as microbial pesticides for controlling insect pests has been limited by the fungi's environmental requirements (often free moisture or high humidity and suitable temperatures) for spore production (sporulation) or spore germination. Species in the fungal genera *Beauvaria*, *Entomophthora*, *Erynia*, *Metarhizium*, and *Nomuraea* are especially prominent pathogens for pest control. *Verticillium lecanii* is applied to control aphids and other homopterous pests in greenhouses. Many insect-pathogenic fungi have never been grown in artificial media, which limits their commercial production to use in controlled applications; however, they may be important natural controls for pest populations.

Nematode Pathogens

Nematodes are small to microscopic, unsegmented worms that occupy a wide variety of habitats in soil, water, plants, and animals. Some species of nematodes are important insect pathogens; others are human or animal pathogens that are transmitted by insect vectors. Nematodes may be either facultative or obligate parasites. Some species of nematodes infect their insect hosts passively through ingestion; others actively seek out and penetrate the host integument to enter the hemocoel.

Mermithid nematodes are striking parasites because of their large size relative to the host insect when they emerge (Fig. 11.3d). Early stages of the nematodes penetrate insects externally or after ingestion by the host. In an acceptable host, the worms increase in size through a series of molts. Their emergence from the insect kills the host. The nematodes form a cyst or resting stage in insects that are not suitable for development. Mermithids are frequently found in aquatic habitats.

Steinernematid and heterorhabditid nematodes are used for biological control because they are highly virulent, have a wide insect host range, can be produced commercially, and retain viability after prolonged storage. These nematodes have an environmentally resistant juvenile stage (dauer stage) that can be applied to soil or locations moist enough for nematode survival and movement. The nematodes enter and penetrate insects through the alimentary tract or enter spiracles and penetrate the tracheae to reach the hemocoel. Once

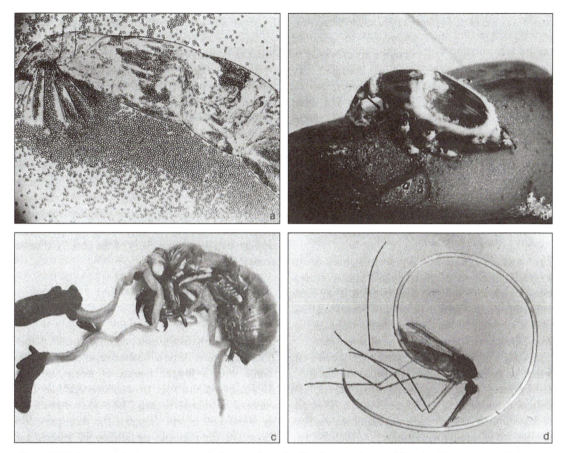

Figure 11.3 Insect pathogens. a, an aquatic larvae of the fly *Chaoborus astictopus* (Chaoboridae), ruptured by infection of the coccidian protozoan, *Adelina* sp.; b, adult weevil *Metamasius hemipterus* (Curculionidae) infected with the fungus *Beauveria bassiana*—mycelia are emerging from between segments; c, Nymph of cicada *Diceroprocta apache* (Cicadidae) infected with the fungus *Cordyceps sobolifera*—note terminal spore-producing structure; d, Adult mosquito, *Anopheles funestus* (Culicidae) with the nematode *Empidomermis cozii* emerging from its body. *(Photos courtesy of and with permission of G. M. Thomas and G. O. Poinar, Jr.)*

in the hemocoel, the nematodes defecate to release bacteria (*Xenorhabdus* spp.) that are symbiotic in the nematode gut. The bacteria rapidly multiply and kill the insect host, while producing a powerful antibiotic that excludes other bacteria from colonizing the dying or dead host. The nematodes consume as food the bacteria and decaying insect host to develop into mature adults, mate, and lay eggs. Juvenile nematodes develop into adults or remain in the environmentally resistant dauer stage for as long as 5 years.

INSECTS AS VECTORS OF MICROBES

Insect vectors transport and transmit microorganisms to other organisms. Transmission may be the consequence of simply transporting a microbe to a site where the microbe can invade a suitable host. For example, numerous flower-visiting species of insects in several orders may be vectors of the bacterium *Erwinia amylovora*, which causes fire blight disease of apple and pear. The vectors carry the bacterium as an external contaminant from flower to flower. More often, insects that are vectors have special physical or physiological characteristics and behaviors that increase the chances of successful transmission of pathogens. The bacterium *Pantoea* (*Erwinia*) *stewartii* causes Stewart's wilt of maize and can be recovered from many types of insects, but only flea beetles (Chrysomelidae, Halticinae) infect plants through their feeding wounds. Similarly, any insect that feeds on the blood of a person with malaria may acquire the malarial parasite, but only certain mosquito (Culicidae) species in the genus *Anopheles* are vectors, that is, capable of transmitting the parasite to another person.

Vector Characteristics

Most ways of characterizing pathogen-vector relationships emphasize how the efficiency of vector transmission changes from species to species and over time. In addition, we may categorize the mechanisms by which insects transmit pathogens. Vector specificity refers to the degree of exclusivity of transmission between a particular pathogen and its vectors. For example, the beet curly top virus has only one known vector, the beet leafhopper, *Circulifer tenellus* (Cicadellidae). In contrast, many different aphid species (Aphididae) can transmit watermelon mosaic virus. Transmission efficiency or vector competence is a measure of how frequently an infectious vector successfully transmits a pathogen to a host. The brown citrus aphid, *Toxoptera citricida* (Aphididae), is a very efficient vector of citrus tristeza virus. In contrast, the melon aphid, *Aphis gossypii*, is a very inefficient vector of this virus; each new infection requires many thousands of aphids. Yet when it is extremely abundant, the melon aphid can be important in the spread of tristeza disease.

The following sections briefly describe how insects serve as vectors of pathogens that cause plant diseases.

Virus Vectors

Most plant viruses have mobile insect, acarine, or nematode vectors because plants are not mobile and viruses do not disperse on their own. Some species of beetles and thrips are important vectors of plant viruses, but the most important groups of insect vectors of plant viruses are in the Hemiptera. The needlelike mouthparts and feeding behavior of sucking insects makes them ideal for transmitting viruses. Among the Hemiptera, aphids (Aphididae), leafhoppers (Cicadellidae), whiteflies (Aleyrodidae), and mealybugs (Pseudococcidae) have the greatest number of species that are plant virus vectors. Sucking, blood-feeding insects such as mosquitoes (Culicidae), black flies (Simuliidae), kissing bugs (Reduviidae), and sucking lice (suborder Anoplura) are important vectors of animal pathogens.

Changes in transmission efficiency over time (Fig. 11.4) are clues to the method or mechanism of transmission and may be features of the transmission process that are extremely important in trying to prevent the spread of particular diseases. The length of time required between a vector's acquiring a pathogen, usually by feeding on an infected host, and its first successful inoculation of another host is the latent period (or extrinsic incubation period), usually estimated as an average or median value. Because not every insect capable of transmitting a pathogen will do so at every opportunity, it is not possible to determine absolutely that there is no latent period, only that it is very short. Another dynamic characteristic of vector transmission of pathogens is how long a vector is able to continue to transmit after becoming infectious. Many aphid-borne viruses are nonpersistent; their aphid vectors are capable of transmitting to only one or a few plants in succession. The maximum efficiency of aphid acquisition of nonpersistently transmitted plant viruses occurs within seconds of feeding on virus-infected plants. Likewise, inoculation efficiency is at a maximum within seconds or minutes of beginning to feed on a new host. Molecules produced from the transcription of viral genes serve as "helper factors" that apparently aid aphid transmission. Viruses with defective helper factor genes are not aphid-transmissible even though they may still be readily transmitted by mechanical inoculation in the laboratory. The latent period is an important consideration in trying to control the spread of vector-borne diseases. Insecticides cannot prevent the spread of nonpersistent, aphid-transmitted viruses because aphids transmit so quickly after acquisition

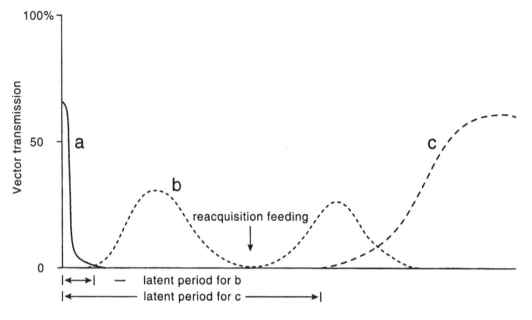

Figure 11.4 Changes in transmission efficiency over time after acquisition: a, nonpersistent (hours) transmission; b, persistent (days), circulative, nonpropagative transmission with the vectors feeding for a second time on an infected host (arrow); c, persistent (weeks), circulative, propagative transmission. The time between acquisition and first transmission is the latent period for b and c.

(no latent period). The key to control is usually to reduce the number of infected plants within the crop to prevent aphids from picking up the virus and transmitting it to nearby plants. Where aphids fly into the crop from a distance, repelling the landing aphids with reflective mulches or interfering with virus transmission with oil sprays on plant surfaces reduces virus spread. For pathogens transmitted only after a considerable latent period, insecticidal control of vectors often can reduce disease spread.

Vectors of Bacteria

Most bacterial plant pathogens do not require insect vectors, but some very important plant diseases do require vectors. Flea beetles (Chrysomelidae) transmit the bacterium *Erwinia stewartii* to maize through feeding wounds, causing Stewart's wilt of maize. Various strains of the bacterium *Xylella fastidiosa* cause debilitating or lethal diseases of grape, peach, citrus, and numerous tree species in the Americas. The bacterium exclusively inhabits the xylem (water-conducting) tissues of plants. Vectors are xylem-feeding insects in several hemipteran families (Cicadellidae: Cicadellinae; Cercopidae, Cicadidae). Both bacterium and vectors are extreme organisms. The xylem-feeding vectors have the highest feeding rates of any terrestrial animal, consuming hundreds of times their weight daily. The bacterium thrives in the high-velocity (10–50 cm/second), nutritionally dilute stream of ingested sap in the vectors' foreguts. Vectors transmit the bacterium after a very short or no latent period but retain their infectivity indefinitely unless they molt. These observations point to the foregut as the critical site for bacterial retention and release during transmission. The bacteria are not merely passive participants in the transmission process. Mutants that lack a signaling system for cell-to-cell bacterial communication can cause plant disease but are not insect transmissible.

Phytoplasmas and spiroplasmas are small bacteria that lack a cell wall (class Mollicutes) and occupy the sugar-rich phloem tissues of plants. These plant pathogens usually complete the complex process required for vector transmission only in certain species of phloem-feeding leafhopper (Cicadellidae), planthopper (Fulgoroidea), or psyllid (Psyllidae) vectors. Transmission requires a latent period of several weeks while the phytoplasmas or spiroplasmas multiply and pass from the gut to the hemocoel and eventually to the vector's salivary secretions, where they enter new plant hosts during the feeding of their insect vectors. Each of these steps represents a hurdle that the pathogen must overcome and results in the high degree of vector specificity and interspecific differences in vector competency of these pathogens.

Insects as Intermediate Hosts of Flatworms and Tapeworms

Insects may serve as intermediate hosts of tapeworms (Cestoda) and other types of helminth worms such as flukes or flatworms (Platyhelminthes). The insects are not truly vectors for these worms because they do not initiate infection of another host. Instead, the tapeworms infect animals that eat the insects containing the worms. For example, larval fleas or lice consume eggs of the dog tapeworm, *Dipylidium caninum*. Dogs or cats acquire the still viable tapeworms by consuming the lice or adult fleas. Flukes in the genus *Prosthogonimus* emerge from snail hosts and seek out dragonfly naiads. The flukes enter the anus of the naiad when it takes in water for respiration. Birds that eat the infected naiads or adult dragonflies become infected with the worms.

Vectors of Fungi

Fungi transmitted by insect vectors cause important plant diseases. Bark beetles (Scolytidae) are the main vectors of fungal plant pathogens in the genera *Ceratocystis* and *Ophiostoma*. Bark beetles such as *Scolytus multistriatus* (Scolytidae) are the principal vectors of the fungus *Ophiostoma ulmi*, which causes Dutch elm disease. The sticky spores of the Dutch elm fungus line the galleries beneath the bark of diseased elms and adhere to the bodies of adult bark beetles that emerge from the galleries. The adults then disperse to new elm trees and introduce spores into feeding wounds made by the beetles on healthy elm twigs. Trees weakened by Dutch elm disease are especially attractive to egg-laying adult beetles and are more susceptible to bark beetle attack, thus accelerating the spread of both the disease-causing fungus and bark beetles. In a similar fashion, various bark beetle species spread fungi of the genus *Ceratocystis* to cause blue stain disease of conifers. Sap beetles (Nitidulidae) pick up spores of the oak wilt fungus, *Ceratocystis fagacerum*, from sap that oozes from fungal cushions found beneath the bark of diseased trees. Subsequently, the beetles may introduce the spores to fresh wounds on uninfected trees. The *Amylosternum* yeasts previously mentioned in the discussion of symbiotic relationships are plant pathogens only in conjunction with the oviposition wounds made by their *Sirex* wasp hosts and vectors. In this situation the vector is also a direct component of the resulting plant disease.

Vectors of Nematodes to Plants

The nematode *Bursaphelenchus xylophilus* causes pine wilt disease by invading the woody tissues of pine trees. The nematodes crawl into the tracheae of adult pine sawyer beetles (Cerambycidae) (Fig. 11.5) while the beetles bore galleries in

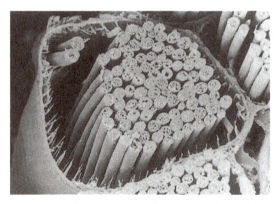

Figure 11.5 The pine wilt nematode, *Bursaphelenchus xylophilus,* in the tracheae of sawyer beetles (Cerambycidae) that bore galleries in the diseased trees. The beetles transport the nematodes to other trees. *Source: Photos courtesy of and with permission of M. Linit.*

nematode-infested trees. After the beetles have moved to different trees to feed or excavate egg-laying tunnels, the nematodes exit their beetle vectors to infect the tree. Other species of wood-boring beetles transport fungal-feeding rhabditoid nematodes externally. The nematodes feed on fungi or decaying matter in the beetle galleries but cause no direct harm to their beetle vectors or plants. Insects are vectors of nematodes that cause some important diseases of humans and animals, as will be discussed in the following chapter on medical and veterinary entomology.

CHAPTER 12

Pest Management

The word "insect" unfortunately conjures up negative images for many people who think of insects as undesirable. Insects are on balance beneficial to humans, but they also compete with or exploit them. Any insect that interferes with human welfare or aesthetics can be considered a pest. Even beneficial insects can be pests in certain circumstances. For example, honey bees provide a valuable service as pollinators (Chaps. 1 and 7), but their stings can provoke severe allergic reactions. The public image of the aggressive Africanized honey bee pursuing and stinging people often looms larger than its service in pollination and honey production. Ladybird beetles (Coccinellidae) are widely recognized for their helpful role as insect predators, but a larval ladybird beetle crawling about fresh produce is automatically considered to be a pest by most shoppers.

Pest management draws upon and integrates many of the facts and principles of insect biology covered in previous chapters. In this chapter we discuss approaches to preventing pest damage to agricultural crops, forests, and commercial products (including structures). Many of the same principles outlined here apply to control of medically important pests. Modern methods of pest management emphasize interactions of insects with the environment and the importance of maintaining environmental quality. Pest management methods, whether based on pesticides or agronomic practices such as plowing or other forms of tillage, can have major impacts on the environment. Pest management uses tools and ideas from ecology, population biology, systematics, and physiology. The evolution of plant attributes that prevent or reduce insect herbivory (Chap. 9) provides a basis for breeding plants that resist insect pests. This is useful for preventing insect pest damage to crops, but as we shall see, this approach is also controversial.

PEST DAMAGE

Direct pests cause immediate injury or disturbance (Fig. 12.1). Biting flies or insects that feed on marketable fruits such as apples or on furniture are examples. Cosmetic damage is an injury to appearance or aesthetic value. Damage that detracts only from the appearance of ornamental plants is cosmetic. Thrips feeding on rinds of oranges or skins of grapes create noticeable scars on the fruit (Fig. 12.2a) but do not alter their nutritional value or flavor. Thrips scarring blemishes grapes as fresh fruit but is inconsequential to grapes used to make raisins or wine. Educating consumers to tolerate inconsequential damage such as thrips scarring of oranges would reduce the economic and environmental costs of preventing such cosmetic damage.

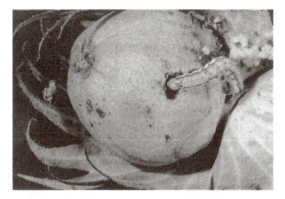

Figure 12.1 Direct pest damage to cotton fruit by the cotton bollworm, *Helicoverpa zea* (Noctuidae).
Source: Photograph courtesy of and with permission of University of California at Davis, IPM Education and Publications.

Figure 12.2 Cosmetic injury. a, Thrips scarring of grape berry; b, Sooty mold on leaf growing on sugary excrement (honeydew) from whiteflies.
Sources: a. Photo by and with permission of J. K. Clark; b. Photo by Alexander Purcell.

Indirect pests create conditions that lead to damage by other factors. Thrips can cause immense damage to fruit or flower crops as vectors of tomato spotted wilt virus even when their numbers are so low that direct injury by the thrips' feeding is not noticeable. Honeydew excreted by scale insects (Coccoidea) or whiteflies (Aleyrodidae) causes indirect cosmetic injury to plants by supporting the unsightly growth of otherwise harmless sooty mold fungi on leaves or fruit (Fig. 12.2b). If the sooty mold layer is thick and widespread on most leaves, this can significantly reduce photosynthesis by shading leaves. Honeydew can cause direct injury if it falls on cotton fibers or some other plant product that directly affects the usefulness of the crop.

Complete or total control of insect pests is usually not attainable except for special cases such as protecting valuable historical documents or artifacts from insect feeding. Biological control of pests by natural enemies or disease requires that some level of the pest population persist or else the natural enemies will starve or emigrate. The concept of an economic injury level (EIL), where the amount of damage equals the cost of preventing damage, guides decisions about what level of damage should trigger control actions (Fig. 12.3). Because pest (and plant) populations are not static but are constantly changing, control action should be taken at an economic threshold of pest density.

This is defined as the level of pest density where action must be taken to prevent the economic injury level from being reached.

The characteristic abundance or general equilibrium level of population density for various insects may vary enormously from one pest species to another. The codling moth *Cydia pomonella* (Tortricoidea) is a serious pest of apples at much lower densities than other common caterpillar pests that attack forage crops such as alfalfa. Much higher densities of caterpillars can be tolerated in alfalfa hay without concern for economic damage. The EIL is inversely related to the unit value of the crop or commodity to be protected. For high-value crops such as apples or other fresh fruits, small percentages of insect infestation can economically

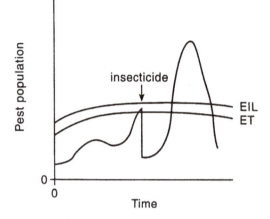

Figure 12.3 Economic injury level (EIL) and economic threshold (ET) for pest density. The pest population (solid line) treated with insecticide (arrow) dropped below the ET, then rebounded above the EIL. The variation of the EIL over time depends on changes in the crop's vulnerability to pests and to other factors.

justify control measures. Only very substantial damage to low-value crops such as hay meets or exceeds the cost of insecticide applications. Changes in the cost of control have the opposite effect. As control measures become more expensive, the EIL increases because a greater amount of damage must be sustained to equal or exceed the cost of control. Various combinations of economic and ecological modeling have produced methods to objectively calculate EILs. In simple general terms, the EIL should equal the cost (*C*) of control on a per production unit basis (for example, $/hectare) divided by the product of the market value ($/kg) of the crop (*V*), the yield loss caused per pest (*D*, in $/insect), and the decimal fraction of pests reduced by the control used (*K*):

$$EIL = C/VDK$$

Changes in the independent variables *V*, *D*, and *K* with time or other variables may further complicate the calculation of EIL. The effects of insect density on yield loss may vary considerably with time. For example, pests that attack fruit may be tolerated in higher levels before bloom time or after a vulnerable stage in crop maturity is past.

The relationship of the typical or characteristic densities of pest populations to the EIL determines whether an insect is never a pest (populations always remain below the EIL), a minor or occasional pest (populations occasionally or rarely exceeding the EIL), or a major pest (populations always near or above the EIL). Pests whose numbers either usually or always exceed the EIL are key pests if pest management programs for a particular crop must always contend with their control. The codling moth is a key pest of apples and pears in North America and Europe because its damage to fruit is always above the EIL if uncontrolled. If insecticides are not used to control it, other control measures such as the application of pheromones to disrupt mating or the application of insect viruses and natural enemies are needed. Using insecticides may disrupt natural enemies that usually keep densities of occasional pests below the EIL. Using non-insecticidal control methods may allow pests that are normally suppressed by insecticides to reach damaging levels.

To use an economic threshold to trigger the timing of pest control actions, pest populations must be monitored by sampling. The reliability and expense of sampling and the predictability of changes in pest population have to be taken into account in devising threshold levels. Economic thresholds also require an understanding of how pest populations relate to damage. The EIL may change as the crop matures. For example, lygus bugs, *Lygus* spp. (Miridae), feeding on the flower buds of cotton plants can exceed the EIL early in the plant's fruiting cycle because their feeding causes the eventual loss of cotton fruits (bolls). Higher densities of damage by lygus bugs feeding later in the season may actually increase cotton yields. They do not feed on fruit buds close to maturity (called bolls) but can eliminate young buds that would not mature before harvest and would divert some of the plant's reserves as they mature.

A BRIEF HISTORY OF PEST CONTROL
In the preinsecticide era, traditional agricultural practices intentionally or unintentionally incorporated a variety of pest control methods. A degree of plant resistance resulted from the greater survival

and seed production of plants that better resisted local pests and diseases and were adapted to local climates, even without conscious selection by farmers. Natural enemies and diseases of pest insects provided a degree of natural control. In regions where the characteristic abundance of key pests always exceeded levels where the crop could be profitably produced, farmers simply did not attempt to grow the crop. Early farmers often had little appreciation or understanding of how insects and mites damaged plants, apart from the most obvious cases of damage to foliage or fruit. As agriculture increasingly applied scientific methods, pests and their damage came under greater scrutiny. The result was new information on pest identification, life histories, natural enemies, and diseases. Some pests could be avoided or their damage minimized by manipulating normal agricultural practices such as the date of planting or harvesting.

The first recorded intentional use of natural enemies for pest control was in 324 BCE in China. Farmers there placed colonies of predaceous ants in citrus orchards to control pests. In this example, native insects were used to control native pests, a practice that has continued to the present.

Scientists realized that exotic insects often became pests or worse pests after arriving in a new area without the natural enemies that had coevolved with them. In their native home of origin, these insects were often held in check by various natural enemies. Freed from regulation by these natural enemies, the exotic insects became abundant enough to reach pest status. Logically, such exotic pests should be controlled by importing specific natural enemies from the region where both the pest and the natural enemies originated. The first spectacular success with this approach, called classical biological control, was the control of cottony cushion scale on citrus in 1889. But to most farmers interested in pest control, biological control was ignored or underutilized for decades afterwards. Early insecticides such as soaps, oils, and arsenic compounds were widely used where pest damage justified the expense.

Beginning in the 1940s, the development of synthetic pesticides rapidly led to widespread and intensive use of insecticides to control agricultural

and medical pests. The new chemical pesticides had many advantages. They were effective and fast-acting, offered a choice of control action for individual farmers, and were inexpensive compared to pest damage. Negative aspects of pesticides that offset these advantages were soon recognized. Insects and mites that had been only occasional or minor pests, or that never were considered to be pests because of their rarity, began to appear in damaging numbers following treatments with some insecticides. Before pesticide treatments, these secondary pests had been limited to nondamaging levels by natural enemies but reached abnormally high populations after pesticide treatment reduced populations of key predators and parasitoids (Fig. 12.4).

A related effect was pest resurgence, the rapid increase of pest populations to even higher than normal levels following pesticide applications (Fig. 12.4). Again, the cause could be attributed to disruption of preexisting levels of natural enemies. In both phenomena, pest insects were less sensitive to pesticides than were natural enemies. In a few cases, notably spider mites (Acarina: Tetranychidae), pesticides even stimulated or increased pest reproduction.

Both pest resurgence and secondary pest outbreaks tended to intensify when the same insecticides were used repeatedly. By comparing the insecticide sensitivity of field populations of pests to the sensitivity of reference populations not exposed to insecticides, entomologists documented the evolution of genetically controlled resistance to insecticides. Resistance is now a major problem for pest management in many crops worldwide. It also is an unambiguous example of evolution by natural selection on observable time scales.

Other problems in using pesticides were recognized later. Investigations of the fate of the insecticide DDD (*d*ichloro*d*iphenyl*d*ichloroethane), chemically similar to the better known insecticide DDT (*d*ichloro*d*iphenyl*t*richloroethane), in the environment showed that concentrations of the insecticide in animal tissues increased at higher levels in the food chain. Clear Lake, California, was treated three times over 3 years with DDD. Chemical analyses detected only trace amounts of the chemical, about 0.01 parts per million (ppm),

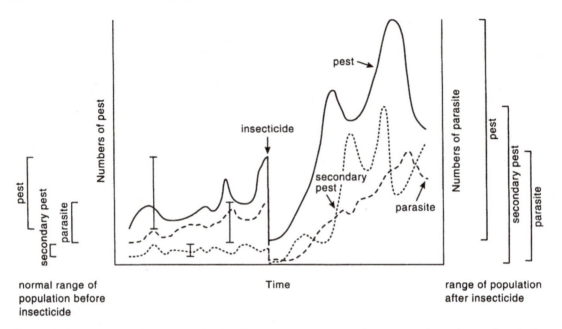

Figure 12.4 Resurgence of pest population (solid line) caused by insecticide treatment (arrow) severely reducing a key parasite population (dashed line). Insects that are habitually below injurious levels (dotted line) can become secondary pests when released from control by natural enemies eliminated by insecticides. Brackets indicate the characteristic abundance (normal range) of the three populations.

in the lake's water immediately after treatment. Algae had about 5 ppm, small fish 40 to 250 ppm, larger fish 340 to 2500 ppm, and fish-eating birds an average of 1600 ppm (Rudd, 1964). This biological amplification was especially prominent in DDT as well as for DDD and chemically related pesticides because it was chemically stable for years, whether in soil, plants, or animals. Furthermore, it was fat soluble, so it tended to accumulate in fatty tissues of animals that ate other plants or animals that had DDT residues. Predators such as fish-eating birds had high levels of DDT because their prey were predatory fish that also had amplified DDT levels. Experiments demonstrated that these high levels of DDT interfered with the reproduction of some birds. The landmark book *Silent Spring* (1962), by Rachel Carson, documented or suggested numerous adverse side effects of pesticides on nontargeted organisms such as pollinating insects, birds, and earthworms. Data increasingly suggested that some pesticides had chronic effects on the health of mammals. The hormonal disruption and

oncogenic (cancer-causing) and teratogenic (causing developmental defects in fetuses) potential of pesticide residues on food had obvious implications for human and animal health.

The heightened public concern and political reactions to pesticides continue to motivate pest management approaches that reduce or eliminate pesticide applications and to more rigorously screen candidate insecticides for safety to humans and animals. Rather than relying solely on insecticides, pest management practices increasingly integrate a variety of control methods for economical pest control while minimizing adverse environmental or social impacts. This approach is called integrated pest management (IPM). The exact definition of IPM varies from one user to another, but the key elements are to consider the pest problem in an ecological context, evaluate all feasible solutions, and balance economic concerns in recognition of possible environmental and health impacts. Every category of pest control method—including pesticides, biological control, and plant resistance—has some advantages and disadvantages. IPM attempts to use a blend of techniques that optimizes the cost-benefit tradeoffs of a combination of methods.

Integrated pest management programs take a broad perspective of how insects, weeds, and diseases fit into the ecology of a cropping system. Fungicides to control plant diseases may upset fungal diseases of insects that would otherwise adequately control a key pest. Weed control may eliminate sources of pollen and nectar required by adult predators or parasitoids or eliminate nonpest insects that serve as prey to increase populations of key predators or parasitoids. Crop yields may be relatively constant over a range of crop densities, so that planting crops at higher initial densities may allow for some plant mortality or damage due to pests without affecting yields. Plowing may kill some important pests but increase soil erosion and release noxious fumes and carbon dioxide from the operation of tractors. Good pest management requires an understanding of the overall ecology of the crop and of adjacent crops or ecosystems from which harmful or beneficial insects may immigrate. Economic consequences of weeds, nematodes, pathogens, and insect pests should be incorporated into management decisions and practices. Agronomic practices such as the selection of the crop variety, fertilization, and irrigation attempt to optimize crop yields and make the growing of the crop as economical as possible by allowing large areas to be planted, maintained, and harvested as a single unit. Unfortunately, a uniform stand of plants of the same age and genotype under optimum growth conditions provides near ideal conditions for the explosive growth of most pest populations. In natural settings, plants can reduce their "apparency," or the ease with which insects can locate them by being widely dispersed among other plants. Small-scale growers can take advantage of this by interplanting compatible crops, but on a larger scale, maintenance and harvesting crops grown in this is too expensive to be economically competitive.

PEST MANAGEMENT METHODS

Integrated pest management aims to integrate a variety of pest control actions to optimize the advantages and reduce the disadvantages of any one method. Some methods are inherently incompatible. For example, insecticides often upset biological control of pests by killing natural enemies. The choice and dosage of insecticide and the timing of application may reduce interference with natural control to acceptable levels. Some pest control methods are complementary, meaning that their effects are compatible or may even may be additive when used together. Other combinations of control methods may be synergistic—the combined effects are greater than the sum of the different methods used separately. As an example, the use of crop varieties that are resistant to pests and biological control of the same pests often reduces pest populations more than does the sum of reductions obtained by using either method alone. Here we briefly describe the major types of pest control methods and the advantages and disadvantages of each method.

Insecticides

Development of synthetic insecticides in the 1940s rapidly led to their widespread use for pest control. Advantages of insecticides are that they effectively and rapidly reduce pest numbers, allow individual choice of insecticide and timing of use, can be applied with relative ease, and provide high economic returns if used properly. For almost 20 years, most other methods of pest control were ignored or neglected, but the serious disadvantages we have already discussed soon became evident. Most species of pest insects have developed genetic resistance to one or more insecticides, requiring more frequent or heavier applications. The consequences of these adverse effects often led to a "pesticide treadmill." This describes a situation where pest resurgence, secondary pest outbreaks, or resistance required greater use of pesticides but the increased use only made the problems worse. The hazards and adverse environmental impacts posed by insecticides require regulation of their use. Moreover, insecticide manufacturers and applicators are subject to civil suits for unsafe use or adverse effects from applications. As a result of regulation and legal concerns, the cost of developing new insecticides continues to climb, allowing only the largest chemical companies to pursue the development of new insecticides. The reduced competition means even greater prices for pesticides. When resistance shortens the marketable life span of a new insecticide, this further discourages commercial efforts to develop new products. These development expenses also mean that only broad-spectrum chemicals that can control a variety of pests on a variety of crops can sell well enough to recover development costs.

Despite the problems posed by insecticides, their use is a cornerstone of modern agriculture because no equally effective alternative methods of control have been found for many pest problems. Many adverse effects of insecticides result from misuse or abuse. The proper choices of insecticide and dosage are critical to pest management programs in order to minimize adverse effects.

Insecticides can be classified in several ways. Depending on the route of entry, insecticides are classified as contact poisons absorbed through the exoskeleton, stomach poisons that act after being ingested, or fumigants that enter the insect through the tracheal system. Insecticides can also be classified by their mode of action. Already mentioned, for example, are nervous system poisons that inhibit acetylcholinesterases (Chap. 6), chitin synthesis inhibitors, and insect growth regulators (Chaps. 4 and 9). The following sections describe some of the distinguishing properties of various categories of insecticides grouped by similarities of origin or chemical structure.

Inorganic Chemicals

The earliest insecticides were inorganic chemicals. Arsenic salts such as lead arsenate and copper acetoarsenite were easily produced with the industrial technology of the nineteenth century and effectively killed a wide range of chewing pests. Because of their direct toxicity to some plants (phytotoxicity), risk as toxicants for humans and animals, and long persistence in the environment, arsenic compounds are rarely used for insect control. The legacies of their overuse many decades ago are persistently high levels of arsenic in some soils, even after more than 100 years. Cryolite (sodium fluoroaluminate) can be mined directly or manufactured.

It is much less toxic to vertebrates than many current insecticides and is selective in that it acts as a stomach poison to kill chewing insects such as caterpillars and grasshoppers. Soaps made as salts of fatty acids and oils of plant, animal, or petroleum origin continue to be used as insecticides. They require direct contact to kill insects and have little residual activity. Silica gels and boric acid are relatively nontoxic to humans but are used to kill household roaches. These materials absorb wax from the cuticle of roaches, killing the insects by increasing water loss from the cuticle.

Botanicals

Plants that contain chemicals to reduce insect herbivory are sources of natural insecticides (Chap. 9). One of the earliest insecticides consisted of pulverized flowers of the pyrethrum daisy, *Pyrethrum cinearaefolium*. The active chemicals in the flower extracts, collectively called pyrethrum, are chemically classified as pyrethrins and cinerins. Insecticides incorporating pyrethrum are still manufactured because of their low mammalian toxicity and because they are made from natural sources rather than synthesized. They provide extremely quick knockdown of insects. For this reason and because of their low risk to vertebrates, they are widely used for household pest control. Piperonyl butoxide is commonly added as a synergist to pyrethroid insecticides. By itself, piperonyl butoxide is scarcely toxic, but in mixtures it multiplies the toxicity of pyrethrins to insects manyfold by inhibiting enzymes that break down the insecticide.

The alkaloid nicotine is a pyrrolidine alkaloid present in high concentrations (2–14 percent) in tobacco species and some other plants. It was first used for aphid control in 1763. It interferes with insect nerve transmission by reducing the action of acetylcholinesterase and disturbing the postsynaptic receptors of the central nervous system. Its dangerous toxicity to mammals has greatly reduced its use. Other insecticides manufactured as natural products from plants are ryanodine, from the roots and stems of the tropical tree *Ryania speciosa*, and rotenones, from a variety of tropical plants that were used originally by native peoples for fish poisons. Botanical insecticides are generally too expensive for large-scale use in agriculture but are applied for rapid pest control by organic farmers or in dairies or other situations where synthetic chemicals are not legally permitted. The tropical neem tree, *Azadirachta indica*, has long been a source of crude insecticides produced from infusions of its fruits soaked in water. Azadirachtin is the principal active component among the numerous anti-feedant, repellent, and insecticidal compounds in neem extracts. This liminoid chemical can kill insects by disrupting normal development but is used mainly for its anti-feedant or insect-repellent properties.

Synthetic Organic Insecticides

These are by far the most widely used chemicals for insect and mite pest control. Most organochlorine compounds such as DDT, chlordane, lindane, and dieldrin have been banned in most developed countries because of the problems of biological amplification and environmental persistence discussed above. Some organochlorine insecticides are reserved for emergency control of medically important vector insects in developing tropical countries. Organochlorine compounds interfere with the sodium channel function of the central nervous system.

Organophosphate (OP) insecticides were developed during World War II as a by-product of nerve gas research. These compounds are derivatives of phosphoric acid. They inhibit the acetylcholinesterase enzyme that breaks down acetylcholine during the process of synaptic nerve transmission. As a result, acetylcholine accumulates at synapses to hinder normal nerve transmission.

Some organophosphates break down rapidly in many environments and do not persist as pollutants or interfere with the activity of natural enemies beyond initial mortality. Plants can take up and translocate some organophosphate compounds throughout the plant. This systemic action is especially effective against sap-feeding insects. The more popular and enduring organophosphate insecticides include chlorpyrifos and malathion.

Carbamate insecticides are esters of carbamic acid. They have short (days) to medium (weeks) persistence after application. They inhibit acetylcholinesterase enzymes, allowing acetylcholine to accumulate and block nerve transmission. Carbaryl is a commonly used carbamate with low mammalian toxicity. Aldicarb is a highly toxic carbamate that is also systemic in plants.

Pyrethroids are synthetic analogs of botanical pyrethrins that have been developed as commercial insecticides. Although they may have chemical structures that appear rather different from natural pyrethrins, the pyrethroids have chemically active sites that are similar in activity to natural pyrethrins. Their mode of action is similar to pyrethrins in disturbing the normal function of sodium channels in depolarization of the nerve cell membrane during nerve signal transmission. Pyrethroids have become widely used because they act very quickly at low dosages and have low mammalian toxicity. In addition, they persist in an active form for longer periods than natural pyrethrins, which break down quickly in sunlight.

Neonicotinoids are relatively recently developed insecticides that act on the central nervous system in a manner similar to nicotine by binding to the postsynaptic nicotinic acetylcholine receptor (Chapter 6). They kill insects at very low dosages and with much less toxicity to mammals. Popular neonicotinoids are highly systemic in plants and especially effective against sucking insects. They can also be used as seed treatments, with the insecticide being translocated systemically to the newly germinated plant.

Insect Growth Regulators

A variety of chemicals that affect insect growth perpetuate immature stages of development, so they are most appropriate for insects such as fleas and mosquitoes, which are only damaging in the adult stage. These chemicals are nontoxic and break down rapidly after application, especially in ultraviolet light. Insect growth regulators are selective in their effects because each analog is only effective against a limited range of insect families. Chitin synthesis inhibitors such as diflubenzuron interfere with development of the cuticle, usually leading to death during or immediately after molting.

Microbial Insecticides

Preparations of insect pathogens can be applied much in the same ways (dusts, sprays, baits, etc.) as chemical insecticides. Preparations of spore-forming bacteria such as *Bacillus thuringiensis* (*Bt*) (Chap. 11) have been widely marketed. The toxicity of *Bt* applications is short-lived, lasting only a few days at most. The numerous strains of *B. thuringiensis* differ in the spectrum of insects that are poisoned. For example, strains effective against larvae of some Lepidoptera are not active against beetle or mosquito larvae. *B. thuringiensis israeliensis* is effective against aquatic fly larvae, including mosquitoes and black flies. Genetic manipulations of *Bt* genes responsible for their toxic proteins can be used to create new toxin combinations tailored for specific pests or to avoid the development of resistance to *Bt*. Insecticidal preparations of *Nosema* pathogens of grasshoppers or *Bacillus popillae* for the Japanese beetle (*Popillia japonica*) are inoculative; they multiply within infected insects and survive for prolonged periods (months to years) before new inoculative applications are needed. The bacterium *Serratia entomophila* is produced commercially and applied to the soil as an inoculative control for root-feeding larvae of the scarab beetle pest, *Costelytra zealandica*. This controls the beetle grubs adequately for at least several years.

Baculoviruses, especially nuclear polyhedrosis viruses, are produced commercially for control of certain species of caterpillars. They are expensive but particularly useful where poisonous insecticides cannot be used because of hazards to people or animals. Baculoviruses only infect arthropods and are considered safe to mammals. Disadvantages are that baculoviruses break down rapidly in ultraviolet light after application and the diseases they cause require several to many days to kill the targeted pest. Many viruses cause the host insect to cease feeding within hours of ingesting the virus, so that the slowness of killing the pest is not a problem. The limited host range of baculovirus insecticides can be a disadvantage if numerous pests need to be controlled, but this may be an advantage if only one pest species needs to be controlled and insecticides would upset natural control of other potential pests.

Fungi have been tried as microbial pesticides since the late 1800s with inconsistent success. Fungal pathogens must have the humidity, surface moisture, and temperature within narrow limits to germinate and infect insects or to sporulate. Commercial preparations of the fungi *Verticillium lecanii* and *Hirsutella thompsonii* are marketed for controlling some pests such as aphids and thrips in greenhouses.

Nematodes of the families Steinernematidae and Heterorhabditidae can be mass-produced along with their accompanying bacteria (Chap. 11) and remain viable in storage for many months. These nematodes attack and kill a wide range of insects but require suitable moisture to survive for long out of soil.

Formulation and Application

Insecticides that have the same active toxicants may perform differently because of other ingredients, called adjuvants, with which they are packaged. A variety of chemicals can be used to dilute the concentration of active toxin in the final mixture. They are usually named on the insecticide's label as "inert ingredients." Dilution makes packaged insecticides safer and easier to ship, store, and apply. Talc, powdered clay, or other finely pulverized carriers provide a surface to which the active insecticide can be adsorbed for application as a dust or wettable powder. Other adjuvants serve various purposes. Emulsifiers disperse water-insoluble insecticides to form an emulsion when added to water so that the mixture can be easily sprayed in large quantities of water. Spreaders reduce the surface tension of droplets on waxy plant surfaces and cause the droplet to spread. Stickers help it to remain firmly attached to plant surfaces. Buffers maintain a desired level of acidity or alkalinity that may be essential for the chemical stability of the insecticide.

The manner in which insecticides are applied may affect their performance. Only a small amount of an insecticide applied to a crop reaches a location that is likely to affect targeted pests, so the method of delivery is critical for effectiveness. Insecticides that kill pests only by contact (e.g., soaps or oils) require thorough coverage and will not affect insects sequestered within plants or under soil. The concentration, droplet-size distribution, and sprayer velocity of the pesticide application can affect the

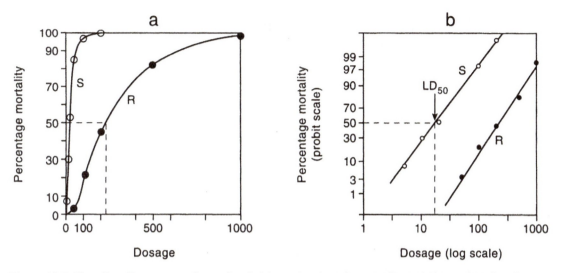

Figure 12.5 Mortality of insects experimentally administered various dosages of insecticides: a, plotted on linear scales; b, mortality plotted on probit scale and dosage plotted on a logarithmic scale. R represents a population resistant to the insecticide; S represents a population 10 times more susceptible. The dose that kills half of the test population (LD_{50}) is shown by the dotted line.

final coverage of the insecticide on the targeted crop. Some insecticides are encapsulated in minute granules for gradual release to soil or plants.

Insecticide Toxicity

Comparing the toxicity of different insecticides to a particular insect or to different populations of insects, as in studies of insecticide resistance, requires standardized measures of toxicity. Individuals within any population of insects will vary in susceptibility to a uniform dosage of any insecticide. The mortality that results from varying the dose of insecticide administered to individual insects produces results such as those depicted in Figure 12.5a. Plotting dosage on a logarithmic scale and mortality on a probit scale (normalized percentages) results in the data approximately fitting a straight line (Fig. 12.5b) to simplify mathematical computations and comparisons. The line that best fits the data points (by linear regression) can be described as an equation from which we can estimate the dosage that kills a given percentage of the test insects. There is no probit for 100 or 0 percent because probits approach these limits asymptotically. The most common statistic from dosage-mortality experiments is the median lethal dosage, or LD_{50}. This

is the estimated dosage that will kill half of the treated insects. The median toxicity is less variable from one experiment to another than are other levels of toxicity, such as the LD_{95} (dosage lethal to 95 percent of test organisms). Where it is difficult to deliver the same dosage per insect, concentrations of insecticide are varied to estimate the LC_{50} (concentration of insecticide that kills 50 percent of the insects). The toxicity must be measured separately for oral (ingestion) or dermal (external) applications.

The relative toxicity of insecticides to mammals is expressed as milligrams of pure insecticide per kilogram of body weight or as an LC_{50}. Hazards of toxicity to humans and animals are estimated from tests using small mammals such as rats or rabbits. Table 12.1 shows the LD_{50} for legally defined toxicity categories and some examples of widely used insecticides in each category.

Insecticide Resistance

The rapid evolution of resistance to insecticides in pest populations is a clear example of Darwin's natural selection or "survival of the fittest." When populations of insects are exposed to an insecticide, usually only a few individuals tolerate the

Table 12.1 Toxicity Categories and Selected Insecticides

Toxicity category	Oral/Dermal LD_{50} (mg toxin/kg body weight)	Required signal words on label	Representative insecticides
Highly toxic	<50 oral and/or <200 dermal	Danger—Poison	Parathion, aldicarb, nicotine sulfate
Moderately toxic	50–500 oral and/or 200–2000 dermal	Warning	Rotenone, diazinon
Slightly toxic	500–5000 oral and/or 2000–20,000 dermal	Caution	Malathion, carbaryl, acephate, chlorpyrifos
Low toxicity	>5000 oral and/or >20,000 dermal	Caution	Permethrin

exposure without dying. If the basis of their tolerance is genetic, a larger percentage of genes for resistance will occur in the next generation. After repeated cycles of selection for resistant individuals, most of the population may have genes for resistance (Fig. 12.6). Not only insects, of course, but all other pest or pathogen groups selected by chemical treatments—mites, nematodes, weeds, fungi, protozoa, and bacteria—have evolved resistance to chemicals. Resistance appears to occur much less frequently in populations of natural enemies. Resistance will develop most rapidly where all individuals in the population are exposed to high doses that leave only a few resistant survivors. If the resistance allele is dominant, resistance may develop rapidly. The frequency of *Anopheles gambiae* mosquitoes in Africa that were homozygous (RR) or heterozygous (Rr) for dieldrin resistance ranged from 0.04 to 6 percent before spraying. After a single spray cycle using dieldrin insecticide, the frequency of individuals with R alleles increased to 86 percent. Resistance alleles that are recessive reduce the speed with which resistance appears in a population under selection. Resistance based on more than one gene (polygenic) will develop more slowly than will monogenic resistance. Resistance may persist for many generations without further selection once it has developed in a population. Reversion, or the decrease in frequency of resistance alleles in the absence of selection, has been documented.

Resistance to DDT was first noted in houseflies in Sweden in 1946, not long after it first began to be used. The numbers of insecticide-resistant arthropod species have steadily grown from fewer than 50 recorded as resistant in 1955 to 481 in 1989 (Georghiou and Lagunes-Tejeda, 1991). The diamondback moth, *Plutella xylostella* (Yponeumeutidae), became resistant to the biological pesticide *Bacillus thuringiensis* within 3 years of heavy use in Hawaii and in Japan. More than half of the resistant pest species have multiple resistance to insecticides in more than one chemical class. This usually reflects the development of resistance to a succession of insecticides as first one, then another insecticide is adopted following the development of resistance to previously used chemicals. Cross-resistance occurs when resistance to one insecticide confers resistance to a related insecticide to which the species has never been exposed. This is because a single mechanism such as a detoxifying enzyme may act against insecticides that share the same mode of action. For example, resistance to one organophosphate compound may confer resistance to another organophosphate. Cross-resistance to insecticides in other chemical groupings also occurs.

Research into the basis of resistance has revealed numerous biochemical, physiological, and behavioral mechanisms. Enzymes responsible for detoxification in resistant insects include various esterases, mixed microsomal oxidases, glutathion transferases, and others. Resistance to pyrethroids may be caused by reduced sensitivity of the sodium channels of nerve axons caused by a reduced number of receptor sites for the insecticide. Altered acetylcholinesterases may be less susceptible to

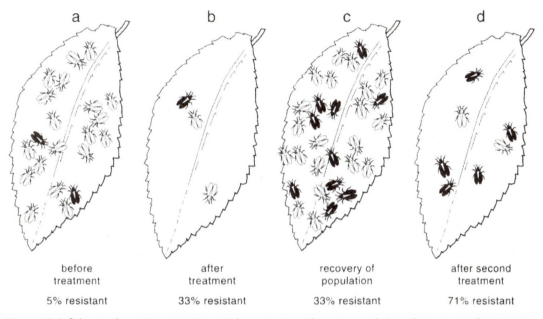

Figure 12.6 Selection for resistance to insecticide: a, starting with a pest population where genetically resistant individuals (dark) are greatly outnumbered by susceptible individuals (light), insecticides differentially kill susceptibles, producing b, a higher proportion of resistant individuals whose offspring, c, are more likely to be genetically resistant; d, after a second treatment, the percentage of resistant individuals increases again.

inhibition by carbamate and organophosphorus insecticides. Resistance to toxins of *Bacillus thuringiensis* appears to be due to changes in receptor sites on midgut cells where the toxins attach. Amplifying the amount of enzymes that break down pesticides by having multiple copies of the gene for the enzyme may increase resistance dramatically (Devonshire and Field, 1991). Resistant *Culex* mosquitoes may have hundreds of copies of an esterase gene. One proposed explanation for the rapid appearance of resistance to insecticides in herbivorous insects is that their natural adaptations to detoxify a variety of plant defensive chemicals also enable them to detoxify synthetic pesticides (Chap. 9). A counterargument to this hypothesis is that mosquitoes and other insects that do not feed on plants are among the best examples of insects to have rapidly become resistant. Resistance is rarely observed in insect predators and parasitoids, which compounds the problems posed by resistance developing in pest species. By intensive selection and breeding in the laboratory, resistance

to specific pesticides has been incorporated into predators and parasitoids. Establishing resistant natural enemies in field populations may reduce the likelihood of pest resurgence and thus reduce the number of pesticide treatments needed (Hoy and Herzog, 1985).

Resistance Management

Resistance to insecticides is one of the most pressing problems in pest management. New chemicals are being commercialized at a slower rate because development and safety testing are so expensive and lengthy. Older insecticides have been rendered obsolete by resistance, and pests are becoming resistant to new insecticides. The observed leveling of the increase in the numbers of resistant species is not cause for optimism. It simply means that most major pests have become resistant. The problem is especially acute in controlling mosquitoes and other insect vectors of human pathogens, where insecticides have been the only effective or most environmentally acceptable control measure.

Preventing or slowing the development of resistance in pest populations depends on understanding the factors that lead to its evolution and modifying practices to avoid or minimize these factors. Ideally, other pest control methods should replace insecticides except for emergencies. Tolerating higher levels of damage, especially where these are largely cosmetic, reduces the need for control. Reducing the number of insecticide applications reduces the number of selection cycles. The simplest and most broadly applicable approach is to eliminate unnecessary applications by basing control decisions on careful monitoring of pest populations and a good understanding of the relationship of pest numbers to crop damage. Economic thresholds are of no value if populations are not monitored regularly. Economic injury levels that are unrealistically low increase insecticide use.

Resistance develops most rapidly where all stages and all members of a population are exposed to an insecticide. For example, insecticides that kill larval mosquitoes often are applied over large areas. This approach exposes most of the population to selective pressure for resistance. Treatments aimed at killing adult mosquitoes near human habitations exert selective pressure for resistance on a much smaller fraction of the mosquito population. The latter method may be as effective as larval treatments in preventing mosquito transmission of human pathogens if humans are the main sources for mosquitoes to acquire the pathogens. Monophagous pests that feed on frequently treated crops, such as the cotton boll weevil, *Anthonomus grandis* (Curculionidae), will develop resistance more quickly than will polyphagous pests. Examples of the latter are grasshoppers that feed and reproduce in surrounding habitats where they are not exposed to insecticides. Insecticide-treated populations that frequently interbreed with unexposed populations will develop resistance more slowly. Short generation times and numerous offspring per generation speed the development of resistance.

Lower doses and chemicals that break down rapidly after use help to avoid resistance. Persistent insecticidal residues can continue to select for resistance long after applications are made. Applying insecticides in a rotation or as a mixture of different chemicals with different modes of action can delay the onset of resistance.

Biological Control

Recall that interspecific competition, predators, or parasitoids are important constraints on the growth of insect populations (Chap. 8). Secondary pests, for example, are organisms that become pests when insecticides eliminate natural enemies. DeBach (1964) defined biological control as "the action of predators, parasites and pathogens in maintaining another organism's density at a lower average than would occur in their absence." Biological control avoids hazards to health or the environment associated with insecticides. Where effective biological control can be established, it is generally more stable and far more economical than chemical methods. As of 1992, 517 projects in 196 islands or countries were declared successful in reducing the population densities of 421 pest species. The most severe shortcoming of biological control may be that it has been difficult to implement in certain situations, such as crops grown for only a few months. Natural enemies may not increase quickly enough to be effective in such rapidly changing environments. Another caution in the use of biological control is to avoid effects on nontargeted species such as native insects that might be driven to extinction by importing certain natural enemies.

Biological control research is not cheap, but the benefits far outweigh the investments in its research and implementation because they may continue for many years without any further inputs. Only a few successful major projects are needed to return more than the amount invested in all biological control projects.

Four approaches are used for biological control: importation of natural enemies to control exotic pests (classical biological control), conservation of existing natural control agents, augmentation of natural enemies by releasing mass-reared natural enemies, and inundation of pest populations by releases of mass-reared agents.

Classical Biological Control

The best known type of biological control is the importation and establishment of natural enemies to control pests that become established in new areas away from their area of origin. The first success with this method controlled the cottony cushion scale, *Icerya purchasi* (Margarodidae), in California in the 1880s by introducing a coccinellid, the vedalia beetle, *Rodolia cardinalis*. A key to this first success was determining that the cottony cushion scale came from Australia and identifying its specific natural enemies. Entomologists found the scale to be rare in Australia and discovered that vedalia beetles and parasitoid flies, *Cryptochetum iceryae* (Cryptochetidae), attacked the scale. These natural enemies were carried to California, reared, and released. The success of this first effort was rapid and spectacular. Both natural enemies still keep the scale well below economic thresholds. The vedalia beetle predominates in warmer areas, the fly in cooler locations.

Classical biological control has become far more complicated today. Molecular as well as traditional methods are used to identify pests and parasitoids. More elaborate steps are taken in quarantine to confirm the identity of natural enemies, their freedom from disease and hyperparasitoids, and their suitability for release. Possible adverse effects on native nonpest insects that might be eliminated by imported natural enemies must be considered. Generally the most host-specific natural enemies are most effective for biological control. A notable exception was control of the endemic coconut moth, *Levuana iridescens* (Zygaenidae), in Fiji by the parasitoid tachinid fly, *Bessa remota*. This fly was imported from Malaysia, where it attacked other moth species. In Fiji, biological control by the tachinid fly made the coconut moth extinct, although the area of origin of the coconut moth is a subject of controversy.

Biological control has succeeded in controlling only some exotic pests, despite major efforts in cases such as the gypsy moth, *Lymantria dispar* (Lymantriidae), in North America, where more than sixty species of natural enemies have been introduced. A continuing controversy is whether importing only one or a few natural enemies is more effective than introducing a larger complex of parasitoids and predators. For example, hyperparasitoids and superparasitoids

may upset effective control of a targeted pest by other parasitoids.

The more detail given to study of the targeted pests and its natural enemies, the greater chances of success with biological control. The walnut aphid, *Chromaphis juglandicola* (Aphididae), is native to the Old World and was not controlled satisfactorily in central California by introductions of the parasitoid wasp, *Trioxys pallidus* (Chalcidoidea), from France. The parasitoids collected from France were effective only in cooler localities near the ocean. Later collections of this species from hot, arid regions of Iran quickly brought the walnut aphid under excellent biological control. The ability of the Iranian strains to withstand the hotter, drier interior areas of California was the key difference.

There are fewer cases of successful biological control of exotic weeds than of insect pests. On the other hand, a higher percentage of biological control attempts against exotic weeds have succeeded. This may be due to the more detailed biological studies prompted by concerns that herbivores imported to control weeds might also damage desirable plants. Exotic plants that have no host-specific herbivores are usually the worst weeds. The first step in classical biological control of weeds is to identify the region where the weed originated and to discover its natural enemies. Next, the weed's ecology, including natural enemies and diseases, must be studied in detail. This includes testing potential biological control agents to ensure that useful plant species are not attacked. Finally, the most promising species of natural enemies are imported and screened under quarantine to ensure that the imported organisms themselves are free of parasitoids and diseases before mass rearing and release. Biological control of weeds has worked best against perennial rangeland or pasture weeds rather than against weeds in frequently disturbed situations such as annual croplands. This is a result of the combined effect of the herbivore on the target weed and competition from surrounding plants. Outstanding examples of biological control of weeds are the control of the rangeland weed *Hypericum perforatum* and the aquatic weed *Salvinia molesta* on two continents. *Hypericum perforatum* is native to Europe, where it is known as St. John's wort. Australian researchers identified several beetle herbivores specific to *H. perforatum* and established them in Australia. Two of these, *Chrysolina hyperici* and *C. quadragemina* (Chrysomelidae), were imported and released in California. Ever since the 1950s, these imported insects have effectively controlled this weed in the western United States (where it is known as Klamath weed).

Salvinia molesta is an aquatic weed that choked lakes and other waterways after its accidental introduction throughout Asia, Africa, and Australasia. Initial attempts at classical biological control failed because of the misidentification of both the weed as a look-alike but nonpest species, *Salvinia auriculata,* and its herbivore weevil, *Cyrtobagous singularis* (Curculionidae). Further research established the true identity of the exotic weed as *Salvinia molesta* but misidentified another weevil, now named *C. salvinae,* as a specific herbivore of the weed. After the taxonomic problems were clarified, *C. salvinae* was finally introduced and established. One more detail, however, remained for the project to be successful. *Salvinia molesta* thrives best on nitrogen-rich waters. Fertilizing the targeted lakes with nitrogen increased the food quality of the plants for weevils. In turn, weevil populations increased to levels that devastated the remaining weeds.

Conservation

A strategy to enhance the effectiveness of established natural enemies is to minimize disturbances to existing natural control. The most common approach is to eliminate or severely restrict the use of insecticides. A second method is to preserve or provide habitats or refuges from which natural enemies can move into nearby crops to attack pests. Leaving strips of alfalfa among cotton fields in California provides reservoirs of predators and parasitoids of key pests. Harvesting only half of the alfalfa strip at any time further ensures a continuing habitat for natural enemies. This method has not been widely adopted because it requires higher costs for cutting alfalfa and closer monitoring of pest and natural enemy populations and because pesticides applied to the cotton

crop are difficult to keep out of adjacent alfalfa. Furthermore, a small amount of damage to cotton from pests that move in from the alfalfa strips must be tolerated, and growers have traditionally tried to ensure extremely low damage levels by using pesticides.

Augmentation

Releasing natural enemies that have already been established may be useful where unusual environmental conditions eliminate or reduce a number of natural enemies below an effective level. Where unusually bad weather has eliminated a natural enemy, inoculative releases of the natural enemy may reestablish it. Augmentative releases also may increase natural enemies at a particularly critical phase of the crop to be protected or during a vulnerable phase of the target pest's life cycle. Inoculative applications of the bacterial pathogen *Bacillus popilliae* against Japanese beetles (*Popillia japonica*) often must be repeated after a few years to achieve effective levels of beetle control (Chap. 11).

Inundative Releases

Releasing large numbers of natural enemies can be similar to the use of insecticides, as discussed in the previous section for microbial pesticides. They are used where establishing stable biological control is not feasible. Greenhouse pests such as aphids or whiteflies (Aleyrodidae) may be controlled by releasing predaceous fungus gnats (Mycetophilidae), beetles (Coccinellidae), parasitoids (Chalcidoidea), or fungal sprays (*Verticillium lecanii*). The rapid, short growing cycle in the highly artificial greenhouse environment makes long-term control with inoculative releases an unrealistic goal. Most microbial pesticides do not remain at effective levels in the environment after application. Viral or bacterial pesticides kill their targeted hosts rapidly, leaving few infected hosts to provide inoculum to infect additional hosts over a prolonged period. Exceptions may be the granulosis viruses of sawflies (Tenthredinidae) and the western grape leaf skeletonizer, *Harissina brillans* (Zygaenidae). These granulosis viruses have effectively controlled their host insects for years after being applied. The viruses appear to be limited to cells lining the midgut, and some virus-infected larvae survive to the adult stage where the virus carries over on eggs as a surface contaminant to the next generation or from one season to the next.

Host Plant Resistance

Plant resistance to insects is a heritable set of characteristics of a plant that make it less susceptible to damage by insects. You may recall three categories that describe mechanisms of plant resistance to insects discussed in Chapter 9: (1) antibiosis adversely affects herbivore longevity or fecundity; (2) antixenosis (formerly called nonpreference) deters insect feeding or oviposition; and (3) tolerance minimizes loss of yield despite insect damage. These are relative properties in comparison to other plant genotypes that have a lesser degree of resistance and are thus said to be more susceptible.

Plant resistance is one of the cheapest and most easily applied methods of pest management. It is compatible or even synergistic with other methods of pest control, except for a few examples where plant resistance also harms important natural enemies. The major direct costs are for breeding and testing new varieties (or cultivars). Indirect costs are the lower yields or quality of some resistant varieties. Plant resistance is especially desirable for crops with a low unit value so that insecticides are too expensive to use. For this reason, plant resistance is used most extensively in grain and forage crops. One major problem is the long time required to develop adequately resistant varieties. For some crop pests, no satisfactory resistant varieties have ever been developed. Another problem is the development of insect populations that overcome plant resistance.

Conventional plant breeding by controlled crosses of selected parents has identified specific genes for resistance to insects. In most cases the specific mechanism of resistance is unknown, but this is no impediment to using plant resistance for pest management. Once a breeder can identify specific genes for resistance (R genes), it is relatively easy to incorporate them into new varieties. A problem is that resistant races of pests may appear after cultivars with the new gene combinations become widely used. Fewer problems with resistant

races or pests have been noted in breeding for plant resistance to insects compared to plant diseases. One example for insects is resistance in wheat to the cecidomyiid pest, the Hessian fly, *Mayetiola destructor*. Populations of the Hessian fly that overcame resistance appeared in response to the widespread deployment of R genes. Such populations with distinct abilities to overcome resistance have been termed biotypes, but the definition or usage of the term biotype varies widely. With respect to plant resistance, biotypes are defined by the plant cultivars the insect does or does not infest. There are now at least ten biotypes of Hessian fly defined by their reactions to combinations of four alleles in wheat for resistance to Hessian fly.

Laboratory experiments with the brown planthopper, *Nilaparvata lugens* (Delphacidae), demonstrated how biotypes might evolve in this pest. The planthopper biotypes were reared on rice cultivars that were resistant at the start of the experiment. After ten to fifteen generations, each biotype was reproducing about as well on the originally resistant cultivars as they did on other cultivars formerly classified as susceptible (Claridge and Den Hollander, 1982). This microevolution of changes of the brown planthopper in response to plant resistance is analogous to laboratory selection for insecticide resistance. In both types of selection, survival increased gradually from one generation to the next after each cycle of selection.

Durability is a major goal for resistance breeding. Repeated selection among high-yielding cultivars for resistance is aimed at achieving long-lasting resistance, defined as resistance that avoids the development of pest biotypes. One of the earliest examples of using plant resistance was the adoption of wild grapevines, *Vitis rupestris* and other *Vitis* spp., from North America as rootstocks to counteract the grape phylloxera, *Daktulosphaira vitifoliae* (Phylloxeridae). This approach was instantly successful and has endured in wild grape selections or in hybrids between wild grapes for at least 140 years in France. Hybrid rootstocks of wild grape species crossed with commercial grape, *Vitis vinifera*, however, have not been durable; biotypes of phylloxera have appeared on hybrids that have *V. vinifera* as a parent. The breakdown of resistance (formation of resistant races of phylloxera) caused losses of over one billion dollars over 10 years in California alone in replacing vines grafted on the failed rootstock varieties of *V. vinifera* hybrids.

Effective plant resistance is a still elusive goal for some major pests of important crops. The long time required for the repeated selection and testing necessary to develop resistant cultivars is the main barrier for breeding resistant fruit trees. Breeding trials may require 3 to 10 years for tree crops to produce fruit for a single selection cycle. A major advantage of plant resistance is that it can be combined with other methods to produce acceptable results where either method alone would not. For example, resistance to an aphid pest of wheat and other grain crops, the greenbug, *Schizaphis graminum* (Aphididae), is adequately controlled by a parasitoid in combination with a resistant variety, but either method alone is not adequate for control (Fig. 12.7). By reducing the rate at which pests can multiply, plant resistance may allow predation and parasitism to adequately control pests.

Even where resistant varieties do not lower pest densities enough to avoid the need for insecticides, they may delay or reduce the number of pesticide applications.

Early human selection that domesticated plants for cultivation intentionally selected for rapid growth, gigantism of fruits, taste, reliable germination, and ease of harvest. Natural selection increased crop resistance to climatic unpredictability, pests, diseases, and weeds with little conscious direction from humans. Breeding governed by understanding of genetic principles rapidly improved yields, adaptability, and resistance of major crops to multiple pests and diseases. About the same time, other developments in soil and fertilizer technology and chemicals for disease and pest control decreased farmers' dependence on inherent plant resistance to pests. The goal of decreasing pesticide use has renewed interest in plant resistance. Wild relatives and progenitors of domesticated plants have provided important sources of resistance genes, but these often have led to impermanent resistance.

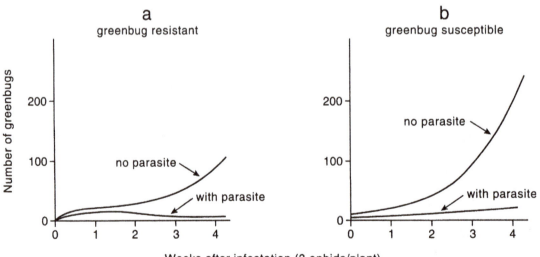

Figure 12.7 Effects on populations of the greenbug, *Schizaphis graminum* (Aphididae), of biological control with a parasitic wasp combined with plant resistance: a, on a greenbug-resistant (Will barley) variety; b, on a susceptible (Rogers barley) variety. The parasite lowers greenbug populations on the resistant variety.
Source: Redrawn from data by Starks et al., 1972.

Cultural Controls

Manipulations of normal agricultural practices to reduce pest damage are classified as cultural control techniques. The most common types of cultural control are tillage, timing of harvest and planting, crop rotation, sanitation, and water management.

Tillage, or cultivation, is the mechanical manipulation of soil, chiefly for weed control, drainage, and disrupting impenetrable soil layers. Plowing is perhaps one of the best known methods of cultivation. It inverts a layer of soil to bury weeds and crop residues, giving crop seedlings a fast start over competing weeds. Its main disadvantages are that plowing, followed by hoeing or disking for weed control, requires large amounts of fuel, may cause compaction layers under the plow bottom, and is environmentally damaging because it increases rates of soil erosion and the release of environmentally damaging greenhouse gases. Shredding and deep plowing to bury cotton plant residues after harvest destroy overwintering larvae of the pink bollworm, *Pectinophora gossypiella* (Gelechiidae), but are useful only where applied in a regionwide program. Otherwise, surviving pink bollworms in unplowed fields will develop into adults and migrate into adjacent fields the following spring. Plowing usually does not kill soil-inhabiting pests that are mobile in soils. The widespread increase in using herbicides (pesticides that control weeds) has largely replaced tillage in some crops, reducing erosion, soil compaction, energy use, and costs. The objections to herbicides are concerns to human and animal health and effects on nontargeted organisms. No-till farming has reduced some pest problems and plant diseases and increased others compared to traditional tillage, requiring changes in pest management tactics.

The timing of planting and harvesting is a long established method for pest management. For example, wheat planted after the Hessian fly's brief period of egg-laying in the autumn escapes most infestation by this pest. Crops vulnerable to some virus diseases can be planted after the main flights of specific insect vectors. Delaying planting of crops until after spring migratory flights of the beet leafhopper, *Circulifer tenellus* (Cicadellidae), reduces losses to the curly top virus transmitted by this insect. Similarly, the timing of harvest can avoid pest injury. Adult moths of the navel orange worm, *Amyelois transitella* (Pyralidae), lay eggs on ripe almonds in trees or on the ground. Damage from larvae that hatch from the eggs increases steadily after the nuts are mature, especially when the nuts are on the ground. Early harvest minimizes damage. Potatoes grown for use as seed are harvested in many regions as soon as possible following increases in flights of the green peach aphid, *Myzus persicae*, which is the main vector of potato leaf roll virus. Infection with the virus greatly reduces the value of a seed crop but can be tolerated at higher levels in main crops grown for food. Early-maturing cultivars of cotton allow the profitable harvest of cotton crop in regions where late season pest explosions lead to the pesticide treadmill syndrome.

Crop rotation varies the place rather than the time that crops are planted. It may be effective against crop-specific pests that have soil-inhabiting stages between crop cycles. For example, Western corn rootworm, *Diabrotica virgifera* (Chrysomelidae, Coleoptera), populations that increase in maize (corn) fields do little or no damage to soybeans that are planted in the same fields the next year because the root-feeding larvae cannot develop on soybeans. Crop rotation is one of the main methods used to prevent rootworm damage. In the midwestern United States, a traditional 3-year rotation of maize alternates either soybeans or a grain crop (oats, wheat, or barley) the next 2 years before replanting to maize. The adult rootworms lay their eggs in soil in maize fields—thus replanting maize in the same field would expose the crop to hatching rootworm larvae. Maize fields planted after a crop rotation have reduced numbers of rootworms because few rootworms develop in soybeans or small grain crops. However, a new variant of the Western corn rootworm has appeared that endangers the effectiveness of using rotations for control. The new "biotype" will fly into soybean fields to lay its eggs, negating the effect of the rotation to soybeans. This new "rotation-resistant" form appeared in the late 1980s in Illinois and has continued to expand its range to other states. This is an example of natural selection prompting behavioral changes in response to human-induced environmental changes. Crop rotation has additional benefits. It may reduce crop-specific pathogens, weeds, and nematodes in the soil and help to maintain soil fertility.

Sanitation is the removal of weeds or crop residues that might lead to pest problems. Burning or shredding pruned branches from fruit trees or grapevines destroys borers (Coleoptera: Bostrichidae, Anobiidae, and others) that overwinter in prunings to emerge during the spring and damage remaining branches. Knocking down and disking under the "sticktight" nuts remaining in winter-dormant almond trees destroys naval orangeworm larvae that overwinter in the old nuts. Some pests build up populations faster in fields with certain weeds. For example, Russian thistles growing in sugar beet fields leads to denser populations of beet leafhopper, *Circulifer tenellus* (Cicadellidae), and beet webworms, *Loxostege sticticalis* (Pyralidae). Weeds next to croplands may harbor stinkbugs (Pentatomidae) or lygus bugs, *Lygus* spp. (Miridae), that move in to damage fruit crops. On the other hand, outlying weeds may be important reservoirs and sources of nectar and pollen for natural enemies. Animal waste management in dairies and poultry farms is an important feature of controlling flies that breed in manure.

Water management by draining or tillage to avoid standing water is a major tactic to reduce mosquito populations by eliminating breeding habitats. The adverse environmental consequences of draining wetlands require new ideas for mosquito control. Flooding dormant grapevines for more than a month during the winter effectively controls the root-inhabiting grape phylloxera, but prolonged flooding cannot be tolerated by most crops.

Pheromones and Attractants

Research on the effects of external chemicals on insect behavior has led to the use of semiochemicals (Chap. 6) for pest management. Chemical attractants are used to monitor pest abundance and activity. Food lures, aggregation pheromones, and sex pheromones attract insects in measurable numbers where densities are too low to monitor by direct sampling methods. Pheromones are especially valuable for detecting initial insect outbreaks

or evaluating the effectiveness of quarantine programs in excluding insects. For some pests, traps baited with sex pheromones can indicate when mating flights begin, helping to determine when to apply control efforts.

Mass-trapping of pests attracted to traps has not been widely successful as a method of pest control. Low-density pests such as tsetse flies, *Glossina* spp. (Glossinidae), can be controlled in limited areas with inexpensive and simple traps baited with attractants. The most attractive components and combinations of chemicals were derived from analyzing the components of cattle breath and their effects on tsetse fly behavior. Other studies have demonstrated the feasibility of mass-trapping with aggregation pheromones to control bark beetles (Scolytidae) (Chap. 6) and sex pheromones for grape berry moth, *Endopiza viteana* (Tortricidae), but cost: benefit ratios have been uneconomic.

Mating disruption of pests using sex pheromones has succeeded in controlling the pink bollworm in cotton and the oriental fruit moth, *Grapholita molesta* (Tortricidae), in peaches and plums. A steadily growing list of pests are being controlled by using sex pheromones to disrupt mating. Synthesized blends of volatile sex pheromones can be delivered by emitters dispersed throughout the crop. In cotton the emitters are segments of thin, hollow fibers dispensed by aircraft or ground equipment. In orchards, the emitters are hollow plastic tubes tied onto branches. High levels of widespread pheromones in the crop inhibit normal mating or reproduction. The exact mechanism of inhibition is unknown. Males may be unable to locate female pheromone plumes within the high background concentrations of sex pheromone. The effects of high-pheromone backgrounds on females are less clear.

Using pheromones avoids the environmental damage and health risks of using toxic chemicals and may enhance biological control. Unfortunately, many important pests do not use pheromones that act at long distances. Moreover, timing and coverage of pheromone application are critical for success, and the method controls only a single pest species. Thus a pest species targeted for control with pheromones is usually a key pest.

Genetic Manipulation

Mass releases of insects that can mate normally but do not reproduce have been used increasingly to attempt to eliminate certain pests from entire regions. The landmark first success of the sterile insect release method was with the screwworm fly, *Cochliomyia hominivorax* (Calliphoridae), a serious pest of livestock. This fly occurs in very low population densities during winter months. Mass-reared flies were sterilized by radiation and the males distributed by aircraft over infested areas to overwhelm natural populations. The populations of screwworm decreased exponentially with each release of the same numbers of sterile flies because the ratio of sterile to fertile flies increased as the wild screwworm fly populations decreased. In contrast, insecticides kill a relatively constant proportion of insects with each treatment. The screwworm fly was eradicated first from the island of Curaçao. Eventually coordinated control and releases of sterile males eliminated the pest from the United States and northern Mexico. The program continues in Central America to keep screwworm flies from reestablishing northward. Sterile releases eradicated an incipient outbreak of screwworm fly in Libya in 1992.

For sterile releases to succeed in eradicating a pest from a region, the number of offspring produced from matings after each release has to drop enough so that the ensuing adult generation decreases in size with each release. The ratio of sterile to fertile insects required to decrease populations depends on many factors. Examples of insect characteristics that tend to facilitate the sterile insect release method are single instead of multiple matings by females (unless the first male's sperm takes precedence over later matings), methods to economically separate mass-reared males

from females, a low rate of increase of natural populations, and competitive mating performance of sterilized compared to normal insects. Lowering the target pest population before release increases the sterile-to-fertile ratio after release. For this reason, sterile insects are first released when populations are just recovering from normal seasonal lows. Often the targeted pest population is reduced by insecticides or other controls before releases. Using combinations of control methods to lower populations, even pests that can reach high peak populations, such as the cotton boll weevil, the codling moth, and fruit flies (Tephritidae), have been eradicated from prescribed regions by using sterile releases. Because the ultimate goal of sterile releases is to eventually eliminate an insect species from a particular region, this approach must be coordinated over an entire region and requires massive investments for research and implementation. As a result, only programs that will produce economic and social benefits at least several times the total costs of control and that lack any other adequate control methods have been attempted.

The release of pest insects altered with introduced genes to decrease targeted insect populations is still a goal for pest management. Natural selection quickly selects against genes that make an insect population less competitive. A more realistic but still speculative goal would be to change the pest status rather than the population density of targeted insects. For example, reducing the ability of key populations of mosquitoes to transmit virus would greatly reduce the damage or risk caused by the mosquitoes without having to change their population size. A mechanism to do this must increase the numbers of insects with replacement genes relative to numbers of wild-type insects.

CHAPTER 13

The Study of Classification

This chapter introduces a brief history of insect classification and describes how insect systematists study the historical genealogical relationships among insects today. The reader should find this a useful primer to read before beginning Chapter 14 which reviews the actual comparative morphological and fossil evidence that underpins the current insect classification.

THE EMERGENCE OF INSECT CLASSIFICATION

Little in the way of a true classification of insects existed before the eighteenth century. Most literature on insects then consisted of whether they were useful or harmful, as well as some early observations of behavior and development. The pathbreaking work of Linnaeus (1707–1778), who established our modern system of binominal nomenclature (genus name + specific epithet, e.g., *Apis mellifera*), began the process of hierarchical classification of plants and animals. During an age in which much of the world was being explored botanically and zoologically following European colonization, Linnaeus eventually described roughly 2000 species of insects. He classified them in seven orders (then called classes) based on wing characteristics. Linnaeus suggested that features of the mouthparts would also be useful for classification, a suggestion that was pursued by his student, J. C. Fabricius (1745–1808), who described about 10,000 species and modernized the concept of what we now recognize as insect orders (still called classes). P. A. Latreille (1762–1833), a contemporary and colleague of Fabricius, integrated many more diverse morphological traits into his comparative study and classification and introduced the concepts of many

insect families, many of which are still recognized today.

Insect classification continued to develop as an organization of diagnostic characters through the nineteenth century and focused mostly on practical aspects of describing and classifying the huge diversity of insects revealed by world exploration. The last individual to attempt to classify all insects by his own efforts was J. O. Westwood (1805–1893); since that time insect systematists have tended to specialize either on subgroups or on higher level phylogeny rather than the details of classification. It would be a truly daunting task today to repeat Westwood's efforts, with roughly a million insect species described and several million more still to be described. Today, literally thousands of taxonomists worldwide describe insect species, making it a challenging task to catalogue and classify even the species already known.

Since Darwin, biological classification has more or less attempted to reflect the phylogeny (ancestry) of organisms. In practice, classification initially did not change significantly from the influence of evolutionary thought until the advent of modern systematic methodologies in the late twentieth century made it possible to infer or estimate phylogenetic relationships with some degree of certainty. Even today, insect classifications are in many cases practical compromises between modern phylogenetic findings and more familiar traditional arrangements. As our understanding of insect phylogeny deepens, we expect to see the classification of insects integrate this understanding more and more thoroughly into a truly evolutionary (phylogenetic) classification.

DESCRIBING AND NAMING NEW SPECIES AND HIGHER TAXA

As scientists began to more thoroughly explore the world and describe many new species of organisms, it became apparent that some sort of nomenclatural rules for establishing official scientific names for the new species and higher taxa were needed. The function of the International Code of Zoological Nomenclature, and the International Commission on Zoological Nomenclature that administers it, is to provide a series of guiding rules for establishing the priority and uniqueness of scientific names for taxa and to promote stability and standardization in the use of names for organisms worldwide. Generally, the first name given to a species (or higher taxon) is given priority unless there are complicating factors; the application of the name is also fixed by requiring that a type specimen be deposited in a permanent collection for future reference. Needless to say, standard taxonomic practices have changed considerably over the past 250 years, so that species described now are subject to much more scrutiny and many more rules than those described in Linnaeus's time.

RECONSTRUCTING INSECT EVOLUTION

Fossils represent our only direct evidence of the identities and morphologies of previously existing but extinct insects. One might assume, then, that they would figure directly in the reconstruction of evolutionary history of extant taxa. While they do play important roles in inference about insect evolution, they tend not to be included directly in analyses of phylogeny, since it is seldom if ever clear whether any given fossil is directly ancestral to specific living insects or, alternatively, represents an extinct side lineage. Instead, fossils provide very strong evidence about how recently certain insect groups and morphological traits could first have appeared on earth. They also provide a valuable perspective on which morphological or ecological traits evolved before or in combination with others and in which kinds of habitat. We will have more to say about insect fossil history in Chapter 14.

With the development of modern statistics early in the twentieth century, the added invention and development of computers in the latter half, and the later advent of powerful molecular biology techniques in the closing decades, there has been an accelerating pace of progress in analyzing evidence for insect relationships. The most current phylogenetic understanding of most of the insect order and family relationships is still under highly active investigation. We can expect classifications to continue to change significantly in the future as our knowledge improves.

Figures 13.1 through 13.3 provide a brief example of how comparative morphological traits of insects can be used to infer phylogenetic relationships (in this case among the holometabolous insect orders). These figures should only be considered as an example of how the basic logic works, not as a comprehensive depiction of how holometabolous insects are actually related. Our modern understanding of insect relationships also takes into account analyses of molecular genetic data and in addition, the comparative morphological evidence for insect relationships is more extensive and complex than shown here (see the chapters on individual insect orders and also Chap. 14).

The study of phylogeny has moved from a once speculative exercise to becoming one of the most vibrant and methodologically complex fields in modern biology. Insect relationships have played major roles in a number of analytical and philosophical controversies in systematics over the last few decades. Every year new findings are reported that affect the higher classification (order and family relationships) of insects, and this edition of this text includes a number of major reclassifications

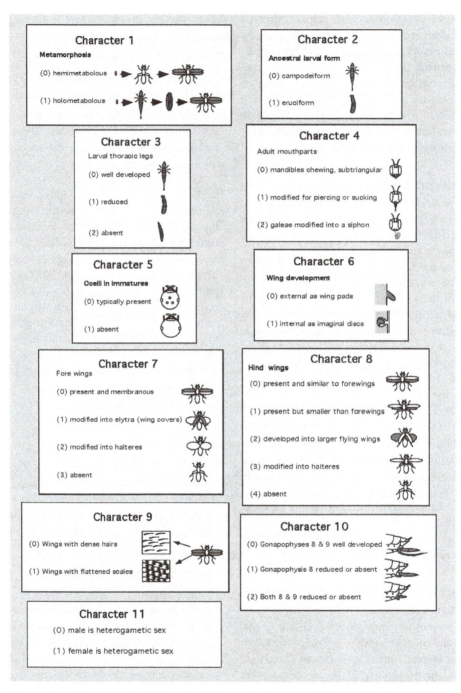

Figure 13.1 Characters for Holometabola. Note that these are highly simplified, intended as an example only. See the chapters for each holometabolous order for more details on higher-level relationships among these orders.

Characters	1	2	3	4	5	6	7	8	9	10	11
Taxa											
Non-endopterygotes	0	0	0	0	0	0	0	0	0	0	0
Neuroptera s. l.	1	0	0	0	1	1	0	0	0	1	0
Coleoptera	1	0	0	0	1	1	1	2	0	1	0
Hymenoptera	1	1	1	0	1	1	0	1	0	0	0
Strepsiptera	1	1	0	0	1	1	2	2	0	?	0
Mecoptera	1	1	1	1	1	1	0	0	0	2	0
Diptera	1	1	2	1	1	1	0	3	0	2	0
Siphonaptera	1	1	2	1	1	1	3	4	0	2	0
Lepidoptera	1	1	1	2	1	1	0	0	1	2	1
Trichoptera	1	1	1	0	1	1	0	0	0	2	1

Figure 13.2 Matrix of coded characters from Fig. 13.1. Analysis leads to 4 trees with 20 steps, one of which is summarized in Fig. 13.3.

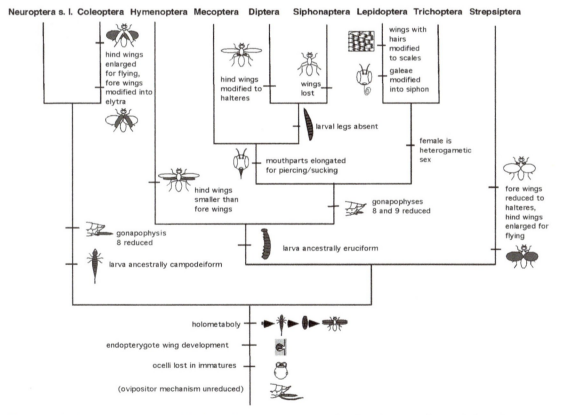

Figure 13.3 Highly simplified tree of Holometabola relationships, with major character changes indicated on branches. See Figs. 13.1 and 13.2 for details, as well as chapters on holometabolous orders.

from earlier editions. The references at the end of this chapter provide introductions to many aspects of current systematics, but the reader is advised that advances in the field may quickly render many of these references out-of-date.

NEW DIRECTIONS IN INSECT SYSTEMATICS

Phylogenetic analysis is not the only area of systematics being revolutionized by computerization and Web-based information sharing. Comparative morphological, genetic, and distributional data for many insects are available on the World Wide Web via a plethora of databases such as GenBank (for protein and DNA sequences), MorphBank and Morphobank (for morphological traits), the Tree of Life Project (overviews of biology of organismal groups), and GBIF (the Global Biodiversity Information Facility). These databases not only provide archival storage for scientific data, they have democratized and standardized many aspects of comparative biology by providing universal worldwide access to information.

The identification of insects has also been enhanced by the introduction of interactive keys, many of them available online. The advantages to such keys are, first, that each feature can be fully illustrated and, second, that the user can identify an insect using the process of elimination, starting with any character or body region, rather than in a set order as with dichotomous printed keys. The current edition of this textbook contains dichotomous keys for the various insect orders and families, but future editions are likely to move completely to interactive keys.

The recent revolutionary advances in systematic methodology, combined with the obvious power of the Web for reaching people worldwide, lead to the inevitable conclusion that systematics of insects and other organisms is moving gradually toward a more Web-based information system. Ultimately, all new descriptions, names, phylogenies, and comparative biological information are likely to be integrated and searchable via computer, rather than being available only through access to thousands of published papers in many journals throughout the world. Current insect taxonomy is not there yet, but no doubt our access to information will change radically within the lifetime of the reader.

CHAPTER 14

An Evolutionary Perspective
of the Insects

The phylum Arthropoda is split into major subdivisions that reflect basic differences in structural organization (Table 14.1). We recognize here five subphyla, one of which, the Trilobita, is entirely extinct. Some controversy still exists over how these five subphyla are related. One longstanding view, based primarily on gross comparative morphology, considers Hexapoda, including the insects, to be most closely related to the Myriapoda and calls this larger lineage the Atelocerata.

In this view, the Atelocerata is characterized by mouthparts consisting of the labrum, a pair of mandibles, a pair of maxillae, and a pair of second maxillae, which are fused into a labium, which

Table 14.1 Conspectus of the Higher Classification of the Phylum Arthropoda

Subphylum Trilobita—trilobites
Subphylum Chelicerata
  Class Merostomata—horseshoe crabs and eurypterids
  Class Pycnogonida—sea spiders
  Class Arachnida—spiders, mites, scorpions, phalangids, etc.
Subphylum Crustacea—crabs, shrimp, lobsters, etc.
Subphylum Myriapoda
    Class Diplopoda—millipedes
    Class Chilopoda—centipedes
    Class Pauropoda—pauropods
    Class Symphyla—garden centipedes
Subphylum Hexapoda
    Informal group Entognatha
      Order Protura
      Order Collembola
      Order Diplura
    Class Insecta—insects

forms the floor of the oral cavity. Although there are differences in detail of the mouthpart structures among the classes of Atelocerata, their close relationship is suggested by a general similarity in body organization. In all Atelocerata, the head bears a single pair of antennae and the mandibles never bear palps, as they do in many Crustacea. The segmental nature of the trunk is fundamentally similar in all Atelocerata, with dorsal terga, lateral pleura, and ventral sterna, the last producing invaginations or apophyses into the hemocoel for the purpose of muscle attachment. The alimentary canal consists of fore-, mid-, and hindgut divisions with a peritrophic membrane and Malpighian tubules. Respiration is by a tracheal system, usually with two spiracles per segment, and circulation is by means of a dorsal vessel with segmental contractile chambers. In Crustacea, Arachnida, and Pycnogonida the alimentary canal typically expands into a large, glandular digestive and storage organ in the trunk region, and respiration is typically by book lungs or gills. In addition, in the last three subphyla the head and thorax often appear to be fused into an integral body region or tagma, the cephalothorax. In the Atelocerata the head is always an independent and freely moving tagma. The differentiation of the arthropod body into distinct regions specialized for different functions is called tagmosis. This is one of the most important developments in the evolution of the Arthropoda and has produced the distinctive body plans of the various arthropod phyla and classes.

Within the Atelocerata two patterns of tagmosis are evident. In the Myriapoda the body is differentiated into head and trunk. In the Hexapoda the trunk is further differentiated into the thorax and abdomen—the former specialized for locomotion, the latter for housing the intestine and reproductive organs. This organizational change led first to the shortening of the body and to the characteristic hexapod gait and eventually to the evolution of wings in pterygote insects. The general organizational features of the hexapod body as proposed by Kukalová-Peck (1987) are shown in Figure 14.1.

An alternative view has recently become increasingly well supported by comparative developmental studies, some newer comparative morphological surveys, and molecular phylogenetics. It holds that the Hexapoda are actually best seen as a subgroup of a larger complex called Pancrustacea, including the classes of the Crustacea. The closest relationship of Hexapoda appears to be with several subgroups of Crustacea, thus, the class Crustacea as traditionally defined may well be paraphyletic unless the Hexapoda are included. A number of aspects of developmental gene expression are shared by these groups; the comparative morphological implications of homology are still being elucidated, especially with respect to fossil taxa such as the enigmatic marine "hexapod" *Devonohexapodus* (Fig. 14.2), which was contemporaneous with early terrestrial hexapod fossils.

Until the last few decades, all six-legged arthropods have been included in the Insecta. However, the hexapodous condition represents an evolutionary grade, and some hexapods lack many features that appear to be "standard" within the true insects. Disregarding the unusual Devonian fossils such

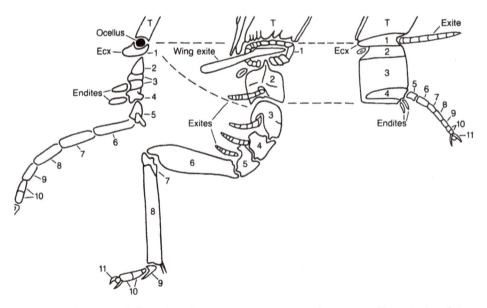

Figure 14.1 Diagrams of a hypothetical, primitive, winged insect, showing serial homologies of the appendage segments on the head, thorax, and abdomen. Some basal leg segments are incorporated into the sides of the body as follows: head (epicoxa); thorax (epicoxa, subcoxa); and abdomen (epicoxa, subcoxa, trochanter). Lobes between some segments of the appendages are called exites (outer lobes) and endites (inner lobes). Leg segments are: 1, epicoxa; 2, subcoxa; 3, coxa; 4, trochanter; 5, prefemur; 6, femur; 7, patella; 8, tibia; 9, basitarsus; 10, tarsus (primitively with two subsegments); 11, posttarsus (primitively with two claws). ECX sidelobe, a lobe of the first leg segment or epicoxa; T, tergum.

Source: Redrawn from Kukalová-Peck, 1987. Reproduced by permission of Canadian Journal of Zoology.

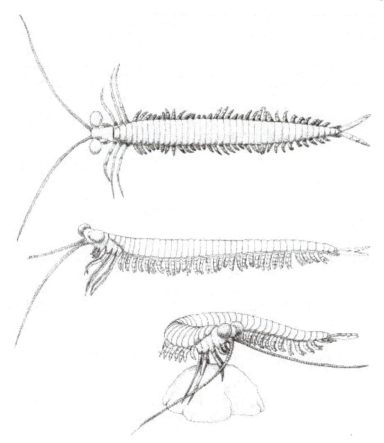

Figure 14.2 *Devonohexapodus bocksbergensis,* a marine hexapodous arthropod from the early Devonian slates of Germany (from Haas et al. 2003; used with permission).

as *Devonohexapodus*, the extant hexapods consist of the orders (sometimes called classes) Protura, Collembola, and Diplura, among which relationships are still uncertain, and the class Insecta. Together these four taxa are referred to as the subphylum Hexapoda; although the monophyly of Hexapoda has been disputed, most studies appear to confirm it.

Separation of the Protura, Diplura, and Collembola from Insecta is strongly indicated by their fundamental differences in head structure. First, all antennal segments of Collembola are musculate (antennae are entirely absent in Protura), while in the Insecta only the basal two segments bear muscle insertions. More importantly, the Protura, Diplura, and Collembola have

the mandibles and maxillae deeply retracted into pouches in the head capsule (Fig. 16.1c), the tips being protruded during feeding. This condition is termed entognathy or endognathy. The entognathous hexapods have monocondylic mandibles, adapted for externally triturating food particles into minute pieces before ingestion. In all Insecta except Archaeognatha, mandibles are dicondylic, or articulated with the head capsule at two points. These mandibles move transversely and are adapted for biting off and grinding food particles. The elongate styliform or bladelike mandibles of the noninsect hexapods lack the basal grinding lobe of Insecta. Differences in mouthpart structure among the entognathous taxa suggest, however, that entognathy may not represent a primitive divergence from the ectognathous hexapods but is a condition that has arisen independently several times in taxa that primarily inhabit soil and litter. Entognathy may be favored in soil interstice habitats because the enclosed mouthparts are protected from abrasion as the animals force their way through constrictions.

Ecologically, as well as in some morphological features, the noninsect hexapods resemble myriapods more than insects. Like myriapods, they are primarily inhabitants of litter and the upper soil layers, requiring high relative humidity or sources of free water. Direct evidence of their former multilegged condition is provided by their styliform, rudimentary abdominal appendages, which function as skids in supporting the abdomen. Sperm transfer is indirect, by means of a spermatophore that is attached to the substrate and later picked up by the female, with or without direct contact with the male. Some of these primitive features are also characteristic of apterygote insects, as discussed below. Diplura have often been grouped with either Collembola or Protura or both as Entognatha because the mouthparts are elongate and partly enclosed by the head capsule and the mandibles are monocondylic. The mandibles of Diplura, however, have a distinct incisor lobe and the maxillae are relatively unmodified compared to those of Protura and Collembola. Based on these characters Diplura have often been considered the entognathous order most closely related to Insecta. In a fossil interpreted as a dipluran from the Pennsylvanian Period of Illinois, the mouthparts are clearly ectognathous and large compound eyes were present. Protura and Collembola differ from Diplura as well as from Insecta in basic features. Protura lack antennae, and they grow by a process of anamorphosis, whereby abdominal segments are added at the time of molting. The newly eclosed nymphs have nine segments, while mature individuals have twelve segments. Anamorphosis is characteristic of myriapods. In contrast, insects (and other hexapods) grow by epimorphosis, in which the number of segments remains constant. Collembola are unique among hexapods in their six-segmented abdomen. Recent molecular studies have clashed strongly with respect to the relationships among the three noninsect

hexapod orders and with respect to which, if any, of them are most closely related to the true Insecta. For convenience, we adopt here an informal grouping of the three orders as "Entognatha," recognizing that it may not be monophyletic.

The great differences among the various entognathous orders indicate a very early divergence. Fossil Collembola are recorded from the Devonian (Shear et al., 1984), but the only other fossil representing this group is the Pennsylvanian dipluran mentioned above. This is probably because the small, soft bodies of Protura and Diplura do not facilitate preservation. The extant orders of the primitive hexapods perhaps represent remnants of a much more diverse Paleozoic fauna.

ORIGIN OF WINGS

Wings have contributed more to the success of insects than any other anatomical structures; yet the historical origin of wings remains largely a mystery. The earliest true insect fossil that has been discovered, from the early Devonian, may have already been winged (based on other correlated features; the part of the body that would have borne wings or not was not preserved). The primitively wingless hexapods, from which we might have gained insight into the origin of flight, are poorly represented in the living fauna. Thus the body structures that developed into wings, the steps in the evolution, and the ecological circumstances that favored wings are debatable. Nevertheless, there are some clues from several lines of evidence.

Preadaptations for Flight

Before discussing theories of wing origin, let us examine certain general deductions about the origin of pterygotes and their early environment. Wings evidently evolved only once, because the veins and articulatory sclerites at the wing base can be homologized among all pterygotes. Furthermore, wings probably evolved after the apterygotes were resistant to desiccation, breathed air by means of tracheae, and were six-legged. These attributes may be considered preadaptations for flight. In other words, these are features that evolved in response to one set of environmental conditions and have provided the basis for the evolution of new features under another set of conditions. Here adaptations to terrestrial life and locomotion have provided the preadaptations for aerial life and locomotion.

Protection against desiccation is obviously needed by small organisms that are freely exposed to the drying effect of air currents. The assumption that the tracheal system preceded wing evolution is supported by the association of that system with wing ontogeny. At one time tracheae were even thought to influence the pattern of venation, but this is incorrect. The veins arise as blood-filled extensions of the hemocoel between the layers of wing epidermis. Nevertheless, tracheae supply oxygen within the wing and presumably antedate the wing in evolutionary origin.

The six-legged condition probably also evolved before the wings. The reasoning is that wings are restricted to the thoracic segments because only those segments had the skeletal and muscular arrangements associated with walking legs. Certain flight muscles were originally leg muscles, and the enlarged pleura that first evolved to support the legs were modified as a stiff fulcrum for the wings. Of the three segments, the mesothorax and metathorax supported the more powerful legs and possibly were also in the most favorable position aerodynamically to evolve the flight mechanism.

Fossils indicate that the early pterygotes may have molted more than once after the wings became functional, thus perpetuating the apterygote pattern of life history in which adults molt repeatedly. Today only the Ephemeroptera have two instars with wings. All other pterygotes possess functional wings only in the final instar. Judging by the occasional failure of living insects to properly shed the cuticle enclosing the wings, one suspects that repeated molting of wings large enough to fly would involve a decided risk during growth. Flight is also expensive energetically and would divert nutritional resources away from growth in the immature stages. The strategy of delaying reproduction and dispersal until the last instar has been successful in the insects. This style of life history combines the advantages of flight and offsets the hazards of wing molting. Flight is advantageous in promoting sexual outcrossing among unrelated mates, thus reducing inbreeding; in dispersing to new habitats; in escaping enemies; and in locating specific feeding and oviposition sites. It is no coincidence that three of the largest, most diverse orders of insects—Lepidoptera, Diptera, and Hymenoptera—are primarily aerial in habit.

Ecological Considerations

Turning to the probable circumstances under which flight evolved, we should note that by the middle of the Devonian Period some plants reached 6 m or more in height. Later, in the Mississippian and Pennsylvanian Periods (=Carboniferous Period), trees of the coal swamps reached 41 m in height. Food in the form of spores, pollen, seeds, or vegetative parts might have provided incentives for early apterygotes to adopt arboreal habits. Arboreal herbivores also may have been joined by predatory relatives and arachnids. Therefore, wings may have arisen first as a device to escape predators, to move from plant to plant, or to sail back to the plant after being dislodged.

The distribution of vegetation and bodies of freshwater was probably patchy and seasonal in the Ordovician and Devonian. Dispersal by air would have been a distinct asset, both for terrestrial forms as well as for those that had become secondarily aquatic. It has been emphasized by Rainey (1965) that dispersal by wind is especially advantageous in arid regions. Thermal updrafts over heated ground lift objects aloft. The prevailing winds tend to converge at low-pressure areas where rainfall is likely. Thus wind movement is toward areas of renewed plant growth and freshwater.

Flight is not essential for passive airborne dispersal, but organisms must be quite small to benefit from passive dispersal. Glick (1939) collected wingless arthropods by airplane at heights up to 4500 m. Included in his catch were mites, spiders, Thysanura, Collembola, Siphonaptera, ants, and immature instars of Hemiptera, Orthoptera, Coleoptera, Lepidoptera, and Diptera. The possession of wings, however, permits insects both large and small to reach altitudes of favorable wind currents and to remain there longer, thus greatly increasing the distances traveled. Wings are, of course, important for local movements because insects can search for mates and select suitable feeding or breeding sites.

An obstacle that all theories of wing origin must overcome is to explain how selection would favor intermediate steps of wing development before the wing could function in flight. To lift the insect into the air and provide a forward thrust, the wings must be thin, stiff, and large relative to the body; articulated at the bases; and moved by muscles in a complex propulsive stroke.

The time from the appearance of fossil land plants in the Devonian Period to the first fossils of winged insects in the Pennsylvanian Period is 100 million years. Perhaps it is not surprising that arthropods could become fully terrestrial and then aerial during that time. A period of much less time in the Cenozoic Era was sufficient to evolve humans from shrewlike prosimians.

Paranotal Origin of Wings

According to this theory, flight first evolved in an arboreal insect. If dislodged by wind or while escaping predators, a wingless insect would fall to the ground unchecked. Although probably not harmed, it would be necessary for the insect to regain its footing before escaping further. The addition of thin, lateral expansions on the terga, called paranotal lobes, might create a stable orientation or attitude during falling so that the insect would land on its feet for a quick getaway. Calculations have shown that some degree of attitudinal control is given when even small, fixed wings are added to models, provided the models are no smaller than 1 cm in length. Paranotal lobes also might have functioned in other ways, such as giving increased lateral protection from predators when the insect was flattened against the substrate.

Larger paranotal lobes would facilitate gliding to the ground or from plant to plant. The development of a suitable basal hinge on the thoracic nota would permit the paranotal lobes to flap when the thorax was deformed by muscular contractions. The final refinements would include modification of the basal hinge, muscle attachments, and nervous system so that the wing angle could be varied during each stroke. The anatomical evidence for this theory is based mainly on the presence of broad thoracic nota in Thysanura and Archeognatha and winglike prothoracic lobes on some fossil Palaeodictyoptera, Ephemeroptera, and Protorthoptera. The prothoracic lobes of some fossils appear to be articulated and to have well-developed venation (Fig. 14.3b).

Exite Origin of Wings

This theory proposed that flight originated in an insect with an aquatic, gill-bearing nymph. The lateral abdominal gills of certain extant Ephemeroptera bear some resemblance to wings, being movable, thin, and membranous and having a ramifying pattern of tracheae. The similarity is quite striking in nymphs of several orders of fossil insects from the Paleozoic (Kukalová-Peck, 1983) (Fig. 14.3a). These fossils have wings on the mesothorax and metathorax that are too small for flight but are curiously held outstretched and curved obliquely backward from the body. The winglets have fairly distinct venation, which seems to be homologous to that of the wings and what appears to be a movable basal articulation.

The similarity of the winglets and abdominal gills in the Paleozoic mayfly nymphs reinforces the idea that the gills and wings are serially homologous. By beating the gills, living mayfly nymphs create currents of water over the body to replenish oxygen. Sclerotized gills or gill plates also function to some extent in locomotion. The articulated winglets of Permian nymphs may have moved in the same fashion and for the same purposes. More powerful strokes might have propelled the nymph forward. Thus the winglets might have functioned like fins, stiffened by venation, and moved in a propulsive stroke by muscles.

In another Paleozoic order, the Palaeodictyoptera, nymphs are believed to have been terrestrial. The flattened abdominal flaps may have functioned as covers for the spiracles, reducing water loss. Flight could have originated when the spiracle covers became enlarged to allow gliding in nymphs which fell from emergent vegetation. Palaeodictyoptera appear to be among the most primitive flying insects, suggesting that the aquatic habits of nymphal Ephemeroptera and modifications of the spiracle covers as gills were probably

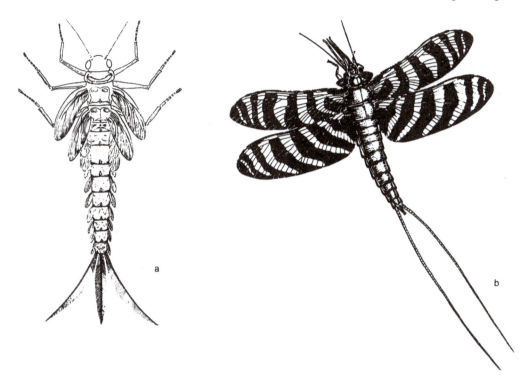

Figure 14.3 Reconstruction of Paleozoic insects by J. Kukalová-Peck: a, naiad of *Protereisma* sp., a mayfly from the Permian Period (290–245 million years ago); b, *Homaloneura joannae*, a palaeodictyopteran from the Pennsylvanian Period (= Upper Carboniferous Period; 323–290 million years ago).
Source: Courtesy of and with permission of J. Kukalová-Peck.

evolved subsequent to the origin of flight. This is important since it much easier to imagine flight arising in a terrestrial, rather than an aquatic ancestor. The prothoracic winglets were placed too far forward to be aerodynamically useful and were fused to the pronotal margin and then lost in modern insects. It has also been proposed that the original function of articulated notal flaps was in courtship display (Alexander and Brown, 1963), in thermoregulation (Kingsolver and Koehl, 1985), in sculling along the surface of water, as in some winter stoneflies (Marden and Kramer, 1994; but see Will, 1995), or a variety of other hypotheses (Kingsolver and Koehl, 1994).

If wings and gills are serially homologous, what familiar structures of the arthropod body do they represent? Opponents of the gill theory, such as Snodgrass (1958), believed that the abdominal gills were serially homologous to the thoracic legs. Since

legs are already present on the thoracic segments, the wings must have evolved from something else, such as paranotal lobes. Wigglesworth (1976), however, argued in favor of the gill theory by homologizing gills with basal lobes of the arthropod limb, called exites. If derived from exites, both wings and legs could be present in the thorax.

Kukalová-Peck (1983, 1987) has shown that exites existed on the leg bases of numerous Paleozoic insects, both winged and wingless. The styli of modern Parainsecta and Apterygota have previously been interpreted as endite lobes, and it appears that early insects had a variety of endite and exite processes on the basal (precoxal and subcoxal) segments of their legs. The fossils illustrated by Kukalová-Peck further show that Paleozoic Thysanura (originally referred to as the independent order Monura; Fig. 14.4) had well-developed abdominal legs homologous with those on the

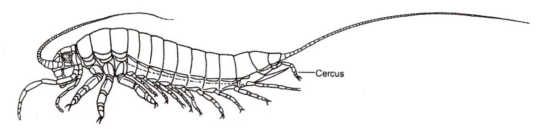

Cercus

Figure 14.4 Reconstruction of a primitive apterous insect, Thysanura (*Dasyleptus*, originally described as a distinct order, Monura), from the Pennsylvanian Period (= Upper Carboniferous; 323–290 million years ago) of Illinois. Length without antennae or terminal filament is about 35 mm. Note the cerci retain the shape of legs. Not shown are the dense bristles that covered the body.
Source: Redrawn from Kukalová-Peck, 1987. Reproduced by permission of Canadian Journal of Zoology.

thorax. The earlier presumption of Manton (1972, 1973) and others that the insect leg is primitively uniramous and thus basically different from the legs of other arthropods appears unfounded. The primitive structure of the myriapod trunk segment probably incorporated both exites and endites of various function (Fig. 14.1), and this is certainly true of most Crustacea, even extant ones.

An important aspect of the exite theory of wing origin is that wings are thereby derived from a structural feature that already had the required basal articulation and musculature. Complex structures are almost always evolved from precursory features that are preadapted for the necessary modifications. If wings were developed from fixed paranotal lobes, it is difficult to explain how the basal hinge and musculature could have appeared completely *de novo*.

The Modern Developmental Perspective on Wing Evolution

As mentioned in Chapter 2 and shown in Figure 2.23, studies of developmental gene expression have confirmed that wings and many external gills share a common underlying developmental mechanism across arthropods. This is not to say necessarily that wings evolved from ancestral gills (although this is possibly true even without clear fossil evidence) but to say instead that both structures have an underlying common developmental basis, which itself may have originated once and then been modified differently in different arthropod and insect groups.

CLASS INSECTA

As restricted here, the class Insecta may be defined as those hexapodous arthropods that at some time in their life show the following features: head, five-segmented with ectognathous mouthparts comprising mandibles without palps; maxillae and labium, both with palps; antennae with intrinsic muscles in the first two segments; thorax, three-segmented, with one pair of legs per segment; legs, six-segmented (coxa, trochanter, femur, tibia, tarsus, pretarsus); abdomen, primitively eleven-segmented with cerci; respiration through a tracheal system with spiracles on the last two thoracic and first eight abdominal segments; excretion by Malpighian tubules; postembryonic development epimorphic. In addition, the vast majority of insects are characterized by the possession of compound eyes and wings in the adult stage, though these are absent in some primitive forms and secondarily lost in many specialized ones.

By conservative estimates the class Insecta encompasses at least 876,000 extant described species (and many more yet to be described), classified in 29 orders and about 939 families (Figure 14.5). These numbers dwarf those in any other class of animals. Indeed, each of the the the four largest orders of insects (Coleoptera, Lepidoptera, Hymenoptera, and Diptera) exceeds the next largest phylum of animals (Mollusca) by a considerable margin. The family Curculionidae (weevils), with approximately 65,000 species, is currently the largest family of animals in terms of described species—larger than all nonarthropod phyla except

Mollusca. Several other families in Coleoptera (especially Staphylinidae), Diptera (several muscoid families), Hymenoptera (Ichneumonidae, Braconidae), Hemiptera (Cicadellidae), and Lepidoptera (Noctuidae) are also likely to have similar species richness. Species known only from fossils represent at least twelve additional orders, which arose and radiated in the late Paleozoic and early Mesozoic Eras.

The morphological and ecological diversity within the Insecta is enormous. As detailed elsewhere in this book, insects occupy essentially every conceivable terrestrial habitat and have extensively colonized freshwaters. Only marine situations are relatively devoid of insects, although several thousand species inhabit the sea/land interface presented by intertidal regions, and a few Hemiptera live on the surface of open seas far from land. Insects include scavengers, herbivores, predators, and parasites. Many species are narrowly specialized for highly restricted foods, substrates, or activity periods, with consequent morphological modifications. The immensity of diversity within the Insecta and the attendant complexity of their phylogenetic relationships have resisted attempts to develop an entirely consistent and comprehensive evolutionary history. Nevertheless, the broad outlines of insect evolution have been repeatedly confirmed by independent lines of investigation, and modern approaches to comparative morphology, developmental biology, and molecular phylogeny are rapidly improving our understanding. Areas of uncertainty or dispute are indicated where appropriate in the following account.

The classification proposed by Handlirsch (1908, 1926) has formed the framework for most subsequent classifications (Brues et al., 1954; Hennig, 1953; MacKerras, 1970; Richards and Davies, 1957; Rohdendorf, 1969; Wille, 1960), although many minor differences are evident. Kristensen (1991) and Kukalová-Peck (1991) provide excellent reviews of ideas regarding hexapod phylogeny based on comparative morphology, while Wheeler et al. (2001), Gaunt and Miles (2002), and Kjer et al. (2006) provide nice summaries of recent findings from molecular data.

The Apterygote Insects

These insects are similar in body organization to pterygote (winged) insects but primitively lack wings. Apterygotes have ectognathous mouthparts and intrinsic musculature in only the basal two antennal segments, and they have compound eyes and ocelli. The long, slender cerci and median filament recur in Ephemeroptera (mayflies), and the body form of apterygotes strongly resembles that of Ephemeroptera nymphs. Like the Entognatha, apterygotes have rudimentary abdominal appendages, employ indirect insemination, and molt throughout life. These are ancestral features that have been retained in both hexapod lines of evolution; they do not indicate close relationship.

Two levels of organization are evident within the apterygotes. In the Archeognatha the elongate monocondylic mandible has the incisor lobe distinct from the molar lobe (Fig. 17.2b), a primitive condition similar to that in many Crustacea. In Thysanura the dicondylic mandible has the incisor and molar lobes much closer, as in pterygote insects. Archeognatha have large, compound eyes, as in Pterygota. In extant Thysanura have the eyes reduced to a few lateral facets, but relatively large compound eyes are present in some fossils. The Archeognatha and Thysanura also show differences in the endocrine system (Watson, 1965), structure of the spermatozoa (Wingstrand, 1973), and abdominal musculature (Birket-Smith, 1974).

The Winged Insects—Pterygota

The Pterygota are primitively winged insects with the mesothoracic and metathoracic segments enlarged and bearing wings, or secondarily wingless; mandibles primitively dicondylic, adapted for chewing, or highly modified; abdomen primitively with eleven segments; the anterior ten segments without appendages and the eleventh segment frequently with cerci.

Pterygota differ biologically from apterygotes in employing direct insemination via copulation and in molting only until sexual maturity. No adult Pterygota possess abdominal appendages similar to those of apterygotes, but the abdominal gills of mayflies are probably exites of abdominal appendages. The abdominal appendages found on

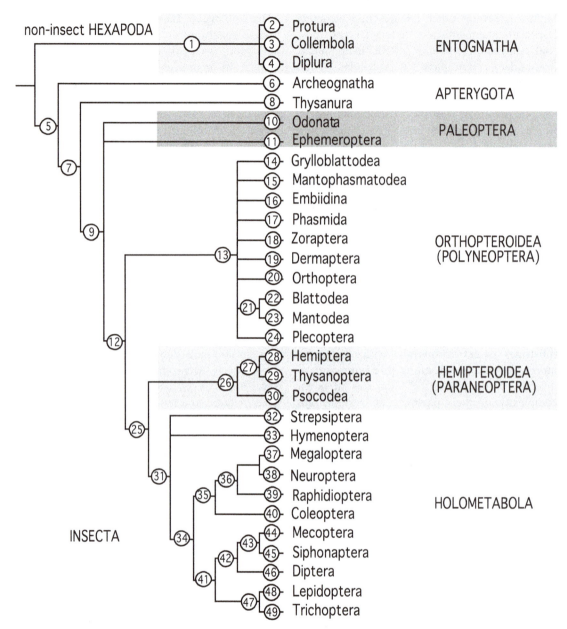

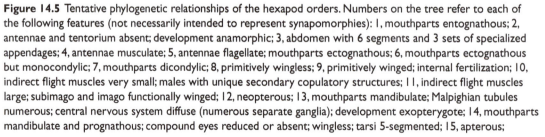

Figure 14.5 Tentative phylogenetic relationships of the hexapod orders. Numbers on the tree refer to each of the following features (not necessarily intended to represent synapomorphies): 1, mouthparts entognathous; 2, antennae and tentorium absent; development anamorphic; 3, abdomen with 6 segments and 3 sets of specialized appendages; 4, antennae musculate; 5, antennae flagellate; mouthparts ectognathous; 6, mouthparts ectognathous but monocondylic; 7, mouthparts dicondylic; 8, primitively wingless; 9, primitively winged; internal fertilization; 10, indirect flight muscles very small; males with unique secondary copulatory structures; 11, indirect flight muscles large; subimago and imago functionally winged; 12, neopterous; 13, mouthparts mandibulate; Malpighian tubules numerous; central nervous system diffuse (numerous separate ganglia); development exopterygote; 14, mouthparts mandibulate and prognathous; compound eyes reduced or absent; wingless; tarsi 5-segmented; 15, apterous;

some pterygote larvae may be secondarily derived structures.

The Paleopterous Insects

Pterygota comprise two major functional divisions, differing in their wing-hinging mechanism and flight musculature. In the Paleoptera the wings cannot be flexed over the back at rest, whereas in the Neoptera wing flexing is allowed by the configuration of axillary sclerites and folds in the wing base. Wing flexing was advantageous for allowing winged insects to reenter the protected environments provided by litter and by subcortical (underbark) and other restricted situations. Some paleopterous insects, such as the suborder Zygoptera (Odonata), simulate wing flexing through the steep backward tilt of the pterothorax (Chap. 19).

The extant orders of Paleoptera exemplify many atypical features for Pterygota. In Odonata

(dragonflies) the male stores the spermatozoa in a secondary copulatory organ on the second abdominal sternite. The female is responsible for completing sperm transfer, as in Apterygota. Possibly this behavior represents a modified indirect fertilization. Furthermore, in Odonata the flight musculature is entirely direct (attached directly to the wing bases) rather than primarily indirect (attached to the thoracic terga) as in Ephemeroptera and nearly all Neoptera. Notably, ancestral features of Ephemeroptera include the paired male external genitalia of some species and abdominal gills, which, in at least some cases, appear to be homologous with the wings. The winged but sexually immature subimaginal instar of mayflies is unique but suggests the indeterminate molting of Apterygota.

Extant Paleoptera are restricted to the two relatively small orders Ephemeroptera and Odonata, both aquatic as naiads. However, during the Paleozoic Era the Paleoptera radiated extensively,

cerci unsegmented; prothoracic tibiae and femora modified for grasping prey; 16, specialized silk-spinning glands in prothoracic tarsi; wings (when present) soft and flexible except during use; 17, slender, often sticklike; pronotum without ventrolateral extension over pleuron; tarsi 5-segmented; 18, superficially resembling Psocoptera but with cerci present, unmodified mandibulate mouthparts; tarsi 2-segmented; 19, forewings short, elytralike, hindwings semicircular with distinctive radiating venation or absent; cerci usually modified as strongly sclerotized forceps; 20, hind legs enlarged for jumping; communication often through sound production; tarsi 1–4 segmented; 21, dorsal longitudinal muscles very small; corporotentorium perforated; immatures terrestrial; eggs in specialized ootheca (lost in subgroup Isoptera); 22, with intestinal symbiotes (primarily protists and bacteria, depending upon subgroup) for food processing; body form rather generalized 23, head small, triangular, highly mobile; prothorax elongate; forelegs raptorial with elongate coxae; 24, dorsal longitudinal muscles large; corporotentorium imperforate; immatures aquatic; ootheca absent; 25, some tendency toward holometaboly (Hemiptera and Thysanoptera); 26, cerci absent; nymphs without ocelli; mouthparts specialized; 27, mandibles and laciniae developed as stylets; 28, mouthparts symmetrical, both mandibles modified as stylets; labium modified as sheath; 29, mouthparts asymmetrical; right mandible absent; labium unmodified; 30, mandible retained and laciniae modified as rods, or all mouthparts highly modified (Anoplura); 31, development fully endopterygote; 32, legs without trochantins; immatures endoparasitic in insects; mesothoracic wings modified into halterelike structures; 33, haplodiploid sex determination; Malpighian tubules numerous; 34, (no obvious common traits not shared with other Holometabola); 35, legs with trochantins; larvae primitively campodeiform; 36, forewings membranous; 37, larvae aquatic; 38, larvae with incomplete intestine and suctorial mouthparts; 39, larvae terrestrial; 40, forewings modified as elytra; 41, adult with rostrate mouthparts; 4 to 6 Malpighian tubules; 42, larvae usually apodous (exception, Mecoptera); labial glands seldom producing silk; wing membrane bare or with few, unmodified setae; 43, fore- and hind wings (if present) subequal in size; wingless forms (Boreidae and Siphonaptera) weakly to strongly laterally flattened; 44, adults 4-winged (except Boreidae); larvae with thoracic legs and usually with abdominal prolegs; 45, adults apterous; ectoparasites living on mammals and birds; 46, adults typically two-winged; hind wings modified into halteres; 47, larvae eruciform, frequently with abdominal prolegs; larval silk production by labial glands; 48, adults with mouthparts mandibulate (rarely) or developed as coiled proboscis; wing membranes covered with scales; larvae almost always terrestrial; 49, adults with mouthparts atrophied; wing membranes covered with hairs; larvae aquatic.

producing a much larger, more diverse fauna than at present. Several major evolutionary lineages existed. The palaeodictyopteroid lineage comprised a very diverse assemblage characterized by prothoracic winglets, long caudal filaments (cerci), and hypognathous mouthparts, which were adapted for piercing and sucking. The clypeus was strongly domed, apparently to house a cibarial pump. Members of this lineage are classified as four (in some classifications more) orders: Palaeodictyoptera, Diaphanopterodea, Megasecoptera, and Permothemistida (Dicliptera). Some of these species had wingspreads of 50 cm or more. It has been suggested that the piercing mouthparts were used to reach into the cone-like fruiting structures of primitive plants such as lycopods and seed ferns. Other species were probably predators, while some very small, delicate Diaphanopterodea had the body form of mosquitoes and may have been ectoparasites. The Permothemistida had the hind wings reduced, as in modern mayflies. Palaeodictyopteroid nymphs were terrestrial, and some were probably capable of flight, judging from the size of their wings (Shear and Kukalová-Peck, 1990). No members of the palaeodictyopteroid assemblage survived later than the Permian Period.

The other major lineage of paleopterous insects includes the extant Ephemeroptera and the dragonflylike Protodonata (griffenflies). Paleozoic mayflies were generally similar to modern forms but often were much larger (to 45 cm wingspan). Some of the fossils show mouthparts that appear to have been functional in the adults. Many of the Paleozoic forms had prothoracic winglets, and in some the fore and hind wings were subequal. Naiads were aquatic. Protodonata were generally similar to Odonata in body form. Like Odonata, they had powerful chewing mandibles and strong legs used for grasping prey. Protodonata lacked the specialized secondary genitalia of dragonflies and differed in several features of wing venation and head structure. This order contained the largest known insects, with wingspans up to 71 cm in *Meganeuropsis permiana*.

Due to conflicts between different sources of evidence, both morphological and molecular, there has been uncertainty about whether the Paleoptera form a monophyletic group or, if not, which of the extant orders, Ephemeroptera or Odonata, is more closely related to the Neoptera. The winged orders of insects appeared to have diverged from each other relatively rapidly over 350 million years ago, so the hierarchical pattern of shared characteristics among orders is very small compared with the differences among them.

Infraclass Neoptera

The Neoptera comprise about 99 percent of all insects. They present an exceedingly diverse assemblage but have in common the ability to flex the wings over the back by means of a pleural muscle inserting on the third axillary sclerite. For almost every other character the Neoptera are variable, or at least subject to exceptions. Neopterous insects may be classified in two major divisions, the possibly paraphyletic Exopterygota and the clearly monophyletic Endopterygota, based on modes of growth and development. In Endopterygota the wings and other presumptive adult structures develop as internal buds, or anlagen, in the immature (larva), which usually differs from the adult in many features. In Exopterygota the immature (nymph) is generally similar to the adult except in the size of the wings, which develop as external pads. Some Hemiptera and Thysanoptera are physiologically holometabolous, with a "pupal" instar in which occurs histolysis of muscles, alimentary canal, and other structures. However, the ontogeny of the wings is exclusively external in these species, which are highly evolved members of their evolutionary lineages. Thus, holometabolism represents an evolutionary grade that has developed independently in Exopterygota and Endopterygota. For these reasons, Exopterygota and Endopterygota are preferable to Hemimetabola and Holometabola as taxonomic names.

Division Exopterygota

The Exopterygota comprise two major lineages, designated here as the superorders Orthopteroidea (also called Polyneoptera) and Hemipteroidea (also called Paraneoptera). These differ in numerous major morphological characteristics, the most important involving the structure of the central nervous system, the number of Malpighian tubules, and wing venation. In all these features, as well as in the structure of the mouthparts, the Hemipteroidea appear evolutionarily derived compared to the more generalized Orthopteroidea. Fossils assignable to both the Orthopteroidea and the Hemipteroidea appear in the Pennsylvanian Period, when they coexisted with the very diverse paleopterous fauna. The division Exopterygota is considered by some authorities as an evolutionary grade rather than a monophyletic lineage, and they reject it as a formal taxonomic name.

Superorder Orthopteroidea (Polyneoptera)

The orthopteroid orders are characterized by chewing mouthparts, long multiarticulate antennae, multiarticulate cerci, and numerous Malpighian tubules, as well as other relatively ancestral insect features. One possible derived orthopteroid feature is the enlarged anal lobe of the hind wing.

Undoubtedly, the orthopteroid insects include the most morphologically ancestral Neoptera. Yet the interrelationships of the orthopteroid orders are poorly understood because of the mosaic distribution of ancestral and derived characteristics and because the groups apparently radiated rapidly very early in insect evolution, causing difficulty for molecular phylogeny. For example, Dermaptera are the only Neoptera that possess (presumably ancestral) paired external genitalia (paired penes of some species), but they display highly derived hind-wing venation (to permit folding beneath the specialized forewings) and strongly sclerotized cerci modified as forceps. Morphological features of Embiidina are strongly specialized for life within self-constructed silken tubes and Phasmatodea for camouflage. In contrast, nearly all features of the Grylloblattodea and Plecoptera are so generalized that their relationships with other orders are difficult to determine.

Not only are the relationships among the various orthopteroid orders difficult to specify, but their relationship to the Hemipteroidea and the Endopterygota remains a matter of contention. The derived characters linking the orthopteroid orders could have evolved independently more than once, and Orthopteroidea may not represent a monophyletic assemblage (Kristensen, 1975, 1991)

Within the Orthopteroidea, the Blattodea (cockroaches and termites) and Mantodea (mantids) form a group of clearly related orders, sometimes combined as the single order Dictyoptera. All Dictyoptera share several unique morphological features (proventricular armature, wing venation pattern, structure of female genitalia, perforated corporotentorium). Phylogenetically the termites are specialized cockroaches, and the mantids might well be considered another specialization from a proto-roach lineage.

Among the remaining Orthopteroidea, Plecoptera are noteworthy for their retained ancestral morphological features, including tracheal gills in nymphs, distinct anapleurite and coxopleurite in some nymphs, and multiarticulate cerci. Orthoptera show a long, separate evolution, with recognizable fossils as early as the Pennsylvanian Period. Their closest affinities are possibly with the Phasmatodea. In the Zoraptera are combined features of both the Orthopteroidea (chewing mouthparts, cerci) and the Hemipteroidea (six Malpighian tubules, concentrated central nervous system, two-segmented tarsi, and reduced wing venation). Based on wing venation they have been placed near the Blattodea (Kukalová-Peck and Peck, 1993), but their relationships remain uncertain; recent molecular studies sometimes place them among the hemipteroids.

Superorder Hemipteroidea (Paraneoptera)

The most important feature in interrelating the hemipteroid orders is the configuration of the mouthparts. With the exception of Zoraptera, all Hemipteroidea have the clypeus enlarged to house the cibarial pumping structure and the lacinia modified as a sclerotized, styliform organ. In other mouthparts much variation exists. In one evolutionary line leading through Psocoptera (book lice) and Phthiraptera (lice), the lacinia is a sturdy rod that is applied to the substrate as a brace to stabilize the head while the mandibles scrape food particles. The hypopharynx and pharynx are modified as a mortar and pestle for crushing. Some Phthiraptera (suborder Anoplura) have the mouthparts further modified to suck blood, but their general similarity in numerous characters to other lice (suborders Amblycera and Ischnocera) and Psocoptera leaves little doubt of their evolutionary origin.

In the evolutionary line leading to Thysanoptera (thrips) and Hemiptera (bugs), both mandibles and maxillae are modified as styliform piercing organs. The putatively more plesiomorphic arrangement occurs in Thysanoptera, where the maxilla and single mandible are relatively short, thick blades that are guided by the conical labrum and labium. As in Hemiptera, the maxillary stylets are attached to the stipes by a special lever arm. In Hemiptera the two mandibular and two maxillary stylets interlock to form a highly efficient piercing-sucking tube, which is enclosed by the troughlike labium. A few Thysanoptera and Hemiptera are physiologically holometabolous, as mentioned above. The exopterygote orders display limited biological diversity compared to the Endopterygota. Most Orthopteroidea live on the surface of ground, on vegetation or in litter, while the great majority of Hemipteroidea inhabit foliage (or freshwater habitats). Specialized modes of life, such as ectoparasitism in Phthiraptera and Hemiptera, or social behavior in termites are exceptional. There are no parasitoids, no endoparasites (see Chap. 10), and very few fully aerial species. Exopterygota scarcely participate in exploiting pollen and nectar—otherwise important factors in the diversification of the Endopterygota. In general, exopterygotes tend to utilize the same food resources in all instars (exceptions include cicadas) and are almost all external feeders. Inability to feed internally has barred them from a multitude of niches occupied by endopterygote larvae. For example, except for termites (Isoptera), almost no exopterygotes bore or mine living or dead wood, foliage, fruits, or seeds. By external feeding only Hemiptera have exploited the vascular region of woody plants, an important habitat for endopterygotes, especially Coleoptera.

Most extant orders of Hemipteroidea are known as fossils from the Permian. Phthiraptera are definitively known only from the Recent Epoch, less than 10,000 years ago, although there are older controversial fossils.

Division Endopterygota

Included here are about 85 percent of the extant species of insects, representing immense taxonomic and biological diversity. While the evolution of holometaboly is not entirely clear, its chief adaptive value seems to be in allowing adults and larvae to utilize different resources. Resource division is usually accompanied by morphological divergence. This divergence is extreme in Diptera and most Hymenoptera, where the legless, grublike larvae frequently lack eyes and have the antennae and mouthparts reduced to papillae. Differences between adults and larvae appear least in Megaloptera and Raphidioptera, which are among the earliest-diverging endopterygote orders. Besides sharing holometabolous development, the Endopterygota also flex the wings along the jugal fold, in contrast to flexion along the anal fold in Exopterygota.

The origin of the Endopterygota is obscure. As mentioned above, holometabolous development has independently evolved in the Hemipteroidea, but there are no other Exopterygota that suggest a tendency toward holometabolism. The most morphologically generalized endopterygotes—Mecoptera, Megaloptera, and Raphidioptera—provide little hint of what their ancestors may have been. Their wing venation has been compared to that of Palaeodictyoptera, but the resemblance is apparently convergent. Venational similarities to Plecoptera (Hamilton, 1972) have also been suggested but are contradicted by phylogenetic results from both comparative morphology and molecular data.

The endopterygote orders form three evolutionary lineages, recognized as the superorders Neuropteroidea, Mecopteroidea, and Hymenopteroidea.

Superorder Neuropteroidea

Morphologically, the most plesiomorphic Neuropteroidea are the Megaloptera and Raphidioptera. The pupae of both groups are capable of limited movements and have functional (decticous) mandibles, which are used in defense. Although larvae of Megaloptera are aquatic and those of Raphidioptera terrestrial, their morphological similarity is so great that they have sometimes been classified as a single order. They are distinct as fossils since the Jurassic Period, however, and are here recognized as separate orders in accordance with recent practice.

Adult Neuroptera are similar to Megaloptera, but the larvae are highly specialized in having the mandibles and maxillae adapted for piercing and sucking and in having the midgut end blindly. In addition, fossil evidence shows a long period of evolution independent of Megaloptera, with great diversification in the Permian and Cretaceous Periods, justifying recognition as a separate order.

Coleoptera and Strepsiptera are so highly specialized that their exact relationships are uncertain. Larvae of some aquatic Coleoptera are exceedingly similar to immature Megaloptera, and fossils such as *Tshekardocoleus* may represent intermediates between Megaloptera and Coleoptera. Mickoleit (1973) has shown that the coleopteran ovipositor is homologous with that in other neuropteroids.

The Strepsiptera are narrowly specialized for endoparasitism and are morphologically distinct from (other) neuropteroids. Like Coleoptera, they fly using the hind wings (posteromotoria). First-instar larvae (triungulins) of Strepsiptera are similar to triungulins of the parasitoid beetle families Meloidae and Rhipiphoridae. Some recent molecular studies have indicated that the Strepsiptera might be related to Diptera, while others indicate that they might represent a quite isolated endopterygote lineage.

Superorder Mecopteroidea

Extant Mecoptera number no more than a few hundred species, but fossils record a large, diverse fauna in the late Paleozoic and Mesozoic Periods. Like Megaloptera, pupae of Mecoptera are decticous. Mecoptera have larvae with compound eyes, a feature shared with some primitive Lepidoptera. Most modern Mecoptera are characterized by the elongation of the lower face and mouthparts, producing a strong resemblance to early lineages of Diptera. Likewise, wing venation in nematoceran flies is similar to that of Mecoptera, and some fossils are difficult to place in either order. Modern Diptera are characterized by the reduction of the

hind wings to halteres, but metathoracic wing pads are visible in some *Drosophila* pupae, and certain fossil Diptera may have been four-winged.

Siphonapteran (flea) adults are extremely specialized as ectoparasites of mammals and birds, but the larvae are similar to larvae of primitive Diptera. The former winged condition of Siphonaptera is revealed by pupal wing pads in some species. Recent molecular studies indicate that Siphonaptera are likely to represent derived Mecoptera, related to the wingless Boreidae ("snow fleas").

Lepidoptera and Trichoptera are generally similar in wing venation, mouthpart morphology, and larval body form, though their larvae are terrestrial and aquatic, respectively. The most basal lineages of Lepidoptera (suborder Zeugloptera) and all Trichoptera have decticous pupal mandibles, and Zeugloptera have compound larval eyes, as in Mecoptera. Lepidoptera are generally characterized by the elongation of the apposed maxillary galeae as a sucking proboscis. In Zeugloptera, however, the maxillae are unmodified, and in Eriocraniidae the proboscis is formed by relatively short galeae, apposed only during feeding.

Superorder Hymenopteroidea

Relationships of Hymenoptera have been controversial. They possess numerous Malpighian tubules, differentiating them from all other Endopterygota, which typically have four or six tubules. Hymenopteran wing venation is highly distinctive, with fusions of many veins to produce a few large cells. The eruciform larvae of primitive Hymenoptera share many characteristics with larvae of Lepidoptera and/or Mecoptera, including abdominal pseudopodial legs, a single tarsal claw, and labial silk glands. On morphological grounds it would appear Hymenoptera diverged relatively early from the mecopteroid lineage, but recent molecular phylogenetic studies suggest that the order diverged from other Holometabola even before the split between the neuropteroid and mecopteroid superorders.

THE GEOLOGICAL RECORD OF INSECTS

Our knowledge of insects that lived in the past is based on traces or remains preserved as fossils in the earth's sedimentary rocks and hardened plant resins (Fig. 14.6). The hard parts of insects are resistant to decay. If the insect is quickly immersed in a protective, fine-grained medium such as mud or volcanic ash, the chances are great that at least its wings will be preserved. Unfortunately, through the ages terrestrial deposits on the surface are usually destroyed by the erosive action of water or wind. Other deposits become too deeply buried to be discovered. Today the places where insect fossils can be found are limited in number. Still, the fossil record of insects is surprisingly extensive, and throughout most of their evolutionary history the diversity of fossilized insects is greater than that of vertebrates (Labandeira and Sepkoski, 1993).

Fossil evidence can be used to establish the first appearance of a taxon and, in the case of extinct taxa, the last appearance. When contemplating the duration of their existence, remember that the organisms must have existed for some period prior to the earliest record we have found. Likewise, taxa now extinct may have persisted for some time after the last fossils.

The chronological sequence in which the insect orders appear in the geological record is of limited value in determining their phylogenetic relationships. For example, members of the extant order Protura, though apparently of ancient origin, are unknown as fossils, as is the suborder Mallophaga of the Phthiraptera. Unfortunately, evidence of the crucial steps leading to the origin of insects has not yet been found in the fossil record. Attention has been focused on strata of the Devonian Period (Figure 14.7) because the first definite land plants have been found here. Fossils named *Eopterum devonicum* and *Eopteridium striatum* in these rocks were thought to be the oldest winged insects. They have now been identified not as insects but as the winglike tails of crustaceans. The first fossil Collembola of Devonian age were also controversial, but recently other fossils of Collembola and possibly of Archeognatha have been found.

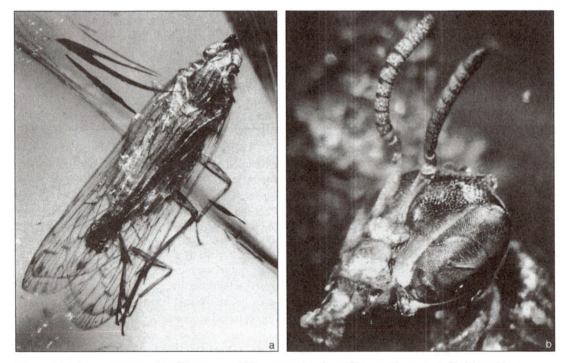

Figure 14.6 Insects preserved in Oligocene and Miocene amber from Chiapas, Mexico: a (left), fulgorid bug; b (right), stingless bee, *Trigona silacea* (Apidae). Note sensilla on antennae.
Source: Photos courtesy of Joseph H. Peck, Jr.; b, reprinted by permission of Journal of Paleontology.

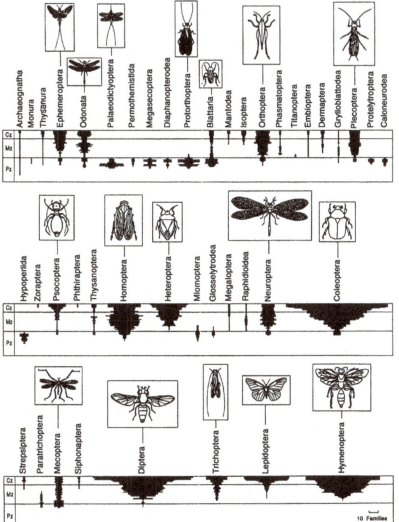

Figure 14.7   Fossil history of insect orders, including extinct ones. Width of black bands indicate relative species richness through time.
Source: Modified, with permission, from Labandeira and Sepkoski, 1993.

The earliest fossil interpreted as a true insect (*Rhyniognatha hirsti*) is only a partial fossil from early Devonian chert deposits.

Most insect fossils are found in sedimentary deposits, either coal beds or fine-grained siltstones. One difficulty with sedimentary fossils is that the anatomical detail necessary to critically analyze small, complex organisms like insects is obscured. Many of the fossils taken from coal beds show remarkable preservation of mouthparts, the wing base, and other important structures; the large size of many Paleozoic insects eases anatomical interpretation. For the great majority of small insects, especially those preserved in siltstones or sandstones, only gross features of anatomy are recognizable.

In the Cretaceous Period many insects are preserved in hardened lumps of fossil plant resin called amber. The brownish, translucent substance often contains organic matter such as pollen, hair, leaves, or insects. Polished pieces are prized as jewelry. The resins do not decay or dissolve but harden with time. Because of their light weight, lumps are carried by stream water to be deposited, eroded, and redeposited several times. Consequently, pieces may be much older than the rock in which they are finally embedded. When found with lignite or coal seams, however, it is likely that the amber has not been carried far from the trees that produced the resin. Fossil inclusions in amber are remarkable for the detail of preservation, including setation, surface sculpturing, mouthparts, genitalia, and sometimes even soft tissues.

Extinct Orders

The most common fossils are wings or fragments of wings. Many extinct orders have been described on the basis of unusual patterns of venation exhibited by fossil wings. As a result, the classification and the phylogenetic relationships of extinct insects are controversial. The first appearance of a modern insect order is also subject to argument because the ancestral insects often do not closely resemble the modern forms. Recent treatments of fossil insects differ in methodology, higher categories, the number of orders, and in the families assigned to the orders. Carpenter's (1992) classification is traditional and evolutionary, while that of Kukalová-Peck (1991) is cladistic. Recent compilations by Rasnitsyn and Quicke (2002) and Grimaldi and Engel (2005) differ strongly in nomenclature as well as classificatory philosophy. Currently, the Grimaldi and Engel (2005) treatment is the most authoritative and readable. Despite some instability in classification of fossil taxa, here we list orders commonly mentioned in scientific literature.

Subclass *Pterygota*

Infraclass Paleoptera

Order Protodonata (= Meganisoptera). Closely related to Odonata, but lacking the nodus, pterostigma, and arculus in the venation. *Meganeura* reached 71 cm in wingspread, but this was exceptional. Most were the size of large dragonflies. Nymphs had a labial mask similar to that of Odonata; presumably they were aquatic predators. Adults were predatory. (Pennsylvanian-Triassic)

Order Palaeodictyoptera. This order (Fig. 14.3b) and the next three are probably related. All possessed cerci (no caudal styles), external ovipositors, and long, piercing beaks of five stylets with the paired mandibles and maxillae and the hypopharynx supported in an elongate, troughlike labium. The beaks were held vertically beneath the head or slightly thrust forward, not directed to the rear as in Hemiptera. The postclypeus was swollen and may have housed a sucking pump. Nymphs of this order and of Megasecoptera and Diaphanopterodea were evidently terrestrial. The wing pads were free and curved obliquely backward. Nymphs had beak s like the adults, and all stages presumably had similar feeding habits. Many probably sucked the juices of plants, but they may also have probed crevices to reach the ovaries of primitive gymnosperm plants. Other species were probably predators, and some small mosquitolike forms may have been ectoparasites of vertebrates. Adults of this order reached 50 cm in wingspread, but most were smaller. Adults may have continued molting, as in Apterygota. The wings were often dark in color with patterns of light spots or transverse bands. The hind wings were variable in size and shape, being larger than, equal to, or smaller than the forewings. The pronotal lobes were large and sometimes exhibited a veined membrane. The abdominal terga had lateral lobes. Cerci were about twice as long as the abdomen. (Pennsylvanian-Permian)

Order Megasecoptera. Distinguished by long cerci and by distinctive wings that were long, nearly equal in size and shape, frequently elongate, and petiolate at the base. Megasecoptera and Diaphanopterodea had simple or branched integumental processes on the body that were densely covered with setae. Some processes were quite long, even longer than the body, and formed fringelike rows on the tergites. (Pennsylvanian-Permian)

Order Diaphanopterodea. Distinguished by a wing-flexing mechanism superficially like that of Neoptera. The nymphs were shrouded in filamentous projections of unknown function. (Pennsylvanian-Permian)

Order Permothemistida (Dicliptera). Small with reduced prothoracic winglets and very small hind wings and short rostra. Nymphs unknown. (Permian)

Infraclass Neoptera, Division Exopterygota

Order Paraplecoptera or Protoperlaria. Believed to be related to Plecoptera and possibly belong within it. The anterior and posterior median veins are fused basally, and the anal region of the hind wing is expanded. Mouthparts were mandibulate, prothoracic winglets were present, and moderately long cerci were present. In her classification, Kukalová-Peck (1991) includes fossils from the Jurassic in this lineage. Carpenter (1992) placed these insects in Protorthoptera. (Pennsylvanian-Jurassic)

Order Titanoptera. These were gigantic orthopteroids with raptorial forelegs and elongate mandibles suggesting predaceous habits. The forewings bore prominent fluted regions, which may have functioned in sound production. Wingspans reached 36 cm. (Triassic)

Order Protorthoptera. This order previously contained a diversity of fossils, most of which have been reclassified as members of other orders of the Orthopteroidea or of other superorders. The Protorthoptera may eventually be eliminated as a formal taxonomic group. Carpenter (1992), however, placed families of the Paraplecoptera and Protoperlaria in the Protorthoptera. According to him, all families but one of uncertain classification were extinct at the end of the Permian. (Pennsylvanian-Permian)

Order Protelytroptera. Small, robust insects superficially resembling beetles, but probably related to Dermaptera. Forewings formed into elytra, the broad hind wings folded beneath. Short cerci were present. (Permian)

Orders Caloneurodea and Glosselytrodea. Small to large insects with domed clypeal region to accommodate sucking pump. The mandibles are variable but often elongate and styliform. Maxillae were sometimes modified as a chisel-shaped organ, as in Psocoptera. Antennae were long, wings subequal. Short, annulate cerci were present. Caloneurodea are likely to be related to modern Orthoptera and Phasmatodea and the extinct Titanoptera. Carpenter (1992) considered the Glosselytrodea to be endopterygotes based on resemblance of some fossils to some neuropteroids, but relationships of the group as a whole are enigmatic. Caloneurodea (Pennsylvanian-Permian); Glosselytrodea (Permian-Jurassic)

Order Miomoptera. Small insects with mandibulate mouthparts, short cerci, and wings of equal size with simplified venation. In some Permian deposits they are more numerous than all other orders. The phylogenetic relationships of the Miomoptera are uncertain. According to some authors, they are endopterygotes and possibly ancestral to the Hymenoptera. (Pennsylvanian-Jurassic)

Fossil History of Insects

Depending on how the orders are classified, the fossil record indicates that at least sixteen orders of hexapods were present before or during the Pennsylvanian: Collembola, Diplura, Archeognatha, Thysanura, Palaeodictyoptera, Megasecoptera, Diaphanopterodea, Protodonata, Ephemeroptera, Paraplecoptera, Protoperlaria, Protorthoptera, Orthoptera, Blattodea, Caloneurodea, and Miomoptera. Considerable insect evolution must have taken place before the Pennsylvanian because both the Paleoptera and Neoptera are represented plus what some authorities believe are the first the Endopterygota, namely, the Miomoptera. A possible endopterygote larva also has been described from the Pennsylvanian. From this early fauna of hexapods, however, only the Collembola, Diplura, Archeognatha, Thysanura, Ephemeroptera, Orthoptera, and Blattodea survived (at least recognizably) until the present. Fossil wings of Blattodea are so common that the period is often called the "Age of Cockroaches." These insects were similar to modern cockroaches except that the wings were different in some features, their ovipositors were much longer, and they probably did not deposit oothecae. The Orthoptera had jumping hind legs, as in modern forms.

During the next period, the Permian, the climates of the world changed drastically. In the Northern Hemisphere increasing aridity is evidenced by great wind-blown deserts and thick salt deposits left by drying seas. Colder climates in the Southern Hemisphere were marked by several periods of glaciation.

Permian limestones of the Wellington Formation in Kansas and Oklahoma have abundant insect fossils. The strata measure some 213 m and include fossils of terrestrial plants, marine arthropods, and salt deposits. This is interpreted as a coastal swamp with intermittent freshwater habitats.

Associated with the changing climate and landscape of the Permian, the insect fauna also underwent major changes in composition. Ten new orders that have survived until the present made their appearance: Plecoptera, Embiidina, Hemiptera, Psocoptera, Thysanoptera, Neuroptera, Megaloptera (including a larva), Mecoptera, Diptera, and Coleoptera. Both the Orthopteroidea and Hemipteroidea are now represented, and the Endopterygota are definitely present as Neuropteroidea and Mecopteroidea. Three additional orders are also new: Permothemistida, Glosselytrodea, and Protelytroptera. With the appearance of new kinds of insects and the survival of a rich fauna from the Pennsylvanian, life of the Permian may have been more diverse in insect orders than even the present.

Permian Ephemeroptera had wings of equal size, and some had conspicuous mandibles. Nymphs are also found in Permian deposits (Fig. 14.3a). The mandibles, known from one specimen, are large and have well-developed teeth. The small wings have distinct veins and curve obliquely backward from the body. Gills were present on the first nine abdominal segments.

The end of the Permian was a time of massive extinction of Palaeozoic hexapods that had thrived for at least 78 million years and probably much longer. Of the fourteen ancient orders present in the Permian, the following seven were extinct by the end of the period: Diaphanopterodea, Palaeodictyoptera, Megasecoptera, Permothemistida, Caloneurodea,

Protoperlaria, and Protorthoptera. The remaining five ancient orders survived into the Mesozoic Era, later also to become extinct by the end of the era: Protodonata (extinct by the end of the Triassic Period); Paraplecoptera, Glosselytrodea, and Miomoptera (Jurassic); and Protelytroptera (Cretaceous).

The Mesozoic Era was a crucial time when modern insects were evolving in association with flowering plants and mammals. After the various extinctions of the ancient orders that arose in the Paleozoic, the Mesozoic fauna took on a relatively familiar and modern appearance. Odonata, Phasmatodea, Grylloblattodea, Trichoptera, and Hymenoptera are first represented in the Triassic Period, plus the huge Titanoptera that existed only in the Triassic. During the Jurassic Period, fossils of Dermaptera, Raphidioptera, and Lepidoptera first appear. Also at this time, existing families of Odonata, Orthoptera, Diptera, and Hymenoptera can be recognized.

Cretaceous amber has been found in Alaska, Canada, Siberia, Lebanon, and New Jersey. The insects thus preserved include the earliest ants (worker caste), earliest stingless bees (worker caste), Strepsiptera, and undisputed Collembola, plus Aphididae, parasitoid Hymenoptera, and Chironomidae. Mesozoic insect fossils have also been found in other types of preservation in Russia and China. Several specimens of Siphonaptera have been described. Isoptera are known from Cretaceous deposits in Labrador. Thus, by the end of the Cretaceous, termites, ants, and bees had evolved eusocial behavior with worker castes and entomophagous parasites and parasitoids were present. Most modern families had evolved, and the fauna had essentially a modern aspect. The Cretaceous insect fauna survived into the Cenozoic Era, apparently without the extinctions suffered by the dinosaurs.

Of the several epochs of the Tertiary Period in the Cenozoic Era, the Oligocene is best known. Amber from the shores of the Baltic Sea is dated from the Eocene and Oligocene. Here for the first time are Mantodea, plus many representatives of modern orders. The species do not exist now, but the insects belong to modern types. The silverfish family Lepidotrichidae was thought extinct, but living relatives of those entrapped in Baltic amber were discovered in 1959 in the coastal forests of northern California. In general, Baltic amber fossils represent distinct genera in modern families.

Oligocene and Miocene amber from Chiapas, Mexico, and from the Dominican Republic has yielded many insects (Fig. 14.6). Included are Zoraptera and stingless bees of the genus *Trigona* that differ only slightly from species living today in Central America. In general, fossils from Mexican and Dominican amber are very similar to modern species, and the greatest significance of these deposits is probably for paleoecology and biogeography.

Another important locality of Oligocene age is the Florissant Fossil Beds National Monument, near Colorado Springs, Colorado. A creek flowing through the area was dammed by lava and mud-flows emanating from volcanoes about 24 km away. Fine ash from the volcanoes settled into the lake and buried a great variety of plants and animals that fell into the still water. Over a hundred species of higher plants and thousands of insect species have been described from these deposits. Half or more belong to genera existing today. Some, however, are no longer present in North America. *Glossina*,

the tsetse fly, is present as a fossil at Florissant but today is found only in tropical Africa.

Calcareous nodules containing insects have been found in the Mojave Desert, California, from deposits of the Miocene Epoch. Apparently a Miocene freshwater lake was surrounded by volcanoes. Many orders of insects are represented, but none for the first time. Some are beautifully preserved by replacement of the original organic matter by colloidal silica.

Asphalt or tar pools in southern California, dating from the Pleistocene Epoch of the Quaternary Period, have yielded abundant insect remains. Inside the skulls of the saber-toothed cat, *Smilodon*, are puparial shells of dipterous larvae that presumably were scavengers on animals trapped in the tar. Adult aquatic insects, such as dragonflies and water beetles, are attracted today to the waterlike reflections of the liquid tar, and entombed. The fossils taken from the asphalt pools without exception represent species that occur in southern California today.

The oldest record of the suborder Anoplura of the Phthiraptera was obtained when lice were found on the carcass of a rodent frozen during the Pleistocene Epoch in Siberia (1.64 million to 10,000 years ago).

CHAPTER 15

Keys to the Orders of Hexapoda

The keys in this book consist of series of numbered dichotomies or two-way choices. Each pair of choices is called a couplet. Each half of a couplet leads to either a subsequent couplet (as indicated by the appropriate number) or the name of a taxon (order, family), which is then identified.

Beginning with the first couplet, one works through the key, comparing the characteristics of the specimen at hand with the dichotomous choices. If a couplet lists more than one feature, the primary, or most diagnostic, character is compared first. Secondary characters that follow should be used to confirm identifications made with the primary character or as alternatives when primary characters are missing or damaged. Numbers in parentheses indicate the couplet immediately preceding, so that the keys can be worked backward or forward. It should be emphasized that any large group of organisms, such as the Insecta, contains exceptional species that do not fit keys. The keys in this book should identify nearly all the adult insects encountered during general collecting in North America. It will not work for the strictly African order Mantophasmatodea.

Once a taxon has been identified as belonging to a particular order, it will be useful to go to the chapter treating that order to confirm the identification. The chapter number for each order is provided at each endpoint in the key.

KEY TO THE CLASSES AND ORDERS OF COMMON HEXAPODOUS ARTHROPODS

Body without wings, or with rudimentary or vestigial wings less than half body length

    Key A

Body with 1 or 2 pairs of wings at least half as long as body (wings may be modified as rigid covers over abdomen, or folded)   **Key B**

KEY A

Wings Absent or Rudimentary

1. Legs absent or reduced to unsegmented papillae shorter than one-fifth body width

    2

    Legs with 4 to 5 distinct segments, almost always terminated by 1 or 3 claws   **3**

2(1). Mouthparts enclosed in a slender, tubular rostrum (Fig. 15.1a); antennae and eyes usually absent; body segmentation indistinct or absent; sessile plant feeders, frequently covered by a waxy or cottony shell (Fig. 32.16d,e) **Hemiptera** (Chap. 32) Mouthparts mandibulate or internal, never enclosed in a tubular rostrum; antennae and eyes present or absent; body segmentation usually distinct; seldom with protective covering **Legless endopterygote larvae, not keyed further**

3(1). Legs terminated by a single claw or without claws **4**

At least middle legs terminated by 2 claws **16**

4(3). Head with large compound eyes almost always present laterally; ocelli frequently present on vertex **13**

Head with compound eyes absent or vestigial, stemmata (lateral ocelli) often present (Fig. 15.1b,d); ocelli absent from vertex **5**

5(4). Mouthparts enclosed in a slender, tubular rostrum; antennae and eyes usually absent; body segmentation frequently indistinct

or absent; plant feeders, frequently with body covered by a waxy or cottony shell (Fig. 32.16d) **Hemiptera** (Chap. 32) Mouthparts mandibulate or concealed in head capsule; body segmentation rarely indistinct **6**

6(5). Antennae with 2 or more segments **8**

Antennae absent **7**

7(6). Tarsus terminating in single claw; lateral ocelli absent; body cylindrical, elongate (Fig. 15.1c); minute, pale arthropods found in soil **Protura** (Chap. 16) Tarsus without claw; stemmata (lateral ocelli) large, usually set in pigmented patches (Fig. 15.1d); body fusiform (Fig. 15.1d); minute insects usually found in flowers, foliage **Strepsiptera** (First instars, Chapter 38)

8(6). Abdomen with 6 segments; segments 1, 3, and 4 usually with medial unpaired appendages (Fig. 16.3)

Collembola (Chap. 16)

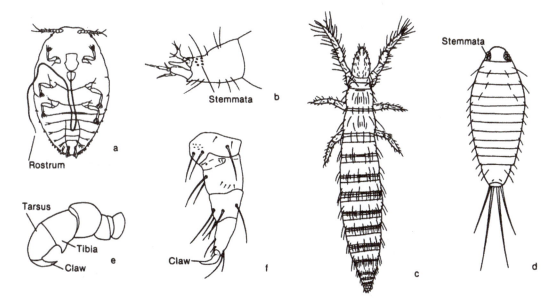

Figure 15.1 Key characters: a, scale insect, Hemiptera (*Parlatorea*); b, head of larval carabid beetle, showing stemmata (simple eyes); c, Protura (Acerentomidae); d, triungulin larva of Strepsiptera (Mengeidae); e, chelate tarsus of Phthiraptera (Anoplura); f, leg of endopterygote larva (Lepidoptera, Noctuidae). *Source: a, from University of California Extension; b, from Melis, 1936; c, from Berlese, 1909.*

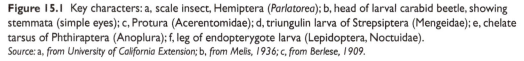

Abdomen with 8 to 11 segments; without appendages or with paired unsegmented appendages on some segments   **9**

9(8). Tarsus and claw chelate (Fig. 15.1e); abdomen without appendages; ectoparasites with flattened body, thick, short, 3- to 5-segmented antennae
    Phthiraptera (Psocodea) (Chap. 31)
    Tarsus and claw usually fused (Fig. 15.1f), very rarely chelate; abdomen with or without appendages; body shape extremely variable   **(Endopterygote orders, larvae)**
                                           10

10(9). Abdomen with unsegmented, paired walking appendages on some preterminal segments; body caterpillar-shaped (Fig. 15.2a)   **11**
    Abdomen without walking appendages on preterminal segments; not usually caterpillarlike   **12**

11(10). Abdominal appendages bearing rows, circles, or patches of short curved spines (crochets, Fig. 15.2b)
    Lepidoptera (Chap. 43)

Abdominal appendages without crochets
    Hymenoptera (Chap. 39)

12(10). Abdomen with ventrally directed hooked prolegs on last segment (Fig. 44.2); thoracic legs with trochanters 2-segmented; aquatic larvae usually dwelling in tubular cases
    Trichoptera (Chap. 44)
    Abdomen without appendages or with dorsal or lateral appendages without terminal hooks; if prolegs are ventrally directed, then terminal hooks are absent
    Coleoptera (Chap. 37)

13(4). Mouthparts enclosed in tubular rostrum (Fig. 15.1a); cerci absent
    Hemiptera (Chap. 32)
    Mouthparts mandibulate, never enclosed in a rostrum; cerci present or absent   **14**

14(13). Terminal abdominal segment with 2 or 3 long filaments; antennae multiarticulate, slender; long-legged aquatic insects, usually with leaflike abdominal gills
    Ephemeroptera (naiads) (Chap. 18)
    Terminal abdominal segment without filaments; terrestrial gills absent   **15**

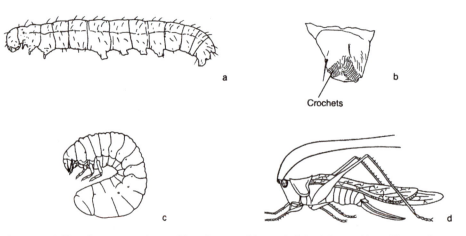

Figure 15.2 Key characters: a, larva of Lepidoptera (Noctuidae); b, abdominal leg of larva of Lepidoptera (Noctuidae); c, scarabaeiform grub of Endopterygota (Coleoptera, Scarabaeidae); d, a grasshopper (Orthoptera).
Source: a, b, from Melis, 1936; c, from Ritcher, 1944; d, from University of California Extension.

15(14). Body caterpillar-shaped (Fig. 15.2a) or grublike (Fig. 15.2c), with short, thick legs; tarsi with 1 segment

> **Mecoptera (larvae)** (Chap. 40)

Body slender elongate, with long, slender legs; tarsi with 4 to 5 segments

> **Mecoptera (adults)** (Chap. 40)

16(3). Anterior legs with first tarsal segment globular, at least 3 times as thick and long as second segment   **Embiidina** (Chap. 26)
Anterior legs with tarsal segments about equal in thickness and length   **17**

17(16). Posterior legs modified for jumping, with femur greatly thickened (Fig. 15.2d)

> **Orthoptera** (Chap. 27)

Posterior legs similar to middle pair   **18**

18(17). Last abdominal segment bearing cerci, either single-segmented (Figs. 15.3b, 15.4a) or multiarticulate (Fig. 15.3a)   **19**
Last abdominal segment without cerci   **32**

19(18). Last abdominal segment with median, multiarticulate filament; short, 1-segmented appendages present on at least abdominal segments 7 to 9   **20**

Last abdominal segment without median filament; appendages absent from segments 7 to 9   **21**

20(19). Head with large, contiguous compound eyes; maxillary palp with 7 segments; body cylindrical, arched

> **Archeognatha** (Chap. 17)

Head with widely separated lateral ocelli or without eyes; maxillary palp with 5 segments; body usually flattened

> **Thysanura** (Chap. 17)

21(19). Mouthparts modified as a rostrum or haustellum (Fig. 15.3c) or projecting beaklike below the head (Fig. 32.2e); labial palp with 2 segments   **22**
Mouthparts short, mandibulate (Fig. 16.3d), never forming a beak or haustellum; labial palp with 3 segments   **23**

22(21). Mouthparts modified as a rostrum or haustellum (Fig. 15.3c); metathorax frequently with halteres; antennae frequently with 5 or fewer segments   **Diptera** (Chap. 41)
Mouthparts with elongate mandibles and maxillae projecting beaklike (Fig. 40.1b);

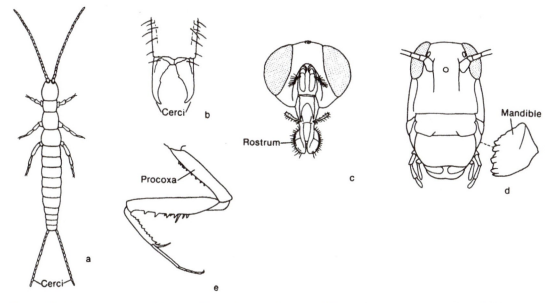

Figure 15.3 Key characters: a, Diplura, Campodeidae; b, forceps of Diplura, Japygidae; c, haustellum of Diptera (Muscidae); d, chewing mouthparts of a grasshopper (Orthoptera); e, raptorial leg of a mantid (Mantodea). *Source: a, d, from University of California Extension; c, modified from Snodgrass, 1953.*

halteres absent; antennae with at least 12 segments   **Mecoptera** (Chap. 40)

23(21). Tarsi with 5 segments   24
         Tarsi with 1 to 4 segments   27

24(23). Anterior legs raptorial (Fig. 15.3e)
                              Mantodea (Chap. 21)
         Anterior legs not raptorial, fitted for walking
                                                        25

25(24). Head prognathous (mouthparts directed anteriorly)   26
         Head hypognathous (mouthparts directed ventrally)   **Blattodea** (Chap. 20)

26(25). Cerci with 5 to 9 segments, long, flexible; body not modified for camouflage
                        Grylloblattodea (Chap. 22)
         Cerci with 1 segment, usually short; body almost always sticklike or leaflike
                          Phasmatodea (Chap. 28)

27(23). Tarsi with 1 to 2 segments   28
         Tarsi with 3 to 4 segments   29

28(27). Cerci either multiarticulate (Fig. 15.3a) or 1- segmented, forceps-shaped (Fig. 15.3b); tarsi with 1 segment; soil or litter dwellers with long, parallel- sided body (Fig. 15.3a)
                              Diplura (Chap. 16)
         Cerci with 1 segment, short, not forceps-shaped, tarsi with 2 segments; body stouter, not parallel-sided   **Zoraptera** (Chap. 29)

29(27). Cerci forceps-shaped, strongly sclerotized (Fig. 15.4a)   **Dermaptera** (Chap. 24)
         Cerci not forceps-shaped   30

30(29). Tarsi with 3 segments   31
         Tarsi with 4 segments
                      Isoptera (Blattodea) (Chap. 20)

31(30). Antennae more than half as long as body; labium small, without movable, apical teeth
                                                        41
         Antennae much less than half as long as body; labium jointed, with large, movable apical teeth (Fig. 20.5).
                              Odonata (Chap. 19)

32(18). Tarsi with 5 segments   36
         Tarsi with 1 to 3 segments   33

33(32). Mouthparts enclosed in a long, slender rostrum projecting beneath the head (Figs. 15.1a, 32.1b); maxillary and labial palps absent
                           Hemiptera (Chap. 32)
         Mouthparts not in the form of a rostrum; maxillary and labial palps usually present
                                                        34

34(33). Antennae longer than head, with at least 5 segments, usually with more than 10 segments   35
         Antennae shorter than head, with 3–7 segments   40

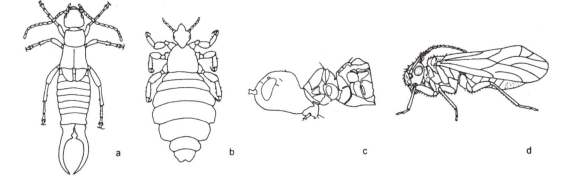

Figure 15.4 Key characters: a, an earwig (Dermaptera); b, a louse (Phthiraptera); c, lateral aspect of the anterior end of a thrips (Thysanoptera); d, a book louse (Psocoptera).
Source: a, b, from University of California Extension; c, from Faure, 1949.

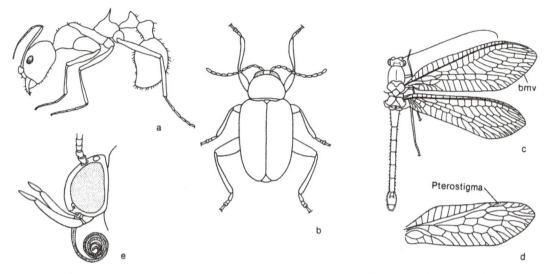

Figure 15.5 Key characters: a, an ant (Hymenoptera, Formicidae); b, a beetle (Coleoptera); c, wings of Neuroptera, showing bifurcating marginal veins (bmv); d, wing of a snakefly (Raphidioptera), showing pterostigma; e, head of a moth, showing coiled proboscis.
Source: a, b, from University of California Extension; e, modified from Snodgrass, 1935.

35(34). Head cone-shaped, directed ventrally or posteriorly (Fig. 15.4c); antennae with 4 to 9 segments; body elongate, slender
                Thysanoptera (Chap. 33)
Head not cone-shaped; antennae almost always with more than 12 segments; body stout (Fig. 15.4d)
           Psocoptera (Psocodea) (Chap. 30)

36(32). Abdomen strongly constricted at base (Fig. 15.5a); antennae frequently elbowed (Fig. 15.5a)   **Hymenoptera** (Chap. 39)
Abdomen not constricted at base; antennae not elbowed   **37**

37(36). Body densely covered with scales or long hairs; mouthparts usually a coiled proboscis (Fig. 15.5e) (sometimes vestigial)
                Lepidoptera (Chap. 43)
Body bare or sparsely covered with hairs, rarely scaled; mouthparts not a coiled proboscis   **38**

38(37). Mouthparts a slender tube-shaped rostrum (Figs. 42.1b, 41.3b) or a haustellum

(Fig. 15.3c); antennae usually with 3 or fewer segments   **39**
Mouthparts mandibulate (Fig. 15.3d), never forming a rostrum or haustellum; antennae almost always with 9 to 11 segments
                Coleoptera (Chap. 37)

39(38). Body strongly flattened laterally; thorax and head usually bearing large flattened, backwardly directed spines
                Siphonaptera (Chap. 42)
Body cylindrical or flattened dorsoventrally; head and thorax not bearing special spines   **Diptera** (Chap. 41)

40(34). Antennae usually concealed in grooves; ectoparasites on birds, mammals, with flattened bodies, reduced eyes and pigmentation (Fig. 15.4b).
          Phthiraptera (Psocodea) (Chap. 32)
Antennae-free; free-living aquatic naiads with long legs, large compound eyes, and darkly pigmented bodies (Fig. 20.6)
                Odonata (Chap. 19)

41(31). Cerci with numerous segments; aquatic naiads with gills usually present on legs, thorax, or abdomen

        Plecoptera (Chap. 25)

Cerci with 1 segment; terrestrial, without gills   **Phasmatodea** (Chap. 28)

KEY B

Wings Present, Functional

1. Mesothoracic wings thick, strongly sclerotized or parchmentlike, at least at base, or vestigial   **2**

Mesothoracic wings membranous, sometimes covered with scales; never vestigial   **10**

2(1). Mesothoracic wings vestigial, scalelike or club-shaped; hind wings large, fan-shaped

      9

Mesothoracic wings covering about one half or more of abdomen; never club-shaped or scalelike   **3**

3(2). Abdomen with large, strongly sclerotized forceps-shaped cerci (Fig. 15.4a); forewings short, leaving at least 3 abdominal segments exposed

      Dermaptera (Chap. 24)

Abdomen with cerci absent or not forceps-shaped; forewings usually covering entire abdomen   **4**

4(3). Mouthparts a slender, elongate rostrum, hinged at the base and projecting beneath head and backward (Figs. 15.1a, 32.1b)

      Hemiptera (Chap. 32)

Mouthparts short, mandibulate (Fig. 15.3d); head sometimes prolonged as a beak (Figs. 37.19b, 40.1b) but never with a hinged base   **5**

5(4). Forewings without venation, usually strongly sclerotized and meeting at midline at rest (Fig. 15.5b); antennae rarely with more than 11 segments, frequently clubbed

      Coleoptera (Chap. 37)

Forewings with extensive, reticulate venation; antennae usually with more than 12 segments, filiform, never clubbed

      6

6(5). Hind legs with femora greatly enlarged for jumping (Fig. 15.2d)

      Orthoptera (Chap. 27)

Hind legs not modified, similar to middle legs   **7**

7(6). Anterior legs raptorial (Fig. 15.3e)

      Mantodea (Chap. 21)

Anterior legs not raptorial, fitted for walking   **8**

8(7). Head prognathous (mouthparts directed anteriorly); body sticklike or leaflike

      Phasmatodea (Chap. 28)

Head hypognathous (mouthparts directed ventrally); body not sticklike or leaflike

      Blattodea (Chap. 20)

9(2). Hind legs with femora enlarged for jumping (Fig. 15.2d); prothorax projecting posteriorly over wings and abdomen

      Orthoptera (Chap. 27)

Hind legs not modified for jumping, similar to middle legs; prothorax small, not projecting over hindbody

      Strepsiptera (Chap. 38)

10(1). One pair of wings present   **11**

Two pairs of wings   **13**

11(10). Abdomen bearing 1 to 3 long filaments on terminal segment; mouthparts vestigial **12**

Abdomen not bearing long filaments; mouthparts rarely vestigial

      Diptera (Chap. 41)

12(11). Antennae long, filiform; wing with a single vein, without cells

      Hemiptera (Chap. 32)

Antennae short, bristle-shaped; wing with closed cells, usually with veins very numerous

      Ephemeroptera (Chap. 18)

13(10). Abdomen bearing 2 to 3 long, terminal filaments   **21**

Abdomen without terminal filaments or with very short processes   **14**

14(13). Tarsi with 1 to 4 segments   **15**

Tarsi with 5 segments   **22**

15(14). Mouthparts a slender, tube-shaped rostrum projecting below the head (Figs. 15.1a, 32.1b)

      Hemiptera (Chap. 32)

Mouthparts short, mandibulate, never an elongate projecting rostrum (Fig. 15.3d)
16

16(15). Forelegs with basal segment of tarsi globular, at least twice as thick as second segment
Embiidina (Chap. 26)
Forelegs with tarsal segments about equal in size
17

17(16). Antennae shorter than head, bristle-shaped; insects at least 1.5 cm long; abdomen long, slender; wings with extensive network of veins
Odonata (Chap. 19)
Antennae longer than head, filiform
18

18(17). Tarsi with 4 segments
Isoptera (Blattodea) (Chap. 20)
Tarsi with 2 to 3 segments
19

19(18). Last segment of abdomen bearing cerci; minute insects occurring in decaying wood
Zoraptera (Chap. 29)
Cerci absent
20

20(19). Wings linear, narrow, with no more than 2 veins; head cone-shaped, directed ventrally or posteriorly (Fig. 33.1a)
Thysanoptera (Chap. 33)
Wings oval with at least 4 longitudinal veins; head not cone-shaped
Psocoptera (Psocodea) (Chap. 30)

21(13). Hind wings larger than forewings; mouthparts mandibulate, almost always well developed
Plecoptera (Chap. 25)
Forewings larger than hind wings, broadly triangular; mouthparts vestigial
Ephemeroptera (Chap. 18)

22(14). Forewings densely covered with hairs or scales
23
Wings bare, or with fringe of marginal hairs
24

23(22). Wings covered with scales; mouthparts usually a coiled proboscis (Fig. 15.5e)
Lepidoptera (Chap. 43)
Forewings covered with hairs; mouthparts mandibulate
Trichoptera (Chap. 44)

24(22). Forewings about 1.5 times longer than hind wings; fore- and hind wings usually markedly different in shape and venation; abdomen usually strongly constricted at base
Hymenoptera (Chap. 39)
Forewings and hind wings similar in size, shape, and venation; abdomen not constricted at base
25

25(24). Head not prolonged ventrally; wings usually with numerous crossveins in costal margin (exception, Coniopterygidae)
26
Head prolonged ventrally, beaklike (Fig. 40.1b); wings with 1 to 3 crossveins in costal margin
Mecoptera (Chap. 40)

26(25). Hind wings broader at base than forewings; veins not bifurcating near wing margin (Fig. 34.1c)
27
Hind-wing breadth at base less than or equal to forewing breadth; at least radial veins bifurcating just before wing margin (Fig. 15.5c) (exception, Coniopterygidae)
Neuroptera (Chap. 36)

27(26). Pronotum quadrate or nearly so; wings without pterostigma
Megaloptera (Chap. 34)
Pronotum at least 3 times longer than wide; wings with pterostigma (Fig. 15.5d)
Raphidioptera (Chap. 35)

CHAPTER 16

The Noninsect Hexapoda:
Protura, Collembola, and Diplura

ORDER PROTURA

Very small to minute entognathous hexapods. Eyes, antennae absent; mandibles, maxillae styliform, retracted into invaginations in cranium; 1-or 2-segmented, nonambulatory appendages present on abdominal segments 1 to 3; cerci absent; postembryonic development anamorphic, with 9 abdominal segments in first instar, 12 segments in adult.

ORDER COLLEMBOLA

Entognathous hexapods characterized by 6-segmented abdomen with specialized, nonambulatory appendages on segments 1, 3, 4; antennae with 4 (rarely 6) musculate segments; cerci absent.

ORDER DIPLURA

Entognathous hexapods with elongate body, short legs, slightly differentiated thoracic and abdominal segments. Eyes absent, antennae multiarticulate, flagellum musculate. Abdomen with 10 or 11 segments, appendages on segments 1 or 2 to 7; cerci developed as multiarticulate, moniliform appendages or as short, sclerotized forceps, bearing terminal gland openings.

The three orders included in this chapter have sometimes been included among the insects but today are generally considered to be distinct enough biologically, morphologically, and phylogenetically to be treated as separate hexapod groups (see also Chap. 16). These three orders have been variously classified together as a group (Entognatha), sister to the true insects, or with Diplura, more closely related to the true insects than Protura and Collembola. Recent phylogenetic analyses utilizing both molecular and morphological data have not fully resolved the relationships among these orders but tend to support the monophyly of the true Insecta as well as a close relationship between these three noninsect hexapods and the true insects.

ORDER PROTURA

Protura are weakly sclerotized, elongate, fusiform arthropods that occur exclusively in moist decaying leaf litter, moss, soil, and similar situations. About 500 species are known; all are easily recognized by the absence of antennae and the characteristic body shape with conical head. Because of their small size and cryptic habits, Protura are seldom encountered, but soil or litter extracts frequently contain large numbers of individuals. Eggs of *Eosentomon*, the only genus observed, are spherical and covered with either large conical tubercles or irregular polygonal plates. First-instar larvae, called prelarvae, eclose with 9 abdominal segments. Uneclosed prelarvae lie in a spiral position inside the eggs, with the head and posterior abdomen adjacent. No other details of embryology are known. The prelarva molts to a larva with 9 abdominal segments, the apical one different from that of the prelarva. The first instar larva molts to a second instar larva with 10 abdominal segments, followed by another molt to produce a subadult with 12 abdominal segments. The tenth segment is added between the eighth and ninth. The eleventh and twelfth segments are added on either side of

the newly acquired tenth. There is no instar with 11 abdominal segments. A final molt produces the sexually mature adult, also with 12 abdominal segments. Occasional molts then occur throughout life.

Among hexapodous arthropods, Protura are unique in lacking antennae in all instars, but pseudoculi (Fig. 16.1a,b), present on the anterior cranium, may represent rudimentary antennae. The forelegs are usually held in front of the body, apparently functioning as sensory organs. Mouthparts of Protura are quite unlike those of insects but resemble those of Collembola. Mandibles and maxillae are in the form of elongate, simple blades,

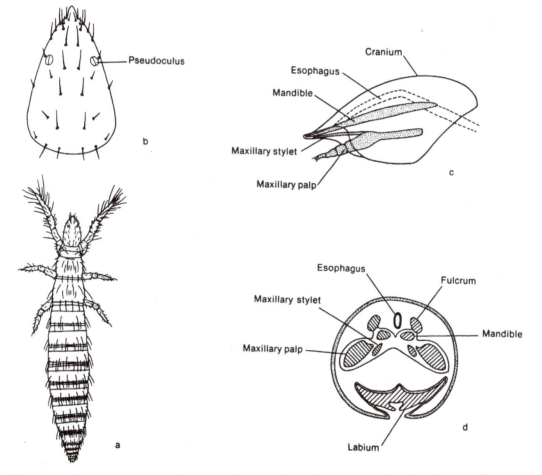

Figure 16.1 Protura: a, Acerentomidae, dorsal (*Acerentus barberi*); b, Acerentomidae, dorsal aspect of head (*Acerentomon*); c, longitudinal section of head, diagrammatic; d, transverse section of head, diagrammatic.
Source: a, from Ewing, 1940; b, from Price, 1959.

apparently used to triturate food particles. At rest the mouthparts are deeply retracted into invaginations in the head (Fig. 16.1c,d), with only the labrum visible externally. The only observation of feeding has been on fungal mycorrhizae associated with tree roots. The Protura have been monographed by Tuxen (1964), whose classification is followed here. Their biology and ecology are summarized by Bernard and Tuxen (1987).

KEY TO THE FAMILIES OF PROTURA

1. Spiracles present on mesothorax and metathorax; tracheae present
   Eosentomidae
   Tracheal system and spiracles absent   **2**
2. Abdominal terga with transverse sutures and laterotergites   **Acerentomidae**
   Abdominal terga without transverse sutures or laterotergites   **Protentomidae**

ORDER COLLEMBOLA (SPRINGTAILS)

The name "springtail" refers to the ability of many species to jump using the abdominal appendages. Springtails range in length from less than 1 mm to more than 10 mm, averaging 2 to 3 mm. About 6000 species occur worldwide.

Several different body plans are characteristic of different families (Figs. 16.2, 16.3), the most distinctive being the compact, globular form in Sminthuridae. Collembola occur in a wide variety of moist situations, especially soil and litter, where they are frequently exceedingly abundant and are among the most important consumers in many soil ecosystems. In Arctic regions Collembola are particularly dominant in soil habitats. Densities of 2 to 5 individuals per cm^3 have been recorded, and under these conditions the soil may consist largely of collembolan fecal pellets. Collembola also inhabit fungi, lower levels of vegetation, the surface of slowly moving or stagnant water, snow or ice fields, and, occasionally, buildings. Several species are abundant on seashores, ranging into the upper intertidal zone; a few forms occur in ant or termite nests. Collembola are readily wind-borne, and many species are extremely widespread, appearing even on coral atolls and oceanic islands. They have also been widely spread by commerce.

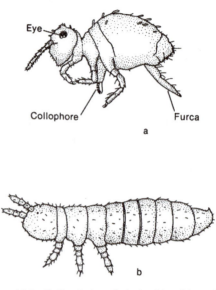

Figure 16.2 Collembola: a, Sminthuridae (*Neosminthurus curvisetis*); b, Onychiuridae (*Onychiurus subtenuis*). *Source: a, from Mills, 1934; b, from Folsom, 1917.*

Mouthparts of Collembola are always entognathous, but the mandibles vary depending on the type of food consumed, rather than being uniformly styliform, as in Protura. Species feeding on plant material have a defined molar surface for grinding. Simple, piercing mandibles are found in species that feed on liquids such as juices of fungi. Collembola share a six-segmented abdomen with early instars of very distantly related myriapods, which add segments after each molt, until the adult segmentation is attained. The arrangement and structure of abdominal appendages (Figs. 16.2, 16.3b) are unique among arthropods. The collophore, used in imbibing free water, is almost always present on the first abdominal segment. The furca, usually 3-segmented, is held under tension beneath the abdomen by the retinaculum when at rest. When released it propels the animal up to 0.2 m through the air.

Most Collembola are bisexual, but parthenogenetic forms are known. Insemination is indirect, males depositing stalked spermatophores on the substrate. The spermatophores are later picked up by the females, usually without direct participation by males. Old spermatophores may be consumed as fresh ones are deposited. In Sminthuridae some species have developed elaborate courtship interactions

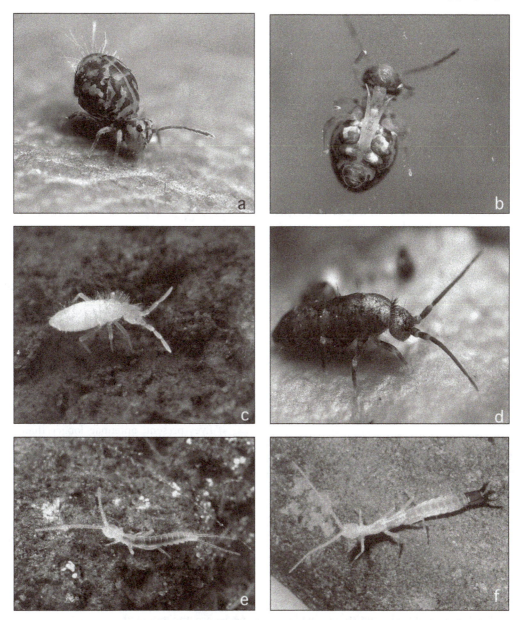

Figure 16.3 Collembola and Diplura. a: Sminthuridae, such as this *Dicyrtomina ornata*, are commonly found on plants in gardens; b: the underside of this sminthurid (photographed through the glass upon which the animal was standing) clearly shows the forked furcula in "cocked" position; c: entomobryids are among the most common arthropods in soil samples; d: unlike the antennae of true insects, segments of collembolan antennae are individually musculated beyond the first two; e: campodeid diplurans are extremely delicate animals, commonly recovered from soil samples; f: this dipluran, *Occasjapyx kofoidi* (Japygidae), was found in Potter Creek Cave in northern California. None of the cave japygids is especially adapted for cave life.

Source: Photos a, b, and d by Brian Valentine, used with permission; c and e by Alex Wild, used with permission.

involving contact between special sense organs on the frons. In *Sminthurides*, which live on the surface of water, the prehensile antennae of males are used to grasp the antennae of females. The much larger females carry the males about, periodically lowering them to the water surface where they deposit a spermatophore, which is then taken up by the female. Eggs, deposited singly or in clutches, are dropped onto the substrate. Those oviposited during unfavorable seasons may diapause for many months before hatching. Eggs of Collembola are atypical among those of hexapodous arthropods, being microlecithal (lacking large yolk reserves) and developing by holoblastic or complete cleavage (some tiny parasitoid Hymenoptera also do this). The first four or five molts are accompanied by minor changes in body proportions, chaetotaxy, and color pattern. Sexual maturity is usually attained in the fifth or sixth instar, but occasional molts occur throughout life, with a total of 50 recorded for some species. Most springtails probably feed on decaying vegetation, but spores, pollen, fungal mycelia, or animal material are included in the diet of various species. Some species attack living plant material, occasionally becoming economically important, and a few are predators on microorganisms such as rotifers and nematodes.

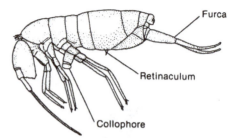

Figure 16.4 Collembola: Entomobryidae, *Tomocerus flavescens*. Entomobryidae usually have the fourth abdominal segment largest; *Tomocerus* is exceptional with the third segment largest.
Source: From Folsom, 1913.

KEY TO THE FAMILIES OF COLLEMBOLA

1. Body elongate, cylindrical or subcylindrical; first 6 postcephalic segments distinct (Figs. 16.4, 16.2b)   2
   Body globular; first 6 postcephalic segments fused (Fig. 16.2a)   6
2(1). Prothorax developed as distinct segment, bearing short setae (Fig. 16.2b)   3
   Prothorax reduced, indistinct, without setae (Fig. 16.4)   5
3(2). Head prognathous; eyes present or absent (as in Fig. 16.2b); third antennal segment often with large, complex sensilla   4
   Head hypognathous; eyes present; third antennal segment without unusually large sensilla   **Poduridae**
4(3). Eyes absent; third antennal segment with large, complex sensilla; body usually white   **Onychiuridae**

Eyes present or absent; third antennal segment with simple sensilla   **Hypogastruridae**
5(2). Body with scales or clavate setae; fourth abdominal segment usually much longer than third   **Entomobryidae**
   Body with simple setae only; third and fourth abdominal segments subequal   **Isotomidae**
6(1). Eyes absent; antennae shorter than head   **Neelidae**
   Eyes present; antennae longer than head   **Sminthuridae**

Sminthuridae (Figs. 16.2a, 16.3a,b). Easily distinguished from other Collembola by the globose body, frequently covered with brightly colored scales. The collophore, retinaculum, and furca are usually present. The few agriculturally important Collembola are sminthurids.

Neelidae. Three genera of very small (<1 mm) collembolans usually found in highly organic situations such as humus or rotten logs. Many of the species are unpigmented.

Onychiuridae (Fig. 16.2b). Usually unpigmented, lacking furca and retinaculum. Mostly inhabitants of soil interstices.

Poduridae. Includes only *Podura aquatica*, which is cosmopolitan on the surface of still freshwater. The body is bluish to reddish in color and about 1.25 mm long.

Hypogastruridae. Body form variable; small, active species with furca and retinaculum present are similar to Onychiuridae; large (to 1 cm long), sluggish species superficially resemble immature Coccinellidae (Coleoptera). The color of hypogastrurids varies from creamy to greenish, bluish, or black. *Anurida maritima* occurs in marine intertidal situations. *Hypogastrura nivicola,* which occurs on snow and ice, is known as the snow flea.

Entomobryidae (Fig. 16.4). A large family, frequently encountered under bark, on the lower leaves of plants, or occasionally in buildings and other relatively dry situations, as well as in leaf litter and soil. Many species are closely covered by overlapping scales. Most entomobryids are easily separated from other families by the enlarged fourth abdominal segment.

Isotomidae. A large group of common species that occur in leaf litter, garden soil, or rotting vegetation. Body shape is similar to that of Entomobryidae, but simple setae are present rather than scales. Isotomids, like entomobryids, are often greenish, bluish, or purplish in color.

ORDER DIPLURA (DIPLURANS)

Extant species of Diplura are functionally entognathous, with the mandibles and maxillae largely enclosed in an invagination on the ventral side of the head (Fig. 16.5b–d). They also differ from typical insects in lacking eyes. In the Pennsylvanian Period fossil *Testajapyx*, the mouthparts appear to be ectognathous and well-developed compound eyes are present (Fig. 16.6; Kukalová-Peck, 1987). In these characters as well as overall morphology, Diplura are the most insectlike of the entognathous hexapods. Unlike in insects, the antenna is musculate rather than flagellate.

Like the other entognathous orders, Diplura inhabit leaf litter, soil interstices, moss, and other

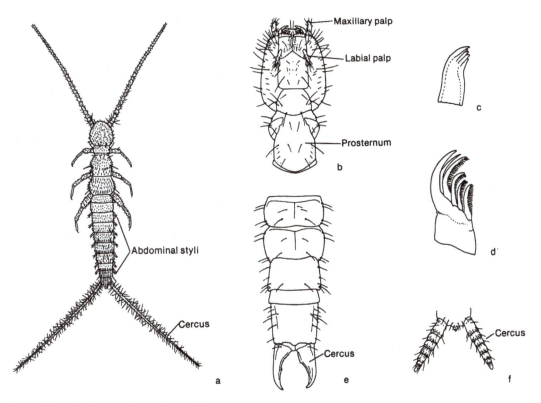

Figure 16.5 Diplura: a, Campodeidae (*Campodea montis*); b, Japygidae (*Occasjapyx kofoidi*), ventral aspect of head; c, same apical portion of mandible; d, same, apical portion of lacinia; e, Japygidae (*Metajapyx steevsi*), abdomen; f, Projapygidae (*Anajapyx menkei*), cerci.
Source: a, from Gardner, 1914; b, c, d, from Silvestri, 1928; e, from Smith and Bolton, 1964; f, from Smith, 1960.

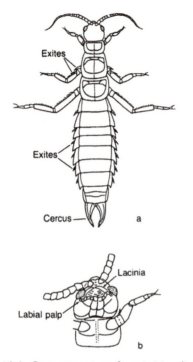

Figure 16.6 Reconstruction of a primitive dipluran, *Testajapyx thomasi*, from the Pennsylvanian Period (= Upper Carboniferous Period; 323–290 million years ago) of Illinois. Length without antennae was 47.5 mm. Note the compound eyes (facets of ommatidia were rounded rather than hexagonal), segmented labial palp, comblike structure of the laciniae, exites on the legs and abdominal segments, and stout cerci. a, Dorsal view of whole insect; b, ventral view of head.
Source: Redrawn from Kukalová-Peck, 1987. Reproduced by permission of Canadian Journal of Zoology.

moist, cool situations. Typically, they are unpigmented and weakly sclerotized organisms, 5 to 15 mm long, but *Anajapyx* (Japygidae) reaches 50 mm in length. Diplura are cosmopolitan, with the Campodeidae best represented in temperate regions, especially in the Northern Hemisphere. The Japygidae are predominantly tropical, subtropical, and south temperate. About 800 species are known.

Known species are bisexual. Insemination is indirect, by means of a stalked spermatophore, which the males attach to the substrate. Early instars are similar to adults except for chaetotaxy, and metamorphosis is essentially absent. Molting continues throughout life, as in the Thysanura, but sexual maturity is attained when the setation is complete. Campodeidae apparently feed on decaying vegetation, while Japygidae are predatory, using the forceps to capture prey. Projapygidae appear to feed chiefly on mites.

Diplura were formerly placed in a single taxon with Thysanura, which are also primitively wingless with multiarticulate antennae and cerci. However, the antennae of Thysanura are flagellate (lacking intrinsic muscles in the flagellum). Imms (1936) has shown that the Diplura are generally similar to the Symphyla in body organization. The Japygidae are unique among hexapodous arthropods in possessing four pairs of thoracic spiracles, with two pairs on each of the mesothorax and metathorax.

KEY TO THE FAMILIES OF DIPLURA

1. Cerci multiarticulate, not sclerotized   2
   Cerci with 1 segment, sclerotized, forcipulate (Fig. 16.5e)   **Japygidae**
2. Cerci less than half length of antennae (Fig. 16.5f), strongly moniliform, and bearing terminal gland openings   **Projapygidae**
   Cerci about as long as antennae, filiform or weakly moniliform, without terminal gland openings (Fig. 16.5a)   **Campodeidae**

Campodeidae (Fig. 16.5a). White, delicate, active species mostly about 3 to 5 mm long. Abundant in leaf litter and soil throughout North America.

Japygidae. Body more compact than in Campodeidae, with terminal abdominal segment and forceps sclerotized. Commonly ranging from 5 to 15 mm in length; in North America most abundant in Southwest and West.

Projapygidae. Body compact, as in Japygidae, but cerci multiarticulate. In North America occurring only in California.

CHAPTER 17

The Apterygote Insects: Archeognatha and Thysanura

ORDER ARCHAEOGNATHA

Elongate, subcylindrical, with arched trunk and 3 multiarticulate terminal filaments; antennae multiarticulate; 3 ocelli and large, dorsally contiguous compound eyes; mouthparts partly retracted into head, mandibles elongate, monocondylic; maxillary palp elongate, 7-segmented; 1-segmented appendages (styli and protrusible vesicles) present on abdominal segments 1 to 9; styli also present on meso- and metathoracic leg bases.

ORDER THYSANURA

Elongate to ovate, flattened insects with 3 multiarticulate terminal abdominal filaments; antennae multiarticulate; eyes of compound type but small, rudimentary, or absent; mouthparts mandibulate; mandibles short, semidicondylic, maxillary palps 5-segmented; 1-segmented appendages absent or present on variable number of abdominal segments.

ORDER ARCHEOGNATHA (MACHILIDS, BRISTLETAILS, MICROCORYPHIANS)

Superficially the Archeognatha are similar to the Thysanura, and the two have sometimes been classified as a single order. Structurally the Archeognatha share certain ancestral features, especially mouthpart structures (Fig. 17.1b), with the entognathous hexapods, especially the Diplura, as well as showing more derived similarities to the Thysanura and Pterygota. The occurrence of archeognathanlike fossils as early as the Devonian (Labandeira et al., 1988) seems to indicate that a much larger fauna of primitive, flightless insects existed during the Paleozoic Era (Kukalova-Peck, 1987; Sharov, 1966). The Archeognatha probably represent a relictual stock of such forms.

In contrast to the entognathous hexapods, the Archeognatha have successfully adapted to a variety of habitats besides soil and litter. Machilids commonly occur in chaparral or dry coniferous woodlands, where they are dependent on localized sources of free water, which they imbibe through the eversible membranous vesicles adjacent to their abdominal appendages. Meinertellids can inhabit extremely xeric habitats, including desert sand dunes, and require little if any free water, but also are known from tropical forest.

Biology and life histories are known in detail for only a few species. Fertilization is indirect, as in the entognathous groups (Chap. 16). During courtship the males of most genera spin a fine thread from the gonapophyses. One end of the thread is attached to the substrate, the other end held at the

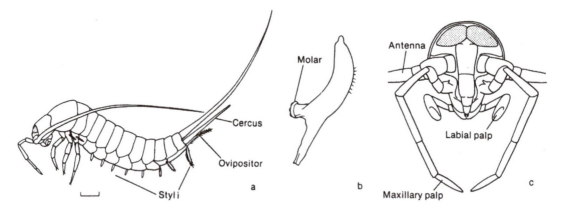

Figure 17.1 Archeognatha, Machilidae: a, *Machilis*, female (scale equals 1 mm); b, mandible of *Machilis*; c, head of *Machilis*.
Source: b, modified from Snodgrass, 1950.

tip of the abdomen (Fig. 17.2). Several spermatophores are deposited on the thread as it is formed. These are later picked up by the female after a series of complicated courtship maneuvers. Considerable variation in the details of courtship have been observed among various genera of Archeognatha (Sturm, 1992). In *Petrobius* the sperm droplet is placed directly on the ovipositor. Females mate during each instar, depositing clutches of up to 30 eggs in crevices.

Immatures resemble adults except for chaetotaxy and the lack of styli on the thorax. Scales appear on the body in the third instar. Food consists of lichens, algae, vegetable detritus, and probably dead arthropods. Sexual maturity is attained after about 8 to 10 molts over a period of up to 2 years. Sexually mature individuals continue to molt periodically throughout life. Life spans of 2 to 4 years have been recorded, with a total of as many as 66 molts. These are active, fast-running insects, capable of random escape jumps by suddenly arching the body and slamming down the abdomen. The abdominal appendages are used as skids to support the abdomen. About 350 species of Archeognatha have been described.

Although wingless both in origin and currently, arboreal bristletails show an ability to control and direct their fall from treetops to sail back spirally towards the tree trunk, from which they can find their way back up to the canopy. This

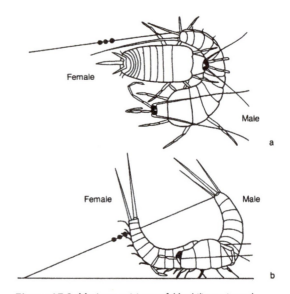

Figure 17.2 Mating positions of *Machilis*; a, viewed from above; b, viewed from the side. The male spins a thread fastened to the ground at one end and held by his genitalia above the female at the other end. Sperm drops (black dots) are suspended on the thread.
Source: Redrawn from Schaller, 1971. Reproduced, with permission, from the Annual Review of Entomology, Volume 16, 1971, by *Annual Reviews Inc.*

gliding ability has been hypothesized as potentially contributing to the evolution of insect flight. The ground-dwelling Archeognatha do not display this ability.

Meinertellidae (Fig. 17.3a). Abdominal sterna small; antennae without scales. Mostly small (<5 mm) diurnal species found in dry regions. Sometimes found in association with ant nests.

Machilidae (Fig. 17.3b). Abdominal sterna relatively large; antennae with scape and pedicel usually scaled. Larger (5–15 mm), mostly crepuscular species. Cosmopolitan. The commonly encountered North American archeognathans are all machilids.

ORDER THYSANURA (SILVERFISH)

In the morphology of the mouthparts, the Thysanura are comparable to primitive pterygote insects, such as mayflies, rather than the superficially similar Archeognatha. Structurally they are more variable than the Archeognatha, ranging from slender, elongate, free-living forms to ovate, short-tailed types that frequent mammal burrows or ant nests. Thysanura are rapid runners but lack the jumping ability of archeognathans. Various species occur in many different habitats, including relatively hot, dry situations, including houses. *Ctenolepisma* is able to extract water through the rectum from atmospheres as low as 60 percent relative humidity (Fig. 17.4).

Life histories of Thysanura are similar to those of Archeognatha, with complex courtship and

Figure 17.3 Archeognatha. a) a meinertellid from California; b) a Californian machilid examines the photographer. Note the large contiguous eyes.

Source: Photos a and b by Alex Wild; used with permission.

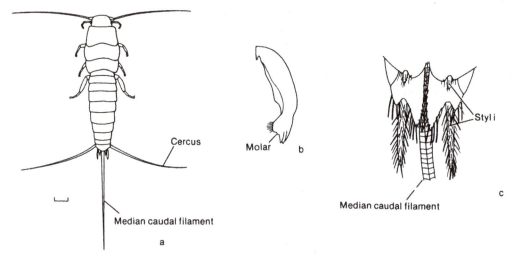

Figure 17.4 Apterygota, Thysanura, Lepismatidae: **a**, *Ctenolepisma lineata* (scale equals 1 mm); **b**, mandible of *Ctenolepisma*; **c**, abdominal apex of *Heterolepisma*.
Source: b, modified from Snodgrass, 1950; c, from Folsom, 1924.

indirect insemination, ametabolous metamorphosis, and surprisingly slow development and long life spans. *Tricholepidion gertschi* (Lepidotrichidae) (Fig. 17.5) has a 2-year cycle. Survival of lepismatids over periods of 3 to 4 years is recorded, during which molting and reproduction continue. As in Archeognatha, mating occurs in each instar after sexual maturity. Most Thysanura are probably omnivorous. The household species occasionally damage books by consuming the starch in their bindings. The order includes about 370 species.

KEY TO THE FAMILIES OF THYSANURA

1. Compound eyes present   2
   Compound eyes absent   **Nicoletiidae**
2. Body hairy; ocelli present; tarsi with 5 segments; exertile vesicles present on abdominal segments 2 to 9

                   Lepidotrichidae

   Body scaly; ocelli absent; tarsi with 3 or 4 segments; exertile vesicles absent

                   Lepismatidae

Nicoletiidae. Mostly commensal forms, occurring in ant or termite nests or mammal burrows or cavernicolous or subterranean. Eyes and ocelli

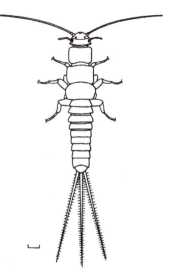

Figure 17.5. Apterygota, Thysanura, Lepidotrichidae, *Tricholepidion gertschi* (scale equals 1 mm).

are absent; body either elongate, slender or broad, oval, with short terminal filaments. In the United States, restricted to the South and Southwest.

Lepidotrichidae. Body slender, elongate, moderate in size (12–15 mm); lateral and median ocelli present. *Tricholepidion* (Fig. 17.5), the only genus, has the body sparsely clothed with simple setae. The family was known only from Oligocene fossils until living insects were discovered in northwestern California in 1959.

Lepismatidae. Body slender, elongate, small to moderate in size; lateral ocelli absent, median ocellus reduced as median frontal organ. The body of lepismatids is always clothed in scales, differentiating them from Lepidotrichidae and most Nicoletiidae. Lepismatids occur in litter, rubbish, in dry areas beneath loose bark, in animal burrows, and many other situations. All the frequently encountered silverfish belong to this family, including common household species such as *Lepismodes inquilinus* (the firebrat) and *Lepisma saccharina* (the silverfish).

CHAPTER 18

Order Ephemeroptera (Mayflies)

ORDER EPHEMEROPTERA

Adult. Delicate paleopterous insects with elongate, subcylindrical body (Fig. 19.1a). Head with short, multiarticulate, filiform or setaceous antennae, 3 ocelli, and large, compound eyes; mouthparts mandibulate, vestigial. Mesothorax enlarged, supporting large, triangular forewings; hind wings small, rounded, or absent; venation extensive, with numerous crossveins (Fig. 19.2). Abdomen slender, elongate with 10 segments; tenth segment with long cerci and usually with median caudal filament.

Naiads. Extremely variable in body form (Fig. 18.3) but almost always with large compound eyes, short multiarticulate antennae; mouthparts mandibulate; tarsi 1-segmented, with single claw; 10-segmented abdomen with cerci and usually median caudal filament; gills present on at least some of anterior 7 abdominal segments.

Subimago. Similar to adult but with duller coloration and relatively shorter appendages.

Some Paleozoic fossils have forewings and hind wings of similar size and venation, but all extant adult and subimaginal mayflies are easily recognized by the large, triangular forewings, which are held vertically above the body at rest, and by the long abdominal filaments. Naiads are usually distinguished by the combination of three terminal filaments and abdominal gills. Forms with two terminal filaments (Fig. 18.3c) may superficially resemble nymphs of Plecoptera. Mayflies are exceedingly abundant in nearly all permanent freshwater habitats, where they constitute a basic food item for nearly all predators. Adults of many species, especially in temperate climates, tend to emerge synchronously, sometimes in extremely large numbers, and may be temporary nuisances near lakes or large streams.

The Ephemeroptera, together with the Odonata, are the only extant insects with a paleopterous wing-folding mechanism. These two relatively small orders, sometimes classified as the infraclass Paleoptera, constitute the remnants of an extensive fauna, which mostly became extinct at the end of the Mesozoic Era (see Chap. 14). In mayflies, although wing-flexing capability is lacking, the dorsal longitudinal thoracic muscles are relatively large, and flight is largely powered by these indirect muscles. In Odonata, flight is powered entirely by direct musculature. The major differences in flight mechanism, as well as the grossly different body plans and biological adaptations, indicate the extreme antiquity of the paleopterous orders and their long period of independent evolution. Due to the difficulty of resolving such ancient relationships using either morphological or molecular evidence, it has remained relatively controversial which order, Ephemeroptera or Odonata, is more closely related to the Neoptera or if they might actually have diverged together from the lineage leading to modern Neoptera.

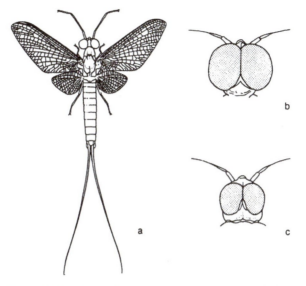

Figure 18.1 a, Adult of *Hexagenia bilineata* (Hexageniidae);
b, *Homoneura dolani*, dorsal aspect of head of male; c, same of
female.
Source: a, from Needham, 1917; b, c, from Edmunds et al., 1958.

Mayflies are remarkable in possessing paired penes and oviducts opening through separate gonopores. Among other insects, the external genitalia are paired only in some male earwigs (Dermaptera). Mandibles of nymphs (Fig. 18.4b) are notable in their general similarity to those of Archeognatha (Fig. 17.1b) but are unique in having three points of articulation, one being a movable slider that affords unusual mobility (Kukalová-Peck, 1985). A second mandibular condyle can be slid along a troughlike socket, enabling the mandible to roll as in Archeognatha (Kukalová-Peck, 1985). Musculature and development of the nymphal gills show similarities to legs, suggesting possible homology with the abdominal appendages of the entognathous hexapods. In most other morphological features, mayflies are highly specialized, as described below.

Mayfly naiads molt to subimagos similar to the adults but sexually immature. Coloration of the subimago is usually subdued relative to the adult, the caudal filaments and tarsi are often shorter, and the wing veins are microsetose (bare in imagos). Compared with adults the subimagos are slow,

clumsy fliers. All mayflies have subimaginal instars; in the old world family Palingeniidae the adult female instar is absent and the subimagos are reproductively functional. In Oligoneuridae the subimaginal cuticle is molted from the body but retained on the wings. Mayflies are the only extant insects in which a functionally winged form molts, but functionally winged immatures may have occurred in several orders which became extinct at the end of the Paleozoic (Kukalová-Peck, l991). Duration of the subimaginal instar is generally correlated with the length of adult life. In species that mate and reproduce the same night, the subimago usually lasts only a few minutes; in other species the adult may live several days, and the subimago usually persists for about 12 to 24 hours. In a few species the imaginal molt is partly or completely suppressed. Adults engage in mating flights, which may involve large, dense swarms, especially in nocturnally active, temperate zone species. The male swarms often occur in characteristic situations. Some species, for example, typically fly over emergent objects in streams, others in forest clearings near water. Time of day of swarming and, in temperate climates, season are

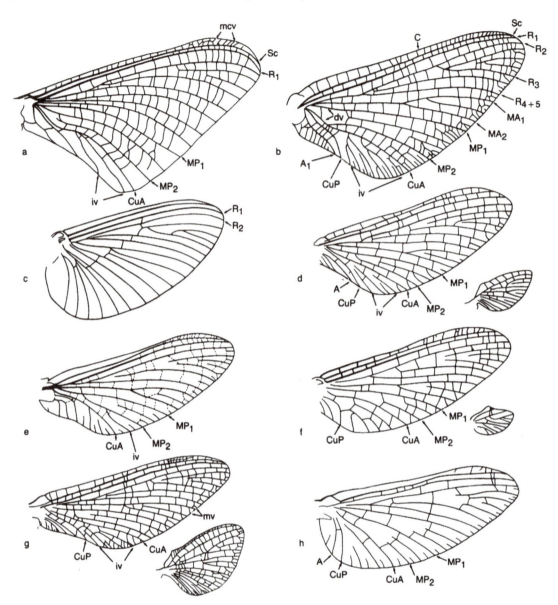

Figure 18.2 Representative mayfly wings: a, Polymitarchidae (*Tortopus primus*); b, Ephemeridae (*Ephemera simulans*); c, Caenidae (*Brachycercus lacustris*); d, Heptageniidae (*Heptagenia maculipennis*); e, Ephemerellidae (*Ephemerella lutulenta*); f, Leptophlebiidae (*Thraulodes speciosus*); g, Ametropodidae (*Siphloplectron basale*); h, Baetidae (*Callibaetis fluctuans*). iv, intercalary veins; mv, marginal veinlets; mcv, marginal crossveins; dv, divergent veins. *Source: From Burks, 1953.*

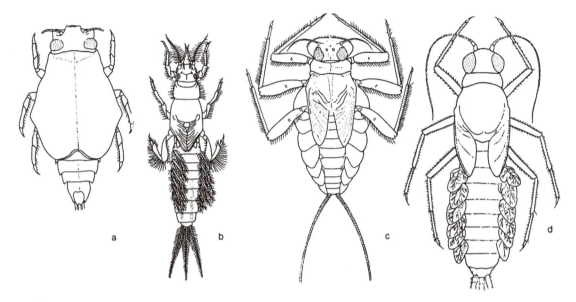

Figure 18.3 Representative mayfly naiads: a, Baetiscidae (*Baetisca columbiana*), caudal filaments removed; b, Ephemeridae (*Ephemera varia*); c, Heptageniidae (*Epeorus longimanus*); d, Baetidae (*Callibaetis coloradensis*), caudal filaments removed.
Source: a, from Edmunds, 1960; b, from Needham, 1920; c, d, from Jensen, 1966.

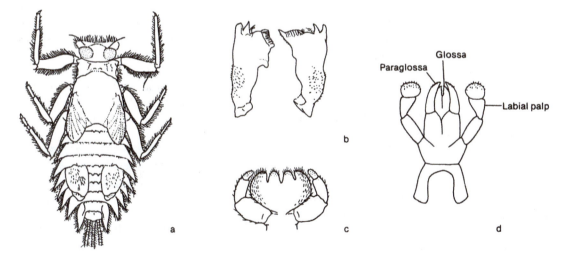

Figure 18.4 Structures of mayfly naiads: a, *Ephemerella hecuba*, dorsal view, caudal filaments removed (Ephemerellidae); b, mandibles, *E. hecuba*; c, labium, *E. hecuba*; d, labium, *Baetis* (Baetidae).
Source: a, b, c, from Allen and Edmunds, 1959; d, from Murphy, 1922.

also characteristic of species to a degree. In tropical areas seasonality is often absent. Nuptial flights or dances frequently entail characteristic movements, especially by males, which rapidly fly upwards several meters, then slowly drift downward. These patterns are assisted by the caudal filaments, which act as a counterbalance during the downward segment of the flight. In a few species, solitary males patrol in search of females. Adults and subimagos have vestigial mouthparts and never feed. The midgut is enlarged as an air sac, increasing the efficiency of flight. The esophagus and hindgut are modified as valves, apparently for regulating and maintaining the amount of air in the midgut.

Before copulation, the male approaches the female from below, grasping her mesothorax with his elongate forelegs, which are provided with a reversible joint at the first tarsal segment. Also correlated with courtship is the enlargement of the eyes in males (Fig. 18.1b,c). In many genera the dorsal facets are larger than the ventral ones; in some the eyes are divided into distinct dorsal and ventral lobes.

Several hundred to several thousand minute eggs are deposited, either in small batches or en masse, depending upon the species. Females may oviposit while in flight or may descend below the water surface. Eggs are quite variable in shape and sculpturing among different genera and families and are typically provided with coiled filaments that unroll when the eggs contact water, acting to attach the eggs to the substrate. Eggs have been characterized for many species and are of some taxonomic importance. Hatching occurs after one to several weeks or may follow diapause of several months. Development of immatures requires one to several years, with up to 27 instars. Most species are scavengers or herbivores; a few predaceous forms are known. Naiad morphology is generally correlated with habitat. Free-swimming, campodeiform, or shrimplike nymphs (Fig. 18.3d) usually occupy small standing bodies of water or sluggish streams. Flattened bodies with widely separated, laterally directed legs (Fig. 18.3a,c) are adaptations for clinging to stones or logs in fast-flowing streams. In burrowing forms, the legs or mandibles (Fig. 18.3b) may be specialized for digging, and the plumose gills for creating a current in the burrow, as in *Ephemera* and *Polymitarcys*. Species that live in water with high sediment content

display a variety of structural modifications to protect the gills. In Tricorythidae, Neoephemeridae, and Caenidae the gills on abdominal segment 2 are enlarged, forming a protective cover over the respiratory gills. In Baetiscidae and Prosopistomatidae the mesonotum is expanded as a carapace over the thorax and anterior abdominal segments and gills. Many Ephemerellidae and Heptageniidae are specialized for living in fast-moving water. In these nymphs the body is broad and flat, the legs robust and directed laterally—adaptations allowing them to cling to stones or submerged logs while presenting a low profile in the fast-moving water. Some species occur in the spaces beneath stones where the water movement is greatly reduced.

Classification of mayflies has been subject to numerous rearrangements, with great differences in the number of families recognized. Molecular data have so far influenced mostly the subordinal classification, which is not yet stable and is not presented here. The relatively conservative classification adopted here follows Edmunds (1959) and Edmunds et al. (1976). Adult mayflies, probably because of their high degree of specialization for a short adult life and aerial copulation, show only minor morphological differences among families. The key to adults is based largely on wing venation and must be used carefully. Nymphs are highly specialized for living on or in different substrates, as reflected in their great morphological divergence, and usually can be easily keyed to family. The general geographic and habitat distribution of nymphs and the numbers of North American species in each family are listed in Table 18.1.

KEY TO THE NORTH AMERICAN FAMILIES OF EPHEMEROPTERA[1]

Adults .

    1. Forewings with veins MP1 and CuA strongly divergent at base; vein MP2 abruptly bent toward MP1 at base (Fig. 18.2a,b)   **2**

---

[1] Names of the veins (Fig. 18.2b) are those of Edmunds et al. (1976); MA, medius anterior; MP, medius posterior; CuA, cubitus anterior (not cubitoanal); CuP, cubitus posterior.

Forewings with veins MP1 and CuA gradu-ally divergent at base; vein MP2 gradually bent toward MP1 at base (Fig. 18.2d,e,h)   **6**

2(1). Basal portion of forewing with costal crossveins faint or incomplete; hind wing with veins MP1 and MP2 separating dis-tally to middle of wing (as in Fig. 18.2f)   **Neoephemeridae**
Forewing with costal crossveins strong, never incomplete; hind wing with veins MP1 and MP2 separating proximally to middle of wing (Fig. 18.2a)   **3**

3(2). Forewings with veins Sc and R1 curving posteriorly around apical margin of wing (Fig. 18.2a); middle and hind legs atrophied to trochanter   **Polymitarcyidae**
Forewings with veins Sc and R1 straight to apex, or slightly curved, ending anterior to apex (Fig. 18.2b); middle and hind legs with femur, tibia normal   **4**

4(3). Apical anterior portion of forewing with dense, interconnecting network of marginal crossveins (Fig. 18.2a); cubital intercalary veins straight, not connected with vein CuA (Fig. 18.2a)   **Polymitarcyidae**
Radial portion of forewing with marginal crossveins sparsely interconnecting, not forming network; cubital intercalary veins sinuous, joining vein CuA (Fig. 18.2b)   **5**

5(4). Vein A1 forked   **Potamanthidae**
Vein A1 not forked (Fig. 18.2b)   **15**

6(1). Lateral ocelli about one-tenth to one-fourth as large as compound eye   **8**
Lateral ocelli about one-half as large as compound eye   **7**

7(6). Forewing with numerous crossveins between veins R1 and R2   **Tricorythidae**
Forewing with 1 or very few crossveins between R1 and R2 (Fig. 18.2c)   **Caenidae**

8(6). Forewing with venation reduced to 7 to 9 longitudinal veins; veins Sc and R1 fused   **Oligoneuriidae**
Forewing with at least 15 longitudinal veins; veins Sc and R1 separate from base to apex (Fig. 18.2d-h)   **9**

9(8). Forewing with cubital intercalary veins present (Fig. 18.2d)   **10**
Forewing with cubital intercalary veins absent   **Baetiscidae**

10(9). Hind leg with tarsus 5-segmented   **Heptageniidae**
Hind leg with tarsus 3- or 4-segmented; basal segment(s) fused with tibia   **11**

11(10). Forewing with 1 or 2 long intercalary veins between veins MP2 and CuA (Fig. 18.2e)   **Ephemerellidae**
Forewing without long intercalaries between veins MP2 and CuA (Fig. 18.2f–h)   **12**

12(11). Forewing with vein CuP abruptly angled near midpoint (Fig. 18.2f)   **Leptophlebiidae**
Forewing with vein CuP straight or evenly curved   **13**

13(12). Forewing with 1 to 2 pairs of long, parallel cubital intercalary veins; marginal veinlets never free (Fig. 18.2g)   **Ametropodidae**
Forewing with several short cubital inter-calary veins, or with marginal veinlets free, detached (Fig. 18.2h)   **14**

14(13). Hind wing very small or absent; forewing with vein MP2 detached from MP1 (Fig. 18.2h)   **Baetidae**
Hind wing large; forewing with vein MP2 basally connected to MP1   **Siphlonuridae**

15(6). Male with pronotum about three times as broad as long; female with caudal filaments shorter than body   **Palingeniidae**
Male with pronotum less than twice as broad as long; female with caudal filaments longer than body   **Ephemeridae**

Naiads

1. Thorax with notum enlarged posteriorly, leaving only 4 or 5 abdominal tergites exposed, and concealing gills (Fig. 18.3a)   **Baetiscidae**
Thorax with notum smaller, leaving at least 7 abdominal tergites exposed; gills exposed (Fig. 18.3b–d)   **2**

2(1). Mandibles produced as long tusks project-ing forward beyond the head (Fig. 18.3b)   **3**

Table 18.1 Characteristics of Families of Mayflies in North America.

Family	Approximate no. of species in North America	Distribution	Habitat	Substrate	Habits
Siphlonuridae	56	Widespread	Streams, rivers, lakes, ponds, swamps, temporary pools	Sandy or rocky or on vegetation or debris	Active swimmers; filter feeders
Oligoneuriidae	30	Southeast to southwest	Moving water, small to large streams, rapids	Shifting sand or under sticks, stones	Burrowing in sand or clinging to substrates; poor swimmers; filter feeders
Heptageniidae	132	Widespread	Mostly in moving water; also in ponds, rock pools, lake margins	Usually on stones; sometimes in vegetation or debris, or under stones, logs	Clinging to substrate with gill holdfast; poor swimmers
Ametropodidae	9	Northeast to southwest	Slow-moving streams	On vegetation, stones or sand	Crawling and swimming
Baetidae	146	Widespread	Streams, rivers, ditches, lakes, ponds	Usually in vegetation; also in debris, under stones	Most actively swimming or climbing about on vegetation; negatively phototactic; herbivorous
Leptophlebiidae	70	Widespread	Most slow moving streams, backwashes, lake margins	Under stones, logs; in accumulations of leaves or other debris	Mostly crawling; negatively phototactic; herbivorous or omnivorous
Ephemerellidae	79	Widespread	Variable	On stones, logs, other objects	Climbing about substrate; poor swimmers
Tricorythidae	22	Widespread	Streams, rivers	On sticks, logs, branches, or on sand, gravel	Crawling about substrate; poor swimmers; herbivorous
Caenidae	18	Widespread	Slowly moving, quiet, or stagnant water	Sand or silt bottoms	Resting on bottom, partly covered by fine sand or silt; herbivorous or omnivorous
Neoephemeridae	4	Southeast to central	Slow to fast moving strams	On or in debris anchored in current; in vegetation	Crawling about substrate; poor swimmers
Potamanthidae	8	East, midwest	Swift streams	On sandy or silty bottoms or under stones	Collectors, filterers
Behningiidae	1	Southeast	Large rivers	Loose sand	Burrowing in sand and swimming; predatory
Ephemeridae	13	Widespread	Streams, rivers, lakes	Sand, silt, or mud	Burrowing in substrate; probably ingest mud
Palingeniidae	2	Central and southeast	Large rivers	Clay banks	Burrowing
Polymitarcyidae	6	Widespread	Large rivers; swift streams	Sand, silt, or mud substrates	In burrows under partly embedded stones or in clay banks; filter feeders
Baetiscidae	12	Widespread	Small to moderate streams; lake margins	Sandy or gravelly substrates	Resting partially buried in sand or silt

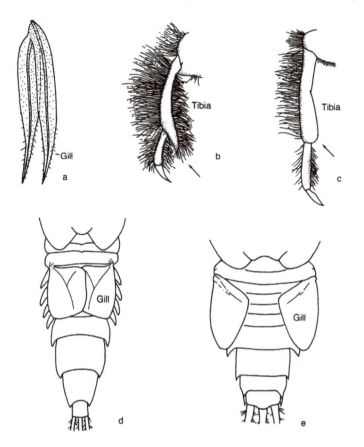

Figure 18.5 Characters of mayfly naiads. a, bifurcate gill of Leptophle-biidae; b, hind tibia of Ephemeridae; c, hind tibia of Polymitarcyidae; d, abdominal gills of Caenidae; e, abdominal gills of Tricorythidae.
Source: a–c, e, adapted from An Introduction to the Aquatic Insects of North America, *3rd ed., by R. W. Merritt and K. W. Cummins. Copyright © 1996 by Kendall/Hunt Publishing Company. Used with permission.*

Mandibles short, without tusks **6**

3(2). Gills with fringed margins (Fig. 18.3b)   **4**
Gills bifurcate, with margins straight, unfringed (Fig. 18.5a)   **Leptophlebiidae**

4(3). Gills dorsal, held above abdomen; fore tibiae broad, flattened   **5**
Gills lateral, held at sides of abdomen; fore tibiae slender, cylindrical   **Potamanthidae**

5(4). Hind tibiae with apex produced into a sharp, angulate point, with a row of spines on the apical margin (Fig. 18.5b)   **17**

Hind tibiae with apex rounded or truncate (Fig. 18.5c)   **Polymitarcyidae**

6(2). Head with dense tufts of setae on anterior corners   **Behningiidae²**
Head without dense tufts of setae   **7**

7(6). Forelegs with dense row of long setae on inner margin; maxillae with tuft of gills at base   **8**
Forelegs without setae arranged in dense row on inner margin; maxillae without gills   **9**

_____

² Known only from naiads in North America.

8(7). Abdomen with dorsal gills on first segment; forelegs with gills on coxae
Siphlonuridae
Abdomen with ventral gills on first segment; forelegs without gills on coxae
Oligoneuriidae

9(7). Abdominal segment 2 with gills subquadrate, meeting at midline of abdomen (Fig. 18.5d)   10
Abdominal segment 2 with gills rounded or triangular, not meeting at midline of abdomen (Fig. 18.5e)   11

10(9). Mesonotum with rounded lobes projecting laterally from anterior corners; metathorax with distinct wing pads
Neoephemeridae
Mesonotum without projecting lateral lobes; metathorax without wing pads   **Caenidae**

11(9). Gills absent from abdominal segment 1, or vestigial, threadlike; sometimes absent from segments 2 or 3   12
Gills present on at least abdominal segments 1 to 5   13

12(11). Gills present on abdominal segment 2, triangular or oval   **Tricorythidae**
Gills absent on abdominal segment 2, sometimes absent from segment 3
Ephemerellidae

13(11). Body cylindrical; eyes or antennae located anteriorly or laterally (Fig. 18.3d)   15
Body dorsoventrally flattened; both eyes and antennae located dorsally (Fig. 18.3c)   14

14(13). Gills with 2 equal filaments; labial palps with 3 segments   **Leptophlebiidae**
Gills unbranched or with slender filament attached near base; labial palps with 2 segments   **Heptageniidae**

15(13). Claws on middle and hind legs long, slender, and without spines; claws on forelegs bifid or with several stout spines
Ametropodidae
Claws on all legs similar in structure   16

16(15). Labium with glossae and paraglossae long, slender (Fig. 18.4d); abdomen with posterolateral angles of segments 8 and 9 prolonged as flattened spines
Siphlonuridae
Labium with glossae and paraglossae short, broad (Fig. 18.4c); abdomen seldom with segments 8 and 9 produced posteriorly (Fig. 18.3d)   **Baetidae**

17(5). Mandibular tusks round in cross section
Ephemeridae
Mandibular tusks with toothed, dorsal keel
Palingeniidae

CHAPTER 19

Order Odonata
(Dragonflies and Damselflies)

ORDER ODONATA

Adult. Medium to large, paleopterous insects with large mobile head, enlarged thorax, two pairs of long, narrow, membranous wings, and slender, elongate abdomen. Head globular, hypognathous, constricted behind into a petiolate neck; compound eyes large, multifaceted; median, and two lateral ocelli present; antennae short, setaceous with 3 to 7 segments; mouthparts mandibulate. Prothorax reduced, mobile; posterior thoracic segments fused into rigid pterothorax with enlarged pleura, reduced sterna, terga; wings similar, with extensive network of veins, apical pterostigma near anterior margin. Abdomen of male with sterna 2 to 3 modified as complex copulatory organs; tergite 10 with unsegmented appendages.

Naiad. Body more robust than in adult, with less mobile head, smaller eyes, and longer antennae; mouthparts mandibulate, with labium modified as jointed, extensile grasping organ; abdomen 10-segmented; respiration in Anisoptera by tracheate lamellae in rectum; Zygoptera with 3 external caudal gills or lamellae.

Their distinctive body configuration makes adult Odonata one of the most readily distinguished groups of insects (Fig. 19.1). From the superficially similar Myrmeleontidae (Neuroptera) they differ in having setaceous antennae (clubbed in Myrmeleontidae) and in numerous other features. The unique prehensile labium, which is thrust rapidly forward to grab prey, differentiates the naiads from all other insects. All Odonata are predatory in all instars—mostly on organisms much smaller than themselves—and are probably important natural control agents of mosquitoes and other aquatic insects. Odonata are commonly associated with aquatic habitats, but as adults many dragonflies are highly vagile, straying many miles into deserts or other arid regions, where they may colonize animal watering troughs or other very limited sources of water. Two broad adaptive modes exist within the order. The more robust dragonflies (Anisoptera) are powerful, active, searching predators. The more frail damselflies (Zygoptera) spend more time perching and darting after suitable prey.

The flight mechanism is in some ways more "primitive" in Odonata than in any other insects. Not only is wing-flexing capability lacking, as in Ephemeroptera, but the flight muscles attach directly to the bases of the wings, with a consequent reduction in the size of the thoracic terga and sterna. The wing venation of extant Odonata is highly modified and very difficult to homologize with other insects. In some Paleozoic fossils, however, the venation is similar to that of contemporary mayflies (Riek and Kukalová-Peck, 1984). Despite

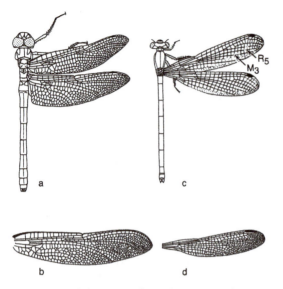

Figure 19.1 Adult dragonflies: a, Anisoptera, Macromiidae (*Macromia magnifica*); b, Petaluridae (*Tanypteryx hageni*); c, Zygoptera, Coenagrionidae (*Argia*); d, Lestidae (*Archilestes*).
Source: Redrawn from Kennedy, 1915, 1917.

their "primitive" thoracic structure, dragonflies evolved independently to become among the most powerful, agile fliers. Aerial efficiency is increased by synchronization of the two pairs of wings, which move slightly out of phase. In the Zygoptera, wing flexing is simulated by the extreme backward tilt of the pterothorax.

Odonata are unique in the development of elaborate secondary male copulatory organs (Fig. 19.3b,c). Prior to courtship the male transfers sperm from the primary genitalia on the ninth abdominal segment to the genital fossa on sternite 3 by curling the abdomen anteroventrally. The cerci are modified as strong clasping organs, used to hold the female by corresponding grooves on the prothorax (Zygoptera) or occipital region of the cranium (Anisoptera). While the courting partners are joined in tandem (Fig. 19.3a), the female bends her abdomen forward to the male genital fossa, where it is grasped by a second set of male claspers. Copulation and insemination then ensue in this "wheel position."

Other adaptive features facilitate efficient predation. In Anisoptera the nearly hemispherical eyes allow almost 360° vision. In Zygoptera the eyes are smaller but are placed at the lateral extremities of the transversely elongate head, maximizing the field of vision. The arrangement of the legs, with the posterior pairs successively longer, creates a basket for sieving prey from the air. The spinose femora and tibiae are useful in holding prey items, and the forward shift of leg attachments allows easy transfer of prey items to the mouth while in flight.

Naiads emerge from the water before the final molt, usually at dawn. Dispersal flights away from the place of emergence succeed metamorphosis in most species. These "maiden flights" may extend over distances of more than a kilometer in the stronger flying Anisoptera or may cover only the few yards to riparian vegetation in the Zygoptera. The first few weeks of adult life, during which sexual maturity and complete coloration develop, may be spent well away from aquatic habitats, but courtship and mating usually occur around water. Male Anisoptera frequently defend territories. Aeshnidae and Corduliidae patrol elongate areas over open water, sometimes flying almost incessantly. Libellulidae, Gomphidae, and Petaluridae usually protect a spot of open water by darting out from a central perch where they spend most of their time. Mid-air clashes are mostly between conspecific males, but other species may be investigated as well. Territoriality is less obvious but also important in Zygoptera (Paulson, 1974).

Both sexes of Odonata are polygamous, and males of some damselflies are able to remove sperm from previous matings before depositing their own sperm (Waage, 1979, 1984). In order to ensure sperm precedence, males of many Odonata maintain the tandem position for long periods after insemination, and in some species the male remains in tandem until the eggs are deposited. In other species the male hovers near the female until after oviposition, driving away competing males. Zygoptera and some Anisoptera (Aeshnidae) oviposit in stems of vegetation; most Anisoptera lack an ovipositor and deposit eggs freely or attach them to submerged vegetation.

Naiads of nearly all Odonata are aquatic, occupying a variety of situations from swift, cold streams to lakes, ponds, and even estuaries.

Respiration is by diffusion through the cuticle, enhanced in Zygoptera by the caudal lamellae, which function as gills. In Anisoptera water is rhythmically pumped in and out of the muscular rectum, where the highly convoluted surface increases gas exchange. Forceful contractions of the rectum are also used to propel the naiad forward in rapid spurts.

Naiads of a few species inhabit unusual situations such as tree holes, and some tropical Zygoptera live in bromeliads or hollow, water-filled, broken stems of plants. Naiads of most Petaluridae dig burrows in bogs. The bottom of the burrow, where daylight hours are spent, is below the water level, but the naiads may venture out of the water while hunting at night. A few other naiads, such as *Megalagrion* (Coenagrionidae) are essentially terrestrial, and many other genera are able to withstand prolonged periods out of water. This relative lack of dependence on water has been interpreted to indicate the Odonata were originally terrestrial as naiads, as in the palaeodictyopteroid orders (see Chap. 14).

Naiads are ambush predators, waiting concealed until prey is close enough to seize or slowly creeping until within range of the prehensile labium. Development ranges from a few months in species that colonize temporary ponds to several years in larger species inhabiting permanent bodies of water. Naiad morphology is broadly correlated with microhabitat. Naiads of Zygoptera are typically slender and shrimplike. They usually frequent aquatic vegetation. Body form in the Anisoptera is variable. Relatively elongate, slender naiads (Aeshnidae, some Gomphidae, Corduliidae, and Libellulidae) mostly frequent vegetation or surfaces of substrates. Burrowing forms tend to be shorter, broader, and often flattened (Petaluridae, Corduligastridae, and most Gomphidae). Long-legged, flattened naiads (most Corduliidae, Libellulidae) usually sprawl on soft bottoms.

The family-level classification of Odonata is relatively stable for extant taxa, although there is some disagreement about relationships of the numerous fossil species. The order and related extinct orders date from the Carboniferous, over 350 mya (million years ago), but the diversification of the modern odonate fauna is likely to be much more recent, as representatives of the extant families do not extend back before the late Jurassic, about 160 mya.

KEY TO THE SUBORDERS OF ODONATA

Adults

1. Hind wings with base broader than in forewings; venation dissimilar (Fig. 19.1a)   **Anisoptera**
   Fore and hind wings similar in shape and venation (Fig. 19.1c,d)   2
2. Eyes separated by more than their width in dorsal view (Fig. 19.1c)   **Zygoptera**
   Eyes separated by less than their width in dorsal view (Oriental region)   **Anisozygoptera**

Naiads

1. Body slender with 3 caudal tracheal gills (Fig. 19.6b)   **Zygoptera**
   Body stout, without external gills (Fig. 19.6a)   2
2. Antennae with 4 or 6 to 7 segments   **Anisoptera**
   Antennae with 5 segments (Oriental region)   **Anisozygoptera**

KEY TO NORTH AMERICAN FAMILIES OF ZYGOPTERA

Adults

1. Wings with 10 or more antenodal crossveins (as in Fig. 19.2b)   **Calopterygidae**
   Wings with 3 or fewer antenodal crossveins 2
2. Veins M_3 and R_5, arising closer to arculus than nodus (Fig. 19.1d)   **Lestidae**
   M_3 and R_5 arising closer to nodus than to arculus (Fig. 19.1c)   **Coenagrionidae**[1]

[1] Protoneuridae, with two uncommon species in southern Texas, will key here. Vein Cu1 is absent or rudimentary and Cu2 short in Protoneuridae. Both veins are well developed in Coenagrionidae.

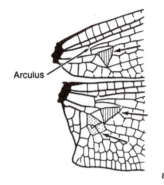

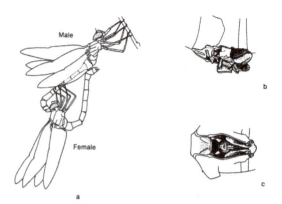

Figure 19.3 Copulation in Odonata: a, tandem pair of *Macromia magnifica* (Macromiidae) in "wheel position"; b, c, secondary male genitalia of *Macromia* (b, lateral aspect; c, ventral aspect).
Source: From Kennedy, 1915.

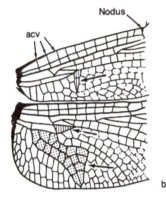

Figure 19.2 Venation at base of wings of dragonflies: a, Gomphidae (*Ophiogomphus*); b, Libellulidae (*Erythemis*). Arrow and stippled pattern indicates anal loops; arrow and vertical lines indicate triangles; acv, antenodal crossveins.
Source: Redrawn from Needham and Westfall, 1955.

Naiads

1. Median and lateral gills similar **2**
   Median gill bladelike, flattened; lateral gills triangular in cross section
   Calopterygidae
2. Median lobe of labium grooved (Fig. 19.5c)
   Lestidae
   Median lobe of labium projecting, not grooved (Fig. 19.5d) **Coenagrionidae[2]**

---

[2] Protoneuridae key here. They differ from Coenagrionidae in having the proximal portion of the gills thickened and darkened.

KEY TO NORTH AMERICAN FAMILIES OF ANISOPTERA

Adults

1. Triangles similar in fore- and hind wings (Fig. 19.2a) **2**
   Triangles dissimilar in fore- and hind wings (Fig. 19.2b) **5**
2(1). Eyes broadly contiguous dorsally (Fig. 19.4a) **Aeshnidae**
   Eyes separated or barely touching dorsally (Fig. 19.4b) **3**
3(2). Labium with median lobe deeply notched **4**
   Labium with median lobe not notched **Gomphidae**
4(3). Pterostigma moderate, intersecting 4 to 5 crossveins **Cordulegastridae**
   Pterostigma elongate, intersecting 6 to 7 crossveins (Fig. 19.1b) **Petaluridae**
5(1). Eyes with posterior margin lobed (Fig. 19.4c) **6**
   Eyes with posterior margin straight (Fig. 19.4d) **Libellulidae**
6(5). Hind wing with anal loop elongate, with two parallel rows of cells (Fig. 19.2b)
   Corduliidae

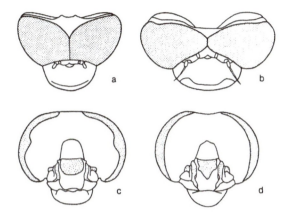

Figure 19.4 Anisoptera heads: a, Aeshnidae (*Anax junius*); b, Cordulegastridae (*Cordulegaster dorsalis*); c, Corduliidae (*Cordulia shurtleffi*); d, Libellulidae (*Libellula nodisticta*); a, b, dorsal views; c, d, posterior views.

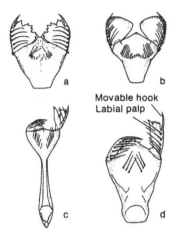

Figure 19.5 Ventral aspects of labia of naiads of Odonata: a, Cordulegastridae (*Cordulegaster*); b, Libellulidae (*Tarnetrum*); c, Lestidae (*Lestes*); d, Coenagrionidae (*Argia*). *Source: a, b from Musser, 1962; c, d, from Kennedy, 1915.*

Hind wing with anal loop short, rounded, with cells not arranged in 2 parallel rows (Fig. 19.2a)   **Macromiidae**

Naiads

1. Labium strongly convex, protecting anteroventrally   **2**
   Labium flat   **5**
2(1). Lateral lobes of labium with large, irregular, interlocking teeth (Fig. 19.5a)   **Cordulegastridae**
   Lateral lobes of labium with regular dentition; teeth usually small, not interlocking (Fig. 19.5b)   **3**
3(2). Head with prominent, pyramidal horn or tubercle; squat naiads with very long legs   **Macromiidae**
   Head bare or sometimes with small tubercle; body form variable   **4**
4(3). Cerci nearly as long as median anal spine (epiproct); lateral spines on abdominal segment 9 less than one-fourth length of segment   **Corduliidae**
   Cerci usually about one-half to three-fourths length of epiproct; if longer, lateral spines of segment 9 more than one-fourth length of segment   **Libellulidae**
5(1). Antennae with 4 segments; fore and middle tarsi with 2 segments   **Gomphidae**

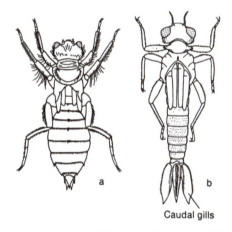

Figure 19.6 Naiads of Odonata: a, Anisoptera, Libellulidae (*Libellula*); b, Zygoptera, Coenagrionidae (*Argia*).

Antennae with 6 or 7 segments; tarsi with 3 segments   **6**
6(5). Antennae with segments hairy, wider than long   **Petaluridae**
   Antennae with segments bare, longer than wide   **Aeshnidae**

Suborder Zygoptera

Coenagrionidae. The dominant family of Zygoptera, especially in temperate regions. Naiads inhabit a variety of situations, especially sluggish or static waters, where they clamber about on vegetation. Protoneuridae occur along streams.

Lestidae. Relatively large damselflies that frequent marshlands and swamps; bodies usually metallic blue, green, or bronze, wings transparent. Eastern North America to the Pacific Coast.

Calopterygidae. Primarily a tropical family, with only two genera in North America; bodies variably colored, wings usually pigmented, at least basally; naiads inhabit riffles in streams and creeks, eastern North America to Pacific Coast.

Suborder Anisoptera

Aeshnidae. Cosmopolitan; possibly the most familiar dragonflies to nonentomologists; adults strong-flying, sometimes migratory, and found far from water. Naiads occupy a variety of habitats, including brackish water.

Gomphidae. Cosmopolitan, medium to large in size, found most abundantly about moving water; the enlarged abdominal apex is a useful diagnostic character for adults; naiads usually burrow or hide in debris.

Petaluridae. Large, dull-colored, clear-winged species that mostly frequent bogs (where the naiads burrow) or, less frequently, streams; a relictual family with only about a dozen species, distributed mostly in temperate regions. Two species in North America.

Cordulegastridae. A small family of medium to large dragonflies, dark brown to black with yellow striping; adults frequent brooks and streams, where the naiads inhabit bottom debris. Holarctic and Indomalaysian, 25 species.

Corduliidae and Macromiidae. Widespread and abundant, especially in north temperate regions; adults mostly metallic blue or green with spots, bands; naiads usually occur in vegetation in a variety of aquatic habitats, including streams, ponds, and swamps.

Libellulidae. A dominant family, especially in the tropics; in cross section the abdomen is triangular; cuticle usually nonmetallic, but often powdered with silvery pruinosity; wings frequently banded or clouded; naiads stout, usually inhabiting vegetation or bottom debris in still water.

Suborder Anisozygoptera. Includes only two species from Japan and India, but Mesozoic fossils are numerous. Wing venation of adults is similar to Zygoptera, while the naiad resembles Anisoptera.

CHAPTER 20

Order Blattodea (Including Former Isoptera): Cockroaches and Termites

ORDER BLATTODEA (EXCEPT ISOPTERA, TREATED SEPARATELY BELOW)

Small to large Exopterygota with broad, flattened body, strongly hypognathous head. Mouthparts mandibulate, generalized; compound eyes large, ocelli developed as two "ocelliform spots"; antennae long, multiarticulate. Prothorax large, mobile, with shield-like notum; meso- and metathorax similar, rectangular, slightly smaller than prothorax. Wings with longitudinal veins usually much branched, connected by numerous cross-veins; forewing sclerotized as a protective tegmen, with anal lobe undifferentiated; hind wing membranous, with large, fan-shaped anal lobe that is folded by longitudinal pleats at rest. Legs robust, long, often spiny; coxae very large, tarsi 5-segmented. Abdomen with 10 distinguishable segments; tergite 10 bearing cerci of 1 to many segments; tergites 5 to 6 and sometimes sternites 6 to 7 bearing openings of scent glands; ovipositor of 3 pairs of small internal valves; male genitalia complex, strongly asymmetrical. Nymphs similar to adults except in development of wings, genitalia, and sometimes color and integumental texture.

ISOPTERA (BLATTODEA)

Neopterous Exopterygota living socially in small to very large colonies. Individuals small to moderate in size, with subcylindrical, weakly sclerotized body. Head hypognathous or prognathous with mandibulate mouthparts, 10 to 32 segmented moniliform antennae; compound eyes and sometimes 2 ocelli present in reproductives, usually absent in soldiers and workers; frons almost always bearing conspicuous frontal pore or fontanelle, through which discharges frontal gland. Thorax with subequal segments, usually with lightly sclerotized tergites, membranous sterna; reproductives with 2 pairs of similar wings with simple, longitudinal venation or extensive network of crossveins; wings broken at humeral suture and lost after dispersal flight; soldiers, workers, and nymphs without wings; legs short, adapted for walking. Abdomen with 10 similar segments and reduced eleventh segment; terminal segment bearing 1 to 8 segmented cerci.

BLATTODEA
(EXCLUDING ISOPTERA)

The insects treated in this chapter and the next are often grouped into a superorder, Dictyoptera, now widely accepted as a monophyletic lineage. In some older classifications, Dictyoptera is even treated formally as an order, grouping cockroaches, termites, and mantids into a morphologically somewhat coherent but biologically diverse assemblage.

Within this assemblage of three orders, interordinal relationships have been controversial. Recently, however, mounting morphological and molecular evidence has converged on a phylogeny that subsumes termites (formerly treated as the order Isoptera) within the cockroaches (Blattodea or Blattaria), as long suspected by some biologists on comparative natural history grounds (see below). Here we treat the termites together with the cockroaches but retain separate diagnoses for the two groups, and the more traditional family-level classifications, until a new consensus emerges concerning which families to recognize. Inward et al. (2007), in formally sinking the Isoptera into the Blattodea, suggested placing all termites in the single family Termitidae. This reasonable but relatively new suggestion has not yet met with wide acceptance among termite specialists.

The Blattodea in the strict sense (without counting the roughly 2300 species of termites) constitutes a relatively small order at present, with about 4000 described species, but comprised the dominant group of Neoptera in the upper Carboniferous. About 30 species of cockroaches are cosmopolitan inhabitants of man's dwellings, placing them among the most familiar insects to nonentomologists. Domiciliary species are commonly regarded with disgust because of their invasions of cupboards and pantries, which they spot with their feces and, depending on the species, may permeate with the musty-smelling secretions from their abdominal glands. Cockroaches have not been definitely implicated in transmission of human diseases, and they neither bite nor sting. Yet they must be considered of potential medical importance because of the large number of pathogenic microorganisms, including bacteria, viruses,

protozoa, and parasitic worms, that have been isolated from their bodies or excrement. It may be pointed out that the same characteristics that adapt species such as *Periplaneta americana* (Blattidae) to life in dwellings also produce a hardy laboratory animal. Cockroaches have been important subjects for the investigation of hormonal control of insect growth and development and especially for the study of insecticidal mode of action and insecticide resistance.

Cockroaches are among the first neopterous insects to appear in the fossil record and are extremely generalized in most morphological features. The Blattodea (and other dictyopterans) share a number of features with Orthoptera, such as mandibulate mouthparts, a common pattern of wing venation, and the presence of cerci. However, they differ from Orthoptera in several profound characters, the most important probably being the mode of oviposition and the type of flight musculature. All modern cockroaches (although not all fossil species) produce characteristic oothecae (Fig. 20.1), consisting of a double-layered wrapper protecting two parallel rows of eggs. This habit extends to the Mantodea, and the egg arrangement occurs in the primitive termite *Mastotermes* (see below), although the oothecal covering is absent.

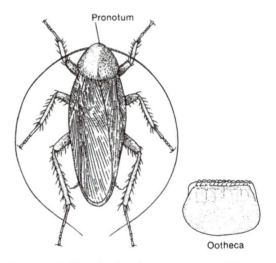

Figure 20.1 Blattidae, *Periplaneta americana*: adult female and ootheca.
Source: From Patton, 1931.

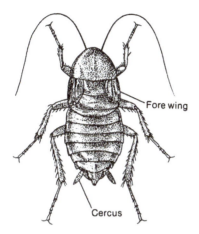

Figure 20.2 Blattidae, *Blatta orientalis*.
Source: From Patton, 1931.

In contrast, Orthoptera lay eggs singly or in pods, never with an ootheca. In most neopterous insects, including Orthoptera, wing movements are powered almost entirely by indirect flight muscles. In Blattodea, Mantodea, and Isoptera the indirect muscles are very small, the downstroke of the wing apparently being effected by basalar and subalar muscles (Snodgrass, 1958). These differences, together with the long period of separate evolution indicated by fossils, suggest that it is appropriate to recognize dictyopterans as a distinct group of orthopteroid insects.

Most of the familiar household cockroaches are nocturnal, positively thigmotactic crevice dwellers. Tropical species, comprising the bulk of the world fauna, include diurnal species, often brightly colored, and sometimes arboreal. North American species of *Arenivaga* (Polyphagidae) are burrowers in sand or litter, and *Attaphila fungicola* (Polyphagidae), less than 3 mm long when mature, is a symbiont of the leaf-cutting ant *Atta fervens* in Texas. Some tropical species are known to be amphibious and some in Australia are cavernicolous. Most species are solitary or simply gregarious, but *Cryptocercus punctulatus* (Cryptocercidae), the wood roach, occurs subsocially in galleries in punky, rotting wood. In this species, as in termites, wood is digested with the aid of symbiotic flagellate protozoa, which are lost

after each molt. The defaunated nymphs become reinfested by consuming fecal pellets of adults or older nymphs.

Sex pheromones have been isolated from several species and are probably present in all. Courtship by males, involving stridulation or posturing, normally precedes copulation. Males also produce a secretion from the abdominal tergites, which induces the female to mount the male just before copulation. After genitalic contact the pair turn tail to tail. Copulation, during which a spermatophore is formed and transmitted by the male, is usually of considerable duration. Most species exhibit both sexes, but some, including *Periplaneta*, may reproduce parthenogenetically.

Oothecae are quite variable in structure, enabling recognition to species of those of household pests. Species also differ in the length of time the ootheca is retained by the mother, in its placement upon deposition, in the number produced by a single female, and in the number of eggs contained. For example, the American roach deposits an average of 50 oothecae per female, each containing about 12 to 14 eggs, while the German roach, *Blattella germanica* (Blattellidae), averages only five oothecae, each containing about 40 eggs. The American roach cements the oothecae to objects in the environment, while the German roach retains each ootheca until shortly before the eggs hatch. Cockroaches that deposit oothecae long before eclosion usually attempt some form of concealment. Some species tuck the ootheca into cracks or crevices, while in others dirt or debris may adhere to its outer surface.

Members of the family Blaberidae and a few Blattellidae are ovoviviparous or viviparous. In most ovoviviparous species an ootheca is formed, then retracted into a brood sac where the eggs develop. The oothecal wall is usually membranous rather than sclerotized. In a few genera no ootheca is formed, the eggs moving directly to the brood sac. In these ovoviviparous species the eggs are produced with sufficient yolk to complete development, needing only water from the mother. The blaberid genus *Diploptera* is strictly viviparous, producing small eggs with little yolk that depend on the mother for nutrition.

Embryonic and nymphal development is relatively slow, ranging from a few months in smaller species to over a year in larger ones. Food of domestic species is extremely generalized, but many wild species are difficult to rear, especially in early instars. Adult roaches are often long-lived, some species surviving 4 years or more under laboratory conditions.

The North American fauna, numbering about 50 species, includes about a dozen important domiciliary species, including *Blattella germanica* (German roach) and *Supella longipalpa* (brown-banded roach), both species in Blattellidae, and *Blatta orientalis* (Oriental roach) and *Periplaneta americana* (American roach), both species in Blattidae (Figs. 20.1, 20.2). Only *Parcoblatta* (Blattellidae) among native roaches is of any economic significance. The pest roach species are often inaccurately named with respect to their geographical origin; for instance, *Periplaneta americana*, the so-called American roach, apparently originated in Africa (Cornwell, 1968).

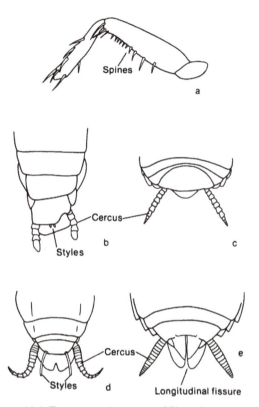

Figure 20.3 Taxonomic characters of Blattodea: a, anterior leg, Blattellidae (*Blattella*); b–e, ventral aspects of external genitalia; b, Blattellidae, male (*Blattella*); c, Blatellidae, female; d, Blattidae, male (*Periplaneta*); e, Blattidae, female.

KEY TO COMMON NORTH AMERICAN FAMILIES OF BLATTODEA

1. Body length less than 3 mm; eyes very small; symbiotic in nests of *Atta* (Formicidae)   **Polyphagidae** (*Attaphila*)
   Body length greater than 3 mm; eyes almost always large; not in ant nests   **2**

2(1). Middle and hind femora with numerous stout spines on posterior margin   **3**
   Middle and hind femora with one or two apical spines or lacking spines (hairs may be present)   **5**

3(2). Fore femora with posterior row of similar spines   **4**
   Fore femora with spines long and stout proximally, shorter and slender distally (Fig. 20.3a)   **Blattellidae**

4(3). Female with last visible sternal plate divided by a longitudinal fissure (Fig. 20.3e); male styli slender, long, straight (Fig. 20.3d)   **Blattidae**

Female with last visible sternite not divided (Fig. 20.3c); male with last sternite and styli usually asymmetrical or unequal in size (Fig. 20.3b)   **Blattellidae**

5(2). Abdomen with seventh tergite and sixth sternite expanded as plates that conceal abdominal apex, including cerci; body elongate, parallel-sided   **Cryptocercidae**
   Abdomen with eighth to tenth segments and cerci exposed; body oval or broadly oval   **6**

6(5). Fore femora with 1 to 3 spines on posterior margin   **Blaberidae**
   Fore femora without spines on posterior margin   **7**

7(6). Frons flat; wings with anal region folded in fanlike pleats at rest; at least 16 mm long; pale green   **Blaberidae**

Frons bulging, swollen; wings with anal region folded flat against preanal portion at rest (females frequently wingless); less than 16 mm long; brown, tan, or gray   **Polyphagidae**

Blattellidae. A large, worldwide family of small to large, mostly slender-bodied, fast-moving roaches. The most common introduced domiciliary pests (*Blattella, Supella*) and most of the roaches commonly encountered outdoors in North America are blattellids.

Blattidae. Primarily tropical roaches of medium to large size, represented in North America almost entirely by introduced domiciliary species (*Periplaneta, Blatta, Neostylopyga*).

Polyphagidae. A relatively small family distributed mostly in arid and semiarid regions. North American genera (*Arenivaga, Eremoblatta, Compsodes*) are almost entirely restricted to the arid Southwest. Males are winged, while females are apterous. *Attaphila texana,* which is commensal in the nests of leafcutter ants, is usually placed in Polyphagidae, sometimes in Blattellidae.

Blaberidae. This large, primarily tropical family contains ovoviviparous and viviparous species exclusively. North American representatives are tropical species that have usually become established only near ports of entry (e.g., *Leucophaea,* the Madeira cockroach, in heated buildings in New York City) or in the subtropical areas of southern states (*Blaberus, Nauphotea, Panchlora*).

Cryptocercidae. This is the phylogenetically crucial family of Blattodea closest to the termites, with two species in China and one in North America. *Cryptocercus punctulata* occurs in moist, rotten wood in the Pacific Northwest and the Appalachians. The roaches occur as family groups of a male, female and their offspring. Adults are apterous.

TERMITES (FORMERLY TREATED AS ORDER ISOPTERA)

Termites (Fig. 20.4) constitute a relatively small but ecologically dominant lineage of social insects, with about 2300 described species. All termites are specialist feeders on cellulose, the most abundant organic compound in terrestrial habitats. Because of this a large part of the stored energy in an

Figure 20.4 Workers of the fungus-growing termite, *Macrotermes* sp. (Termitidae), foraging on leaf mulch in Tanzania.

ecosystem is available to them. Wood of living trees is usually avoided, but sound or decaying deadwood, twigs, fresh and dead grass, leaves, seeds, humus, and dung are consumed. Termites may be pests of agricultural crops and forestry. In tropical regions, termites are ecological analogs of earthworms in the temperate zone, playing a crucial role in soil development, drainage, and erosion.

North American species are of retiring habits, seldom venturing out of their concealed galleries, with the exception of the reproductives, which disperse aerially prior to founding new colonies. Yet most nonentomologists are familiar with termites because of their nearly worldwide distribution and especially because of their proclivity to invade and consume manmade wooden or paper objects. They also damage dried food, fabrics, rubber, hides, wool, linoleum, and insulation materials. Termites are economically important everywhere except extremely cold regions, and in parts of the

tropics and subtropics may be among the most destructive arthropods. In northern Australia *Mastotermes* (Mastotermitidae) is considered the single most important insect pest. These termites form immense underground colonies from which they rapidly invade manmade structures as well as dead and living trees and crop plants. Aside from their economic importance, termites are of general biological interest because of their highly developed social behavior (see also Chap. 7).

Termites are highly adapted for living in densely populated colonies, apparently having evolved by the early Mesozoic Era or even earlier. Yet many characteristics, especially of the earliest diverging families, are shared with cockroaches (Blattodea). In particular, the termite family Mastotermitidae shares many ancestral similarities with the roach family Cryptocercidae (Table 20.1). The Mastotermitidae is now represented by a single species, *Mastotermes darwiniensis*, in northern

Table 20.1 Comparison of Selected Characters of "Isoptera" and Blattodea s. s.

	"Isoptera"		Blattodea	
	Isoptera except Mastotermitidae	Mastotermitidae	Cryptocercidae, Panesthiidae	Blattodea except Cryptocercidae
No. of antennal segments	10-22	20-32	>32	>32
No. of tarsal segments	3-4	5	5	5
Wing venation	Reticulate in Kalotermitidae, Hodotermitidae, reduced in Termitidae	Reticulate	Reticulate	Reticulate
Hind wing anal lobe	Absent	Present	Present	Present
Humeral suture	Present in both wings	Present in forewings; absent in hind wings	Wings absent in Cryptocercidae; sutures present in Panesthiidae	Absent in both wings
Female external genitalia	Absent or vestigial	Short ovipositor with 3 pairs of valves	Short ovipositor with 3 pairs of valves	Short ovipositor with 3 pairs of valves
Egg deposition	Laid separately	Laid in pods	Contained in oothecae	Contained in oothecae
Intestinal symbiotes	Flagellate protozoa (spirochaetes in Termitidae)	Flagellate protozoa	Flagellate protozoa in Cryptocercidae; amoebae in Panesthiidae	Bacteria
Bacteriocytes	Absent	Present	Present	Present
Social organization	Eusocial	Eusocial	Primitively subsocial	Nonsocial

Australia, but *Mastotermes*-like fossils from the Mesozoic are widespread. The most convincing similarities, perhaps, are the shared features of wing structure, egg deposition, and intestinal symbionts. The woodroach, *Cryptocercus*, harbors 21 to 22 species of hypermastigote and flagellate protozoa. All of these are members of families that occur in primitive families of termites, including the shared genus *Trichonympha*. The nymphs of *Cryptocercus*, which are not unlike termite workers in general appearance, occur gregariously with adults in galleries in rotting wood. Another similarity (not listed in Table 20.1) is the presence of relatively large direct pleurosternal muscles and only small, poorly developed indirect dorsal longitudinal flight muscles in both roaches and termites.

Termites, like social Hymenoptera, are characterized by the development of morphologically differentiated castes, which perform different biological functions. Least modified morphologically are the reproductives (Fig. 20.5), which have relatively large eyes, unmodified mandibulate mouthparts, and a sclerotized body with distinct thoracic and abdominal tergites. Primary reproductives always bear wings initially, later breaking them at a basal fracture zone (humeral suture) by pressing the extended wings firmly against the substrate. Males do not change in external appearance after shedding their wings, but the abdomens of females gradually become swollen, or physogastric, from the enlargement of the ovaries. Physogastry is especially evident in the more specialized termites of the families Termitidae and Rhinotermitidae, in which the abdominal sclerites are eventually isolated as small islands on the bloated, membranous abdomen. Many species have the capacity to develop substitute kings and queens. These are neotenic individuals that never develop wings, do not become so physogastric as the primary reproductives, and differ in other minor morphological features.

The worker caste (Fig. 20.6a) is distinguished mostly by regressive characteristics. Compound eyes are usually absent, the number of antennal segments is usually smaller than in other castes, and the body is very lightly sclerotized except in a few species that actively forage outside their galleries. In Kalotermitidae and Hodotermitidae (such as *Zootermopsis* in North America), the worker caste is absent, being functionally replaced by nymphal reproductives, or pseudergates. Workers or pseudergates are the most abundant caste, performing

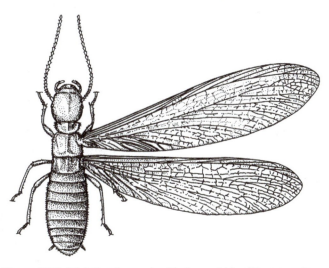

Figure 20.5 Adult female termite, Hodotermitidae (*Zootermopsis nevadensis*).
Source: From Banks, 1920.

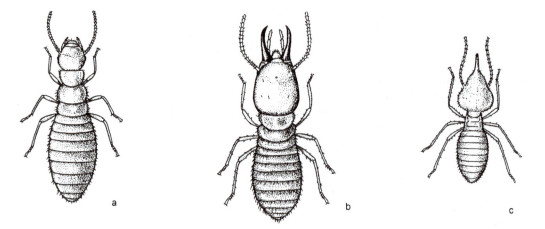

Figure 20.6 Variation in body form of termites: a, worker, Rhinotermitidae (*Prorhinotermes simplex*); b, soldier, Rhinotermitidae (*Prorhinotermes simplex*); c, soldier, Termitidae (*Nasutitermes costaricensis*). *Source: From Banks, 1920.*

foraging, care of young nymphs, feeding of soldiers and reproductives, and nest construction and maintenance.

Soldiers (Figs. 20.6b,c) are characterized by specialization of the head for defense of the colony, especially against ants. The mandibles are usually enlarged as long, slender, scissorlike blades, sometimes with grotesque serrations, ridging, or twisting. In some Termitidae the mandibles of soldiers are atrophied; the head is drawn into a nozzle-shaped projection bearing the opening of the frontal gland. These nasute soldiers deter invaders by spraying or exuding repellent or entangling substances. The proportion of soldiers varies from none in some species of Termitidae and Kalotermitidae to over 15 percent in *Nasutitermes* (Termitidae). In some species two classes of soldiers may occur, differing in structure as well as size.

Social organization in Isoptera is extremely complex and variable (see also Chap. 7). In general, the more biologically "primitive" species (Kalotermitidae, Hodotermitidae) excavate relatively small nests directly in logs, timbers, or other food material. For example, in *Incisitermes minor* (Kalotermitidae), which infests dry wood in California, mature colonies number only a few thousand individuals, including about 8 to 10 percent soldiers. These primitive termites do not construct

passageways or forage away from the nest, thereby limiting the potential size and age of the colony by the size of the initial food source. *Mastotermes*, generally considered the earliest distinct lineage of living termite, is exceptional in constructing massive, densely populated subterranean colonies.

Most members of the families Rhinotermitidae and Termitidae construct subterranean galleries from which the workers forage for food. The size of such colonies varies considerably, but frequently they are very large, containing millions of individuals. In many species the mature termitarium includes a portion that extends above the ground surface as a dome or spire that may reach 7 m in height. Such structures, which may be inhabited for 50 years or more, consist of an ensheathing hard layer of soil particles cemented together with excrement or glandular secretions, enclosing a honeycomb of living chambers (Fig. 20.7). Covered passageways are extended to food sources up to 100 m from the termitarium. In *Macrotermes, Nasutitermes* (Termitidae), and a few other genera, workers forage in the open at night or on cloudy days, collecting grass, pine needles, and similar materials. These are cut into short lengths and carried to storage chambers in the nest. Some members of the Termitidae construct enclosed, carton-type nests on trees, without any direct contact with the

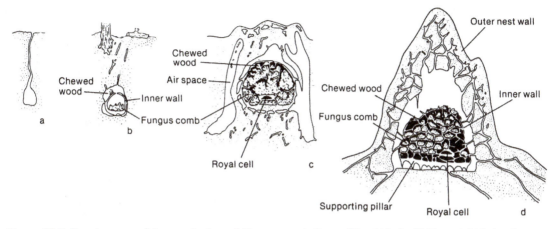

Figure 20.7 Development of the termitarium of *Macrotermes bellicosus* (Termitidae) of Africa; a, initial chamber excavated by founding couple; b, c, intermediate stages of development; d, structure of mature nest. *Source: From Grassé and Noirot, 1958..*

ground, and a few species ("dry wood termites") live in and consume the seasoned timbers in buildings without any contact with the soil. In general, however, some connection with the ground is necessary for the colony to obtain moisture.

Termitaria not only provide protection from predators but stabilize the temperature and humidity within the nest. The Old World subfamily Macrotermitinae (Termitidae) has capitalized on the homeostatic climate within the nest to grow fungi (*Termitomyces*), which are used as food, in a manner analogous to fungus culture by New World attine ants. The fungus is grown on spongy combs constructed of the termites' feces. The fungal combs are continually eaten, together with fresh woody material, and replaced. Lignin in the feces is broken down by the fungus, freeing additional cellulose, which the termites are able to utilize. In other species cellulose is obtained from various sources, including wood or bark, dry grass or leaves, or highly organic soils. The dry dung of ungulates commonly harbors termites, particularly in arid regions. Actual digestion of the cellulose is accomplished by the microorganisms in the gut.

Worldwide, most of the economically important termites are members of the Termitidae. In parts of Africa and Asia, damage to neglected structures occurs with remarkable rapidity. In

temperate regions, where Termitidae are rare (14 species in North America), Rhinotermitidae cause more damage, principally by invading foundation timbers from their subterranean nests.

As in social Hymenoptera, the life history of termites involves dispersal of reproductives from established colonies. Dispersal occurs at characteristic seasons for each species and sometimes at characteristic times of day. For example, *Incisitermes minor*, discussed earlier, flies during warm, dry periods, whereas *Reticulitermes hesperus* (Rhinotermitidae) flies after the first fall rains in California, usually early in the morning. Dispersal flights of a species are often simultaneous over a district and cover variable distances. A few species are efficient colonizers of islands and other isolated habitats, but sexuals of most species probably fly no more than a few hundred meters. After returning to earth, the unmated males and females dehisce their wings, then join in pairs, which almost always exhibit tandem running, the male following the female (Chap. 7). Founding pairs of reproductives initiate colonies in suitable locations in soil, logs, or other substrates. Colony growth is slow at first. After the first year a typical colony may number only a dozen workers and the primary reproductives. The first soldiers appear in 2 or 3 years, and winged dispersal castes are not produced for 4 years or more. Eventually, in species with large colonies egg production is almost continuous, and large, physogastric queens may lay 2000 to 3000 eggs per day. Individuals of all castes develop slowly and survive for considerable periods. The ultimate instar is usually achieved after 7 to 10 molts, requiring a year or more. Workers and soldiers live up to 4 years; reproductives may survive 15 years or more.

Termite colonies and those of ants are exploited by a variety of other organisms. Specialist vertebrate predators include Central and South American anteaters and armadillos; African aardvarks, aardwolves, bat-eared foxes, meerkats, and pangolins; and Australian echidnas and marsupial anteaters. Termites also are eagerly eaten by generalist predators such as birds, reptiles, frogs, and other insects, especially ants. Predators gorge themselves during the mass emergence of the reproductives from a colony, leaving few survivors to establish new nests. Humans also collect and eat the workers and reproductives.

Staphylinid beetles are particularly well represented among insect associates or termitophiles (Kistner, 1979, 1982). Some termitophiles consume eggs or young nymphs of their host, while others are nest scavengers. The latter frequently show considerable structural modification, with special secretory areas that are attractive to the termite or with the body modified to resemble the contour or texture of the host.

Knowledge of the diversity of Isoptera has attained a relatively mature state. New genera are infrequently discovered, and no new species have been described from North America since 1941. The family Termitidae is dominant, with about 1400 species, mostly tropical in distribution. Only 14 Termitidae occur in North America, all restricted to the Southwest. The Kalotermitidae contains 16 North American members, including several economically important species. Hodotermitidae are represented by three species and Rhinotermitidae by nine species, including *Reticulitermes*, the most destructive genus in North America. In most of the northern and central United States only a single species of *Reticulitermes* is present. Details of the type of damage caused by various termites and

methods of control are found in texts dealing with economically important insects.

Workers are extremely similar and difficult to distinguish and are not keyed here. All termites should be stored in liquid preservatives to prevent shriveling of their bodies. Separation of termite families is often difficult without a good stereomicroscope for seeing whether a fontanelle is present or not.

KEY TO NORTH AMERICAN FAMILIES OF ISOPTERA

Reproductives (Wings or Wing Bases Present)

1. Head with median dorsal pore or fontanelle; wings with 2 strong and several faint veins, without crossveining **2**
   Head without fontanelle; wings with at least 3 strong veins, usually with extensive crossveining **3**
2(1). Pronotum flat; portion of dehisced wing longer than pronotum **Rhinotermitidae**
   Pronotum concave, saddle-shaped; basal portion of dehisced wing shorter than pronotum **Termitidae**
3(1). Head with ocelli **Kalotermitidae**
   Head without ocelli **Hodotermitidae**[1]

Soldiers

1. Head produced anteriorly into a hornlike projection (Fig. 20.6c); mandibles vestigial **Termitidae**
   Head without hornlike projection; mandibles large **2**
2(1). Mandibles with one or more large, marginal teeth **4**
   Mandibles without marginal teeth **3**
3(2). Head longer than broad, with median dorsal pore or fontanelle Rhinotermitidae
   Head about as long as broad, without fontanelle Kalotermitidae
4(2). Mandibles with 1 prominent marginal tooth; fontanelle present; head narrowed anteriorly Termitidae

CHAPTER 21

Order Mantodea (Mantids)

ORDER MANTODEA

Moderate-sized to very large Exopterygota adapted for predation. Head small, triangular, freely mobile on slender neck; mouthparts mandibulate; eyes large, lateral; ocelli usually 3 or none; antennae slender, filiform, multiarticulate. Prothorax narrow, elongate, strongly sclerotized; meso- and metathorax short, subquadrate; forewings narrow, usually cornified with reduced anal region; hind wings much broader, membranous, with large anal region; venation of largely unbranched longitudinal veins, usually with numerous crossveins; wing folding of longitudinal pleats; wings frequently reduced, especially in females; forelegs raptorial, with very long coxae; middle and hind legs long, slender, adapted for walking; tarsi 5-segmented. Abdomen with 10 visible segments, weakly sclerotized, slightly flattened dorsoventrally; segment 10 with variably segmented cerci; ovipositor short, mostly internal; male genitalia bilaterally asymmetrical. Nymphs similar to adults except in development of wings and genitalia.

Mantids are primarily tropical in distribution but are among the more familiar insects to nonentomologists everywhere because of the large size and striking appearance of many species. The common mantid pose, with the anterior part of the body elevated and the forelegs held together as if in prayer, has inspired folk beliefs since ancient times and the common name, praying mantis. In reality, this is the posture assumed for ambushing prey.

Mantids show a high morphological resemblance to cockroaches (Blattodea). Important similarities include the strong development of the direct pleurosternal flight muscles (see Chap. 20) and relatively small size of the indirect longitudinal muscles, 5-segmented tarsi, multisegmented cerci, and strongly asymmetrical male genitalia. The eggs are always enclosed in an ootheca, which differs in details but is generally similar to that of cockroaches. These features strongly differentiate roaches and mantids from Orthoptera, with which

they are sometimes classified as a single order. The features that distinguish mantids from roaches are nearly all related to their predatory mode of life. Most obvious is the modification of the forelegs as strong grappling hooks for snaring arthropod prey (Fig. 21.1a). The elongate thorax and coxae increase the reach, while the sharp femoral and tibial teeth secure struggling victims. The long, slender posterior pairs of legs elevate the body, allowing unimpeded vision. The large eyes are situated far apart, as in other visually orienting predators such as damselflies, and the mobile head can be cocked in any direction. The cryptic coloration, usually in shades of green, brown, or gray, is a further adaptation for the sedentary mode of predation practiced by these insects.

In temperate regions adult mantids appear late in summer, when they are seen on vegetation, tree trunks, and other perches. As predators they are generalists, consuming any arthropod small enough

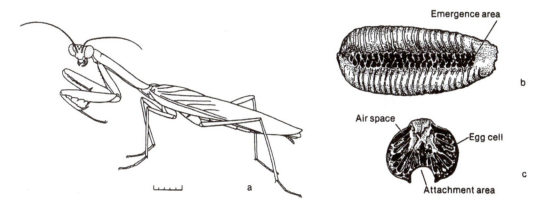

Figure 21.1 Mantodea: a, *Stagmomantis carolina*, male (scale equals 5 mm); b, ootheca, dorsal (diagrammatic); c, ootheca, transverse section (diagrammatic).
Source: b, c from Breland and Dobson, 1947.

to subdue, including smaller mantids. Larger species have occasionally been seen feeding on small frogs or lizards. Large mantids may be quite pugnacious, readily unfurling the hind wings, which may be brightly colored, and adopting threatening postures. All mantids are "sit-and-wait" predators that rely on camouflage to get within reach of prey, as mentioned above. Tropical species may be gaudily colored with foliate appendages and specially textured or sculptured cuticle creating a strong overall resemblance to flowers, leaves, bark, or other plant parts.

Sex pheromones have been found in females of some species. Copulation may be preceded by courtship, allowing the male to reduce the chances of attack as he approaches the larger female. During copulation the male sits astride the female with his abdomen curled ventrally. Sperm is transferred by a spermatophore. In many species the female attacks and partly or wholly consumes the male during or immediately after copulation, but this may be partly an artifact caused by confined spaces of captivity.

Gravid females become heavily swollen by the large mass of developing eggs. At the time of oviposition the female constructs a characteristic ootheca (Fig. 21.1b,c) from the frothy secretion from the accessory glands. This secretion quickly hardens into a protective envelope around the eggs, which are aligned in rows, four abreast. Structural

details of oothecae, as well as site of attachment, vary greatly among species. Each female normally lays several to many egg masses. Females of a few species protect the ootheca until the nymphs have emerged. In temperate regions, overwintering usually occurs in the egg stage.

As in cockroaches, the first-instar nymph is a nonfeeding stage that quickly molts. Succeeding instars vary in number among different species from 3 or 4 to 12. There is usually a single generation per year.

Although they are commonly considered beneficial insects and their oothecae are often sold in nurseries, mantids are generalist predators that consume both desirable and undesirable species alike. In any case, natural populations seldom if ever become dense enough to have much effect on most pests.

A single family, Mantidae, occurs in North America. The most familiar species is the Carolina mantid, *Stagmomantis carolina*. Other conspicuous forms include the Chinese mantid, *Tenodera aridifolia*, and the praying mantid, *Mantis religiosa*, both introduced into Eastern United States. In the western states the small, wingless gray or brown species of *Litaneutria* are not infrequently encountered on the ground or on low vegetation. The world fauna comprises approximately 2000 species in eight families.

CHAPTER 22

Order Grylloblattodea
(Grylloblattids)

ORDER GRYLLOBLATTODEA

Elongate, cylindrical, or slightly flattened Exopterygota of moderate size. Antennae multiarticulate, filiform, about as long as thorax; compound eyes reduced or absent, ocelli absent; mouthparts mandibulate, prognathous. Prothorax subquadrate, larger than mesothorax and metathorax; wings absent; legs long, slender, adapted for walking, tarsi 5-segmented. Abdomen 10-segmented, with filiform, 8- or 9-segmented cerci; female with straight, short ovipositor. Nymphs similar to adults.

Grylloblattids comprise one of the smallest orders of insects, with 26 species in 5 genera of the family Grylloblattidae occurring in boreal areas of North America and eastern Asia. Collections have been made from as low as 300 m elevation in central Japan and 500 m in northern California, but most specimens are found at high altitudes under cool, moist conditions, frequently near snowfields from central California to western Canada and Wyoming and in Japan, China, Korea, and Siberia. Species in Japan and Korea have been found in caves. Activity occurs nocturnally or diurnally at low temperatures, and grylloblattids scavenge on the surface of snow or ice, as well as around snowmelt or moist surfaces in or near rockslides or under rotting logs. Jarvis and Whiting (2006) state that *Grylloblatta* is so strongly cryophilic that it can be killed simply by the heat of a collector's hand. It is possible that many grylloblattids are under long-term threat from global warming.

Food includes both plant and animal material. Prior to mating the male chases the female, then positions himself on her right side during copulation. The large black eggs are deposited beneath stones, in rotten wood or in moist soil, hatching only after a prolonged diapause of several months to about a year. Development is slow, with the 8 recorded nymphal stadia requiring up to 7 years for completion. As development proceeds segments are added to the antennae and cerci, and later instars gradually darken.

These generalized insects (Fig. 22.1) are of special interest because they combine features of the Orthoptera (external ovipositor, structure of the

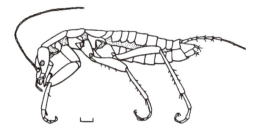

Figure 22.1 Grylloblattidae: *Grylloblatta campodei-formis*, male (scale equals 1 mm).

tentorium) and the Blattodea (5-segmented tarsi, multiarticulate cerci, asymmetrical male genitalia). A close relationship to the recently described order Mantophasmatodea has also been proposed but has not been strongly supported. It has been suggested that the presently reduced fauna represents the remnants of an evolutionary line that produced both the blattoid and orthopteroid orders, but some embryological and comparative anatomical evidence also has also suggested that grylloblattids are a highly specialized lineage of Orthoptera (Ando, 1982). The relationships among the polyneopteran orders have continued to be enigmatic, with different data sets providing evidence for conflicting arrangements. Resolution is made more difficult by the great ages of the orders and the availability of a number of fossils of extinct lineages of uncertain placement. The Mesozoic *Blattogryllus* is the only unquestionable fossil of this order, but the Permian *Tillyardembia* has also been proposed as a possible relative (Grimaldi and Engel, 2005).

CHAPTER 23

Mantophasmatodea
(Heelwalkers or Gladiators)

ORDER MANTOPHASMATODEA

Moderate-sized to small (roughly 2 cm) Exopterygota weakly adapted for predation. Head hypognathous; mouthparts mandibulate; eyes large, lateral; ocelli absent; antennae slender, filiform, multiarticulate. Thorax with each tergum slightly overlapping the following, more or less equal in length; prothoracic pleuron large and fully exposed; legs with coxae elongate, all tarsi 5-segmented with proximal 3 tarsomeres synscleritous but divisions visible ventrally; forelegs not clearly raptorial. Abdomen with 10 visible segments, weakly sclerotized, slightly flattened dorsoventrally; segment 10 with short, 1-segmented cerci; ovipositor short but weakly projecting externally; male genitalia bilaterally partially asymmetrical. Nymphs similar to adults except in development of genitalia; adults short-lived.

Mantophasmatodea is the most recently described order of insects (Klass et al., 2002), having recently been rediscovered alive and in collections in southern Africa after being previously known only from Baltic amber fossils. Despite the relatively unmodified front legs, heelwalkers are predaceous. The common name "heelwalkers" arises from their habit of walking with the pad of the tarsal tip upraised. Two living genera are known (in the single family Mantophasmatidae), containing only a handful of species, but additional species apparently remain to be described from southern Africa.

As the name Mantophasmatodea implies, these insects somewhat resemble a cross between a mantid and phasmid, although in many other general aspects a better morphological resemblance could be claimed with the Grylloblattodea. Early attempts to place the order relative to other orthopteroids using molecular data did suggest a close relationship with Grylloblattodea, but more recent analyses also indicate a possible relationship with Phasmatodea.

CHAPTER 24

Order Dermaptera (Earwigs)

ORDER DERMAPTERA

Neopterous Exopterygota of medium size; body elongate, slightly flattened in dorsoventral plane, with leathery cuticle. Head prognathous; compound eyes moderate or absent, ocelli absent; antennae multiarticulate, filiform; mouthparts mandibulate, adapted for chewing. Prothorax mobile, with large notum (except in wingless forms); cervical region of several separate sclerites; mesothorax small, mesothoracic wings short, elytriform, or absent; metathorax large, wings semicircular with distinctive radiating venation, or absent; legs adapted for walking, tarsi with three segments. Abdomen with 8 segments in female, 10 in male; cerci modified as strongly sclerotized forceps, except in parasitic forms.

The forciculate cerci, terminating the flexible, telescoping abdomen, distinguish earwigs from all insects except Japygidae (Diplura), which are entognathous and blind and which differ in numerous other features (the forceplike appendages of male Meropeidae (Mecoptera) are actually genitalic structures rather than cerci). In the elytriform forewings Dermaptera resemble staphylinid beetles (Coleoptera), which, however, never have forceps. About 2000 species occur in all parts of the world but are most diverse in tropical and subtropical regions. In most temperate regions the familiar earwigs represent a few cosmopolitan, synanthropic species. Although they frequently become extremely abundant, most are of no more than nuisance importance. Earwigs have unmodified, mandibulate mouthparts and unspecialized legs and are generalized in most aspects of their internal anatomy. The retention of three pairs of cervical sclerites is probably an ancestral feature. In the putatively more basal lineages male genitalia are distally paired, as in Ephemeroptera, the only other insects with paired external genitalia.

With a few exceptions the Dermaptera are quite uniform in most anatomical and biological features, being rather narrowly adapted for a cryptic life. For example, entry into crevices, abandoned lepidopteran leaf shelters or leaf rolls, or cavities under stones, or logs, where nearly all Dermaptera take refuge by day, is facilitated by the smooth, leathery cuticle and the specialized wing structure. In the flying wings, the sclerotized, platelike preanal veins are crowded into the wing base. The membranous anal portion, with numerous, radiating veins, folds fanwise and transversely beneath the abbreviated mesothoracic wings, with only a sclerotized apical portion exposed. The mesothoracic wings are sclerotized, without defined veins. At rest they are held closed by rows of bristles on the metanotum.

Forceps are used in defense, prey capture, courtship, and also in wing folding. In most Dermaptera forceps are single-segmented throughout life but are usually straighter, less strongly sclerotized, and setose in early instars. In *Diplatys* and *Bormansia* (suborder Forficulina), the cerci are multiarticulate until the last nymphal instar, in which they are unsegmented but straight, becoming forcipulate only after the ultimate molt. In *Arixenia* and *Hemimerus*, which are commensals or ectoparasites of vertebrates, the cerci are never forcipulate. The forceps of males are often larger and stouter than those of females and sometimes more strongly curved or asymmetrical. Polymorphism in the size of the forceps, as well as in wing development is common.

Biological information is available primarily for a few cosmopolitan species but probably applies to most of the order. Most earwigs appear to be omnivorous, but some, such as *Anisolabis*, are reported to be predatory. Nearly all are nocturnal, retreating to protected shelters during the day. Nymphs and adults differ little in biological requirements. The order is notable for the development of maternal care of the young. Females oviposit clutches of 15 to 80 ovoid, delicate eggs in self-constructed burrows or retreats and then protect the eggs (Fig. 7.1) and first-instar nymphs for several weeks, after which the mother becomes cannibalistic. Nymphs pass through 4 or 5 instars in various species. *Arixenia* and *Hemimerus* are parasites or nest associates of mammals. Both show reduction of sense organs, flattening of the body, and other morphological modifications associated with parasitism. In both genera, embryonic development is internal, with first-instar nymphs appearing viviparously.

The relationships of Dermaptera to other polyneopteran insect orders are not clear; some molecular data have been interpreted as supporting a sister-group relationship with the Zoraptera, while morphological data have been presented to suggest a close relationship with Embiidina, but neither hypothesis yet has fully convincing support. Traditionally, the 11 families in the order have been classified into 3 suborders—the free-living earwigs belonging to the Forficulina suborder, those associated with mammals to the Arixeniina (from bats), and Hemimerina (from rats). The molecular phylogenetic study of Jarvis et al. (2005) found that Hemimerina actually belong within the Forficulina, but their study did not include the Arixeniina and no new subordinal classification was proposed. The traditional classification is retained below with the understanding that it needs reevaluation.

KEY TO THE SUBORDERS AND NORTH AMERICAN FAMILIES OF DERMAPTERA

1. Eyes vestigial or absent; cerci not forcipulate suborders **Arixeniina** and **Hemimerina**
    Eyes well developed; cerci forcipulate, at least in adults suborder **Forficulina**   **2**
2. Second tarsal segment produced longitudinally or laterally (Fig. 24.1d,e)   **3**
    Second tarsal segment cylindrical, not produced (Fig. 24.1f)   **4**
3. Second tarsal segment expanded laterally (Fig. 24.1d)   **Forficulidae**
    Second tarsal segment produced longitudinally below third (Fig. 24.1e), not expanded laterally   **Chelisochidae**
4. Tarsus with padlike arolium beneath claws   **Pygidicranidae**
    Tarsus without arolium   **5**
5. Antennae with 25–30 segments   **Labiduridae**
    Antennae with 10–24 segments   **6**
6. Forewings meeting along entire inner margin   **Labiidae**
    Forewings rounded, inner margins separated at base, or absent   **Carcinophoridae**

Pygidicranidae. This is considered the earliest diverging family of earwigs, with primitively paired, subequal penis lobes on the male both flexed cephalad. Pygidicranidae is primarily a tropical family, with a single adventitious representative, *Pyragropsis buscki*, in southern Florida.

Labiduridae. Represented in North America by only a single introduced species, *Labidura riparia*. Labidurids have paired penis lobes, one pointing caudad, one cephalad. They are medium to large (45 mm), often apterous earwigs that occur in many habitats.

Carcinophoridae. This large, widespread family includes both winged and wingless species and predators as well as omnivores. Males have paired penes as in Labiduridae. Carcinophoridae is represented in North America by two genera, *Euborellia* and *Anisolabis*. *Anisolabis* has been introduced to both the Atlantic and Pacific coasts,

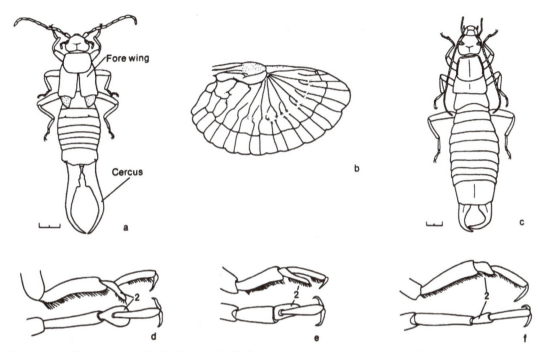

Figure 24.1 Dermaptera: a, Forficulidae, adult (*Forficula auricularia*; scale equals 2 mm); b, wing, Forficulidae (*Doru linearis*); c, Labiduridae (*Anisolabis maritimus*, male; scale equals 2 mm); d, tarsus, Forficulidae (*Forficula*); e, tarsus, Chelisochidae (*Chelisoches*); f, tarsus, Labiduridae (*Labidura*).

where it is restricted to coastal strand situations, including the upper intertidal. At least some species of *Euborellia* are introduced. They are usually encountered in debris or litter. The largest known earwig (55 mm long) is an Australian carcinophorid.

Labiidae. A large, heterogeneous assemblage of primarily winged species distributed mostly in the old world. *Labia* and *Prolabia* are common introduced representatives in North America. *Vostox*, encountered about rotting cacti and other succulent vegetation in the southwestern

states, is extremely active and flies readily when disturbed.

Chelisochidae. Members of this small family are mostly paleotropical. One species, *Chelisoches morio*, has been introduced into California.

Forficulidae. A large, cosmopolitan family of winged species. *Forficula auricularia* (Fig. 24.1a) has been introduced into nearly all temperate regions, including North America, and is often a nuisance because of its large numbers. A single, median penis lobe is present in Forficulidae, Labiidae, and Chelisochidae.

CHAPTER 25

Order Plecoptera (Stoneflies)

(Revised with extensive contributions from Ed DeWalt, Illinois Natural History Survey)

ORDER PLECOPTERA

Neopterous Exopterygota with aquatic nymphs, terrestrial adults. Adults with head hypognathous or prognathous with mandibulate mouthparts, bulging, lateral compound eyes, 2 to 3 ocelli, long, filamentous antennae with numerous segments. Thorax with 3 subequal segments composed of little modified tergal, pleural, and sternal sclerites; wings membranous; forewing narrower than hind wing, with anal region much reduced; venation of longitudinal veins with complex network of crossveins, or with crossveining absent, wing folding of longitudinal pleats; legs robust, long, with 3-segmented tarsi bearing 2 terminal claws. Abdomen weakly sclerotized, cylindrical or flattened, with 10 similar segments and cerci of 1 to many segments. Thorax and/or abdomen often with remnants of nymphal gills. Nymphs similar to adults, but with gills, if present, usually larger or more extensive, cerci longer, developing wingpads present later in development and genitalia undeveloped.

Plecoptera are rather obscure as adults, appearing typically from late winter to midsummer; immatures can be found throughout the year. The nymphs, which are familiar to fishermen and anyone who has turned stones in cool brooks, are important in food chains in aquatic ecosystems.

Stoneflies are extremely generalized in most features of their morphology. Their unmodified, mandibulate mouthparts (Fig. 25.1), with 2-lobed maxilla and 4-lobed labium, each with well-developed palps, are similar to those of Orthoptera. The multisegmented antennae and cerci, unspecialized cursorial legs, and (in primitive forms) complex network of wing veins are shared, at least in part, with a few other insects such as Ephemeroptera and Grylloblattodea. Plecoptera are notable for the putatively ancestral structure of the thoracic pleura, which are not divided into episternum and epimeron in the prothorax. The single, crescent-shaped pleurite is suggestive of the subcoxal arc of primitive insects such as silverfish. Venation is most primitive in the family Eustheniidae, restricted to Australia, New Zealand, and South America, but Pteronarcyidae of the North American fauna show a relatively complete venation (Fig. 25.3a). In other families, especially Nemouridae, venation is variably reduced (Fig. 25.2b–d).

Nymphs (Fig. 25.3b) of most stoneflies respire through the general body surface, but many bear

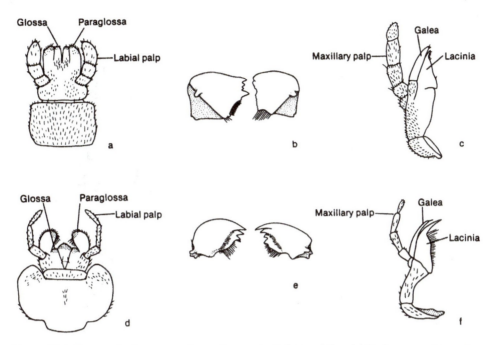

Figure 25.1 Taxonomic characters of stonefly nymphs (left to right): a, d, labia; b, e, mandibles; c, f, maxillae. a–c, Nemouridae (*Amphinemura sp.*); d–f, Perlidae (*Perlinella drymo*).
Source: From Claassen, 1931.

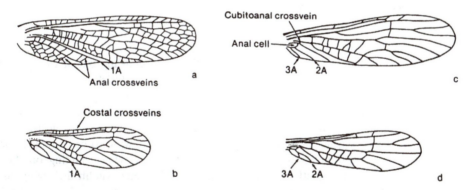

Figure 25.2 Wings of stoneflies: a, Pteronarcyidae (*Pteronarcys proteus*); b, Peltoperlidae (*Tallaperla maria*); c, Perlodidae (*Isoperla bilineata*); d, Chloroperlidae (*Utaperla sopladora*).
Source: a, c, from Needham and Claassen, 1925; b, d, from Ricker, 1943, 1952.

external tracheated gills. The gills vary from strap-like, unbranched filaments to profusely branched tufts and may occur on the head, thorax, or abdomen. Differences in the distribution and structure of gills are of uncertain functional significance but are extremely useful in identification.

Adult stoneflies usually remain near the streams or lakes from which they emerged. Many species rest exposed on stones or vegetation, but others hide in cracks and crevices and may be encountered in such cryptic places as beneath slightly loose bark. Most species are weak, fluttery fliers, and many will

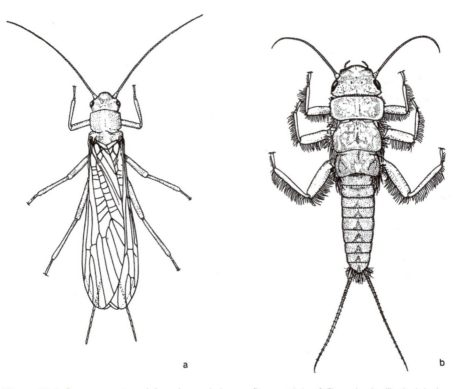

Figure 25.3 Representative adult and nymphal stoneflies: a, adult of *Clioperla clio* (Perlodidae); b, naiad of *Hesperoperla pacifica* (Perlidae).
Source: a, from Frison, 1935; b, from Claassen, 1931.

run rather than fly to elude predators. Most stoneflies are diurnal, but some members of the superfamily group Systellognatha are nocturnal. Within the superfamily group Euholognatha it appears that most adults feed on algae, lichen, rotten wood, or decaying leaves. At least one genus in the family Taeniopterygidae has been implicated in the damage of fruit tree blossoms. Systellognathan adults often do not feed as adults, but some within the families Chloroperlidae and Perlodidae consume pollen and possibly other food sources, presumably to aid in egg development.

Emergence is highly seasonal for most species, and since the adult life span is usually only a few weeks, the composition of stonefly faunas is highly characteristic of time and location. Many species of euholognathan stoneflies within the families Capniidae and Taeniopterygidae are so-called "winter stoneflies" because they emerge during winter

months, often crawling out of streams across ice and snow (Ross and Ricker, 1971). These species are of comparatively low vagility due to low air temperatures limiting the usefulness of wings and due to the males of many capniid species being brachypterous (short-winged) or even apterous. Many winter stoneflies, while standing on the surface tension of the water, will lift their wings and "sail" across brooks and streams (Marden and Kramer, 1994). The entire life cycle of *Utacapnia tahoensis*, apterous as adults, is passed underwater in Lake Tahoe in California and Nevada (Nebeker and Gaufin, 1965).

Over 100 species of the suborder Arctoperlaria (includes both Euholognatha and Systellognatha) occurring in North America utilize "drumming" as a vibrational mate-finding behavior. Males initiate drumming as a series of single beats, grouped beats, or as a rub using either their abdomen or specialized structures of the abdomen. Usually only unmated females answer, the result of which is often a duet that helps the male find the stationary female. Drumming appears to be absent in the other suborder, Anarctoperlaria, a taxon that has a Gondwanan distribution. Copulation, which may occur several times, takes place on substrates, with the male astride the female.

Eggs are extruded in batches from the posterior of the eighth abdominal sternum. Multiple egg batches are normal with total number of eggs extruded being a few hundred to over 1000. Eggs are later dropped into water during flight, washed off by dipping the abdomen just beneath the surface, or deposited when the female crawls into the water, as in wingless species. The eggs of euholognathan species are membranous and are enclosed in a sticky matrix that anchors them to the substrate. Eggs of systellognathan species have sclerous chorions, often with intricate, species-specific surface sculpturing. Eggs in batches scatter upon entry in the water column. Eclosion usually occurs after 2 to 3 weeks, or even several months. Species which inhabit seasonal streams may feature prolonged diapause so that the eggs hatch only under favorable conditions.

Nymphal development rates vary tremendously. Some with short egg development times grow continuously throughout the year. Others with lengthy egg diapause and a one-year life cycle must finish growth within only a few months. Other species may combine a 9-month egg diapause with near-continuous but slow nymphal growth for up to 3 years, producing a 4-year life cycle. From 12 to 24 instars, differing in the number of gills and segments in the antennae and cerci, are recorded for North American species; up to 33 instars have been recorded in other Northern Hemisphere species. Stoneflies are most diverse in cool streams with moderate or swift current but also can be found in the wave-swept shores of cool, oligotrophic lakes; there has been significant radiation of species into clean, warm water. Stony, irregular bottoms support the most varied and most dense fauna. In general, stonefly nymphs require well-oxygenated water and are among the first aquatic life to disappear in polluted streams. In New Zealand and nearby subantarctic islands a number of species have terrestrial naiads that inhabit cool, wet situations.

Most stonefly nymphs are primarily herbivorous, consuming submerged leaves, algae, diatoms, and similar items. Systellognathan nymphs are generally predators, but some in this superfamily group may feed as primary consumers early in their nymphal development, then shift to primarily animal matter later in development. These general feeding differences are reflected in mouthpart structure. Herbivorous species (euholognathans) have short, stout mandibles with a pronounced molar lobe (Fig. 25.1b); predatory species have elongate, sickle-shaped mandibles with the molar lobe reduced or absent (Fig. 25.1e). A few days before adult emergence feeding ceases, and shortly before the last molt, the nymph crawls onto an emergent stone, log, or vegetation, often several feet from the water surface, splits the dorsal suture on the thorax and abdomen, and crawls out of the exuvium. Transformation is usually accomplished within 15 to 30 minutes.

About 3700 species of Plecoptera have been described worldwide, with about 700 species in North America north of Mexico. The order is most abundant and diverse in cool temperate regions and in North America is best represented in four megadiverse regions: the southern Appalachians, interior highlands (Ozark Plateau), Rocky Mountains, and the Coastal, Cascade, and Sierra mountain ranges. The Great Plains and northern transcontinental regions are areas where some species of eastern and western origins meet but are of lower diversity overall. The classification adopted here follows that of Zwick (2000).

Preservation of nymphs and adults will depend upon the desired uses. General collection and morphological examination will benefit from collection in 80 percent ethyl alcohol (EtOH), 5 percent formalin, or Kahle's fluid. Specimens for use in molecular studies should be preserved in 95 to 100 percent EtOH. The exuviae of the last nymphal instars may be softened in warm water to relax the gills and are nearly as useful as nymphs for identification. Nymphs are often easily reared to adulthood in shallow, cool water that is well oxygenated. Nymphs with black or full wingpads are the best to rear because they require little or no feeding. Herbivorous nymphs may be reared on a diet of stream-conditioned tree leaves. Nymphs are superficially similar to those of Ephemeroptera, but mayflies most often appear three-tailed (addition of medial caudal filament), have a single claw and tarsal segment, and have a variety of platelike or forked gills on the abdomen.

KEY TO NORTH AMERICAN FAMILIES OF PLECOPTERA

Adults

1. Labium with glossae and paraglossae of approximately equal length (Fig. 25.1a)   **2**
   Labium with glossae much shorter than paraglossae (Fig. 25.1d)   **7**
2(1). Basal segment of tarsus less than half the length of apical segment; sides of thorax with remnants of gills   **3**
   Basal segment of tarsus more than half the length of apical; thorax without gill remnants   **4**
3(2). Gill remnants present on first 2 or 3 abdominal segments; 3 ocelli   **Pteronarcyidae**
   Abdomen without gill remnants; 2 ocelli   **Peltoperlidae**
4(2). Tarsus with second segment much shorter than first   **5**
   Tarsus with first and second segments subequal   **Taeniopterygidae**
5(4). Cerci with 1 segment   **6**
   Cerci with at least 4 segments   **Capniidae**
6(5). Wings flat over dorsum at rest; labial palp with apical segment globular, much larger than basal segments   **Nemouridae**
   Wings at rest rolled down around sides of abdomen; labial palps with apical and preapical segments subequal   **Leuctridae**
7(1). Thoracic sterna or pleura with profusely branched gill remnants at corners; forewing with cubitoanal crossvein intersecting anal cell (Fig. 25.2b) or distad from it by no more than its own length   **Perlidae**

Thoracic segments without gills or with unbranched, straplike gills; forewing with cubitoanal crossvein distad from anal cell by its own length or more (Fig. 25.2c) or absent   **8**
8(7). Forewing with veins 2A and 3A arising from anal cell separately (Fig. 25.2c); gill remnants sometimes present on thorax or abdomen   **Perlodidae**
   Forewing with veins 2A and 3A arising from anal cell as single vein (Fig. 25.2d); gill remnants absent   **Chloroperlidae**

Nymphs

1. Abdomen with tufts of branched gills present on first 2 or 3 abdominal segments   **Pteronarcyidae**
   Abdomen without gills   **2**
2(1). Thorax with sternites produced posteriorly and overlapping; head flattened; body often cockroachlike   **Peltoperlidae**
   Thorax without overlapping sternites; head not flattened; body not cockroachlike   **3**
3(2). Labium with glossae and paraglossae about equal in length (Fig. 25.1a); dorsal surface of body usually unpatterned   **4**
   Labium with glossae much shorter than paraglossae (Fig. 25.1d); dorsum usually patterned   **7**
4(3). Hind legs extending beyond abdominal apex; wing pads oblique to longitudinal body axis   **5**
   Hind legs not extending as far as abdominal apex; wing pads parallel to longitudinal body axis   **6**
5(4). Tarsus with first and second segments subequal   **Taeniopterygidae**
   Tarsus with second segment much shorter than first   **Nemouridae**
6(4). Abdominal segments 8 and 9 annular, without membranous, ventrolateral fold separating tergal and sternal regions   **Leuctridae**
   Abdominal segments 8 and 9 with membranous lateral folds separating tergal and

CHAPTER 26

Order Embiidina
(Embioptera, Webspinners, Embiids)

ORDER EMBIIDINA

Small to moderate-sized Exopterygota, with subcylindrical, elongate body modified for living in tubular galleries. Head prognathous with mandibulate mouthparts, 12 to 32 segmented antennae, kidney-shaped eyes; ocelli absent. Prothorax small, narrower than head; mesothorax and metathorax broad, flattened, and usually bearing wings in male but narrower, elongate, without wings in female; wings long, narrow, with similar, simple venation; legs short, robust; tarsi 3-segmented; forelegs with basal tarsomere enlarged, globular; hind legs with femur enlarged. Abdomen with 10 subequal segments, 2-segmented cerci on segment 10. Nymphs similar to adults except in development of wings and genitalia.

Embiidina are monotonously similar in body form but vary in size and coloration. Males of some species are attracted to lights, but most are obscure because of their small size, dull color, and retiring habits. This predominantly tropical order is poorly represented in the temperate zones and except for introduced species is absent from islands, presumably because of the lack of wings in females. Although the world fauna included fewer than 200 named species in 1991, it seems likely that as many as 2000 species remain undescribed (Ross, 1991). The United States fauna comprises eight native species and three others introduced from the Old World.

With the exception of dispersal by adult males, all instars of all species of Embiidina exclusively inhabit self-constructed silken galleries. This narrow specialization for a single niche is strongly reflected in the overall morphological uniformity of embiids. The most obvious modification involves the fore tarsi (Fig. 26.1b), which contain the silk-producing glands. Each swollen basitarsomere usually contains up to about 75 secretory units, exceptionally as many as 200, each connected to a hollow spinning bristle by a minute cuticular tubule. Each secretory unit consists of a syncytial layer of cytoplasm enclosing a storage chamber for the liquid silk. Silk production is involuntary, and since each tarsal bristle extrudes an individual fiber, a broad swath of silk is available continuously. The forelegs are rapidly shuttled back and forth, and the body is rotated as galleries are extended onto new food sources. Additional silk is continually added to old galleries, which eventually become opaque from multiple laminations of silk. As individuals grow they progressively construct larger tunnels, so that an active embiid colony contains a maze of passageways of various sizes and ages.

Rapid backward movement within the narrow passageways is aided by the highly tactile cerci

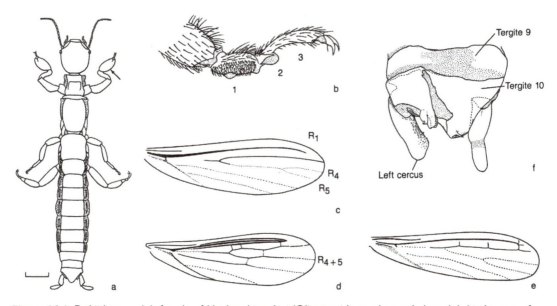

Figure 26.1 Embiidina: a, adult female of *Haploembia solieri* (Oligotomidae; scale equals 1 mm); b, hind tarsus of *H. solieri*; c, forewing of Teratembiidae (*Oligembia brevicauda*); d, forewing of Anisembiidae (*Anisembia schwarzi*); e, forewing of Teratembiidae (*Teratembia geniculata*); f, cerci of Anisembiidae (*Chelicera galapagensis*). *Source: b–f, from Ross, 1940, 1952, 1966.*

and strongly developed tibial depressor muscles of the swollen hind femora. In alate individuals, the wings are soft and flexible, allowing them to flex quickly over the head during backward motion. During flight the wings are stiffened by inflation of a blood sinus formed by the principal veins, especially the radial. The remaining veins are relatively weak, with few branches and with almost no venational differences between forewings and hind wings. In most other morphological features, embiids are somewhat similar to earwigs (Dermaptera) and stoneflies (Plecoptera), indicating a very early divergence from the orthopteroid lineage (the earliest known fossil is from mid-Cretaceous amber, but the order must have been present long before then). Molecular studies have also yet to converge on an answer to which order is most closely related to Embiidina.

Embiidina are strongly gregarious, a typical colony consisting of one or more adult females with offspring of various numbers and sizes. A colony is usually centered around a secure retreat, such as a crevice beneath a stone or a crack in a log. Passageways are extended to close-by leaf litter, bark, lichen, moss, or other vegetable material that is suitable as food. In humid tropical habitats webways may run exposed over bark or other irregular surfaces, sometimes covering considerable areas. In North America, colonies are normally relatively small and concealed beneath stones, logs, or other shelters. In arid regions the galleries extend down cracks into cooler soil regions. Embiids are easily cultured in small containers of leaves or other habitat material.

Eggs are laid within the galleries and are sometimes covered by masticated cellulose. Eggs and newly emerged nymphs are briefly tended by the mother. There are four nymphal instars. The sexes are similar until wing buds appear in the last two instars of males. Asymmetry in male external genitalia appears only after the last molt. At maturity males leave the home colony for a short dispersal flight before entering another colony. They do not feed, and after mating they live only a short time, frequently being consumed by the females. The elongate, prognathous mandibles of male embiids are used to grasp the head of the female during mating. The highly complex, asymmetrical male cerci (Fig. 26.1f) of some species function as clasping organs during copulation. Parthenogenesis is recorded in a few species, and wingless males occur in others, especially those inhabiting arid regions.

In North America the order is restricted to the southern and southwestern states. The best differentiating characters are found in the wings, genitalia, and mouthparts of adult males. The following key is based on the work of Ross (1940, 1944) and Davis (1940).

KEY TO THE NORTH AMERICAN FAMILIES OF EMBIIDINA

Adult Males

1. Mandibles with apical teeth; left cercus with 2 segments; inner apical surface of cercus smooth   **2**

    Mandibles without apical teeth; left cercus with segments fused into a single member; inner apical surface of cercus lobed and minutely spiculate (Fig. 26.1f)

                                   Anisembiidae

2(1). Wings with R_{4+5} unbranched (as in Fig. 26.1d)   **Oligotomidae**[1]

    Wings with R_{4+5} 2-branched (Fig. 26.1c,e)

                                   Teratembiidae

---

[1] *Haploembia solieri*, common in the Southwest, has both parthenogenetic and bisexual races. In all instars the hind basitarsi have two ventral papillae. One papilla is present in all other North American embiids.

CHAPTER 27

Order Orthoptera (Grasshoppers, Crickets, etc.)

ORDER ORTHOPTERA

Usually, medium to large Exopterygota with subcylindrical, elongate body; hind legs almost always enlarged for jumping. Head hypognathous, with compound eyes; ocelli present or absent; antennae multiarticulate; mouthparts mandibulate. Prothorax large, with shieldlike pronotum curving ventrally over pleural region; mesothorax small, with wings narrow, cornified; metathorax large, wings broad, with straight, longitudinal veins, numerous crossveins; wing folding by longitudinal pleats between veins; tarsi with 1 to 4 segments. Abdomen with first 8 or 9 segments annular, terminal 2 to 3 segments reduced; cerci with 1 segment. Nymphs similar to adults except in development of wings and genitalia.

Many Orthoptera, especially tropical species, are bizarrely modified to mimic living or dead vegetation or other insects. In subterranean or myrmecophilous forms wing loss may be accompanied by lack of differentiation of thoracic and abdominal segments and body proportions may be quite different than in surface dwelling types. Yet nearly all Orthoptera are readily distinguished from other insects by the enlarged hind femora and characteristic grasshopperlike shape. The Orthoptera are easily the dominant group of mandibulate hemimetabolous insects, with about 25,000 species distributed throughout the world. Most species are terrestrial herbivores, but various adaptations for predation or for subaquatic or subterranean life characterize many genera, tribes, or families. While Orthoptera clearly are related to other orthopteroid insects in some way (see Chap. 14), the exact relationships are still unclear, despite recent efforts being devoted to molecular phylogenetic studies of their higher classification.

A majority of Orthoptera conform to two broad adaptive modes, which correspond taxonomically to suborders. The Caelifera include mostly alert, diurnal forms with acute vision and hearing, which rely on jumping to escape from predators. Antennae and legs are relatively short, allowing unimpeded, rapid movement (Fig. 27.1). These species, exemplified by the Acrididae (grasshoppers and locusts), are predominantly terrestrial (as opposed to arboreal). They are especially adapted to exposed situations in open habitats, such as savannah woodlands, steppes, and deserts. The great majority are herbivores; many have specialized dentition for consuming the tough foliage of monocotyledonous plants.

In contrast to the Caelifera, the Ensifera (katydids, crickets) tend to rely on crypsis to avoid predators. Most species are nocturnal, either retreating to protected niches during the day or escaping notice through camouflage or mimicry. Antennae and legs are long, sometimes extremely so; tactile responses are highly developed, though vision and

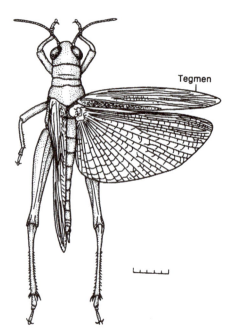

Figure 27.1 A grasshopper, *Melanoplus cinereus*, Acrididae, dorsal aspect. Scale equals 5 mm.

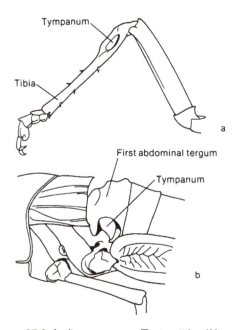

Figure 27.2 Auditory organs: a, Tettigoniidae (*Microcentrum californicum*); b, Acrididae (*Dissosteira carolina*).

hearing are also important. Movements tend to be slow and careful but become abrupt and vigorous upon physical contact. Many Ensifera are omnivorous or predatory. Phytophagous species usually feed on stems or foliage of dicotyledonous plants, rather than grasses.

Nearly all Orthoptera defend themselves by kicking with the hind legs, which are usually sharply spurred on the tibiae. Most species will attempt to bite if held, and the larger ones can readily break the skin. Other defensive mechanisms include regurgitation of the crop contents or release of secretions from integumental defensive glands, often present in species with aposematic coloration. If seized by a predator the hind legs are often autotomized at the femur; one-legged grasshoppers testify to the efficiency of this mechanism.

Hearing is acutely developed in most Orthoptera, and auditory stimuli are important in courtship in many species. In the Caelifera, auditory organs (if present) are on the first abdominal tergite, beneath the folded wings (Fig. 27.2b). Sound is produced by rubbing the hind femora against the lateral, vertical part of the forewings. In the Ensifera, auditory organs are located on the anterior tibiae (Fig. 27.2a). Stridulation is usually accomplished by vibrating the horizontal, overlapping parts of the forewings over one another (Fig. 6.8d,e). The wingless Stenopelmatidae tap the abdomen against the substrate to produce drumming patterns used in courtship.

The songs of Orthoptera are almost always species specific and are often produced at a characteristic time of day. In general, grasshoppers stridulate during the day, katydids at night, and crickets during both day and night. In some species singing may be synchronized and very loud. The crackling sounds produced by many acridids during flight may be designed to confuse birds.

Intraspecific polymorphism is common in Orthoptera, especially in the Caelifera. In the extreme case, distinct differences in behavior and physiology, as well as in color pattern and body proportions, may occur in response to population fluctuations and ecological conditions. Such phase transformation is very pronounced in periodically migratory grasshoppers such as the desert locust (*Schistocerca gregaria*) and occurs in species from all geographic regions (see Chap. 4). More typically, polymorphism involves only color, cuticular sculpturing, or wing length. The function in many aspects is obscure.

Nymphs are structurally similar to adults, except in the less developed wings and genitalia, but may differ strikingly in coloration. Many common katydids, for example, are brightly banded as early instar nymphs but solid green as adults. Immatures of apterous species may be very difficult to distinguish from adults. A peculiarity shared only with Odonata is that the orientation of nymphal wing pads changes during development. In early instars the costal margin is lateral. Later, usually at the penultimate molt, the costal margin is turned inward; subsequent reversal to the lateral position occurs at the adult molt.

Sound communication is a frequent adjunct to courtship (see Chap. 6), and stridulatory mechanisms are frequently restricted to males. Copulation occurs with the male astride the female and may endure for several hours or longer. Orthopteran eggs are typically elongate or oval and relatively large. Eggs are usually inserted singly into plant tissue or soil substrates in the Ensifera. Burrowing species deposit the eggs in the burrow. In the Caelifera clutches of 10 to several hundred eggs are buried in shallow pits excavated by the ovipositor valves. Each pod of eggs is surrounded by a foamy waterproof matrix. Egg diapause is widespread in Orthoptera.

KEY TO THE NORTH AMERICAN FAMILIES OF ORTHOPTERA

1. Antennae usually with many more than 30 segments (suborder Ensifera)   2
   Antennae usually with many fewer than 30 segments (suborder Caelifera)   8
2(1). Forelegs modified for digging, with femur and tibia broad and flat and tibia and tarsus strongly toothed   **Gryllotalpidae**
   Forelegs normal, adapted for walking   3
3(2). Hind coxae nearly contiguous ventrally; eyes very small; body dorsoventrally flattened   **Myrmecophilidae**

Hind coxae widely separated ventrally; eyes large; body cylindrical or subcylindrical   4
4(3). Tarsi with 4 segments, at least on middle and posterior legs   5
   Tarsi with 3 segments   **Gryllidae**
5(4). Anterior tibiae with auditory organs (Fig. 27.2a), without ventral, articulated spines   **Tettigoniidae**
   Anterior tibiae without auditory organs, usually with ventral, articulated spines   6
6(5). Antennal bases separated by at least the length of the first antennal segment   7
   Antennal bases contiguous or nearly so   **Rhaphidophoridae**
7(6). Tibiae narrow; tibial spurs small, inconspicuous   **Gryllacrididae**
   Tibiae robust, flattened or club shaped; tibial spurs large, flattened   **Stenopelmatidae**
8(1). Tarsi with 3 segments   9
   Tarsi with 1 or 2 segments   **Tridactylidae**
9(8). Pronotum extended backwards to cover wings   **Tetrigidae**
   Pronotum not prolonged backwards over wings and abdomen   10
10(9). Antenna longer than fore femur   11
   Antenna shorter than fore femur   **Eumastacidae**
11(10). Antennae at least half as long as body, longer than body in males   **Tanaoceridae**
   Antennae less than half the length of body   **Acrididae**

Suborder Ensifera

Tettigoniidae (katydids). This variable family comprises approximately 5000 species predominantly from the tropics. Most katydids are arboreal, winged insects, but apterous species, such as the Mormon cricket (*Anabrus simplex*), are common. Feeding habits of Tettigoniidae are diverse. While the majority are herbivorous, others are detritivores or predators. A few, including the Mormon cricket, are crop pests. Large, tropical Tettigoniidae may live several years and are often strikingly mimetic. Different species mimic ants and wasps as well as living or dead leaves, flowers, or lichen. Those resembling other insects often display unusual mimetic behavior. Most familiar North American

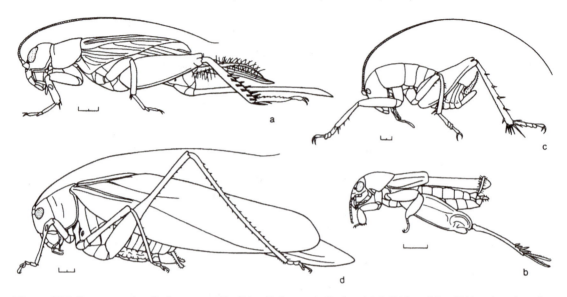

Figure 27.3 Representative Orthoptera: a, Gryllidae (*Acheta assimilis*, female); b, Tridactylidae (*Tridactylus minutus*); c, Rhaphidophoridae (*Ceuthophilus californicus*); d, Tettigoniidae (*Microcentrum californicum*, male). Scales equal 2 mm, except for b, which equals 1 mm.

species are relatively unmodified, green or brown, annual katydids (Fig. 27.3d).

Gryllidae (crickets; Fig. 27.3a). Approximately 1200 species, distributed worldwide. Most species live on the ground under logs or stones or in dense vegetation, but others are arboreal or subterranean. Gryllidae are often brachypterous, apterous, or polymorphic for wing length. Most species are omnivorous.

Gryllotalpidae (mole crickets). About 50 species worldwide, seven occurring in the United States. The antennae are much shorter than the body and the forelegs flat and platelike, adapted for burrowing in moist soil.

Myrmecophilidae. Small, flattened yellowish or brownish crickets that inhabit ant nests, feeding on secretions produced by the ants. Worldwide in distribution, with about 50 described species. Five of the six North American myrmecophilids occur in the west, where they are common in the nests of many different ants.

Rhaphidophoridae (cave crickets, camel crickets). North American representatives are apterous (Fig. 27.3c), but winged species occur in other

areas. Most species are at least partly subterranean, and many are restricted to caves, tree holes, animal burrows, and similar situations. Most appear to be opportunistic scavengers; some are predaceous. There are about 600 species worldwide.

Stenopelmatidae (Jerusalem crickets, potato bugs). Stenopelmatidae number about 200 species of winged or wingless crickets worldwide. The North American fauna includes a few genera of flightless, heavy-bodied crickets, which are found under stones or logs or burrowing in moist soil. They are restricted to the western states, especially of Pacific North America. The commonly encountered *Stenopelmatus*, lacking an obvious ovipositor, is also distinguished by the bicolored, banded abdomen. *Stenopelmatus* species usually turn onto their backs when molested, kicking the legs and moving the mandibles. They bite painfully if handled carelessly.

Gryllacrididae (leaf-rolling crickets). The single North American species, confined to the southeastern states, constructs shelters of rolled leaves tied with silk produced by glands that open into the mouth. Gryllacridids in other regions also use the secretions to stabilize burrows in the ground.

Suborder Caelifera

Acrididae (grasshoppers). The largest family of Orthoptera, with over 10,000 species. The great majority are typical "grasshoppers," with minor structural variation from the acridid body plan. Many species are habitat specific, and many have restricted diets, while others may be diversely adapted or polyphagous. Brachypterous species occur in several different lineages. Such familiar Orthoptera as migratory locusts, lubber grasshoppers, and band-winged grasshoppers are all acridids.

Tanaoceridae (long-horned grasshoppers). This family is restricted to the deserts of southwestern North America, where the few nocturnal species are found on sage brush and other shrubs.

Eumastacidae (monkey grasshoppers). Worldwide Eumastacidae numbers more than 300 species of small to medium, wingless grasshoppers that occur mostly in tropical or warm temperate regions. The 12 North American species are southwestern, occupying a variety of habitats from deserts to high mountains, where they are found at night on the tops of shrubby plants. In life the hind legs are held at right angles to the body.

Tetrigidae (pygmy locusts). A well-defined group of about 1000 species, mostly tropical or subtropical. Few are more than 15 mm long. Polymorphisms involving wing and pronotal development occur in many species, and many are grotesquely flattened and contorted. Many species are associated with moist situations and are proficient swimmers. Rain forest species are often encountered on mossy tree trunks. Tetrigids feed on algae or detritus.

Tridactylidae (pygmy mole crickets; Fig. 27.3b). At least 75 species of this specialized family occur worldwide. All frequent margins of ponds or streams. Eggs are laid in burrows in moist sand. Tridactylids are excellent swimmers and powerful jumpers. Food is apparently algae.

CHAPTER 28

Order Phasmatodea (Stick Insects)

ORDER PHASMATODEA

Cylindrical, elongate or flattened, leaflike Exopterygota, mostly of moderate to large size. Antennae multiarticulate, filiform or moniliform; compound eyes present, ocelli present in some winged males; mouthparts mandibulate, prognathous. Prothorax short, mesothorax and metathorax often elongate (winged species) or short (wingless species); pronotum without ventrolateral extensions covering pleural region; mesothoracic wings narrow, cornified; metathoracic wings broad, with straight longitudinal veins, numerous crossveins, anterior margin cornified, wing folding longitudinal, between veins; wing reduction or aptery common. Legs usually elongate, slender; tarsi with 5 segments. Abdomen with 10 (occasionally 11) apparent segments; terminal 2 or 3 sternites concealed beneath subgenital plate; cerci with 1 segment, external or concealed. Nymphs similar to adults except in development of wings and terminalia.

Approximately 3000 species of phasmatids are known, predominantly from the tropics, especially in the Oriental region. Walkingsticks appear to be relatively rare insects, even considering the extraordinary difficulty of finding them, and seldom become economically important, although in the southern and midwestern United States they can become ecologically significant in the crowns of trees. Superficially they are similar to some Orthoptera but differ in having asymmetrical male genitalia (symmetrical in Orthoptera), in lacking wing pad reversal in the nymphs, and in proventricular structure. As with many other orthopteroids, the exact relationship of walkingsticks to other insects is somewhat obscure, but the relationships within the order have received some phylogenetic attention. Recently, controversy erupted over whether wings could reappear (after being lost) in the evolution of the group, as the current phylogeny appears to suggest.

Nearly all phasmatids are modified to mimic vegetation of various sorts. Many species strikingly resemble either leaves or twigs, but others are generally similar to foliage without clearly conforming to either morphological type. Integumental sculpturing is diversely modified as thorns, tubercles, ridges, and other processes or textures that heighten the similarity to vegetation. Body form ranges from extremely slender to foliate, or only the appendages may be foliate. Resemblance to plant parts is further enhanced by posturing, frequently in grotesque, asymmetrical positions. Such postures may be maintained even when the insects are disturbed, or they may drop from their purchase and feign death. A few species, such as *Carausius morosus*, are able to effect color changes by altering the dispersal of pigment granules in the epidermis. A few others, such as *Anisomorpha buprestoides* (Fig. 28.1b) of the southeastern United States, are aposematically colored. These active, agile insects

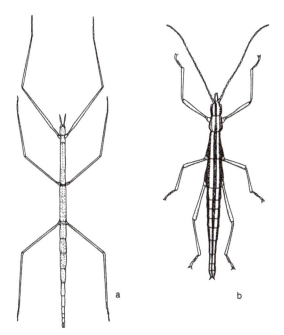

All Phasmatodea are herbivores, typically leading solitary, sluggish lives on vegetation, especially trees and shrubs. Males are usually distinctly smaller than females. During copulation, which may be very prolonged, the male sits astride the female. In *Timema* the male spends nearly the entire adult instar riding on the larger female. Eggs, numbering about 100 to 1200 in different species, are deposited singly, either glued to vegetation or dropped to the ground. Females of *Anisomorpha* dig shallow pits in the sandy substrate, deposit a few eggs, and cover them. Phasmatid eggs are distinctive, with the thick, tough chorion often sculptured or patterned so that superficially the eggs often resemble seeds. Egg diapause is common and may endure more than a single season. Newly hatched nymphs climb onto convenient vegetation. They are generally similar to adults, both in appearance and behavior, but in a few species high population densities cause brightly colored nymphs and adults to develop, in a manner analogous to phase polymorphism in locusts. Unlike locusts there is no modification of behavior and no increase in gregariousness.

Two families, both widespread, are recognized here. The Phylliidae include many leaflike species. There are no North American representatives. The Phasmatidae include species of diverse body form. Most North American species are sticklike, but *Anisomorpha* (southeastern states) and *Timema* (California) are fusiform with relatively short legs. *Timema* is distinguished from all other genera in having the tarsi 3 segmented.

Figure 28.1 Representative Phasmatodea, Phasmatidae: a, *Parabacillus coloradus*; b, *Anisomorpha buprestoides*. Source: a, from Henderson and Levi, 1938; b, from Caudell, 1903.

secrete large quantities of a defensive substance (anisomorphal) in the prothoracic glands, readily spraying or oozing the milky product upon very slight disturbance. Yet another common defensive mechanism is limb autotomy, which takes place between the femur and trochanter. Partial leg regeneration may occur in nymphs.

CHAPTER 29

Order Zoraptera

ORDER ZORAPTERA

Minute to small Exopterygota with mandibulate, chewing mouthparts, 9-segmented moniliform antennae, compound eyes, and ocelli present or absent. Prothorax large, subquadrate; mesothorax, and metathorax subquadrate or transverse; wings with large pterostigma and reduced venation or absent; legs adapted for walking, with 2-segmented tarsi. Abdomen with 11 similar segments, unsegmented cerci on segment 11. Nymphs similar to adults except in development of wings and/or genitalia.

These obscure insects resemble termites in general appearance and frequently occur about decaying wood. The order is known from the single family Zorotypidae, which includes about 30 species, usually placed in a single genus, *Zorotypus*. The species are distributed sporadically in most parts of the world except Australia.

Their unspecialized, chewing mouthparts and the presence of cerci have appeared to indicate a relationship of the Zoraptera with the orthopteroid orders, but their exact phylogenetic placement within the Orthopteroidea, or alternatively within the Hemipteroidea, is not resolved. The male genitalia, which are asymmetrical, suggest affinities with cockroaches and termites. The Malpighian tubules number only six, a feature common to the hemipteroid orders. Termites, however, are unusual among the orthopteroids in also having reduced numbers of Malpighian tubules (6–9). The central nervous system of Zoraptera is concentrated into two large thoracic and two abdominal ganglia, recalling the condition in the hemipteroid orders, and the wings, when present, have the venation extremely reduced, somewhat as in Psocoptera or

Thysanoptera. As in hemipteroids, the hind wings are smaller than the forewings (Fig. 29.1b).

Zoraptera occur in moist leaf litter, in rotting wood or sawdust, and in or about termite colonies. Species that have been studied are polymorphic, one adult form being largely unpigmented and lacking wings, compound eyes, and ocelli (Fig. 29.1c). Winged forms are more darkly pigmented, with small compound eyes and three ocelli (Fig. 29.1a). The significance of polymorphism in Zoraptera is uncertain, but the winged forms probably function in dispersal and may develop when the substrate becomes unsuitable because of drying or other deterioration. Food apparently consists mostly of fungi, though arthropod fragments have also been found in the alimentary canal.

In the only species observed the female climbs onto the back of the male after a courtship dance. After coupling the female moves forward, turning the male onto its back and dragging it about in this position (Fig. 29.2). Eggs of Zoraptera are ovoid, without remarkable sculpturing or micropylar structures. Eclosion is by means of an egg burster on the head of the embryonic nymph. Polymorphism is evident in the nymphs (Fig. 29.1d), which develop

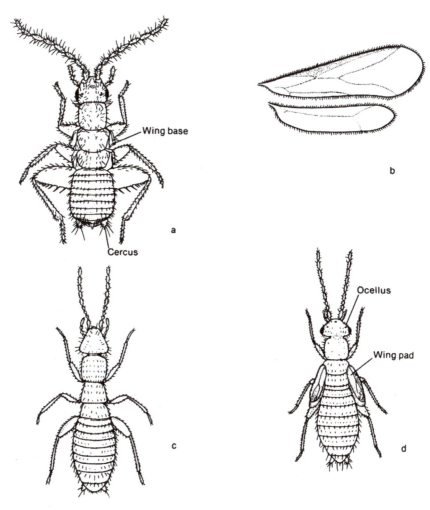

Figure 29.1 Zoraptera, Zorotypidae: a, dealate adult of *Zorotypus mexicanus*; b, wings of *Zorotypus hubbardi*; c, d apterous adult and nymph of alate form of *Zorotypus hubbardi*.
Source: a, from Bolivar y Pieltain, 1940; b–d, from Caudell, 1920.

Female   Male

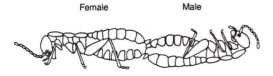

Figure 29.2 Mating position of *Zorotypus hubbardi* (Zorotypidae); male on right, female on left.
Soruce: Redrawn after Shetlar, 1978.

external wing pads and have eyes and ocelli if destined to produce winged adults. The number of nymphal instars and other details of the life history and biology are unknown.

Two species occur in North America. *Zorotypus hubbardi* is widespread in the southeastern United States. *Zorotypus snyderi* is known from southernmost Florida and Jamaica.

CHAPTER 30

Order Psocoptera (Psocids, Bark Lice, and Book Lice)*

ORDER PSOCOPTERA

Minute to small Exopterygota with stout, soft body and large mobile head with swollen clypeus. Mandibles of chewing type, maxillae with lacinia modified as a slender rod, detached from stipes; compound eyes usually large, convex; 3 ocelli present in winged, absent in wingless forms; antennae filiform, usually with 13 segments. Prothorax small, collarlike, especially in winged forms; mesothorax and metathorax subequal, sometimes with fused terga in wingless forms; forewings about 1.3 to 1.5 times longer than hind wings, with simplified venation; few crossveins or closed cells, pterostigma present; legs slender, adapted for walking, tarsi with 2 to 3 segments. Abdomen 9-segmented, without cerci. Nymphs similar to adults, except in the smaller number of antennal segments and ocelli, and undeveloped wings.

Because of their small size and retiring habits, Psocoptera frequently escape notice. However, they are exceedingly common animals in many habitats and number over 5000 species worldwide. Most psocids feed on algae, fungus, and/or lichens and live on bark, foliage, or stone surfaces. Some psocids are scavengers, and they often occur in buildings, where they occasionally cause damage to stored materials. For example, *Liposcelis*, the common book louse, sometimes infests insect collections or herbarium specimens and the bindings of books, as well as invading dry, starchy foodstuffs. In recent years, several species of *Liposcelis* have become serious pests of these stored foods. Heavy infestation destroys the product because of contamination with the bodies and feces of these insects. The most important pest species in North America are *Liposcelis decolor* and *L. entomophila*.

Psocids are among the more generalized of hemipteroid insects (Fig. 30.1), with chewing mandibles bearing well-developed molar and incisor lobes. The maxilla is specialized in having the detached lacinia modified as a stout, heavily sclerotized rod (Fig. 30.2b). Other maxillary structures are unmodified. The hypopharynx bears dorsally a concave sitophore sclerite, which is directly opposed by a cuticular knoblike structure on the dorsal wall of the cibarium (Fig. 30.2a). During feeding the rodlike laciniae are apparently shoved against the substrate as a brace to steady the head while lichen, fungal hyphae, algae, or detritus is scraped up with the mandibles. The

*New text by Emilie Bess and Edward Mockford; illustrated keys by Edward Mockford

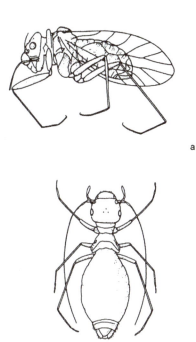

Figure 30.1 Representative adult Psocoptera: a, *Speleketor flocki*; b, *Psyllipsocus ramburi*.
Source: From Gurney, 1943.

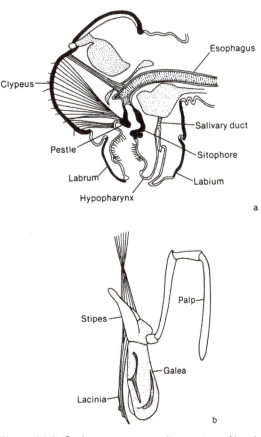

Figure 30.2 Oral structures: a, median section of head of *Ectopsocus briggsi*; b, maxilla of *Speleketor flocki*.
Source: Modified from Snodgrass, 1944.

sitophore apparatus provides an auxiliary grinding and crushing device. The hypopharynx also functions in taking up water vapor from moist atmospheres. The lacinia and sitophore in parasitic lice (Phthiraptera) are highly similar to the structures described above. Similarity in these characters and in the structure of the female genitalia provides strong evidence for a close relationship between Psocoptera and Phthiraptera and suggests that the two groups belong in a single order, Psocodea. Studies using both morphological and molecular data support the presence of three monophyletic groups within Psocodea: suborder Psocomorpha, suborder Trogiomorpha, and a third suborder containing parasitic lice and troctomoph psocids. Molecular studies indicate that parasitic habit of lice evolved within the troctomoph psocids. The psocid family Liposcelididae is the closest relative of the Amblycera parasitic lice, while the non-Amblyceran lice form a monophyletic group nested within the other troctomorph families.

Psocids are frequently gregarious, mixed groups of winged or wingless adults and immatures of various sizes congregating in areas that are damp but not wet. Caves, hollow stumps, lichen-covered rock faces, and similar protected situations are favored by some species, and these frequently enter man-made structures. Other species frequent the foliage, twigs, and branches of trees and shrubs—hence the name "bark lice." Fungi, especially bracket fungi on trees, leaf litter, and bird, mammal, wasp, and ant nests provide suitable habitats for certain forms. Many species are rapid, nimble runners, often moving backward or forward with equal ease. Many psocids have the ability to spin silk from labial glands. Some species use silk only as a dense cover for egg masses (families Caeciliusidae, Ectopsocidae, some Lachesillidae). Others spin a few strands of silk on the undersurfaces of leaves, seemingly providing a focus for aggregation of a few individuals (families Pseudocaeciliidae, Ectopsocidae, some Archipsocidae). Still others spin dense silken tents (family Archipsocidae). When these insects are abundant, entire branches or large areas of tree trunks may be covered by their harmless, gossamer webs, which protect the psocid colonies from marauding ants, spiders, and opilionids.

Polymorphism affecting adult (versus nymphal) characters is of common occurrence and may involve either sex or both. Short-winged and wingless forms lack ocelli, usually have smaller dorsal lobes of the meso- and metathorax, often retain nymphal truncated or apically expanded setae on many body surfaces, and may lack certain specialized adult sense organs (tarsal ctenidiobothria, paraproctal sensory field). Unlike the seasonal changes in Homoptera, polymorphism in Psocoptera is not obviously related to abiotic environmental factors. In some colonial species, long-winged forms appear only at high population density and disperse soon after reaching maturity.

Parthenogenesis is frequent, males being extremely rare or unknown in some species; in others both asexual and sexual races are known. In still others, sexual and asexual species pairs or complexes have been found. The asexual forms are often much more widespread than the sexual. In *Liposcelis bostrychophila*, the asexual form is the host of a microorganism that may control its sexuality. The *Trichadenotecnum alexanderae* species complex, consisting of a single sexual and four parthenogenetic species, has been investigated in considerable depth using a variety of procedures.

Copulation and precopulatory behavior varies with different taxa. In suborder Trogiomorpha, copulation is preceded by almost no "courtship" activity. Upon contact, the male backs under the female from in front, achieves copulatory attachment, and turns immediately through 180° to assume an end-to-end position for a long (1–2 hours) copulatory

period. In *Liposcelis* (suborder Troctomorpha), antennal vibrations alternating between the sexes occur before the male either runs over the female from back to front and pushes backward under her to achieve copulation or turns around after facing her and pushes under her from in front. In suborder Psocomorpha, two forms of courtship and copulatory activity exist. In family Archipsocidae, the female takes an active part in initiating copulation, touching the male's back with the fore tarsi, then mounting the male's back with fore- and middle legs. The male pushes backward with some bodily jerking to achieve genital contact. The two sexes remain facing the same direction throughout the brief (several seconds) copulatory period. In most other Psocomorph groups, the male, upon perceiving a suitable female, runs about flitting or rapidly vibrating its wings, then runs over the female from back to front and pushes under her from in front to achieve copulatory position, with both sexes facing the same direction. Partial or complete reversal of direction may occur before the two separate. Rapid copulation (a few seconds) usually involves placement of a spermatophore in the female's genital chamber, while slow copulation (1 hour or more) usually does not, but some exceptions are known.

Oviposition varies considerably among taxa. In family Lepidopsocidae, eggs are placed singly in crevices in the substrate, thus protecting them from grazing by the parent and conspecifics and, perhaps to some extent, from egg parasitoids. In *Liposcelis*, eggs are laid singly, bare, on open substrate, glued to a surface. Their small size probably excludes parasitoids, and adults and nymphs do not graze the eggs. In *Archipsocus*, eggs are laid in loose groups on the substrate in the communal web and are partially covered with faeces and other debris. In *Archipsocus gurneyi*, females stand guard over the egg mass, with several females guarding the probably communal mass. In Caeciliusidae and Ectopsocidae, eggs are deposited in groups on the substrate, and the egg cluster is covered with dense webbing. In families Mesopsocidae, Psocidae, and Myopsocidae, eggs are deposited in groups on the substrate and the group is covered with a cement of fine particles of food and bark passed rapidly through the female's digestive tract and secreted as a liquid paste, which rapidly hardens. The female often sculpts the surface in a particular design with the tip of her abdomen. Interestingly, the smallest known insect, the wasp *Dichomorpha echmepterygis* (Mymaridae), is an egg parasitoid of the bark louse *Echmepteryx hageni* (Lepidopsocidae).

Hatching from the egg involves rhythmic pressure of the head of the pronymph against the upper egg chorion. The undersurface of the pronymphal head being pushed against the chorion bears a sclerotized ridge with denticles or a knifelike edge that cuts a slit in the chorion. The pronymph pushes partway through this slit, then molts into a first instar, freeing itself completely from the egg.

The number of nymphal instars may vary relative to the state of wing development of the adult. The very small, apterous males of *Embidopsocus* undergo 3 nymphal instars, while the larger but also apterous females undergo 4. Females of *Liposcelis* (all apterous) also undergo 4 nymphal instars. The relatively small, micropterous males of *Archipsocus* undergo from 4 to 6, but usually 5 nymphal instars. Fully winged psocids usually undergo 6 nymphal instars. In all cases, the first instar has fewer antennal segments than the later instars. The first instar of *Liposcelis* has 9 antennal segments. With the first molt, each of the 6 basal flagellar segments divides into 2, thus producing the full complement of 15 segments. In the Psocomorpha, the first instar has 8 antennal segments, and in the same manner as in *Liposcelis*, the full complement of 13 is produced with the first molt.

Most psocids are solitary or assemble in loose groups on a food supply. Some species aggregate in small numbers on the undersurfaces of leaves, especially leathery leaves such as American holly (*Ilex americana*) and live oaks (*Quercus* spp.). These species usually spin a few strands of silk over the leaf surface, thus marking the assembly area, and then forage for food out from the aggregation leaf. The large species of the tribe Cerastipsocini (Psocidae) aggregates as nymphs in large herds, often numbering several hundred individuals. Maintenance of the herd is controlled by an aggregation pheromone. The herd disperses soon after adults reach full coloration after the final molt. In the archipsocid genera *Archipsocus* and *Archipsocopsis*, nymphs and adults spin communal webs on tree trunks and branches, and several

generations may persist under the same web, gradually expanding the web as the population grows. Feeding occurs only under the web, and the web expands to incorporate new food supply. All males and most females are short-winged and leave the web only to add to its surface. Long-winged females appear when the population in the web becomes high. These females disperse, but they copulate before leaving the web, so that much inbreeding occurs.

Collecting methods for psocids include searching tree trunks, beating branches and foliage, sifting litter, sweeping grasses, and the use of light traps, Malaise traps, and soil and litter extraction traps. Tree trunk searching is most productive in areas of sparse trees, where much light reaches the trunks. Here food is often abundant on the trunks, while in dense forest it is not. In tropical forests, large buttress-based trees are important havens for a diversity of psocids. Beating over a white sheet or a black umbrella is a very effective collecting method, especially for collecting inhabitants of dead branches, hanging dead leaves, conifer foliage, and palm leaves. Collecting in ground litter offers some difficulty in that psocids congregate in favorable areas, leaving large areas sterile, and favorable areas cannot be judged easily. Small patches of well-drained litter on hillsides in open woodland are usually more productive than abundant litter in deep forests. Light trapping and Malaise trapping can be very productive, but no habitat data come with the specimens.

Rearing of psocids involves careful control of relative humidity and maintenance of an abundant food supply. A relative humidity of 80 percent at an ambient temperature of 20°C is suitable for most psocid rearing, and these conditions are easily maintained in a closed container over a saturated solution of KCl, which regulates humidity. The psocids may be kept in shell vials plugged with cotton. Bark-inhabiting species should be provided with pieces of bark or twigs bearing the same food (algae or lichens) on which they were feeding in the field. The culture vials should be large enough to allow introduction of fresh bark or twigs when the food supply is exhausted on the old ones. The psocids usually transfer themselves to the new food supply. Gentle urging with a camel's hair brush is sometimes necessary. Dead leaf inhabitants are somewhat more difficult in that it is not so easy to see when the food supply is gone. Dead palm and oak leaves often have a growth of small black rusts. When the leaf surface looks clean, the food supply is gone. Stored product psocids can usually be reared on the stored product in which they were found.

KEY TO SUBORDERS OF PSOCOPTERA (ADULTS)

1. With >18 flagellomeres; hypopharyngeal filaments separate throughout their length (Fig. 30.3a); labial palpus with minute basal segment and rounded distal segment (Fig. 30.3b); tarsi three-segmented
Suborder Trogiomorpha

With < 18 flagellomeres (usually 11–13); hypopharyngeal filaments fused on midline at least part of their length (Fig. 30.3c); labial palpus as above or with no basal segment; tarsi two- or three-segmented   **2**

2. With 13 flagellomeres (rarely fewer); at least some flagellomeres annulated with cuticular sculpture (Fig. 30.3d). Labial palpi usually with minute basal segment and rounded distal segment. Tarsi three-segmented (rarely two-segmented). Forewing of macropterous forms lacking sclerotized pterostigma   **Suborder Troctomorpha**

With 11 flagellomeres (rarely fewer); no flagellomeres annulated with cuticular sculpture but sometimes with reticulate sculpture pattern (Fig. 30.3e). Labial pal-

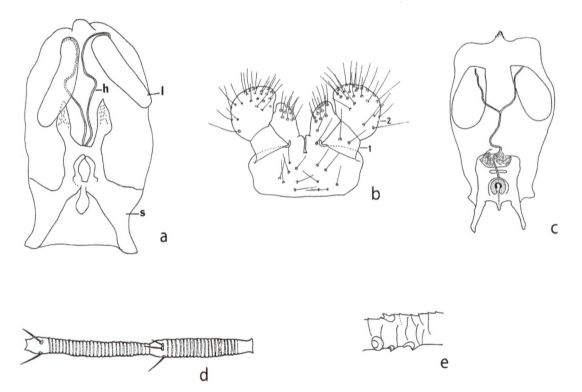

Figure 30.3 a, *Echmepteryx madagascariensis* (family Lepidopsocidae), hypopharynx (s, sitophore sclerite; l, lingual sclerite; h, hypopharyngeal filament); b, *E. madagascariensis*, labium with palpal segments numbered; c, *Peripsocus madidus* (family Peripsocidae), hypopharynx (filaments partially fused); d, *Liposcelis* sp. (family Liposcelididae), two flagellomeres with annulate sculpture; e, *Hemipsocus* sp. (family Hemipsocidae), flagellomere with reticulate sculpture.

pus consisting only of single rounded or triangular segment, without basal segment. Forewings of macropterous forms with sclerotized pterostigma or forewing membrane densely hairy
Suborder Psocomorpha

Suborder Trogiomorpha: Key to Families

1. Spur sensillum always present on 2nd maxillary palpomere (P2) (Fig. 30.4a).

Forewing, when fully developed, with veins Cu2 and IA ending separately on wing margin. Ovipositor valvulae: always with a long, setose outer valvula (v3) sometimes also with a short, slender inner valvula (v1, Fig. 30.4b) **Infraorder Atropetae**   **2**

Spur sensillum of P2 present or absent. Forewing, when fully developed, with veins Cu2 and IA ending together on wing margin (the nodulus).

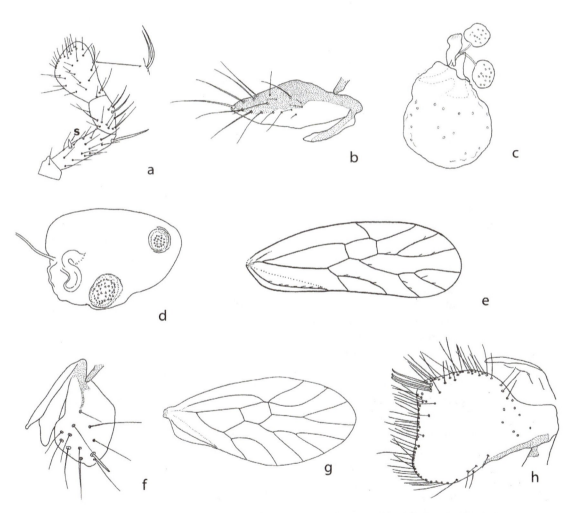

Figure 30.4 a, *Echmepteryx madagascariensis*, maxillary palpus (s = spur sensillum); b, *Soa flaviterminata* (family Lepidopsocidae), ovipositor valvulae; c, *Psoquilla infuscata* (family Psoquillidae), spermatheca; d, *Lepinotus inquilinus* (Family Trogiidae), spermathecal sac; e, *Psyllipsocus oculatus*, forewing; f, *Psocathropos lachlani* (family Psyllipsocidae), ovipositor valvulae; g, *Speleketor flocki* (family Prionoglarididae), forewing; h, *S. flocki*, ovipositor valvulae.

Ovipositor valvulae: v3 setose, never much longer than broad; usually two other valvulae present, both slender and bare: a median (v2) and an inner (v1)

Infraorder Psocatropetae   4

2(1). Body and forewings covered with scales or dense setae. Wings usually pointed apically; if rounded, the scale-covered forewing broad, nearly half as wide as long, or wings greatly reduced, rounded apically, but covered with scales   **Lepidopsocidae**
Wings rounded apically both when fully developed and when reduced. Body and forewings never covered with scales or dense setae   **3**

3(2). Wings with visible veins even when much reduced. Ovipositor consisting entirely of a pair of v3s. Spermathecal sac with a protruding beak and a conspicuous pair of accessory bodies near its opening (Fig. 30.4c)   **Psoquillidae**
Wings greatly reduced (rarely absent), lacking visible veins. Spermathecal accessory bodies usually consisting of two denticulate plaques (maculae) attached to the wall of the sac (Fig.30.4d)   **Trogiidae**

4(1). Wings fully developed or much reduced; when fully developed, basal segment of Sc short, ending blindly (Fig. 30.4e). Ovipositor valvulae: v3 approximately twice as broad as v1 and v2 together, but not forming a large, rounded plate (Fig. 30.4f)   **Psyllipsocidae**
Wings fully developed; basal segment of vein Sc in forewing a long arc joining vein R (Fig. 30.4g). Ovipositor valvulae: v3 developed as a large, rounded plate with or without a small, accessory valvula on its median margin (Fig. 30.4h)   **Prionoglarididae**

Suborder Trogiomorpha

This suborder contains about 360 described species. Most of them are tropical, but several are widespread household inquilines. There are also some early Cretaceous fossils that may represent the most early-derived psocids.

Archaeatropidae

This family is known only from early Cretaceous (114 mya) amber fossils from northern Spain. It is of much interest in demonstrating that some of the specialized characters of Infraorder Atropetae (the major division of Suborder Trogiomorpha), such as the elongate third valvulae, were already established at that time.

Empheriidae

This extinct family is known from five genera, two in early Cretaceous amber from northern Spain, one in later Cretaceous amber from Siberia, and two in the mid-Eocene Baltic amber (ca. 44 mya).

Lepidopsocidae

This group of slightly over 200 described species is largely tropical, with important centers of diversity in the islands of the Pacific and Indian Oceans and the Greater Antilles. Only six species appear to be native to North America, and the entire fauna of the western Palaearctic appears to have been introduced. These insects are often found on palm, banana, and *Heliconia* leaves, especially dead hanging leaves. They run rapidly when disturbed, and winged adults take flight readily. *Echmepteryx hageni* is common on tree trunks and branches in eastern United States.

Trogiidae

This is a small family of 55 described species in nine genera. Each of the genera *Cerobasis*, *Lepinotus*, and *Trogium* includes one or more household inquiline species. Females of *Lepinotus* and *Trogium* produce a sound audible to the human ear by tapping the subgenital plate on a resounding surface. In addition to the modern genera, the genus *Eolepinotus* was described from upper Cretaceous Siberian amber (ca. 80 mya).

Psoquillidae

This is a small family of 28 described species in six genera. Some species of the genera *Psoquilla* and *Rhyopsocus* occur at least temporarily as household inquilines in areas of warm climate. These insects may be found in hanging dead leaves and under loose bark in tropical and subtropical areas.

Psyllipsocidae

This is a small family with 47 named species in five genera. *Psyllipsocus ramburii* occurs in caves and basements throughout Europe and North America. *Psocathropos lachlani* is a common household inquiline in areas of warm climate. Several species of *Dorypteryx* are household inquilines in various areas of the world. In addition to the modern genera, the genus *Khatangia* is known from Cretaceous Siberian amber.

Prionoglarididae

This is a small family of 13 named species in five genera. Most members are cave dwellers, showing a spectacular array of primitive characters and unique apomorphies. The family is distributed spottily across North America, Africa, and Eurasia, suggesting great antiquity (as do also some of the plesiomoprhic characters), possibly predating the breakup of Pangaea.

Suborder Troctomorpha: Key to Families

1. Small forms, usually less than 2 mm in length. Wings, when present and fully developed (females only), without scales, with not more than 2 M branches in forewing, and vein IIA never present. Coxal organ absent or represented by a slight bulge in cuticle
   Infraorder Nanopsocetae   2
   Larger forms, usually 2.5–5 mm in length. Wings present in both sexes; forewings covered with scales or not. In forewing, M 3-branched, vein IIA usually present. Coxal organ usually represented by mirror and a small rasp (Fig. 30.5a)
   Infraorder Amphientometae   4
2(1). Body flattened. Coxae of opposite sides widely separated by broad sternal plates. Forewings, when present (some females), with two parallel veins running most of the length of the wing (Fig. 30.5b)
   Family Liposcelididae
   Body not flattened. Coxae of opposite sides only narrowly separated. Forewings, when present (some females), with several branching veins occupying main body of wing   3
3(2). Compound eyes reduced to 7–10 units and situated forward from hind margin of head. Females with elytriform forewings (Fig. 30.5c) and no hind wings. Males micropterous or apterous; phallosome with basal struts joined anteriorly on midline and extending forward from junction as a short apodeme (Fig.30.5d)
   Family Sphaeropsocidae
   Compound eyes with many facets, situated touching hind margin of head. Macropterous or apterous, rarely micropterous. Macropterous females with forewings membranous, folding flat over back at rest; hind wings present, of same texture and orientation as forewings. Males apterous; phallosome with basal struts fused antero-medially but with no anterior apodeme (Fig. 30.5e)   **Family Pachytroctidae**
4(1). Forewing membranous or coriaceous, but never covered with scales. In forewing, basal region of Sc curving posteriorly and joining R stem
   Superfamily Electrentomoidea 5
   Membrane of forewing covered with dense scales and setae. In forewing, basal region of Sc directed antero-distally and usually reaching anterior wing margin
   Superfamily Amphientomoidea (Family Amphientomidae)
5(4). Front femur with row of denticles or spines   6
   Front femur lacking denticles or spines   7
6(5). Outer cusp of lacinial tip elongate (Fig. 30.5f). Flagellum 11- to 12-segmented. M in hind wing branched
   Family Compsocidae
   Outer cusp of lacinial tip short (Fig. 30.5g). Flagellum 12- to 13-segmented. M in hind wing simple   **Family Protroctopsocidae**
7(6). Pterostigma closed basally. Flagellum 13-segmented
   Family Manicapsocidae
   Pterostigma open basally. Flagellum 9- to 11-segmented   8

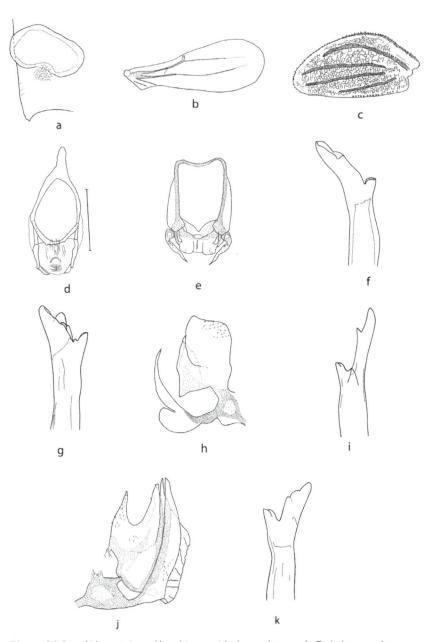

Figure 30.5 a, *Lithoseopsis* sp. (Amphientomidae), coxal organ; b, *Embidopsocus* luteus (Liposcelididae), forewing; c, *Sphaeropsocopsis argentina* (Sphaeropsocidae), forewing; d, *Prosphaeropsocus pallidus* (Sphaeropsocidae), phallosome; e, *Nanopsocus oceanicus* (Pachytroctidae), phallosome; f, *Compsocus elegans* (Compsocidae), lacinial tip; g, *Protroc-topsocus enigmaticus* (Protroctopsocidae), lacinial tip; h, *Musapsocus huastecanus* (Musap-socidae), ovipositor valvulae; i, *Musapsocus tabascensis*, lacinial tip; j, *Troctopsocus bicolor* (Troctopsocidae), ovipositor valvulae; k, *Troctopsocus separatus*, lacinial tip.

8(7). V3 with apex only shallowly indented if at all (Fig. 30.5h). Outer cusp of lacinial tip elongate (Fig. 30.5i)
Family Musapsocidae
V3 with apex deeply indented (Fig. 30.5j). Outer cusp of lacinial tip short (Fig. 30.5k)
Family Troctopsocidae

Suborder Troctomorpha

This suborder includes some 500 described species in nine families. Many species await description. Some of the smallest psocids are included here.

Electrentomidae

This is a small group of 10 named species, entirely tropical and south-temperate in distribution. The genus *Epitroctes* is found on trunks of large trees in tropical forests of Mexico, Central America, and Trinidad. The genera *Electrentomum* and *Parelectrentomum* are fossils in Eocene Baltic amber.

Compsocidae

This family contains only two genera, each with a single species, found on trunks of tropical forest trees in southern Mexico and Central America.

Protroctopsocidae

This is a small family of four described species in three genera, two of them found in the Mediterranean region of Europe and the third in the eastern mountains of Mexico. They are inhabitants of ground litter.

Troctopsocidae

This is a small family of 22 described species in six genera found in the American and Asian tropics. They occur on ferns and other low vegetation on the edges of tropical forest stands, where their numbers are usually low.

Musapsocidae

This is a small family of two genera and nine species restricted to tropical America. These psocids inhabit foliage of palms, heliconias, and banana.

Amphientomidae

This is a family of moderate size, with 150 named species in 20 genera. The greatest diversity is found in southeastern Asia, the Sunda Archipelago, and Madagascar, with relatively poor representation in West Africa and tropical America. Very few species reach Europe and North America. These insects are found on rock outcrops and trunks of large forest trees. They run rapidly when disturbed.

Liposcelididae

This is a relatively large family, with nearly 200 described species in nine genera. Members of subfamily Embidopsocinae are dimorphic, with wingless males and fully winged and wingless females. Most of them live under loose bark. In subfamily Liposcelidinae, the large genus *Liposcelis* (ca. 125 named species) includes some serious pests of stored grain and several common household pests. Most species of *Liposcelis* occur under bark of trees and in ground litter.

Pachytroctidae

This is a family of moderate size, with some 90 named species in 10 genera. Most taxa are tropical, and only five species reach the United States. One of these, *Nanopsocus oceanicus*, has been distributed in human commerce and is found occasionally as an inquiline in human dwellings. Many of the species are found on leaves of palms, bananas, and other tropical monocots. Dimorphism is common, and males are generally wingless.

Sphaeropsocidae

This is a small family, with 21 named species in eight genera. An early fossil, *Sphaeropsocites lebanensis*, from lower Cretaceous Lebanese amber (ca. 125 mya) shows the great antiquity of the family and also demonstrates that the elytriform forewings with reduced venation, characteristic of females of the entire family, were already established at that time. Most of the modern species occur in ground litter and soil. *Badonnelia titei* is inquiline in human dwellings in northern Europe and Alaska.

Suborder Psocomorpha: Key to Families

1. Forewings, when fully developed, with membrane densely setose and venation indistinct (Fig. 30.6a). Brachypterous and micropterous forms with body densely setose, pronotum only slightly smaller than meso- and metanotum. Colonial forms living under webbing
Infraorder Archipsocetae (Family Archipsocidae)
Forewing setae largely confined to veins and margin. Venation of forewing distinct. Brachypterous and micropterous forms with body not densely setose (see also separate key to micropterous and apterous forms). Pronotum generally much smaller than meso- and metanotum **2**

2(1). With the combination of characters: mandibles elongate and hollowed-out posteriorly to accommodate the galeae (Fig. 30.6b) and labrum with a pair of sclerotized ridges running at least part-way through its length (Fig. 30.6c)
Infraorder Epipsocetae   3
Not with the above combination of characters. Labrum never with a pair of sclerotized ridges running through its length (although sometimes with a pair of pigmented bands). Mandibles hollowed-out posteriorly or not **7**

3(2). Labral sclerites reaching hind margin of labrum and each continuing in an arc to side of labrum (as in Fig. 30.6c) **4**
Labral sclerites not reaching hind margin of labrum (Fig. 30.6d) **5**

4(3). Tarsi 3-segmented; vein IIA present in forewing; antennal scape completely sclerotized
Cladiopsocidae
Tarsi 2-segmented; vein IIA absent in forewing; antennal scape membranous on most of ventral surface (Fig. 30.6e)
Epipsocidae,
incl. Neurostigmatidae

5(3). Tarsi 3-segmented; some genera with multiple branches of Rs and M in forewing but pterostigma without spur on hind margin **Ptiloneuridae**
Tarsi 2-segmented; forewing usually without multiple branches of Rs and M but

some forms with pterostigmal spur also having these branches **6**

6(5). Pterostigmal spur vein present in forewing (Fig. 30.6f); pulvillus straight and pointed at tip **Spurostigmatidae**
Pterostigmal spur vein absent. Pulvillus bent near base, knobbed at tip (Fig. 30.6g)
Dolabellopsocidae

7(2). Meso-precoxal bridges wide; meso-trochantins narrow throughout (Fig. 30.6h). Vein Cu1a in forewing usually not joined to M **8**
Meso-precoxal bridges narrow; meso-trochantins wide basally (Fig. 30.6i). Vein Cu1a in forewing usually joined to M either directly or by a crossvein
Infraorder Psocetae   22
Note: This couplet does not work for micropterous and apterous specimens. See separate key to these forms of suborder Psocomorpha.

8(7). Third ovipositor valvulae greatly reduced or plastered to body wall or absent (Fig. 30.6j). Mandibles elongate and hollowed-out posteriorly, thus accommodating bulbous galea (Fig. 30.7a) (exceptions: *Notiopsocus* and *Pronotiopsocus*, with 7 sensilla in distal row of labrum)
Infraorder Caecilietae   9
Third ovipositor valvulae well-developed, usually setose (Fig. 30.7b).
Mandibles short, not hollowed posteriorly; galeae somewhat flattish.
Infraorder Homilopsocidea   13

9(8). Abdomen with 1–3 ventral eversible vesicles (Fig. 30.7c). Labrum
decidedly flat, producing with large galeae an open-mouth appearance; pulvilli broad throughout **Superfamily Caeciliusoidea**
10
Abdomen without ventral eversible vesicles. Labrum at least to some extent curved around mandibles. Pulvilli narrowing before broad apex, or slender, or absent
Superfamily Asiopsocoidea (Asiopsocidae)

10(9). In forewing, pterostigma-R2 + 3 crossvein and M-Cu1a crossvein present (Fig. 30.7d).

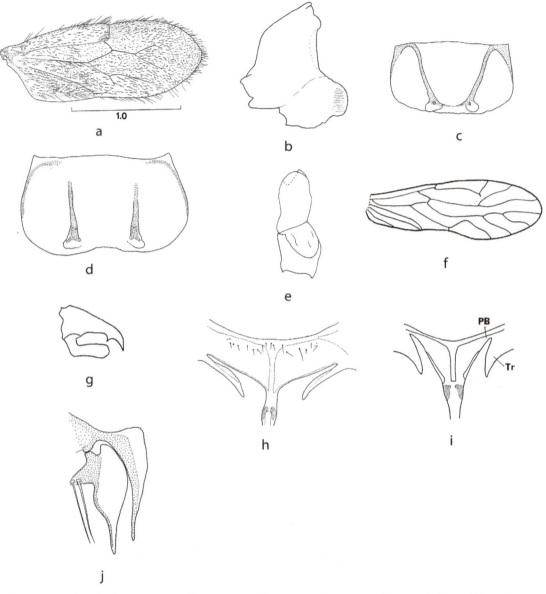

Figure 30.6 a, *Pseudarchipsocus guajiro* (Archipsocidae), forewing; b, *Epipsocus* sp. (Epipsocidae), mandible; c, *Bertkauia lepicidinaria* (Epipsocidae), labrum; d, *Loneura* sp. (Ptiloneuridae), labrum; e, *Bertkauia lepicidinaria*, antennal scape and pedicel; f, *Spurostigma epirotica* (Spurostigmatidae), forewing; g, *Dolabellopsocus* sp. (Dolabellopsocidae), pretarsal claw; h, *Mesopsocus laticeps* (Mesopsocidae), meso-precoxal bridges and meso-trochantins; i, *Trichadenotecnum alexanderae* (Psocidae), meso-precoxal bridges and meso-trochantins; j, *Xanthocaecilius granulosus* (Caeciliusidae), ovipositor valvulae.

Hind wing margin without setae or with a few restricted to cell R3
      Family Stenopsocidae

In forewing, pterostigma-R2 + 3 crossvein absent; Cu1a usually free from M, rarely fused to M   11

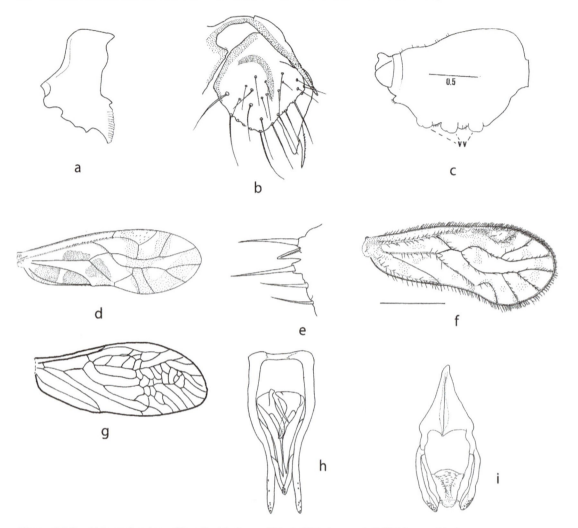

Figure 30.7 a, *Valenzuela micans* (Caeciliusidae), mandible; b, *Trichopsocus dalii* (Trichopsocidae), ovipositor valvulae; c, *Xanthocaecilius* sp. female, abdomen, showing ventral eversible vesicles; d, *Graphopsocus cruciatus* (Stenopsocidae), forewing; e, *Teliapsocus conterminus* (Dasydemellidae), female, spines of paraproct edge; f, *Philotarsus parviceps* (Philotarsidae), female, forewing; g, *Calopsocus* sp. (Calopsocidae), forewing; h, *Heterocaecilius* sp. (Pseudocaeciliidae), phallosome; i, *Philotarsus kwakiutl*, phallosome.

11(10). Ciliation of hind wing margin restricted to cell R3 or none. Spine of free margin of paraproct relatively large (Fig. 30.7e)
Dasydemellidae
Hind wing margin ciliated except basal two-thirds of front margin. Spine of free margin of paraproct smaller or absent   12

12(11). Setae of veins in distal half of forewing on both dorsal and ventral surfaces. P4 = P2 in length   **Amphipsocidae**

Setae on veins in distal half of forewing only on dorsal surface. P4 < P2 in length
Caeciliusidae

13(8). Setae of median cell margins of forewing in two series forming crossing pairs (Fig. 30.7f)   14
Crossing pairs of setae absent on forewing margin   16

14(13). Hind margin of head in lateral view acute-angled. Forewings broad, L/W index ca. 2.

Forewings often with extra veins, sometimes forming reticulate pattern (Fig. 30.7g)   **Calopsocidae**
Hind margin of head rounded in lateral view. Forewings of normal width, L/W index ca. 3.1. Forewings without extra veins   15

15(14). Five distal inner labral sensilla. External parameres (Fig. 30.7h) generally much longer than aedeagal arch (the arch sometimes absent). Primarily leaf-inhabiting forms
Pseudocaeciliidae
Nine distal inner labral sensilla. External parameres only slightly longer to shorter than aedeagal arch (Fig. 30.7i). Bark-inhabiting forms
Philotarsidae

16(13). In forewing, vein Cu1a absent   17
In forewing, vein Cu1a present   18

17(16). Vein R1 in forewing (closing pterostigma posteriorly) parallel to wing margin most of its length (Fig. 30.8a). Vertex bearing numerous setae. Male endophallus complex, asymmetrical (Fig. 30.8b)
Ectopsocidae
Vein R1 in forewing (closing pterostigma) closer to wing margin at base than near apex of pterostigma (Fig. 30.8c). Vertex bare, or with few very short setae. Male endophallus a pair of symmetrical sclerites or a median symmetrical fork (Fig. 30.8d)
Peripsocidae

18(16). Female abdominal terga well-sclerotized and fused together. Forewing membrane setose. M-Cu stem in forewing with 2 ranks of setae   **Bryopsocidae**
Not with above characters   19

19(18). V2 lacking distal process or V2 absent. Almost all setae of V3 on or near outer surface. Subgenital plate rounded distally
Lachesillidae
V2 with distal process. Setae of V3 usually scattered over surface. Subgenital plate with at least slight indication of median distal protuberance   20

20(19). Wings lacking setae. Aedeagal arch rounded distally (Fig. 30.8e). Subgenital plate with single median extension (egg guide, Fig. 30.8f)   **Mesopsocidae**
Wings at least slightly setose on veins and margin. Aedeagal arch (when present) acute-angled or with pointed process distally. Subgenital plate variable but usually not as above   21

21(20). Wings lightly ciliated; ciliation of hind wing restricted to margin in cells R1 and R3. Robust forms (some apterous) inhabiting primarily tree branches   **Elipsocidae**
Wings well ciliated; ciliation of hind wing on margin extending from cell R1 around to cell Cu2. Delicate forms (all macropterous) inhabiting foliage of trees and shrubs
Trichopsocidae

22(7). M in forewing 2-branched and joined to Cu1a directly or by a crossvein (Fig. 30.8g). Subgenital plate rounded distally, lacking an egg guide   **Hemipsocidae**
M in forewing 3-branched. Subgenital plate with a median egg guide, sometimes developed as a separate sclerite   23

23(22). Cu1a in forewing forming a cubital loop not joined to M in any way. Tarsi 3-segmented
Psilopsocidae
Cu1a in forewing either joining directly to M or joined by a crossvein. Tarsi either 2- or 3-segmented   24

24(23). Tarsi 3-segmented. Forewings heavily blotched with brown, the margins with alternating brown and colorless banding (Fig. 30.8h)   **Myopsocidae**
Tarsi 2-segmented. Forewing markings variable, sometimes with extensive brown blotching but margins never with alternating brown and colorless banding (Fig. 30.8i)   **Psocidae**

Family Key to Micropterous and Apterous Forms of Suborder Psocomorpha

1. Body densely setose. Pronotum only slightly smaller than meso- and metanota. Colonial forms living under webbing
Archipsocidae
Body not densely setose. Pronotum usually smaller than meso- or metanotum.

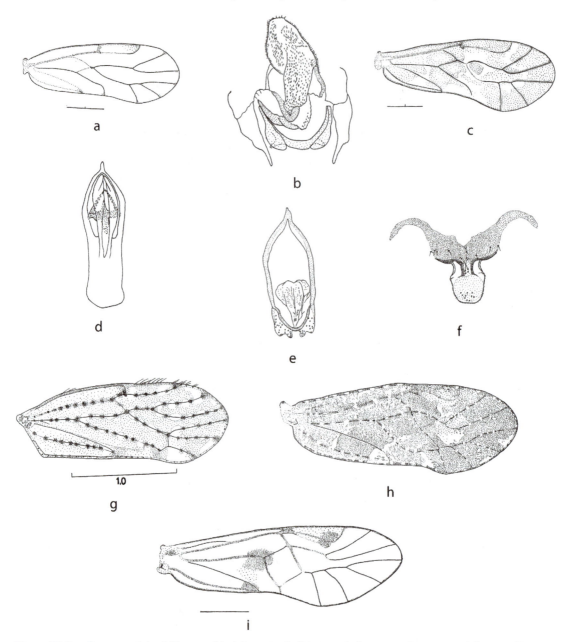

Figure 30.8 a, *Ectopsocus briggsi* (Ectopsocidae), forewing; b, *E. briggsi*, phallosome; c, *Peripsocus subfasciatus* (Peripsocidae), female, forewing; d, *Peripsocus madidus*, phallosome; e, *Mesopsocus unipunctatus* (Mesopsocidae), phallosome; f, *M. unipunctatus*, subgenital plate; g, *Hemipsocus pretiosus* (Hemipsocidae), forewing; h, *Lichenomima coloradensis* (Myopsocidae), forewing; i, *Psocus leidyi* (Psocidae), forewing.

Free-living forms or single individuals under a small web   2

2(1). Mandibles elongate and hollowed-out posteriorly to accommodate the galea. Labrum with a pair of sclerotized ridges running through its length   **Family Epipsocidae**
Without the above combination of characters   3

3(2). Large females (>3 mm body length) with a single median protuberance (egg guide) on subgenital plate   **4**

Smaller forms of both sexes or subgenital plate not as above   **5**

4(3). With three tarsomeres

Mesopsocidae

With two tarsomeres   **Psocidae**

5(3). Females, with at least one ventral abdominal vesicle, often 2, occasionally 3. Third valvula reduced or plastered to body wall **6**

Forms of either sex, lacking ventral abdominal vesicles   **7**

6(5). Large females (ca. 3 mm) with lacinial tip broad and nondenticulate

Dasydemellidae (genus *Ptenopsila*)

Smaller females with lacinial tip slender or denticulate   **Caeciliusidae**

7(5). Females, with subgenital plate rounded distally and ovipositor of a single valvula on each side or none   **8**

Forms of either sex, but females with subgenital plate having 1 or 2 distal protuberances and ovipositor of 3 valvulae each side

10

8(7). Tarsi 3-articulate. Ovipositor valvulae absent. Preclunial abdominal segments fused together as a hard carapace

Family Elipsocidae (females of genus *Lesneia*)

Tarsi 2-articulate. Ovipositor valvulae present. Abdomen not as above   **9**

9(8). Pretarsal claw with a preapical denticle (Fig. 30.9a)   **Lachesillidae**

Pretarsal claw lacking a preapical denticle (Fig. 30.9b) **Asiopsocidae** (females of genera *Asiopsocus* and *Pronotiopsocus*)

10(7). Females, with 3 tarsomeres; 2 distal protuberances (sometimes very shallow) on subgenital plate   **Elipsocidae** (genera *Elipsocus, Hemineura, Roesleria*)

Forms of either sex. Females with two tarsomeres or with three tarsomeres and a single median protuberance on subgenital plate   **11**

11(10). Small females (ca. 1.5 mm) with single median protuberance on subgenital plate, 2 or 3 tarsomeres, large spine on paraproct (Fig. 30.9c)

Family Mesopsocidae (genera *Psoculus* and *Psoculidus* with 3 tarsomeres, *Palmicola* with 2 tarsomeres)

Somewhat larger females with 2 tarsomeres, lacking large spine on paraproct, or males

12

12(11). Females with V3 decidedly slender (Fig. 30.9d). Males with clunial comb of elongate denticles (Fig.30.9e)

Family Ectopsocidae

Females with V3 normal or males without clunial co   **13**

13(12). Forms of both sexes with stout spinulate or knobbed setae on the

body **Family Elipsocidae** (genus *Paedomorpha* with 2 tarsomeres, genus *Nepiomorpha* with 3 tarsomeres)

Lacking stout spinulate or knobbed setae on the body   **14**

14(13). Small (ca. 1 mm) apterous males

Family Asiopsocidae (genus *Notiopsocus*)

Females, 1.5–2 mm in length   **15**

15(14). Subgenital plate with 2 distal protuberances (Fig. 30.9f) **Family Elipsocidae** (genus *Reuterella*) Subgenital plate with a single distal protuberance (as in Fig. 30.9g)

Family Peripsocidae

Suborder Psocomorpha

Psocomorpha is the largest of the three suborders of Psocoptera, including 24 families and some 3600 species. Most psocomorphs are fully winged and hold their wings tentlike over the abdomen when at rest, but wing polymorphism is widespread within species. This suborder includes the largest bark lice, and body size ranges from 1 to12 mm. They generally live on tree bark, foliage, stone surfaces, or leaf litter and feed on lichens, algae, and fungus. Psocomorphs exhibit a range of interesting behaviors, including web-spinning, aggregation, subsociality, nesting, and sound production by stridulation. The earliest known fossil of Psocomorpha is a Psocidae wing that originated during the Late Jurassic (ca. 152 mya) in Karatau, Kazakhstan. A variety of Cretaceous amber fossils from families Elipsocidae, Lachesillidae, and Psocidae have also been found.

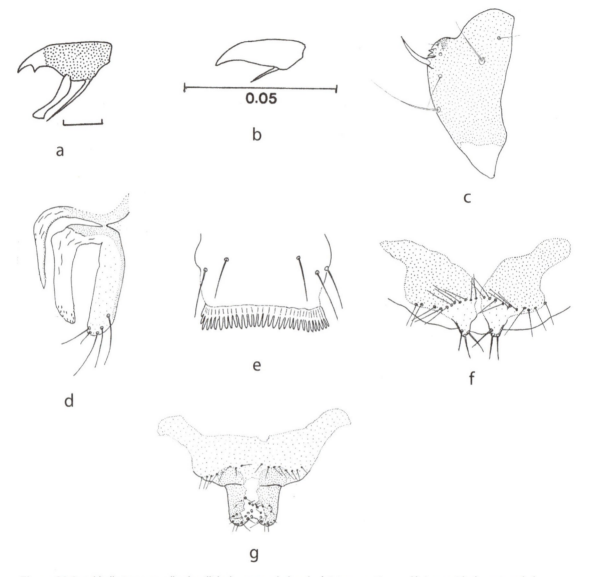

Figure 30.9 a, *Nadleria gamma* (Lachesillidae), pretarsal claw; b, *Asiopsocus spinosus* (Asiopsocidae), pretarsal claw; c, *Palmicola* sp. (Mesopsocidae), female, paraproct; d, *Ectopsocus pumilis*, ovipositor valvulae; e, *Ectopsocus vachoni*, male, clunial comb; f, *Reuterella helvimacula* (Elipsocidae), subgenital plate; g, *Peripsocus subfasciatus*, subgenital plate.

Amphipsocidae

This family includes some 170 species in 17 genera with the center of diversity in Asia and Africa. One species, *Polypsocus corruptus*, is known from North America. Amphipsocids are large-sized bark lice, 4 to 6 mm in length. They are known to inhabit broad-leaved, evergreen, and rain forest trees.

Archipsocidae

This family contains about 80 species in 5 genera distributed worldwide, with the greatest diversity in Central and South America. Seven species are known from North America in the genera *Archipsocus*, *Archipsocopsis*, and *Pararchipsocus*. These are small to medium-sized bark lice, 1.5 to 2.5 mm in length. Body colors are red- or orange-brown, sometimes pale, with creamy white wings. Almost all body surfaces are densely covered with long hairs. Archipsocids live on the surface of bark, often in groups under dense webs.

Asiopsocidae
This family includes 14 species in 3 genera with worldwide distribution. Three species are known from North America: *Asiopsocus sonorensis* from the southwestern United States, *Notiopsocus* sp. from southern Florida, and *Pronotipsocus* sp. from south-central Florida. Asiopsocids are small to medium-sized bark lice, 2 to 3 mm in length, generally brown in color. Asiopsocids inhabit bark, and sexual dimorphism is common in this family, with one sex lacking wings when the other has full wings.

Bryopsocidae
This family includes two species in the genus *Bryopsocus*, known only from damp forests in New Zealand. These medium-sized bark lice (2–3 mm in length) are the only psocids known to inhabit moss. Adults can have full-length or shortened wings, and females have a distinctive sclerotized area on the rear dorsal surface of the abdomen.

Calopsocidae
This family includes some 34 species in 8 genera distributed in tropical regions of Southeast Asia, Australia, and the Pacific Islands. Calopsocids are medium to large-sized bark lice, 4 to 6 mm in length. Forewings have numerous secondary veins, creating a lacy look. Body colors are generally light to dark brown, but some are very brightly colored. Calopsocids inhabit green leaves, including palm fronds.

Cladiopsocidae
This family includes 2 genera. *Cladiopsocus* with 16 species is known from Angola and from Central and South America. *Spurostigma* with 11 species is known from Central and South America. Body color is brown. These bark lice inhabit both wet and dry vegetation in tropical regions, but little is known about their habitat preferences.

Caeciliusidae (formerly Caeciliidae)
This family includes includes some 600 described species in more than 30 genera, distributed worldwide, with the highest diversity in Asia and Africa. There are about 35 species of the family in North America. Caeciliusids are small to medium-sized bark lice, approximately 3 mm in length, frequently yellow in color, but many are brown or red. Caeciliusids are leaf-inhabiting bark lice that live on conifers and broadleaf trees. Most are long-winged, but some of the few ground litter species have both long- and short-winged females. This is the largest family of leaf-inhabiting bark lice.

Dasydemellidae
This family includes 30 species in 4 genera distributed in North and South America and in Asia. One species, *Teliapsocus conterminus*, is known from North America and is widespread throughout the United States and southern Canada, where it is found in the foliage of a variety of trees and occasionally in ground litter. These are fairly large bark lice, 4 to 6 mm in length.

Dolabellopsocidae
This family includes 37 species in 3 genera. *Dolabellopsocus* and *Isthmopsocus* are distributed in Central and South America. *Auroropsocus* is found in China and India. They inhabit trees and shrubs in wet and dry habitats, including palms, ferns, heliconia, and broadleaf trees.

Ectopsocidae
This family includes some 200 species in 6 genera distributed worldwide, with the highest diversity in Asia. Fourteen species are known from North America in the genera *Ectopsocus* and *Ectopsocopsis*. These are small-sized bark lice, 1.5 to 2.5 mm in length, with robust bodies. Body colors are brown; wings are clear with or without markings. Ectopsocids generally hold their wings horizontal over their abdomen at rest and have a

distinctive rectangular-shaped pterostigma. They inhabit dead leaves on tree branches and leaf litter.

Elipsocidae
This family includes some 130 species in 23 genera distributed worldwide. Nine species in 5 genera are known from North America. These are small to medium-sized bark lice 1 to 4.5 mm in length, with robust bodies. Body colors are brown or blackish. Elipsocids inhabit bark, stone surfaces, dead leaves, and leaf litter. Some species are sexually dimorphic (different body forms in male and female) or polymorphic (different body forms within a sex).

Epipsocidae
This family includes more than 140 species in some 20 genera worldwide. The family is mainly tropical with major radiations in South America, Africa, and Malaysia. Two species are known from North America in the genus *Bertkauia*. Epipsocids are medium to large-sized bark lice, most are 5 mm in length or smaller, but some are as large as 10 mm. They live on bark, in foliage, in leaf litter, and on shaded rocks.

Hemipsocidae
This family includes nearly 50 species in 4 genera worldwide. The family is found primarily in the tropics. Most diversity in Hemipsocidae has been found in China, where the family is most actively studied. Two species are known from North America: *Hemipsocus choroticus* from Florida and coastal regions of Texas and North Carolina and *H. pretiosus* from Florida. Hemipsocids live in dead foliage, and, when disturbed, they tend to run rapidly or tumble in a characteristic manner.

Lachesillidae
This family includes more than 270 species in 20 genera distributed worldwide, with the greatest diversity in Central and South America. About 50 species are known from North America, most in the genus *Lachesilla*. These are medium-sized bark lice, 2 to 4 mm in length, with somewhat slender bodies. Body colors are brown; wings are usually clear, with or without markings. Lachesillids inhabit leaf litter, persistent dead leaves, bark,

grasses, and foliage of conifers, and a few species have been found in stored food.

Mesopsocidae
This family includes some 74 species in 12 genera distributed worldwide, with the highest diversity in tropical Africa and Eurasia. Three species are known from North America in the genus *Mesopsocus*. Mesopsocids are medium- to large-sized bark lice, 3 to 4.5 mm in length. Sexual dimorphism is common in this family: males are slender and winged; females are stout with short wings or wingless. Mesopsocids inhabit bark and stone surfaces, and one species has been associated with termite nests.

Myopsocidae
This family includes some 200 species in 7 genera with worldwide distribution. Most species are tropical. Six species are known from North America in the genera *Lichenomima* and *Myopsocus*. Myopsocids are fairly large bark lice, 3 to 10 mm in length, with robust bodies and distinct, heavily patterned wings. Body colors are dark or mottled. Myopsocids can be found living on tree bark and on shaded stone outcrops and cement structures.

Neurostigmatidae
This family includes a single genus, *Neurostigma*, which includes 8 species known from Central and South America. They are known to inhabit bark.

Peripsocidae
This family includes some 235 species in 12 genera distributed worldwide, with the highest diversity in Asia. Thirteen species are known from North America in the genera *Peripsocus* and *Kaestneriella*. Peripsocids are medium-sized bark lice 2 to 4 mm in length, with robust bodies. The body color is dusky brown. Peripsocids inhabit rock surfaces and the bark of trees and shrubs. Some species are sexually dimorphic (different body forms in male and female) or polymorphic (different body forms within a sex).

Philotarsidae
This family includes some 115 species in 7 genera distributed worldwide, with the highest diversity

in Australia. Eight species are known from North America in the genera *Aaroniella, Garcialdretia*, and *Philotarsus*. These are medium-sized bark lice, 2 to 3.5 mm in length, with robust bodies. Body colors are brown or gray. Philotarsids inhabit bark and rock surfaces; some species live under dense webbing, alone or in small groups.

Pseudocaeciliidae

This family includes some 300 species in 23 genera distributed worldwide, with the highest diversity in Asia. Two species are known from North America: *Pseudocaecilius citricola* from Florida and Texas and *P. tahitiensis* from Florida and several regions of Mexico. These are medium-sized bark lice, 2 to 3 mm in length, with robust bodies. Body colors are brown or yellow. Pseudocaeciliids inhabit leaves and can often be found in clusters of dead leaves on tree branches. Some species live under sparse webbing, alone or in small groups.

Psilopsocidae

This family includes a single genus, *Psilopsocus*, with 7 species, found only in Southeast Asia and Australia. Psilopsocids are medium- to large-sized bark lice 3.5 to 6 mm in length, with dark brown bodies. Adults live on the surface of bark and have distinctive spotted wings. The nymphs bore into wood, inhabiting burrows in the tips of twigs. In nymphs, the distal part of the abdomen is sclerotized and creates a "plug" used to block the entrance of the burrow. These are the only known wood-boring psocids.

Psocidae

This is the largest family of bark lice, containing over 900 species in more than 60 genera worldwide. These are large-sized bark lice, 5 to 12 mm in length, with diverse morphology. Most species live on bark, but a few are ground- or rock-dwelling. Some species are gregarious as nymphs and young adults, occurring in large groups. Psocidae includes the largest species of bark louse, in the South American genus *Thyrsophorus*.

Ptiloneuridae

This family includes some 60 species in 9 genera known primarily from Central and South America. Some species of the genus *Loneura* have recently been described from the southwestern United States and Mexico. These are medium- to large-sized bark lice, 4 to 6 mm in length, and body colors are generally brown or reddish brown. Ptiloneurids are known to inhabit forests and rock faces.

Stenopsocidae

This family contains some 100 species in 4 genera distributed worldwide, with the highest diversity in Asia. A few species of *Graphopsocus* are known from North America. These are medium- to large-sized bark lice, 3.5 to 5 mm in length. Body colors are brown and yellowish brown. Stenopsocids are known to inhabit damp forests, where they can be collected from green foliage.

Trichopsocidae

This family contains 11 species in the genus *Trichopsocus* distributed worldwide, but absent from Asia. Two species are known from North America; both are introduced. These are medium-sized bark lice, 2 to 2.5 mm in length, with slender bodies. The body color is yellow. Trichopsocids inhabit leaves on living and dead trees.

CHAPTER 31

Order Phthiraptera (Lice)

ORDER PHTHIRAPTERA

Small, wingless Exopterygota with the body strongly modified for ectoparasitism on vertebrates. Eyes reduced or absent, ocelli absent; antennae short, thick, 3- to 5-segmented; mouthparts mandibulate or highly modified for piercing and sucking. Thorax with segments poorly separated or fused, or with prothorax free, mobile; legs usually short, stout, often with tarsi modified, cheliform for clinging to host's pelage or plumage. Abdomen with 8 to 10 distinct segments, cerci absent. Nymphs similar to adults.

All members of this specialized order, numbering slightly over 3000 species, are permanent ectoparasites of birds and mammals, including domestic stock, fowls, and humans. Most species are scavengers of dried skin or other cutaneous debris. A relatively small number feed on blood, sebum, and other tissue fluids. All species can cause general irritation and may prove debilitating in severe infestations. The Anoplura are of special importance because of their ability to transmit epidemic typhus and other rickettsial diseases to humans; fortunately, the prevalence of these diseases has decreased greatly since the advent of insecticides of low mammalian toxicity.

Lice are highly adapted to survive on the bodies of their hosts. The cuticle is tough and elastic, providing hydrostatic protection for the internal organs. Sensory structures are reduced or absent; antennae are short and can be tucked against the head; the body is usually streamlined or flattened. The tarsi are frequently shaped to cling to the pelage or plumage of the host, and in mammal-infesting species the tarsus and its single claw form a clamp for grasping hairs (Fig. 31.4b–d). Most of these modifications are characteristic of ectoparasitic insects from diverse orders and are evident in parasitic crustaceans as well.

Recent phylogenetic studies show that the suborders of Phthiraptera are clearly derived from within the Psocoptera, perhaps more than once independently. It is likely the two orders will be formally merged into a single order Psocodea, with both free-living and parasitic members. The biting lice (suborders Amblycera and Ischnocera) closely resemble Psocoptera in mouthpart configuration. As in psocids, the maxillary laciniae are sclerotized rods, detached from the stipes, and the hypopharynx and cibarium are specialized as a sitophore apparatus (Fig. 31.1c,d). As in Psocoptera the hypopharynx also functions in uptake of atmospheric water vapor (Rudolph, 1983). The biting lice have normal mandibles, which are used to clip off the host's skin debris or feathers or, in some species, to pierce or lacerate the host integument. In the sucking lice (suborder Anoplura), the functional mouthparts consist of three slender stylets of uncertain homology (Fig. 31.2a–c). At rest these are withdrawn into a pouch in the head capsule. The oral opening bears small denticles, which are anchored in the host's integument while the stylets are inserted. In the small suborder Rhynchophthirina, whose members are parasites of elephants and warthogs, the very small mandibles are attached apically on an elongate rostrum with the cutting edge external and the abductor muscles greatly enlarged (Figs. 31.3).

Phthiraptera are dependent on their hosts for food, shelter, and oviposition sites; most of them survive only a short time if isolated from hosts. All orders of birds and most orders of mammals serve as hosts. Lice are usually quite host-specific, and many species are further restricted to limited parts of the host's body. For example, some bird lice occupy the hollow quills of the large primary wing feathers, while others infest the chin, nape, or other parts of the body that are inaccessible to preening. Many mammal-infesting lice range widely over their hosts, whereas certain others are restricted to localized sites such as legs, tail, head, flippers, or even to the interdigital fossae of the feet, where their hosts have difficulty grooming. The species inhabiting humans occupy head or body hair, including pubic hair. Anoplura associated with marine mammals are unusual in being capable of surviving on land for long periods while their hosts remain at sea. The number of species of lice occupying a host species varies from 1 to at least 15, but averages about 2 to 6 for birds and 1 to 3 for mammals. The numbers of individuals on a host may vary from none to hundreds or even thousands.

Most lice are bisexual, but males are frequently less common than females; a few species are known to be parthenogenetic. The number of eggs per individual varies from a few dozen to over 300 in different species, but usually only a few eggs are deposited each day. In most species the eggs are glued to hairs or feathers or left in places inaccessible to preening and grooming, such as the interior of feather shafts. There are three nymphal instars, each usually molting after about 1 week.

No single classification of lice is generally accepted (see Haub, 1980; Kim and Ludwig, 1982). The suborders Amblycera, Ischnocera, and Rhynchophthirina are sometimes considered a single order, Mallophaga, coordinate with the order Anoplura, which classification has the disadvantage of concealing the structural diversity within the Mallophaga. The most recent authoritative classifications do not use the term Mallophaga.

Several family classifications of Phthiraptera have been proposed. Those of Clay (1970) and Keler (1969) are adopted here. Identification of families

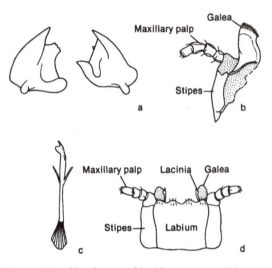

Figure 31.1 Mouthparts of Amblycera: a, mandibles of *Laemobothrion*; b, maxilla of *L. gypsis*; c, maxillary fork (lacinia) of *Ancistrona vagelli*; d, labium and maxillae of same.
Source: Modified from Snodgrass, 1944.

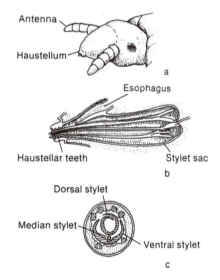

Figure 31.2 Mouthparts of Anoplura, diagrammatic: a, head with stylets protruding from oral opening; b, median section through head; c, transverse section near oral opening.
Source: Redrawn from Furman and Catts, 1961.

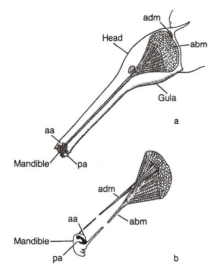

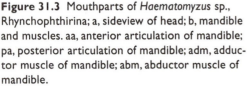

Figure 31.3 Mouthparts of *Haematomyzus* sp., Rhynchophthirina; a, sideview of head; b, mandible and muscles. aa, anterior articulation of mandible; pa, posterior articulation of mandible; adm, adductor muscle of mandible; abm, abductor muscle of mandible.
Source: Redrawn after Haub, 1972.

is frequently difficult without properly prepared slides. The key below includes suborders. Keys to families and genera appear in most medical entomology textbooks.

KEY TO THE SUBORDERS OF PHTHIRAPTERA

1. Head prolonged anteriorly as a rostrum, with mandibles at apex
   Suborder Rhynchophthirina
   Head not prolonged as a rostrum; small mandibles articulated ventrally on head capsule, or absent 2

2(1). Head relatively small, narrower than prothorax, sometimes fused with thorax; mandibles absent
   Suborder Anoplura (Fig. 31.4c,d)
   Head relatively large, wider than prothorax, free, mobile; mandibles adapted for chewing 3

3(2). Antennae capitate, 4-segmented; head with grooves for receiving antennae; mandibles horizontal; maxillary palpi present
   Suborder Amblycera (Fig. 31.4b)
   Antennae filiform, 3 to 5-segmented; head without antennal grooves; mandibles vertical; maxillary palpi absent
   Suborder Ischnocera (Fig. 31.4a)

Suborder Rhynchophthirina. A single family, Haematomyzidae, with two species restricted to elephants and the warthog, respectively. In the reduced mandible and vestigial maxillae and labium, as well as other characteristics, Rhynchophthirina are somewhat intermediate between the Amblycera and Ischnocera and the Anoplura and may not represent a distinct lineage.

Suborders Amblycera and Ischnocera. The largest and most diverse group of lice, with about 2500 species, mostly associated with birds. The two suborders differ in biological as well as morphological characteristics. The Amblycera have horizontally biting mandibles, which in most species are used to scrape off feathers, loose skin, or other cutaneous debris. In a few genera the mandibles are modified for piercing integument. The mammal-infesting species (families Boopidae, Trimenoponidae, and Gyropidae) are almost entirely confined to marsupials and rodents from the Neotropics and Australasian region, but *Heterodoxus* infests dogs in many areas. Two other families (Menoponidae and Laemobothridae) occur on restricted groups of birds. The Ischnocera have vertically biting mandibles. The bird-infesting lice of this suborder (family Philopteridae) feed largely on feathers. This is the largest family, with specialized lineages infesting a diverse array of birds ranging from songbirds to marine and freshwater divers and waders, raptors and galliform birds, including domestic fowl. Mammal-infesting species of Philopteridae infest placental mammals exclusively, feeding on dead or living skin. Bovicolidae infest a great diversity of ungulates and include species of economic importance. Trichodectidae infest carnivores and rodents and some primates. Finally, two families have very restricted hosts (Dasygonidae on hyraxes; Trichophilopteridae on lemurs).

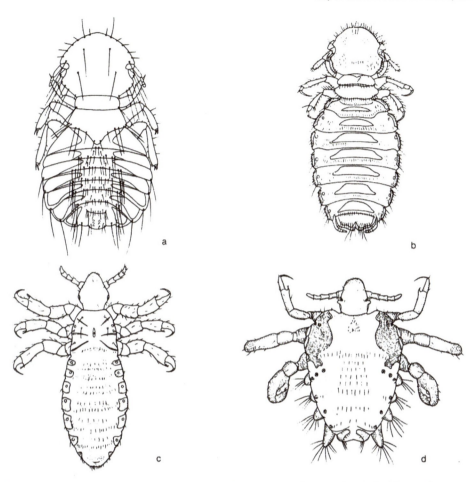

Figure 31.4 Representative lice: a, Ischnocera, *Goniodes gigas*, from poultry; b, Amblycera, from goat; c, d, Anoplura (c, the human body louse, *Pediculus humanus*; d, the pubic louse, *Pthirus pubis*). *Source: a, redrawn from Emerson, 1956; b, redrawn from Patton, 1931; c, courtesy of D. Furman; d, redrawn from Ferris, 1951.*

Suborder Anoplura. About 250 bloodsucking species confined to placental mammals. The family Pediculidae specializes on primates. Included are two species that infest humans—*Pediculus humanus,* the body and head louse (Fig. 31.4c), and *Pthirus pubis*, the pubic or crab louse (Fig. 31.4d).

The largest family of Anoplura is Hopleuridae, infesting rodents. The families Haematopinidae and Linognathidae occur mainly on pigs, ungulates, and hyraxes and include economically important species. Echinophthiriidae are restricted to Pinnipedia (seals, etc.) and Neolinognathidae to insectivores.

CHAPTER 32

Order Hemiptera (Bugs, Leafhoppers, etc.)

ORDER HEMIPTERA

Minute to large Exopterygota with styliform (needlelike) suctorial mouthparts (Fig. 32.1). Head capsule strongly sclerotized, seldom with separate sclerites delimited by sutures; compound eyes usually large, ocelli present or absent; antennae short to long, filiform or setaceous, with 10 or fewer segments (exception, numerous segments in male Coccoidea); mandibles and maxillae extremely elongate, needlelike piercing stylets, ensheathed in elongate, tubular labium, which is jointed or unjointed; palps absent. Prothorax large, distinct, except in wingless forms; mesothorax represented dorsally by scutellum, usually large, triangular; metathorax small. Forewings larger than hind wings,; flexed flat or obliquely over abdomen, usually without folds in wing membrane; legs usually adapted for walking, forelegs raptorial in many predatory taxa (Fig. 32.2a–c); tarsi with 3 or fewer segments, frequently with arolia, pulvilli, or other pretarsal structures. Abdomen commonly with anterior 1 to 2 segments reduced or absent, posterior 1 to 2 fused, reduced, associated with genitalia; cerci always absent.

Hemiptera is the fifth largest insect order and the largest and most diverse order of hemimetabolous (Exopterygota) insects, both in terms of numbers of species (more than 80,000 described) and types of habitats occupied. They are all fluid feeders, mostly on plants. Hemipterans ingest plant fluids or macerated cell contents from a wide variety of plant tissues: vascular fluids (phloem and xylem sap), leaf and stem cells of numerous types, seeds, flowers, fruits, fungal tissues, and algae. There is some evidence that the order may have arisen from predators. In any case, it would take little adaptive change from feeding on the body fluids of arthropod prey to feeding on vertebrate blood, and the family Reduviidae contains both predators of arthropods and feeders on vertebrates.

The higher classification of the Hemiptera is controversial and confusing. The current order Hemiptera was long considered as two separate orders: Hemiptera and Homoptera. After evidence was produced that the Homoptera was paraphyletic, the two groups then were combined into the order Hemiptera with the suborders Heteroptera and Homoptera (as in second edition of this book). More recent molecular and some morphological data now classify the Hemiptera as having three or four suborders. We adopt the classification of Hemiptera as containing four suborders: Heteroptera (true bugs), Coleorrhyncha, Sternorrhyncha (aphids, scales, whiteflies, and psyllids), and Auchenorrhyncha (leafhoppers, cicadas, planthoppers, and allied groups). Further

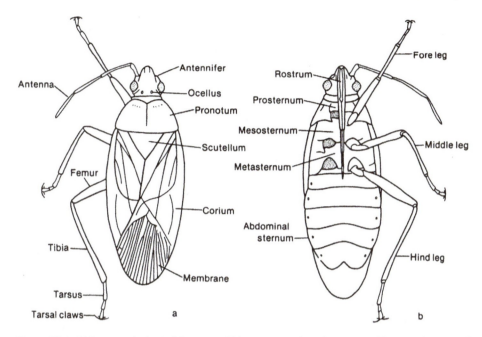

Figure 32.1 Major morphological features of Hemiptera: a, dorsal aspect, and b, ventral aspect of *Boisea trivittata* (Coreidae).

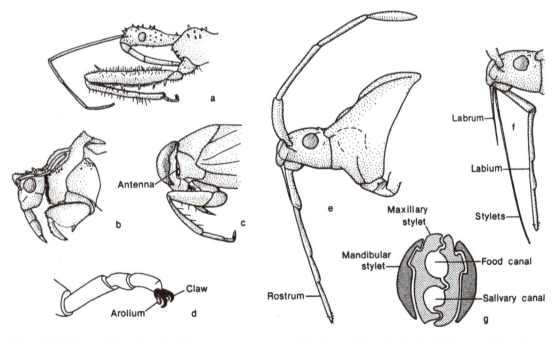

Figure 32.2 Legs and mouthparts of Heteroptera: a, raptorial forelegs of *Sinea* (Reduviidae); b, same of *Phymata granulosa* (Reduviidae); c, same of *Notonecta* (Notonectidae); d, arolium of pretarsus; e, rostrum of *Anasa tristis* (Coreidae); f, rostrum with labium removed to show feeding stylets; g, cross section of feeding stylets.
Source: g, redrawn from Tower, 1914.

changes are proposed. The Auchenorryncha may be paraphyletic, which would require two new suborders, Cicadomorpha and Fulgoromorpha, which we treat as infraorders. Some group the Heteroptera and Coleorrhyncha into the suborder Prosorrhyncha, but this has not been widely accepted. The Coleorrhyncha are a small group (about 12 species) in a single family, Peloridiidae, of moss and liverwort feeders found in southern South America, New Zealand, and Australia (formerly Gondwana). Morphological features similar to many fossil Hemiptera and their geographical distribution strongly implicate Peloridiidae (Fig. 32.3) as a relictual group, possibly the most primitive extant Heteroptera. The most primitive Heteroptera comprise two of the seven heteropteran infraorders: Enicocephalomorpha and Dipsocoromorpha. These are small, cryptic insects, mostly thought to be insect predators; altogether they have fewer than 800 described species worldwide.

Numerous fossils ascribed to the Hemiptera are known from the Permian Period, and it seems certain that the order arose earlier, in the Carboniferous Period. Most of the large group of fossils that appears to precede the hemipteroid orders (see Chap. 16) show that these early hemipteroids had evolved the suctorial mouthparts characteristic of modern forms, making them admirably preadapted to feeding on the vascular tissue of higher plants, which diversified during the Mesozoic Era. Plant fluids do not provide a nutritionally complete diet, and Hemiptera that feed mostly on plant sap or on vertebrate blood support symbiotic bacteria or fungi (previously thought to be yeasts) that serve to synthesize essential amino acids and vitamins (Chap. 11). In Heteroptera the symbiotes may occur freely in the alimentary canal or may be housed in special gut diverticula. In Auchenorrhyncha and Sternorrhyncha they are intracellular inhabitants of special host cells called mycetocytes (for fungal symbiotes) or bacteriocytes. The symbiosis is usually obligatory, and the microorganisms are transmitted transovarially from one generation to the next.

Suctorial plant feeding is the only the major feeding niche that endopterygote insects have failed to invade. It seems likely that their complete exclusion has been due to the prior occupation by the Hemiptera, which differentiated before the large orders of Endopterygota.

Specialized styliform mouthparts are the most obvious feature shared by all Hemiptera. The mandibles and maxillae have become highly specialized as a two-channeled piercing tube for delivering salivary secretions and taking up food (Fig. 32.2e–g). Proximally, the maxillary and mandibular stylets are hinged to a short arm, on which are inserted the protractor and retractor muscles. This specialized structure differentiates the Hemiptera and Thysanoptera from the other hemipteroid insects, in which the muscles insert directly on the stylets. The labium is developed as a stout, usually jointed sheath that surrounds and supports the stylets and is normally the only portion of the rostrum visible externally Fig. 32.2f). Most Hemiptera probe only a short distance into the host tissue, and the stylets are not much longer than the rostrum. Penetration is accomplished by repeated contraction and relaxation of the protractor muscles, necessitated by the small scope of movement allowed by the protractor levers. In Aradidae, which feed on fungal mycelia, which are usually long and sinuous, the stylets are commonly longer than the body and are coiled in pouches within the head capsule. In Psylloidea and Coccoidea, the long stylets are coiled in the crumena, a pouch in the neck or anterior thorax. In predatory forms such as Reduviidae and Belostomatidae, where quick penetration of a

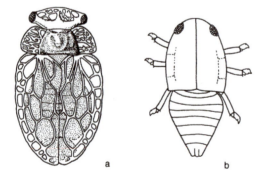

Figure 32.3 Peloridiidae; a, adult; b, nymph.
Source: a, redrawn after Myers and China, 1929; b, redrawn from Helmsing and China, 1937.

potentially dangerous prey is important, the protractor muscles are very long, allowing the stylets
to be inserted with a single muscle contraction.

The alimentary canal of Hemiptera is modified for uptake of liquid food. Salivary glands are
universally present and are usually large relative
to the body size and differentiated into distinct
cell types, each of which produces and stores different salivary components. Extraoral digestion
is apparently widespread, and relatively large
quantities of saliva are injected into the host's
tissue, causing local necrosis in plants or, more
rarely, systemic disturbances. Injection of saliva
may also be an important avenue for the transmission of microorganisms, especially circulative
plant viruses. In predatory species, saliva is highly
toxic and paralytic, enabling relatively large prey
to be quickly subdued. In many phytophagous
Heteroptera, especially Lygaeidae, Pyrrhocoridae,
and Pentatomidae, the midgut is differentiated as
several distinct regions. These may include a dis-
tensible, anterior crop and a variable number of
posterior caecae. The number and configuration
of the caecae, which contain distinctive bacteria
that are thought to be symbiotes, are characteristic
of some families or genera.

Most Homoptera, with the major exceptions of
the Fulgoroidea and some Sternorrhyncha, have
the tubular posterior midgut coiled closely adjacent to the dilated, croplike anterior midgut. The
Malpighian tubules may also be involved, forming
a complex organ called a filter chamber (Chap. 5;
Fig. 5.8c), whose function is apparently to shunt
excessive amounts of water to the hindgut. In
Diaspididae (Coccoidea) no direct connection
between midgut and hindgut exists, excess liquid being resorbed by the enlarged Malpighian
tubules. In many Homoptera and some Heteroptera
(Tingidae), surplus fluids containing sugars, amino
acids, and other small molecules are excreted as
honeydew, which forms the basis for commensal
relationships with ants.

Flight in Hemiptera is powered primarily by
the mesothoracic wings, which are usually longer than the metathoracic pair, in contrast to the
orthopteroid orders, in which the anterior pair of
wings functions primarily as a protective cover.

Venation in Hemiptera is frequently reduced,
especially in the smaller Homoptera, where only
a few longitudinal veins support the wing membrane. In Heteroptera the wings usually conform
closely to the outline of the body, and wing structure is frequently highly specialized. The anterior
basal region, or corium, which covers the abdomen laterally (Fig. 32.4a), is usually more heavily
sclerotized. The corium may bear a distal line of
weakness, the cuneal fracture, which allows the
apex (membrane) of the wing to bend downward
over the abdomen in repose. The arrangement of
veins and closed cells in the corium and membrane is characteristic of families and is of great
taxonomic utility.

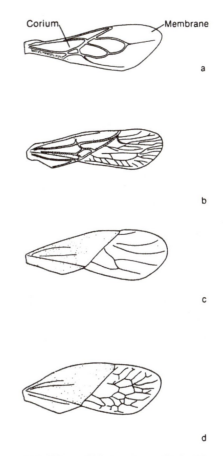

Figure 32.4 Wings of Heteroptera: a, Reduviidae
(*Sinea diadema*); b, Nabidae (*Nabus alternatus*); c,
Lygaeidae (*Lygaeus kalmii*); d, Largidae (*Largus*).

Hemiptera commonly have pretarsal structures (Fig. 32.2d) that facilitate gripping the smooth, exposed surfaces of plants or holding prey. Most Heteroptera secrete aromatic compounds, apparently defensive in nature, although some have sex or aggregation attractants. Many Heteroptera that are well endowed with defensive secretions are aposematically colored, usually in red, orange, or yellow and black. Many Reduviidae are also brightly colored, as an advertisement of their potent bites rather than defensive secretions.

Sound production is widespread in Hemiptera (see also Chap. 6). Some Auchenorrhyncha such as the Cicadomorpha produce sounds by vibrating a pair of taut membranes, the tymbals, in the base of the abdomen (Fig. 6.8a,c). Other Auchenorrhyncha such as the Fulgoromorpha produce sounds by vibrating ("drumming") the abdomen on a plant substrate to communicate to potential mates. Heteropteran sound production may use one of several different stridulatory mechanisms. Reduviidae stridulate by rubbing the rostrum over the striate walls of the prosternal furrow, producing an audible rasping noise. Corixidae stridulate by drawing the anterior femora over the clypeal margin, and some Pentatomidae rub a comb on the inner surface of each wing over a striate area at the base of the abdomen. Other mechanisms involving the coxae and the abdominal venter occur in other families; clearly, acoustical communication has evolved independently several times in Hemiptera.

Most Hemiptera are hemimetabolous. Nymphs differ from adults in such details as number of antennal segments, presence of ocelli, and chaetotaxy, as well as wing development. Morphological differences among instars are usually gradual, but color changes may be abrupt and striking. For example, *Largus cinctus* (Largidae) is bright metallic blue with a red dorsal spot as a nymph but molts to a brownish or blackish adult with a dull orange stripe on each flank. In some Sternorrhyncha developmental differences are as extreme as in the Endopterygota (see below).

Reproduction is usually bisexual, with one to several generations per year. In several families of Sternorrhyncha, however, differences among instars may be extreme, and cyclical or permanent parthenogenesis is common. The increased diversity in development and reproduction is discussed below. Most Hemiptera are oviparous, ovipositing either on or in plant tissue or on soil, stones, or other substrates. Eggs of Sternorrhyncha and Auchenorrhyncha are usually simple, ovoid structures, whereas those of Heteroptera are greatly differentiated among various families, frequently with opercula, spines, or appendages (Cobben, 1968).

KEY TO THE SUBORDERS OF HEMIPTERA

1. Mesothoracic wings usually folded flat over abdomen; basal portion of wing thickened, apical portion membranous; rostrum arising anteriorly on head (Fig. 32.2e), with sclerotized gula, antennae 4- or 5-segmented

   Suborders Heteroptera and Coleorrhyncha

   Mesothoracic wings almost always held rooflike over abdomen, uniformly membranous (but some adults may be wingless); rostrum arising near posterior margin of head, without gula.   **2**

2. Antennae filiform with more than 4 segments, tarsi 2-segmented

   Suborder Sternorrhyncha

   Antennae 3-segmented (usually) and setaceous or short; tarsi 3-segmented

   Suborder Auchenorrhyncha

Suborder Heteroptera (True Bugs)

Head horizontal, with base of rostrum clearly distinct from prosternum, although the distal segments may extend between the anterior coxae; antennae with 4 or 5 (occasionally fewer) segments; prothorax usually large, trapezoidal or rounded, frequently longer than broad; tarsi usually with 3 segments.

Although the large majority of Heteroptera feed on plants and some are ectoparasites or scavengers, they are believed to have been primitively predaceous. Heteroptera include a few viviparous species, but apparently none is parthenogenetic, in sharp contrast to the rest of the order. Phytophagous Heteroptera include many species that feed on higher plants (Miridae, Coreidae, Pentatomidae, Tingidae, etc.) and some that are restricted to particular plant parts. For example, many Lygaeidae, Pyrrhocoridae, and Largidae feed preferentially or exclusively on seeds, sometimes from a very narrow host range. Many Cydnidae are subterranean, apparently feeding on roots. Other Heteroptera limit their feeding to certain plant taxa. For example, certain Miridae are restricted to ferns, others to specific genera or families of Angiosperms. Peloridiidae, possibly the most primitive extant Heteroptera, feed only on liverworts and mosses. Aradidae feed exclusively on wood-rotting fungi.

Among terrestrial forms, the Reduviidae, Nabidae, and Anthocoridae are almost entirely predaceous, and many Miridae, Pentatomoidea, and Lygaeoidea, as well as occasional members of other families, are predatory. In addition, many more species are facultatively predatory. Modes of predation range from "sit-and-wait" types (many Reduviidae) to active searchers, such as *Geocoris* (Lygaeoidea, Geocoridae). Some species, especially in the Pentatomidae, are host-specific, but most heteropteran predators prey on all suitably sized organisms in a given habitat. All aquatic Heteroptera are predators except some Corixidae, which feed on algae and diatoms. The numerous amphibious families are either scavengers or predators, mostly of much smaller, weaker organisms.

Two families of Heteroptera are exclusively ectoparasitic. Cimicidae inhabit bird and mammal nests, including houses, and suck blood from the occupants. Polyctenidae suck blood from bats. In addition, all members of the reduviid subfamily Triatominae are ectoparasites of mammals and birds and vectors of trypanosomes that cause animal and human disease.

Heteroptera are among the most successful insect colonizers of marine habitats. *Halobates* (Gerridae) occurs in all tropical seas, commonly hundreds or thousands of kilometers from land. *Rheumatobates* occurs in brackish mangrove swamps and *Hermatobates* around coral reefs. Many Saldidae frequent seacoasts, and some, such as *Aepophilus*, are intertidal, enduring long periods of submergence during high water.

The Heteroptera includes seven infraorders. Gerromorpha and Nepomorpha are aquatic; Leptopodomorpha is semi-aquatic. The remaining four are terrestrial: the most primitive are Enicocephalomorpha and Dipsocoromorpha, followed by Cimicomorpha and Pentatomomorpha. Approximately 89 families of Heteroptera are recognized, with a total of over 43,000 species described (Henry, 2009). This more than doubles the number of families noted in the second edition of this book. The following key includes the major families found in North America or important worldwide and largely follows the classification scheme of Henry (2009) and Schuh and Slater (1995).

Key to the Families of the Suborder Heteroptera

1. Antennae very short, inserted ventrally, with no more than 1 or 2 segments visible from above (aquatic or subaquatic species) **2**

   Antennae much longer than head, usually inserted dorsally, and with at least 3 segments visible from above **9**

2(1). Anterior tarsus with a single flattened, paddle-shaped segment (Fig. 32.10c); rostrum short, with 1 or 2 segments

   Corixidae

   Anterior tarsus usually with 2 or 3 segments, never paddle-shaped; rostrum long, with 3 or 4 segments **3**

3(2). Ocelli present **4**
   Ocelli absent **5**

4(3). Antennae retractable into grooves beneath eyes; anterior femora thickened, raptorial

   Gelastocoridae

   Head without grooves for antennae; anterior femora slender **Ochteridae**

5(3). Abdomen with slender caudal respiratory appendages at least one-fourth length of body (Fig. 32.10b); tarsi with 1 segment

   Nepidae

Abdomen with respiratory appendages short or absent; tarsi with 2 segments   **6**

6(5). Body oval, flattened; anterior legs raptorial, with femora thickened and grooved to receive tibiae   **7**

Body strongly convex, arched dorsally, usually elongate; anterior femora not raptorial   **8**

7(6). Eyes bulging, prominent; wing membrane with reticulate veins; abdomen with short, flat apical breathing appendages   **Belostomatidae**

Eyes molded to outline of head and thorax, not prominent; wing membrane without veins; abdomen without breathing appendages   **Naucoridae**

8(6). Hind legs with tibia and tarsus flattened, oarlike, without tarsal claws; body length greater than 5 mm   **Notonectidae**

Hind legs with tibia and tarsus cylindrical, with 2 tarsal claws; body length less than 4 mm   **Pleidae**

9(1). Eyes present   **10**

Eyes absent; body flattened; forewings short, hind wings absent   **Polyctenidae**

10(9). Anterior legs with claws attached before apex of tarsi; apical tarsal segment longitudinally cleft   **11**

Anterior legs with claws attached apically; apical tarsal segment not cleft   **12**

11(10). Middle coxae close to hind coxae, remote from fore coxae; hind femora much longer than abdomen   **Gerridae**

Middle coxae about equidistant from fore and hind coxae; hind femora about as long as abdomen   **Veliidae**

12(10). Head linear, as long as thorax including scutellum; eyes distant from base of head   **Hydrometridae**

Head shorter (usually much shorter) than thorax, including scutellum; eyes usually near base of head   **13**

13(12). Scutellum concealed by pronotum; ocelli absent; tarsi with 2 segments; hemelytra frequently with raised, reticulate venation (Fig. 32.5d)   **51**

Scutellum usually visible, frequently large; if concealed, ocelli present or tarsi with 3 segments   **14**

14(13). Body greatly flattened dorsoventrally, with texture and appearance of tree bark; forewings in repose usually much narrower than abdomen (Fig. 32.7c)   **Aradidae**

Body not textured like bark, usually not greatly flattened; forewings in repose usually about same width as abdomen   **15**

15(14). Antennae with basal 2 segments short, thick; apical segments slender, filamentous   **16**

Antennae with segments of subequal length and diameter or gradually tapering, or with only first segment short, thickened   **17**

16(15). Pronotum, wings, and head, including eyes, with long erect spines   **Leptopodidae**

Body not spiny   **Ceratocombidae, Dipsocoridae, Schizopteridae**

17(15). Wings reduced to leathery flaps no longer than first abdominal segment, without veins; body ovoid, flattened, leathery (Fig. 32.7b)   **Cimicidae**

Wings usually large; if reduced, usually extending beyond first abdominal segment and with veins; body elongate, thick   **18**

18(17). Forewings with cuneus present (Fig. 32.5a,b)   **19**

Forewings without cuneus   **21**

19(18). Ocelli present; forewing with 0 to 1 basal cells in membrane (Fig. 32.5b)   **20**

Ocelli absent; forewing with 2 to 3 closed cells in membrane near cuneus (Fig. 32.5a)   **Miridae**

20(19). Rostrum with 4 segments; tarsi with 2 segments   **Microphysidae**

Rostrum and tarsi each with 3 segments **48**

21(18). Antennae with 4 segments   **25**

Antennae with 5 segments   **22**

22(21). Scutellum large, triangular or U-shaped, reaching wing membrane (Fig. 32.7d)   **23**

Scutellum smaller, not reaching wing membrane (as in Fig. 32.6a-d)   **24**

23(22). Tibiae set with stout spines; apices of hind coxae fringed with stiff setae   **Cydnidae**

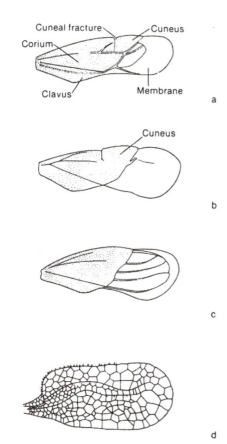

Figure 32.5 Wings of Heteroptera: a, Miridae (*Irbisia mollipes*); b, Anthocoridae (*Tetraphelps*); c, Saldidae (*Salda abdominalis*); d, Tingidae (*Corythuca obliqua*).

Tibiae not set with spines; hind coxae without fringe of stiff setae   52

24(22).Anterior femora thickened, raptorial (Fig. 32.2a); antennae with second segment about one-fourth length of third   **Nabidae**
Anterior femora slender; antennae with second segments about one-half length of third   **Hebridae**

25(21).Ocelli present   26
Ocelli absent   43

26(25).Middle and hind tarsi with 1 or 2 segments; size minute to small   27
Middle and hind tarsi with 3 segments; size small to large   30

27(26).Forewings entirely membranous; forelegs thickened, raptorial; body slender, elongate, linear   **Enicocephalidae**
Forewings with base thickened, coriaceous; forelegs slender; body not linear   28

28(27).Rostrum with 4 segments; corium shorter than wing   29
Rostrum with 3 segments; wing with corium extending to apex (1 species, Florida)   **Thaumastocoridae**

29(28).Body covered with velvety hairs; femora extending far beyond body margins   **Hebridae**
Body coarsely punctuate on dorsum; never with velvety hair; femora extending about to body margins   **Piesmatidae**

30(26).Rostrum with 3 segments   31
Rostrum with 4 segments   32

31(30).Ocelli set between eyes; forewings with 4 to 5 elongate, parallel cells in membrane (Fig. 32.5c); head broader than long   47
Ocelli set behind eyes; forewings usually with 1 or 2 large cells in membrane (Fig. 32.4a); head longer than broad   **Reduviidae** (incl. **Phymatidae**)

32(30).Rostrum with basal segment projecting, not capable of being apposed to ventral surface of head; tarsi without arolia   **Nabidae**
Rostrum capable of being apposed to ventral surface of head and pronotum; tarsi with arolia (Fig. 32.2d)   33

33(32).Pronotum with posterior median lobe that covers scutellum; similar to Microveliidae in general appearance   **Macroveliidae**
Pronotum variable in shape, but without lobe covering scutellum   32

34(33).Forewings with 6 or fewer main longitudinal veins; if wings abbreviated, antennifers (tubercles on which antennae are inserted) located ventrolaterally   35
Forewings with at least 7 (usually many) main longitudinal veins (Fig. 32.6d); antennifers located dorsolaterally   49

35(34).Body elongate, slender, linear; antennae and legs extremely long, threadlike, with

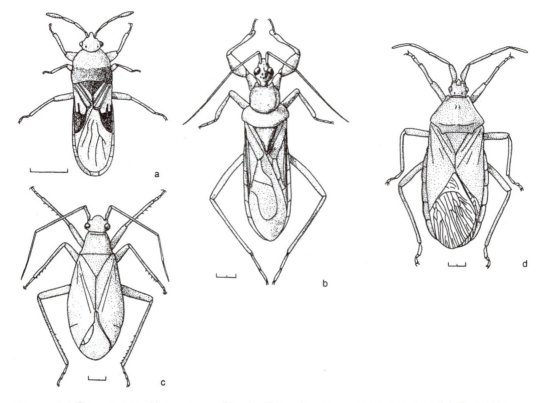

Figure 32.6 Representative Heteroptera: a, Blissidae *(Blissus leucopterus*, scale equals 1 mm); b, Reduviidae *(Triatoma protracta*; scale equals 2 mm); c, Miridae *(Irbisia mollipes*; scale equals 1 mm); d, Coreidae *(Anasa tristis*; scale equals 2 mm).

tips of femora and antennal scape slightly swollen (Fig. 32.7a) **Berytidae**
Body usually stout; legs and antennae not extremely long and slender **36**

36(35). Forelegs with enlarged femora; suture separating segments 3 and 4 curving forward and not reaching margin of abdomen **Rhyparochromidae**
Not as above **37**

37(36). Spiracles on abdominal sterna segments 2–7 are dorsal; Y-shaped pattern on scutellum **Lygaeidae**
At least one pair of spiracles on abdominal segments 2–7 is ventral **38**

38(37). Spiracles of abdominal segment 7 are ventral, others dorsal **39**
At least spiracles on abdominal segments 6 and 7 are ventral **41**

39(38). Hemelytra are coarsely punctuate **40**
Hemelytra are not or only faintly punctate **Blissidae**

40(39). Posterior tip of scutellum bilobed; eyes stalked **Ninidae**
Not as above; usually brownish yellow and smaller than 10 mm **Cymidae**

41(38). Spiracles on abdominal segments 3–7 are ventral **42**
Spiracles on abdominal segments 2–5 are dorsal; prominent, kidney-shaped compound eyes **Geocoridae**

42(41). Lateral margin of pronotum laminated **Artheneidae**
Lateral margin of pronotum rounded **Oxycarenidae**

43(25). Tarsi with 3 segments **45**
Tarsi with 2 segments **44**

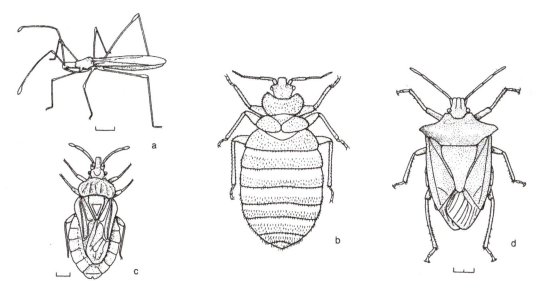

Figure 32.7 Representative Heteroptera: a, Berytidae (*Metapterus banksi*; scale equals 2 mm); b, Cimicidae (*Cimex lectularius*); c, Aradidae (*Aradus proboscideus*; scale equals 1 mm); d, Pentatomidae (*Euschistus conspersus*; scale equals 2 mm).

44(43). Tarsi with 2 segments; rostrum retractable into groove between fore coxae
Reduviidae
Tarsi with 1 segment; rostrum not retractable into groove
Nabidae

45(43). Anterior femora swollen, bearing several teeth; forewing with 4–5 longitudinal veins in membrane (Fig. 32.4c)
Lygaeidae
Anterior femora slender, usually without teeth; forewing with numerous longitudinal veins in membrane (Fig. 32.4d)   **46**

46(45). Pronotum with upturned lateral margins
Pyrrhocoridae
Pronotum with lateral margins rounded
Largidae

47(31). Hemelytra with 4–5 closed cells in membrane (Fig. 32.5c)   **Saldidae**
Hemelytra without veins
Mesoveliidae

48(20). First and second abdominal terga with laterotergites; remaining tergites form a single plate   **Lasiochilidae**

All abdominal terga have laterotergites
Anthocoridae

49(34). Scent glands open between mid and hind coxae   **50**
No visible scent gland opening on lateral thorax; usually pale-colored
Rhopalidae

50(49). Head almost as wide and as long as anterior pronotum   **Alydidae**
Head narrower and shorter than pronotum
Coreidae

51(13). Basal forewings heavily punctate or with numerous wing cells with raised (reticulate) veins; usually 4 or fewer antennal segments; 2–8 mm long   **Tingidae**
Scutellum does not cover corium, 5-segmented antennae; large mesosternal keel
Acanthosomatidae

52(23). Shieldlike shape; scutellum does not cover most of abdomen
Pentatomidae
Oval shape; strongly convex; scutellum covers abdomen and most of hemelytra
Scutelleridae

Infraorder Enicocephalomorpha (>400 species)

Enicocephalidae. The more prominent of the two families in the small infraorder. Small, reduviid-like bugs with enlarged head, globose behind eyes, antennae and rostrum with four segments, tarsi with 1 to 2 segments; forewings entirely membranous. Most species frequent damp leaf litter or other debris and are predaceous. Adults form dense aerial swarms at dusk in the autumn. Widespread but uncommon.

Infraorder Dipsocoromorpha (~340 spp.)

Ceratocombidae (~50 spp.) have long, slender antennae; bristlelike setae on the antennae, head and tibiae; and short, distinct fracture midway on the costa of the hemelytra. Dipsocoridae (~50 spp.) also have long, slender antennae with bristlelike setae, but with a cuneuslike break about a third of the wing length from the wing's apex. These small predators occur mostly along streams, marshes and lakes. The Schizopteridae (>230 spp.) are minute (2 mm), strongly convex, beetlelike bugs. Their hind legs give them an impressive jumping ability.

Infraorder Gerromorpha (>2100 spp.)

Gerroidea (3 families). Gerridae (water striders): Slender, elongate insects specialized for skating on the surface film of still or moving water, using the greatly elongate middle and hind legs; rostrum with four segments. Gerrids are most abundant on still or slowly moving water, where they scavenge and attack small arthropods. *Halobates* occurs in marine situations, including the open seas at great distances from land, and other genera inhabit estuaries or other saline situations. Veliidae: Similar in appearance and habits to Gerridae but with three-segmented rostrum and shorter middle and hind legs. *Velia* commonly frequents margins of quiet pools in streams; *Rhagovelia* (Fig. 32.9) inhabits riffles in fast-flowing streams, using the fanlike semicircle of setae on the hind tarsi to swim against the current. Hermatobatidae (seabugs; not keyed): Small (3–4 mm) predators with large eyes and strong claws found on rocks in intertidal zones.

Hebroidea. Hebridae: Small, stout-bodied, predaceous bugs; antennae with four or five segments, ocelli present, rostrum with three visible segments, retractable into groove between coxae. Hebrids are covered with hydrofuge pile, which allows them to crawl beneath the water surface without wetting their cuticle. Widespread in marshes and at stream margins but uncommon. Macroveliidae: A small family confined to the New World. A single species occurs in western North America, commonly about the margins of springs, seeps, and streams, sometimes in cavities beneath logs or large stones or in other protected sites. Unlike Veliidae and Mesoveliidae, these bugs are incapable of skating on the surface film of water.

Mesovelioidea. Mesoveliidae (pond treaders): Similar to Veliidae but with claws inserted apically. Mesoveliids are predators and scavengers frequenting seeps, margins of quiet pools along streams, or ponds and damp leaf litter. In this family, as well as the Veliidae and Gerridae, wing polymorphism is common, and wingless, short-winged, or fully winged individuals may be encountered in the same population.

Hydrometroidea. Hydrometridae: Extremely elongate, linear bugs with threadlike legs and antennae. They are sluggish, slow-moving predators, inhabiting quiet pools, marshes, and swamps, mostly in tropical regions.

Infraorder Nepomorpha (>2300 spp.)

Nepoidea (2 families). Nepidae (water scorpions; Fig. 32.10b): Elongate, cylindrical or flattened with long, slender legs; anterior legs raptorial; posterior pairs adapted for walking. Antennae with 3 segments, tarsi with 1 segment, without claws on anterior legs. Nepids are sluggish predators resembling sticks or leaves. They ambush prey from camouflaged positions on aquatic vegetation or debris and are not highly adept swimmers. They are widespread, but more abundant in tropical regions. Belostomatidae (giant water bugs; Fig. 32.10a): Broad, flattened, oval bugs with the middle and hind legs flattened, fringed for swimming, the anterior pair raptorial; antennae with 4 segments.

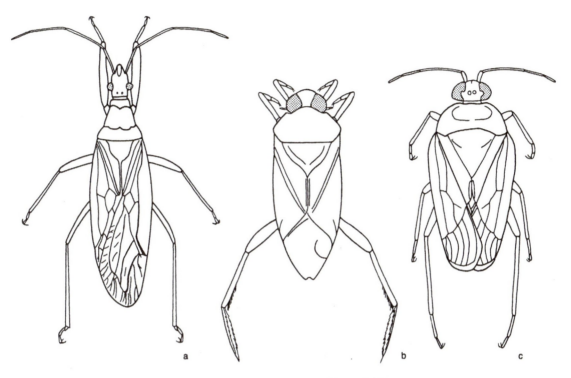

Figure 32.8   Representative Heteroptera: a, Nabidae; b, Notonectidae; c, Saldidae.

Belostomatids are moderate- to large-sized (up to 100 mm long), free-swimming predators that kill small vertebrates as well as other arthropods. They have a painful bite if mishandled. Females of *Belostoma* and *Abedus* have the peculiar habit of attaching their eggs to the backs of males, where they are carried until hatching. Widely distributed.

Naucoroidea *(3 families).*   Naucoridae: Similar to belostomatids in general appearance, but more rounded and streamlined in dorsal aspect; anterior femora extremely thick; middle and posterior legs cylindrical, spiny. Naucorids frequent vegetation, bottom debris, undersides of stones, and similar concealed situations, in both running and still water, where they prey on a variety of small arthropods. Respiration is by means of a gas bubble, periodically renewed at the surface, or in some species by a plastron, which occurs in no other Hemiptera and enables these bugs to remain permanently submerged. Mostly tropical

in distribution. North American species are found mostly in the Southwest.

Corixoidea.   Corixidae (water boatmen; Fig. 32.10c): Flattened, streamlined swimming bugs with forelegs and hind legs oarlike, without tarsal claws. Rostrum short, concealed; antennae with 3 or 4 segments. Corixidae occur in almost all still water, including salt ponds and brackish pools; they are less abundant in streams. Corixids include a few predators, but most use the paddlelike forelegs to scoop up bottom detritus containing algae, diatoms, and other microorganisms, from which they suck the protoplasm.

Ochteroidea *(2 families).*   Gelastocoridae (Fig. 32.10d): Squat, stout-bodied, with prominent, bulging eyes and raptorial forelegs. Gelastocorids are commonly associated with sandy or muddy banks. They actively search for prey, which they capture by sudden, short hops. As in damselflies, the widely spaced, hemispherical eyes allow

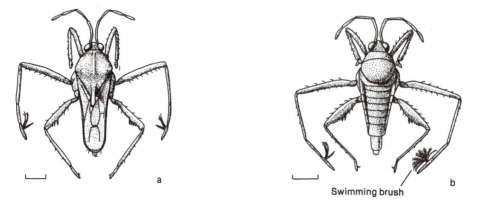

Figure 32.9 Veliidae (*Rhagovelia distincta*): a, winged form; b, apterous form. Scale equals 1 mm.

vision in all directions and probably provide binocular depth perception. Ochteridae: Similar to Gelastocoridae, but with longer antennae, long, 4-segmented rostrum, and anterior legs similar to posterior pairs. Ochterids are widespread but infrequently encountered. They occupy the same situations as saldids and gelastocorids.

Notonectoidea (3 families). Notonectidae (back swimmers; Figs. 8.6d, 32.8b): Stout, wedge-shaped bugs that swim on their backs, using oarlike hind legs. Ocelli absent, antennae with 4 segments. Notonectids are free-swimming predators in open water or small pools or along stream and lake margins. Their bite is painful. Pleidae: These insects are distinguished from notonectids by having three antennal segments and by the partial fusion of the head and thorax. Both families are widely distributed, but pleids are much less commonly encountered.

Infraorder Leptopodomorpha (>380 spp.)

Leptopodoidea (2 families). Leptopodidae: These insects are native to tropical and subtropical regions of the Eastern Hemisphere, where they mostly inhabit dry areas along streams. One species, *Patapius spinosus*, introduced to California, occurs under debris in arid regions.

Saldidoidea (2 families). Saldidae (shore bugs; Fig. 32.8c): Oval, with broad head and large, prominent eyes. Ocelli present; 3-segmented rostrum

projecting below head, even in repose; tarsi with 2 segments; forewing with 4 or 5 long, parallel-sided cells in membrane (Fig. 32.5c). Active predatory bugs capable of rapid running, jumping, or flying. Saldids are widely distributed in wet habitats such as marshes and stream margins, including saline situations, but may sometimes be found far from water, especially during dry seasons.

Infraorder Cimicomorpha (~20,500 spp.)

Cimicoidea (5 families). Cimicidae (bedbugs; Fig. 32.7b): Oval, flattened ectoparasites, with forewings always reduced to short scalelike flaps. Ocelli absent, rostrum retractable into ventral groove, tarsi with 3 segments. Unlike lice, bedbugs contact their hosts only for periodic feeding, spending most of their time in nest material or adjoining cracks or crevices. Most are parasites of birds, some attack cave-dwelling bats, and one species, *Cimex lectularius*, is associated with humans. Polyctenidae: Highly specialized parasites of bats, mostly in tropical and subtropical regions. They are highly modified for moving through the fur of their hosts. All species are viviparous, with only two nymphal instars. Two species occur in southwestern United States. Anthocoridae (minute pirate bugs): Compact, short-legged 1 to 5 mm long. Ocelli present, antennae with four segments, rostrum with 3 segments; tarsi with 3 segments; forewings with incomplete cuneus (Fig. 32.5b). Most species are predators of foliage-inhabiting

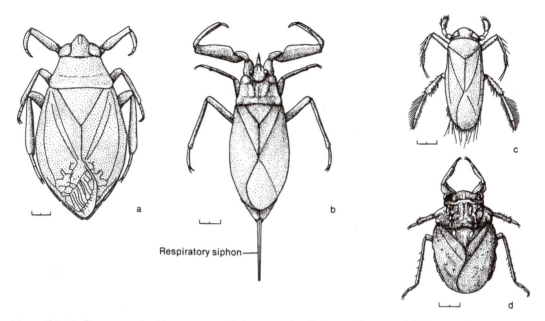

Figure 32.10 Representative Heteroptera: a, Belostomatidae (*Belostoma flumineum*); b, Nepidae (*Nepa apiculata*); c, Corixidae (*Hesperocorixa laevigata*); d, Gelastocoridae (*Gelastocoris oculatus*). Scales equal 2 mm.

insects, but others occur on bark, in leaf litter, on fungus, and even in stored products, where they attack mites and small insects or insect eggs. Over 300 species, widespread. Lasiochilidae: Size 3 to 4 mm; general appearance similar to Anthocoridae but with dorsal lateral tergites on first and second abdominal segments. About 60 species.

Microphysoidea. Similar in appearance to Anthocoridae, but tarsi with 2 segments, rostrum with 4 segments; size 2 to 3 mm. Most species, including the single North American representative, *Mallochiola gagates,* inhabit moss and lichen on or about dead trees, and are probably predators.

Miroidea (3 families). Miridae (plant bugs; Figs. 32.1, 32.6c): Delicate, oval or elongate bugs, mostly 2 to 10 mm long. Ocelli absent, antennae and rostrum with 4 segments; tarsi with 3 segments; forewings with distinct corium, clavus, and (usually) a large cuneus (Fig. 32.5a). This is one of the largest heteropteran families: more than 10,000 species in more than 1300 genera. Most are phytophagous, as suggested by the common name of the family,

feeding on nearly all higher plants, with a wide range in host specificity. *Lygus* is among a number of economically important species, feeding on many crops. Many mirids, however, are predaceous, feeding on mites, aphids, and other small arthropods.

Thaumastocoridae is a small family of small, flattened bugs with broad head, eyes frequently on short, lateral stalks; tibiae usually with membranous apical appendage. Predominantly a tropical family with few New World species, of which a few are important pests of palms.

Tingidae (lace bugs). Small, dorsally flattened species with raised, reticulate venation (Fig. 32.5d); usually pale-colored. Ocelli absent, antennae with 4 segments, tarsi with 2 segments; pronotum usually covering scutellum, often crested or hood-shaped. Tingids are gregarious, phytophagous bugs, which usually feed on the undersides of leaves. Honeydew is secreted, and some species are tended by ants; a few form galls. Nymphs of tingids are often strikingly different from adults, with spines or other outgrowths from the dorsum. Over 2150 species, widely distributed, with about 150 in North America.

Reduvioidea (2 families). Reduviidae (assassin bugs; Fig. 32.6b): Small to large; ocelli present (rarely absent); antennae with 4 or 6 to 8 segments; rostrum with 3 segments, fitting into intercoxal stridulatory groove on prosternum; tarsi usually with 3 segments. Reduviidae is the second largest heteropteran family, with almost 7000 described species. These vary enormously in body shape, size, color, and cuticular sculpturing, ranging from the gnatlike to squat, robust types. Nearly all reduviids are predatory, but *Triatoma, Rhodnius,* and a few others are ectoparasites of mammals and birds and vectors of *Trypanosoma cruzi,* which causes Chagas' disease in humans. The forelegs are raptorial in all predaceous reduviids and may be greatly enlarged and specialized, as in *Phymata* (Fig. 32.2b), which ambushes prey from hiding places on flowers. Several of the larger reduviids, sometimes called kissing bugs, freely deliver painful bites if not handled cautiously.

Naboidea (2 families). Nabidae (damsel bugs; Fig. 32.8a): Superficially similar to Reduviidae, but without prosternal groove for rostrum. Ocelli present or absent, antennae with 4 or 5 segments, rostrum with 4 segments. These predaceous bugs commonly inhabit vegetation and may be of some importance in controlling crop pests. Insemination in some genera is directly into the hemocoel (traumatic), as in Cimicidae. About 400 species, widely distributed.

Infraorder Pentatomomorpha (>16,200 spp.)

Aradoidea. Aradidae (flat bugs; Fig. 32.7c). Strongly flattened, oval in dorsal silhouette, with rough gray, black, or brown cuticle; mostly 5 to 10 mm long. Ocelli absent, antennae with 4 segments, usually thick, cylindrical; tarsi with 2 segments. Aradids feed on wood-rotting fungi, about which large numbers of all instars often congregate, either on the outer surface of trees or beneath loose bark. The feeding stylets are extremely elongate, frequently longer than the entire body, apparently as an adaptation for penetrating and following fungal mycelia. The elongate stylets also allow the feeding

insects to remain camouflaged on bark-adjoining fungal masses. Abbreviation or absence of wings is common in aradids, either in both sexes or in females only.

Pentatomoidea. (includes 16 families, many of which consist of few species.) Cydnidae (burrowing bugs and Negro bugs): Broadly oval, brown or black bugs 3 to 10 mm long, sometimes with orange or yellow bars on the forewings. Ocelli present, antennae with 5 segments, tarsi with 3 segments; forelegs frequently modified for digging. Most species are probably phytophagous, feeding on either roots or foliage. As in pentatomids, thoracic scent glands are present. Pentatomidae (stink bugs; Fig. 32.7d): Broad, usually triangular or trapezoidal in shape. Ocelli present, antennae with 5 segments; tarsi with 3 segments and arolia. Most are phytophagous, including economically important pests on rice and crucifers; members of the subfamily Asopinae are predators, mostly of lepidopterous larvae. All instars produce aromatic repugnant secretions released through openings in the thorax (adults) or abdomen (nymphs). Some species are able to spray these secretions several centimeters. Pentatomids include a variety of brightly metallic-colored or grotesquely sculptured members, particularly in the tropics, where the majority of the approximately 4700 described species occur. Scutelleridae (shield bugs) are thick-bodied bugs with a large scutellum that covers most of the abdomen. A *Eurygaster* species is a major grain pest in Asia. Acanthosomatidae is a small family (about 200 spp.) differing from other Pentatomoidea by having the combination of 2-segmented tarsi and a large mesosternal keel.

Coreoidea (5 families). Coreidae (squash bugs, etc., Fig. 32.6d) range from small to large and are robust, elongate bugs with a trapezoidal prothorax. Ocelli present, antennae with 4 segments, inserted dorsolaterally on head, tarsi with 3 segments and arolia; wing membrane with numerous veins (Fig. 32.6d). All species are phytophagous, mostly attacking growing shoots. A few of the almost 2000 species worldwide, such as *Anasa tristis* in North America, are economically important. Most produce repulsive secretions, which some large species can spray over short distances. In Rhopalidae the openings of the scent glands are slitlike or concealed. The box elder bug, *Boisea trivittata*, is a common species in most of North America. In Alydidae the body is narrow and the head is nearly as broad as the prothorax (much broader in the other subfamilies), having about 250 species worldwide.

Lygaeoidea *(16 families).* This superfamily contains 15 former subfamilies or tribes that have recently been elevated to family status from what was formerly within the family Lygaeidae (seed bugs, and the classification probably is still in flux). Oval or elongate bugs, mostly 2 to 15 mm in length. Ocelli present, antennae with 4 segments (with few exceptions), inserted ventrolaterally on head; tarsi with 3 segments and arolia, wing with 4 to 5 veins in membrane (Fig. 32.4c). *Lygaeus, Oncopeltus,* and other large, bright orange or red and black "milkweed bugs" are the most commonly noticed members of what is currently considered in Lygaeidae. The great majority of species are small, usually dull-colored, frequently ground- or litter-dwelling insects. Many feed on seeds. Loss or shortening of wings is frequent in ground-dwelling lygaeids, which may mimic ants. Geocoridae, commonly known as big-eyed bugs, are significant predators in some crops such as alfalfa and cotton. The chinch bugs, *Blissus* spp. (Blissidae, Fig. 32.6a), feed on stems and roots of grasses, including lawns, pastures, and grain crops. Cymidae are small (<8 mm), brownish bugs that feed mainly on sedges, as do also members of the small family Ninidae. The Artheneidae and Oxycarenidae are small families with few species in North America. The largest lygaeoid family, Rhyparochromidae, with about 50 Nearctic species and almost 2000 worldwide, are mostly seed feeders.

Other prominent families now considered in Lygaeoidea are Piesmatidae (ash-gray lace bugs) and Berytidae (stilt bugs). The small, oval Piesmatidae resemble Tingidae in the lacelike texture of the dorsum, but with the scutellum exposed; ocelli present, antennae with 4 segments; tarsi with 2 segments and pulvilli. They are phytophagous, widespread but uncommon. The Berytidae (Fig. 32.7a) are slender, elongate bugs.

All are primarily phytophagous but apparently require animal food such as insect eggs in order to mature.

Pyrrhocoroidea *(2 families).* Pyrrhocoridae (cotton stainers) and Largidae (bordered plant bugs). Oval or elongate bugs, mostly more than 5 mm in length. Ocelli absent, antennae with 4 segments, tarsi with 3 segments and arolia; wing with anastomosing veins in membrane (Fig. 32.4d). Most species are bright orange, red, or yellow and black. Feeding is predominantly on fruiting structures of plants, and several species are economically important. Worldwide in distribution, but mostly tropical.

SUBORDERS AUCHENORRHYNCHA AND STERNORRHYNCHA

Suborder Auchenorrhyncha (Leafhoppers, Cicadas, Planthoppers, etc.)

Head bent toward venter, with rostrum closely appressed to prosternal region, its base extending between anterior coxae; antennae usually with 2 to 10 segments; prothorax usually small (exception, large in Membracidae), transverse or quadrangular; tarsi with 3 or fewer segments.

All Auchenorrhyncha and Sternorrhyncha are phytophagous but highly diversified in body form and in details of life history. Most species can be grouped into one of two broad categories on the basis of morphological and biological features. The superfamilies Cicadoidea, Fulgoroidea, Cercopoidea, and Cicadelloidea comprise the suborder Auchenorrhyncha. In this group the antennae are very short and bristlelike, the rostrum is clearly articulated with the head, and the ovipositor is bladelike for inserting the eggs into plant tissue. The Auchenorrhyncha are mostly active insects that live exposed on plant surfaces and jump to escape predators. Exceptions include cicadas, which cannot jump but readily fly as adults, and nymphal cercopids, which live immersed in a self-produced coating of froth as immatures and have practically lost the ability to jump. The Auchenorrhyncha have simple life cycles, but wing dimorphism and seasonal differences in markings are not uncommon. Immatures generally resemble adults, except for the absence of wings, although nymphs of cicadas are specialized for subterranean life and are superficially quite different from adults. Nearly all Auchenorrhyncha are bisexual, and they never have alternating generations as in the Sternorrhyncha, discussed below. Males of many species produce sound from a pair of tymbals in the base of the abdomen. In cicadas the tymbals are associated with a resonating chamber. The chorusing noise produced by large aggregations of individuals may be almost deafening and probably functions in confusing birds and other predators, as well as in courtship. Sounds produced by cicadellids are too feeble to be heard by humans without amplification. Their acoustical repertoires include territorial and courtship calls.

Suborder Sternorrhyncha (Aphids, Whiteflies, Psyllids, scales)

The superfamilies Psylloidea, Aphidoidea, Aleyrodoidea, and Coccoidea constitute the suborder Sternorrhyncha. They have relatively long, filiform or filamentous antennae, not bristlelike as in the Auchenorrhyncha. The rostrum is inserted on the extreme posteroventral part of the head and is usually closely appressed to the thoracic sterna, from which it may appear to arise. Tarsi are 1- or 2-segmented. The ovipositor is usually undeveloped, and eggs are deposited on plant surfaces. Adult psyllids and aleyrodids are active jumping or flying insects, but the Sternorrhyncha are predominantly nonvagile, specialized for rapid feeding and reproduction. Parthenogenesis is widespread and in Aphidoidea may be alternated with sexual reproduction and shifts in plant hosts at certain times of year (Chap. 4). Many Sternorrhyncha are notable for the tendency of the immature and adult stages to diverge morphologically. In aleyrodids and male coccoids the degree of differentiation is equivalent to that in endopterygote insects. The nonmotile nymphs have greatly reduced appendages and sense organs and are essentially a feeding stage. The motile adult is preceded by a nonfeeding "pupal" stage, and physiologically these insects are holometabolous.

Many Sternorrhyncha feed on plant phloem sap (food-conducting vascular tissue), and a few Auchenorrhyncha form commensal relationships with ants, which are attracted to their copious exudates of honeydew and which to some extent protect these sap-feeders from predators and move them about their host plants. Ant-tended hemipterans exude honeydew, rather than ejecting it, as do many free-living forms, and some ant-tended species have special devices for holding droplets of honeydew until it is taken by ants. Ant-tended Homoptera tend to live in clusters. Alarm pheromones cause them to aggregate rather than disperse, and they allow themselves to be carried about by ants without struggling. Such relationships are most specialized with some aphids. For example, the corn root aphid, *Anuraphis maidiradicis*, is tended by ants (chiefly *Lasius*) in all instars. The eggs are collected in the fall and overwintered in the ant nests. The first-instar aphids are carried to the spring host (smartweed). Later generations are transferred to the summer host (corn). In many Sternorrhyncha dermal glands secrete a protective cover of wax either in the form of characteristic filaments or plates or as a powdery deposit. In some Coccoidea the eggs are enmeshed in a fluffy envelope of wax filaments. The scales of diaspidids and other armored scales also contain wax.

Aphidoidea and Coccoidea require slide mounting or other special preparation for reliable identification to family. For that reason families of these groups do not appear in the following key. Nearly all the commonly encountered Aphidoidea are members of the family Aphididae. Most Coccoidea encountered by nonspecialists are members of Diaspididae (armored scales) and Pseudococcidae (mealy bugs). Pseudococcidae (mealybugs) are mobile, with externally visible body segments, in all instars.

KEY TO SUPERFAMILIES AND MAJOR DIVISIONS OF THE SUBORDERS STERNORRHYNCHA AND AUCHENORRHYNCHA

(Adults)

1. Tarsi with 1 or 2 segments, or legs absent; antennae filiform, not hairlike, and with

more than 3 segments
(suborder **Sternorrhyncha**)   2
Tarsi with 3 segments, antennae usually
3-segmented
(suborder **Auchenorrhyncha**)   5

2(1). Tarsi with 1 segment and 1 claw or legs vestigial or absent; females larviform or gall-
like or scalelike (Fig. 32.16c–e), males with
mesothoracic wings and atrophied mouthparts (Fig. 32.16b)   **Coccoidea**
Tarsi with 2 segments and 2 claws; body
insectlike, with recognizable legs, antennae, and mouthparts; wings 4 in number, if
present   3

3(2). Tarsal segments subequal in size and shape;
antennae with 7 to 10 segments   4
Tarsi with first segment small, triangular, or
absent; antennae with 1 to 6, usually with 5
or 6 segments   **Aphidoidea**

4(3). Antennae with 7 segments; wings membranous, without sclerotized leading edge
(Fig. 32.14e); body powdered with white
wax   **Aleyrodoidea (Aleyrodidae)**
Antennae usually with 10 segments;
wings with sclerotized leading edge;
body usually not powdered with wax
            Psylloidea (Psyllidae)

5(1). Antennae inserted on frons, between eyes
(Fig. 32.13a,b); pedicel subequal to scape in
diameter; mesocoxae short, contiguous or
close together   6
Antennae inserted laterally, beneath eyes
(Fig. 32.13c,d); pedicel enlarged, often globular; mesocoxae elongate, distant
                        Fulgoroidea

6(5). Head with 3 ocelli; anterior femora thickened; body more than 15 mm long
            Cicadoidea (Cicadidae)
Head with 2 ocelli, or ocelli absent; anterior
femora slender; body rarely more than 10
mm long   7

7(6). Metatibiae with 1 or 2 large spurs on shaft
(Fig. 32.11a); metacoxae short, conical
            Cercopoidea (Cercopidae)
Metatibiae with numerous small spines on
shaft (Fig. 32.11b); metacoxae transverse,
platelike   **(Cicadelloidea)**   8

8(7). Scutellum exposed as a triangular sclerite
(Fig. 32.13b)   9
Scutellum hidden by pronotum, which
extends backward usually about half the
length of wings   10

9(8). Hind tibia with one or several rows of small
spines   **Cicadellidae**
Hind tibia with one or two spines on shaft
and circlet of spines at apex, or lacking
spines   **Aetalionidae**

10(8). Pronotum extended backward as a narrow,
straplike process that exposes the scutellum laterally   **Aetalionidae**
Pronotum extended broadly backward to
completely conceal the scutellum (Fig.
32.13a)   **Membracidae**

Cicadoidea (cicadas; Fig. 32.12). A single family, **Cicadidae**, of moderate to very large insects
with membranous, transparent wings, sometimes
banded or spotted. Adult cicadas are familiar
because of the high-volume sounds produced by

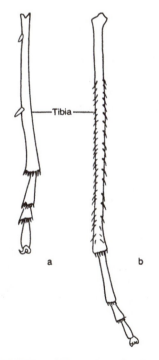

Figure 32.11 Tibiae of Homoptera showing spurs and spines: a, Cercopidae; b, Cicadellidae.

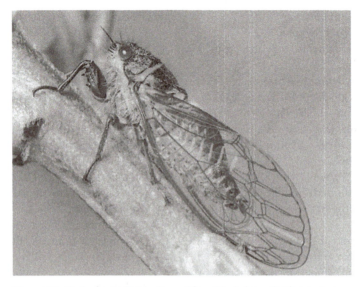

Figure 32.12 A cicada, *Platypedia* sp. (Cicadidae), from California. Immature cicadas develop underground, sucking sap of roots.

the males, often singing in groups (see Chap. 6; Fig. 6.8a,b). Females use their sharp ovipositors to create a slit in twigs to insert their eggs, sometimes causing economic damage. Newly eclosed nymphs drop to the soil, where they lead subterranean lives feeding on xylem sap in roots. Most species require several years to mature, and often all of the members of a single generation appear simultaneously. Usually, a generation matures each year, but in the "periodical cicadas" (*Magicicada*) of eastern North America, generations are restricted to certain years at a specific location. These species require either 13 or 17 years to mature and consequently may unexpectedly appear after several years' absence. At most localities the individuals emerging in any given year consist of two or three morphologically similar species with synchronized life cycles. The ecological and evolutionary significance of these synchronized mass emergences restricted to certain years is not entirely understood but may be a means of alternately "swamping" and starving potential predators.

Fulgoroidea (fulgorids; Fig. 32.13c,d). Clypeus separated from frons by distinct frontoclypeal ridge, not extending posteriorly between eyes;

antennae usually with second segment enlarged, with conspicuous sensory organs; 2 ocelli (occasionally 3) on sides of head near eyes; tegulae present on mesothorax.

Fulgoroidea range from small to large in size and occur in a diverse and sometimes bizarre array of body forms. Many species (Cixiidae, Delphacidae, Achilidae) superficially resemble Cicadellidae. Others have broad, mothlike wings (Flatidae) or have the wings reduced and strap-shaped (Dictyopharidae, Issidae). Exaggeration of the proportions of various parts of the body, including head, wings, and legs, occurs in various species for as yet undetermined reasons.

Most fulgoroids are phloem sap feeders, but some (Achilidae, Derbidae) feed on fungus. Most species probably spend their entire lives on foliage, but most Cixiidae apparently live underground as nymphs, feeding on grass roots, or in ant nests. Fulgoroids are primarily a tropical group, and most of the relatively few economically important species occur on sugar cane, coffee, and similar crops.

Recognition of fulgoroid families is rendered difficult by the great structural variation within many families, together with a repeated tendency toward convergence in external appearance, especially in the numerous flightless species from different families. About 20 families, many obscure or principally tropical, are recognized by most authorities. About half of these occur in North America. The interested student is referred to Kramer (1950) and Nault and Rodriguez (1985) for more complete discussions of the superfamily.

Cercopoidea *(spittlebugs or froghoppers).* Adults superficially similar to Cicadellidae; best distinguished by the tibial characters given in the key. The soft, whitish nymphs of Cercopidae live within frothy masses of "spittle," which is produced by blowing air through a viscous mucosaccharide secreted by the Malpighian tubules and released through the anus. Air is expelled by contracting a cavity formed by the ventral extension of the abdominal tergites beneath the abdomen. The spittle serves both as a protective device and as a means of reducing evaporation (Marshall, 1966). Nymphs of some Australasian species (family Machaerotidae) construct hardened tubes or shells from the anal secretion, attaching these to twigs or branches. The tube in turn is filled with spittle. Cercopids are cosmopolitan, but most species are tropical.

Cicadelloidea *(leafhoppers and treehoppers; Fig. 32.13a,b).* Distinguished by the tibial and coxal characters listed in the key. Cicadelloids also differ from other superfamilies in venational characters and in having the tentorium reduced.

Cicadelloids feed in the phloem, xylem, or parenchyma tissues, almost always of higher plants. Many species are cryptically colored, and some, especially membracids, are structurally modified to resemble thorns or other plant parts. Membracid nymphs are often gregarious and may be tended by ants. The great majority of Cicadelloids have simple life cycles with many generations per year. Cicadellidae use a wide array of host plants, including herbs, grasses, shrubs, and trees. In contrast, Membracidae and Aetalionidae feed mostly on woody vegetation. Adults, especially of Cicadellidae, are highly mobile and may migrate long distances, particularly when local food sources (such as ephemeral native plants)

wither. Cicadellids are important vectors of plant pathogens, especially viruses and bacteria. The dominant family Cicadellidae has over 20,000 species worldwide, and some species are among the most common insects collected by sweeping low vegetation. Membracids include more than 3200 species and Aetalionidae only about 50, with 3 species in North America.

Psylloidea *(psyllids; Fig. 32.14a,b).* Minute to small Homoptera with long, bristlelike antennae, 3 ocelli; forewings with venation reduced by fusion of R, M, and Cu veins; metacoxae enlarged, elongate.

Psyllids may superficially resemble minute cicadas. As adults they are highly active, readily jumping and flying if disturbed. Nymphs are sluggish and are morphologically distinct from adults, with squat, flattened bodies. Many Australian species produce protective polysaccharide shells or lerps; North American species may be covered with honeydew but never produce hardened coverings. They mature on foliage, usually of woody plants. A few species cause the growth of galls. The life cycle is simple, with one to several bisexual generations per year. Some psyllids are important vectors of plant pathogens. Psyllidae, is the largest family.

Aleyrodoidea (whiteflies; Fig. 32.14c-e). A single family, **Aleyrodidae**. Adults similar in general size and body form to Psyllidae; antennae long, bristlelike; wings opaque with dusting of white wax, venation greatly reduced, with 2 or 3 longitudinal veins, no clavus; all instars with characteristic vasiform orifice on last abdominal tergite (Fig. 32.14d).

These minute insects are phloem feeders, usually on the undersides of the leaves of angiosperms. A few are known from ferns, none from gymnosperms. Eggs, attached by short pedicels to the undersurfaces of leaves, are frequently arranged in arcs or circles. First instars are flattened, ovoid, with short legs. The second and third instars (larvae) are sessile, with atrophied legs. The fourth instar ("pupa") is similar to the larva but feeds only briefly, then becomes inactive.

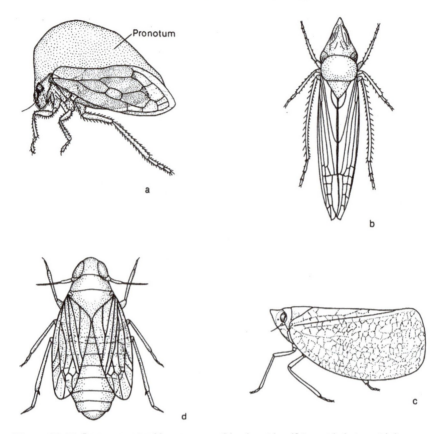

Figure 32.13 Representative Homoptera: a, Membracidae (*Stictocephala inermis*); b, Cicadellidae (*Draeculacephala mollipes*); c, d, Fulgoroidea; c, *Acanalonia bivittata* (Acanaloniidae); d, *Delphacodes campestris* (Delphacidae).
Source: a, from Yothers, 1934; b–d, from Osborn, 1938.

Before the final molt, the developing adult organs become visible through the pupal cuticle. The pupal stage is frequently ornamented by wax plates or filaments in highly individualized patterns, and the taxonomy of Aleyrodidae is based mainly on these structures. Reproduction may be either sexual or asexual; in some species unfertilized eggs produce males, but parthenogenetic races of only females are known. Aleyrodids are primarily tropical, and the most economically important species in North America attack citrus or greenhouse plants.

Aphidoidea (aphids; plant lice; Fig. 32.15). Soft-bodied, globular or flattened insects with abdominal segmentation indistinct; either winged or apterous as adults, forewings with 3 to 5 longitudinal veins and distinct clavus; cornicles usually present but sometimes reduced, cone-shaped or ring-shaped.

The Aphidoidea are remarkable for the development of highly complex life cycles, which may involve cyclical parthenogenesis, cyclomorphosis (seasonal change in body form), and alternation of hosts. These are apparently adaptations to survive in seasonal climates, and in highly temperate areas all three phenomena may occur in a single species (see discussion of aphid cycles in Chap. 4). Parthenogenetic and viviparous typically occur in the spring and summer, while in autumn, winged adults parthenogenetically produce males and females. These, in turn, produce 1 or a few over-wintering eggs.

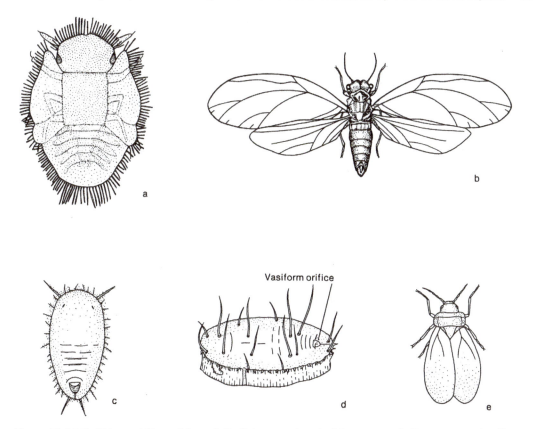

Figure 32.14 Psyllidae and Aleyrodidae: a, b, Psyllidae, nymph and adult, respectively (*Paratriazoa cockerelli*). c–e, Aleyrodidae, nymph, pupa, and adult, respectively (*Trialeurodes vaporariorum*).
Source: a, b, redrawn from Knowlton and Janes, 1931; c–e, redrawn from Lloyd, 1922.

Taxonomy of Aphidoidea is unsettled, with 1 to 9 families recognized by different specialists. The more important differences among families involve features of the life cycle, rather than morphological characters. Aphididae is the dominant family, with over 3000 species, the great majority from north temperate areas and very few from the tropics. Two other small families are commonly recognized: Adelgidae cause conelike galls on conifers, with at least one generation occurring on spruce (*Picea*); Phylloxeridae occur on oak (*Quercus*), grape (*Vitus*), walnut (*Juglans*), and other plants, usually forming galls on at least one part of the plant. All but a few species have a single host. In the latter two families the cornicles are reduced or vestigial. Aphidoidea are of exceptional economic importance. Besides the direct effects caused by sucking plant sap, their saliva may be toxic to plants. Aphids make up the largest group of plant virus vectors. One species alone, *Myzus persicae*, the green peach aphid, vectors more than 100 different viruses.

Coccoidea (scale insects; mealybugs; Fig. 32.16). Minute to small, sexually dimorphic Homoptera; mature females with ovoid to globular body, frequently with segmentation indistinct, eyes small, arranged in groups of simple lenses; antennae 1- to 9-segmented or vestigial; legs segmented, with single tarsal claw, or atrophied; rostrum 1- to 2-segmented, short; stylets extremely long, coiled in body; body frequently protected by waxy, cottony or hard, scalelike covering. Mature male usually gnatlike, with long, multiarticulate

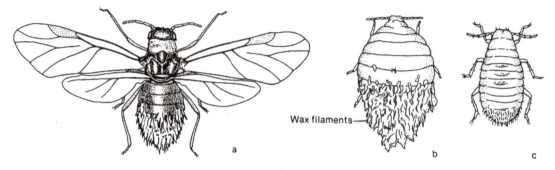

Figure 32.15 Aphidoidea, Aphididae (*Eriosoma lanigerum*, the woolly apple aphid). a, Winged female; b, apterous female; c, male.
Source: Redrawn from Baker, 1915.

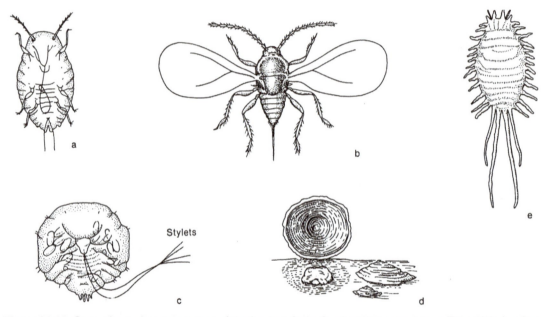

Figure 32.16 Coccoidea: a–d, various instars of the San José Scale, *Quadraspidiotus perniciosus* (Diaspididae); a, first-instar nymph or crawler; b, adult male; c, adult female, ventral, extracted from scale revealing eggs inside body; d, scales; e, *Pseudococcus longispinus* (Pseudococcidae).
Source: a–d, redrawn from U.S. Department of Agriculture; e, redrawn from Green, 1896–1922.

antennae, lateral eyes, enlarged mesothorax bearing 1 pair of wings with 1 or 2 longitudinal veins; hind wings reduced to halteres.

Morphologically Coccoidea are among the most highly modified insects, being specialized for a sessile or nearly sessile life attached to their host plants. The first instars, or crawlers (Fig. 32.16a), are always motile and function in dispersal. Pseudococcidae (Fig. 32.16e) (mealybugs) possess functional legs in all instars, but in other families of scale insects at least 1 instar is nonmotile. The crawlers can survive up to several days without feeding, during which period surprisingly large distances may be traversed. If a spot suitable for feeding is located, the first instar ("crawler") inserts its mouthparts to thereafter permanently remain attached in place. The legs, antennae, and eyes atrophy, and the developing scale secretes a protective covering, or scale. In a few species no scale is produced, protection being afforded instead by greatly thickened cuticle. Males pass through one more instar than females. The last two larval instars in males display external wing pads and other adult features, much as in endopterygote pupae, and the last instar does not feed. Adult males do not feed and usually survive only a few days. Reproduction is oviparous or viviparous. Oviparous forms usually cover the eggs with waxen oothecae or retain them beneath the mother's scale or body. Parthenogenesis occurs in various families. Some Margarodidae are hermaphroditic (have both male and female sex organs).

Various scale insects have one or more instars specialized beyond the description given above. In Margarodidae intermediate instars of females are legless "cysts," capable of withstanding periods of up to several years without food or water. Some members of the Eriococcidae form galls, within which the insects mature. In some species the male and female galls are strongly differentiated, while in others the male remains within the parent gall. First instars of some species, such as the coccid, *Walkeriana*, bear elongate filaments enabling them to travel on breezes, in the manner of a dandelion seed, and first instar females of *Cystococcus* (Eriococcidae) cling to adult males as a means of dispersal.

It should be pointed out that although many coccoids are serious economic pests, several useful products, such as cochineal and other dyes, lac (used in shellac), and wax, have been derived from certain species, and *Dactylopius* (Eriococcidae) has been used in biological control of prickly pear cactus (*Opuntia*).

Many species of armored scales (Diaspididae) may be recognized by the characteristic scales they produce (Fig. 32.16d), but family and species classification and identification of most Coccoidea are extremely difficult, requiring removal of the scale, clearing, staining, and mounting on slides. About a dozen families are recognized by most authorities. For identification of the common economic species, the reader is referred to Metcalf and Metcalf (1993). The most complete taxonomic treatments of North American species are those of Ferris (1937–1955) and McKenzie (1967).

CHAPTER 33

Order Thysanoptera (Thrips)

ORDER THYSANOPTERA

Minute to small Exopterygota with slender, elongate body. Head elongate, hypognathous or opisthognathous; mouthparts asymmetrical, with maxillae and left mandible modified as piercing stylets; eyes prominent, round or kidney-shaped, with large, round facets; antennae 4- to 9-segmented, inserted anteriorly. Thoracic segments subequal or prothorax largest; mesothorax and metathorax fused; wings narrow, straplike, fringed with long setae; legs short, adapted for walking; tarsi 1- or 2-segmented with eversible, glandular vesicle at apex. Abdomen 11-segmented, terminating in a tubular apex or with a large, serrate ovipositor.

These highly distinctive insects are remarkable for their ubiquity, inhabiting all sorts of vegetation, as well as leaf litter, fungus, and the subcortical region of decaying trees. About 5000 species are known worldwide, with temperate and subtropical grasslands and woodlands supporting a particularly diverse fauna. Most foliage-inhabiting thrips imbibe the contents of cells pierced by the oral stylets. The resulting air-filled cells produce a characteristic mottling or silvering of the affected plant surfaces. Many phytophagous species preferentially infest the reproductive structures of plants, and the most serious economic damage involves buds, flowers, and young fruits. Only a few species transmit plant viruses, the most serious being tomato spotted wilt virus, which can infect a wide range of plants in addition to members of the tomato family (Solanaceae). Thrips vectors typically breed on weeds from which they carry the virus to crop plants when weeds are mown or removed. Adult thrips can transmit virus only if they acquire it as nymphs feeding on virus-infected plants. Some thrips feed on aphids or other small, soft-bodied arthropods, especially eggs of mites and Lepidoptera. A few species feed on insects

as ectoparasites. Some feed on fungal spores, and a few form galls. Flower-inhabiting thrips commonly suck the liquid contents from pollen grains, and the larger Tubulifera can ingest small pollen grains and fungal spores. Thrips as pollinators are often overlooked, despite their often reaching large populations in flowers. Because of their small size, thrips carry far fewer pollen grains than do larger flower-visiting insects, but their huge numbers in flowers may offset this disadvantage. Some plants are specialized for thrip pollination.

Thrips' bites of humans may cause an itching sensation. The unique, asymmetrical mouthparts (Fig. 33.1d) of thrips share several general similarities with the specialized styliform mouthparts of Hemiptera. The single, awl-shaped mandible and the maxillary stylets (Fig. 33.1b,c) slide through a conical guide formed by the labrum, labium, and external maxillary plates. Internally the maxillary stylets are attached to the maxillary plates by a slender lever bearing the attachments of the protractor and retractor muscles, as in the Hemiptera. In feeding, the oral cone is pressed against the surface and the stylets are punched a short distance into the tissues

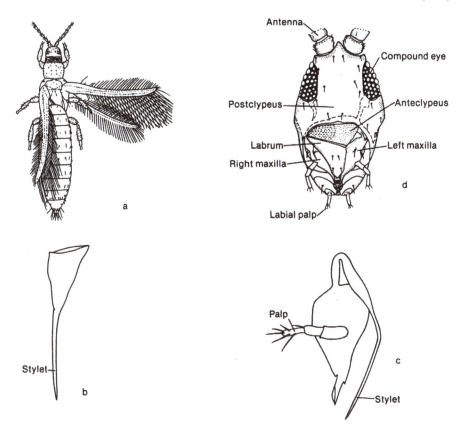

Figure 33.1 Thysanoptera: a, *Anaphothrips striata* (Thripidae), dorsal aspect; b, mandible, same; c, maxilla, same; d, frontal view of head, *Chirothrips hamatus* (Thripidae).
Source: a–c, redrawn from Hinds, 1900; d, redrawn from Jones, 1954.

and saliva injected. The exuding fluids are sucked into the alimentary canal by a cibarial pump.

Fringed wings are a specialization that appears repeatedly in very small insects, frequently in conjunction with reduction of the area of the wing membrane. Winged thrips are competent fliers, but their small size means that air movements may govern the direction and distance of long-range movements. Mass flights are recorded for several species. Wingless species often climb to the tops of grass stems and are readily dispersed by wind. The tarsus bears an adhesive vesicle that protrudes during walking, enabling thrips to cling to very smooth surfaces. These vesicles, apparently an adaptation for clinging to exposed plant surfaces, are unique to thrips, but analogous structures occur in a variety of phytophagous species in different orders.

Most thrips are bisexual, but parthenogenesis is common, and males are rare or unknown in some species. Polymorphism is also frequent. Most evident is the degree of development of wings, which may be present, reduced, or absent, especially in males.

Female thrips are diploid; males are produced from unfertilized eggs and are haploid. This type of parthenogenetic reproduction, termed arrhenotoky, is nearly universal in Hymenoptera (Chap. 44). Eggs are large in relation to body size. In different species, numbers of eggs vary from as few as two to several hundred. The Phlaeothripidae is the sole family in the suborder Tubulifera, in which the ovipositor is not developed, and eggs are deposited in cracks, crevices, under bark, or exposed on foliage surfaces. In the suborder Terebrantia, the serrate, bladelike ovipositor is used to insert the eggs into plant tissue. A few species are ovoviviparous. Eggs hatch after about 2 to 20 days, depending on temperature. The newly emerged nymphs resemble adults except in the absence of wings and in the reduced number of antennal segments. Two feeding instars are followed by a "prepupal" or "propupal" instar and one "pupal" instar in the Terebrantia, or by one prepupal and two pupal instars in the Tubulifera (Fig. 33.2). During these transformational stages the mouthparts are nonfunctional and the insects are capable of only slow, restricted movement. The pupal stages are usually passed in a cell prepared by the previous instar from soil or litter particles, but some species construct a silk cocoon or transform naked in cracks or crevices or on foliage. The prepupal and pupal stages are analogous to the pupae of endopterygote insects, and histolysis of various organs, especially head muscles, has been demonstrated in various species. Unlike metamorphosis in endopterygotes, external growth of wing pads in thrips occurs in more than one instar. Furthermore, their specialized mouthparts, reduced wing venation, and the organization of internal organs indicate that thrips represent a specialized derivative of hemipteroid insects, rather than primitive Endopterygota.

Many thrips pass through numerous generations annually, requiring as few as 10 days to develop from egg to adult. Others with a single generation spend most of the year in pupal diapause. Nymphs, pupae, or adults may overwinter in various species. Adults are commonly dark or cryptically colored, but larvae are often bright red or yellow because of their colored tissues showing through the transparent cuticle.

Many thrips aggregate on plants or fungi but are not social. However, a few Australian species in the Phlaeothripidae are eusocial. The female adult foundresses of these eusocial thrips have offspring that have reduced reproduction and act to defend the colony within a leaf gall. Later generations have normal wings and reproduction to disperse to new locations and establish new colonies. Social interactions among generations have been described in a species from Panama (Kiester and Strates, 1984), in which nymphs engage in communal foraging.

Characters useful for identifications of thrips include antennal characters that often require a compound microscope to see in slide-mounted specimens. The following key does not include the rare families Uzelothripidae or Fauriellidae.

KEY TO THE FAMILIES OF THYSANOPTERA

1. Terminal abdominal segment cylindrical, tubular (Fig. 33.3a); forewing (if present) with membrane smooth, glabrous (suborder **Tubulifera**)   **Phlaeothripidae**

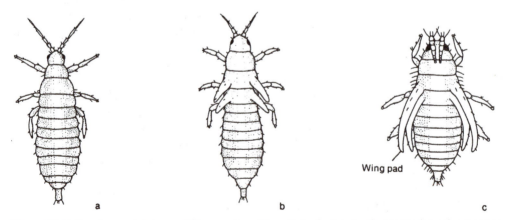

Figure 33.2 Developmental stages of Thysanoptera (*Hercothrips fasciatus*): a, last-instar larva; b, prepupa; c, pupa.
Source: Redrawn from Bailey, 1932.

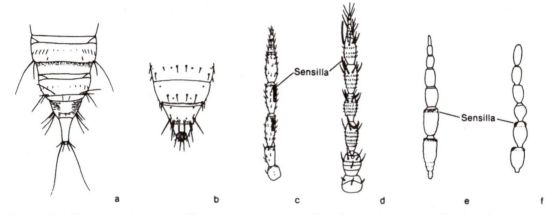

Figure 33.3 Taxonomic characters of Thysanoptera: a, suborder Tubulifera, abdominal apex (*Hoplandothrips irretius*, Phlaeothripidae); b, suborder Terebrantia, abdominal apex (*Anaphothrips striata*, Thripidae); c–f, antennae, showing arrangement of sensilla: c, Aeolothripidae (*Aeolothrips scabiosatibia*); d, Thripidae (*Anaphothrips striata*); e, Heterothripidae (*Heterothrips salicis*); f, Merothripidae (*Merothrips morgani*).

Terminal abdominal segment rounded or conical (Fig. 33.3b); forewing (if present) with membrane bearing minute hairs (microtrichia) (suborder **Terebrantia**)   **2**

2(1). Antennae with 9 segments and longitudinal patches of sensilla on segments 3 to 4 (Fig. 33.3c); females with ovipositor curved upward   **Aeolothripidae**
Antennae with round or transverse patches of sensilla on segments 3 to 4 (Fig. 33.3d–f); ovipositor curved downward or reduced   **3**

3(2). Antennae with sensilla on segments 3 to 4 arranged as simple or forked cones (Fig. 33.3d)   **5**
Antennae with sensilla on segments 3 to 4 as continuous or interrupted bands encircling apices of segments (Fig. 33.3e,f)   **4**

4(3). Antennae with 8 segments   **Merothripidae**
Antennae with 9 segments   **6**

5(3). Third and fourth antennal segments with conelike sensilla   **Adiheterothripidae**
Third and fourth antennal segments have simple hairlike or forked sensilla **Thripidae**

6(4). All antennal segments have transverse rows of small hairs (microtrichia)
Melanthripidae
Antennal segments without transverse rows of microtrichia   **Heterothripidae**

Most common thrips, including those of economic significance, are in the families Thripidae and Phlaeothripidae. Aeolothripidae are sometimes abundant on flowers or in litter. Most species have banded or maculated wings. Merothripidae and Heterothripidae are poorly represented in North America, with a few species occurring in litter or on fungus and in flowers, respectively. Adiheterothripidae has only two known species in western North America and several others from the Mediterranean region to India.

CHAPTER 34

Order Megaloptera (Alderflies and Dobsonflies)

ORDER MEGALOPTERA

Moderate-sized to large Endopterygota. Adults with strong mandibles, maxillae, and labium, large compound eyes, ocelli present or absent. Thorax with 3 subequal segments; mesothorax and metathorax with large, subequal wings; metathoracic wings with pleated anal region folded over abdomen at rest. Larva with large head, large prognathous mandibles, large sclerotized prothorax, smaller sclerotized mesothorax and metathorax; abdomen elongate, with segmented or unsegmented filaments on segments 1 to 7 or 8, sometimes with tufts of gills at the filament bases; terminal segment bearing elongate medial appendage or paired prolegs. Pupa decticous, motile.

Megaloptera are often regarded as among the most primitive endopterygote insects. Recent phylogenetic studies indicate that while it is clear that Megaloptera, Raphidioptera, and Neuroptera form a monophyletic lineage (the Neuropterida), monophyly of the Megaloptera itself is not so clear, and it might well represent several early-diverging lineages of the other Neuropterida. There are about 300 species worldwide, distributed mostly in temperate regions.

Megaloptera are morphologically similar to Neuroptera as adults (Fig. 34.1a) and in some classifications are included as a suborder of Neuroptera. The most reliable differentiating feature is the presence of an anal fold in the hind wing (Fig. 34.1c) (no folding in Neuroptera). In contrast, larvae of the two groups are well differentiated. Whereas nearly all neuropterous larvae have specialized suctorial mouthparts (Fig. 36.1c), in the Megaloptera mandibles and maxillae are unmodified and adapted for chewing (Fig. 34.2). The alimentary canal is complete in Megaloptera (ending in the blind midgut in Neuroptera), and pupation occurs in an earthen cell (within a silk cocoon in Neuroptera). In general structure (Fig. 34.2), larvae of Megaloptera are more similar to adephagous beetle larvae than to Neuroptera, and separation from Gyrinidae and Dytiscidae is sometimes difficult.

Dobsonflies and alderflies are familiar insects around permanent water, being especially prevalent around cool, well-oxygenated streams. Adults are weak fliers, seldom straying more than a few hundred meters from water. Apparently they do not feed, despite their well-developed mandibles, which are greatly enlarged in males of some Corydalidae (Fig. 34.1b). Mating occurs on streamside vegetation or on the ground. Single females may produce as many as several thousand eggs, and communal egg deposition sites on vegetation or other objects overhanging water may be used by many individuals, resulting in large egg masses. Eggs hatch in 1 to 4 weeks, the newly emerged larvae dropping into the water.

Larvae shelter beneath stones, sunken vegetation, or in other protected situations. All types of

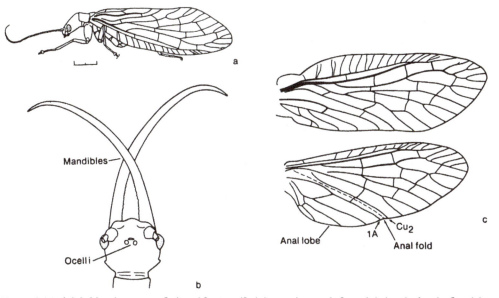

Figure 34.1 Adult Megaloptera: a, *Sialis californicus* (Sialidae; scale equals 2 mm); b, head of male *Corydalus*, antennae removed (Corydalidae); c, wings, *Sialis californicus*.

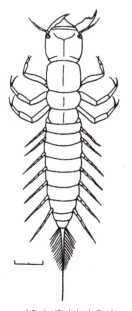

Figure 34.2 Larva of *Sialis* (Sialidae). Scale equals 2 mm.

aquatic situations are utilized, but different species have distinct preferences for temperature, substrate, and other characteristics. Sialidae require permanent aquatic habitats, but some Corydalidae

occupy intermittent streams in western North America, where they burrow into the substrate while the stream is dry. All megalopteran larvae are predaceous, attacking any aquatic organisms small enough to subdue, and in turn are consumed by fish, frogs, and other predators. The tactile filaments along the sides of the abdomen are used in detecting prey and avoiding enemies. They may be lost without permanent detriment to the larva. There are numerous larval instars, and full growth apparently requires a year in Sialidae and up to 5 years in Corydalidae. The mature larvae leave the water to excavate a pupal chamber in moist sand or soft earth beneath stones or driftwood, where they may remain for several months in the larval stage before pupating. Pupae retain full mobility and can defend themselves with the mandibles.

KEY TO THE FAMILIES OF MEGALOPTERA

1. Adults with 3 ocelli (Fig. 34.1b), fourth tarsal segment cylindrical, wingspan 45 to 100 mm; larvae with 8 pairs of unsegmented

abdominal filaments, prolegs on last abdominal segment **Corydalidae**
Adults without ocelli, fourth tarsal segment bilobed, wingspan 20 to 40 mm; larvae with 7 pairs of segmented abdominal filaments, terminal filament on last abdominal segment (Fig. 34.2) **Sialidae**

Corydalidae. Widespread, with species throughout North America. Larvae occur predominantly in running water, especially fast-flowing streams.

The wings are folded flat over the back at rest. Adults are partly nocturnal, commonly appearing about lights.

Sialidae. Widespread in North America, most diverse in the west. The wings are held at a steep angle over the back at rest, in the manner of lacewings. Larvae occur in ponds and sluggish watercourses, as well as in swiftly flowing streams. Adults diurnal.

CHAPTER 35

Order Raphidioptera (Snakeflies)

ORDER RAPHIDIOPTERA

Medium-sized Endopterygota. Adult with strong mandibles, maxillae, and labium, prominent compound eyes, ocelli present or absent. Prothorax slender, elongate, neck-like; mesothorax and metathorax subequal; wings similar in size and venation, with large pterostigma; female with long, slender ovipositor. Larva with large head, strong prognathous mandibles; prothorax sclerotized, slightly larger than membranous meso-thorax and metathorax; abdomen fleshy, cylindrical. Pupa decticous, motile.

In most features snakeflies resemble Megaloptera (Fig. 35.1). They are sometimes considered specialized terrestrial members of that order. Unlike the Megaloptera, which do not feed as adults, snakeflies are voracious predators throughout life, and the elongate head capsule and prothorax

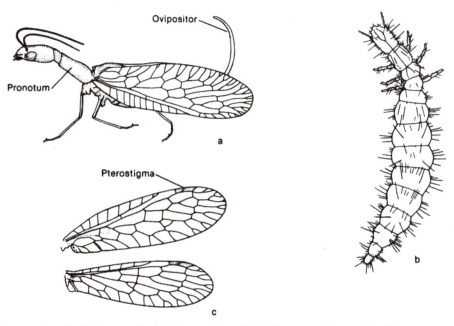

Figure 35.1 Raphidioptera, Raphidiidae: a, adult of *Agulla bracteata*; b, larva of *Agulla bracteata*; c, wings of *Agulla astuta*.
Source: Redrawn from Woglum and McGregor, 1958, 1959.

are adaptations to increase mobility of the head, which is used to strike the prey in a snakelike fashion. Fossils similar to modern Raphidioptera date from the Jurassic Period, suggesting a very early divergence from the Megaloptera. The world fauna numbers about 175 species in two families, Raphidiidae and Inocellidae.

Raphidioptera predominantly inhabit woodlands, where the adults are encountered on foliage, flowers, tree trunks, or similar places. The male copulates from beneath the female, following a long courtship. Up to 800 eggs are inserted into crevices in bark or rotting wood in small groups or in clusters up to about 100. Larvae of Raphidiidae commonly occur under loose bark, in porous rotten wood, leaf litter, and similar places; those of Inocellidae have been found only under loose bark. In addition to their legs, larvae use a pygidial adhesive organ to nimbly move in any direction and to maintain a purchase on vertical surfaces. As in Megaloptera, pupation is in a rough cell constructed by the last-instar larva, without a cocoon. The pupa retains full mobility and frequently moves to a second location before adult emergence. The life cycle normally requires 2 years.

Snakeflies occur throughout the temperate areas of the Northern Hemisphere. Species in the southern extremes of this range in Mexico, northern Africa, and northern India inhabit montane areas. Both families occur in North America, restricted to the region west of the Rocky Mountains. In Raphidiidae (Fig. 35.1) ocelli are present, antennal segments are basally constricted, and antennae are usually less than half as long as the body. In Inocellidae ocelli are absent, antennal segments are cylindrical, and antennae are usually about as long as the body.

CHAPTER 36

Order Neuroptera (Lacewings, Antlions, etc.)

ORDER NEUROPTERA

Small to large Endopterygota. Adult with chewing mouthparts, large lateral eyes, ocelli present or absent; antennae multiarticulate, usually filiform or moniliform. Mesothorax and metathorax similar in structure, with subequal wings, usually with similar venation. Abdomen without cerci. Larvae with clearly defined head capsule; mandibles and maxillae usually elongate, slender, modified for sucking; thoracic segments with walking legs with 1-segmented tarsus usually bearing 2 claws; abdomen frequently bearing adhesive disks on last 2 segments, without cerci. Pupa exarate, decticous, enclosed in silken cocoon.

The Neuroptera comprise a small but highly variable order of predominantly predatory insects that display a mixture of primitive and specialized features. Permian fossils include several families of Neuroptera, and by the mid-Mesozoic Era forms similar to most modern families had appeared. Neuroptera occur in all parts of the world, but many families show relict distributions. The fauna of Australia is especially rich, including several endemic families as well as a great variety of primitive forms from other families. Ithonidae are restricted to North America and Australia, and Polystoechotidae to North America and southern South America. Most North American Neuroptera are green or various shades of gray or brown, sometimes with dark wing maculations; while sometimes large in size, they are mostly rather obscure insects. Many Australian and Asian species are strikingly colored or patterned. The order numbers about 5000 species worldwide. The families Chrysopidae and Myrmeleontidae are dominant, each with about 2000 species.

Adult Neuroptera are soft-bodied insects of generalized body plan (Fig. 36.1a). The mouthparts are adapted for chewing, with strong mandibles and maxillae, and small labium. The thoracic structure and wing venation are especially generalized, compared with those of most other Endopterygota. In most species the thoracic segments are subequal, with the prothorax freely movable. Mantispidae, which have enlarged, raptorial forelegs, also have the prothorax elongate, further increasing their striking range. The wings (Fig. 36.5) are similar in size, shape, and venation, except in Coniopterygidae, in which the hind wings may be much reduced, and in Nemopteridae, in which the hind wings are very long and slender, usually expanded at the apex. In Ithonidae, as well as several exotic families, characteristic sensory structures (nygmata) occur in the wing membrane. Wing coupling occurs in several families, especially those with strong flying members. In Hemerobiidae coupling is by a bristlelike frenulum on the base of the hind wing. Similar coupling

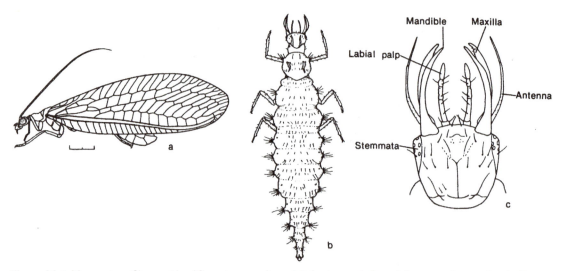

Figure 36.1 Neuroptera, Chrysopidae (*Chrysopa carnea*); a, adult (scale equals 2 mm); b, mature larva; c, head of larva. *Source: b, redrawn from Tauber, 1974; c, redrawn from Withycombe, 1924.*

occurs in some Chrysopidae and Mantispidae. The wings of Coniopterygidae are coupled by hamuli-like setae on the costal margin of the hind wings. Abdominal scent glands, usually situated on the first or fifth to seventh abdominal tergites, occur in several families, including Myrmeleontidae and Ascalaphidae. In Chrysopidae glands occur on the prothorax and offer some protection against vertebrate and invertebrate predators.

Immatures of Neuroptera (Figs. 36.1b, 36.2) are all predatory, frequently specialized for particular types of prey. A peculiarity of all species is the blindly ending midgut, a feature that also occurs in most Hymenoptera. During pupation the alimentary canal becomes complete, and the accumulated fecal material is voided shortly after adult emergence.

In Polystoechotidae and Ithonidae larval mandibles and maxillae are short and blunt. The larva of *Polystoechotes* is somewhat similar in general morphology to larvae of Megaloptera (Withycombe, 1925), and Polystoechotidae are probably among the most morphologically generalized extant Neuroptera. Larvae of Ithonidae are scarabaeiform in general shape, living in soil. It is unclear whether they are strictly carnivorous or feed at least in part on decaying plant material. All other neuropteran larvae are predatory with specialized mouthparts. The mandibles are elongate, with a deep mesal groove that is covered by the maxilla to form an efficient piercing and sucking apparatus (Fig. 36.3a). Configuration of the larval body is related to mode of life. Chrysopidae (Fig. 36.1b), Hemerobiidae, and Coniopterygidae are freely roving predators of small, soft-bodied arthropods, especially Homoptera and mites. Their legs are relatively long with large empodia; the abdomen is slender, tapering, and prehensile, with a terminal adhesive disk, adaptations for clinging to vegetation. Larvae of Osmylidae (Old World and South America) are similar in body form to Berothidae. They occur in forested situations, sometimes near streams, and may be subaquatic. In contrast, in the Myrmeleontidae (Fig. 36.2) and Ascalaphidae, the abdomen and thorax are either globular or broadly flattened and are relatively inflexible. The legs are short and in Myrmeleontidae are adapted for backward movement and for digging. The head is broad and flat, with enormous, sickle-shaped jaws. These larvae are sedentary predators that ambush prey from concealed positions. Some Myrmeleontidae construct conical pitfalls in dry, friable soil, the larva buried at the bottom with open jaws (Fig. 36.3b), while most either lie in ambush at the surface or

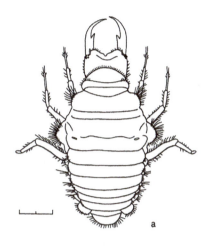
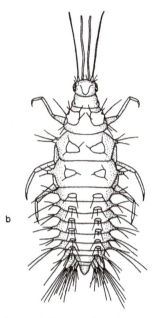

Figure 36.2 Larvae of Neuroptera: a, Myrmeleontidae (*Vella*; scale equals 2 mm); b, Sisyridae (*Climacia areolaris*).
Source: b, redrawn from Parfin and Gurney, 1956.

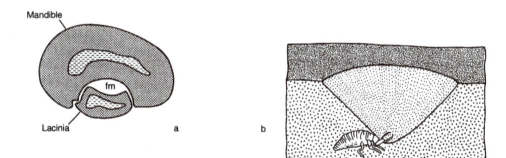

Figure 36.3 Larvae of ant lions, Myrmeleontidae: a, cross section of mandible; b, diagram of a larva in sand pit. The larva seizes small insects that fall into the pit and are unable to climb out.
Source: Modified from Scholtz and Holm, 1985.

actively hunt on or just beneath the surface. Walking is either backward or forward depending on the species, but it is backward in most pit-building forms. Ascalaphid larvae sit motionless in litter or on vegetation, snapping the jaws closed on any suitably sized prey. Larvae that frequent exposed situations on vegetation sometimes bear specialized hooked or spatulate setae which anchor a camouflaging layer of debris, including the dry husks of prey. This habit is shared with some chrysopid larvae. Larvae of Nemopteridae have a short bulbous abdomen, similar to that of Myrmeleontidae, but the prothorax forms a neck up to several times as long as the abdomen. These larvae frequent rock crevices and caves in the Old world and South America. The family Psychopsidae (Old World) is related to Chrysopidae, but the larvae have become adapted as ambush predators living under *Eucalyptus* bark and resemble myrmeleontid larvae in body form, with the head grossly enlarged and the body obese.

Larvae of Sisyridae (Fig. 36.2b) are superficially similar to those of Chrysopidae and Hemerobiidae but are specialized predators of freshwater sponges. They are distinguished by their straight, needlelike mandibles and jointed abdominal gills. Mantispid larvae are specialized parasitoids of spider egg masses or of larvae of vespid or sphecoid wasps or bees. The motile first instar, which is similar to the larvae of chrysopids, searches for a suitable host, sometimes attaching to adult hosts as a means of locating the immatures. After entering the spider egg sac or hymenopteran brood cell the mantispid larva molts to the grublike second and third instars, with short, ineffective legs and reduced head capsule. All feeding and development occur in a single host cell or egg sac. Larvae of Berothidae are predators of immature ants and termites, in whose nests they live, using neurotoxic chemicals to overcome their much larger hosts (Brushwein, 1987; Johnson and Hagen, 1981). Eggs of berothids are placed on termite-infested logs. Like mantispids, larvae of berothids are hypermetamorphic, with the motile first instar entering the termite nest and the later, obese instars feeding and completing development.

Many adult Neuroptera are predators of a variety of insects; other species imbibe nectar, pollen, or the honeydew secreted by homopterous insects. Adults are relatively short-lived except for those that diapause. Most of them are weak, erratic fliers, but flight is strong and direct in many Ascalaphidae. Eggs are deposited in soil in ground-dwelling forms or on foliage in arboreal types, being sessile (Hemerobiidae, Dilaridae, Coniopterygidae) or attached by a short or long stalk (Mantispidae, Chrysopidae, Berothidae). Three larval instars are typical of most families, with four or five instars reported in Ithonidae. The length of larval life is correlated with type of prey and prey availability. In those species that consume relatively abundant prey, such as Homoptera, full growth may be attained after only a few weeks. In

Myrmeleontidae and Ascalaphidae, where hunting success is highly unpredictable, larvae are capable of ingesting very large meals, then fasting for many months if necessary. In such forms larval life may last several years.

All Neuroptera pupate within shelters of silk produced by the Malpighian tubules and spun out through the anus. In Hemerobiidae the cocoon is usually wispy, but Chrysopidae and soil-dwelling types may incorporate sand or debris to form a tough protective covering. In some families the cocoon is double-layered. The movable pupal mandibles are sharply toothed, serving to open the cocoon. It may be noted that pupal mandibles in Trichoptera and a few primitive Lepidoptera have the same function. Pupae of Neuroptera usually emerge from the cocoon and may move about before eclosing.

KEY TO NORTH AMERICAN FAMILIES OF NEUROPTERA

1. Wings with few veins and fewer than 10 closed cells (Fig. 36.4a); small to minute insects covered with whitish exudate **Coniopterygidae**
Wings with numerous veins; more than 10 closed cells (usually very numerous); body without whitish exudate   **2**

2(1). Forelegs raptorial (Fig. 36.5b); prothorax elongate   **Mantispidae**
Forelegs not raptorial; prothorax subquadrate or slightly longer than broad   **3**

3(2). Antennae filiform, moniliform, or pectinate (see Fig. 2.7)   **5**
Antennae gradually or abruptly clubbed   **4**

4(3). Antennae nearly as long as body, with abrupt knob at apex   **Ascalaphidae**
Antennae about as long as head and thorax, gradually thickened (Fig. 36.5a)   **Myrmeleontidae**

5(3). Wings with sensory spots (nygmata) in membrane between branches of radial vein; hind wing abruptly narrowed at base   **Ithonidae**
Wings without sensory spots; hind wing gradually narrowed at base (Fig. 36.4b–d)   **6**

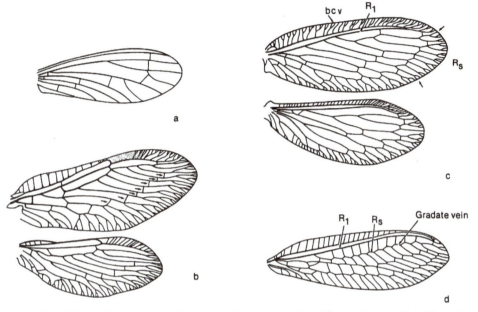

Figure 36.4 Wings of Neuroptera: a, forewing of Coniopterygidae (*Conwentzia hageni*); b, Mantispi-dae (*Plega signata*); c, Hemerobiidae (*Hemerobius ovalis*); d, forewing of Chrysopidae (*Chrysopa carnea*). bcv, bifurcate costal crossveins; extent of Rs marked by marginal arrows; gradate vein in b marked by arrows.
Source: a, redrawn from Banks, 1907.

6(5). Antennae pectinate (see Fig. 2.7) (male), or with ovipositor exserted, as long as abdomen (female)   **Dilaridae**
Antennae filiform or moniliform; ovipositor internal, not visible   **7**

7(6). Wings with numerous crossveins between veins R_1 and R_s (Fig. 36.4d) or R_s arising from 2 to 3 stems (Fig. 36.4c)   **8**
Wings with 1 to 4 crossveins between veins R_1 and R_s (as in Fig.36.4b)   **9**

8(7). Forewing with some costal crossveins bifurcate (Fig. 36.4c); body and wings usually brown   **Hemerobiidae**
Forewing without bifurcate costal crossveins (Fig. 36.4d); body and wings green or brown   **Chrysopidae**

9(7). Forewing with about 15 parallel branches of radial sector, wingspan at least 40 mm   **Polystoechotidae**

Forewing with 4 to 7 branches of radial sector (as in Fig. 36.4b); wingspan less than 30 mm   **10**

10(9). Forewing with some costal crossveins bifurcate (as in Fig. 36.4b,c); gradate vein present (as in Fig. 36.4b)   **Berothidae**
Forewing without bifurcate crossveins (Fig. 36.4d); gradate vein absent   **Sisyridae**

Coniopterygidae. Small or minute insects, superficially resembling whiteflies (Aleyrodidae) because of the reduced wing venation (Fig. 36.4a) and the whitish powder that covers the body and wings. Coniopterygids are common, widespread insects but usually escape notice because of their small size. Adults may feign death when disturbed.

Ithonidae *and* **Polystoechotidae.** A single, rare ithonid, *Oliarces*, occurs in desert areas of southern

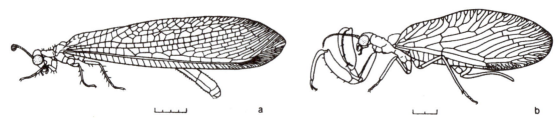

Figure 36.5 Adult Neuroptera: a, Myrmeleontidae (*Myrmeleon immaculatus*; scale equals 4 mm); b, Mantispidae (*Plega dactylota*, female; scale equals 2 mm)

California. Adults sometimes emerge in large evening swarms in spring or summer. The two North American polystoechotids are more widespread but also rare. Adults of both families are relatively large (30–70 mm) with broad wings and robust body.

Dilaridae. Small lacewings (wingspan 6–12 mm) whose larvae hunt on tree trunks. Widely distributed, with 2 uncommon species of *Nallachius* in eastern North America. Females oviposit in bark crevices of dead trees, especially oak and tulip trees, where the larvae are probably predators of beetle larvae.

Berothidae. Small lacewings, superficially similar to Hemerobiidae, usually distinguished by scalloped outer wing margins. The North American fauna includes fewer than 15 uncommonly encountered species.

Sisyridae. Widespread but seldom encountered, being restricted by the occurrence of freshwater sponges, their only hosts. Eggs are deposited on foliage overhanging water, the newly emerged larvae dropping into the water. Pupation occurs on emergent vegetation or on other objects near the water. At least one of several species occurs in most regions of North America.

Hemerobiidae *and* **Chrysopidae** (lacewings; Fig. 36.1). The superficially similar adults are reliably distinguished by the presence or absence of bifurcate costal crossveins. Wings of Hemerobiidae are frequently hairy, those of Chrysopidae bare. These two families represent one of the dominant groups of Neuroptera, with many North American species.

Mantispidae (Fig. 36.5b). Distinguished from all other Neuroptera by the enlarged raptorial forelegs. Nocturnal species are gray or brown; diurnal species include bright green forms as well as others patterned in red, yellow, and black, apparently mimicking vespid wasps. Mantispidae are general predators, most abundant in tropical and subtropical regions.

Myrmeleontidae *and* **Ascalaphidae**. Differentiated from other Neuroptera by the elongate body form, similar to that of Odonata. Ascalaphidae are not familiar insects in most of North America, but Myrmeleontidae (ant lions, Figs. 36.2a, 36.3b, 36.5a) are among the most common Neuroptera, especially in arid regions, where large numbers of the nocturnal adults often collect about lights.

CHAPTER 37

Order Coleoptera (Beetles)

ORDER COLEOPTERA

Adult. Minute to large Endopterygota. Compound eyes present (absent in many specialized forms), ocelli almost always absent; antennae almost always with 11 or fewer segments; mouthparts adapted for chewing, with recognizable mandibles, maxillae, and labium. Prothorax large, mobile, with ventral extensions of notum usually extending to vicinity of coxae; mesothorax small, fused with metathorax ventrally; mesonotum visible externally as scutellum; mesothoracic wings developed as sclerotized, rigid, elytra with specialized longitudinal venation; metathorax large (small in flightless forms), with strongly sclerotized sternum, weakly sclerotized tergum. Abdomen usually with sterna strongly sclerotized, terga membranous or weakly sclerotized with 2 or more terminal segments usually reduced; cerci absent.

Larva. Body form extremely variable but almost always with sclerotized head capsule, antennae with 2 to 4 segments, mouthparts mandibulate; thoracic segments with 4 or 5 segmented legs bearing 1 or 2 terminal claws; abdomen without legs, but frequently bearing segmented or unsegmented, sclerotized urogomphi on terminal segments. Wood-boring or other internally feeding larvae sometimes have legs, sensory organs, and head capsule reduced.

Pupa. Adecticous, exarate, similar to adult in general form or rarely obtect.

The Coleoptera, with more than 350,000 described species, constitutes the largest order of insects in terms of its described fauna. Five families of beetles number over 20,000 species, with weevils (Curculionidae) topping the list with at least 60,000 members, substantially more than most individual phyla of animals. As might be expected in such a large assemblage, Coleoptera are exceedingly variable ecologically and biologically. The majority of beetles are terrestrial herbivores, but several entire families and portions of others are predatory, frequently with highly specialized host ranges or life cycles. Either larvae or adults or both may be aquatic, with numerous freshwater and a few marine (intertidal) species known. In addition, beetles have exploited an extraordinarily diverse array of narrow, specialized niches, such as seed predation; boring in leaves, stems, or other restricted plant parts; formation of galls or tumors on plants; life inside ant or termite colonies; and many others. Certain relatively specialized habitats have been extensively utilized, often to the near exclusion of other insects: (1) the subcortical region of woody plants; (2) leaf litter and the top layers of soil; and (3) fungi, including fungi in soil, subcortical habitats, dung, and carrion, as well as the fruiting bodies of mushrooms and shelf fungi. One major source of food that has been rarely exploited by Coleoptera is nectar from flowering plants, probably because beetles are relatively inefficient fliers.

To a major extent, the evolutionary success of Coleoptera seems to stem from the protection from physical trauma provided by the high degree of sclerotization and body compaction. The most obviously affected structures are the forewings or elytra, which are modified as protective covers for the hind body (Fig. 37.1). Elytra are almost always present, even in wingless forms, but may be abbreviated (Staphylinidae and others), in which case the abdominal terga are sclerotized. In most beetles the elytra form hard sheaths, molded to the shape of the abdomen and interlocking with one another along the midline at rest. The movement

of the elytra is limited to opening and closing at times of flight. In flightless forms the elytra often interlock immovably along the midline and may dovetail with the abdominal sterna laterally.

Except during flight, the metathoracic wings are concealed beneath the elytra. This is accomplished by a complex series of foldings in the wing membrane (Figs. 37.1c, 37.3b), conforming the size and shape of the wing to the space available in the subelytral cavity. Wing venation is specialized to allow the necessary folding, with the few closed cells in the unfolded anal portion of the wing (Fig. 37.1a). Reduction in venation is common in species of small body size, which often have only a few longitudinal veins and no closed cells. Species of very small body size may also show reduction in the area of wing membrane, compensated by a fringe of long setae on the posterior margin, as in the family Ptiliidae. Flightlessness is relatively common in Coleoptera and may only entail degeneration of the

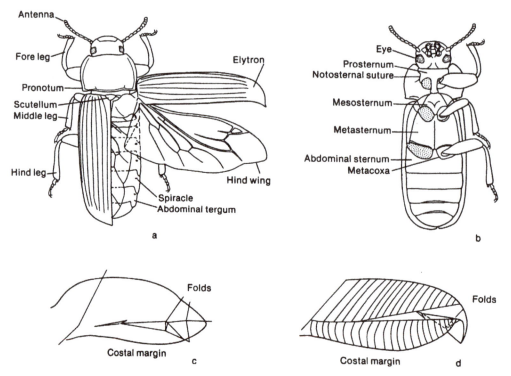

Figure 37.1 Anatomy of Coleoptera: a, dorsal aspect of *Tenebrio molitor* (Tenebrionidae), right elytron and hind wing extended; b, ventral aspect of same; c, diagram of folds in hind wing of same; d, folding of wing of same.

flight muscles following an initial dispersal flight, as in *Helichus* (Dryopidae), or may involve permanent shortening or complete loss of the wings and permanent atrophy of the associated muscles. In lineages that have long been flightless, the pterothorax becomes shortened, the metatergum is greatly reduced or lost, and the elytra are usually immovably attached along the midline and interlocked with the abdominal sternites, as in many desert dwelling Curculionidae and Tenebrionidae. Wing polymorphism (reviewed by Thayer, 1992) is also of common occurrence in Coleoptera.

In various families body compaction may be increased in a variety of ways. Commonly the head is more or less deeply retracted into the thorax. In many families the coxae are enclosed in cavities in the sternal sclerites, reducing mobility but increasing the strength of the coxal articulations. Antennae and the distal joints of legs may also be withdrawn into concavities, and in strongly modified species the heavily armored body is practically without projections when the head and appendages are retracted.

The spiracles open into the subelytral cavity, with the exception of the mesothoracic pair, which is concealed between the prothorax and mesothorax. The enclosed position of the spiracles creates a moisture gradient, greatly enhancing water retention, and Coleoptera are usually a conspicuous element of arid habitats. Arid adapted forms are often flightless, with elytra tightly interlocking with the abdominal sternites, further decreasing water loss. Spiracles may be provided with special closing mechanisms, and Malpighian tubules may be cryptonephric, additional means of increasing water retention efficiency.

In general, body form in Coleoptera reflects the mode of life. Subcortical species are often dorsoventrally flattened, sometimes extremely so, allowing entry between tightly adherent bark and wood. In these species the legs are laterally directed and are usually short. Species that burrow through substrates tend to be very compact and either circular or oval in cross section. Swimming forms are streamlined.

Many Coleoptera, especially those that lead exposed lives, store defensive substances that are self-produced or obtained from plants. These may be contained in the hemolymph (Buprestidae, Coccinellidae, Endomychidae, Lycidae, Lampyridae, Meloidae, Oedemeridae, some Chrysomelidae) or held in reservoirs and forcibly ejected when needed (Carabidae, Rhysodidae, Cicindelidae, Dytiscidae, Gyrinidae, Silphidae, Staphylinidae, Tenebrionidae, some Cerambycidae). Compounds produced range from alkaloids to various organic acids, aldehydes, and quinones. The bombardier beetle, *Brachinus* (Carabidae), releases a gaseous mixture of quinones and steam, produced by an exothermic reaction in two-chambered reservoirs located at the abdominal apex. Larvae of several of these families also produce defensive chemicals, sometimes different from those of the adults. The variation in structure and location, as well as the diversity of substances produced, makes it clear that chemical defensive mechanisms have arisen numerous times independently in Coleoptera.

Coleoptera include the most familiar terrestrial bioluminescent organisms (fireflies, glowworms), in most of which luminescence is used in mate location and courtship when one sex is flightless (see Cantharoidea, below).

LIFE HISTORY

Sexes are generally similar except for minor differences in size or development of sense organs or appendages. Striking exceptions include many Lampyridae (fireflies) and related families, in which females are wingless and frequently larviform. In some Scarabaeidae (June beetles) and Lucanidae (stag beetles), the males bear elaborate horns or greatly enlarged mandibles, which are used in jousting over females in some species. Female-produced pheromones have been isolated from many beetles and are probably produced by the great majority. The pheromones have the usual function of attracting males, whose antennae are frequently larger or more elaborate than those of females, apparently to enhance pheromone detection.

Copulation almost always proceeds with the male mounted dorsally on the female and may be preceded by specialized courtship behavior, such as stridulation (Scolytidae) or palpation and antennation (Meloidae). Eggs are usually simple and ovoid, with relatively thin, unsculptured chorion, with oviposition on or close by the larval food. Eggs are laid singly or in small batches in many Coleoptera, but egg masses are deposited by some (e.g., Coccinellidae). Numbers of eggs vary from one or two produced at a time in Ptiliidae and some other very small forms to over 1000 in Meloidae. Ovoviviparity occurs sporadically in several families, including Carabidae, Staphylinidae, Tenebrionidae, and Chrysomelidae. Parental care is not typical of Coleoptera, but pairs of sexton beetles (Silphidae) and dung-feeding Scarabaeidae prepare special larval chambers, stock them with appropriate food, and sometimes remain with the larvae until growth is partly completed. Adult Scolytidae excavate oviposition galleries in dead or living trees and remain for some time with their larvae, as do a few Staphylinidae and Carabidae in burrows excavated in soil or sand.

Immatures are usually of limited mobility, but many predatory larvae are active hunters; in parasitoid Coleoptera the highly mobile first instars (triungulins) function in locating the host. Body form among larvae reflects mode of life and is at least as variable as in adults. However, most coleopterous larvae can be classified as (1) campodeiform (Fig. 37.5b): active, predatory, with prognathous head, long legs, and elongate urogomphi (Carabidae; Staphylinidae); (2) eruciform (Fig. 37.17h): caterpillarlike, with cylindrical body, short legs, and hypognathous head (Coccinellidae; Erotylidae); (3) scarabaeiform (Fig. 37.9d): obese, C-shaped body with moderate legs; burrowing in soil or rotting wood (Scarabaeidae; Lucanidae); (4) cucujiform (Fig. 37.17i): flattened or subcylindrical body, prognathous head, moderate legs, laterally directed; usually with well-developed urogomphi of diverse configuration (Cucujidae, Nitidulidae, Colydiidae); (5) apodous (Fig. 37.19e): grublike, with legs, eyes, antennae, and urogomphi reduced or absent (Curculionidae, Bruchidae); or (6) elateriform (Fig. 37.12b): elongate, cylindrical, with short legs and tough leathery cuticle; urogomphi present or absent; usually soil-inhabiting (Elateridae, Tenebrionidae). Normally, all larval instars, which vary greatly in number, are similar,

but hypermetamorphosis is the rule among parasitoid forms, with an initial campodeiform instar and later instars grublike. It should be stressed that many other, often highly distinctive body forms are associated with specialized modes of life and that the larval types described above correspond to general modes of adaptation, not evolutionary relationships.

Pupation usually occurs in a cell constructed by the larva in the feeding substrate or nearby. Cocoons are uncommon but are formed by some Curculionidae as well as a few other families from silk produced by the Malpighian tubules. Limited pupal mobility is enabled by abdominal movements. Various parts of the body frequently bear tubercles, knobs, or setae characteristic of the pupa. Adjacent abdominal segments sometimes have opposable "gin traps," pointed sclerotized prominences supposedly used in defense. Pupae that occur on exposed surfaces (e.g., Coccinellidae) often remain within the tough larval cuticle. The pupal stage usually lasts 2 to 3 weeks; pupal diapause occurs in forms that pupate in protected cells (e.g., Cerambycidae) but seems to be relatively uncommon in Coleoptera, which diapause more frequently as larvae or adults.

The economic importance of Coleoptera proceeds almost entirely from their consumption of crops and other products valued by humans. Scolytidae are probably the most important pests in coniferous forests, and, aside from termites, beetles (Anobiidae, Ptinidae, Dermestidae) are the most important pests in wooden structures. Coleoptera also include a multitude of species (many families) that feed on stored food products, and some of these will also consume woolen articles, preserved insects, or herbarium specimens. Beetles are essentially of no medical importance.

The four suborders of Coleoptera are relatively well established and have been confirmed as lineages by recent comparative molecular and morphological phylogenetic studies. Most current classificatory changes involve establishing the limits of large families with respect to some closely related, and possibly subordinate, small families.

KEY TO THE SUBORDERS OF COLEOPTERA

1. Pronotum with notopleural sutures (Fig. 37.2)   **2**
   Pronotum without notopleural sutures (Fig. 37.1b)[1]   **Polyphaga**
2. Metacoxae dividing first visible abdominal sternite (Fig. 37.2)   **Adephaga**
   Metacoxae not dividing first visible abdominal sternite (Fig. 37.1b)   **3**
3. Antennae clubbed (Fig. 37.3c; two small families in North America, Sphaeriidae and Hydroscaphidae)   **Myxophaga**
   Antennae filiform (Fig. 37.6a; two small families in North America, Cupedidae and Micromalthidae)   **Archostemata**

Suborder Archostemata

Archostemata, including only the families Cupedidae, Micromalthidae, and Omatidae, are most notably distinguished from other beetles by having the wing membrane spirally rolled in repose, rather than folded (Fig. 37.3b). The presence of distinct pleural sclerites in the prothorax (Cupedidae) and closed cells in the cubitomedian area of the wing are primitive features shared with the Myxophaga and Adephaga.

Cupedidae are wood-inhabiting insects of moderate size (5–15 mm) that bore in decaying trunks of both conifers and angiosperms. They are seldom encountered but are locally abundant. Males of some species are attracted to chlorine bleaches, which apparently mimic a sex pheromone. Adults are characterized by an elongate, parallel-sided body, with reticulately sculptured elytra (Fig. 37.3a). Larvae are eruciform, with five-segmented legs with one or two claws. Only 25 species are known; five occur in North America. The earliest fossils definitely assignable to the Coleoptera, impressions of elytra from Permian beds in Asia, are very similar to elytra of extant Cupedidae. The genus *Tshekardocoleus*, described by Rohdendorf

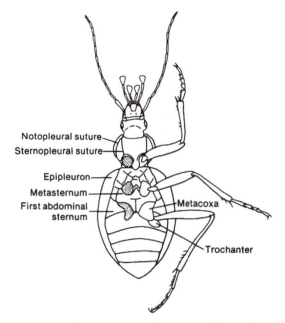

Figure 37.2 Ventral aspect of *Scaphinotus* (Carabidae); scale equals 2 mm.

(1944), appears to have venation intermediate between that of Coleoptera and Megaloptera.

The Micromalthidae are minute beetles (1–2 mm) known from the single species *Micromalthus debilis*, which occurs in rotting wood in eastern North America and has been introduced to several other continents. The life history of *Micromalthus* is extremely complex, with at least five distinct larval forms, including one that is neotenic and another that is ectoparasitic upon the mother (Pringle, 1938). Females are diploid, males haploid.

Omatidae includes three Old World genera.

Suborder Myxophaga

Four small families are included in the Myxophaga. The wings are folded basally, but rolled apically, and contain a closed cell (oblongum) in the cubitomedian area of the wing; the prothoracic notopleural sutures are distinct, and the mandible has a distinct molar lobe. These features suggest relationships to both Archostemata and Polyphaga. Larvae are ovate with broad, deflexed head, five-segmented legs with single claws, and tracheal gills on the abdomen. All species that have been studied are aquatic, apparently feeding on algae.

---

[1] *Micromalthus*, with one rare species in eastern North America, lacks notopleural sutures. It will key to Staphylinidae.

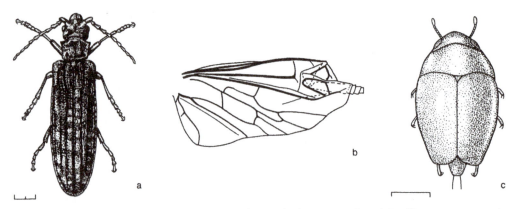

Figure 37.3 Coleoptera, Archostemata and Myxophaga; a, Archostemata, Cupedidae (*Priacma serrata*; scale equals 2 mm); b, rolled wing of *P. serrata*; c, Myxophaga, Hydroscaphidae (*Hydroscapha natans*; scale equals 0.25 mm).

Two families occur in North America. Hydroscaphidae inhabit margins of small sluggish streams and springs, including hot springs, in western North America. Adults are 1 to 2 mm long, broad and flattened, with truncate elytra exposing two or three abdominal tergites (Fig. 37.3c). Pupae are pharate in the last larval cuticle and float on the surface film. Five species are known, with *Hydroscapha natans* the single North American representative. Microsporidae (= Sphaeriidae) are minute (0.5–1 mm long), highly convex beetles that occur in the interstices of wet gravel and debris along stream margins. The elytra completely cover the abdomen, differentiating microsporids from hydroscaphids. There are two North American species, 11 worldwide. Torridincolidae, with several genera in tropical regions of the New and Old Worlds, inhabit splash zones around cataracts and similar situations.

Suborder Adephaga

These predominantly predatory beetles represent one of the two major evolutionary lines of Coleoptera, with two large, diverse families and several smaller, more specialized ones. Adephaga have distinct pleural sclerites in the prothorax (Fig. 37.2) and usually have a distinct oblongum in the wing but fold the wing membrane in repose. They are distinguished from all other Coleoptera by the immovable metacoxae, which are fused to the first abdominal sternite, and by the enlarged trochanters of the hind legs (Fig. 37.2). In all Adephaga the molar lobe of the mandible is reduced or absent, a common modification in predatory species of Polyphaga as well. Other useful distinguishing features of Adephaga include large, round (not emarginate) eyes; long, filiform or moniliform antennae; abdomen with six visible sternites; tarsal segmentation 5-5-5. Larvae are campodeiform, with five-segmented legs with two claws.

The dominant family is the Carabidae (ground beetles; Fig. 37.5a), with about 20,000 species distributed throughout the world. Carabidae are most abundant in moist situations and tend to be nocturnal, but there are many exceptions. Size ranges from about 1 to 50 mm. Adults and larvae of most species are probably nonspecific predators, but diets are highly restricted in others (e.g., *Brachinus*, which attacks hydrophilid prepupae or pupae). About 2500 North American species are known.

Cicindelidae and Rhysodidae are closely related to Carabidae, where they are sometimes placed as subfamilies, although Rhysodidae superficially appear quite different. Cicindelidae (tiger beetles) are fast-running, mostly diurnal predators, which are among the most agile flying beetles. Larvae are sessile predators that construct deep burrows in friable soil, usually near water. Rhysodidae are sluggish, heavily sclerotized beetles that inhabit moist, decaying wood.

Haliplidae inhabit slowly moving or standing water, especially the shallows along streams or pond margins. The strongly convex adults are distinguished by the large metacoxal plates, which conceal the bases of the hind legs. Larvae feed exclusively on algae. Adults mingle this diet with occasional small arthropod prey.

The Dytiscidae, Gyrinidae, and Amphizoidae are aquatic predators. Dytiscidae (Fig. 37.5b,c) is the dominant family, with about 4000 species (330 in North America) occupying all freshwater habitats. Adults and larvae must periodically surface to renew their air supply, somewhat restricting them to shallow water. Adults carry an air bubble beneath the elytra, renewing it by breaking the surface film with the abdominal apex. Body size ranges from about 1 mm to more than 25 mm. Adults and larvae are predatory, the larger species being able to kill tadpoles and small fish as well as invertebrates. Adults have typical chewing mouthparts, while the larval mandibles are deeply grooved or hollow, allowing suctorial feeding.

Gyrinidae (whirligigs) are specialized for surface swimming, with the eyes divided, allowing simultaneous submarine and aerial vision. Adults tend to be highly gregarious, often congregating in dense swarms in eddies or quiet pools, where they gyrate rapidly and erratically about on the water surface, sometimes diving to escape disturbances from above. Their specialized, club-shaped antennae are sensitive to surface film disturbances, enabling the beetles to avoid one another and to locate the floating insects on which they prey. Larvae frequent aquatic vegetation.

Amphizoidae include only five species, all but one from western North America. A single species occurs in Asia. Adults and larvae inhabit floating debris trapped in eddies or entangled in shoreline vegetation, mostly along cold, fast-flowing streams, but they are unable to swim.

KEY TO THE FAMILIES OF THE SUBORDER ADEPHAGA

1. Metacoxae covering no more than first visible abdominal sternite   **2**
Metacoxae expanded as plates covering at least first 3 sternites   **Haliplidae**

2(1). Hind coxae greatly enlarged, extending laterally to meet elytra (Fig. 37.4)   **3**

Hind coxae not extending laterally as far as elytra (Fig. 37.2)   **5**

3(2). Legs with long hairs fringing tibiae and tarsi (Fig. 37.5d)   **4**
Legs without long fringing hairs   **Amphizoidae**

4(3). Eyes divided into dorsal and ventral lobes   **Gyrinidae**
Eyes rounded or emarginate, not divided   **Dytiscidae** (including **Noteridae**)

5(2). Metasternum without transverse suture in front of metacoxae; antennae filiform or serrate (Fig. 37.6a,e)   **6**
Metasternum with transverse suture just before metacoxae; antennae thick, moniliform (Fig. 37.6b)   **Rhysodidae**

6(5). Clypeus extending laterally in front of antennal insertions   **Cicindelidae**
Clypeus not extending laterally as far as antennal insertions   **Carabidae**

Suborder Polyphaga

This is an exceedingly diverse assemblage, including more than 90 percent of the Coleoptera. The wings lack a closed cell (oblongum) in the cubitomedian field and are always folded (not rolled) in repose. The pronotum extends ventrally to the vicinity of the lateral coxal articulation, with the reduced pleural sclerites internal. Larvae are

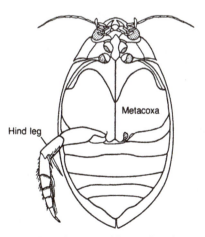

Figure 37.4 Ventral aspect of Dytiscidae, showing large hind coxae.

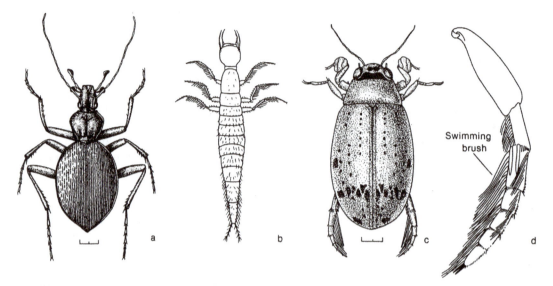

Figure 37.5 Representative Adephaga: a, Carabidae, adult (*Scaphinotus*); b, Dytiscidae, larva (*Hydaticus*); c, Dytiscidae (*Eretes sticticus*); d, hind leg of *E. sticticus*. Scales equal 2 mm.
Source: b, redrawn from Böving and Craighead, 1931.

variable but have legs with four segments and a single claw (or legs vestigial or absent).

The Polyphaga are divided into numerous superfamilies. Most of these are relatively homogeneous in terms of both morphological and ecological characteristics, the major exceptions being the Cucujoidea and Tenebrionoidea. The following key is modified from that of Britton (1970); the taxonomic arrangement is essentially that of Crowson (1960). The most important taxonomic characters include the configuration of the antennae (Fig. 37.6), the number of abdominal segments, and tarsal structure. The numbers of tarsal segments on the fore-, middle-, and hind legs, respectively, are indicated by three-digit formulas in the key (e.g., tarsal segmentation 5-5-4). The procoxal cavities are considered open if not enclosed posteriorly by the prothorax (Fig. 37.16g); they are closed if a sclerotized portion of the notum extends medially behind them to the procoxal process (Fig. 37.16c). It should be stressed that keying members of the Polyphaga requires great care because of the many exceptional genera. For a more detailed treatment of North American Coleoptera adults, Arnett (1960) is the most complete reference, although the higher classification does not reflect many recent changes.

KEY TO THE SUPERFAMILIES OF THE SUBORDER POLYPHAGA

1. Antennae with terminal 3 to 7 segments enlarged and flattened as a discrete, lamellate, one-sided club (Fig. 37.6i, 37.9b) **Scarabaeoidea**
   Antennae variable, but not lamellate, or if rarely lamellate, more than 7 segments involved **2**

2(1). Antennae with segment 6 (occasionally segment 4 or 5) cup-shaped, transverse (Fig. 37.7b); maxillary palpi frequently as long as antennae or longer (Fig. 37.7b) **Hydrophiloidea**
   Antennae with segment 6 unmodified; maxillary palpi almost never elongate **3**

3(2). Elytra truncate, exposing 2 or more abdominal tergites, which are sclerotized (Fig. 37.8a–d) **4**
   Elytra usually covering entire abdomen, or rarely exposing 1 or 2 tergites **10**

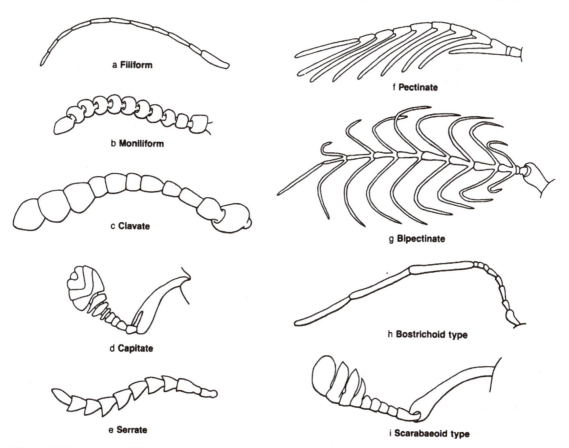

Figure 37.6 Antennae of Coleoptera.

4(3). Wings at rest extending beyond elytra, without transverse folds   **5**

Wings at rest folded transversely beneath elytra or absent   **6**

5(4). Maxillary palpi flabellate; eyes nearly contiguous dorsally   **15**

Maxillary palpi filiform; eyes widely separated   **27**

6(4). Claws dentate (Figs. 37.15c, 37.16d) or appendiculate (Fig. 37.16e)   **10**

Claws unmodified   **7**

7(6). Antennae elbowed with last 3 segments forming capitate club (Fig. 37.6d)   **Histeroidea**

Antennae not elbowed   **8**

8(7). Abdomen with 6 or 7 visible sternites; antennae usually clavate or filiform (Fig. 37.6a,c)
    Staphylinoidea

Abdomen with 5 visible sternites; antennae with capitate club (Fig. 37.14b)   **9**

9(8). Abdomen with 3 or more tergites exposed   **Staphylinoidea**

Abdomen with 2 or fewer tergites exposed   **10**

10(3,6,9). Abdomen with 6 or 7 visible sternites   **11**

Abdomen with 3 to 5 visible sternites   **16**

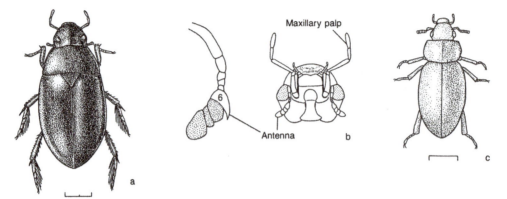

Figure 37.7 Hydrophiloidea: a, Hydrophilidae (*Tropisternus ellipticus*; scale equals 2 mm); b, ventral aspect of head of *Tropisternus* with cup-shaped sixth antennal segment marked; c, Limnebiidae (*Ochthebius rectus*; scale equals 0.5 mm)

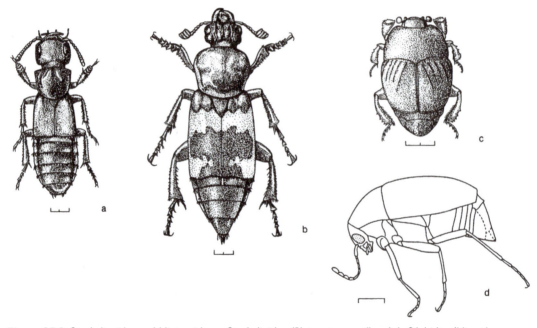

Figure 37.8 Staphylinoidea and Histeroidea: a, Staphylinidae (*Platycratus maxillosus*); b, Silphidae (*Nicrophorus marginatus*); c, Histeridae (*Saprinus lugens*); d, Scaphidiidae (*Scaphisoma quadriguttatum*). Scales equal 2 mm, except for d, which equals 0.5 mm.

11(10). Tarsal segmentation 3-3-3
      Cucujoidea (Coccinellidae)
      Tarsal segmentation 5-5-5   **12**

12(11). Antennae with distinct, usually, capitate club (Fig. 37.14b)   **13**
      Antennae filiform, serrate, or pectinate (Fig. 37.6a,e,f)   **15**

13(12). Tarsi with segment 4 deeply bilobed; claws frequently toothed (Fig. 37.15c)
      Cleroidea
      Tarsi with segment 4 not bilobed; claws simple   **14**

14(13). Antennal club with 3 segments
      Staphylinoidea

Antennal club with 5 segments **Hydrophiloidea**

15(5,12). Tarsi filiform (Fig. 37.16f), slender, at least as long as tibiae **Lymexyloidea**
Tarsi stout, not filiform, shorter than tibia **Cantharoidea**

16(10). Metacoxae with posterior face vertical, concavely excavated for reception of femora (Fig. 37.11b)   **17**
Metacoxae flat or convex, without excavated posterior face (Fig. 37.14a)   **23**

17(16). Procoxae globular round or oval (Fig. 37.16g), not projecting from coxal cavities   **18**
Procoxae transversely or dorsoventrally elongate (Fig. 37.16c), sometimes projecting from coxal cavities (Fig.37.8d) **19**

18(17). Abdomen with first 2 sternites fused, sutures obscured   **Buprestoidea**
Abdomen with all sternites free, separated by distinct sutures   **Elateroidea**

19(17). Antennae with last 3 to 5 segments differentiated as strong, capitate club (Fig. 37.12d), or greatly elongate (Fig. 37.6h)**20**
Antennae filiform, serrate, or clavate, but not with last 3 segments strongly differentiated   **21**

20(19). Antennae with last 3 segments greatly elongate   **Bostrichoidea**
Antennae with last 3 to 5 segments forming short, capitate club (Fig. 37.14c,d) **Dermestoidea**

21(19). Tarsi with next to last segment deeply bilobed (Fig. 37.10e)   **Dascilloidea**
Tarsi with next to last segment not bilobed; tarsi usually filiform (Fig. 37.16f)   **22**

22(21). Head with clypeus distinctly separated from frons by fine suture   **Dryopoidea**
Head with clypeus fused with frons   **Byrrhoidea**

23(16). Tarsal segmentation 5–5-5   **24**
Tarsal segmentation 5–5-4, 4-4-4, or 4-4-3 **27**

24(23). Prothorax usually hoodlike, produced forward over head (Fig. 37.14a); legs usually with trochanters elongate; antennal insertions usually very close   **Bostrichoidea**

Prothorax not hoodlike; trochanters small; antennal insertions distant   **25**

25(24). Tarsi with large, bisetose empodium (Fig. 37.15f); procoxae conical, projecting from cavities; body usually with numerous, erect bristles   **Cleroidea**
Tarsi with empodium absent or small, inconspicuous (Fig. 37.15c); procoxae not projecting; body usually without erect bristles   **26**

26(25). Tarsi with fourth segment minute, concealed in groove of bilobed third segment so that tarsi appear 4- segmented (Fig. 37.18b)   **30**
Tarsi with fourth segment subequal to third; if small, not concealed in deep groove of third segment   **Cucujoidea**

27(5,23). Tarsal segmentation 5–5-4   **Tenebrionoidea**
Tarsal segmentation 4-4-4 or 4-4-3   **28**

28(27). Tarsal segments not lobed beneath**34**
Tarsi with 1 to 3 basal segments lobed beneath (Figs. 37.10e, 37.15c)   **29**

29(28). Metasternum without transverse suture   **Cleroidea**
Metasternum with transverse suture just before metacoxae   **30**

30(26,29). Head prolonged as a beak on which the antennae are inserted (Fig. 37.19a,b)   **Curculionoidea**
Head not prolonged as a beak; antennae inserted on frons near eyes (Fig. 37.19d)   **31**

31(30). Antennae filiform, serrate, or occasionally clavate   **Chrysomeloidea**
Antennae strongly capitate   **32**

32(31). Antennae elbowed (Fig. 37.6d)   **Curculionoidea**
Antennae straight   **33**

33(32). Body densely covered by scales or hairs **Curculionoidea (Anthribidae)**
Body glabrous or sparsely pubescent

Cucujoidea (Erotylidae, Languriidae)

34(28). Abdominal sternites all freely movable; antennal insertions usually exposed   **Cucujoidea**
Abdominal with basal 3 or 4 sternites connate   **Tenebrionoidea**

Superfamily Hydrophiloidea *(Fig. 37.7).* Predominantly aquatic beetles, characterized by short antennae, with elongate scape and compact club with 3 to 5 segments. In many species the maxillary palpi are elongate, frequently longer than the antennae (Fig. 37.5b).

Hydrophilidae (Fig. 37.7a,b) occupy all aquatic situations. The most familiar species are relatively large, free-swimming forms. Others burrow in bottom debris or hide in aquatic vegetation. Many hydrophilids inhabit the littoral zone, especially along gravelly margins, and do not swim. The numerous members of the subfamily Sphaeridiinae inhabit dung or moist humus in terrestrial habitats, and a few species breed in piles of marine algae that have been stranded on beaches. Aquatic species obtain air by breaking the surface film with an antenna, which communicates with the subelytra air reservoir by an air film held on a plastron on the ventral surface of the body. Hydrophilids are mostly scavengers as adults, predatory as larvae.

Hydraenidae (Fig. 37.7c) are small beetles (1–2 mm) that occur in a variety of littoral habitats, including seeps, wet gravel, or coarse sand banks, splash zones, and marine rock pools. Adults are aquatic, using an antenna to break the surface film to obtain air, while larvae frequent adjacent, moist, terrestrial habitats. Georyssidae creep about mud or sand bars along streams. Adults coat the elytra with a camouflaging of mud. Both families are widespread within their restricted habitats.

KEY TO THE FAMILIES OF HYDROPHILOIDEA

1. Abdomen with 5 visible sternites; antennae with club 3-segmented   **2**
Abdomen with 6 to 7 visible sternites; antennae with club 5-segmented   **Hydraenidae** (including Limnebiidae)

2. Tarsal segmentation 4-4-4   **Georyssidae**
Tarsal segmentation 5-5-5   **Hydrophilidae**

***Superfamily* Histeroidea** *(Fig. 37.8c).* Adults distinguished by compact body, frequently with retractile appendages; antennae elbowed with abrupt, capitate club. Appendages frequently bear enlarged, spinose setae in characteristic patterns, and the cuticle is typically very hard, black, and polished.
Histeridae are cosmopolitan predators, abundant about carrion and dung, where they are most frequently encountered. Numerous species burrow in sand, many occur beneath bark of dead trees, and others are restricted to mammal burrows or ant or termite nests. Many of the free-living species are predators of fly larvae or pupae. Species frequenting dung or living in sand are usually globular, while those in subcortical habitats are either flattened or cylindrical and elongate, the latter an adaptation for moving through burrows made by scolytids or other beetles. Sphaeritidae, with three known species, are occasionally found in association with decaying plant and animal products. Their biology is unknown. Sphaeritids are distinguished from histerids in having the antennae straight (not elbowed) and the front tibiae simple (not dentate).

***Superfamily* Staphylinoidea** *(Fig. 37.8a,b,d).* Adults with wing venation reduced by loss of the mediocubital loop and with only 4 Malpighian tubules. Larvae usually with articulated urogomphi and 4-segmented legs with single claws. In external features the Staphylinoidea are extremely variable and difficult to recognize. Most have the elytra abbreviated, exposing 3 or more abdominal tergites, which are sclerotized, but there are many exceptions.

Staphylinoid beetles are found mainly in moist habitats, frequently in association with fungi or in leaf litter. Ptiliidae, Pselaphidae, and Scydmaenidae are largely restricted to fungus–leaf mold environments, but in all these families a few species are known to be commensals in ant or termite nests, and others may be encountered beneath bark.
Silphidae, as well as some Leiodidae and many Staphylinidae, are associated with carrion, either as scavengers or as predators, especially of Diptera.

Nicrophorus (Fig. 37.8b) (Silphidae) is well known for burying corpses of small vertebrates, which are then prepared as food for the grublike, sedentary larvae. Many Leiodidae breed in fungi or slime molds, but others feed on carrion or inhabit litter. A few occur in ant or termite nests. Leptinidae are all specialized as nest inhabitants or ectoparasites of mammals. Adult leptinids have reduced eyes and antennae, flattened body with leathery cuticle, and mouthparts adapted for feeding on cutaneous debris. The best known species, *Platypsylla castoris*, inhabits the lodges and fur of beavers. Scaphidiidae (sometimes considered a subfamily of Staphylinidae) are distinguished by the short, convex body, elytra that largely cover the abdomen, and the pointed, conical abdominal apex. Adults and larvae feed on basidiomycete fungi or slime molds.

The Staphylinidae is a very large, diverse group, encompassing all the modes of life described above. Perhaps the largest number of species is associated with fungi, including molds and rusts. Many are found near water, and staphylinids are the most common marine Coleoptera. Very few staphylinoid beetles are phytophagous; although none is presently of economic significance, many Staphylinidae are predaceous, especially on Diptera, and may become important as biological control agents.

A number of the families keyed here are suspected of belonging to lineages actually within Staphylinidae and are then treated as subfamilies. The classification of Staphylinoidea is receiving significant attention at present; the interested student should consult the primary literature for updates.

KEY TO THE FAMILIES OF STAPHYLINOIDEA

1. Elytra truncate, exposing at least 3 abdominal tergites   **2**
   Elytra exposing no more than 2 tergites, frequently covering
   entire abdomen   **3**
2(1). Abdomen swollen, at least twice width of pronotum; tarsi with 1 claw or with claws unequal   **Pselaphidae**

Abdomen slender, less than 1.5 times width of pronotum; tarsi with claws of equal size
                                   Staphylinidae (Fig. 37.8a)
3(1). Last abdominal tergite conical, as long as 3 preceding segments (Fig. 37.8d); first abdominal sternite as long as 2 to 4 combined   **Scaphidiidae**
   Last abdominal tergite about equal to preceding segment; first sternite about equal to second   **4**
4(3). Metacoxae with posterior surface concave, excavated for reception of femora; length less than 1.5 mm
              Ptiliidae (including **Limulodidae**)
   Metacoxae with posterior surface flat or convex; length greater than 1.5 mm   **5**
5(4). Procoxae conical, prominent (Fig. 37.8d) **6**
   Procoxae globular, not projecting from cavities   **Leptinidae**
6(5). Metacoxae contiguous or approximate   **7**
   Metacoxae widely separated **Scydmaenidae**
7(6). Tibial spurs small, indistinct   **8**
   Tibial spurs large, conspicuous; body frequently convex, capable of being rolled into a ball   **Leiodidae** (incl. **Leptodiridae**)
8(7). Antennae clavate with segment 8 smaller than 7 or 9 (Fig. 37.6d)   **Leiodidae**
   Antennae clavate or capitate; segments 7 to 9 subequal
   Silphidae (including **Agyrtidae**) (Fig. 39.8b)

Superfamily **Scarabaeoidea** *(Fig. 37.9).* Distinguished from nearly all other Coleoptera by short antennae with asymmetrical, lamellate club (Fig. 37.9b). Most scarabaeoid beetles are stout-bodied, with the head sunk deeply into the prothorax and anterior tibiae expanded and serrate for digging. Larvae are stout, with large, hypognathous head, well-developed legs, and large, soft abdomen, usually curled ventrally beneath the forebody (Fig. 37.9d).

Lucanidae (stag beetles) and Passalidae inhabit decaying wood, where larvae and adults are usually associated in common galleries. The mandibles of most male Lucanidae vary allometrically, being similar to those of the female in small individuals and disproportionately enlarged in large individuals.

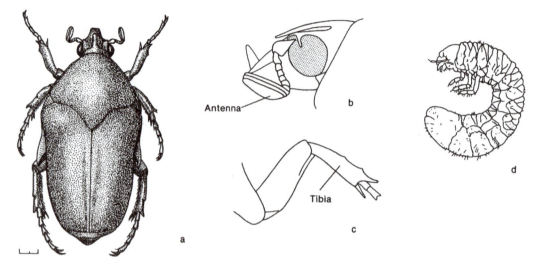

Figure 37.9 Scarabaeoidea, Scarabaeidae: a, *Cotinis nitida* (scale equals 2 mm); b, antenna and c, foreleg of *Cotinis nitida*; d, larva of *Popillia japonica*.
Source: d, redrawn from Böving and Craighead, 1931.

They are apparently used in agonistic encounters between males in some species. Passalids are flattened and heavily sclerotized, with striate elytra. Larvae, which always occur with adults, stridulate by rubbing the tiny, digitate hind leg over the microscopically ridged mesocoxa. Adults apparently macerate food for the larvae. The only U.S. species occur in the South and Southeast.

Scarabaeidae (chafers, June beetles, etc.) is the dominant family. Some scarabaeid larvae feed on rotting wood, but mammal dung, humus, carrion, and roots of living vegetation are consumed by others. Most of the dung-feeders excavate a burrow and larval feeding chamber beneath the dung pile, but species of *Canthon* form a mass of dung into a ball and roll it away before excavating the burrow. *Trox* are found about dry animal carcasses, where they feed on hair, dried skin, etc. The larvae inhabit burrows beneath the food, coming to the surface at night to feed. Adults of many species, such as the Japanese beetle (*Popillia japonica*), attack living foliage, especially of the Rosaceae. A few species are specialized as predators or commensals in ant or termite nests. Scarabs range in size from burrowing species less than 2 mm long to giants over 10 cm in length, including the bulkiest extant insects. Males of many

species, especially from the tropics, bear horns on the head or prothorax or have the legs modified in various ways. Most species are brown or black, but many are brightly pastel or metallic, often in striking patterns.

KEY TO THE FAMILIES OF SCARABAEOIDEA

1. Lamellae of antennal club thick, rounded, not capable of close apposition (Fig. 37.6i) **2**
   Lamellae of antennal club flattened, apposable (Fig. 37.9b)
   Scarabaeidae (including **Trogidae, Geotrupidae**)
2(1). Antennae elbowed, first segment longer than next 5 segments combined (Fig. 37.6i)
   Lucanidae
   Antennae straight, first segment shorter than next 4 combined **Passalidae**

Superfamily Dascilloidea (Fig. 37.10). Antennae filiform or serrate, procoxae conical, projecting from cavities; procoxal cavities open; intercoxal process narrow; metacoxae posteriorly concave for reception of femora; tarsal segmentation 5-5-5; 5 visible abdominal sternites.

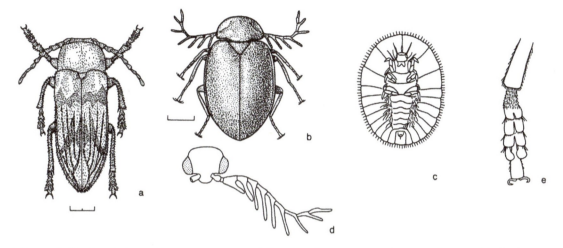

Figure 37.10 Dascilloidea and Dryopoidea: a, Dascillidae (*Dascillus davidsoni*; scale equals 2 mm); b, Psephenidae, adult (*Eubrianax edwardsi*; scale equals 1 mm); c, Psephenidae, larva (*Eubrianax*); d, antenna of adult *Eubrianax*; e, tarsus of *Dascillus* (Dascillidae).

The Dascilloidea contains a diverse assemblage of beetles from a few small families. Adults of most tend to occur in moist situations, frequently around water, but only the larvae of Scirtidae (= Helodidae) are aquatic, sometimes living in streams, but usually in tree holes (water filled cavities in trunks or large branches of trees). Their multiarticulate antennae are unique among immatures of endopterygote insects. Adults of some scirtids have the hind femora enlarged for jumping. Larvae of Dascillidae are scarabaeiform and occur in soil, feeding on roots (*Dascillus*). Adults are usually seen clinging to the tops of grass stems.

Rhipiceridae adults (Fig. 37.12c) are robust, elongate, brown or black beetles 16 to 24 mm long. Their scarabaeiform larvae are internal parasitoids of cicada nymphs. Eucinetidae occur under bark in association with fungi or slime molds. Adults are flattened, ovoid beetles with the head deflexed and the hind femora enlarged for jumping. The Clambidae, included in the Dascilloidea because of similarities of the larvae to that of *Eucinetus* (Crowson, 1960), are minute beetles with 4–4–4 tarsal segmentation and clubbed antennae. Adults are capable of rolling into a ball. The life histories and biology of clambids are almost unknown.

KEY TO THE FAMILIES OF DASCILLOIDEA

1. Length less than 1 mm; antennae clubbed **Clambidae**
   Length greater than 3 mm; antennae filiform, serrate (Fig. 37.6a,e), or flabellate (Fig. 37.12c) **2**
2(1). Antennae filiform or serrate **3**
   Antennae flabellate **Rhipiceridae**
3(2). Metacoxae broad, platelike, oblique **Eucinetidae**
   Metacoxae narrow, transverse **4**
4(3). Tarsi with segments 2- to 4-lobed ventrally (Fig. 37.10e) **Dascillidae**
   Tarsi with only segment 4-lobed **Scirtidae**

Superfamily Byrrhoidea. Strongly convex beetles with the head strongly deflexed, the clypeus fused with the frons, antennae filiform; procoxae transverse, intercoxal process broad, procoxal cavities open; tarsal segmentation 5-5-5; 5 visible abdominal sternites.

The single family, **Byrrhidae**, includes about 40 North American species. They are small to moderate in size. Adults are convex with retractile appendages, accounting for the common name, "pill beetles." Adults and larvae are herbivorous, many feeding in moss or liverworts. Larvae are scarabaeiform or onisciform.

Superfamily Dryopoidea (Fig. 37.10). Procoxal cavities open posteriorly, prosternal process fitting into concavity in mesosternum; metacoxae concavely excavate posteriorly for reception of femora, tarsal claws very elongate; abdomen with 5 visible sternites. Larvae eruciform, with retractable anal gills, or flattened, onisciform (Psephenidae [Fig. 37.10c]).

Dryopoidea are aquatic or subaquatic in some stage of their lives, with the exception of Callirhipidae and Artematopidae and Chelonariidae. Adult **Artematopidae** are encountered on foliage, and the larvae live in moss. Adults and larvae of **Callirhipidae** occur under the bark or in the wood of dead logs. The single North American species, *Zenoa picea*, is in Florida. The predominantly tropical **Chelonariidae,** represented in North America by a single species, *Chelonarium lecontei*, are myrmecophilous. The strongly sclerotized, oval adults have the head retracted into the underside of the pronotum and grooves on the venter for receiving the retractile appendages.

Heteroceridae and **Limnichidae** occur along stream margins. Adults and larvae of heterocerids burrow in mud or sand, feeding on algae, diatoms, and detritus. Limnichids are most frequently encountered on mud or sand banks or on low vegetation near water, feeding on detritus. A few species are intertidal.

Dryopidae are unusual among insects in having terrestrial larvae and aquatic adults. The larvae burrow in damp sand or soil. Newly emerged adults fly to streams, where they cling to the undersides of submerged stones or logs, often in association with elmid adults and psephenid larvae (water pennies). In contrast to the Hydrophiloidea, aquatic dryopoids respire with tracheal gills (larvae) or cuticular plastrons (adults) and may remain submerged indefinitely without replenishing their air supply. Most of these beetles are infrequently collected, but many are common stream inhabitants.

Psephenidae (water pennies) are most commonly encountered as the extremely flattened, disc-shaped larvae that occur under stones, submerged logs, and the like in flowing water. Larvae move to crevices or overhangs just above the water line and pupate pharate in the larval cuticle. Adults are short-lived, brownish or blackish beetles usually seen on streamside rocks or vegetation. Males usually have pectinate antennae. Most adults and all larvae of Elmidae are aquatic, most frequently inhabiting lotic situations with rocky bottoms, where they feed on decaying vegetation. Larvae leave the water to pupate. Ptilodactylidae are small to medium-sized brownish or blackish beetles typically found on vegetation near water. Antennae of females are serrate, those of males usually pectinate. Larvae mostly tunnel through very moist, rotten wood or live in very moist leaf litter, but *Anchytarsus* is aquatic, feeding on submerged wood. Only a few genera and species of ptilodactylids occur in North America, but the family is abundant in moist tropical environments.

KEY TO THE FAMILIES OF DRYOPOIDEA

1. Procoxae conical, projecting from cavities   **2**
   Procoxae transverse or rounded, not projecting   **6**
2(1). Maxillary palpi with second segment as long as next 2 segments combined; abdomen usually with 6 to 7 visible sternites   **Psephenidae**
   Maxillary palpi with second segment subequal to third; abdomen with 5 visible sternites   **3**
3(2). Antennae flabellate (as in Fig. 37.12c), inserted on a strong protuberance on the frons   **Callirhipidae**
   Antennae filiform, serrate, or clavate (Fig. 37.6a,d,e), frequently concealed within prosternal cavity; frons without antennal protuberance   **4**
4(3). Posterior margin of pronotum wrinkled   **5**
   Posterior margin of pronotum not wrinkled   **Artematopidae**

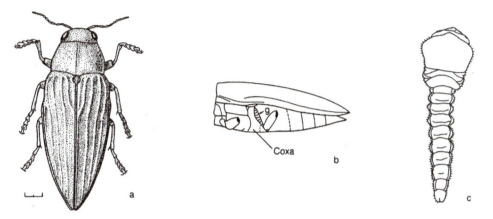

Figure 37.11 Buprestoidea, Buprestidae: a, *Buprestis aurulenta* (scale equals 2 mm); b, metacoxal region of *Buprestis*; c, dorsal aspect of larva of *Chrysobothris*. g, groove in coxa for reception of femur.
Source: c, redrawn from Böving and Craighead, 1930.

5(4). Ventral surface of body with grooves for retraction of legs and antennae
                                    Chelonariidae
Ventral surface of body without grooves for retraction of appendages   **Ptilodactylidae**
6(1). Tarsi with 4 segments   **Heteroceridae**
Tarsi with 5 segments   7
7(6). Tarsi with last segment longer than preceding segments combined   8
Tarsi with last segment shorter than preceding segments combined   **Limnichidae**
8(7). Antennae filiform, elongate   **Elmidae**
Antennae short, with stout, pectinate club
                                      Dryopidae

Superfamily Buprestoidea (Fig. 37.11). Fore coxae globular, procoxal cavities open posteriorly, prosternal process articulated with mesosternum; metasternum with transverse suture; metacoxae grooved for reception of femora; tarsal segmentation 5–5–5; abdomen with 5 visible sternites, the first 2 rigidly fused.

The single family Buprestidae is characterized by the closely articulated, heavily sclerotized body with short serrate antennae and short legs. Buprestids are active, diurnal insects found either about their host plants or on flowers or foliage, on which most adults feed. Unlike in nearly all other Coleoptera, the elytra are lifted slightly but remain closed during flight, with the wings protruding below emarginations just behind the humeri. Many adult buprestids are metallic bronze or green; *Acmaeodera*, marked with contrasting black, yellow, and red, mimic bees and wasps in flight (Silberglied and Eisner, 1968). Larvae are narrowly adapted for boring in tissues of perennial plants. The head and thorax are flattened (hence the name "flat-headed borers"), legs are absent, and the abdomen is long, slender, and unsclerotized. Most buprestids bore in wood of trees or shrubs, but some mine the pithy stems of rosaceous plants, others mine leaves or form galls, and a few genera have larvae that feed externally on roots.

Superfamily Elateroidea (Fig. 37.12). Fore coxae globular or rounded, cavities open posteriorly, prosternal process elongate, articulated with mesosternum; metasternum without transverse suture; metacoxae grooved for reception of femora; tarsal segmentation 5–5–5; 5 visible abdominal sternites, all freely articulated. Larvae mostly elongate, cylindrical or subcylindrical, coriaceous, prognathous with well-developed legs; urogomphi frequently present.
This superfamily includes one large, familiar family, the Elateridae (click beetles), and several small, obscure families. Most elateroid beetles are

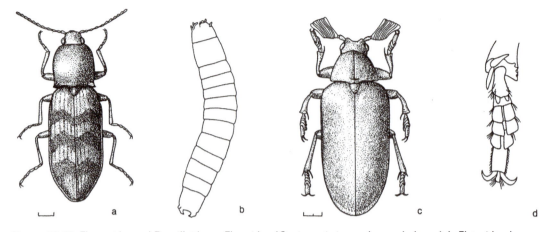

Figure 37.12 Elateroidea and Dascilloidea: a, Elateridae (*Ctenicera tigrina*; scale equals 1 mm); b, Elateridae, larva (*Cryptohypnus*); c, Rhipiceridae (*Sandalus niger*, male; scale equals 2 mm); d, tarsus of *Sandalus* (Rhipiceridae). *Source: b, redrawn from Böving and Craighead, 1930.*

associated with wood, at least as larvae, but the family Elateridae has many soil-dwelling forms. Many of these wireworms were important pests of cereal and root crops before the advent of soil insecticides. Other elaterid larvae are predatory or omnivorous. Adult elaterids are distinguished from all other beetles by the loosely articulated prothorax and the modification of the prosternal process and mesosternum into a jumping mechanism. The beetle first deflexes the prothorax, then suddenly straightens the body, producing an audible click, suggesting the name "click beetles." Small click beetles are able to propel the body 30 cm or more into the air, but in larger species jumps are much shorter and the click may function primarily to startle predators. Some Throscidae and Eucnemidae are also able to click. Most adult elaterids are small to medium-sized, brownish or blackish beetles, but *Alaus*, up to 40 mm long, is marked with a pair of prominent, eyelike spots on the pronotum, apparently as a protective startle mechanism against vertebrate predators. The tropical genus *Pyrophorus*, with a few species in the southern United States, bears a pair of luminescent spots on the pronotum. The larvae, called railroad worms, luminesce through a large transparent spot on the head and a row of smaller spots on each side of the body.

Most Cebrionidae, occasionally seen in the southwestern United States, develop in soil.

Selenodon and Scaptolenus are termitophilous and fly only during rain. Throscidae are small beetles with the body form of buprestids. Their biology is poorly known. Eucnemidae is a diverse but rare family whose larvae live in wood or bark. The larvae have highly modified mouthparts and an enlarged thorax with the legs reduced or absent so that they superficially resemble dipterous larvae.

KEY TO THE FAMILIES OF ELATEROIDEA[1]

1. Head with labrum freely articulated; antennae inserted close to eyes   **2**
   Head with labrum fused to clypeus, indistinguishable; antennae closer to mandibles than eyes   **Eucnemidae**
2(1). Mandibles large, prominent, sickle-shaped   **Cebrionidae**
   Mandibles small, not noticeably protruding from mouth.   **3**
3(2). Prothorax loosely joined to mesothorax   **Elateridae**

---

[1] Perothopidae and Cerophytidae, each represented by a single rare species in North America, are not included in the key.

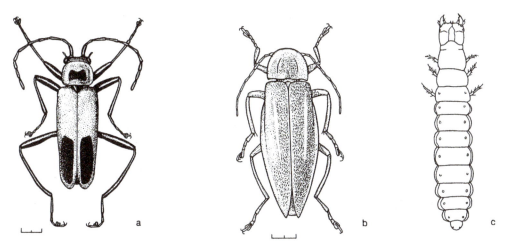

Figure 37.13 Cantharoidea: a, Cantharidae (*Chauliognathus pennsylvanicus*); b, Lampyridae (*Photurus pennsylvanicus*); c, larva of *Chauliognathus*. Scales equal 2 mm.
Source: c, redrawn from Böving and Craighead, 1930.

Prothorax tightly joined to mesothorax; prosternal process fused with mesosternum   **Throscidae**

Superfamily Cantharoidea *(Fig. 37.13).* Procoxae prominent, conical, projecting; procoxal cavities open posteriorly, intercoxal process reduced, seldom reaching mesosternum; tarsal segmentation 5–5–5; abdomen almost always with 6 to 7 visible sternites. Larvae eruciform (Fig. 37.13c) or highly modified, with suctorial mandibles.

The cantharoid beetles are characterized by their leathery, flexible integument and loosely articulated, flattened bodies. Antennae are usually serrate, less commonly pectinate or flabellate. The elytra are approximated in repose but not interlocked as in many other beetles. Larvae are eruciform, with small head and soft abdomen, or have the body segments platelike and usually flattened dorsally. Some of the abdominal segments may be strongly differentiated from the thoracic region, as in Lycidae and Brachypsectridae, resulting in distinctive "trilobite larvae." Most cantharoid beetles are predators, at least as larvae, which have the mandibles medially channeled for uptake of liquid food. Cantharidae (Fig. 37.13a) and Brachypsectridae are nonspecific predators of small, soft-bodied arthropods. Adult cantharids

are common on flowers and foliage; the larvae live in leaf litter. Brachypsectrid larvae are locally abundant in dry areas beneath loose bark; the adults are short-lived and rarely seen. Lampyridae (Fig. 37.13b) feed almost exclusively on terrestrial snails, and some Phengodidae attack only millipedes, which are immobilized by toxin produced by oral glands. Lycidae larvae feed on fluids associated with rotten wood, often in aggregations of dozens of individuals. Adults are usually encountered on flowers or foliage. The biology of Telegeusidae is unknown.

The hemolymph of cantharoid beetles contains various antifeedant compounds, and cantharids also release defensive secretions from segmental glands. The bright, conspicuous coloration of these diurnal insects is aposematic, and many members of Cantharidae, Lycidae, and Lampyridae serve as models for mimicking insects ranging from Hemiptera to Lepidoptera as well as other Coleoptera, especially Cerambycidae. Many Lampyridae luminesce from abdominal organs and use flash patterns in communication between the sexes, which are sometimes strongly dimorphic, with wingless, larviform females. Where several species occur together at the same time, they differ in color, pattern, or duration of their signals. Females of some species of *Photuris*, however, have more complex flashing repertoires, using one set of signals to attract conspecific males and mimicking the flashing patterns of other fireflies in order to attract males, which are then eaten. Female *P. versicolor* prey on at least 11 other species. Males of some *Photuris* have also been observed to mimic the signals of males of other species and may be predators as well. Some larval and adult female phengodids also luminesce, but communication between the sexes is by pheromones.

KEY TO THE FAMILIES OF CANTHAROIDEA

1. Abdomen with 5 sternites visible **Brachypsectridae**

Abdomen with 6 to 7 sternites visible **2**

2(1). Elytra with reticulate sculpture; mesocoxae distant **Lycidae**

Elytra striate, punctate, or smooth, mesocoxae usually contiguous **3**

3(2). Antennae inserted laterally in front of eyes; maxillary palpi greatly elongate **Telegeusidae**

Antennae inserted on frons between eyes; maxillary palpi normal **4**

4(3). Head retracted beneath pronotum, concealed in dorsal aspect (Fig. 37.13b) **Lampyridae**

Head fully visible in dorsal aspect **5**

5(4). Antennae with 12 segments, usually flabellate, pectinate (Fig. 37.6f) or bipectinate (Fig. 37.6g) **Phengodidae**

Antennae with 11 segments, usually filiform or serrate **Cantharidae**

Superfamily Dermestoidea (Fig. 37.14c,d). Antennae abruptly capitate, procoxae prominent, projecting or transverse; metacoxae excavated for reception of femora; tarsal segmentation 5–5–5, with simple tarsomeres; abdomen with 5 visible sternites. Larvae eruciform, usually with long spinose setae on the tergites.

These small to moderate-sized beetles are compact, oval, and usually densely covered with setae or appressed scales. Many species bear a single median ocellus. Typically the posterior margin of the pronotum is medially produced, concealing the scutellum. Three families are included: Dermestidae, Derodontidae, and Nosodendridae.

As larvae, Dermestidae (Fig. 37.14c,d) are scavengers of both plant and animal material but are especially abundant on drying skin, feathers, fur, and other proteinaceous substances. The smaller species frequently occur in old nests of Hymenoptera, feeding on cast skins, pollen, and dead insects. Several species are pests on stored products, woolen or fur articles, and in museum collections of insects and plants. Females of *Thylodrius contractus* are larviform, and a few species are myrmecophilous. Adults of many dermestids are common on flowers.

Adult and larval Nosodendridae occupy slime fluxes (flows of fermenting sap) on trees, both conifers and angiosperms, probably feeding on fungi. For example, *Nosodendron californicum* is common in sap flows on white fir, *Abies concolor*.

Derodontidae are rarely collected beetles that mostly inhabit the fruiting bodies of slime molds and wood-rotting fungi, frequently under the bark of dead trees. Members of the genus *Laricobius* feed on Adelgidae (Homoptera).

KEY TO THE FAMILIES OF DERMESTOIDEA

1. Head with pair of ocelli near inner margins of compound eyes **Derodontidae**

Head with single median ocellus or without ocelli **2**

2(1). Forelegs with tibiae flat, serrate; median ocellus absent **Nosodendridae**

Forelegs with tibiae cylindrical or oval in cross section, with straight margins; median ocellus frequently present **Dermestidae**

Superfamily Bostrichoidea (Fig. 37.14a,b). Antennae filiform, frequently with last 3 segments enlarged or elongate; head usually deflexed, frequently concealed dorsally by hoodlike prothorax,

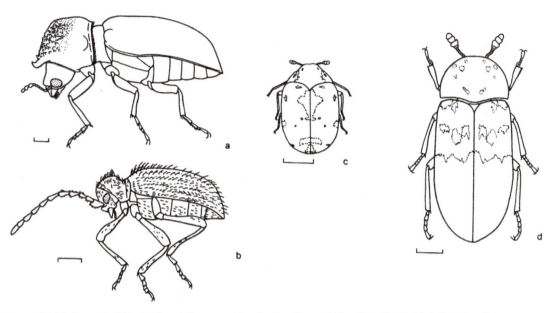

Figure 37.14 Bostrichoidea (a, b) and Dermestoidea (c, d): a, Bostrichidae (*Apatides fortis*); b, Ptinidae (*Ptinus clavipes*); c, Dermestidae (*Anthrenus lepidus*); d, Dermestidae (*Dermestes lardarius*). Scales equal 1 mm, except for b, which equals 0.5 mm.

tarsal segmentation 5–5–5; abdomen with 5 visible sternites. Larvae weakly sclerotized, glabrous, with reduced head capsule and legs; urogomphi absent; body stout, C-shaped, weakly sclerotized.

The beetles of this superfamily mostly feed on dead plant material. Bostrichidae, Lyctidae, and some Anobiidae bore in dead wood of both dicots and monocots such as bamboo. Bostrichids mostly attack recently felled trees, while anobiids and lyctids may enter seasoned timbers, usually affecting primarily the sapwood. *Rhyzopertha dominica* (Bostrichidae) is a cosmopolitan pest of stored grain. Ptinids and some anobiids are more polyphagous, consuming a variety of plant and animal products, including leather, seeds, tobacco, fish meal, spices, certain drugs, and especially stored cereal products. Anobiids also feed in fungi, including puffballs. A significant number of ptinids are commensals in ant nests, and many others frequent bird or mammal nests. Some ptinids and anobiids form cocoons from chitin filaments derived from the peritrophic membrane.

KEY TO THE FAMILIES OF BOSTRICHOIDEA

1. Metacoxae with posterior face excavated, concave, for reception of femora   **Anobiidae**
   Metacoxae with posterior face flat or slightly convex   **2**
2(1). Antennae filiform, inserted in close proximity between eyes (Fig. 37.14b)   **Ptinidae**
   Antennae clubbed or with 2 to 3 terminal segments longer than preceding 7 segments (Fig. 37.6h)   **3**
3(2). Prothorax hoodlike, concealing head in dorsal aspect (Fig. 37.14a), or, if not hoodlike, first abdominal sternite subequal to second   **Bostrichidae**
   Prothorax not hoodlike; first abdominal sternite nearly as long as second and third combined   **Lyctidae**

Superfamily Cleroidea *(Fig. 37.15).* Head usually prognathous, with antennae clubbed or filiform;

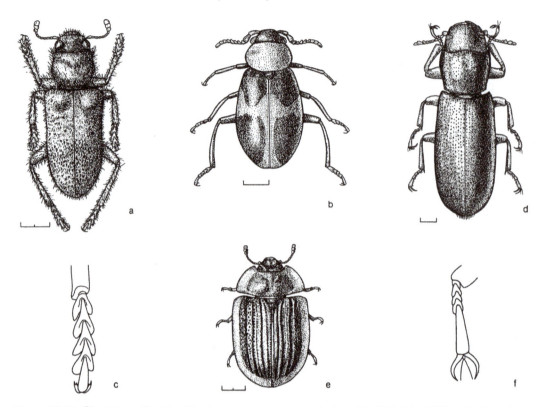

Figure 37.15 Cleroidea: a, Cleridae (*Enoclerus moestus*; scale equals 2 mm); b, Melyridae (*Collops histrio*; scale equals 1 mm); c, tarsus of *Enoclerus* (Cleridae); d, Trogossitidae (*Temnochila virescens*; scale equals 1 mm); e, Trogossitidae (*Ostoma pippingskoeldi*; scale equals 2 mm); f, tarsus of *Ostoma* (Trogossitidae).

procoxae transverse; metacoxae usually extending laterally beyond metasternum, prominent or flat, but not posteriorly concave for reception of femora; tarsal segmentation 5-5-5, with large bisetose empodium frequently present (Fig. 37.16f); abdomen with 5 (rarely 6) visible sternites. Larvae eruciform, with prognathous head, sclerotized prothorax, well-developed legs, urogomphi arising from sclerotized tergal plate.

Cleroids are nearly all predatory, at least as larvae. Exceptions include the Corynetinae (Cleridae) and several Trogossitidae (*Ostoma, Calitys, Thymalus*). Most Corynetinae feed on carrion, including preserved meat, hence the name "ham beetles," but some species are predators of *Dermestes* larvae. *Ostoma* (Fig. 37.15e), *Calitys*, and *Thymalus* are fungus feeders. Many clerids and trogossitids, such as *Temnochila* (Fig. 37.15d), are broad-spectrum predators, while

others specialize on one or a few prey organisms. For example, some species of *Enoclerus* (Cleridae; Fig. 37.15a) locate scolytid (bark beetle) burrows by following sex pheromone gradients produced by the female bark beetles, while *Trichodes* prey on bee larvae in their galleries or on grasshopper eggs. Several small trogossitids have cylindrical bodies, allowing them to move through the galleries of bark beetle prey. Adults of *Eronyxa* (Trogossitidae) are commonly found on flowers feeding on pollen; the larvae are predators of soft-bodied insects such as coccids. Larvae of Melyridae are predators in diverse situations, including beneath bark, in litter or soil, in galls and termite nests, and in the refuse chambers of harvester ants. *Endeodes* inhabits the upper intertidal zone along the Pacific coast. Some Melyridae (*Collops*; Fig. 37.15b) are of minor benefit as predators of aphids and other soft-bodied insects on crops; the adults of many species also feed on pollen.

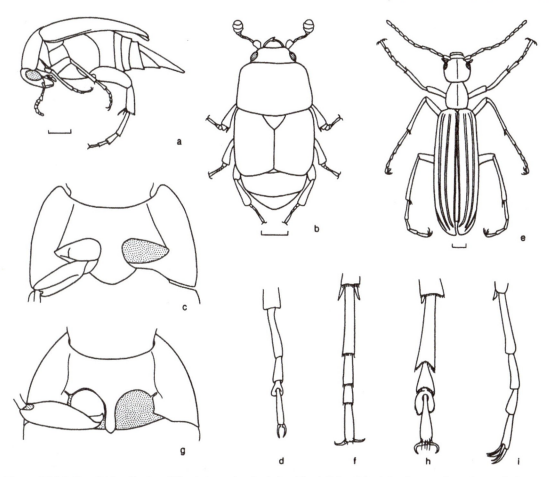

Figure 37.16 Cucujoidea (b, c) and Tenebrionoidea (a, d–i): a, Mordellidae (*Mordella albosuturalis*; scale equals 1 mm); b, Nitidulidae (*Carpophilus hemipterus*; scale equals 0.5 mm); c, ventral aspect of prothorax of *Carpophilus* (Nitidulidae); d, tarsus of *Stereopalpus* (Pedilidae); e, Meloidae (*Pleuropompha tricostata*; scale equals 2 mm); f, tarsus of Tenebrionidae (*Telacis opaca*); g, ventral aspect of prothorax of *Synchroa* (Melandryidae); h, tarsus of *Ditylus quadricollis* (Oedemeridae); i, tarsus of *Pleuropompha tricostata* (Meloidae).

KEY TO THE FAMILIES OF CLEROIDEA

1. Procoxae prominent, conical, projecting (as in Fig. 37.8d)   2
   Procoxae transversely elongate (as in Fig. 37.16c), not projecting   **Trogossitidae**
2(1). Antennae usually clubbed; tarsi with segments 1 to 4 ventrally lobed (Fig. 37.15c)   **Cleridae**
   Antennae usually filiform, rarely with loose 2- to 3-segmented clubs; tarsi simple, filiform, or with fourth segment lobed   **Melyridae**

Superfamily Lymexyloidea. Elongate, with 5 to 7 visible abdominal sternites; procoxal cavities confluent, mesocoxal cavities contiguous; antennae very short; tarsi at least as long as tibiae, with 5–5–5 segmentation.

The extremely slender, elongate larvae of the single family, Lymexylidae, tunnel through hard wood of moribund or dead trees or sawn lumber, apparently feeding on fungus. Adult males are remarkable for the elaborate, flabellate maxillary palpi. The elytra of *Atractocerus* (Neotropical) are extremely short, almost completely exposing the longitudinally folded wings.

Superfamily Cucujoidea *(Figs. 37.16, 37.17).* Antennae filiform, moniliform, or clubbed, very rarely serrate; metacoxae flat or convex, lacking concave posterior face for reception of femora; tarsal segmentation usually 5-5-5, occasionally 4-4-4 or 3-3-3; abdomen almost always with 5 sternites, rarely with 6 or 7. Larvae eruciform or highly modified; head capsule sclerotized, thoracic legs well developed. Abdomen usually with urogomphi on one or more terminal segments.

The Cucujoidea, containing about 25 families, is exceedingly diverse morphologically and ecologically, yet the number of species is relatively low, including less than 5 percent of all the Coleoptera. Coccinellidae and Nitidulidae are the largest families, with about 4500 and 3000 species, respectively; the other families are mostly small and frequently quite restricted biologically to narrow, specialized niches.

Habits and habitats of Cucujoidea are extremely variable, but the majority frequent the subcortical region of dead trees and shrubs. This habitat is extremely rich in fungi, molds, and other saprophytic organisms, and many Cucujoidea are fungivorous. For example, Erotylidae and Endomychidae are strictly dependent on fungi, mostly those associated with rotting wood. Many other families are less obviously associated with fungi but probably feed on molds, yeasts, or other saprophytes, at least as larvae (e.g., Lathridiidae, Rhizophagidae, Cryptophagidae, Phalacridae, some Cucujidae, Nitidulidae). Both adult and larval Sphindidae feed on slime mold spores. The biology of many families, though clearly associated with the subcortical habitat, is almost unknown. Several families, such as Nitidulidae and Cucujidae, include predatory forms. *Oryzaephilus* and several other cucujids are pests in stored grains, while many nitidulids feed in fruiting structures of higher plants. Coccinellidae are familiar as predators of aphids and other soft-bodied arthropods but also include fungus-feeding forms *(Psyllobora* and relatives) and plant-feeders (*Epilachna* and relatives). Coccinellidae are unusual among Cucujoidea in that the larvae feed exposed on plant surfaces. Classification of the Cucujoidea is extremely complex, and identification is difficult without a reference collection. Many families are only rarely encountered or are poorly differentiated. The following key includes only the more familiar groups and should be used with care. For identification of the smaller, more obscure families, Arnett (1960) and Lawrence and Britton (1991, 1994) are the most useful references.

KEY TO THE COMMON FAMILIES OF CUCUJOIDEA

1. Tarsal segmentation 5-5-5   **8**
Tarsal segmentation 4-4-4 or less   **2**

2(1). All tarsi with 4 segments   **4**
Tarsi with 3 segments or fewer   **3**

3(2). Elytra truncate, exposing terminal abdominal segment (Fig. 37.16b)   **4**
Elytra covering entire abdomen   **5**

4(2,3). Procoxae transversely elongate (Fig. 37.16c)   **Nitidulidae** (Fig. 37.16b)
Procoxae globular, round (Fig. 37.16g)   **Rhizophagidae**

5(3). Tarsi with second segment slender   **7**
Tarsi with second segment dilated (Fig. 37.17c)   **6**

6(5). First abdominal sternite with impressed lines parallel to coxal cavities   **Coccinellidae** (Fig. 37.17a)
First abdominal sternite without impressed lines   **Endomychidae**

7(5). Antennae filiform; body strongly flattened   **Cucujidae** (Fig. 37.17g)
Antennae capitate, with basal segment (scape) usually large, globular (Fig. 37.17f); minute to small in size   **Lathridiidae**

8(1). Antennae filiform or moniliform (Fig. 37.6a,b); body usually strongly flattened   **Cucujidae** (Fig. 37.17g)
Antennae clubbed (Figs. 37.6c, 37.16b); body usually convex   **9**

9(8). Epipleura continuing to elytral apices   **11**

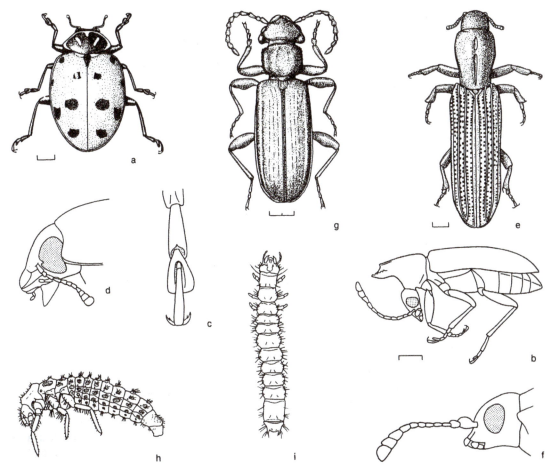

Figure 37.17 Cucujoidea (a, c–d, f–i) and Tenebrionoidea (b, e): a, Coccinellidae (*Hippodamia convergens*; scale equals 1 mm); b, Anthicidae (*Notoxus monodon*; scale equals 0.5mm); c, d, tarsus and antenna of *Hippodamia* (Coccinellidae); e, Colydiidae (*Deretaphrus oregonensis*; scale equals 1 mm); f, antenna of *Enicmus* (Lathridiidae); g, Cucujidae (*Cucujus clavipes*; scale equals 2 mm); h, larva of Coccinellidae (*Coccinella*); i, larva of Cucujidae (*Cucujus*). *Source: h, i, redrawn after Böving and Craighead, 1930.*

Epipleura terminating before elytral apices
                                    10

10(9). Antennae with 10 segments (10 and 11 fused)   **Rhizophagidae**[1]
Antennae with 11 segments
                    Nitidulidae (Fig. 37.16b)

11(9). Body convex, short, hemispherical; claws appendiculate or toothed (Figs. 37.16i, 37.17c)   **Phalacridae**

Body elongate, ovoid; claws simple   **12**

12(11). Body with upper surface smooth, without setae   **13**
Body with upper surface pubescent
                            Cryptophagidae

13(12). Procoxal cavities closed posteriorly (Fig. 37.16c)   **Erotylidae**
Procoxal cavities open posteriorly (Fig. 37.16g)   **Languriidae**

Superfamily Tenebrionoidea (Figs. 37.16a,d–i; 37.17b,e). Antennae filiform, moniliform, clubbed

---

[1] Males of Rhizophagidae have 5-5-4 tarsal configuration.

or serrate; metacoxae flat or convex, without concave posterior face for reception of femora; tarsal segmentation 5-5-4, reduced to 4-4-4 in a few lineages; abdomen with 5 or 6 sternites. Larvae eruciform or highly modified, especially in parasitoid forms; head capsule sclerotized, thoracic legs almost always well developed, abdomen with or without terminal urogomphi.

Various authorities recognize 25 to 30 families of Tenebrionoidea, but in number of species they constitute only about 7 to 8 percent of the Coleoptera, and Tenebrionidae is the only major family, with about 20,000 species. Most families of Tenebrionoidea have some members that inhabit rotting wood, either as larvae and often as adults, and many of the smaller families are more or less restricted to this habitat. Most of these species are probably feeding on fungi, which may be the primitive food for the entire superfamily. All species of Ciidae feed in fruiting bodies of basidiomycete fungi; Mycetophagidae and many Melandryidae and Zopheridae also feed directly on or in fungi. The most successful families by numbers of species and individuals, however, are those that are not primarily restricted to fungal feeding. Tenebrionidae includes several mycetophagous lineages and others that are associated with rotting wood, but the larger part of the family comprises species that have become specialized for a ground-dwelling existence in arid habitats. Many of these arid-adapted forms are large, conspicuous ambulatory beetles, and a few are important pests in field crops, especially in the Old World. Meloidae (Fig. 37.16e) consists entirely of parasitoid forms, the larvae feeding either in the buried egg masses of grasshoppers or on immature bees and wasps in their brood cells. The first instars are vagile triungulins, which seek the host upon which later sessile instars will feed. Those species with hymenopteran hosts are usually phoretic on adult bees or wasps as a means of entering their nests. Anthicidae adults (Fig. 37.17b) are encountered in various habitats (flowers or foliage; soil surfaces; under bark); larvae have been found in leaf litter. Colydiidae (Fig. 37.17e) have diverse feeding habits, including rotting wood, fungi, and lichens; many species inhabit leaf litter, and a few are predators of wood-boring

beetles. Larvae of Oedemeridae and Mordellidae inhabit punky rotten wood.

KEY TO THE COMMON FAMILIES OF TENEBRIONOIDEA

1. Procoxal cavities closed posteriorly by a process of the hypomeron (Fig. 37.16c) **9**
Procoxal cavities open posteriorly (Fig. 37.16g) **2**

2(1). Head abruptly constricted behind eyes, exserted **6**
Head gradually narrowed behind eyes, usually retracted into prothorax **3**

3(2). Tarsi with fourth segment dilated, ventrally pubescent (Fig. 37.16h) **Oedemeridae**
Tarsi with no dilated segments **4**

4(3). Mesocoxal cavities closed laterally by mesosternum (as in Fig. 37.14a,b) **Cryptophagidae**
Mesocoxal cavities closed laterally by mesepimeron (Fig. 37.1b) **5**

5(4). Body very strongly flattened **Cucujidae** (Fig. 37.17g)
Body rounded in transverse section **Melandryidae**

6(2). Prothorax with sharply defined lateral margins; last abdominal tergite conical, projecting beyond elytral apex **Mordellidae** (Fig. 37.16a)
Prothorax rounded, without distinct lateral margins **7**

7(6). Base of pronotum narrower than elytra **8**
Base of pronotum at least as wide as elytra **Rhipiphoridae**

8(7). Tarsal claws toothed or serrate or bifid (Fig. 37.16i) **Meloidae** (Fig. 37.16e)
Tarsal claws simple, not toothed or serrate **Anthicidae and Pedilidae** (Fig. 37.17b)

9(1). Tarsal formula 5–5-4 **10**
Tarsal formula 4-4-4 **Colydiidae** (including Cerylonidae) (Fig. 37.17e)

10(9). Prosternal process expanded laterally behind coxae, closing cavities **Zopheridae**
Prosternal process expanded only slightly behind coxae; cavities closed by

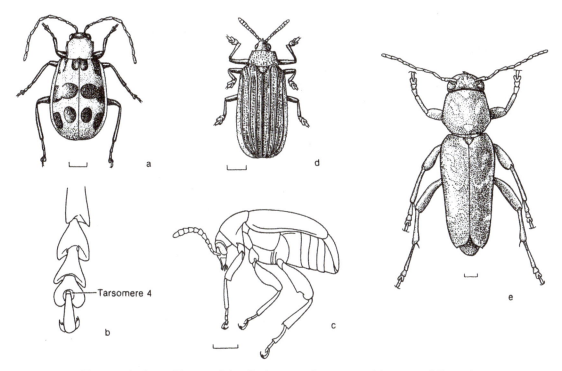

Figure 37.18 Chrysomeloidea: a, Chrysomelidae (*Diabrotica undecimpunctata*); b, tarsus of *Chrysochus* (Chrysomelidae); c, Bruchidae (*Bruchus rufimanus*); d, Chrysomelidae (*Odontata dorsalis*); e, Cerambycidae (*Xylotrechus nauticus*). Scales equal 1 mm.

medial extension of postcoxal process of hypomeron   **11**

11(10). Tarsal claws simple   **12**

Tarsal claws pectinate

Tenebrionidae (Alleculinae)

12(11). Apical antennal segment 2 to 4 times length of preceding segments

Tenebrionidae (Lagriinae)

Apical antennal segment about as long as preceding segments

Tenebrionidae (Fig. 37.1a)

Superfamily Chrysomeloidea *(Fig. 37.18).* Antennae filiform, serrate, or clavate, rarely capitate; metacoxae flat or convex, without posterior concavity for retraction of femora; tarsal segmentation 5-5-5, but with fourth segment almost always extremely small, concealed in bilobed third segment; abdomen with 5 visible sternites. Larvae eruciform, with 3 segmented antennae and well-developed legs (externally feeding forms) or with antennae and legs reduced (borers, internally feeding forms).

Nearly all the 40,000 members of this superfamily are phytophagous. As a group they feed on all parts of higher plants, from the roots, on the one hand, to the flowers and seeds, on the other. Most species of higher plants are probably utilized by at least one of these beetles, and the economic importance of chrysomeloid beetles is almost entirely due to their consumption of plant parts. Chrysomelidae such as *Diabrotica* (cucumber beetles, corn root worms; Fig. 37.18a), *Haltica* and *Epitrix* (flea beetles), and *Leptinotarsa* (Colorado potato beetle) are significant pests on many crops. A few chrysomelids, such as *Diabrotica*, have been implicated in transmission of plant viruses. Cerambycidae (Fig. 37.18e) include many minor forest pests. The apple tree borer, *Saperda candida*, infests the living trunks and large branches of several rosaceous trees.

Chrysomelidae, popularly called "leaf beetles," actually attack most parts of plants. The majority of species are external feeders on leaves, flowers, buds, or stems, but many occur in the soil as larvae, feeding on roots or underground stems. Members of the predominantly tropical subfamily Hispinae are leaf miners, and species of Cryptocephalinae specialize on dead leaves. Chrysomelids also feed on aquatic plants (*Donacia*), form galls, and live as commensals in ant nests. Adults are extraordinarily variable in body form, ranging from disc-shaped (subfamily Cassidinae) to cylindrical types (subfamilies Cryptocephalinae, Chlamisinae). All species of Alticinae (flea beetles) have the hind femora enlarged for jumping. The hind femora are also greatly enlarged in the tropical subfamily Sagrinae, which do not jump.

Cerambycidae as larvae are primarily borers in woody plants, either living or dead. They attack trunks, branches, stems, and roots. Some feed on stems or roots of herbs, a few mine in leaf petioles; there appear to be no true leaf miners. Adults may be either nocturnal or diurnal. The diurnal forms often visit flowers to feed on pollen, and some of these resemble Hymenoptera, particularly in flight. Other cerambycids mimic lycids and lampyrids. *Parandra* and relatives are unusual in having 5 subequal tarsomeres. Bruchidae (bean weevils, Fig. 37.18c; often included in Chrysomelidae) are borers in seeds, mostly of Leguminosae and Palmaceae, but many other plant families are attacked. The great majority oviposit in immature seed pods or capsules, but *Acanthoscelides obtectus*, the bean weevil, and a few other species attack dried peas and beans and are important pests of stored products.

KEY TO THE FAMILIES OF CHRYSOMELOIDEA

1. Antennae usually inserted on prominent tubercles, capable of being reflexed backward over body, and frequently very elongate **Cerambycidae**

Antennae not usually inserted on tubercles, usually less than half length of body **2**

2(1). Head usually deeply retracted into prothorax; fore coxae usually distinctly separated **Chrysomelidae**

Head prominent, not deeply set in prothorax; fore coxae usually contiguous **Bruchidae**

Superfamily Curculionoidea (Fig. 37.19). Antennae elbowed and capitate, or occasionally filiform, serrate, or clavate; oral region more or less prolonged as a distinct rostrum; prothorax without distinct lateral margins; metacoxae flat or convex, without posterior concavity for reception of femora; tarsal segmentation 5-5-5, with fourth segment minute, hidden in bilobed third; abdomen with 5 visible sternites. Larvae apodous (with reduced legs in a few primitive genera), with papilliform, 1 or 2-segmented antennae; no urogomphi.

The Curculionoidea, comprising over 65,000 species, is easily the largest superfamily group of Coleoptera and the most important from an economic standpoint. The evolutionary success of these beetles apparently stems at least in part from the elongation of the stomal region of the head into a rostrum, which is used in preparing oviposition holes as well as in feeding. The rostrum is most specialized in the Curculionidae, where the labrum is absent and maxillae and labium are reduced and partly concealed. The most specialized members of Curculionoidea are the weevils (Curculionidae) and bark beetles (Scolytidae). Both families are characterized by capitate, elbowed antennae. In the remaining curculionoid families, one or both of the antennal specializations is lacking. The number of families recognized in this superfamily varies by treatment and is currently being reevaluated.

The Curculionidae (weevils; Fig. 37.19a,b), with an estimated 60,000 described species, are probably the most morphologically and biologically derived family of Coleoptera. Weevils are typically found on foliage or flowers as adults, but some species are ground dwellers, some burrow in sand dunes, and a few are aquatic or marine. Larvae are

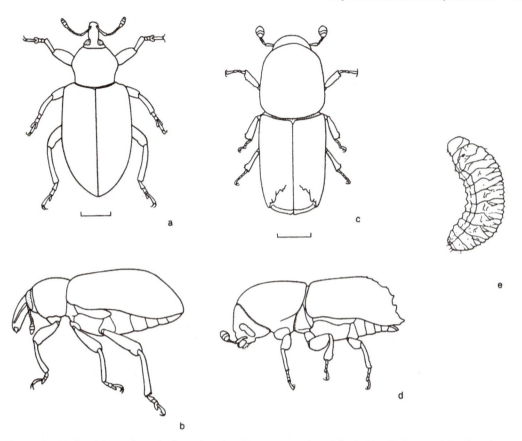

Figure 37.19 Curculionoidea: a, b, Curculionidae (*Hypera postica*); c, d, Scolytidae (*Ips*); a, c, adults, dorsal aspect; b, d, adults, lateral aspect; e, larva of Curculionidae (*Naupactus*). Scales equal 1 mm.
Source: e, redrawn from Böving and Craighead, 1930.

mostly internal feeders or subterranean and are rarely encountered without searching, but some, such as *Hypera,* feed externally on foliage.

Weevils include many serious economic pests, such as the alfalfa and clover weevils (*Hypera*), the cotton-boll weevil (*Anthonomus grandis*), Fuller's rose weevil (*Pantomorus godmani*; infests many ornamentals), strawberry weevils (*Brachyrhinus*; infests roots of ornamentals and row crops), and the plum curculio (*Conotrachelus nenuphar*; infests fruit of many orchard crops). Species of *Sitophilus* are important pests of stored whole grains.

Brentidae and Anthribidae are relatively small families, closely related to Curculionidae. Most Brentidae are elongate, cylindrical beetles that feed in decaying wood as larvae, the adults usually being encountered under bark. Anthribidae feed in fungi, decaying wood, or, occasionally, seeds or fruit. Males of some species have antennae considerably longer than the body. The small, compact, black adults of Apionidae are extremely common on flowers.

Nemonychidae infest the fruiting structures of gymnosperms and a few primitive angiosperms. Adults are not uncommon on Pinaceae when the staminate cones are mature. Attelabidae (including Rhynchitidae), with about 60 species in North America, mostly feed in rose hips, acorns, or other fruiting structures of plants. *Attelabus* females form a leaf roll into which they oviposit and in which the larva feeds. Ithyceridae includes a single species that is associated with oak, hickory, and beech in eastern North America.

The Scolytidae (bark beetles; Fig. 37.19c,d) attack woody plants by boring into the cambial tissue, where adults feed and prepare oviposition galleries. The adult beetles transport yeast spores in special mycangia, inoculating the walls of the galleries. The growths that result are consumed by both adults and larvae. Most species infest necrotic or freshly killed plants, but some invade apparently healthy trees and are the world's major forest pests. In North America species of *Ips* and *Dendroctonus* cause severe damage on conifers. Morphologically, bark beetles are virtually indistinguishable from some weevils, and it may be pointed out that Cossoninae (Curculionidae) are biologically very similar to bark beetles. The two groups are treated separately here because of the massive amount of applied literature keyed to traditional classifications. Many authors now consider Scolytidae as a subfamily of Curculionidae.

KEY TO THE NORTH AMERICAN FAMILIES OF CURCULIONOIDEA[1]

1. Antenna straight; scape about as long as succeeding segments.   3
   Antenna elbowed; scape at least as long as next several segments combined (Fig. 37.19a)   2

2(1). Antennae inserted distant from eyes (Fig. 37.19b); rostrum short to very long
                        Curculionidae
   Antennae inserted contiguous to eyes (Fig. 37.19d); rostrum short
                Scolytidae (including Platypodidae)

3(1). Maxillary palp segmented; tarsal claws usually toothed or bifid   7
   Maxillary palp fused into single, rigid article; gular sutures fused or absent   4

4(3). Antenna with 11 segments   5
   Antenna with 10 segments   **Apionidae**

5(4). Antenna with compact club, segments appearing partly fused   **Ithyceridae**
   Antenna with loose club, segments appearing separate   6

6(5). Body elongate, cylindrical (more than three times as long as wide); rostrum long, narrow   **Brentidae**
   Body about twice as long as wide, narrowed anteriorly; rostrum long, narrow or short, broad
           Attelabidae (including Rhynchitidae)

7(3). Gular sutures separated; pronotum without lateralcarinae   **Nemonychidae**
   Gular sutures absent; pronotum with short to complete lateral carinae   **Anthribidae**

---

[1] Oxycorinidae, with one species in southern Florida, is not included.

CHAPTER 38

Order Strepsiptera

ORDER STREPSIPTERA

Minute to small Endopterygota specialized for endoparasitism. Males (Fig. 38.1a) highly distinctive, with large, transverse head; antennae with 4 to 7 pectinate segments; bulging eyes with large facets; ocelli absent. Prothorax and mesothorax very small, mesothorax with wings reduced, elytralike, without veins, functioning as halteres; metathorax very large, bearing broad, fan-shaped wings with a few thick, radiating veins; legs usually without tarsal claws, first two pairs without trochanters, hind pair without coxae. Abdomen cylindrical, tapering, concealed basally by large postscutellum. Females wingless, coccidlike (Mengenillidae) (Fig. 38.2c) or larviform, grublike. First-instar larva (Fig. 38.1b,c) (triungulin) spindle-shaped, lacking antennae, mandibles; 3 pairs thoracic legs without trochanters; abdomen with bristlelike caudal styles. Later instars grublike (Fig. 38.1d), without appendages or distinct mouthparts and with alimentary canal ending blindly. Pupa exarate, adecticous, pharate in puparium formed by cuticle of last larval instar.

While numerous members of several orders live as parasitoids, Strepsiptera are among the very few insects that internally parasitize other arthropods without killing them. They infest or stylopize a variety of insects, including Thysanura, Blattodea, Mantodea, Orthoptera, Hemiptera, Diptera, and Hymenoptera. Of these, Hemiptera (especially auchenorrhynchous Homoptera) and Hymenoptera are the most common hosts. Host specificity of Strepsiptera varies. Stylopidae that parasitize Hymenoptera tend to be narrowly host-specific, whereas Halictophagidae, which parasitize auchenorrhynchous Homoptera, often have host ranges that include many species in several genera. In the great majority of species both sexes use the same host, but in the family Myrmecolacidae males parasitize ants while females parasitize Orthoptera and Mantodea. Unlike parasitoid Hymenoptera and Diptera, which eventually kill their host

and usually consume most of its internal organs, Strepsiptera mature within active, living insects. The females of the great majority of species never leave the host's body.

Morphologically Strepsiptera are superficially more similar to Coleoptera than to any other order of insects. The antennae and elytriform forewings resemble the same structures in parasitoid beetles (Rhipiphoridae), and the first-instar larvae are similar to triungulins of Meloidae and Rhipiphoridae. Furthermore, subsequent larval instars of parasitoid beetles are grublike, as in Strepsiptera. However, the reduced prothorax and enlarged postscutellum are not features that occur in Coleoptera. Strepsiptera are unique in lacking trochanters in all instars and, unlike Coleoptera, lack a gula. Moreover, purported homologies in wing venation of Strepsiptera and Coleoptera have been disputed; instead, a common ancestry with

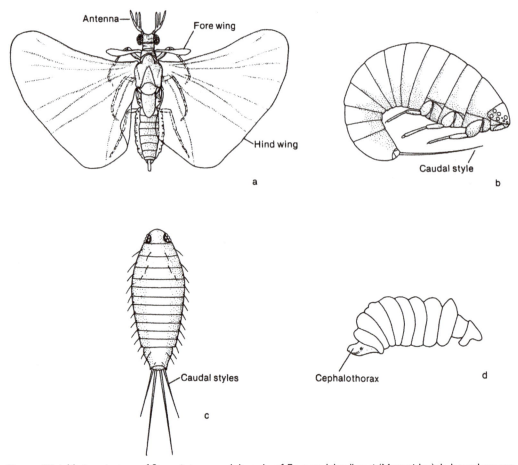

Figure 38.1 Various instars of Strepsiptera: a, adult male of *Eoxenos laboulbenei* (Mengeidae); b, lateral aspect of first-instar larva or triungulin of *Halictophagus tettigometrae* (Halictophagidae); c, dorsal aspect of first instar of *Eoxenos* (Mengeidae); d, lateral aspect of third-instar larva of *Halictophagus*.
Source: a, c, redrawn from Parker and Smith, 1934; b, d, redrawn from Silvestri, 1940.

Diptera was proposed and initially supported by molecular data. Elenchidae, a derived family, is known from Cretaceous amber from Lebanon, suggesting a possible divergence from Coleoptera or some other endopterygote order as early as the Carboniferous. As a further bit of conflicting evidence, however, the pupal stage of some Strepsiptera is preceded by pharate instars with wing rudiments, suggesting that they may not be true endopterygotes at all (Kristensen, 1991). Recent molecular phylogenetic studies employing six nuclear genes appear to confirm the earlier proposed relationship with Coleoptera, and other

developmental data suggest that the relationship with Diptera may not be tenable. It is likely the correct placement will be resolved when full genomes are available for all holometabolous insect orders, as is developing rapidly.

The life cycle is strongly modified for endoparasitism. Males spend their life of 1 or 2 days seeking females, which in typical species remain inside the host with only the anterior end or cephalothorax exposed. Virgin females secrete a sex pheromone from a gland on the cephalothorax, which provides an efficient means of collecting. Males will be attracted for many days unless copulation

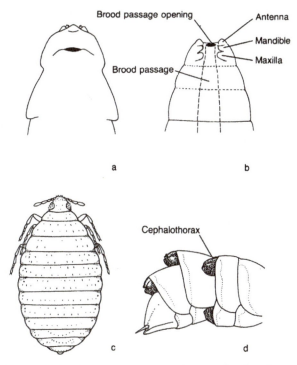

Figure 38.2 a, b, Cephalothoraces of mature females of: a, Halictophagidae (*Halictophagus tettigometrae*); b, Corioxenidae (*Triozocera mexicana*); c, mature female, Mengeidae (*Eoxenos laboulbenei*); d, puparia of Stylopidae (*Xenos vesparea*) in the abdomen of a vespid wasp (*Polistes nympha*).
Source: b, redrawn from Silvestri, 1940; c, redrawn from Parker and Smith, 1940; d, redrawn from Szekessy, 1960.

is allowed, whereupon pheromone production ceases. Numerous eggs are produced (as many as 75,000 have been recorded), being retained within the female until hatching. The minute triungulins emerge to seek appropriate hosts, which are usually immature instars of the proper species. Movement is by crawling or by springing, which is effected by curving the abdomen ventrally, then suddenly snapping it backward. If an appropriate host is encountered, the triungulin immediately enters its body, aided by an oral secretion that softens the host cuticle. The triungulin soon molts, producing the legless, grublike second instar; subsequent growth and development occur within the host body cavity. There are three grublike feeding instars, with the cuticle from each molt remaining as a persistent

sheath around the larva. The larval alimentary canal ends blindly, probably to prevent the feces from entering and injuring the host (Hymenoptera and Neuroptera show an analogous adaptation). The fully grown last instar larva extrudes its anterior end through the intersegmental membrane of the host abdomen (Fig. 38.2d), and pupation occurs in this position. Pupation usually occurs in adult hosts but sometimes occurs in immatures of exopterygote hosts. In males the protruding head becomes tanned, forming the cephalotheca, which the adult later pushes aside in order to emerge. The neotenic, grublike adult female remains pharate within the saclike cuticle of the last larval instar, through which respiration and food absorption continue to take place. The anterior portion of the body, equivalent to the head, prothorax, and mesothorax, is extruded as the cephalothorax. The brood passage provides an opening through which copulation occurs and through which the emerging triungulins later escape. The actual genital openings are pores on abdominal segments two to five.

The typical life cycle described above differs in certain species. In the suborder Mengenillidia, with about 20 species in three genera in the family Mengenillidae, the females, though wingless, have well-developed legs and live free of their hosts, which are Lepismatidae (Thysanura). Last-instar larvae of both sexes leave the host and pupate externally. The Mengenillidia undoubtedly represent the most primitive Strepsiptera, although they occur relatively late (Eocene) in the fossil record. Extant species are known from the Mediterranean region, tropical Africa, Asia, and Australia.

In Stylopidae, which parasitize bees and wasps, the triungulins are deposited on flowers. When adults of suitable host species visit the flowers, the triungulins usually cling to the bee or wasp, or, as in *Stylops pacifica* (see Chap. 11, Fig. 11.7), the triungulins are ingested with nectar. Once in the host's nest, the triungulins disembark (or are regurgitated) and parasitize the host's eggs or larvae. It may be noted that many meloid and rhipiphorid triungulins are likewise phoretic on adults of their hosts. Stylopidae is the largest family of Strepsiptera.

Stylopized hosts frequently differ from unparasitized members of the same species in morphological details. Especially affected are secondary sexual features, which may be modified to resemble homologous structures of the opposite sex. Internal organs, especially the gonads, are stunted, suggesting the term "parasitic castration." However, at least some male hosts appear to be capable of reproduction. In cases where the strepsipteran pupates in an immature host, the host does not molt to the adult stage. It is known that host life expectancy is sometimes increased by stylopization, but it is not clear whether differential survival has a physiological basis or is mediated by behavioral changes that decrease predation.

About 525 species are known worldwide; several times that many probably remain undescribed.

KEY TO NORTH AMERICAN FAMILIES OF STREPSIPTERA

1. Males with tarsi 5-segmented, the terminal segment narrow with minute claws; females with brood passage opening terminally (Fig. 38.2b)
   Corioxenidae
   Males with tarsi 2- to 4-segmented, without claws; females with brood passage opening preterminally (Fig. 38.2a) **2**
2(1). Males with tarsi 4-segmented; parasites of Hymenoptera **Stylopidae**
   Males with tarsi 2- or 3-segmented; parasites of Hemiptera or Orthoptera **3**
3(2). Males with tarsi 2-segmented; parasites of Fulgoroidea **Elenchidae**
   Males with tarsi 3-segmented; parasites of Homoptera, Orthoptera
   Halictophagidae

Corioxenidae. In the United States occurring uncommonly in the southern states. Hosts are Heteroptera.

Stylopidae. Common parasites of bees (Andrenidae, Halictidae, Colletidae) and wasps (Sphecidae, Vespidae). Over 150 species occur in North America.

Halictophagidae. In North America locally common parasites of Homoptera; pygmy grasshoppers (Tetrigidae) are also infested, and exotic species parasitize Tridactylidae, Blattodea, Heteroptera, and Diptera.

Elenchidae. Similar to Halictophagidae. Elenchidae parasitize Fulgoroidea, especially Delphacidae in North America; uncommon.

CHAPTER 39

Order Hymenoptera
(Bees, Wasps, Ants, etc.)

ORDER HYMENOPTERA

Adult. Minute to large Endopterygota, usually with 2 pairs of wings. Head free, mobile, with mandibles adapted for chewing; galeae or labium frequently elongate, suited for imbibing liquids; antennae multiarticulate, usually filiform or moniliform; compound eyes large, lateral; 3 ocelli usually present. Mesothorax enlarged, fused with small, collarlike pronotum and small metathorax; wings membranous, stiff, hyaline, with relatively few veins, which usually delimit a few large, closed cells; forewings about 1.5 to 2 times as long as hind wings; wing coupling effected by hamuli (hooklike setae on leading edge of hind wing). Abdomen with first segment inflexibly joined with metathorax, usually incorporated into thorax as propodeum, second abdominal segment usually constricted as narrow petiole, followed by swollen gaster, female with ovipositor modified as slicing or piercing organ for inserting eggs into tissue or adapted as sting.

Larva. Eruciform, with distinct head capsule, chewing mandibles, 3 pairs thoracic legs and 6 to 8 pairs abdominal prolegs; or apodous, grublike, often with head capsule and mouthparts reduced or vestigial.

Pupa. Adecticous, usually exarate; frequently enclosed in silken cocoon secreted from labial glands.

Certain Hymenoptera, such as honey bees, ants, and some of the larger wasps, are among the most familiar insects. In contrast to most other orders, many Hymenoptera are beneficial to agriculture through either pollination of crops or destruction of phytophagous insects (Fig. 39.1). Parasitoid species have proved extremely useful in biological control, especially of introduced pests, and have been transported extensively about the earth. *Apis mellifera*, the honey bee, among the most thoroughly domesticated insects, is extremely important as a

pollinator, and its products, honey and wax, are still commercially important, despite competition from alternative materials. Approximately 20,000 species of other bees worldwide also provide crucial and often rather specific pollination services to a huge variety of flowering plants, including many not able to be pollinated by honey bees. Bee and wasp stings, annoyances from ants, and occasional damage by sawflies, gall wasps, or other phytophagous species are of only nuisance significance when compared with the positive attributes

Figure 39.1 Diversity of Hymenoptera: a, tenthredinid sawflies mating on a leaf in an Ecuadorean montane forest; b, diprionid sawfly larva wandering to a pupation site in coastal Maine conifer forest; c, *Hartigia maculata*, a cephid stemboring sawfly on a rose leaf in Illinois; d, an ichneumonid (Anomaloninae: Gravenhorstiini) searching leaves in coastal scrub in Cornwall; e, a *Microplitis* parasitoid wasp (Braconidae) exploring a hairy sage leaf in Illinois; f, a female *Gasteruption* wasp (Gasteruptiidae) stops to groom herself on a stone wall in Cornwall; g, a female tiphiid wasp (*Myzinum* sp.) nectaring on goldenrod; h, A leafcutter bee (Megachilidae) with a section of leaf she has cut to line her nest cells.

Source: All photos by J. B. Whitfield, except e by Won Young Choi. Used with permission.

of this order. Hymenoptera currently comprise about 115,000 described species, but this number is certain to increase to several times this number as the numerous small, parasitoid species are discovered. Some tropical biodiversity specialists have even suggested that when the full fauna is known, Hymenoptera might well be the most species-rich insect order, surpassing even Coleoptera.

The most important adaptive features in Hymenoptera involve the mouthparts, thoracic and wing structure, and abdomen, including the ovipositor, all these structures apparently being modified in relation to predation or parasitism. All Hymenoptera possess well-developed mandibles, which in many species are used for tasks other than feeding and may be modified for specific functions. For example, some megachilid bees and some ants cut foliage with the sharp scissorslike mandibles, while in the soldier castes of many ants the mandibles are designed for offense or defense. In most species, however, multiple functions are served. For instance, a worker ant may dig, transport food or soil particles, subdue or manipulate prey, defend the colony, or tend immatures—using the mandibles as a primary tool. Many Hymenoptera, especially many stinging wasps and all bees, have evolved the ability to ingest liquid food, primarily nectar, through a proboscis formed by the glossae (part of the labium) and the maxillary palpi and

galeae (Fig. 39.3b). In apid bees the proboscis may be longer than the body.

As in other aerial insects, the hymenopterous thorax is strongly modified for efficient flight. The mesothorax is much the largest segment and in most species is solidly fused to the reduced pronotum and metathorax (Fig. 39.2). The head and forelegs are attached to the propleura, which are usually flexibly articulated to the pronotum and mesopleura. In general, the prothorax is best developed in ambulatory species such as ants or in those that use the forelegs to dig (Pompilidae, Scoliidae). In most strong fliers the pronotum is much smaller than the mesothorax and in bees and many wasps may appear as a narrow collar in dorsal aspect. The metanotum is represented by a narrow sclerite situated between the mesopostnotum and the first abdominal tergite, which is inflexibly incorporated into the thorax as the propodeum. The propodeum may be recognized by the presence of the first abdominal spiracle (Fig. 39.2). In apocritan (waisted) hymenopterans, the metapostnotum or second phragma is an internal cowling situated just beneath the propodeum. The dorsal longitudinal flight muscles attach to the large second phragma, which is held close to the propodeum by a very short muscle. The functional significance of this complex arrangement is obscure. The pleural sclerites are usually large, distinct plates, bearing

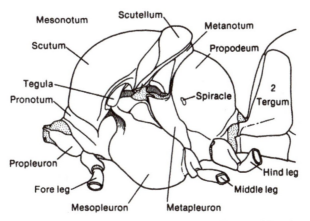

Figure 39.2 Lateral aspect of pterothorax of *Apis mellifera*, the honey bee.
Source: Modified from Snodgrass, 1942.

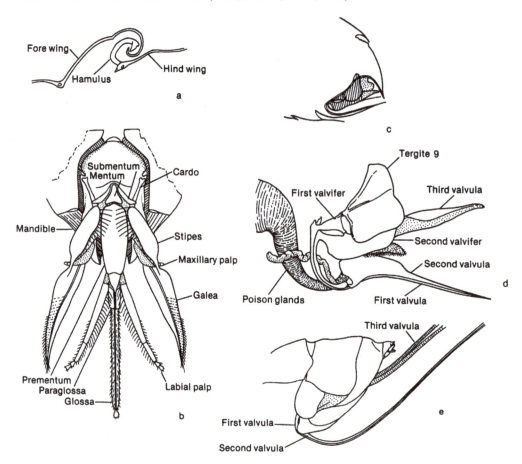

Figure 39.3 Structural features of Hymenoptera: a, diagrammatic cross section of wings of *Apis mellifera* (Apidae), showing wing coupling; b, mouthparts of *Apis mellifera*; c, medial section through abdomen of *Apis mellifera*, showing sting in repose; d, dissected sting of *Apis mellifera*; e, end of abdomen and ovipositor of *Megarhyssa atrata* (Ichneumonidae).
Source: Modified from Snodgrass, 1933, 1942.

the mesocoxal and metacoxal attachments. In general the extensive variation in size and shape of the various thoracic sclerites is distinctive of families and superfamilies and is of some use in classification.

The hind wings are always coupled to the much larger forewings by a series of hooked setae (hamuli), which extend forward from the costal margin of the hind wing and catch beneath a ventral fold of the anal margin of the forewing (Fig. 39.7 a,c,d). The hamuli number from two or three in small species such as Chalcidoidea to many in the larger

wasps where wing coupling may be very tenacious. Longitudinal wing folding is rare in Hymenoptera but occurs in several families (Vespidae, some Gasteruptiidae) in which wing coupling is so strong that the wings do not separate when held over the back. Simplification of wing venation by reduction is typical of Hymenoptera, being broadly correlated with decrease in body size. Very small species, such as some Chalcidoidea, may retain only a single longitudinal vein in the forewing, with the hind wing veinless (Fig. 39.13e). In general, venation is highly variable and characteristic of various families or superfamilies, but aerodynamic significance of the different configurations is not obvious.

The morphological specializations perhaps most responsible for the success of the Hymenoptera involve the abdomen. In the earliest diverging lineages of phytophagous insects known as sawflies, the abdomen is broadly and relatively inflexibly joined to the thorax. The ovipositor is bladelike or sawlike in most of these families, adapted for slicing plant tissues. In the apocritan Hymenoptera, the abdomen is strongly constricted at the second segment, greatly increasing its mobility (Fig. 39.2). The ovipositor is always developed as a piercing organ, which can be inserted deeply into host (usually animal) tissues. Although highly specialized for slicing or piercing tissues, the hymenopteran ovipositor shows obvious homologies to the ovipositors of even more ancient insect groups such as Orthoptera and even Thysanura (compare Figs. 39.3 and 2.16). Morphologically, the ovipositor consists of the modified appendages of abdominal segments eight and nine. Each appendage consists of a short triangular or trapezoidal gonocoxite or valvifer, which rocks about its articulation with the abdominal sternite (Fig. 39.3c–e). Each gonocoxite bears a bladelike gonapophysis or valve. Rapid, repeated rocking of the gonocoxites is translated to linear movement of the gonapophyses, which generally bear barbs that assist their penetration. Gonocoxite nine bears an additional apophysis, the gonostyle or ovipositor sheath. The gonostyles, which are frequently fused, do not enter the oviposition substrate but provide a tubular support for the more delicate gonapophyses. The lumen of the ovipositor is very narrow, and eggs of parasitoid Hymenoptera are capable of great deformation, regaining their normal elongate oval shape once inside the host. The gonapophyses vary from short, relatively stout blades to slender, flexible stylets several times the body length, enabling the wasps to reach hosts concealed in cocoons or in galleries in fruiting structures or wood. Insertion of such elongate ovipositors may require bending into a loop, steadied in some species by the hind coxae (Fig. 39.4a). Many parasitoid species lap up hemolymph exuding from oviposition wounds. If the host is secreted in a cocoon or gallery, the female may secrete material that hardens around the ovipositor, forming a tube that functions as a drinking straw when the ovipositor is withdrawn (Fig. 39.4b). In stinging wasps and bees the first and second gonapophyses are further specialized as a stinger (Fig. 39.3d), discharging only venom, the eggs issuing at the base. In most wasps the stinger is used to immobilize prey, but in Vespidae (paper wasps, yellow jackets, hornets), many ants, and many Apoidea (bees), its primary function is defense against vertebrates, and the venoms produce intense, instantaneous pain. Their stings provide efficient protection against many predators, and bees and wasps are probably the most familiar examples of warning coloration. Stinging

Figure 39.4 Postures of female wasps during oviposition. a, Ichneumonidae; b, Chalcididae.
Source: Redrawn after Askew, 1971.

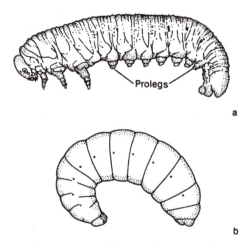

Figure 39.5 Larvae of Hymenoptera: a, eruciform larva of Symphyta (Diprionidae, *Neodiprion lecontei*); b, apodous larva of Apocrita (Apidae, *Xylocopa virginica*). Source: a, redrawn from Middleton, 1921; b, redrawn from Stephen et al, 1969.

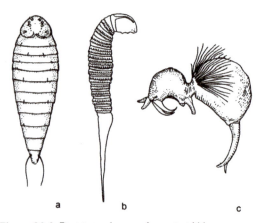

Figure 39.6 First-instar larvae of parasitoid Hymenoptera: a, planidium larva of Chalcidoidea (Perilampidae, *Perilampus hyalinus*); b, caudate larva of Ichneumonoidea (Ichneumonidae, *Cremastus flavoorbitalis*); c, eucoiliform larva of Proctotrupoidea (Scelionidae, *Scelio fulgidus*). Source: a, redrawn from Smith, 1912; b, redrawn from Bradley, 1934; c, redrawn from Noble, 1938.

Hymenoptera are widely mimicked by insects of many other orders and occasionally by spiders.

In general, larvae of Hymenoptera conform to two body plans. Larvae of many sawflies are highly similar to lepidopterous caterpillars and, like most caterpillars, are external foliage feeders (Fig. 39.5a). The large, chewing mandibles are set in a strongly sclerotized head capsule. The thick, cylindrical trunk bears appendages on most abdominal segments, as well as on the thoracic segments. The abdominal prolegs, unlike those of lepidopterous larvae, lack grasping spines or crochets. Other useful differentiating characters include the presence of a single stemma (six stemmata usually present in lepidopterous larvae) and the reduction of the antennae to papillae (antennae usually 3-segmented in lepidopterous larvae).

The families Siricidae and Cephidae, as well as certain Tenthredinidae, are internal plant feeders that lack abdominal legs and have the thoracic legs reduced. All larvae of the Apocrita ("waisted Hymenoptera") are apodous, with soft, maggot-like bodies and sometimes with the head capsule unsclerotized or retracted into the thorax (Fig. 39.5b). These larvae all inhabit protected environments (usually the body cavity of a host insect or a

cell prepared by the mother); hence the structural simplification. The midgut is not connected to the hindgut in most apocritan larvae. This is evidently an adaptation for avoiding fecal contamination of the host body or larval cell. At the last larval molt, the alimentary canal becomes complete, and the accumulated fecal material, or meconium, is voided just before pupation. Often the meconium can be found alongside the last larval skin inside the cocoon or pupal chamber.

Various parasitoid Hymenoptera are hypermetamorphic, the first instar, or occasionally more than one instar, being differentiated from the later maggotlike stages. For example, in some chalcidoid wasps that do not deposit their eggs in a host, the first instar is a mobile planidium (Fig. 39.6a). Planidia resemble the first-instar triungulin larvae of Coleoptera (Meloidae, Rhipiphoridae) and Strepsiptera, and perform the same function—host location. Locomotion is accomplished with the movable ventral extensions of the segmental sclerites, as legs are absent. Caudate larvae, found in some Ichneumonidae and Braconidae, and nauplioid larvae (Fig. 39.6b,c), in some Proctotrupoidea, hatch internally in the host. They may function in eliminating supernumerary parasites, as suggested by the large, sickle-shaped mandibles of the nauplioid type. A variety of other first-instar forms, sometimes bizarrely spined or shaped and of uncertain function, occur in different families of parasitoids.

Adult Hymenoptera, with the notable exception of some nocturnal ichneumonids, bees, and a few others, are largely sunshine-loving creatures that frequent flowers, sunlit vegetation, or open soil. However, as parasitoids and predators, they have followed other insects into almost every niche imaginable. As adults, many Hymenoptera consume nectar, pollen, and honeydew produced by Hemiptera, but many sawflies (Tenthredinidae) are predators of small, soft-bodied insects, and some parasitoid forms imbibe hemolymph, which oozes from wounds made during oviposition.

Mating may occur in the air (as in honey bees, ants) or on some substrate. Males seek out females about their emergence sites or in areas that the females will visit, such as stands of blooming plants. In some species, males establish territories including emergence sites and defend them against other males. In some Tiphiidae and Mutillidae the male is often much larger than the wingless female and carries her about during copulation. In Hymenoptera, males (if present) are typically produced parthenogenetically from unfertilized eggs and are haploid, while the diploid females are produced from fertilized eggs. This mode of reproduction is called arrhenotoky. Sperm are stored in a spermatheca, where they may remain viable for long periods, their release being controlled by the female. Female control of fertilization allows for production of unusual sex ratios, which may be extreme in social species, where relatively small numbers of males are produced only once or a few times a year. Various species of Hymenoptera are entirely parthenogenetic (thelytokous), being known only from diploid females. Many Cynipidae (gall wasps) are cyclically parthenogenetic, with generations of bisexual males and females alternating with generations of asexual females.

Arrhenotoky allows recombination in each generation of females, promoting heterozygosity, while causing elimination of recessive lethal alleles in each generation of males. Arrhenotoky also provides a mechanism influencing interactions within societies in bees and wasps, because of the asymmetrical degrees of relatedness, with sisters sharing more genes than do daughters and mothers (see Chap. 7). Thelytoky is of particular advantage in populations that are highly dispersed, as in colonizing species, since male–female encounters may be rare. Thelytoky would also seem to be advantageous in hyperparasitoid species, where populations would normally be sparse.

Several distinct types of life history occur within the order. As mentioned above, most or all members of the sawfly lineages are strictly herbivorous as larvae, and this mode of feeding is apparently ancestral for the order. The vast majority of apocritans are zoophagous as larvae. The majority of species are parasitoids, living internally or sometimes externally on a host that is almost always selected by the mother. The host is always at least as large as the parasitoid, so that a single victim is sufficient to allow a single parasitoid, or multiple parasitoids, to develop. An interesting offshoot of the parasitoid habit is inquilinism. Inquilines inhabit galls or nests of other species, consuming the food intended for the original occupant, which is incidentally killed and eaten as well, usually as an egg or early-instar larva. Several mostly parasitoid families are known to include inquiline species.

Parasitoidism is practiced by about half the families of Hymenoptera, including some eminently successful taxa such as Ichneumonidae, Braconidae, and most families of Chalcidoidea. Most orders of insects as well as spiders and ticks are attacked. All instars, including eggs, are utilized as hosts. In typical life histories, larval development of the parasite begins soon after oviposition, and in some species eggs hatch during oviposition. Eggs of some species are deposited into immature hosts (eggs or early-instar larvae), where they may eclose, but larval development is postponed, sometimes as much as a year, until the host is nearly full-grown. Such delayed parasitoidism occurs in Ichneumonoidea and especially in the Proctotrupoidea. Many chalcidoids, as well as occasional members of other superfamilies, are secondary parasitoids (hyperparasitoids). The host is another parasitoid rather than a primary victim.

Secondary and even tertiary levels of hyperparasitoidism are known. Most hyperparasitoid species oviposit into insects that already have the required primary parasitoid. A few oviposit speculatively into unparasitized individuals, their eggs developing only if the host is subsequently attacked by the proper primary parasitoid. Some species can complete development as either primary or secondary parasitoids. Most parasitoids develop alone in a host (solitary parasitoids), and many species chemically mark hosts in which they have oviposited. In cases where additional eggs are laid, only a single parasitoid larva survives to maturity. Other parasitoid species, especially in Ichneumonoidea and Chalcidoidea, live gregariously, many individuals in or on a single host. The gregarious condition may result from multiple oviposition or from polyembryonic development of a single egg. Typically eggs are deposited in or on hosts or very close to them. Some species of Chalcidoidea and Ichneumonoidea oviposit on vegetation. The motile first-instar larvae (planidia; Fig. 39.6a) search for a suitable host to which they attach.

Hymenopteran parasitoids also differ in whether they develop externally on the host (ectoparasitoids) or internally (endoparasitoids) and whether they tend to quickly paralyze and feed upon a moribund host (idiobionts) or develop slowly, with more intimate and prolonged nutritional, endocrine, developmental, and immunological interactions with the host (koinobionts). (See Chap. 10 for additional biological details.)

Most of the stinging wasps are predators that immobilize or kill one or more prey, which are installed in a specially constructed cell in which the larva matures. The Pompilidae (spider wasps), Bethylidae, Scoliidae, and Tiphiidae clearly bridge the difference between parasitoidism and typical predation. For example, some pompilids oviposit on a temporarily paralyzed host, which may regain partial motility before being consumed. Some species attack spiders in their burrows, where wasp larval development occurs. The basic behavioral sequence of hunting for prey, stinging, and oviposition is repeated by an individual wasp many times, just as in parasitoids. Other pompilids prepare their own burrows after subduing the prey.

The immobilized prey must be concealed during gallery preparation to prevent it form being stolen by other wasps or scavengers such as ants. All Pompilidae and many Sphecidae utilize prey that are too large to be carried in flight. In the more typical mode of predation, most highly developed in the superfamilies Vespoidea and Sphecoidea, prey size is reduced so that numerous prey items may be quickly flown to the larval gallery, which is prepared before provisioning begins. Many of these more specialized hunters temporarily close the gallery between provisioning trips and permanently close it after provisioning is complete, excluding parasitoids and scavengers. Many wasps have evolved specialized methods of rapid transport of prey, as by removing less nutritious body parts such as antennae and legs, or by holding the prey on the barbed stinger so that the legs are immediately available to open the gallery. Behavior in hunting wasps has undergone two other specializations. First, each burrow leads to several cells, each containing a developing larva. Each cell is individually closed after provisioning is complete, and the nest is closed after all cells have been provisioned. Second, many Sphecidae and Vespidae progressively provision their larvae, adding prey items as required. This contrasts with mass provisioning, in which the required amount of prey is provided before an egg is laid. The behavior of progressively provisioning sphecids and vespids, where a female tends many cells with larvae at different developmental stages, is probably the most complex of all nonsocial insects. Progressive provisioning has the advantage of allowing the mother to periodically inspect the cell for parasitoids or scavengers or to remove spoiled provisions and, through the increased contact between mother and offspring, was probably prerequisite to the development of sociality.

Bees are secondarily phytophagous. Like the morphologically similar Sphecidae, bees construct larval galleries, but they provision their larvae with pollen and nectar, rather than with paralyzed insects. The families Megachilidae and Apidae include some *cleptoparasitic* species, which appropriate the nests and provisions of phytophagous species. Bees are notable, along with the ants and vespid wasps, in having evolved complex societies. In all these taxa, sociality involves repression of reproduction in most female individuals, which function in obtaining food and maintaining and defending the colony. Small numbers of males are produced periodically, functioning only in reproduction. Sociality is discussed in greater detail in Chapter 7.

Due in part to similarities of sawfly larvae to some caterpillars and scorpionfly larvae, Hymenoptera were traditionally thought to be the sister group to superorder Mecopteroidea. Recent comparative genomic data suggest instead that Hymenoptera represent the earliest diverging extant Endopterygota, as some earlier fossil specialists had suggested. At the time of this writing, the complete genomes of the honey bee, two ant species, and three *Nasonia* (Chalcidoidea: Pteromalidae) had been sequenced, with others well under way.

Modern classifications of Hymenoptera have eliminated the use of the traditional suborders Symphyta (sawflies, horntails, and woodwasps) and Apocrita (the "waisted" Hymenoptera), although the terms symphytan and apocritan are still used as informal categories. While the Apocrita do indeed appear to be monophyletic, the Symphyta are now well known to comprise a paraphyletic grade of superfamilies; thus the superfamilies themselves are now the most commonly recognized groups beneath the order level. Phylogenetic relationships among larger lineages of Hymenoptera are still incompletely understood, but the superfamily and family classification used here is relatively standard. There is some recent evidence that the Vespoidea, as defined since the mid-1970s, and Siricoidea, as defined in the key below, may not be monophyletic.

KEY TO THE NORTH AMERICAN SUPERFAMILIES OF HYMENOPTERA

1. Abdomen with base broadly attached to thorax (Fig. 39.9); thoracic nota almost always with cenchri (Fig. 39.9b); hind wings almost always with 3 basal cells (Fig. 39.7b–d) (previously called a suborder,

Symphyta, which is now known not to be monophyletic)   **2**

Abdomen with base narrowly constricted at attachment to thorax (Fig. 39.11c,d); thoracic nota without cenchri; hind wings with 2 basal cells or fewer (Fig. 39.11e) (often called suborder Apocrita, a monophyletic group)   **7**

2(1). Antennae inserted beneath a transverse ridge below the eyes; wings with veins strong, thick in basal half, faint apically (Fig. 39.10)   **Orussoidea**

Antennae inserted between the eyes; veins equally developed throughout wing   **3**

3(2). Anterior tibia with 1 apical spur   **4**

Anterior tibia with 2 apical spurs   **5**

4(3). Pronotum much wider than long (Fig. 39.8a,b), often with posterior margin strongly incurved; abdomen cylindrical   **Siricoidea**

Pronotum longer than wide or subquadrate (Fig. 39.9c); abdomen slightly compressed in lateral plane   **Cephoidea**

5(3). Costal cell divided by intercostal vein, or if fused with subcostal vein, forewing with 3 marginal cells (Fig. 39.7d)   **6**

Costal cell without intercostal vein (Fig. 39.7c); forewing with 1 or 2 marginal cells   **Tenthredinoidea**

6 (5). Antenna with first flagellar segment longer than all other (usually few) flagellar segments combined (Fig. 39.7d)   **Xyeloidea**

Antenna with first flagellar segment much shorter than other flagellar segments combined (can still be the longest segment)   **Megalodontoidea**

7(1). Abdomen with basal 1 or 2 segments differentiated as a distinct node or nodes (Fig. 39.16b); minute to moderate-sized insects, usually with antennae elbowed and wings absent   **Vespoidea** (ants)

Abdomen with basal segments not modified as a node; antennae elbowed or straight; wings usually present   **8**

8(7). Hind leg with trochanter 1-segmented   **9**

Hind leg with trochanter 2-segmented   **17**

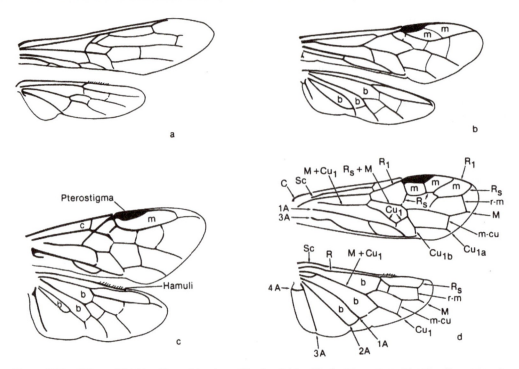

Figure 39.7 a, Wings of Siricidae (*Tremex*); b, wings of Tenthredinidae (*Tenthredo*); c, wings of Argidae (*Arge dulciana*); d, wings of Xyelidae (*Macroxylea ferruginea*). b, basal cell; c, costal cell; m, marginal cell; Sc, intercostal vein (subcosta).

9(8). Hind tibia with inner spur hooked, toothed, or with comb of hairs on inner surface (Fig. 39.19d) (or one or both spurs absent) **10**
Hind tibia with spur or spurs simple, without teeth or hairs (Fig.39.11b) **14**

10(9). Hind wings without closed cells **Chrysidoidea**
Hind wings with at least 1 closed cell **11**

11(10). Pronotum with dorsal lobes extending posteriorly to tegulae (small sclerites at wing bases) (Fig. 39.16a,c); ventral lobes short, not covering anterior thoracic spiracle **13**
Pronotum with dorsal lobes short, not reaching tegulae (Figs. 39.2, 39.17a); rounded ventral lobes covering anterior thoracic spiracles **12**

12(11). Body with some hairs branched; hind tarsi with basal segment wider than following segments (Fig. 39.18b) **Apoidea** (bees)

Body without branched hairs; hind tarsi without basal segments broadened (Fig.39.17a,b) **Apoidea** (sphecoid wasps)

13(11). Eyes deeply notched, or if round, then antennae clavate
Vespoidea (paper and potter wasps)
Eyes round or oval; antennae filiform
Vespoidea (spider wasps)

14(9). Pronotum large, extending posteriorly as far as tegulae (small sclerites at wing bases) (Figs. 39.14a, 39.15a) **15**
Pronotum small, not extending posteriorly as far as tegulae (Fig. 39.14b) **Chrysidoidea**

15(14). Hind wings without closed cells; venation almost always much reduced (Fig. 39.11 c,d) **16**
Hind wings with at least 1 closed basal cell; forewings usually with extensive venation (Fig. 39.15c) **Vespoidea** ("scolioid wasps" in older classifications)

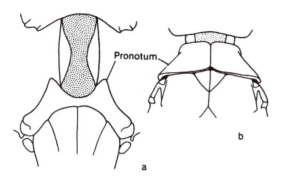

Figure 39.8 Taxonomic characters of Hymenoptera, Symphyta: a, b, dorsal aspects of pronota of Xiphydriidae and Anaxyelidae, respectively.

16(15). Abdomen attached high on propodeum, remote from hind coxae (Fig. 39.11d)
                                                                    Evanioidea
Abdomen attached near base of propodeum, contiguous with or very close to hind coxae (Fig. 39.12a)   **Proctotrupoidea** (some)

17(8). Head with deep groove below base of antennae and circle of teeth on vertex (Fig. 39.11a)   **Stephanoidea**
Head without groove below antennal base or circle of teeth on vertex   **18**

18(17). Costal vein present between wing base and pterostigma (Fig. 39.11d), forewing typically with pterostigma   **20**
Costal vein absent in proximal portion of wing (Fig. 39.12c); forewing typically without pterostigma   **19**

19(18). Antennae elbowed, with first segment as long as succeeding 2 to 3 segments; abdomen cylindrical, subcylindrical or dorsoventrally flattened   **20**
Antennae filiform, with first segment subequal to second, or with second segment small and first and third segments subequal; abdomen laterally compressed
                                                                    Cynipoidea

20(18,19). Pronotum extending posteriorly as far as tegulae (Fig. 39.12a)   **21**
Pronotum short, not approaching tegulae (Fig. 39.13a–d)   **Chalcidoidea**

21(20). Abdomen attached high on thorax, remote from coxae (Fig. 39.11d)   **Evanioidea**

Abdomen attached low on thorax, close to coxae (Fig. 39.11a)   **22**

22(21). Hind wings without closed cells (Fig. 39.12a)   **23**
Hind wings with at least 1 closed cell   **25**

23(22). Fore tibiae with 2 apical spurs
                                                                    Ceraphronoidea
Fore tibiae with 1 apical spur   **24**

24 (23) Antennae inserted on a shelf-like prominence between the eyes, remote from clypeus: midtibiae with 2 spurs
                                      Proctotrupoidea (Diapriidae)
Antennae not inserted on a shelflike prominence, typically arising low on face close to clypeus   **Platygastroidea**

25(22). Forewing with costal cell reduced or absent (Fig. 39.11b,e,f); posterior lobe of pronotum not edged by hairs   **Ichneumonoidea**
Forewing with distinct costal cell; posterior lobe of pronotum edged by short, close-set hairs   **Trigonalyoidea**

The "Symphytan" Superfamilies Abdomen with base broadly attached to thorax, but with first abdominal tergite distinct, not incorporated as part of thorax; metanotum bearing raised, roughened, knoblike cenchri (Fig. 39.9b) (absent in Cephidae); hind wings usually with 3 basal cells (Fig. 39.7c,d) (except in Orussoidea, Fig. 39.10). Larvae eruciform, caterpillarlike, with large, round head capsule and 6 to 8 pairs of abdominal legs (external feeders, Fig. 39.5a), or abdominal legs absent (some internal feeders).

The paraphyletic Symphyta include the most generalized Hymenoptera, in terms of both many morphological characteristics and biological adaptations. Larvae are herbivorous, with the exception of the Orussoidea. In number of species, only the superfamily Tenthredinoidea, whose members are largely adapted as external foliage feeders, is dominant today.

Superfamily Xyeloidea. Small to moderate sawflies with the wing venation very generalized (Fig. 39.7d) and the ovipositor short or rudimentary.

The earliest known fossil hymenopteran, *Liadoxyela praecox,* from the Lower Triassic

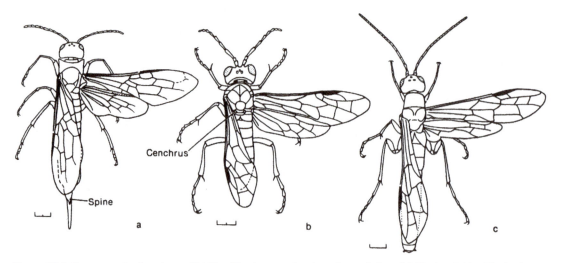

Figure 39.9 Representative Symphyta: a, Siricidae (*Sirex juvencus*, female; scale equals 2 mm); b, Tenthredinidae (*Tenthredo*; scale equals 2 mm); c, Cephidae (*Cephus clavatus*; scale equals 1 mm).

Period, is very similar to existing xyelids, which have the most generalized wing venation in the order. The superfamily contains one family in North America. Xyelidae are distinguished by the elongate third antennal segment, longer than the remaining segments combined. The larvae, which have prolegs on all abdominal segments, feed on staminate (male) cones of pines (*Xyela*), needle buds of firs (*Macroxyela*), or foliage of other trees such as walnut, pecan, and elm. Mature larvae drop from the trees to pupate in earthen cells, where they remain until the following year. Adults appear very early in the spring and may visit early flowers, especially those with massed blooms, such as chokecherry.

*Superfamily **Megalodontoidea**.* Pamphiliidae have at least 13 antennal segments of approximately equal length. Larvae, which lack abdominal legs and have the thoracic legs greatly reduced, move by a wriggling, peristaltic action. They tie or roll leaves of various trees (both Angiosperms and Gymnosperms), some species feeding gregariously beneath self-constructed webs, eventually forming large, unsightly tangles of silk, partly consumed vegetation, frass, and cast skins. Pamphiliid larvae depend entirely on the silk scaffolding they produce for locomotion and must extend the webbing in order to reach new foliage. Pupation is in cells in the soil. A second family, Megalodontidae, is restricted to temperate regions of Eurasia.

*Superfamily **Tenthredinoidea**.* Small to moderate, mostly stout-bodied sawflies with the ovipositor blades with sawtooth serrations for slitting vegetation.

Tenthredinoidea is the dominant superfamily of symphytan Hymenoptera. In north temperate regions the family Tenthredinidae (Fig. 39.9b), with over 4000 described species, many abundant and widespread, includes nearly all commonly collected sawflies. Sawfly larvae are characteristically external foliage feeders. Some consume herbs or occasionally ferns, but trees and shrubs are especially favored. Externally feeding larvae are usually caterpillarlike in general appearance, with six to eight pairs of abdominal legs, which lack crochets (Fig. 39.5a). Sawfly larvae are often camouflaged in shades of green, sometimes with contrasting tubercles or spines in red, white, or other colors. Many species evert from the abdominal sternites glandular invaginations that resemble the osmeteria of Lepidoptera. Some tenthredinid and cimbicid larvae are covered with a whitish powder, and some tenthredinids are sluglike and covered with slimy secretions. Internally feeding larvae, which include leaf miners (Argidae, Tenthredinidae) and gall formers (Tenthredinidae), may lack abdominal legs. Sawflies typically have a single generation per year. Eggs are deposited in slits made in leaves or needles, and mature larvae usually leave the host plant to pupate in a cocoon or a cell in soil or leaf litter. Adults are commonly found on flowers and may imbibe nectar or eat pollen, but many Tenthredinidae are predatory.

The most important families are Tenthredinidae and Diprionidae, each with a number of economically significant species. Among the more familiar tenthredinid pests are *Caliroa cerasi*, the pear and cherry slug, *Endelymyia aethiops*, the rose slug, and *Nematus ribesii*, which consumes currant leaves. Many other genera may defoliate or cause cosmetic damage on a wide variety of trees and shrubs. Tenthredinidae feed on a great variety of other plants, including ferns, sedges, grasses, and horsetails (*Equisetum*), but seldom cause serious damage. More important, especially in North America, are forest pests of conifers, which include *Pristiphora erichsonii* (Tenthredinidae) on larch and various species of *Diprion* and *Neodiprion* (Diprionidae) on spruce and pine. Larvae of Diprionidae feed exclusively on conifers (Pinaceae and Cupressaceae), almost always as external foliage feeders. They are usually gregarious, at least in early instars, and often aposematically colored. When disturbed they synchronously rear up with a jerky motion and may regurgitate resinous substances.

Argidae is the second largest family of Symphyta, with over 800 species, the great majority tropical. About 60 species occur in North America, the larvae feeding on a variety of shrubs, trees, and less frequently on herbs. The 12 species of North American Cimbicidae feed on foliage of trees, especially of the families Rosaceae, Salicaceae (willows), and Betulaceae (birches). The family Pergidae is largely restricted to South America and Australia, where the other families of Tenthredinoidea are relatively uncommon. The few North American species are leaf skeletonizers on oak, walnut, and a few other hardwoods.

KEY TO THE NORTH AMERICAN FAMILIES OF TENTHREDINOIDEA

1. Antennae with 6 or more segments   **2**
   Antennae with 3 segments; apical segment elongate, sometimes bifurcate   **Argidae**
2(1). Antennae with apical club; abdomen with separate pleural sclerites   **Cimbicidae**
   Antennae filiform or pectinate; abdomen without pleural sclerites   **3**
3(2). Antennae with 6 segments   **Pergidae**
   Antennae with 7 or more segments   **4**
4(3). Antennae with 7 to 10 segments
       Tenthredinidae
   Antennae with 13 or more segments
       Diprionidae

Superfamily **Cephoidea** *(stem sawflies)*. Very slender, elongate insects with the abdomen slightly compressed in the lateral plane.

The single family Cephidae includes delicate wasps, either black or patterned in black and yellow (Fig. 39.9c). Adults are encountered on vegetation or flowers. Larvae of most species bore the stems of grasses, including small grains, and *Cephus cinctus* is a significant pest on wheat. Rose, blackberry, gooseberry, and other shrubs with pithy stems are attacked by a few species. Pupation occurs within the larval gallery. The lack of cenchri and the relatively great constriction of the abdomen of adults was once suggested to indicate that the cephoid line of evolution may have given rise to the Apocrita. While this relationship is also reinforced by the legless, internally feeding larvae, it is now contradicted by much morphological and molecular data.

Superfamily **Siricoidea** *(woodwasps, horntails)*. Moderate or large wasps with robust, cylindrical body; anterior tibia with single apical spur.

The Siricoidea are associated with wood, where the short-legged larvae excavate tunnels. Females possess a cylindrical ovipositor, which is used to bore oviposition holes into living or felled trees. The most familiar insects in this superfamily are the Siricidae (Fig. 39.9a), whose larvae bore into the heartwood of various trees. The larvae actually feed on symbiotic fungi, which are transmitted by the mother at the time of egg laying, and do not digest wood. Siricids normally attack damaged or dead trees but occasionally cause minor damage. Pupation in Siricoidea is inside the larval gallery. Anaxyelidae includes a single species, *Syntexis libocedrii*, whose larvae tunnel incense cedar (*Calocedrus*), western juniper, and western red cedar (*Thuja*) in the Pacific Coast states. Females oviposit in trunks still warm and smoking from forest fires. Larvae of Xiphydriidae bore in angiosperm trees, including birch, maple, oak, elm, and poplar. A symbiotic fungus, on which they presumably feed, is associated with their galleries. The wasps are fairly common in the Eastern states. There is some morphological and molecular evidence that Anaxyelidae and Xiphydriidae may be more closely related to Orussidae and Apocrita than Siricidae and that the Siricoidea should be split accordingly.

KEY TO THE FAMILIES OF SIRICOIDEA

1. Abdomen with last segment modified as short, upturned prong or spine (Fig. 39.9a); mesonotum bearing 2 diagonal grooves converging toward scutellum   **Siricidae**
   Abdomen with last segment not modified as a prong or spine; mesonotum without grooves   **2**
2(1). Pronotum with posterior margin strongly incurved (Fig. 39.8a)

     Xiphydriidae
   Pronotum with posterior margin nearly straight (Fig. 39.8b)   **Anaxyelidae**

*Superfamily **Orussoidea** (Fig. 39.10).* A single family, Orussidae, of uncommon wasps whose biology and evolutionary relationships appear to be linked with the Apocrita.

Adults are similar to other Symphyta in having a broad thoracic-abdominal articulation. Unlike Symphyta, orussids have the wing venation

reduced, and the very long ovipositor is coiled like a watchspring inside the abdomen. The insertion of the antennae beneath a ridge below the eyes is unique among Symphyta. The apodous larvae occur in galleries of buprestid beetles, siricid woodwasps, or other wood-boring insects and appear to be parasitoids upon these associates.

The "Apocritan" Hymenoptera Appear to comprise a monophyletic lineage, in contrast to the above "Symphytan" lineages. Represent numerically the vast majority of the Hymenoptera. Share abdomen with first segment incorporated into thorax as propodeum, second segment narrowly constricted as petiole, bearing the enlarged gaster (Fig. 39.11c,d); thorax without cenchri; hind wings with no more than two basal cells. Larvae lacking legs, usually with head capsule unsclerotized; maxillae and antennae reduced to papillae (Fig. 39.5b).

Apocrita comprise all the diversified groups of dominant Hymenoptera, with the exception of the symphytan superfamily Tenthredinoidea. Parasitoidism, a specialized form of predation, is the ancestral mode of larval feeding. In some derived members of the superfamilies Vespoidea and Apoidea, typical predation is the dominant mode of feeding. Several groups of Apocrita, most notably the bees (superfamily Apoidea) and gall wasps (Cynipoidea), have reverted to phytophagy,

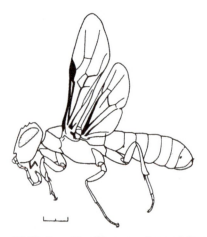

Figure 39.10 Orussidae (*Orussus*; scale equals 2 mm).

but of relatively specialized types. Those families of Apocrita that have the ovipositor developed as a stinger collectively constitute the Aculeata, a monophyletic group including the bees, ants, and most of the familiar wasps.

*Superfamily **Stephanoidea.*** Small to moderate parasitoids with trochanters 2-segmented; forewing with 3 to 6 closed cells, pterostigma and costal cell present, hind wing without closed cells; ovipositor long, external.

The single family, Stephanidae, is represented by six uncommon species in North America. Superficially stephanids (Fig. 39.11a) resemble certain ichneumonids, but they are distinguished by the crown of teeth on the head, by the swollen hind femora, and by the brush of preening hairs on the hind tibial apex. The subantennal grooves, the crown of teeth on the cranium, and the reduced wing venation suggest relationships to the Orussidae. Stephanids are usually encountered on trunks or large branches of dead, frequently charred trees, searching for their hosts, which are the larvae of various wood-boring beetles and siricids. The long ovipositor is used to parasitize the victims in their galleries. Pupation is within the host gallery without a cocoon.

*Superfamily **Trigonalyoidea.*** Stout-bodied, moderate-sized, wasplike parasitoids with long antennae with at least 14 segments; forewing with at least 6 closed cells, costal cell, and pterostigma present; hind wing with closed basal cells; trochanters 2-segmented; ovipositor short, internal.

The single family, Trigonalyidae, includes rare species that deposit the eggs in slits cut in developing foliage by the sharp, pointed apical sternite. Eggs hatch when consumed by caterpillars of Lepidoptera or sawflies, and the latter are hosts for a few species. Most species are hyperparasites. They hatch in the guts of caterpillars and enter the hemocoels but do not develop further unless the caterpillar is secondarily parasitized by certain ichneumonid wasps or tachinid flies which become the trigonalyid's host. Some species parasitize larvae of vespoid wasps, being consumed with caterpillars provided as food by the adult wasps.

The first three instars feed internally, the last two externally.

In body form and wing venation trigonalyids are similar to aculeate wasps, especially in Apoidea, but they share the long, multiarticulate antennae, 2-segmented trochanters and unmodified hind tibiae with the parasitoid families. Because of this combination of characteristics, trigonalyids have often been considered an extremely primitive family, but a relatively basal phylogenetic position has not been confirmed.

*Superfamily **Ichneumonoidea.*** Minute to large parasitoids with 2-segmented trochanters, antennae with at least 13 segments (usually more than 16); forewing with at least 3 closed cells, costal cell absent, pterostigma present or absent; hind wing with 2 closed cells, or occasionally without closed cells. Ovipositor short, internal or long, external.

Ichneumonoidea constitute one of the large diversified superfamilies of parasitoid Hymenoptera, the other being the Chalcidoidea. The family Ichneumonidae has often been stated to contain conservatively 60,000 species, but it is probably more like 100,000 to 200,000, mostly undescribed, a number approached perhaps only in the largest families of Coleoptera. Most Ichneumonoidea have the abdominal cuticle soft and flexible, at least ventrally, unlike most other Apocrita. The flexible abdomen apparently assists in oviposition. The larvae may develop either internally (endoparasitoids) or externally (ectoparasitoids), but pupation is nearly always outside the host in a silken cocoon. Most species are primary parasitoids, but many hyperparasitoids are recorded, especially in the Ichneumonidae. As with many parasitoids, ichneumonoids are often highly host-specific, except in some ectoparasitoids.

The superfamily contains two extant families. Ichneumonidae include mostly moderate-sized to large species that have two recurrent veins in the forewings (Fig 39.11b,e). They are parasitoids of immature endopterygote insects, especially Lepidoptera and symphytan Hymenoptera, and also attack spiders, spider egg sacs, and pseudoscorpions. The eggs may be laid externally on

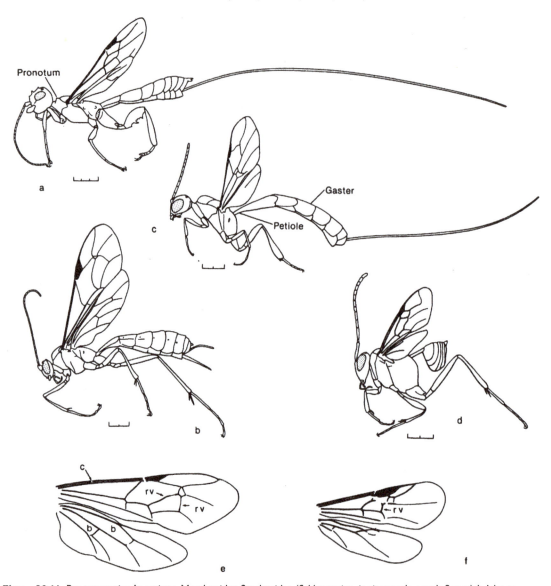

Figure 39.11 Representative Apocrita: a, Megalyroidea, Stephanidae (*Schlettererius cinctipes*; scale equals 3 mm); b, Ichneumonoidea, Ichneumonidae (*Netelia leo*; scale equals 2 mm); c, Evanioidea, Gasteruptiidae (*Rhydinofoenus occidentalis*; scale equals 2 mm); d, Evanioidea, Evaniidae (*Evania appendigaster*; scale equals 2 mm); e, wings of Ichneumonidae (*Catadelphus atrax*); f, wings of Braconidae (*Cremnops haematodes*). rv, recurrent veins; c, costal cell; b, basal cell.

the host or placed inside the host, which may be paralyzed by injected venom. A few species leave the eggs apart from the larval host, which is then located by the motile first instar larva. Larvae are mostly solitary parasitoids, but some are gregarious. Ichneumonidae are especially abundant in moist situations with rank vegetation, where caterpillars are abundant. However, they occur in a wide spectrum of habitats, including arid situations. Some diurnal species are strikingly similar in color or pattern to stinging wasps, though very few ichneumonids are able to sting humans. *Rhyssa* and related genera are remarkable for the extraordinarily long ovipositor (Fig. 39.4a), which is used to reach wood-boring Coleoptera or Hymenoptera deep in their galleries. Ovipositors of some species may be up to 15 cm long, about three times the body length.

Braconidae, with a single recurrent vein in the forewing (Fig. 39.11f), are generally smaller than Ichneumonidae but may be even more species-rich. Braconids are parasitoids of immature and occasionally adult or pupal endopterygote insects, as well as various Homoptera. The great majority are primary parasitoids, but a few hyperparasitize other ichneumonids and tachinid flies, and a very few are secondarily phytophagous. Some braconids attack hosts only slightly larger than themselves. For example, *Praon* and *Aphidius* utilize aphids. Other species may parasitize hosts hundreds of times their own bulk, such as the 2- to 3-mm-long *Microctonus*, which oviposit in adult tenebrionid beetles as large as 50 mm long. *Aphidius* and a few other genera pupate in a thin cocoon within the mummified remains of their host, but most spin a cocoon externally, often attached to the host's body. *Praon* spins a cocoon beneath the husks of consumed aphids, while *Cotesia* cocoons attached to caterpillars are a familiar sight in eastern North America.

*Superfamily **Evanioidea**.* Small to moderate-sized parasitoids with the abdomen attached high on the propodeum, remote from the hind coxae (Fig. 39.11c,d); hind trochanters 1- or 2-segmented; forewing with pterostigma, costal cell, and several other closed cells; hind wing with only narrow costal cell closed; ovipositor short, internal or long, external.

The insects placed in this superfamily are widespread but seldom very common, except in tropical forests. All are parasitoids with restricted host ranges. For example, Evaniidae (Fig. 39.11d) oviposit in cockroach oothecae. The larva consumes the cockroach eggs, then pupates within the ootheca without a cocoon. Aulacidae are internal parasitoids of xiphydriid wood wasps or of Cerambycidae or Buprestidae (Coleoptera). The larvae do not complete their development for about a year, when

the host has reached the last instar. Pupation is in the host gallery. Gasteruptiidae (Fig. 39.11c) oviposit in the cells of solitary bees or of sphecid or vespid wasps. Their larvae consume the host egg or young larva, but the host's provisions form the bulk of their food. Pupation is in a cocoon in the host cell.

KEY TO THE FAMILIES OF EVANIOIDEA

1. Abdomen with gaster subquadrate, laterally compressed, attached to a slender petiole (Fig. 39.11d) **Evaniidae**
Abdomen with gaster elongate, gradually enlarged, round in cross section, petiole not distinct from gaster (Fig. 39.11c) **2**
2(1). Abdomen clavate; forewing with 2 recurrent veins (as in Fig. 39.11e) **Aulacidae**
Abdomen cylindrical, slender to apex (Fig. 39.11c); forewing with 0 or 1 recurrent veins (as in Fig. 39.11f) **Gasteruptiidae**

*Superfamily **Proctotrupoidea**.* Small to moderate-sized parasitoids of extremely variable body configuration; trochanters 1- or 2-segmented; forewings usually without closed cells or with narrow costal cell, or with 2 to 5 closed cells in a few primitive species, pterostigma present or absent; hind wing almost always without closed cells.

The Proctotrupoidea constitute a large, diverse assemblage of parasitoids whose biologies and relationships are often poorly known. They are also one of the more ancient extant lineages of Hymenoptera with two families known from Jurassic fossils. Heloridae resemble braconids in the general body shape and wing venation. Similar venation is found in Vanhorniidae and Roproniidae, but in the other families venation is reduced, frequently to a single vein. Body size is extremely variable, reaching 60 mm in Pelecinidae, but the great majority of proctotrupoids are small or minute, and many can superficially resemble Chalcidoidea, except in the large prothorax. The Diapriidae and Proctotrupidae are diverse families of several hundred to several thousand species worldwide, while the Heloridae, Pelecinidae, Roproniidae, and Vanhorniidae are rare, relictual taxa of one to a few species each.

Because of their small size and specialized habits, most Proctotrupoidea are unfamiliar to many biologists, but numerous individuals of many species may frequently be found by using flight intercept traps or soil extraction methods (see Chapter 45).

Proctotrupoidea are internal parasitoids of a variety of insects, including larval Chrysopidae (Heloridae), coleopterous larvae (Heloridae, Proctotrupidae, Vanhorniidae, Pelecinidae), sawflies (Roproniidae), and dipterous larvae (many Diapriidae). Some adult Proctotrupoidea (especially Diapriidae) are short-winged or wingless, antlike inhabitants of leaf litter. Winged species are commonly encountered on foliage. More than 2500 world species have been recognized, and it is certain that many remain undescribed.

KAY TO THE NORTH AMERICAN FAMILIES OF PROCTOTRUPOIDEA

1. Antennae elbowed, with first segment usually longer than succeeding 3 to 5 segments **Diapriidae**

   Antennae not elbowed, first segment about as long as third **2**

2(1). Body length less than 20 mm; hind leg with basal tarsomere longer than succeeding segments **3**

   Body length 20 to 70 mm; hind leg with basal tarsomere shorter than succeeding segments **Pelecinidae**

3(2). Antennae with 14 or 16 segments **4**

   Antennae with 13 segments **5**

4(3). Mandibles with tips pointing out, not meeting when closed **Vanhorniidae**

   Mandibles with tips pointing inward, meeting or overlapping when closed **Proctotrupidae**

5(3). Antennae with 14 segments; abdomen strongly flattened in lateral plane **Roproniidae**

   Antennae with 13 segments; abdomen slightly wider than high **Heloridae**

Superfamily Platygastroidea. This superfamily was only relatively recently separated from Proctotrupoidea, but in many analyses it shows a closer relationship with the Chalcidoidea and/or Cynipoidea. There are two families. Scelionidae are parasitoids of the eggs of numerous orders of insects and of spiders. Females of the genus *Mantibaria* have the unusual habit of riding about on adult mantids, then ovipositing in freshly laid egg masses before the oothecal covering has hardened. Some Platygastridae parasitize the eggs of Coleoptera, Homoptera, and Diptera (especially Cecidomyiidae) and probably other insects, while others are parasitoids of immatures of Coccoidea and Aleyrodoidea (Homoptera). Recent analyses suggest that Scelionidae and Platygastridae may eventually be combined into a single family.

KEY TO THE NORTH AMERICAN FAMILIES OF PLATYGASTROIDEA

1. Antennae with 10 or fewer segments, forewing venation consisting of short submarginal vein **Platygastridae**

   Antennae with 11 to 12 segments (10 segments rarely present); forewing venation consisting of marginal vein and short vein extending beyond pterostigma **Scelionidae**

Superfamily Ceraphronoidea. As with Platygastroidea, these tiny wasps were previously lumped in with Proctoctrupoidea but now appear to be rather distantly related, based on both comparative morphology and molecular data. They are unique among apocritan wasps in possessing two spurs on the fore tibia. There are two relatively common but inconspicuous families. Ceraphronidae and Megaspilidae are internal parasitoids of a variety of insects (Coccoidea, Aphidoidea, Neuroptera, Thysanoptera, Mecoptera, and Diptera). In addition, many species are hyperparasitoids of other Hymenoptera.

KEY TO THE NORTH AMERICAN FAMILIES OF CERAPHRONOIDEA

1. Forewing typically with large pterostigma; mesothoracic tibia with 2 spurs; male and female antennae with 9 flagellomeres **Megaspilidae**

Forewing with linear pterostigma; meso-thoracic tibia with 1 spur; female antenna with 8 flagellomeres; male with 9

Ceraphronidae

*Superfamily **Cynipoidea**.* Small to moderate-sized insects with the abdomen usually compressed in the lateral plane and the tergites extending ventrally to conceal the sternites (Fig. 39.12b); hind trochanters 1- or 2-segmented; forewing with marginal vein and pterostigma lacking, usually without closed cells; hind wing almost always with a single vein; ovipositor almost always short, internal.

The great majority of these wasps may be recognized by the reduced forewing venation with a characteristic marginal cell in the distal third (Fig. 39.12c). They differ from Chalcidoidea and most Proctotrupoidea in having filiform rather than elbowed antennae.

Ibaliidae, comprising an early-diverging lineage of Cynipoidea, are endoparasitoids of siricid and anaxyelid wood wasps. The larva initially feeds internally, then externally in later instars. Figitidae are internal parasitoids of dipterous larvae and puparia and of hemerobiid lacewings. Eucoilidae, the largest family of Cynipoidea, are internal parasitoids of calyptrate Diptera, the adults emerging from the puparia. The family is mainly tropical, but adults may often be found around fly-infested dung pats. Liopteridae are uncommon wasps whose biologies are almost unknown. A relatively small number of Cynipidae are parasitoids of Diptera or hyperparasitoids of Hymenoptera (Braconidae, Aphelinidae) attacking aphids. The large subfamily Cynipinae, comprising about 600 North American species, all inhabit galls, either as the primary causative agent or as inquilines. Family classification of Cynipoidea has been in some flux over the years, and the classification used in the key below is relatively conservative.

Galls vary enormously in size, shape, color, and texture, as well as in internal structure, and are frequently diagnostic of the species of wasp that formed them. Typically an external epidermis, which may be extremely hard or tough, encloses a thick layer of spongy parenchyma tissue. At the center of the gall is usually a hard shell lined with nutritive tissue, within which the larva feeds. A single gall may contain one to many larval cells and may occur on any part of the host plant. About 75 to 85 percent of all cynipid species attack oaks (*Quercus*) and closely related trees and shrubs; another 7 percent attack the genus *Rosa* (roses); the remainder utilize a variety of other plants, especially Compositae (sunflower family). Inquilines oviposit into galls formed by other species, frequently causing modifications of the original structure. The original occupant is either killed outright or starved by appropriation of food. Cynipoidea pupate inside the galls or hosts, usually without a cocoon.

In most Cynipidae, sexual generations, occurring in the warm part of the year, alternate with asexual (parthenogenetic) generations, which appear during winter. The individuals representing the sexual and asexual generations are often strikingly different in appearance and usually produce dissimilar galls on different parts of the host. Consequently, alternate generations have often been described as different species, which may be associated only as the biological relationships are discovered.

KEY TO THE FAMILIES OF CYNIPOIDEA

1. Abdomen with fourth, fifth, or sixth segment of metasoma larger than other segments; body length, 5 to 15 mm   **2**
   Abdomen with second or third segment of metasoma larger than other segments; body length seldom greater than 6 mm   **3**
2(1). Metasoma with segments strongly compressed in lateral plane, with sharp dorsal edge   **Ibaliidae**
   Metasoma not strongly compressed; dorsal edge rounded   **Liopteridae**
3(1). Scutellum raised into round or teardrop-shaped plate; head and mesoscutum smooth   **Eucoilidae**
   Scutellum without raised plate; head and mesonotum usually roughly textured   **4**
4(3). First segment of metasoma shorter than second   **Figitidae**

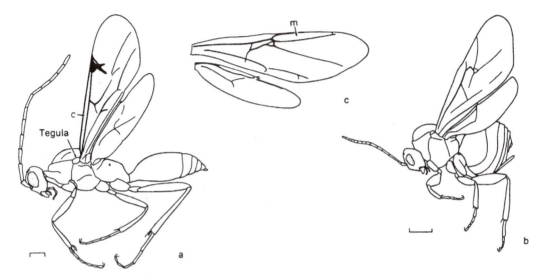

Figure 39.12 Proctotrupoidea and Cynipoidea: a, Proctotrupoidea, Proctotrupidae (*Proctotrupes*; scale equals 0.5 mm); b, Cynipoidea, Cynipidae (*Andricus californicus*; scale equals 1 mm); c, wings of Ibaliidae (Cynipoidea). c, costal cell; m, marginal cell (radial cell).

First segment of metasoma much longer than second   **5**

5(4). First segment of metasoma at least half as long as entire abdomen excluding petiole **Cynipidae**

First segment of metasoma less than half as long as entire abdomen **Figitidae**

Superfamily ***Chalcidoidea.*** Minute to small or occasionally moderate-sized wasps with the antennae elbowed (exception, Eucharitidae), trochanters 1- or 2-segmented; wings with venation reduced to a single vein, without closed cells; pronotum with posterior lobe usually rounded, not contiguous with tegulae; ovipositor short, internal or exserted, elongate.

The Chalcidoidea, which possibly comprise the largest superfamily of Hymenoptera, are remarkable for their great variety of body configurations (Fig. 39.13a–d), which are diagnostic for some families (e.g., Leucospidae, Chalcididae, and Agaonidae) and may verge on the fantastic or capricious. In most species the cuticle is strongly sclerotized and coarsely sculptured; many are iridescent blue or green. Chalcidoidea range to about 30 mm in length, but the great majority are less than 3 mm and many less than 1 mm long. Nearly all Chalcidoidea may be recognized by the combination of reduced wing venation (Fig. 39.13e) and short pronotum (Fig. 39.13a).

Some Eurytomidae, Pteromalidae, Torymidae, Perilampidae, and Eulophidae are phytophagous, either tunneling in seeds or forming galls and a few species, such as *Bruchophagus platyptera*, which infests clover seeds, are of minor economic importance. Members of the family Agaonidae engage in an obligatory symbiotic relationship with species of figs (*Ficus*), whose flowers they both gall and pollinate. Certain commercial varieties of figs, such as Smyrna, are pollinated only by the wasps, which have been transported about the world for this purpose. The great majority of Chalcidoidea, however, are internal or external parasitoids of insects of all orders, but especially Homoptera and immature stages of Endopterygota. Coleoptera, Diptera, and other orders are included in the host range, but Lepidoptera are especially favored among the Endopterygota. Eggs, larvae, and pupae are utilized by various species, and some families attack only a single developmental stage (e.g., Trichogrammatidae, Mymaridae on eggs).

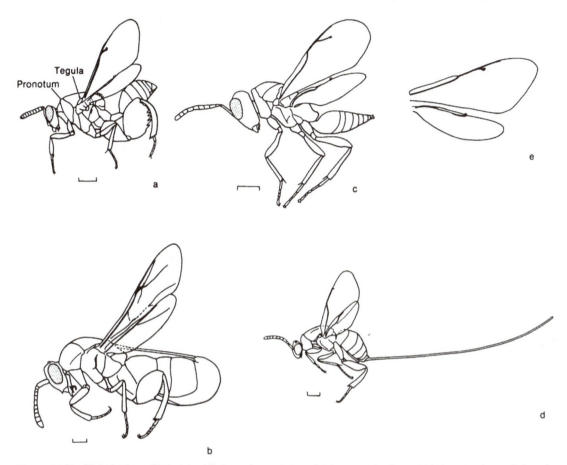

Figure 39.13 Chalcidoidea: a, Chalcididae (*Chalcis*; scale equals 1 mm); b, Leucospidae (*Leucospis birkmani*; scale equals 1 mm); c, Pteromalidae (*Pteromalus vanessa*; scale equals 0.5 mm); d, Torymidae (*Torymus californicus*; scale equals 1 mm); e, wings of Chalcididae (*Chalcis divisa*).

Numerous species are hyperparasitoids of primary parasitoids, and care must be exercised in deducing host relationships.

Many chalcidoids are hypermetamorphic, with the first one or two larval stages specialized for host location or elimination of competitors within the host (Fig. 39.6). Later instars are always maggotlike. Parthenogenesis occurs in various families, and polyembryony has been described in many species. Both features are common components of the life cycles of hymenopterous parasitoids. Pupation occurs within the host remains or externally in soil or litter or on plants. Very few chalcidoids spin a cocoon.

Different authorities recognize from 9 to 22 families of Chalcidoidea. A few are easy to identify, but most families are difficult to separate with existing keys. An additional difficulty is the small or minute size of most species, necessitating excellent optical equipment to see the taxonomic characters. Interested students are referred to Grissell and Schauff (1990), Gauld and Bolton (1988), Goulet and Huber (1993), and Gibson et al. (1997) for identification.

Superfamily **Chrysidoidea.** Small to moderate-sized (1–10 mm) wasps with hind trochanters 1-segmented; forewings with at least 2 to 4 closed cells (veins may be faint), including costal cell (Fig. 39.14c); hind wings without closed cells; ovipositor internal.

Bethylidae are common inhabitants of leaf litter, tree trunks, and fungi, where they are parasitoids of larvae of Lepidoptera and Coleoptera. While Chrysidoidea do not prepare special nests for larval development, some species of Bethylidae move the immobilized prey to a sheltered spot before oviposition. Unlike the fossorial wasps (Scolioidea, Pompiloidea, Sphecoidea), they deposit more than one egg on a single host. Except for their reduced wing venation, Bethylidae are morphologically similar to Tiphiidae and in general are intermediate between typical parasitoids such as the Proctotrupoidea and the primitive stinging wasps.

Dryinidae are internal parasitoids of Homoptera; females are apterous with the protarsus modified as a pincer for clasping the host. The larvae develop internally, but later instars form a characteristic sac bulging from the host abdomen. The relatively rare Sclerogibbidae are external parasitoids of Embioptera, and the few rearing records of Embolemidae suggest that they attack nymphs of fulgoroid hemipterans. Chrysidoidea include many wingless species and others in which only the females are wingless. Sexual dimorphism in other characteristics may be extreme, and sexes may be difficult to associate.

Chrysididae (Fig. 39.14b) are readily recognized by the metallic green and blue colors and the concave ventral surface of the abdomen. Chrysidids roll the heavily sclerotized body into a ball when molested, the head and thoracic venter being concealed in the abdominal concavity. Chrysididae typically attack larvae of bees and wasps in their cells, pupating within a cocoon in the host cell. The most familiar species attack mud daubers (*Sceliphron*, Sphecidae) and vespids that reuse mud dauber nests. Adult chrysidids are most commonly encountered on the surface of

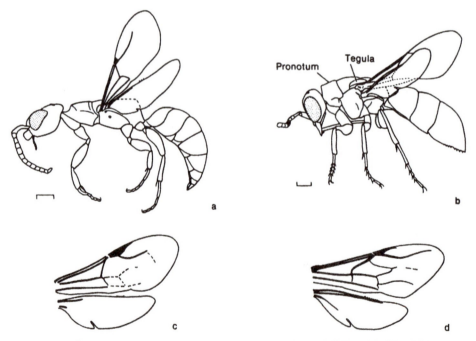

Figure 39.14 Chrysidoidea: a, Bethylidae (*Anisepyris williamsi*; scale equals 0.5 mm); b, Chrysididae (*Parnopes edwardsi*; scale equals 1 mm); c, wings of Bethylidae (*Pristocera armi*); d, wings of Chrysididae (*Chrysis*).

soil or tree trunks or occasionally on flowers. One subfamily contains wasps that attack the eggs of Phasmatodea.

KEY TO COMMON NORTH AMERICAN FAMILIES OF CHRYSIDOIDEA

1. Antennae with 11 to 13 segments; fore tarsi unmodified   **2**
    Antennae with 10 segments; females with fore tarsi claw-shaped   **Dryinidae**
2(1). Abdomen with 7 to 8 segments in metasoma; pronotum extending posteriorly to tegulae (Fig. 39.14a)   **Bethylidae**
    Abdomen with 3 to 6 visible segments in metasoma; pronotum not reaching tegulae   **Chrysididae**

Superfamily **Vespoidea.** This is a large assemblage of previous wasp superfamilies and ants brought together into a single superfamily by the classic study of Brothers (1975). More recent comprehensive molecular phylogenetic studies suggest that Vespoidea, in this enlarged sense, is not monophyletic, but a new classification has not yet stabilized. Below, we treat several previously recognized superfamilies as informal subgroups of Vespoidea: the "scolioid" wasps, the "pompiloid" (spider) wasps and relatives, and the "formicoid" ants.

The "scolioid" wasps are all parasitoids, biologically similar to Bethylidae. Morphologically they are diverse, variably showing similarities to Chrysidoidea, ants, the spider wasps and relatives, and paper wasps. We have not provided a detailed key to families below; due to strong sexual dimorphism, and repeated shifts in the family definitions, separating the families is a complex task. The reader is referred to the excellent treatment by Brothers in Goulet and Huber (1993).

Most scolioid wasps are highly fossorial, with strong forelegs and large prothorax extending posteriorly to the region of the wing bases (Fig. 39.15a). Sexual dimorphism is frequently marked, especially by winglessness in females, which may be very difficult to associate with the winged males. In contrast to many other Hymenoptera, the females are often smaller than the males, probably an adaptation that allows increased dispersal of the flightless females, which are carried about by the males during a nuptial flight.

Tiphiidae and Scoliidae mostly attack ground-dwelling beetle larvae. Scarabaeidae are the most common hosts, but Tenebrionidae and Cicindelidae (tiger beetles), as well as Orthoptera and bees and wasps, are utilized by Tiphiidae. Mutillidae (velvet ants), whose powerfully stinging females are always wingless, are mostly parasitoids of bees and wasps. Most diurnal species of velvet ants are brilliantly patterned in red or orange and black pile, rendering them highly conspicuous as they rapidly run about the ground searching for host nests. Nocturnal species are tan, brown, or black, resembling tiphiids. Sapygidae are parasitoids of bees and wasps. Bradynobaenidae have seldom been reared but have been recorded as ectoparasitoids of Solifugae (Chelicerata).

"Pompiloid" Vespoidea. Small to very large wasps with large, mobile prothorax, hind trochanters 1-segmented; forewings with 7 to 10 closed cells, including costal cell, hind wing with at least one closed cell; ovipositor developed as internal sting.

This informal grouping comprises two families of greatly unequal size and abundance. Rhopalosomatidae, represented by only two rare species in North America, parasitize Gryllidae. The larvae cause a hernialike pouch in the abdominal wall of the host, much in the manner of Dryinidae. Pompilidae (spider wasps, Fig. 39.16a), characterized by the presence of a horizontal furrow on the mesopleuron (absent in Rhopalosomatidae), all utilize spiders as larval food. Pompilids provide each larva with a single prey item, which is necessarily larger than the wasp. A wide variety of spiders is attacked, including tarantulas, which serve as prey for some *Pepsis*.

Because of the large body size of their prey, pompilids usually construct a burrow near the site of attack and immobilization. The paralyzed spider is frequently concealed to avoid attack by parasites and scavengers, then quickly moved a short distance into the completed larval gallery. Other species simply conceal the prey in an appropriate crevice or in the host's own tunnel, and some only temporarily paralyze the spider, which regains activity before being killed by the maturing wasp larva. Such species are essentially parasitoid, showing no important biological differences from Bethylidae or Tiphiidae. More specialized pompilids resemble primitive sphecids in preparing a larval gallery before prey acquisition, while a number of species oviposit on prey of other pompilids. The family is cosmopolitan but is especially diverse in tropical and subtropical regions.

*The **Ants** ("Formicoid" Vespoidea).* Minute to moderate sized Hymenoptera with basal 1 or 2 segments of abdominal gaster modified as a nodelike petiole (Fig. 39.16b); prothorax large, mobile; antennae almost always strongly elbowed.

Ants are among the most ubiquitous and familiar insects. Currently about 12,000 species are known in roughly 300 genera. The greatest variety exists in the tropics, but ants are found north to the Arctic tree line, south to the southern tips of the continents, and on all but the highest mountains. Most oceanic islands within these latitudinal extremes are also inhabited by ants. Wilson (1971) estimates that 1 percent of all insects are ants (i.e., 10^{15} individuals). Several species have become closely associated with humans, occupying their dwellings and other structures throughout the world, including metropolitan areas. Despite great variation in size

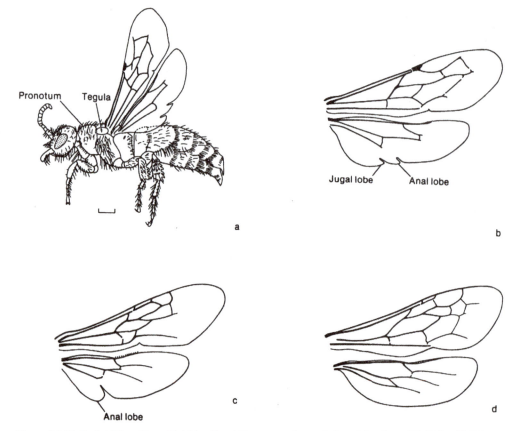

Figure 39.15 Scolioid Vespoidea: a, Tiphiidae (*Paratiphia verna*; scale equals 1 mm); b, wings of Tiphiidae (*Tiphia*); c, wings of Scoliidae (*Scolia nobilitata*); d, wings of Mutillidae (*Timulla vagans*).

and habits, practically all ants are recognizable by the petiolate abdomen and elbowed antennae and are classified as a single family, Formicidae.

Without exception, ants are social insects. Colonies consist of one or a few sexual females, or queens, specialized for reproduction, and a variable number of apterous, neuter females that function as workers or soldiers. Males are normally produced once or a few times a year, surviving only long enough to mate. Males and females are usually larger than nonsexual castes and have the thorax enlarged with the wing muscles. Female sexuals dehisce their wings after a nuptial flight;

males are permanently winged but die shortly after mating. Worker and soldier castes are always wingless. Soldiers, which function in colony defense, are larger than workers, often with the head disproportionately enlarged. Colony organization and structure, which are extremely complex, are discussed in detail in Chapter 7, but it may be mentioned here that colony formation, which is carried out by single females, is critical in the perpetuation of most species.

Ant colonies, which may persist for many years, vary from assemblages of a single queen and a few dozen individuals to multitudes of hundreds of

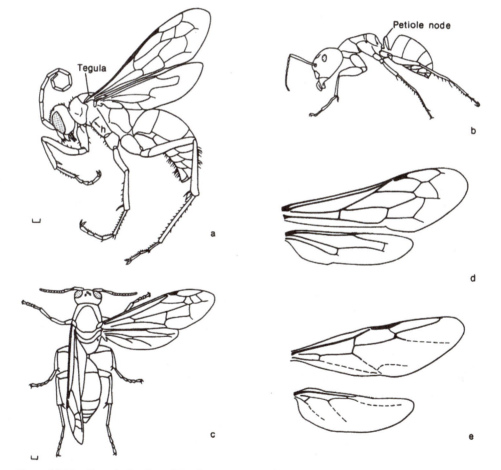

Figure 39.16 a, Pompiloidea, Pompilidae (*Priocnemoides unifasciatus*); b, Formicoidea, Formicidae (*Camponotus*); c, Vespoidea, Vespidae (*Polistes metricus*); d, Vespidae, wings (*Pseudomasaris vespoides*); e, Formicidae, wings (*Camponotus morosus*). Scales equal 1 mm. h, horizontal furrow on mesopleuron.

thousands of individuals with numerous queens. Typically, colonies are established in fixed locations. Subterranean galleries are most common, but many species tunnel dead wood or twigs, and *Oecophylla* species use silk produced by their own larvae to construct nests of living leaves of trees and shrubs. A few tropical species have evolved highly specialized symbiotic relationships with plants. Especially interesting is the relationship between ants of the genus *Pseudomyrmex* and certain species of acacia trees. The large swollen thorns of bull's horn acacia and other species are inhabited by various species of *Pseudomyrmex*. Extrafloral nectaries and Beltian bodies on the leaves supply the ants with sugar and protein. The ants aggressively protect the trees from vertebrate and invertebrate predators (Janzen, 1966, 1967a,b). Army ants do not occupy a fixed colony but lead a nomadic existence, bivouacking in litter or dense vegetation.

A number of species of ants, especially in the genus *Formica*, enslave other species. For example, *F. sanguinea* workers attack the nests of other species of *Formica*, robbing the pupae and carrying them back to their own nest. Still other species live as inquilines in the nests of other ants, a relationship sometimes termed social parasitism.

Ants include herbivores, scavengers, and predators. Most ants are occasionally opportunistic predators, but army ants are exclusively predatory, including a wide variety of small invertebrates in their diet, and a few ants are specific predators of spider eggs, termites, or other arthropods. Many species tend aphids or membracid nymphs, from which they obtain honeydew, and several, such as the argentine ant (*Linepithema humile*), are pests in gardens for this reason. Harvester ants specialize on seeds, and leaf-cutting ants (tribe Attini) cut out pieces of leaves and flowers on which are cultured fungi. The ants eat and replace the gardens in a manner analogous to the fungal gardening of some Termitidae (Isoptera). Ants are sufficiently abundant that certain birds, mammals, and reptiles (horned lizards, anteaters) are specialized as predators on them. Ant colonies are widely exploited by myrmecophilous insects, especially beetles, which may act as predators of the ant brood or as nest scavengers. The more specialized of these myrmecophilous insects may be morphologically modified to provide a tactile resemblance to their hosts.

Paper, Potter and Mason Wasps (Vespoidea in the narrow sense). Small to large wasps with the prothorax with large, triangular lobes extending posteriorly to the tegulae (Fig. 39.16c); hind trochanters 1-segmented; forewings with at least 7 closed cells, including costal cell (Fig. 39.16d); hind wing with at least 2 closed cells; ovipositor developed as internal sting.

The single family, Vespidae, includes diverse wasps that are mostly predatory, usually on caterpillars, occasionally on larvae of sawflies, phytophagous beetles, or other soft-bodied insects. Species of *Vespula* (yellow jackets and hornets) are also aggressive scavengers of all sorts of proteinaceous material, often becoming pests about refuse sites, picnic areas, or pet-feeding locations. Members of the subfamily Masarinae, which provision their nests with pollen and nectar, are locally common in the Western states. Most vespids may be recognized by the emarginate eyes, which partially surround the antennal bases.

Vespids include many solitary forms such as potter wasps (subfamily Eumeninae), but the familiar species, including yellow jackets, hornets, and paper wasps, form annual colonies. Vespid nests may be suspended from branches, rafters, or other aerial structures, may be concealed beneath stones or in hollow logs, or may be constructed in cavities thath the wasps excavate in the ground. A few solitary species use mud as a building material, but nests of social species are commonly constructed of masticated wood or paper. Colonies range in size from a few cells arranged in a single, open comb to multicombed structures enclosed in tough, external jackets of paper. Different species display all stages of intermediacy between solitary and social organization.

Vespids are commonly colored in contrasting patterns of black and yellow, in advertisement of their severe stings. The potter wasps (subfamily Eumeninae) paralyze their caterpillar prey by stinging, in the manner of the sphecoid wasps.

Most social Vespidae, however, kill the prey with the mandibles. Stinging is largely used to deter mammalian predators, which are very sensitive to the pain-producing substances in vespid venoms. The large number of insects that mimic vespids attests to the efficacy of their stings.

Superfamily Apoidea.

This superfamily consists of two functionally different groups that were previously classified as different superfamilies: the predatory apoid wasps (previously called Sphecoidea) and the phytophagous bees (Apoidea in the strict sense). Much phylogenetic evidence now shows that Apoidea in the broader sense is a monophyletic lineage, with the bees arising from within one of the apoid wasp lineages. We treat the apoid wasps and the bees separately below for practicality.

The Apoid Wasps, Small to large wasps with the prothorax collarlike with rounded posterior lobes that do not reach the tegulae (Fig. 39.17a); hind trochanters 1-segmented; forewings with at least 7 closed cells, including costal cell; hind wings with at least 2 closed cells (Fig. 39.17c,d); ovipositor developed as internal sting.

The apoid wasps are commonly called "solitary wasps" or "digger wasps," although some nest in twigs, old beetle burrows, or self-constructed mud cells. These wasps are treated either as a single large family, Sphecidae, including an astonishing diversity of body form and size, as well as behavior, or as a series of families (discussed below as subfamilies).

A few sphecids (subfamily Nyssoninae) are inquilines in the nests of bees or other sphecids, but predation on other insects is much the dominant mode of life. The adult female wasps construct nests containing one or more cells in which the larvae will mature while feeding on prey provided by the mother. A great variety of insects is utilized as prey, and host specificity varies greatly among different species. The most primitive sphecids use a single prey item for each larva. Because of the large prey size, such wasps either drag the prey over the ground or transport it with a series of short flights initiated from trees, shrubs, or other prominences.

Except in the construction of a special larval gallery, these primitive sphecids are not greatly different from bethylids, tiphiids, or other parasitoid wasps.

In the great majority of sphecids, prey size is much reduced, allowing rapid transport by air, so that such abundant insects as aphids, leafhoppers, and thrips may be utilized. Provisioning follows two modes. Mass-provisioning species accumulate the total number of prey required before oviposition. Normally only a single larval cell is tended at one time. Progressive-provisioning species oviposit on one of the first prey items acquired or in an unprovisioned cell, later supplying additional food as needed by the growing larva. In the most advanced species several larvae are tended simultaneously. The resultant close contact between the adult females and their offspring is probably requisite to true sociality (eusociality, occurring in Vespoidea and Apoidea) (see Chap. 7). Communal nesting occurs in several subfamilies of Sphecidae, and division of labor among communal members has been recorded in the subfamilies Pemphredoninae, Philanthinae, and Crabroninae. Eusociality, with reproductive division of labor, has been well documented in *Microstigmus comes* (Pemphredoninae) and probably occurs in other sphecids as well.

The Bees (Apoidea in the strict sense). Small to large Hymenoptera with branched, plumose hairs on at least part of the body; prothorax with rounded posterior lobes that do not reach the tegulae; hind trochanters 1-segmented; forewings with at least 7 closed cells; hind wings with at least 2 closed cells; ovipositor developed as internal sting.

In general morphological characters, bees are very similar to apoid wasps. The most reliable differentiating features are the presence of branched hairs on the body and the specialization of the hind basitarsus as a part of the pollen-collecting apparatus (Fig. 39.19c,d). A few genera, such as *Hylaeus* (Colletidae), carry pollen in the crop. In these and in cleptoparasitic (using the galleries and provisions of other bees) genera of other families, the body is relatively hairless and the tarsi are unmodified, rendering separation from some sphecids difficult.

In terms of nest architecture and behavior, bees (except the social species) are not greatly different

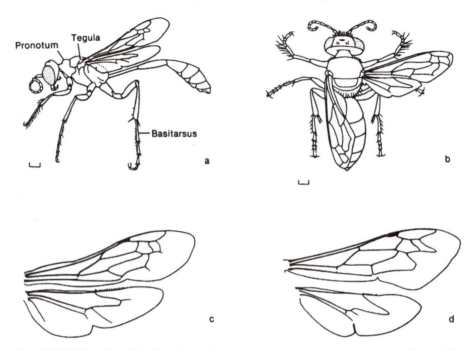

Figure 39.17 Sphecoidea, Sphecidae: a, *Ammophila cleopatra*; b, *Bembex americanus*; c, wings of *Ammophila*; d, wings of *Tachytes*. Scales equal 1 mm.

from solitary wasps, with the major difference that most bees provision their larval cells with nectar and pollen rather than animal material. Their lower position in food chains (as primary rather than secondary or tertiary consumers) is probably responsible for the great abundance of bees in most habitats. As pollinators bees are more important than any other group of insects. The maxillae and labium are modified as a proboscis for sucking nectar and may be very elongate in some Apidae. Scopae, or pollen-collecting structures, may occur on the hind tarsi and tibiae [Apidae (Fig. 46.18c,d), Andrenidae, Halictidae, Melittidae] or on the venter of the abdomen (Megachilidae) of females. Scopae are lacking only in some Colletidae (*Hylaeus*) and in cleptoparasitic or inquiline species.

The honey bee (*Apis mellifera*) is the best known pollinator. Commercial hives are seasonally rotated among many plantings, including fruit trees, alfalfa, and certain row crops. Honey bees visit a wide variety of noncultivated plants as well. Wild bees are more efficient pollinators than honey bees for

some crops. For example, alkali bees (*Nomia*) and leafcutter bees (*Megachile*) are utilized for alfalfa pollination. Pollination of many uncultivated plants is accomplished largely by wild bees, which may be highly specialized for one plant species or a group of closely related species (oligolectic). For example, certain species of *Andrena* (Andrenidae) collect pollen only from night-blooming species of *Oenothera* (Onagraceae), which they visit at night or very early in the morning, before other pollinators have had an opportunity. Polylectic species of bees collect pollen from many plant species, but frequently only a single species is visited on sequential trips for pollen. For example, in the highly polylectic honey bee, any individual bee normally visits only one type of flower until that source of pollen and nectar is exhausted, then an alternative flower type will be selected.

Eusociality has apparently evolved independently several times in bees (see Chap. 7). The largest and most complex colonies are maintained by honey bees (*Apis*) and some tropical stingless bees

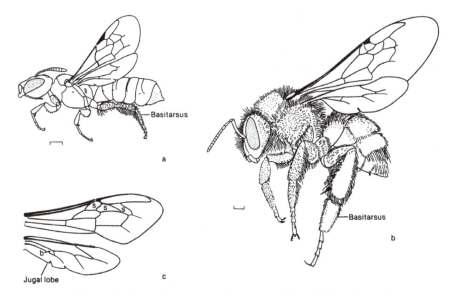

Figure 39.18 Apoidea, Apidae: a, *Ceratina strenua* (scale equals 0.5 mm); b, *Bombus pensylvanicus* (scale equals 1 mm); c, *Apis mellifera*, wings. b, Basal cell in hind wings; s, submarginal cells.

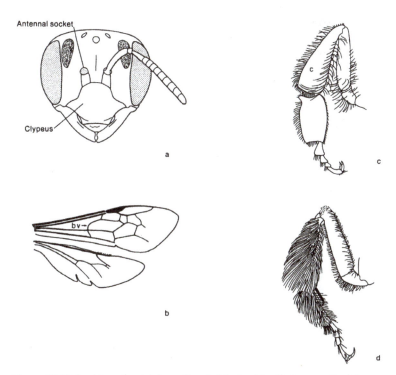

Figure 39.19 Apoidea: a, frontal view of head of Andrenidae (*Andrena gardineri*); b, wings of Halictidae, bv, basal vein (*Halictus ligatus*); c, hind tarsus of Apidae, showing corbicula (c) (*Apis mellifera*); d, hind tarsus of noncorbiculate Apidae, showing scopa (*Anthophora curta*).

(Apidae, Meliponinae), but many other apids, such as bumblebees, which are closely related to the stingless bees, as well as some halictids, form small annual colonies. As in other social Hymenoptera, these are female societies, with the male bees, or drones, appearing only once a year in relatively small numbers. Besides these eusocial forms, many bees, especially Apidae and Halictidae, are presocial, showing various degrees of altruistic or cooperative behavior. The biology of social Hymenoptera is extensively discussed by Wilson (1971), Michener (1974), and Ross and Matthews (1991).

KEY TO THE NORTH AMERICAN FAMILIES OF APOID BEES

1. Face with 2 vertical sutures between each antennal socket and clypeus (Fig. 39.19a)
Andrenidae
Face with a single suture between each antennal socket and clypeus **2**
2(1). Hind wing with jugal lobe as long as basal cell or longer (Fig. 39.19b) **3**
Hind wing with jugal lobe shorter than basal cell (Fig. 39.18c) or lacking (Fig. 39.18b) **4**
3(2). Forewing with the basal vein (basal sector of M vein) strongly arched (Fig. 39.19b)
Halictidae
Forewing with the basal vein straight or feebly curved (Fig. 39.18c) **Colletidae**
4(2). Labial palpi with all segments cylindrical, subequal in length **Melittidae**
Labial palpi with 2 basal segments flattened, at least twice as long as distal segments (Fig. 39.3b) **5**
5(4). Forewing usually with 3 submarginal cells (Fig. 39.18c); labrum usually broader than long; subantennal suture extending from center of antennal socket; females with scopa on hind legs (Fig. 39.19c,d)
Apidae
Forewing with 2 submarginal cells; labrum longer than broad; subantennal suture extending from lateral margins of antennal sockets; females with scopa on venter of abdomen **Megachilidae**

CHAPTER 40

Order Mecoptera (Scorpionflies)

ORDER MECOPTERA

Small to medium-sized Endopterygota. Adult with slender elongate body; elongate hypognathous head capsule; mandibles and maxillae slender, elongate apically serrate; labium elongate, with fleshy 1- to 3-segmented palps. Thoracic segments subequal or prothorax reduced, mesothorax and metathorax fused; wings narrow, elongate, subequal, crossveins numerous. Abdomen cylindrical, elongate, with 1- or 2-segmented cerci. Larvae eruciform or scarabaeiform with sclerotized head capsule, mandibulate mouthparts, lateral compound eyes; thoracic segments subequal, prothorax with distinct notal sclerite; thoracic legs short with fused tibia and tarsus and single claw; abdomen usually with prolegs on segments 1 to 8, segment 10 modified as a suction disk or with a pair of hooks (*Nannochorista*). Pupa decticous, exarate, nonmotile.

In their simple mandibulate mouthparts and membranous wings of similar shape and venation, scorpionflies are similar to the primitive neuropteroid orders Neuroptera and Megaloptera. As in those orders, mecopteran fossils first appear in the Permian Period. Before the end of the Paleozoic Era, a diverse fauna, including three suborders and numerous families, had developed. Most of these taxa, including a fourth suborder that appeared in the Mesozoic Era, are now extinct. Mecoptera are presently widespread, especially in humid temperate and subtropical regions, but a high degree of endemism characterizes local regions. The Australian fauna is highly distinct, with many primitive species and one endemic family, Choristidae. Five families occur in North America, including Meropeidae, otherwise known only from Australia. There are about 500 species worldwide, most in the families Bittacidae and Panorpidae.

Despite the superficial similarity of Mecoptera to Neuroptera, the structural details of the mouthparts of adult scorpionflies indicate a relationship with Diptera and Siphonaptera. The slender mandibles and maxillae and elongate face of bittacids (Fig. 40.1b) are not greatly different from the homologous structures in primitive Diptera (compare with Fig. 41.3a,b), and the labial palps of *Chorista* (Choristidae) are enlarged and fleshy, suggesting a primitive labellum. In addition, the fossils *Permotanyderus* and *Choristotanyderus* are essentially intermediate between Mecoptera and Diptera. In contrast, the larvae of some Mecoptera are similar to those of Lepidoptera and Trichoptera, a relationship that is also indicated by certain fossils.

Most Mecoptera prefer moist conditions, especially in forests, where the adults are to be found on rank vegetation or flying about. In species that inhabit arid or hot regions, such as *Apterobittacus* in California, adult activity is usually restricted to the cool, wet part of the year. Most scorpionflies are winged, but various genera are brachypterous [*Boreus*, Boreidae (Fig. 40.2a)] or completely wingless [*Apterobittacus*, Bittacidae (Fig. 40.2b)]. *Boreus* is a winter form, most commonly found about moss, which is used as adult and larval food. *Apterobicttacus* frequents low vegetation, clambering nimbly about with the long legs. In *Brachypanorpa* (Panorpodidae) females are brachypterous, males fully winged.

Bittacidae are predatory as adults, but most Mecoptera, such as *Boreus*, mentioned above, are apparently herbivores or scavengers. In Bittacidae the single large tarsal claws fold bladelike against the terminal tarsomeres (Fig. 40.2c) and are used to snare small, soft-bodied insects. Panorpidae occasionally visit flowers, probably for nectar, but feed chiefly on dead insects and snails, which they locate on the ground or low vegetation. *Panorpa* also steal prey items from spiderwebs and are able to move about in webs with some alacrity.

Larvae of Panorpidae and Bittacidae are caterpillarlike with prolegs on the first eight abdominal segments. Larvae of Nannochoristidae (Australia and Chile) are elongate and slender, and those of Boreidae and Panorpodidae are scarabaeiform, all without abdominal legs.

Life histories have been described for only a few species. Copulation takes place on vegetation or the soil. In *Panorpa* and many Bittacidae mating is accompanied by complex courtship behavior involving a sex attractant secreted by abdominal glands of males. During copulation males of some species present the female with a previously acquired food item, which she partially consumes. Males of *Panorpa* may also produce a salivary mass, which hardens and is then consumed by the female during copulation, which may last several hours in some species. In Boreidae classical courtship is lacking; the male uses his shortened wings to position the female and hold her on his back during copulation. Egg deposition is variable. Panorpidae and Choristidae oviposit small or large batches of delicate, smooth eggs in crevices, while Bittacidae scatter their tough, leathery, polygonal or spherical eggs on the soil. *Nannochorista* oviposits along stream margins. Eggs of most Mecoptera absorb water from the substrate and increase substantially in size subsequent to oviposition. Larvae emerge after a brief period or, as in *Harpobittacus* and *Apterobittacus*, after a diapause of several months. Habits of immatures are poorly known; most larvae apparently occur in moist litter, consuming dead insects or plant material. The larva of *Nannochorista* is aquatic and predatory; *Boreus*

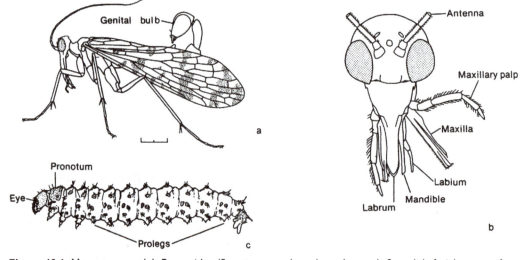

Figure 40.1 Mecoptera: a, adult Panorpidae (*Panorpa anomala*, male; scale equals 2 mm); b, facial aspect of head and mouthparts of *Bittacus chlorostigma*, antennae partly removed (Bittacidae); c, larva of *Panorpa nuptialis* (Panorpidae).
Source: c, redrawn from Byers, 1963.

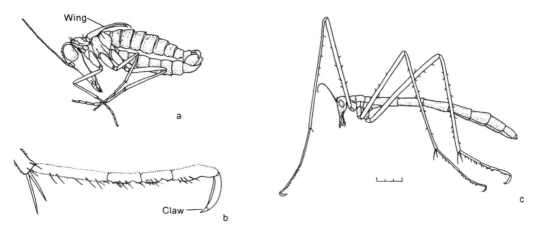

Figure 40.2 Mecoptera: a, Boreidae (*Boreus brevicaudus*); b, tarsus of *Apterobittacus*; c, Bittacidae (*Apterobittacus apterus*; scale equals 3 mm).
Source: a, redrawn from Byers, 1961.

larvae feed on moss and *Brachypanorpa* on herbaceous vegetation. Where known, pupation occurs in a cell in soil, decaying wood, or litter, without a cocoon. Prepupal diapause of several months is recorded in *Harpobittacus* and in *Panorpa*.

Mecoptera are a distinctive group of insects, almost always classified as a single order. It has been suggested, however, that the family Nannochoristidae (Australia, New Zealand, South America) should be considered a separate order, Nannomecoptera, based on differences in adult mouthpart and larval leg structure. Nannomecoptera has been proposed as the sister taxon to Diptera or Siphonaptera. Recent molecular studies are inconclusive about the monophyly of the Mecoptera but confirm the close relationshiop with Diptera and (especially) Siphonaptera.

KEY TO NORTH AMERICAN FAMILIES MECOPTERA

1. Tarsi with single large claw about as long as apical tarsomere (Fig. 40.2b,c)   **Bittacidae**
    Tarsi with claws paired, usually much shorter than apical tarsomere   **2**
2(1). Ocelli absent or very small; wings with reticulate crossveining in costal space, or rudimentary, without veins   **3**

Ocelli present; wings with a few simple crossveins in costal space (wings may be reduced, but venation is preserved)   **4**
3(2). Wings large, broad with complex venation; body dorsoventrally flattened
                                             Meropeidae
    Wings scalelike or bristlelike; body cylindrical or laterally flattened (Fig. 40.2a)
                                               Boreidae
4(2). Rostrum more than twice as long as width at base; wings fully developed in males and females   **Panorpidae**
    Rostrum slightly longer than width at base; wings fully developed in males, reduced in females   **Panorpodidae**

Bittacidae *(Fig. 40.2b)*. Slender-bodied, long-legged Mecoptera resembling craneflies. Widespread and locally common in woodlands throughout the world, with 9 species in North America. Adults commonly hang by their forelegs from vegetation, hence the name "hangflies" or "hanging flies."

Panorpidae *(Fig. 40.1a,c)*. The most abundant and diverse family in the Northern Hemisphere, with about 40 species of *Panorpa* in Eastern North America. The common name, "scorpion fly," refers to the enlarged genital bulb of males.

Panorpodidae. A single genus, *Brachypanorpa*, with two species in the Appalachian Mountains and two in the Pacific Northwest. Males are normally winged, females brachypterous.

Meropeidae. A single species, *Merope tuber*, in the Eastern states, and one species in Australia. Body relatively robust, flattened.

Boreidae *(Fig. 40.2a)*. Restricted to the Northern Hemisphere, with about a dozen North American species, mostly in western mountain ranges. The dark colored adults, about 4 mm long, sometimes crawl about on the surface of snow, hence the name "snow fleas."

CHAPTER 41

Order Diptera (Flies)

ORDER DIPTERA

Adult. Minute to moderately large Endopterygota, usually with functional wings on the mesothorax. Head free, mobile, usually with large compound eyes, 3 ocelli; antennae multiarticulate, filiform or moniliform, or reduced, with terminal segment bristle-shaped; mouthparts diversely modified, usually as a proboscis for imbibing liquid food. Prothorax reduced to a small, collarlike region, usually immovably fused to the mesothorax; mesonotum large, convex, usually divided into prescutum, scutum, and scutellum by transverse sutures; metathorax reduced to a narrow, transverse band or lateral plates bearing halteres and metathoracic spiracles; wings with few or numerous longitudinal veins, few closed cells; legs usually with pulvilli, empodia, or other pretarsal structures.

Larva. Usually maggotlike or vermiform, without legs, but occasionally with unsegmented pseudopods on various segments, especially in Nematocera; head capsule sclerotized as discrete tagma or undifferentiated; eyes and antennae reduced or absent. Thorax usually undifferentiated from abdominal region; abdomen almost always without cerci or urogomphi.

Pupa. Adecticous; obtect (Nematocera, Orthorrhapha) or exarate, enclosed in sclerotized puparium (Cyclorrhapha).

The Diptera comprise approximately 150,000 described species, but many more remain to be described, so that the total species richness of the order is likely to rival Coleoptera and Hymenoptera. The flies must be considered one of the dominant groups of insects because of their extreme abundance over a great variety of ecological situations.

Diptera include predators, parasites, and parasitoids, but the majority of species are saprophytic, and Diptera are usually the dominant invertebrate consumers of decaying vegetation or decomposing animal products. Flies are common visitors to flowers, which provide the only food for adults of some families (e.g., Bombyliidae, Conopidae,

Acroceridae, Apioceridae). However, few Diptera have engaged in the intimate type of symbiotic relationship displayed by the angiosperm plants and Hymenoptera, and flies appear to be of sporadic importance in pollination except in Arctic and high montane regions. Almost no adult Diptera directly damage living plants, except by oviposition.

Larvae occur predominantly in moist or subaquatic situations, less frequently in either dry or strictly aquatic habitats. Many species of arid or subarid regions exist in moist microhabitats, such as animal burrows, rotting succulent vegetation, or feces, and many are specialized for quick exploitation of ephemeral sources of food and water. In such species the larval growth phase may span 1 to 2 weeks or less. Immatures of several families are internal parasites of vertebrates, the only significant group of insects to acquire this habit. A relatively small number of fly species attack living plants, as larval miners or borers in various plant parts, or by causing galls.

The chief economic importance of Diptera is due to their role as vectors for virulent diseases of humans and domestic animals. Malaria, yellow fever, and encephalitis, all transmitted exclusively by mosquitoes, have been three of the most persistent and debilitating human diseases, particularly in tropical climates. Houseflies (Fig. 41.1), as well as other species, mechanically vector numerous microorganisms, especially those causing dysenteries and other enteric ailments. With notable exceptions such as the Hessian fly [*Mayetiola destructor* (Cecidomyiidae)], root maggots (Psilidae), and fruit flies (Tephritidae, Otitidae), Diptera cause relatively minor damage to crops, as would be expected from their predominance as decomposers rather than herbivores. In addition, many phytophagous Diptera, such as fruit flies (Tephritidae), tend to be highly host-specific and susceptible to control by cultural practices such as staggering crops, isolating plantings, and removing crop debris.

MORPHOLOGICAL ADAPTATIONS

The most conspicuous characteristic of the body plan of adult Diptera is the high degree of

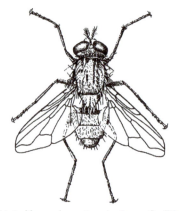

Figure 41.1 *Musca domestica*, the housefly (Muscidae). *Source: Redrawn from U.S. Department of Agriculture.*

adaptation for rapid, efficient flight. For most flies, flight is required for location of food or other resources, escape from predators, and successful reproduction. Diptera, along with Hymenoptera and Lepidoptera, are notable for the strong partitioning of resources between immatures and adults. For most species, highly developed flying ability has facilitated such partitioning.

Obvious adaptations to aerial life include foreshortening and streamlining of the body and shortening of the antennae, as in most Cyclorrhapha. More profound modifications involve structural changes in the thorax. In all Diptera, the wing-bearing mesothorax has become enormously enlarged, while the prothorax and metathorax are reduced to narrow, collarlike regions (Fig. 41.2). The prothorax, marked by the attachments of the forelegs, is largest in Nematocera; in most Brachycera it is represented by narrow pleural sclerites and dorsolateral prominences, or humeri. The metathorax is represented in most species by small lateral sclerites supporting the halteres. The metapleura and sterna are closely associated with the mesothorax and difficult to distinguish.

The architecture of the mesothorax is complicated by the development of secondary sutures, as well as by distortion caused by enlargement, especially of the notal region. Both the notum and postnotum are usually distinct (Fig. 41.2a,b). The notum is divided into prescutum, scutum, and scutellum by transverse sutures, which are represented internally by ridges, providing braces against the stress of the highly developed flight muscles. In Nematocera the postnotum is a relatively large, dorsal sclerite (Fig. 41.2b). In Brachycera the scutellum is prominent, becoming especially enlarged in Cyclorrhapha, where it may be divided into distinct scutellum and subscutellum (Figs. 41.2c, 41.18). In these flies the postnotum is represented by small, lateral plates. Enlargement of the scutellum probably provides additional area for muscle attachment, which is further increased in most Diptera by large, laminate phragmata, which project ventrally into the thoracic cavity (Fig. 41.2d). The prephragma is often small or ridge-shaped, but the postphragma (postnotum) is almost always a prominent, convex

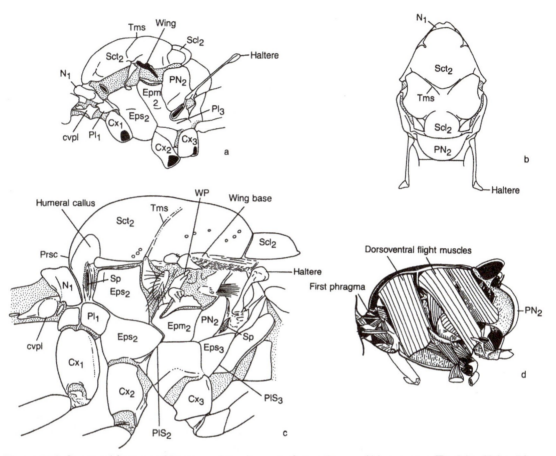

Figure 41.2 Structural features of Diptera: a, lateral aspect of pterothorax of Nematocera (Tipulidae, *Holorusia*); b, dorsal aspect of same; c, lateral aspect of pterothorax of Brachycera (Asilidae, *Mallophora*); d, longitudinal section through pterothorax of *Tabanus* (Tabanidae). cvpl, cervical plates; Cx, coxa; Epm, epimeron; Eps, episternum; N, notum; Pl, pleuron; PlS, pleural suture; PN, postnotum; Scl, scutellum; Sct, scutum; Sp, spiracle; Tms, transmesonotal suture; WP, pleural wing process.
Source: c, redrawn from Bonhag, 1949; d, modified from Cole, 1969.

plate, sometimes nearly separating the internal thoracic and abdominal spaces.

The pleural region of the mesothorax is exceedingly variable and complex, frequently with secondary sutures, incorporation of portions of the coxae, and radical shifts in size, shape, or position of certain sclerites. The principal landmarks are (1) the pleural wing suture, straight in Nematocera (Fig. 41.2a) but sinuous in most Brachycera (Fig. 41.2c); and (2) the metathoracic spiracle, located at the anterior dorsal margin of the metapleuron, usually just below the haltere. The homology of

certain structures and their functional significance are obscure, but the topographic features illustrated in Figure 41.2 are important for identification of many taxa.

The thoracic specializations of Diptera are correlated, of course, with two-winged flight. Whereas the wings of most Hymenoptera and Lepidoptera are coupled during flight, functioning as a single airfoil, the hind wings of Diptera are transformed into halteres (Fig. 41.2), which are vibrated rapidly during flight, apparently functioning as gyroscopic balancing organs. In some Brachycera (the

Calyptratae) the halteres are enclosed in cavities covered by calypters, foldings of the trailing edge of the wing base (Fig. 41.18d). Venation is highly variable. Some Nematocera have extensive, complex venation, not greatly different from that in Mecoptera (Figs. 41.6a, 41.7a). In most families, some reduction in the number of veins and closed cells is apparent (Figs. 41.10, 41.14), often accompanied by a crowding of veins (usually the costal, subcostal, and first radial) into the leading edge of the wings, a strengthening configuration that occurs repeatedly in insects with a rapid wingbeat. Also correlated with the development of more efficient flight is an increasing consolidation of the nervous system. In Nematocera several abdominal and three thoracic ganglia are usually present. In Brachycera these numbers are variously reduced, and in the most specialized Cyclorrhapha all are incorporated into a single, large ganglionic mass in the thorax.

Diptera all imbibe liquid food, but unlike the Hemiptera, which are narrowly specialized for piercing and sucking, flies have evolved a variety of modes of feeding, and some Syrphidae and Bombyliidae ingest finely particulate material such as pollen along with liquids. In certain Nematocera, mouthparts (Fig. 41.3a) are similar to those of Mecoptera, with bladelike mandibles, maxillae with lacinia and multiarticulate palpi, and elongate labium with one- or two-segmented fleshy palpi or labellar lobes. This configuration appears to be primitive. In Tabanidae and some bloodsucking Nematocera, the bladelike mandibles, maxillae (laciniae), labrum, and hypopharynx are enclosed in the stout U-shaped labium, which acts as a guide for the cutting mouthparts. The labellum bears numerous minute surface channels, or pseudotracheae, for sponging up oozing blood by capillarity. In Culicidae the mandibles and laciniae are fine, styliform blades, and the labrum and hypopharynx are delicate tubes for uptake of food and delivery of saliva (Fig. 41.3b). All are ensheathed by the flexible labium and form a piercing structure analogous to that of Hemiptera. In predatory Brachycera (Asilidae, Empididae), the labial sheath is rigidly sclerotized, and the laciniae are stout blades used to pierce the integument of prey. Mandibles are absent, and the labellum vestigial. The salivary glands have become specialized in various bloodsucking families to deliver anticoagulants, or in predators to produce paralytic toxins, which are delivered through the hypopharynx.

In many flies, including nearly all Cyclorrhapha, the mouthparts show a highly derived configuration, consisting of the labrum, labium, and hypopharynx combined to form a thick, fleshy rostrum, or proboscis, terminated by a greatly enlarged labellum (Fig. 41.3c). Soluble constituents of solid foods are dissolved in saliva or regurgitated gut contents before being swallowed. Predation and parasitism in these specialized forms, which have lost the laciniae, are accomplished by the secondary development of sharp, sclerotized projections of the pseudotracheae or preoral region of the labellum, which are used to rasp or pierce the host integument. In the highly specialized tsetse fly (*Glossina*), the rostrum has secondarily become slender and elongate, with highly reduced labella. The stout, elongate maxillary palpi are closely applied to the rostrum, providing additional rigidity.

Most Diptera have one or sometimes three diverticula of the foregut, which function as a crop. Liquid meals, large in volume, may be temporarily held in the diverticula, to be gradually released into the intestine.

A peculiarity of the order is the temporary or permanent rotation of the male genitalia 90 to 360° from their original position. Permanent rotation, termed torsion, occurs during pupal development. When rotation proceeds through 180° ("inversion") the anus comes to lie below the genitalia. If rotation have proceeded through 360° ("circumversion") the original dorsal position of the anus is restored, but the sperm duct is twisted around the hind gut. The function of torsion is not understood, but the degree of twisting is of taxonomic significance, as explained below. The male terminalia are also often bent anteriorly or laterally (flexion). In Cyclorrhapha both torsion and flexion are apparent and are associated with marked asymmetry of the external genitalia. In various taxa rotation and flexion may be in part permanent and in part temporary at the time of mating.

Diptera couple with the male and female initially facing the same direction, usually with the

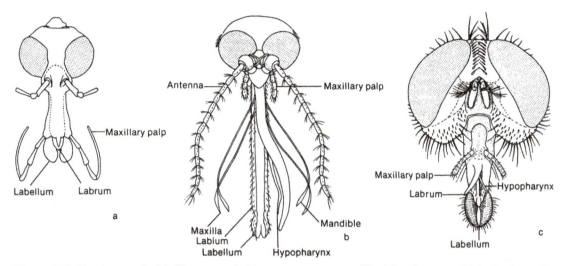

Figure 41.3 Mouthparts of adult Diptera: a, primitive nematoceran type (Tipulidae, *Ctenacroscelis*); b, piercing-sucking type (Culicidae, *Culex*); c, sponging type (Muscidae, *Musca*).
Source: Redrawn after Crampton and Snodgrass.

male dorsal. After coupling, however, the pair turn tail to tail and in flies without flexion or rotation the male is turned onto its back. Rotation and flexion are apparently adaptations that allow the male to remain upright in the tail-to-tail position and have apparently occurred independently many times in Diptera, producing the complexity of male genitalic orientation. Circumversion, however, appears to have occurred a single time and is one of the defining characters of the Cyclorrhapha.

Diptera are remarkable for the profound morphological difference between adults and immatures. Larvae usually have three thoracic and eight or nine discernible abdominal segments, but these may be fused or secondarily increased by subdivision, so that the original segmentation is obscured. In Bibionidae 10 pairs of spiracles are present (prothorax, metathorax, eight abdominal segments), but in other Diptera the number is reduced. Often only the anterior and posterior spiracles are retained in an amphipneustic arrangement or only the anterior or posterior are present (propneustic and metapneustic, respectively). Metapneustic spiracles may be associated with a breathing siphon, as in Culicidae. In several families with aquatic larvae, such as Blephariceridae

and Simuliidae, all spiracles have been lost, producing an apneustic arrangement. In various families the configuration of functional spiracles changes in different instars. All dipterous larvae are without true legs, and the great majority lack trunk appendages entirely. Leglessness among insect larvae is mostly restricted to borers in plant tissues (Coleoptera, Lepidoptera, Hymenoptera), parasitoids (Hymenoptera, Coleoptera), and those inhabiting other protected microenvironments with assured food supply (nest-making Hymenoptera). Reduction of locomotory and cranial structures in fly larvae is probably related to their widespread occurrence in rich nutrient sources, where it is advantageous to feed and grow as rapidly as possible. Under such circumstances extreme divergence between adults and larvae is possible, the larvae being specialized as a feeding stage, the adults as a reproductive and dispersal stage. Divergence not only allows specialization for rapid exploitation of temporary food sources but also decreases competition between larvae and adults.

Larvae of the major taxonomic groups of Diptera differ in important morphological features. In most Nematocera the eucephalic head capsule is sclerotized, with functional mandibles,

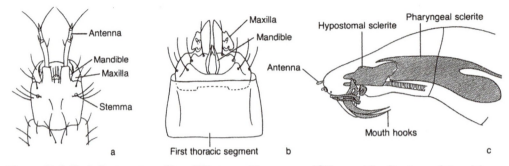

Figure 41.4 Cephalic structure of larval Diptera: a, Nematocera (Chironomidae, *Tanytarsus fatigans*); b, Brachycera, Orthorrhapha (Asilidae, *Dasyllis*), c, Brachycera, Cyclorrhapha (Anthomyiidae, *Spilogona riparia*). *Source: a, redrawn from Branch, 1923; b, redrawn from Malloch, 1917; c, redrawn from Keilin, 1917.*

maxillae, stemmata, and antennae (Fig. 41.4a). Mandibles move in a transverse plane, as in most mandibulate adult insects. In Orthorrhapha and some Nematocera, such as Tipulidae, the head capsule is unsclerotized posteriorly (hemicephalic) (Fig. 41.4b) and usually retracted within the thorax, which is tough and leathery. Mandibles move in a more nearly vertical plane. Cyclorrhapha are acephalic, without a differentiated head. Mouthparts consist of a pair of protrusible, curved "mouthhooks" and associated internal sclerites (Fig. 41.4c). To a certain extent larval morphology corresponds to habitat. For example, free-swimming nematoceran larvae frequently have pseudopodial appendages, sometimes jointed and bearing terminal bristles or hairs (Fig. 41.5a,b). Internal parasitoids and larvae inhabiting rich food sources, such as rotting flesh or dung, usually lack appendages (Fig. 41.5c).

LIFE HISTORY

Diptera are predominantly bisexual, with most cases of parthenogenesis recorded from the Nematocera. Eggs are usually small and ovoid, with thin, unsculptured chorion, and are typically deposited singly or in small masses on or near the larval food. Eggs usually hatch quickly, and egg diapause is infrequent. Cyclorrhapha include several ovoviviparous (larviparous) forms, as well as others that oviposit eggs that hatch immediately after deposition (ovolarviparous). Considerable larval development may occur in the mother's oviducts, and in Glossinidae (tsetse flies), Hippoboscidae, Streblidae, and Nycteribiidae

the larvae feed to maturity on oviduct secretions, then pupating almost immediately after deposition. Larval development in many families is extremely rapid, with many generations per year, but many aquatic species and free-living predators, such as Therevidae and Asilidae, have a single generation each year. Four larval feeding instars are typical, more being recorded for some species and only three in Cyclorrhapha, where the fourth instar has become specialized to produce the protective puparium that encloses the pupa. Aquatic families include some that use dissolved oxygen for respiration (Chironomidae, Simuliidae, Blephariceridae), but most obtain air by either rising periodically to the surface (Culicidae) or extending a breathing siphon to the surface (Syrphidae). Some mosquito larvae have spiracles modified to pierce the tissues of aquatic plants in order to obtain the contained air. Many endoparasitic larvae maintain an opening through the host's integument for breathing (Tachinidae, Oestridae), while others use dissolved oxygen.

Pupation occurs either in the larval substrate or a short distance away where drier conditions prevail. Obtect pupae frequently bear tubercles, horns, or other prominences on various parts of the body. In most cases these are probably used by the pharate adult to work its way through loose substrates before emerging from the pupal skin. In forms with coarctate pupae the puparium is burst along a transverse line of weakness by expansion of an eversible elastic sac, or ptilinum, which is extruded through the ptilinal suture, just above the antennal bases of the pharate adult. In the mature adult this

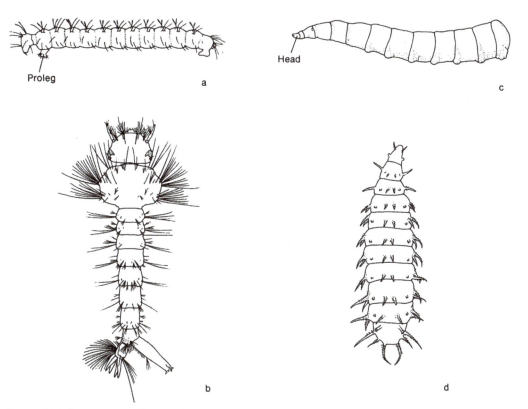

Figure 41.5 Representative Diptera larvae: a, Nematocera (Ceratopogonidae, *Forcipomyia specularis*); b, Nematocera (Culicidae, *Aedes stimulans*); c, Brachycera, Cyclorrhapha (Muscidae, *Musca domestica*); d, Brachycera, Cyclorrhapha (Anthomyiidae, *Fannia canicularis*).
Source: a, redrawn from Malloch, 1917; b, redrawn from Matheson, 1945; c, redrawn from Patton, 1931; d, redrawn from Detweiler, 1929.

suture is permanently sealed. Orthorrhapha and Cyclorrhapha pupate naked; many Nematocera spin silk cocoons. Adult emergence in aquatic forms usually occurs at the surface (Culicidae, Chironomidae, Tabanidae, etc.), but in cases in which the puparium is attached to the substrate, the adult emerges under water and floats quickly to the surface (Simuliidae, Blephariceridae).

Adult Diptera characteristically have rather short life spans of a few days to a few weeks, and some ephemeral forms do not feed. Courtship and mating commonly take place in the air, at least in part. Aerial swarming behavior, in which males form dense, dancing masses, often near characteristic landmarks or over water, precedes mating in many Nematocera and some Brachycera, notably Empididae. A female entering the swarm pairs with a male; usually the coupled flies then leave the swarm. Elaborate courtship behavior typifies various families. Empidid males frequently present females with corpses of prey. Many drosophilids and tephritids show complex premating behavioral displays involving posturing or visual presentation of the pictured wings by the males. In both families courtship differences are important isolating mechanisms among species.

CLASSIFICATION

Diptera are clearly members of the "mecopteroid" lineage of holometabolous insects, most closely related to either Mecoptera or Siphonaptera. Some authors think that Mecoptera may be paraphyletic with respect to both of the other two orders, but so far no convincing evidence has been presented. Some excellent phylogenetic studies on the higher level phylogeny and timing of evolutionary diversification in Diptera have appeared recently. For the most part these confirm earlier suspicions about the major patterns of fly evolution.

There has been continuing proliferation of family names in the Diptera, often as designations for one or a few small genera that did not entirely conform to larger, more well-known and accepted families. There have also been proposals for revolutionary reclassifications of all or part of the order (Griffiths, 1972). The classification presented here is conservative on both counts. Most of the family names omitted from the key apply to very uncommonly encountered insects, often with extremely specialized habits. Exhaustive treatments of North American Diptera include Curran's (1934), Cole's (1969), Griffiths' (1980–) and that of McAlpine et al. (1981–89). Larvae are treated in Stehr (1991) and Ferrar (1988).

Although flies are predominantly aerial as adults, about 20 families include members that are brachypterous or apterous. The following keys do not accommodate these flightless forms.

KEY TO THE SUBORDERS OF DIPTERA

1. Antennae with at least 6 segments, usually filiform or moniliform and frequently longer than thorax; maxillary palpi with 3 to 5 segments; pleural suture straight (Fig. 41.2a) **Nematocera**
   Antennae with fewer than 6 segments,[1] terminal segment usually either elongate or with bristlelike style or arista; maxillary palpi with 1 to 2 segments; pleural suture with two right-angle bends (Fig. 41.2c) **Brachycera**

---

[1] The terminal segment is sometimes annulate, giving appearance of more than 6 segments. The annulations are shallower and less well defined than the true divisions between the basal segments.

Suborder Nematocera

Adults: Antennae with 6 to 14 or more segments, usually filiform or moniliform, frequently longer than head and thorax, never with style or arista; maxillary palpi usually elongate, drooping, with 3 to 5 segments; mesothorax with pleural suture straight or slightly curved (Fig. 41.2a); wings usually without discal cell, and with veins Cu_2 and 2A never intersecting, usually diverging (Fig. 41.7a). Larvae: Head capsule sclerotized, with labium, maxillae, and horizontally biting, toothed mandibles. Pupa: Obtect, free (not enclosed in larval cuticle).

The flies comprising the suborder Nematocera are mostly frail-bodied, gnatlike insects with long, slender legs and thin, delicate cuticle. Body size is commonly about 1 cm or less, with the notable exception of some Tipulidae, which have wingspans greater than 100 mm. The great majority of Nematocera are dull yellowish, brown, or black, usually nocturnal insects.

Structurally, Nematocera include the most primitive Diptera. In adult Blephariceridae, which are predatory, the mandibles are stout, bladelike cutting organs. The maxillary palpi are long, with four segments, and the labium is elongate, not distally enlarged or specially textured as a labellum. Such mouthparts, which superficially are not greatly different from those of Bittacidae (Mecoptera) (see Fig. 42.1b), except in the development of the maxillae, may represent the primitive condition in Diptera. In Simuliidae, Culicidae, Ceratopogonidae, and Psychodidae, the mouthparts are only slightly modified from this condition and are used to puncture the integument, usually of vertebrate hosts. Specialization for ectoparasitism on vertebrates probably represents the initial adaptive radiation of the Diptera. In nonbiting Nematocera, the mouthparts are either vestigial (Deuterophlebiidae, Cecidomyiidae, Chironomidae) or adapted for imbibing nectar or other free liquids (Tipulidae, Scatopsidae, Bibionidae, Anisopodidae), with reduced mandibles, maxillae, and labellum.

Relatively large, articulated prothoracic sclerites are present in Tipulidae, Tanyderidae, Bibionidae, and Scatopsidae; a distinct metanotal sclerite is present in Psychodidae. Wing venation is highly variable. Four radial branches are characteristic of several families, and five radials are present in Tanyderidae, together with four branches of the media (Fig. 41.6a). This is probably the most generalized venation in the order; most families show fusion or loss of certain veins (Figs. 41.6b, 41.11, 41.14), and in Cecidomyiidae and Simuliidae all but two or three anterior veins are extremely weak or absent. Foldings or creases in the wing membrane form a faint network between the normal veins in Blephariceridae and Deuterophlebiidae (Fig. 41.6d). This peculiarity is apparently an adaptation to allow the fully hardened, folded wings to be used immediately after the adults emerge from the torrents and cataracts where pupation occurs.

Adult Nematocera are predominantly nocturnal or, if diurnally active, usually restrict their activity to deep shade or to periods of overcast skies. The major exception is the family Bibionidae, which consists of diurnal, flower-visiting members. Food habits of most adults are obscure, but members of families such as Scatopsidae, Anisopodidae, Tipulidae, Sciaridae, and Mycetophilidae probably imbibe liquid decay products from plant or animal material or nectar or honeydew from plants. Most Chironomidae, Deuterophlebiidae, and Cecidomyiidae, as well as some members of other families do not feed as adults. Some Blephariceridae prey on soft-bodied flying insects, including other flies and mayflies. Ceratopogonidae, commonly known for their habit of sucking blood from vertebrates, are predominantly parasites of other insects, usually much larger than themselves. Female Simuliidae (blackflies) and Culicidae (mosquitoes) use vertebrate blood as food, attacking reptiles, birds, and mammals. In many species a blood meal is required for production of fertile eggs. Members of both families transmit serious diseases of humans (see Chap. 11), constituting the chief economic importance of nematocerous flies.

Swarming behavior occurs in many families, being especially pronounced in Chironomidae, Bibionidae, and Mycetophilidae. Eggs are deposited on or near the larval substrate, usually singly, occasionally in adhering masses. Egg rafts, formed by one to many females, are characteristic of many mosquitoes. The great majority of Nematocera reproduce bisexually, but pedogenetic parthenogenesis occurs in *Miastor* (Cecidomyiidae), in which larviform females reproduce ovoviviparously. *Miastor* inhabits the fermenting phloem beneath bark of dead trees. Parthenogenesis is apparently a mechanism for rapidly exploiting this rich food source. When the tree begins to dry out, winged, bisexual adults appear.

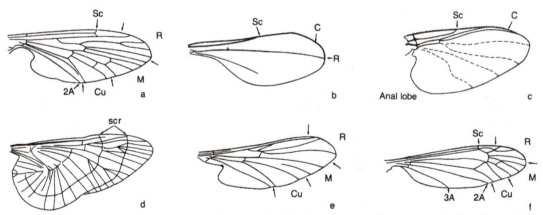

Figure 41.6 Wings of Nematocera: a, Tanyderidae (*Protoplasa*); b, Cecidomyiidae (*Contarinia sorghicola*); c, Simuliidae (*Simulium argus*); d, Deuterophlebiidae (*Deuterophlebia*); e, Blephariceridae (*Agathon comstocki*); f, Tipulidae (*Nephrotoma wulpiana*). sc, secondary creases; arrows at margin of wings (between veins) delimit branches of major veins.
Source: a, redrawn from Alexander; d, modified from Cole, 1969.

Larvae of Nematocera include aquatic, subaquatic, and terrestrial forms. Blephariceridae and Deuterophlebiidae occupy torrents and cataracts, clinging to stones with suckers and grazing on diatoms and algae. Simuliidae occur primarily in moving water, including rapids, attaching either to stones or larger aquatic organisms with the single sucker on the abdominal apex. All pupate near the site of larval attachment. Chironomidae and Culicidae occur in a diverse array of aquatic habitats. Chironomid larvae are usually bottom dwellers, frequently forming tubes in mud, silt, or under stones in still or moving water. Both well-oxygenated and stagnant waters are occupied, and many species have oxygen-binding pigments, including hemoglobin in the blood. Most chironomid larvae are detritivores, but the entire subfamily Tanypodinae is predaceous. *Clunio* larvae (Chironomidae) inhabit intertidal cracks and crevices throughout the world. Emergence of the terrestrial adults is usually synchronized with full moons, ensuring favorable low tides. *Pontomyia*, inhabiting coral reefs in the Indo-Pacific region, is marine in all instars. The highly modified males superficially resemble Veliidae (Hemiptera). They skate about on the surface of tide pools, carrying the vermiform females.

Culicid larvae, or mosquito wrigglers, are primarily inhabitants of standing water, but backwaters or eddies in slowly moving streams, rivers, or ditches offer suitable habitats as well. Many species are specialized for utilizing very small, enclosed bodies of water, such as rot holes in trees, potholes in rocks, crab burrows, or the water entrapped by bromeliads, pitcher plants, or certain other plants. Rainwater held by discarded cans and bottles provides the preferred larval habitat for some species, which have, in effect, become domesticated. Mosquito larvae are mostly filter feeders on plankton; a few species, including the entire subfamily Chaoborinae (phantom midges), are predatory.

Most dipterous larvae, aquatic or not, require atmospheric oxygen for respiration. The major exceptions are the Blephariceridae, Deuterophlebiidae, and Chironomidae, which respire either through tracheal gills or, occasionally, with

plastrons. Some mosquito larvae can respire slowly under water and may overwinter in this way, but most other "aquatic" Nematocera will drown if long separated from the air. Most of these subaquatic larvae inhabit littoral regions of streams or ponds, semiliquid decaying vegetation or sewage, or other wet situations. A few, such as Ptychopteridae, extend a flexible breathing tube to the surface for air, an adaptation encountered again in some syrphid larvae, but most have only short breathing tubes or none. Many Tipulidae and Ceratopogonidae, as well as the Psychodidae, Anisopodidae, Thaumaleidae, Tanyderidae, and Trichoceridae, can properly be classified as subaquatic. Within their semiliquid habitats these larvae function mostly as decomposers, less frequently as herbivores or predators.

The families Scatopsidae, Bibionidae, Sciaridae, and Mycetophilidae, as well as some Ceratopogonidae and Tipulidae, are best considered terrestrial, though the larval medium is usually moist. Larvae in this category are predominantly decomposers, differing chiefly in the type of material consumed. Mycetophilidae and Sciaridae are commonly associated with fungi, ranging from molds to mushrooms and bracket fungi; Scatopsidae frequent dung, and Bibionidae, moist humus. Bibionid larvae may feed on the roots of plants, occasionally becoming economically important. In all these families exceptional members are predatory. For example, the luminescent "glowworms" of Waitomo Cave in New Zealand are predatory mycetophilid larvae, which dangle mucilaginous filaments to entangle their prey. The mycetophilid genus *Planivora* is an internal parasite of terrestrial planarians.

The Cecidomyiidae cannot be accommodated in a simple classification by larval habitat. These minute, fragile flies are best known as the causative agents of galls, which the larvae produce on a wide variety of herbaceous plants. Other members mine leaves or stems and include agriculturally important species, such as the Hessian fly (*Mayetiola*). A few are predators or parasitoids of aphids, coccids, and other soft-bodied insects, and some live in forest litter and soil, probably feeding on fungi.

In Simuliidae and Mycetophilidae pupation occurs within silken cocoons, but the obtect pupae of most families are naked. Blephariceridae,

Deuterophlebiidae, and Simuliidae, which often pupate in waterfalls or rapids, surface, and immediately take flight without an initial period of wing expansion and hardening of the cuticle. Culicids are exceptional in remaining fully vagile throughout the pupal period.

KEY TO THE FAMILIES OF THE SUBORDER NEMATOCERA

1. Wings about 3 times or less as long as broad, with fringe of marginal setae less than half as long as wing width   **2**

   Wings at least 5 times as long as broad, with fringe of marginal setae more than half as long as wing width

   Nymphomyiidae

2. Wings with secondary network of fine creases or folds in addition to venation (Fig. 41.6d); mesonotum with transverse suture straight

   Deuterophlebiidae and **Blephariceridae**

   Wings without secondary network of creases or folds; mesonotum with transverse suture straight or V-shaped   **3**

3(2). Mesonotum with transverse suture V-shaped, extending to vicinity of scutellum (Fig. 41.2b)   **4**

Mesonotum with transverse suture straight or weakly curved   **6**

4(3). Head with 2 to 3 ocelli   **Trichoceridae**

   Head without ocelli   **5**

5(4). Wing with 2 anal veins reaching margin (Fig. 41.6f); antennae usually with 12 to 13 segments   **Tipulidae**

   Wing with 1 anal vein reaching margin (Fig. 41.6a); antennae with 15 to 25 segments   **Ptychopteridae** and **Tanyderidae**

6(3). Ocelli present   **8**

   Ocelli absent   **13**

7(6). Costal vein ending at or near wing tip (Fig. 41.6c–f)   **8**

   Costal vein weakened behind wing tip but continuing around posterior margin of wing (Fig. 41.6b); wings with 3 to 6 longitudinal veins, usually without crossveins; minute flies with long, many-segmented antennae

   Cecidomyiidae (Fig. 41.8b)

8(7). Wing with discal cell (M_2) present (Fig. 41.7a)   **9**

   Wing without discal cell (Fig. 41.7b)   **10**

9(8). Wing with cell M_3 open externally (Fig. 41.7a)

   Anisopodidae (incl. **Pachyneuridae**)

   Wing with cell M_3 closed (Fig. 41.11a) (aberrant orthorrhaphous Brachycera)

   Xylophagidae

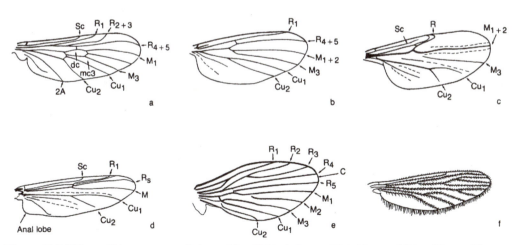

Figure 41.7 Wings of Nematocera: a, Anisopodidae (*Sylvicola fenestralis*); b, Mycetophilidae (*Mycetophila fungorum*); c, Ceratopogonidae (*Culicoides variipennis*); d, Chironomidae; e, Psychodidae (*Pericoma truncata*); f, Culicidae (*Culex tarsalis*). dc, discal cell or cell M_2; mc3, cell M_3.

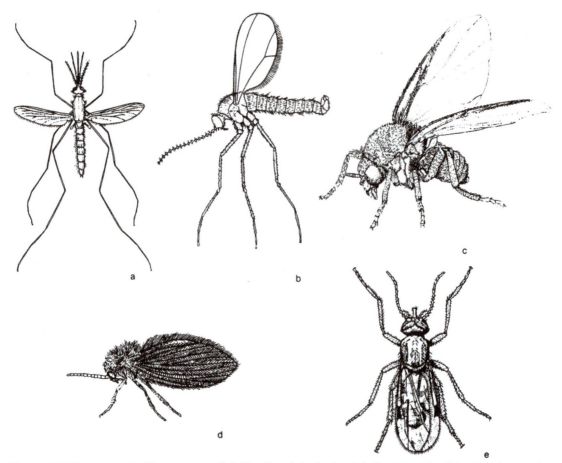

Figure 41.8 Representative Nematocera: a, Culicidae (*Anopheles freeborni*); b, Cecidomyiidae (*Mayetiola destructor*); c, Simuliidae (*Prosimulium mixtum*); d, Psychodidae (*Psychoda*); e, Ceratopogonidae (*Culicoides variipennis*). *Source: a, redrawn from Wilson, 1904; c, redrawn from Peterson, 1970; b, d, redrawn from Cole, 1969; e, redrawn from Hope, 1932.*

10(8). Tibiae without apical spurs (tibial apex may be pointed, spurlike); wings with anterior veins usually much thicker than posterior veins **18**

Tibia with articulated spurs near apex, wings usually with all veins of about same thickness **11**

11(10). Tarsi with broad, fleshy pads (pulvilli) beneath claws (Fig. 41.9a) **Bibionidae**

Tarsi without pulvilli **12**

12(11). Eyes meeting above antennal bases (Fig. 41.9b); coxae moderate in length (Fig. 41.8b); occasionally elongate **Sciaridae**

Eyes separated dorsally (Fig. 41.9c); coxae nearly always greatly elongate **Mycetophilidae[1]**

13(6). Costal vein ending at or near wing tip (Fig. 41.7c,d) **14**

Costal vein weakened behind wing tip, but continuing around posterior margin (Fig. 41.7e) **16**

14(13). Wings broad, oval, with large anal lobe; anterior veins always much stronger than

---

[1] A few Anisopodidae and Pachyneuridae will key out here.

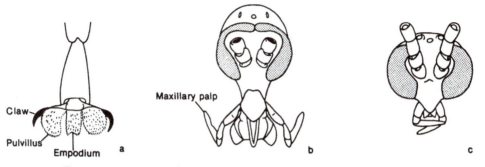

Figure 41.9 Taxonomic characters of Nematocera: a, tarsus of Bibiidae; b, frontal view of head of Sciaridae; c, frontal view of head of Mycetophilidae.

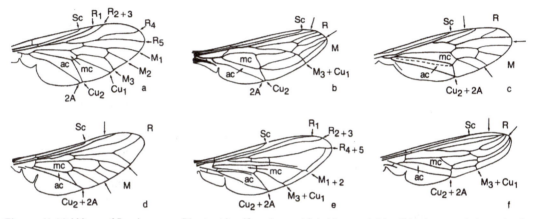

Figure 41.10 Wings of Brachycera: a, Rhagionidae (*Symphoromyia*); b, Nemestrinidae (*Neorhyncocephalus sackeni*); c, Xylophagidae (*Xylophagus decorus*); d, Tabanidae (*Tabanus laticeps*); e, Syrphidae (*Syrphus opinator*); f, Apioceridae (*Rhaphiomidas maehleri*). ac anal cell; mc, median cell; marginal arrows delimit branches of major veins.

faint posterior veins (Fig. 41.6c)

Simuliidae (Fig. 41.8c)

Wings usually lanceolate, elongate, with anal lobe small or absent (Fig. 41.7c,d); venation variable   **15**

15(14). Wing with radial veins intersecting costal margin near midpoint (Fig. 41.7c); radial veins usually stronger than posterior veins; mouthparts of piercing type

Ceratopogonidae (Fig. 41.8e)

Wing with radial veins extending nearly to apex (Fig. 41.7d); radial and posterior veins similar; mouthparts without piercing mandibles

Chironomidae

16(13). Wings broad, with pointed tip; membrane usually densely hairy; radial vein

usually with 5 branches (Fig. 41.7e)

Psychodidae (Fig. 41.8d)

Wings usually narrow, lanceolate, and bare (veins may bear short hairs); radial vein usually with 3 to 4 or fewer branches   **17**

17(16). Wings with 7 or fewer longitudinal veins reaching margin (Fig. 41.6b); minute, frail flies some

Cecidomyiidae and **Thaumaleidae**

Wings with at least 9 longitudinal veins reaching margin (Fig. 41.7f)

Culicidae (including **Chaoboridae, Dixidae** (Fig. 41.8a)

18(10). Wing with costal vein ending much before apex; costal and radial veins darkly pigmented, prominent, others faint

Scatopsidae (including **Synneuridae**)

Wing with costal vein continuing to apex; anterior and posterior veins of about equal prominence **Axymyiidae**

Suborder Brachycera

Adults: Mostly stouter bodied than Nematocera, with antennae short, with terminal style or arista; maxillary palpi usually short, with 1 to 2 porrect segments; pleural suture bent around sternopleural sclerite (Fig. 41.2c); wings with discal cell present (Orthorrhapha) or absent (Cyclorrhapha), anal cell narrowed before wing margin (Figs. 41.10, 41.11, 41.14). Larvae: Head capsule incomplete or vestigial, usually retractable into thorax; mandibles biting in vertical plane (Fig. 41.4b), or mouthparts represented by specialized "mouthhooks" (Fig. 41.4c).

Brachycera comprise the great majority of flies, including many familiar groups such as horse-flies, house flies, and blowflies. The suborder consists of two groups of families, termed divisions. The taxonomic divisions of Diptera are groups of families and should not be confused with the divisions Exopterygota and Endopterygota. The taxonomic category series is used to designate major subdivisions of the Cyclorrhapha, which are difficult to separate by adult features but are well differentiated by larval and pupal characteristics. The division Orthorrhapha is characterized by hemicephalic larvae with vertically biting mandibles. The pupa is obtect (exception: exarate, coarctate in Stratiomyidae). In the division Cyclorrhapha, larvae are acephalic, without true mandibles. The pupa is coarctate, enclosed in a hardened puparium formed from the cuticle of the last larval instar. Cyclorrhapha are further subdivided as the series Schizophora, characterized by the presence of the ptilinal suture just above the antennae. In the series Aschiza the ptilinum and ptilinal suture are absent, although pupation is within a puparium. In the following key the Aschiza segregate with the Orthorrhapha, with which they share several adult features.

KEY TO THE MAJOR DIVISIONS OF THE SUBORDER BRACHYCERA

1. Head with U-shaped ptilinal suture just above antennae (Fig. 41.15a); antennae usually with 3 segments, terminal segment with arista or style attached dorsally before apex (Fig. 41.15a–h)
Cyclorrhapha, Series **Schizophora**
Head without U-shaped ptilinal suture; antennae 3- to 5-segmented, terminal segments elongate or with style or arista attached terminally (Fig. 41.13c) **Orthorrhapha** and **Cyclorrhapha**, Series **Aschiza**

Division Orthorrhapha

Differentiated from Cyclorrhapha primarily by the obtect pupae (pupae exarate, enclosed in puparium in Cyclorrhapha). Exception: Stratiomyidae, with pupa obtect, enclosed in unmodified larval cuticle.

The Orthorrhapha comprise an extremely diverse assemblage of flies, including several of the largest families. Body form varies from rather frail, slender types similar to Nematocera (e.g., Rhagionidae) to those with stout, compact shape resembling that of Cyclorrhapha (e.g., Tabanidae and Dolichopodidae). Many Empididae and Dolichopodidae are minute, but body size in Orthorrhapha averages larger than in Nematocera or Cyclorrhapha. Members of the Asilidae (robber flies), Mydidae, and Apioceridae include the largest flies, with body lengths up to 60 mm. Most Orthorrhapha are active diurnal insects, often conspicuous on flowers or foliage.

Most of their morphological features suggest that the Orthorrhapha are intermediate between the Nematocera and Cyclorrhapha, from which some Orthorrhapha cannot be distinguished on the basis of adult characteristics. For example, antennae in some Coenomyiidae have up to ten segments and in some Rhagionidae have eight segments, recalling the multiarticulate antennae of Nematocera. In contrast, Dolichopodidae and many Empididae, as well as some members of several other families, have short antennae with three to four segments with a terminal arista, as in some Cyclorrhapha. This intermediacy involves all the characteristics used to separate the Cyclorrhapha and Nematocera, including mouthpart structure, thoracic structure and wing venation, genitalic features, and internal organ systems.

The Rhagionidae seem to represent the least specialized Orthorrhapha. Many of these usually soft-bodied flies have piercing-sucking mouthparts with well-developed, bladelike mandibles and maxillae. Similar mouthparts occur in Tabanidae (horseflies). In the predatory Asilidae and Empididae, at least the mandibles are absent, and usually only the hypopharynx is used to pierce the prey. Many Orthorrhapha [Bombyliidae (bee flies), Mydidae, Apioceridae, Acroceridae, Nemestrinidae] have mouthparts adapted for sucking up nectar through an elongate proboscis formed by the labium. In a few Empididae and in Dolichopodidae the labium is produced as a short, flexible proboscis tipped by a fleshy labellum. These flies are predatory, the prey being pierced by cuticular pseudotracheal teeth along the medial groove of the labellum.

The wing venation is very generalized in several families, notably the Rhagionidae (Fig. 41.10a), with the radial vein four-branched and the medial vein four-branched. In a number of families one or more branches of the radial or medial veins are looped forward to intersect the wing margin before its apex (Fig. 41.10b,f) In several of these families (e.g., Bombyliidae, Asilidae, Apioceridae, Mydidae) venation is highly distinctive and useful in identification. In many Empididae and Dolichopodidae the venation is essentially the same as in the Cyclorrhapha, with the radius three-branched, the median two-branched, and the and R_1 extremely short (Fig. 41.11g). Calypters are typically small or absent in Orthorrhapha but are large and conspicuous in Tabanidae and Acroceridae.

Two features in which Orthorrhapha are clearly intermediate between the more generalized Nematocera and the Cyclorrhapha are (1) torsion of the external genitalia in males and (2) the degree of consolidation of the central nervous system. In most Nematocera rotation is not evident, but torsion of 90 to 180°, frequently accompanied by asymmetry of genitalic structures, is typical of most Orthorrhapha. In Dolichopodidae and Cyclorrhapha, torsion (here termed circumversion) has proceeded through 360°, so that the original dorsoventral relationships are restored. Male terminalia of Cyclorrhapha are also usually folded ventrally or ventrolaterally and markedly asymmetrical, making the structures difficult to homologize with those in other Diptera.

The consolidation of the central nervous system is probably related to flying ability. Nematocera, which are often weak, periodic fliers, have three thoracic and six to seven abdominal ganglia. In Orthorrhapha a variable amount of condensation has occurred, and in Cyclorrhapha a single large thoracic ganglion or one thoracic and one abdominal ganglion remain.

Biology

Parasitic blood feeding, of common occurrence in adult Nematocera, reappears in many Tabanidae and in a few Rhagionidae (*Symphoromyia*). In these flies, as in Nematocera, only females are bloodsuckers, opening a wound with the blade-shaped mandibles and lapping up the oozing blood rather than inserting the entire piercing apparatus as in mosquitoes. Females of some species and males have the mandibles atrophied and feed on nectar. Predation, encountered in such Nematocerous families as Ceratopogonidae and Blephariceridae, is universal among Asilidae and widespread among Empididae and Dolichopodidae. Asilidae are aggressive, strong-flying predators, attacking any insects small enough to subdue, including wasps and bees, large Orthoptera, and beetles. The legs are long and robust, frequently bearing strong bristles useful in holding struggling prey. The posterior pairs are successively longer, and together the legs form a basket much as in the Odonata. The strong bristles about the eyes and face probably prevent thrashing prey from injuring the fly's head. Prey items are clasped tightly against the robberfly's body, punctured by the stout, sharp oral stylets and injected with rapidly acting toxic saliva. Many collectors have learned that the pain from a robberfly's bite is as instantaneous and severe as that of stinging Hymenoptera. A number of Asilidae strikingly resemble bumblebees and wasps, doubtless gaining protection from potential predators, but possibly using the resemblance to facilitate predation on flower visiting insects. One group of asilids (Leptogastrinae) superficially resemble ichneumonid wasps, with long slender body and legs. They pick small insects off surfaces, and some species pluck spiders from their webs. Empididae and Dolichopodidae never

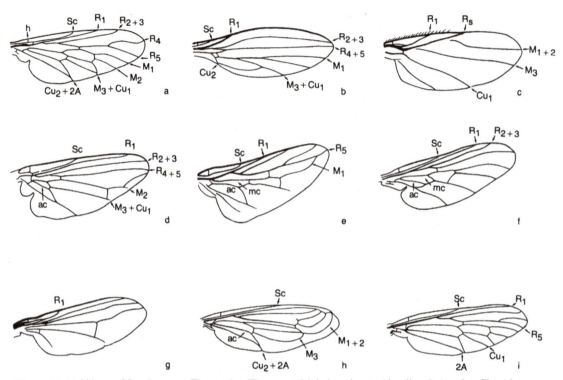

Figure 41.11 Wings of Brachycera: a, Therevidae (*Thereva vialis*); b, Lonchopteridae (*Lonchoptera*); c, Phoridae (*Chaetoneurophora variabilis*); d, Platypezidae (*Calotarsa insignis*); e, Scenopinidae (*Metatrichia bulbosa*); f, Empididae (*Rhamphomyia*); g, Dolichopodidae (*Tachytrechus angustipennis*); h, Mydidae (*Nemomydas pantherinus*); i, Asilidae (*Cyrtopogon montanus*). ac, anal cell; mc, median cell.

achieve the large size of Asilidae and tend to inhabit shadier or moister situations, frequently about seeps or running or standing water. Most members of both families are nimble runners, and many run down their prey rather than pursuing it in the air. Empididae commonly use prey items in courtship, as discussed below.

Adults of Acroceridae, Bombyliidae, Mydidae, Nemestrinidae, Apioceridae, and occasional species of several other families of Orthorrhapha feed on nectar, and some Bombyliidae eat pollen. Some of these nectar-feeding species are equipped with elongate proboscises and hover above flowers while feeding. Other species have only a short proboscis or labellum and alight on the flower. In several families, notably Therevidae, Scenopinidae, Coenomyiidae, and Xylophagidae, adult food is unknown, and some of these may not feed as adults. Saprophagy, extremely widespread

among Cyclorrhapha, is uncommon among adult Orthorrhapha.

Most Orthorrhapha are solitary, but male swarming is highly developed in Empididae and occurs in some Rhagionidae. Individual females enter the swarms of males before pairing takes place. As in Nematocera the swarms are often located above characteristic objects such as shrubs, sunlit spots in forests, or water. During courtship, males of many empidids present females with prey, among which smaller nematocerous Diptera are especially favored. In some species courtship behavior has become ritualized to the extent that males (1) present the female with the prey plastered to the side of a silken or frothy balloon; (2) present the balloon with a nonprey item (small stick, sand grain); or (3) present the balloon alone or present a nonprey item alone. Both sexes of empidids also take nectar, but in many species the only proteinaceous food is that received from males during courtship. Courtship in Dolichopodidae often involves signaling by the males, using the brushes and flanges on the tibiae and tarsi.

Most Orthorrhapha oviposit moderate numbers of eggs on or near the larval substrate. Bombyliidae, Acroceridae, and Nemestrinidae, whose larvae are internal parasitoids of other insects, oviposit extremely large numbers of eggs (up to 900 recorded in Acroceridae and over 4000 in Nemestrinidae) in habitats where the active first-instar larvae are likely to encounter hosts. Most Orthorrhapha apparently deposit the eggs singly or in small batches, but horseflies form tiered egg masses on emergent aquatic vegetation or other objects near aquatic habitats. *Atherix* (Rhagionidae) forms communal egg masses on twigs, small branches, culverts, or other objects overhanging water. The flies die in place after oviposition, their bodies combining with the eggs to form encrustations that were formerly used as food by some Native Americans.

Larvae of Orthorrhapha occupy a broad array of habitats, ranging from aquatic to terrestrial. Aquatic larvae (some Rhagionidae, Stratiomyidae, Tabanidae, Empididae, and Dolichopodidae) require atmospheric oxygen, restricting them to shallow water. Most are inhabitants of seeps, marshes, or the littoral regions of lakes or streams, rather than of open water situations. Empididae, Dolichopodidae, Asilidae, and Xylophagidae usually inhabit moist soil, leaf litter, or rotten wood. Therevidae, Apioceridae, and Mydidae frequent rotten wood, soil, or sand, frequently in quite arid circumstances. Scenopenid larvae are most frequently encountered in carpets, where they feed on moth larvae and other insects.

Larvae of Orthorrhapha are commonly predaceous. For example, therevid larvae, which mostly inhabit loose sandy soils, are general predators of other arthropods, especially the larvae of Elateridae, Scarabaeidae, and Tenebrionidae, and mydids also feed on immature Coleoptera, as far as is known. Biologies of Rhagionidae, Apioceridae, and Xylophagidae are very poorly known, but most are believed to be predators as immatures. The larvae of *Vermileo* (Rhagionidae) construct steep-walled pits in dry, friable soil and trap small arthropods, much in the manner of antlions (Neuroptera, Myrmeleontidae). Several families include both predatory and herbivorous species (Stratiomyidae, Tabanidae) or scavengers (Empididae, Dolichopodidae, Xylomyidae). Tabanid larvae mostly inhabit aquatic or very moist soils, feeding on diverse arthropod prey. Large larvae of *Tabanus* are known to kill newly emerged toads as well. Larvae of Dolichopodidae and Empididae are believed to be generalized predators. Dolichopodids inhabit moist soil, rotting wood, and similar situations. Empididae include both aquatic species and others that live in moist soil.

Larvae of Acroceridae, Nemestrinidae, and Bombyliidae are all internal parasitoids of other arthropods. Acroceridae utilize spiders exclusively; nemestrinid larvae feed on grasshoppers and larvae of scarab beetles. Bombyliidae feed on grasshopper egg masses and are parasitoids of the larvae or pupae of a variety of immature endopterygote insect hosts, including Hymenoptera, Lepidoptera, Diptera, Coleoptera, and Neuroptera. First instars of these flies are vagile, planidiumlike larvae that locate the host upon which the later maggotlike instars feed.

Pupation occurs in or near the larval substrate. The conspicuous, posteriorly directed spines of many orthorrhaphous pupae aid the pharate adult in working its way to the surface of loose sand or soil, from which the cast pupal skins may frequently be seen protruding. Stratiomyidae are unusual in having the exarate pupa enclosed in the cuticle of the last larval instar, which is not modified as a puparium as in the Cyclorrhapha. This arrangement probably functions in floating the pupa in the stagnant waters where many Stratiomyidae develop and is apparently not homologous to the cyclorrhaphous puparium.

The families of the Cyclorrhapha Aschiza display diverse biological adaptations. Lonchopteridae are apparently nectar or honeydew feeders as adults, while the larvae are saprophytic in soil, decaying leaves, and similar situations. Platypezid adults imbibe honeydew or various exudates from leaves in forest understory. The males aggregate in mating swarms in forest openings. The larvae feed in the fruiting bodies of various soft or hard fungi and pupate in the soil. Phoridae are scavengers in many situations, including faces, decomposing vegetation, fungi, and the nests of bees, wasps, termites, and rodents. Other species feed in corpses (including those of humans) or parasitize insects, spiders, and millipedes.

Syrphidae are familiar as adults, which often mimic stinging Hymenoptera and frequently hover motionless above vegetation or flowers. Most of the adults imbibe nectar and eat pollen; the larvae have diverse habits. Most familiar are the taxa that feed on aphids and other Homoptera. Related forms are predators of Thysanoptera and immature Coleoptera and Lepidoptera. Other groups live in decaying succulent plants, especially cacti, or infest the bulbs of monocotyledonous plants, where they are of some economic importance. Still other species inhabit moist feces or water with high organic content, such as that of tree holes, cesspools, or sewage. These larvae are known as rat-tailed maggots after their elongate breathing siphons.

Pipunculidae, which superficially resemble syrphids with enormous heads, as larvae are endoparasites of Homoptera, especially Cicadellidae and Delphacidae. The adult females capture immature hosts and oviposit into their hemocoels. The mature larva leaves the host to pupate in the soil.

KEY TO THE FAMILIES OF THE DIVISIONS ORTHORRHAPHA AND CYCLORRHAPHA (SERIES ASCHIZA)

1. Tarsi with 3 subequal pads below claws (Fig. 41.9a)   **2**
   Tarsi with 2 or fewer pads below claws (Fig. 41.13a)   **8**
2(1). Head very small, thorax and abdomen inflated; calypters very large **Acroceridae**
   Body not so shaped; calypters small or vestigial   **3**
3(2). Antennae with third segment annulate (Fig. 41.13b)   **5**
   Antennae with third segment not annulate; usually elongate or with terminal bristle (Fig. 41.13c)   **4**

4(3). Wings with branches of median vein parallel to posterior margin (Fig. 41.10b)
      Nemestrinidae
   Wings with branches of median vein intersecting posterior margin (Fig. 41.10a)
      some **Rhagionidae**[1]
5(3). Calypters large, conspicuous   **6**
   Calypters small or vestigial   **7**
6(5). Anal cell open (as in Fig. 41.10a); eyes densely hairy   some **Rhagionidae**[1]
   Anal cell closed at or before wing margin (Fig. 41.10c,d); eyes not usually hairy
      Tabanidae (Fig. 41.12a)
7(5). Tibial spurs present at least on middle legs
      20
   Tibial spurs absent   **Stratiomyidae**
8(1). Head hemispherical, larger than thorax, with antennae usually inserted below midline of long, narrow face   **Pipunculidae**
   Head much smaller than thorax, usually with broad face; antennae inserted above midline of head   **9**
9(8). Cu_2 long, reaching wing margin or joining 2A near margin; anal cell much longer than median cell (Fig. 41.10e)   **10**
   Cu_2 absent (Fig. 42.11c) or, if present, joining 2A more than a quarter of its length in from margin (Fig. 41.11d,e,f); anal cell subequal to median cell (Fig. 41.11d,e,f)   **15**
10(9). Medial veins looped forward with M_{1+2} joining R_{4+5} before wing margin (Fig. 41.16a); antennae usually with dorsal bristle
      Syrphidae (Fig. 41.16a)
   Medial veins not joining R_{4+5} (Fig. 41.10f) (but radial veins may unite before wing margin); antennae usually with terminal style **11**
11(10). Vertex concave between eyes (Fig. 41.13d)
      12
   Vertex flat or convex between eyes (Fig. 41.15a–c)   **13**
12(11). Head with 3 ocelli; antennae usually with 3 segments   **Asilidae** (Fig. 41.12b)
   Head with 1 or no ocelli; antennae with 4 segments   **Mydidae**
13(11). Veins R_5 and M_1 looped forward, terminating anterior to wing apex (Fig. 41.10f)
      Apioceridae

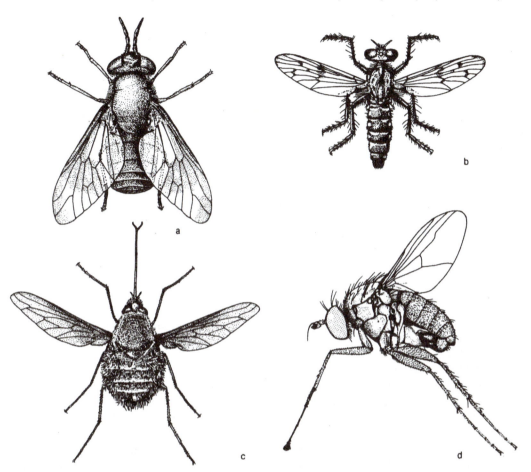

Figure 41.12 Representative Brachycera, Orthorrhapha: a, Tabanidae (*Chrysops proclivus*); b, Asilidae (*Metapogon pictus*); c, Bombyliidae (*Bombylius lancifer*); d, Dolichopodidae (male *Dolichopus*). *Source: From Cole, 1969.*

Veins R_5 and M_1 intersecting wing margin posterior to apex (Fig. 41.11a)   **14**

14(13). Wings with 3 median and cubital cells (Fig. 41.12c); usually scaly or hairy flies with long proboscis

Bombyliidae (including Hilarimorphidae) (Fig. 41.12c)

Wings with 4 median and cubital cells (Fig. 41.11a); body bare or pubescent; proboscis short   **Therevidae**

15(9). Wings with pointed apex; crossveins restricted to extreme basal portion of wing (Fig. 41.11b)   **Lonchopteridae**

Wings with rounded apex; distal crossveins present or absent   **16**

16(15). Radial veins much thicker than medial and cubital veins, which are without crossveins (Fig. 41.11c); antennae appearing 1-segmented   **Phoridae**

Veins all of approximately equal thickness; mediocubital region of wing with at least 1 crossvein; antennae with 2 or more obvious segments   **17**

17(16). Hind tibiae and tarsi dilated, flattened; wings with anal cell pointed distally (Fig. 41.11d)   **Platypezidae**

Hind tibiae and tarsi cylindrical or rounded; if dilated, anal cell not pointed   **18**

18(17). Vein M_1 curved anteriorly, intersecting wing margin before apex (Fig. 41.11e); antennae without arista or style   **Scenopinidae**

Vein M_1 straight or curved posteriorly, intersecting wing margin behind apex

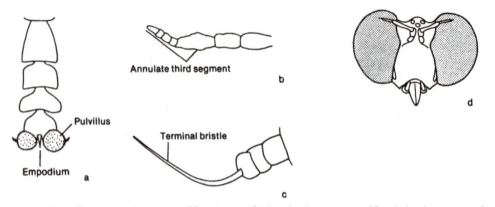

Figure 41.13 Taxonomic characters of Brachycera, Orthorrhapha: a, tarsus of Syrphidae; b, antenna of Tabanidae; c, antenna of Rhagionidae; d, head of Asilidae.
Source: b, redrawn from Brennan, 1935.

(Fig. 41.11f); arista or style frequently present **19**

19(18). Wings with r-m crossvein very close to base; R₁ usually joining C in basal half of wing (Fig. 41.11g); calypters usually conspicuous, fringed; proboscis usually terminating as a fleshy labellum.
Dolichopodidae (Fig. 41.12d)
Wings with r-m crossvein located more than one-fourth distance from base to apex; R₁ usually joining C in distal half of wing (Fig. 41.11f); calypters small; proboscis usually a pointed beak **Empididae**

20(7). Costa ending near apex of wing **21**
Costa continuing around posterior margin of wing **22**

21(20). Foretibia with ventral apical spur some **Xylophagidae[2]**
Foretibia without ventral apical spur **Xylomyidae**

22(20). Clypeus convex, exposed some **Rhagionidae[1]**
Clypeus recessed in deep facial groove and flattened some **Xylophagidae[2]**

[1] Including Vermelionidae, Athericidae, Pelechorhynchidae.
[2] Including Coenomyiidae.

Division Cyclorrhapha

Differing from Orthorrhapha chiefly in having the pupae coarctate and enclosed in a puparium formed by the modified cuticle of the last-instar larva.

The Cyclorrhapha comprise a vast diversity of mostly small, compactly built flies. The great majority of species emerge by pushing off the end of the puparium by inflating the ptilinum. The ptilinal opening is represented by a faint crescentic scar just above the antennal sockets in the fully hardened flies, hence the series name Schizophora. A few families, comprising the series Aschiza, considered in the previous section, lack the ptilinum.

The series Schizophora may be further divided into sections (groups of families). The section Calyptratae includes flies with relatively large calypters (muscids, tachinids, etc.; Fig. 41.18d). The section Acalyptratae contains flies with the calypters small or absent. Included here are numerous families of small flies that are usually difficult to identify.

Mouthparts are much reduced in all Cyclorrhapha, with mandibles absent and maxillae represented by palpi. The labium is modified as a labellum that bears sharp pseudotracheal or prestomal teeth in predatory or parasitic forms. The head typically bears bristles in characteristic positions, which are of great taxonomic utility.

Wing venation is reduced, leaving no more than three radial and three medial branches. The costa is

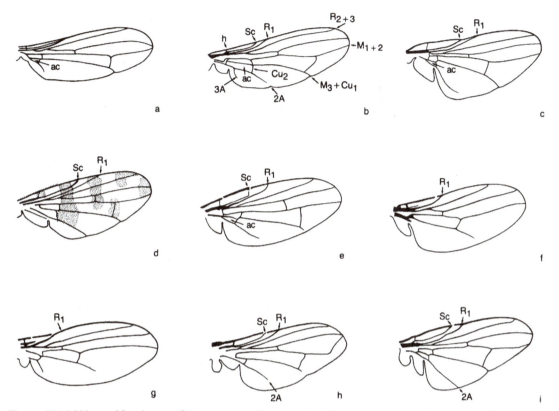

Figure 41.14 Wings of Brachycera, Cyclorrhapha: a, Micropezidae (*Compsobata mima*); b, Sciomyzidae (*Sepedon pacifica*); c, Chamaemyiidae (*Leucopis bivitta*); d, Otitidae (*Ceroxys latiusculus*); e, Tephritidae (*Trupanea nigricornis*); f, Chloropidae (*Thaumatomyia pulla*); g, Agromyzidae (*Cerodontha dorsalis*); h, Muscidae (*Musca domestica*); i, Anthomyiidae. ac, anal cell.

often broken at one or more places; in several families one or more radial or medial veins are looped anteriorly, sometimes joining the next anterior vein. Such anomalies produce highly characteristic venational patterns, which are very useful in identifying many families.

The abdomen of Cyclorrhapha is characteristically divided into an anterior preabdomen and the postabdomen, which is flexed anteroventrally beneath the preabdomen in males. Male genitalia are rotated 360°. This extreme degree of torsion, termed circumversion, differentiates Cyclorrhapha from most Orthorrhapha.

Biology

The most common feeding mode of cyclorrhaphous larvae is decomposition; these larvae are the primary insect consumers of most decaying plant and animal material (Table 41.1). High vagility of adults allows prompt location and exploitation of temporary food sources. High mobility may also be used to exclude competitors. An extreme example is the horn fly (*Haematobia irritans*), which resides on cattle, immediately ovipositing in freshly dropped dung before other insects are attracted.

The variety of feeding substrates of saprophagous Cyclorrhapha is very great. Some species are quite catholic, but most restrict themselves to either plant or animal material, and many are highly specific. For example, Neriidae (in North America) are limited to rotting cacti, Coelopidae to decaying seaweed, and many species of other families have narrow substrate ranges, including such items as fermenting tree sap, frass in burrows

Table 41.1 Biological Characteristics of Common Families of Cyclorrhapha Acalyptratae

Family	Larval habitat/habits	Larval food habits	Adult habitat	Adult food habits
Agromyzidae	Miners in fruits, seeds, twigs, stems, roots and especially leaves; mostly monophagous or oligophagous	Plant tissues	Near host plants	Probably nectar, honeydew, and various plant or animal exudates
Anthomyzidae	Borers in grass stems	Phytophagous or saprophagous	Moist, grassy areas	Unknown
Chamaemyiidae	Predators of aphids, adelgids, coccids, and scale insects	On plants near prey	Near larval hosts	Probably honeydew and nectar
Chloropidae	Mostly borers in grass stems or decaying plants; some in insect frass, in galls, in fungi or predators on insect larvae or eggs. One species is a cutaneous parasite of frogs	Mostly phytophagous or saprophagous	Common on low plants; some attracted to animal exudates	Mostly nectar; some feed on animal exudates
Clusiidae	Rotting wood and termite galleries	Probably saprophagous	On decaying tree trunks	Nectar, sap, rotting plants
Coelopidae	In rotting seaweed	Rotting seaweed	Usually along sea coasts	Rotting seaweed
Conopidae	Internal parasitoids of aculeate wasps, cockroaches, and calyptrate flies	Host tissues	On flowers or near larval hosts	Nectar
Dryomyzidae (including Helcomyzidae)	Carrion, dung, decaying fungi and seaweed; predators of barnacles	Saprophagous or zoophagous	Abundant in many habitats	Unknown
Ephydridae	Aquatic or semi-aquatic, especially in shallow water near shore lines; a few saprophagous in excrement, etc. or leaf miners, parasitoids or predators	Mostly microphagous filter feeders on bacteria, algae, yeasts	Abundant near shorelines	Mostly microphagous on algae, bacteria, etc.; some scavengers on dead insects; a few predators
Heleomyzidae	Many types of decaying plant or animal material, including dung, carcasses, fungi	General saprophages	Near larval food and in moist, shady areas	Unknown
Lauxaniidae	Decaying leaf litter or other plant material; dung	Microphagous on fungi associated with decay	Low vegetation in moist, shady areas	Plant exudates; occasionally nectar
Loncheidae	Decaying flowers, fruits, tubers, succulent plants, or subcortical region of decaying trees	General saprophages	Near larval habitats	Unknown
Micropezidae	Decaying plant matter; dung; tree holes	Saprophagous	Marshes, moist woodlands	Unknown
Milichiidae	Decaying plants; dung; fungi in leaf cutting ant nests	Saprophagous, coprophagous	Many on flowers; some commensal with large predatory insects (Asilidae, Reduviidae)	Nectar; commensals feed on juices from host prey
Neriidae	Rotting succulent plants or fermenting sap	Saprophagous	On rotting succulent plants	Exudates from plant decay?

Table 41.1 *(Continued)*

Family	Larval habitat/habits	Larval food habits	Adult habitat	Adult food habits
Otitidae	Dung, decaying bulbs, tubers, roots, stems; subcortical region of decaying trees; occasionally in live plants	Mostly saprophagous	Near larval habitats; on flowers	Nectar; probably decay exudates
Piophilidae	Protein-rich decaying plants or animals, including meat, bones, hides, fungi, cheese; one species ectoparasitic on nesting birds	Saprophagous or blood feeding	Near larval habitats	Probably decay exudates
Psilidae	Roots and stems of plants	Phytophagous	Moist or wooded areas	Unknown
Pyrgotidae	Internal parasitoids of adult Scarabaeidae	Entomophagous	Nocturnal near hosts; oviposit into host while in flight	Unknown
Sciomyzidae	Associated with freshwater or terrestrial mollusks as predators, parasitoids, or scavengers	Zoophagous	Moist areas near larval hosts or prey	Unknown?
Sepsidae	Dung; sometimes in carcasses, sewage sludge	Coprophagous or saprophagous on protein-rich material	Usually near larval food	Decay exudates
Sphaeroceridae	Dung, carcasses, decaying plants, fungi, seaweed, leaf litter; dung balls of Scarabaeidae	Saprophagous	Moist to wet habitats near larval food	Probably decay exudates
Tanypezidae	Unknown; probably decaying plants	Unknown	Deciduous woods	Unknown
Tephritidae	Flowers, fruits, seeds, stems, leaves, and roots, especially of dicots; some make galls	Phytophagous	Usually near host plants	Nectar or plant exudates
Tethinidae	Mostly unknown; probably in soil or algal masses; some probably in bird dung	Unknown	Mostly marine shorelines and saline areas	Mostly unknown; some feed on nectar

of wood-boring beetles, or the debris that collects in bird nests or animal burrows. Many of these saprophagous species are probably feeding largely on fungi or bacteria, and some will mature on agar bacterial cultures. It may be noted that *Drosophila melanogaster*, an extremely important experimental animal for research in genetics and nutrition, is a decomposer of decaying fruits in nature, accounting for its easy maintenance in confined spaces.

The saprophytic Cyclorrhapha are of considerable economic significance because of their frequent close associations with humans or domesticated animals. Muscidae and Anthomyiidae (as

well as several less significant families) commonly breed in animal dung, often reaching epidemic numbers in stock-raising areas. Many of these species are mechanical vectors of communicable diseases, ranging from dysenteries to poliomyelitis. A few species such as the horn fly (*Haematobia irritans*), the stable fly (*Stomoxys calcitrans*), and the tsetse flies (*Glossina*) have secondarily acquired bloodsucking habits. *Glossina* are intermediate hosts for several trypanosomiases, including sleeping sickness of humans.

The Calliphoridae and Sarcophagidae are known as blowflies and flesh flies, from their habit of ovipositing on carrion. Some species, especially Calliphoridae, may oviposit in wounds of living animals, including humans. Some of these, such as *Cochliomyia hominovorax* and *Lucilia cuprina*, cause serious livestock losses, and the former has been the subject of intense control efforts using sterile-male techniques (see Chap. 12).

Oestridae (including the subfamilies Gasterophilinae, Hypodermatinae, and Cuterebrinae), the botflies and warbles, are almost unique among insects in parasitizing mammals internally. *Gasterophilus* (Gasterophilinae) larvae attach to the stomach lining of horses, donkeys, rhinoceroses, and elephants. *Hypoderma* (Hypodermatinae) larvae live in cutaneous cysts or warbles in cattle, breathing through a small hole maintained in the skin. Most larvae of the subfamilies Oestrinae (primarily old world) and Cuterebrinae (New World) inhabit the nasal or respiratory passages of rodents, lagomorphs, monkeys, and ungulates. The tropical *Dermatobia hominis* (Cuterebrinae), attacking humans as well as livestock, is unusual in that the adult females attach their eggs to mosquitoes or other flies that they capture. When the mosquito later feeds on a warm-blooded animal, the host's body heat stimulates the egg to hatch. The larva then drops to the host and penetrates the skin. Other species oviposit directly on the host's body, usually cementing the eggs to hairs. It seems likely that these parasitic species evolved from flesh flies, and some authorities classify *Dermatobia* as a calliphorid.

Adults of three specialized families of Cyclorrhapha are ectoparasitic. Hippoboscidae are widespread on birds, with a lesser number infesting mammals, especially ungulates, but also dogs. Nycteribiidae and Streblidae are restricted to bats and almost never encountered away from their hosts. All these flies are superficially similar to lice, with flattened bodies and tough, leathery cuticle; many species dehisce their wings after reaching a host.

Braulidae live in the hives of honey bees. Adults are apterous and are usually found on the queen or worker bees. The larvae tunnel in the wax combs.

Parasitoid Cyclorrhapha attack a variety of insect orders, as well as spiders; Sciomyzidae are unusual in attacking freshwater and terrestrial snails. Parasitoid species oviposit directly on or in the host (Chloropidae, Pyrgotidae, Tachinidae) or in locations where the first-instar larvae will encounter the host (Tachinidae, Sarcophagidae). Internal parasitoids frequently tap the host tracheal system, emphasizing the dependence of most flies on atmospheric respiration.

Relatively few Cyclorrhapha are predatory. Carnivorous species include Chamaemyiidae and many Syrphidae (discussed in the previous section; larvae predatory on Homoptera), some Sciomyzidae (larvae predatory on snails), as well as occasional members of many other families. Predatory adults (some Anthomyiidae) usually attack organisms much smaller than themselves.

Eggs of most Cyclorrhapha are deposited directly into appropriate food sources, hatching after a brief time. Most Sarcophagidae, many Calliphoridae, as well as occasional other species, are viviparous or ovoviviparous, depositing larvae onto decomposing organic material. Larvae of *Glossina* (Glossinidae), Hippoboscidae, Nycteribiidae, and Streblidae are retained internally by the female until fully mature. Larval nourishment is by means of a placentalike organ in the common oviduct. The newly deposited larvae pupate immediately.

Cyclorrhapha pass through three functional instars, the fourth instar being suppressed in conjunction with the development of the puparium. Many species pupate in the larval food, protected by the hard, rigid puparial shell, which deters cannibalism as well as offering protection against predation, parasitism, and physical trauma. In contrast to most cocoons and other pupation shelters, the puparium requires no external support or material in its construction; it is probably responsible in large part for the success of the Cyclorrhapha.

KEY TO THE COMMON FAMILIES OF THE DIVISION CYCLORRHAPHA (SERIES SCHIZOPHORA)

1. Body flattened, louselike, with hind coxae widely separated or mesonotum reduced, similar to abdominal segments Hippoboscidae, **Braulidae, Nycteribiidae, Streblidae**

Body not louselike; middle and hind coxae closely approximated   2

2(1). Mouthparts vestigial or lacking; mouth cavity small (Fig. 41.18a) pubescent flies at least 15 mm long **Oestridae (including Cuterebridae, Hypodermatidae, Gasterophilidae)**
Mouthparts functional, usually with large labellum, and labial palps protruding from broad mouth cavity (Fig. 41.3c); usually less than 15 mm long   3

3(2). Mesothorax with complete transverse suture (sometimes faint) anterior to wing attachments (Figs. 41.18b,c; 41.17d); lower calypter usually large; greater ampulla present below wing base (Fig. 41.18d,e) (section **Calyptratae**)   29
Mesothorax with transverse suture present at lateral margins or absent (Fig. 41.17a); lower calypter small or vestigial; greater ampulla usually absent (section **Acalyptratae**)   4

4(3). Vein Sc reaching costal margin of wing (Fig. 41.14a–d), separate from R_1 except at extreme base; anal cell present (Fig. 41.14a–d)   5
Vein Sc not reaching costal margin, frequently fused with R_1 for most of its length (Fig. 41.14f,g); anal cell present (Fig. 41.14a–d) or absent (Fig. 41.14f)   21

5(4). Mesonotum strongly flattened; head closely appressed to thorax, legs bristly   **Coelopidae**
Mesonotum convex, if flattened, head loosely attached to thorax, legs not conspicuously bristly   6

6(5). Head with at least a pair of stout bristles (oral vibrissae) just in front of mouth cavity (Fig. 41.15a,b)   7
Head without oral vibrissae (as in Fig. 41.15g)   11

7(6). Palpi vestigial; metathoracic spiracle with at least 1 peripheral bristle; abdomen usually narrowed at base   **Sepsidae** (Fig. 41.17a)
Palpi functional, carried lateral to rostrum   8

8(7). Head with postvertical bristles parallel or divergent (sometimes absent) (Fig. 41.15b–d)   9

Head with postvertical bristles converging distally; ocellar bristles between median and lateral ocelli (Fig. 41.15a); tibiae with dorsal spines near apex; costal vein spinose   **Heleomyzidae**

9(8). Antennae with subapical arista on third segment; eyes round; head with 1 to 4 pairs of fronto-orbital bristles (Fig. 41.15b)   **Clusiidae**
Antennae with arista attached near base of third segment; eyes elongate, oval; head with 0 to 2 pairs of fronto-orbital bristles   10

10(9). Antennae with third segment at least 2 to 3 times as long as broad (Fig. 41.15c); vein 2A slightly sinuous, reaching wing margin; tibiae usually without preapical bristles; usually stout black or blue-black flies less than 5 mm long   **Lonchaeidae**
Antennae with third segment no more than twice as long as broad (as in Fig. 41.16b); vein 2A straight or curved, not reaching wing margin; costal vein never spinose   **Piophilidae**

11(6). Wings with vein M_{1+2} curving anteriorly, approaching or joining R_{4+5} (Fig. 41.14a); legs often long, stiltlike   12
Wings with vein M_{1+2} not closely approaching R_{4+5} (Fig. 41.14b), or legs not long, stiltlike   13

12(11). Proboscis long, slender, jointed, and bent anteriorly (Fig. 41.17b); body stout or slender   **Conopidae** (Fig. 41.17b)
Proboscis short, thick, not jointed as above; body slender; legs long, slender   **Neriidae, Micropezidae, Tanypezidae**

13(11). Some or all tibiae with dorsal bristles near apex; ovipositor short, membranous, retractile   14
Tibia usually without preapical bristles; if present, ovipositor long, sclerotized   16

14(13). Head with postvertical bristles parallel or diverging (rarely absent) (Fig. 41.15d); antennae without dorsal bristle on second segment   15
Head with postvertical bristles converging (Fig. 41.15a); antennae with dorsal bristle on second segment   **Lauxaniidae**

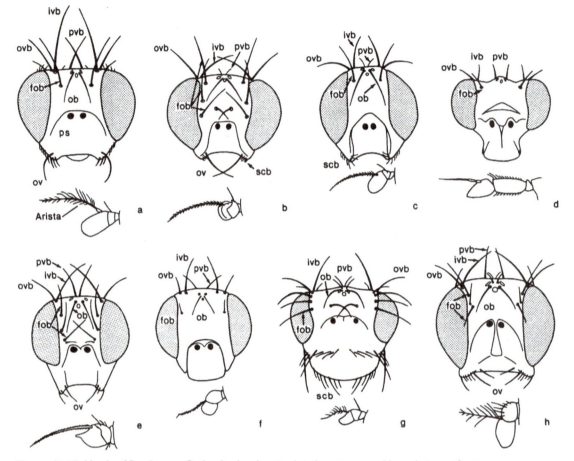

Figure 41.15 Heads of Brachycera, Cyclorrhapha, showing bristle patterns and lateral views of antennae; a, Heleomyzidae (*Suillia limbata*); b, Clusiidae (*Clusia occidentalis*); c, Lonchaeidae (*Earomyia brevistylata*); d, Sciomyzidae (*Sepedon pacifica*); e, Agromyzidae (*Cerodontha dorsalis*); f, Psilidae (*Psila microcera*); g, Ephydridae (*Paracoenia bisetosa*); h, Drosophilidae (*Drosophila simulans*). fob, fronto-orbital bristles; ivb, inner vertical bristles; ob, ocellar bristles; ov, oral vibrissae; ovb, outer vertical bristles; ps, ptilinal suture; pvb, postvertical bristles; scb, subcranial bristles.

15(14). Femora without bristles; vein R_1 intersecting costa beyond middle of wing
Dryomyzidae (including **Helcomyzidae**)
Femora with fine setae and stout bristles, usually with bristle near middle of anterior surface of mid-femur; vein R_1 intersecting costa near middle of wing (Fig. 41.14b)
Sciomyzidae

16(13). Costa with break near intersection of Sc (Fig. 41.14e)   19
Costa not broken near intersection of Sc (Fig. 41.14c,d)   17

17(16). Vein Cu_2 straight, meeting 2A at about a right angle (Fig. 41.14c); vein R_1 bare; ovipositor short, membranous; small gray flies
Chamaemyiidae
Vein Cu_2 angulate, meeting 2A at an acute angle (Fig. 41.14d); vein R_1 usually bearing setae; ovipositor sclerotized, usually projecting   18

18(17). Ocelli absent; base of ovipositor conical
Pyrgotidae
Ocelli present; ovipositor flattened; wings usually banded or spotted   **Otitidae** (Fig. 40.17c)

19(16). Vein Sc bent apically toward costa at nearly a right angle, usually not reaching costa (Fig. 41.14e); wings frequently patterned   **Tephritidae**

Vein Sc bent toward costa at much less than a right angle, usually meeting costa at about 45° angle (Fig. 41.14d)   **20**

20(19). Femora thickened, spiny, abdomen with bristles on second segment   **Otitidae** (Fig. 41.17c)

Femora slender, not spiny; abdomen without bristles on second segment   **Lonchaeidae**

21(4). Vein Sc bent apically toward costa at nearly a right angle, usually not reaching costa (Fig. 41.14e); Cu_2 angulate, meeting 2A at an acute angle   **Tephritidae**

Vein Sc bent toward costa at much less than a right angle, usually meeting costa at about 45° (Fig. 41.14f); Cu_2 straight, meeting 2A at about a right angle   **22**

22(21). Hind legs with 2 basal tarsal segments thickened, larger than succeeding segments; first segment shorter than second   **Sphaeroceridae**

Hind legs with basal tarsal segments not markedly thicker than succeeding segments; first segment usually longer than second   **23**

23(22). Costal vein with single break near Sc or R_1 (Fig. 41.14f)   **24**

Costal vein with breaks near Sc and near humeral crossvein (Fig. 41.14e)   **27**

24(23). Wings with anal cell closed or nearly closed (Fig. 41.24d,e); antenna with arista covered with short hairs   **25**

Anal cell absent (Fig. 41.14f); arista variable; ocellar triangle large, conspicuous   **Chloropidae**

25(24). Head with postvertical bristles converging apically (Fig. 41.15a)   **Tethinidae, Anthomyzidae**

Head with postvertical bristles diverging (Fig. 41.15c) or absent   **26**

26(25). Head with pair of stout bristles (oral vibrissae) just in front of mouth cavity (Fig. 41.15e); postvertical bristles diverging;

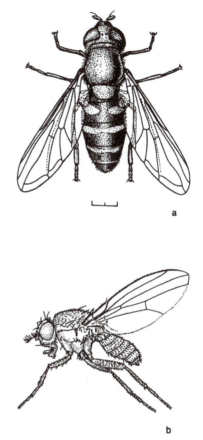

Figure 41.16 Representative Brachycera, Cyclorrhapha: a, Syrphidae (*Syrphus opinator*, scale equals 2 mm); b, Drosophilidae (*Drosophila busckii*). *Source: b, redrawn from Cole, 1969.*

small gray or black flies with ovipositor sclerotized, not retractable   **Agromyzidae**

Head without oral vibrissae; postvertical bristles frequently absent (Fig. 41.15f); small brownish or yellowish flies with very strong break at basal third of costa   **Psilidae**

27(23). Head with postvertical bristles diverging (Fig. 41.15g); pair of stout bristles (oral vibrissae) usually absent in front of mouth cavity (Fig. 41.15g); anal cell absent   **Ephydridae**

Head with postvertical bristles parallel or converging (Fig. 41.15h); (occasionally

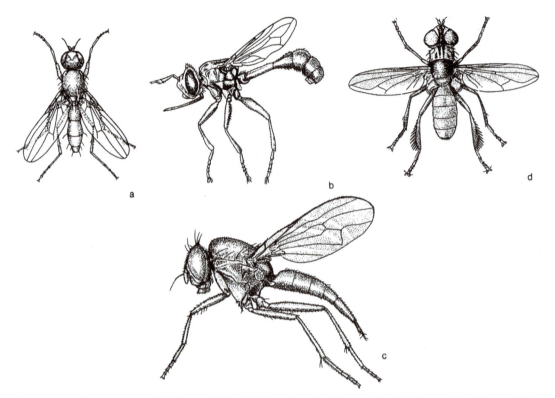

Figure 41.17 Representative Brachycera, Cyclorrhapha: a, Sepsidae (*Sepsis violacea*); b, Conopidae (*Physocephala texana*); c, Otitidae (*Tritoxa flexa*); d, Tachinidae (*Trichopoda plumipes*). *Source: Redrawn from Cole, 1969.*

absent); oral vibrissae usually present; anal cell present or absent   **28**

28(27). At least 1 pair of fronto-orbital bristles converging toward midline of head (Fig. 41.15b,e); antennae with arista usually bare
Milichiidae

Fronto-orbital bristles not directed medially (Fig. 41.15g,h); arista plumose
Drosophilidae (Fig. 41.16b) (including **Camillidae, Curtonotidae, Diastatidae**)

29(3). Meropleuron with row or cluster of large hypopleural bristles below metathoracic spiracle (Fig. 40.18e); wing with vein M_1 most always abruptly curving forward (Figs. 41.14h, 41.17d)   **30**

Meropleuron usually without bristles; if bristled, the pteropleuron lacks bristles

and vein M_1, is straight or gently curved (Fig. 41.14i)   **32**

30(29). Thorax with postscutellum large, convex, transverse (Fig. 41.18d,e); abdominal terga usually bearing strong bristles medially
Tachinidae (Fig. 41.17d)

Thorax with postscutellum small; abdominal terga rarely with medial bristles   **31**

31(30). Body usually metallic green or blue; arista almost always plumose along entire length; mesothorax with 2 or 3 notopleural bristles (Fig. 41.18b)

Calliphoridae

Body not metallic, usually black or gray with dark longitudinal stripes on thorax; arista usually plumose in basal half; mesothorax with 4 notopleural bristles (Fig. 41.18c)   **Sarcophagidae**

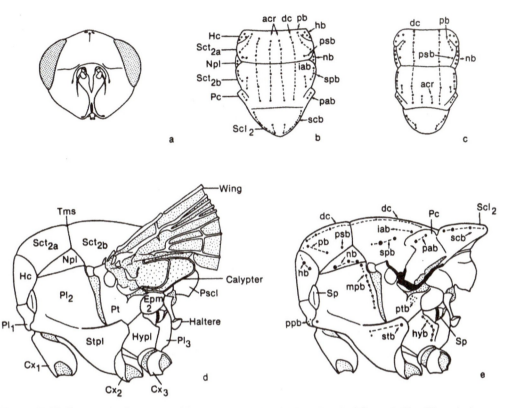

Figure 41.18 Taxonomic characters of Brachycera, Cyclorrhapha: a, head of Cuterebridae (*Cuterebra*); b, dorsal view of pterothorax of Calliphoridae; c, dorsal view of pterothorax of Sarcophagidae; d, e, lateral view of thorax of Calliphoridae (*Phormia regina*). acr, acrostichal bristles; Cx, coxa; dc, dorsocentral bristles; Epm, epimeron; hb, humeral bristles; Hc, humeral callus; hyb, hypopleural bristles; Hypl, hypopleuron; iab, intra-alar bristles; mpb, mesopleural bristles; nb, notopleural bristles; Npl, notopleuron; N, notum; pab, postalar bristles; Pc, postalar callus; pb, posthumeral bristles; Pl, pleuron; psb, prescutal bristles; Pscl, postscutellum; ppb, propleural bristle; ptb, pteropleural bristles; Pt, pteropleuron; scb, scutellar bristles; Scl, scutellum; Sp, spiracle; stpl, sternopleural bristles; Stpl, sternopleuron; spb, supra-alar bristles; Tms, transmesonotal suture.

32(29). Vein Cu$_2$+2A not reaching wing margin (Fig. 41.14h); lower calypter longer than upper calypter **Muscidae**	Vein Cu$_2$+2A reaching wing margin (Fig. 41.14i); lower calypter subequal to or smaller than upper calypter **Anthomyiidae**

CHAPTER 42

Order Siphonaptera (Fleas)

ORDER SIPHONAPTERA
Minute to small, wingless Endopterygota with laterally flattened body. Adults with hypognathous head capsule, 3-segmented antennae, frequently with small lateral ocelli (no compound eyes or dorsal ocelli); labrum and laciniae slender, bladelike; maxillary and labial palps usually 4- to 5-segmented, mandibles absent. Thoracic segments subequal or metanotum slightly enlarged; legs stout with large coxae; hind legs enlarged, modified for jumping. Abdomen 10-segmented, tenth tergite bearing a complex sensilium and associated guard hairs. Larva apodous, vermiform, usually with distinct head capsule, 1-segmented antennae, mandibulate mouthparts. Pupa adecticous, exarate, in wispy cocoon.

About 2400 species of fleas occur throughout the world, parasitizing mostly mammals (94 percent), but about 130 species (6 percent) infest various birds, including domestic fowls. The species that infest birds appear to be specialized derivatives of lineages that otherwise occur on mammals. Unlike lice, which depend on the host for shelter, warmth, and oviposition sites as well as food, fleas are intermittently parasitic. Larvae are mostly nest inhabitants rather than parasites, but larvae of a few species are ectoparasites, and one is an endoparasite. Adults frequently leave the host for various periods of time, and those associated with migratory birds survive in their nests from season to season. Aquatic as well as terrestrial mammals may serve as hosts, but fleas are not known from seals and other marine mammals that do not occupy nests. The chigoe flea, *Tunga penetrans*, of tropical regions, is exceptional in that the females permanently attach to the skin, usually of the feet, of various mammals including humans. Fertilized females burrow into the flesh and distend to the size of a small pea. Host specificity varies greatly among fleas, with up to 35 hosts

known for some species. The number of flea species supported by a given host is likewise highly variable, with 22 species recorded from the rat, *Rattus fuscipes*. Species with broad host ranges readily change host species, and many fleas with restricted host ranges will temporarily infest unsuitable hosts, particularly if hungry. Transfers among individuals of the same species may be frequent, especially among colonial or gregarious hosts. Dead animals are quickly deserted, and in general, high vagility and infidelity to a single host are responsible for the importance of fleas in transmitting bubonic plague from rodents to humans (see Chap. 11). Fleas are also important vectors of murine typhus (*Rickettsia*) and tularemia (*Francisella*) and may serve as intermediate hosts of tapeworms.

Adult fleas are so highly modified that their relationship to other orders of insects is uncertain. However, the absence of mandibles and presence of suctorial mouthparts suggest a possible origin from Diptera or Mecoptera, and flea larvae strongly resemble immatures of the dipterous suborder Nematocera. Initially, some molecular phylogenetic

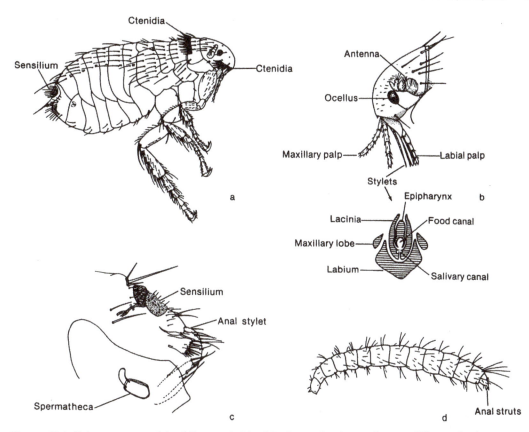

Figure 42.1 Siphonaptera: a, adult of *Ctenocephalides felis*, the cat flea; b, mouthparts of *Xenopsylla cheopis*; c, abdominal sensilium of *Megabothris*; d, larva of *Xenopsylla cheopis*.
Source: a, courtesy of D. P. Furman; b, c, redrawn from Furman and Catts, 1961; d, redrawn from Patton, 1931.

studies suggested that the Siphonaptera arose from within the Mecoptera, near the Boreidae ("snow fleas"), but larger more recent data sets have tended to show a sister-group relationship with Mecoptera, rather than an origin within that order.

Adult morphology (Fig. 42.1) is strongly correlated with the ectoparasitic life. The laterally compressed body and short, retractable antennae allow easy passage through the host's pelage or plumage, and the wedge-shaped coxae appear to be adapted for pushing through hair. Various parts of the body, especially the pronotum, bear combs of stout, backwardly directed, modified setae, or **ctenidia**, which further aid forward progress. Smaller bristles on various parts of the body serve the same function. In general, fleas avoid the

preening or scratching of the host by rapidly running over the host's body. As in most other insect ectoparasites of vertebrates, the cuticle is tough and leathery, protecting the internal organs by a hydrostatic cushion provided by the enclosed hemolymph. The jumping hind legs are used to escape or to mount a potential host. The jumping mechanism (Figs. 42.2a,b) functions after the manner of the release mechanism of a camera shutter. Prior to jumping the hind leg is rotated dorsally and anteriorly at the coxo-trochanteral joint by contraction of the trochanter levator muscle. Rotation continues until the tendon of the large notum-trochanter muscle passes backward across the pivot between the trochanter and coxa (Fig. 42.2a). This position causes the force of the notum-trochanter and coxal

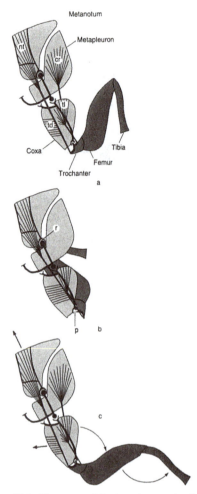

Figure 42.2 Diagrams of the jumping mechanism of a flea, showing part of the hind leg (excluding the tarsus) and musculature as seen from inside the metathorax and with the head to the left; especially note the position of the trochanter relative to the trochanteral-coxal pivot: a, mechanism with leg being drawn up into the cocked position; b, mechanism during the jump, with leg extended by muscular contractions and the rebound of the compressed resilin block. Thick arrows show directions of forces from contracting muscles; thin arrows show direction of movement of the leg; thick lines on sclerites are strong ridges; ntm, notum-trochanter muscle; tdm, trochanter depressor muscle; tlm, trochanter levator muscle; crm, coxal remotor muscle; p, trochanteral-coxal pivot; r, resilin block between pleuron and notum.
Source: Modified from Bennet-Clark and Lucey, 1967.

remotor muscles, which are now contracted, to be transmitted through the sclerotized ridges on the coxa and pleuron to the resilin block, which is situated between the pleuron and notum. With the leg in this cocked position the flea is ready to leap. The jump is initiated by contraction of the trochanter depressor muscle, which pulls the tendon of the notum-trochanter muscle anteriorly across the trochanteral-coxal pivot. The combined force of the contracted muscles and the compressed resilin block causes the hind leg to rotate rapidly downward, striking the substrate and causing the flea to be propelled into the air.

Adult lifespans as long as a year have been recorded, and *Pulex irritans*, the human flea, has survived over 4 months without feeding. Mating and oviposition take place on the host or within its nest or burrow. In species associated with solitary animals, reproduction may be synchronized with hormonal changes in the host, so that a fresh generation of fleas is available to infest the young animals leaving the nest. Blood meals are probably necessary for egg maturation in all species.

Pulex irritans may produce over 400 eggs per individual, and cat fleas up to 1000 eggs, but eggs are laid a few at a time. Eclosion ensues after a few days or weeks, and larval development is complete after about 1 to 3 weeks under favorable conditions such as the warm, humid environs of a mammal nest. Under harsher conditions, especially low temperatures, larval growth may require up to 6 months. Larvae are scavengers and predators, feeding on dried flea excrement, small arthropods, and other debris that collects in animal nests. In houses the cracks and crevices in floors covered by carpets are favorable developmental sites. The third-instar larva spins a thin cocoon, to which dust and debris adhere. Pupation requires a few weeks in most species. Newly emerged adults frequently remain in the cocoon for some time, especially if no host is available. In some species massive emergence of such quiescent adults is apparently stimulated by mechanical disturbances such as movements of a potential host (e.g., swallows returning to their nest in spring or vacationers to an empty house).

CHAPTER 43

Order Lepidoptera
(Butterflies and Moths)

ORDER LEPIDOPTERA

Adult. Minute to large Endopterygota with body and wings covered with scales or hairs, mouthparts almost always modified as a proboscis; head hypognathous with large lateral eyes, frequently with ocelli and/or chaetosemata, antennae with numerous segments; mouthparts consisting of elongate galeae, maxillary palpi (frequently very small), large labial palpi. Prothorax usually small, collarlike, with patagia dorsolaterally; mesothorax large with clearly defined scutum and scutellum, with large lateral tegulae covering wing bases; metathorax usually much reduced, with lateral scutal sclerites. Wings covered with scales (flattened macrotrichia); hind wing coupled to forewing by frenulum, jugum, or basal overlap; venation mostly of longitudinal veins with few crossveins, usually with large discal cell in basal half of wing. Legs adapted for walking, usually with 5 tarsomeres; prothoracic legs occasionally much reduced, almost always with epiphysis on tibia. Abdomen with 10 segments, first segment reduced, with vestigial sternum; segments 9 and 10 strongly modified in relation to genitalia.

Larva. Usually eruciform with sclerotized head capsule; hypognathous, mandibulate mouthparts, short, 3-segmented antennae and usually 6 lateral ocelli; thorax with short, 5-segmented legs with single tarsal claws; abdomen 10-segmented, with short fleshy prolegs on some or all of segments 3 to 6 and 10. Larval morphology often highly modified in leaf mining forms.

Pupa. Rarely decticous, exarate; typically adecticous, obtect, with body parts fused, except for 2 or more free abdominal segments, which remain movable; pupa usually enclosed in silken cocoon or shelter.

Members of the order Lepidoptera have received more attention from naturalists and entomologists than any other group of insects. Most of this effort has been directed toward the butterflies and showier moths, and knowledge of the smaller, more obscure species that constitute the great majority of the order is largely lacking.

Biologically, as well as morphologically, the Lepidoptera are perhaps the most uniform of the large holometabolous orders. Adults, with few exceptions, feed on nectar, honeydew, fermenting sap, or other similar products or have the mouthparts atrophied. Larvae of nearly all species feed on foliage or fruiting structures of angiosperms and

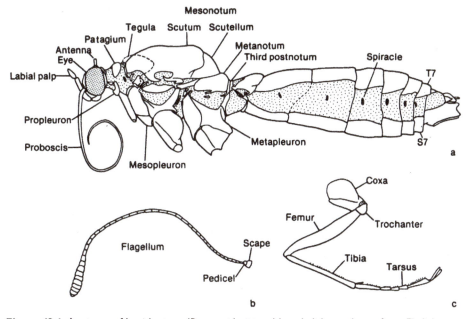

Figure 43.1 Anatomy of Lepidoptera (*Danaus plexippus*, Nymphalidae, redrawn from Ehrlich, 1958): a, lateral view of body; b, antenna; c, mesothoracic leg; T, tergum; S, sternum.

gymnosperms; larval feeding is almost entirely responsible for the great economic importance of the order. It seems likely that nearly all higher plants are host to at least one species of Lepidoptera. In general, Lepidoptera are the dominant insect herbivores in most habitats, having extensively invaded all available niches except that of sucking plant sap. In North America larvae of about 175 species feed on fungi or detritus.

Certain Lepidoptera have been important in the development of ideas in evolutionary biology and ecology. The butterfly *Papilio dardanus*, whose females mimic danaine nymphalids of several species, is one of a few organisms for which the genetic basis of polymorphism is understood. Complex mimetic relationships have been intensively investigated in the monarch (*Danaus plexippus*), queen (*Danaus gilippus*), and viceroy (*Limenitis archippus*). The monarch is also famous for its migratory behavior, documented by Urquhart (1987). Sex-limited genes were first described in the genus *Colias*. The evolution of concealing coloration has been intensively studied by Kettlewell and others

in the geometrid moth *Biston betularia*, and studies of Ehrlich and others (Ehrlich et al., 1975) on the population biology and adaptive ecology of *Euphydryas editha* (Nymphalidae) are important in understanding how populations differentiate. These are just a few examples. In general, butterflies have been favorite subjects for much recent experimental and analytical work, especially regarding plant herbivore coevolution, geographic variation, and population structure.

A distinctive morphological feature of most Lepidoptera is the body covering of scalelike macrotrichia. In most species these are of dull brown or gray hues in mottled patterns, which help to conceal the resting insects. In butterflies the scales produce bright contrasting patterns, which may function aposematically or may provide characteristic patterns allowing easy recognition between the sexes. Scales are easily detached and probably enable these insects to slip from the grasp of many potential predators. In some of the larger moths scales may provide an insulation that allows the internal body to reach the elevated temperatures at which flight occurs more quickly. Body temperatures 10 to 18°C above ambient have been recorded in large Hepialidae and Sphingidae.

The majority of Lepidoptera have the mouthparts adapted as a long coiled proboscis through which liquid food is ingested (Fig. 43.2c). The proboscis consists of the greatly elongate maxillary galeae, structured as a series of sclerotized rings connected by membranes and two sets of muscles, which provide the flexibility required for coiling and uncoiling. The two galeae are tightly interlocked by hooks and spines, forming a double-walled tube. Small maxillary palpi are usually present at the base of the proboscis, and large three- to four-segmented labial palpi are present on all but some nonfeeding forms. Other oral

structures are absent in typical Lepidoptera except as vestiges, but functional mandibles, accompanied by maxillae with both laciniae and unmodified galeae, occur in Micropterygidae and a few other families (Fig. 43.2a,b). Reduced, toothless mandibles also occur in Eriocraniidae, in combination with a short proboscis whose galeae are held together only while feeding.

As in most four-winged insects adapted to an aerial existence, Lepidoptera have evolved mechanisms to synchronize their wing beats. The simplest type of wing coupling consists of the jugum, a lobe projecting from the posterior base of the forewing and overlying the hind wing (Fig. 43.3b). In most species coupling is achieved by one or more stout bristles (the frenulum) which arise from the base of the hind wing and interlock with a cuticular flap or row of setae (the retinaculum) on the underside of the forewing (Fig. 43.3a). In Papilionoidea and a few other Ditrysia, coupling is amplexiform. In

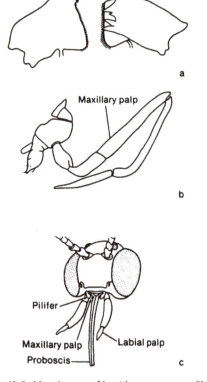

Figure 43.2 Mouthparts of Lepidoptera: a, mandibles of Zeugloptera (Micropterygidae, *Palaeomicroides*); b, maxilla of same; c, head of moth, *Synanthedon*.
Source: a, b, redrawn from Issiki, 1931; c, modified from Snodgrass, 1935.

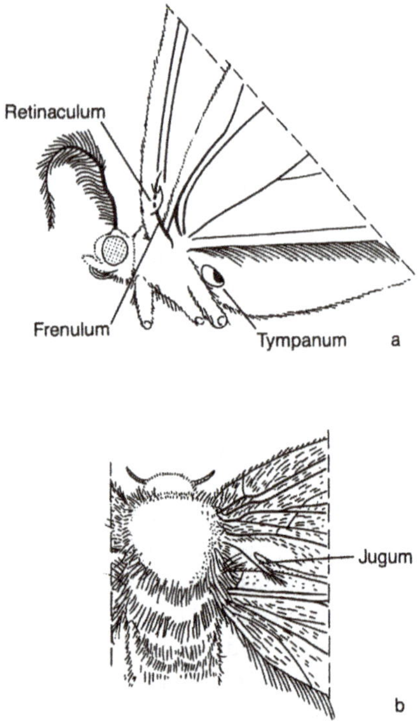

Figure 43.3 Wing coupling mechanisms of Lepidoptera: a, frenulum (Geometridae, *Neoalcis californica*); b, jugum (Hepialidae, *Hepialus*).

these species the forewing, especially in the basal region, broadly overlaps the hind wing, which may bear special humeral veins that stiffen the region of overlap (Fig. 43.10e)

Female Lepidoptera are remarkable for the complexity of the internal reproductive system. Distinct organizational plans typify the suborders Monotrysia and Ditrysia. In all Lepidoptera, sperm are transmitted as a spermatophore, which is stored in the bursa copulatrix. In Monotrysia (Fig. 43.4a) the bursa and common oviduct are confluent and may join the alimentary canal in a cloaca. Eggs are fertilized from the bursa copulatrix as they pass down the common oviduct. The small suborders Zeugloptera and Dacnonypha also have the monotrysian arrangement. In Ditrysia (Fig. 43.4b) the bursa, common oviduct, and rectum open independently. The bursal opening is situated medially on sternite 8, which may be highly modified, along with segments 9 and 10. As in the Monotrysia, spermatophores are stored in the bursa, but the spermatozoa migrate via the ductus seminalis to the spermatheca, from whence the eggs are fertilized. In the suborder Exoporia the openings of the bursa and common oviduct are situated close together on segments 9 and 10, and connected by a groove along which the sperm migrate to the spermatheca. In Trichoptera, the sister taxon to Lepidoptera, the female reproductive tract is of the same general plan as in other insects, with the ovaries and bursa copulatrix branching from the common oviduct and the single genital opening functioning in both copulation and oviposition. Ontogeny of the ditrysian structures during metamorphosis suggests that a single opening on the eighth sternite represents the primitive configuration in Lepidoptera, the common oviduct secondarily extending posteriad to the region of the anus (Dodson, 1937).

The body plan shared by nearly all Lepidopterous larvae, or caterpillars, comprises an elongate trunk with large, sclerotized head capsule, three pairs of thoracic legs, and up to five pairs of unjointed abdominal appendages or prolegs (Fig. 43.5). The prolegs bear terminal patches of short hooked spines, or crochets, which assist the larvae in clinging to the exposed surfaces of vegetation or to silk stands attached to the substrate. Analogous tarsal modifications occur in a variety of other insects that frequent smooth surfaces of leaves. Abdominal segments 3 to 6 and 10 typically bear prolegs, but in many families, such as Geometridae, the number is reduced. Crochets may be arranged in single or multiple circles or in rows or bands. They may be of uniform size, may alternate large and small, or may be larger at the ends of the rows. They are lacking only in a few burrowing forms and in highly specialized leaf mining families such as Nepticulidae, Heliozelidae, and Eriocraniidae (Fig. 43.5c,d) and are consequently of great taxonomic utility. Besides reduction or loss of abdominal prolegs, leaf mining caterpillars often lack thoracic legs and have highly modified, prognathous mouthparts, and flattened bodies. Prolegs and sometimes thoracic legs are

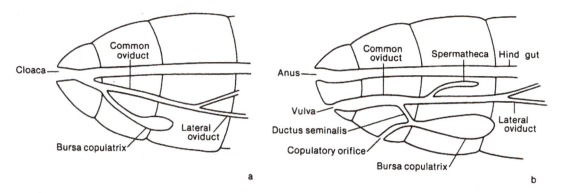

Figure 43.4 Reproductive systems of Lepidoptera (diagrammatic); a, monotrysian system; b, ditrysian system.

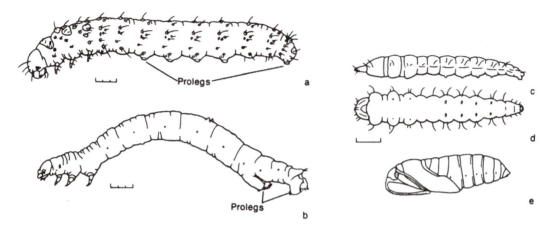

Figure 43.5 Larvae and pupae of Lepidoptera: a, larva of Hepialidae (*Hepialus*; scale equals 3 mm); b, larva of Geometridae (*Synaxis pallulata*; scale equals 3 mm); c, d, lateral and ventral views, respectively, of a leaf-mining larva (Tischeriidae, *Tischeria*; scale equals 1 mm); e, pupa of Sphingidae (*Manduca*).

occasionally lost or reduced in caterpillars that bore in woody substrates.

Caterpillars characteristically bear setae on the head and trunk. Nearly all larvae emerge with a complement of primary setae, which are relatively constant in number and position across families. The primary setae are usually long, tactile structures. Internally feeding forms may have only these primary setate, but in later instars additional setae usually appear, either in addition to or replacing the primary setae. Secondary setae are frequently grouped on elevations, processes, or flat or sclerotized plates and may occur on almost any part of the body, including the mandibles. Most often they are simple, but plumose, spatulate, or knobbed secondary setae occur in many taxa. Often the stiff, bristly secondary setae serve a mechanical protective function, as in many Arctiidae, but the dehiscent setae of caterpillars of several families (Limacodidae, Megalopygidae, some Saturniidae and Lymantriidae) are strongly urticating.

Silk production is universal among lepidopterous larvae, being utilized in the construction of a wide variety of larval shelters or galleries as well as in formation of the pupal shelter or cocoon. Many caterpillars also attach silk strands to the substrate to assist locomotion, and some lower themselves on silk lines to escape predators. Silk is produced by large paired glands, usually extending beside the gut posteriorly into the abdomen and opening through a median spinneret on the labium. The size and configuration of the glands are extremely variable, but the evolutionary significance of this variation is not known.

Pupae are exarate in many primitive forms, including Zeugloptera and Dacnonypha. Obtect pupae (Fig. 43.5e;), with the appendages variably fused to the body, are characteristic of the more derived families, where pupal appendages may be received in grooves in the body wall, and the body regions may be ill defined because of loss of external sutures. Rudiments of mandibles, maxillae, or other mouthparts are present in many families; functional mandibles, occurring in the suborders Zeugloptera and Dacnonypha, are used to open the cocoon just before adult emergence. Members of other suborders escape through a weakened area left by the larva at the time of cocoon formation, or the newly emerged adult regurgitates a liquid that dissolves the silk. At least one pupal abdominal joint remains flexible, and wriggling of the abdomen is the only means of motility in most pupae. Segment 10 may be modified as a cremaster, a series of hooked setae used to anchor pupae of some specialized families to the cocoon or to silk strands attached to the substrate.

BIOLOGY

Monarch butterflies migrate in autumn to overwintering sites, then disperse and reproduce the following spring, and long adult diapause is known in some moths. However, most adult Lepidoptera are rather short-lived. Many species with atrophied mouthparts do not feed. Others imbibe nectar, honeydew, or exudates from fermenting fruit or sap, while a very few are known to suck blood or pierce fruit, using the proboscis to penetrate the skin. Both sexes of nearly all species possess functional wings, but many female Psychidae and Lymantriidae, as well as occasional species of other families are flightless, especially those that are active in winter or at high elevations. Brachyptery in males is limited to a very few species, mostly on oceanic islands. Activity is predominantly nocturnal or at dusk and dawn, but nearly all members of the Papilionoidea (butterflies) are sunshine-loving, and certain families or genera of moths are diurnal. In general, nocturnal species tend toward brown or gray and white mottled color patterns, which are important in concealing the insects on their daytime perches. Tree trunks or branches, lichen, and other irregular surfaces are favorite spots for daytime sheltering, and the scale or hair tufts on many moths enhance their camouflage. At least some moths are able to select backgrounds matching their own ground color (Kettlewell, 1973). The brightly colored hind wings of some nocturnal species are covered by the forewings at rest, being suddenly revealed if the moth is disturbed. Such hind wing patterns sometimes strikingly resemble eyes of vertebrates and are apparently used to startle birds or other potential predators. Aposematism and mimicry are perhaps best understood in the monarch, queen, and viceroy but occur very widely among other butterflies, especially in the families Papilionidae and Nymphalidae, particularly in the tropics (see Papilionoidea, below). Diurnal moths often display aposematic color patterns, sometimes mimicking bees, wasps, or other distasteful Lepidoptera, but the chemicals involved and other details have been little studied.

Lepidoptera almost exclusively have both sexes. Female sex pheromones serving to attract males have been isolated from a great number of species, especially moths, and are probably present in most species. Pheromones may also be produced by males, as in the family Hepialidae, where males may form leks to which the females are attracted. Pheromones may be effective over relatively short distances, as in many of the smaller moths, or may attract males from distances of hundreds of meters, as in Saturniidae. Many butterflies apparently rely on visual cues for initial recognition, frequently followed by behavioral sequences or release of short-range pheromones by males. Copulation takes place on the ground or other perches.

Eggs are characteristically deposited on or near the host plant, either singly or in masses. In Eriocraniidae, Incurvariidae, and a few other families where ovipositors are present, eggs are inserted into plant tissue, and in some Hepialidae they are scattered during flight. The number of eggs per individual varies from about a dozen to more than 18,000 in some Hepialidae. Eggs are extremely variable, ranging from cylindrical or barrel-shaped to flat wafers. The chorion may be smooth and featureless or regularly or irregularly sculptured with ribs, tubercles, or pits. Eggs may be laid singly or in small groups or deposited in large masses, usually on or near the larval food. Egg diapause is a common feature of lepidopteran life cycles. A few European Psychidae, including one species (*Apterona helix*) introduced into North America, are parthenogenetic.

Larvae are predominantly herbivores, but scavengers (many Tineidae, Oecophoridae) and a few predators (a few Lycaenidae, Cosmopterygidae, Noctuidae, Geometridae) are known, feeding on eggs of spiders or other Lepidoptera or on scale insects. Predatory Lycaenidae are symbiotic with ants, feeding on their larvae. Hawaiian species of *Eupithecia* (Geometridae) have become specialized to feed on small flies, especially Drosophilidae. The caterpillars poise on twigs, which they grasp with the terminal prolegs, holding the forebody outstretched, then make sudden striking motions to grasp flies with their enlarged forelegs. Larvae of Epipyropidae are unique in living externally on the bodies of leafhoppers, especially Fulgoroidea, feeding on hemolymph exuding through punctures made by the caterpillar's slender, sharply pointed mandibles. The caterpillars maintain their purchase on the host by clinging with the crochets to lines of silk they attach to the host's body.

Roots, stalks and stems, bark and wood, leaves, buds and flowers, fruits, and seeds are consumed by phytophagous forms. In general, larvae of specialized families of the suborder Ditrysia live in exposed situations and are frequently colored or patterned to resemble their host plants or substrates on which they rest. Larvae feeding exposed often show additional defensive mechanisms. For example, papilionid caterpillars evert bright-colored, aromatic osmeteria. Sphingid caterpillars often rear up and strike at disturbances with the fore part of the body, at the same time regurgitating copious amounts of gut contents. Other defensive mechanisms include dense body coverings of stiff, spiny setae, sometimes urticarious, and contrastingly colored spots concealed in folds in the body wall and revealed by arching the back. Many caterpillars, such as those of the monarch butterfly, sequester toxic compounds from their food, and are aposematically patterned to warn predators of their inedibility. Many caterpillars attempt to escape predators or parasites by feigning death, dropping off their perches or rapidly lowering themselves by silk threads.

Larvae of primitive families of Ditrysia as well as those of the suborders Zeugloptera, Dacnonypha, and Monotrysia frequently feed in leaf mines, tunnels, galls, self-constructed shelters, or other concealed situations. Fossil leaves with mines show that the leaf mining habit was established by the Cretaceous Period. Shelters commonly consist of tied or rolled leaves or twigs, parchment like shields or covers, or silken webs. Most species are solitary, but Lasiocampidae (tent caterpillars), among others, are highly gregarious. Many externally feeding caterpillars show variable color patterns, depending on population density, usually becoming darker or contrastingly striped in denser populations. Host specificity varies from acceptance of a single part of a single plant species, as in many leaf miners, to extreme polyphagy, as in

many Noctuidae and Geometridae. Lepidoptera are heavily parasitized by Diptera and Hymenoptera, probably because of the exposed position of many of the larvae.

Pupation frequently occurs on the host plant, usually in a cocoon formed by the prepupal larva. Leaf miners often pupate within the larval mines, and borers in the larval tunnels, with or without a silken shelter. Noctuidae, Sphingidae, and some others pupate naked in soil or litter, and most Papilionoidea (butterflies) and occasional members of other groups hang exposed as a chrysalis on plants or other erect objects.

CLASSIFICATION

A confusing variety of classifications has been proposed historically. Older arrangements recognize the Frenatae and Jugatae as suborders, based on the type of wing coupling. The terms macrolepidoptera and microlepidoptera refer primarily to body size, which is without phylogenetic significance, and the suborders Rhopalocera and Heterocera refer to antennal shape (clubbed or not clubbed). The classification presented here emphasizes the important structural features of pupal and adult mouthparts and of the female reproductive system and generally follows that of Hodges et al. (1983). The number of families recognized is conservative, many of the small, obscure families being combined with more generally accepted taxa.

In general, the phylogeny of the earlier-diverging lineages of Lepidoptera is relatively well understood, and comparative morphological and molecular phylogenetic studies agree on the relationships. The vast majority of lepidopteran species belong to the Ditrysia, which has not stabilized with respect to its higher classification, although most families treated below are broadly recognized.

KEY TO THE NORTH AMERICAN SUBORDERS OF LEPIDOPTERA

1. Forewings and hind wings ovate, similar in venation, size, and shape (Fig. 43.6a); hind wings with at least 10 veins reaching margin

2

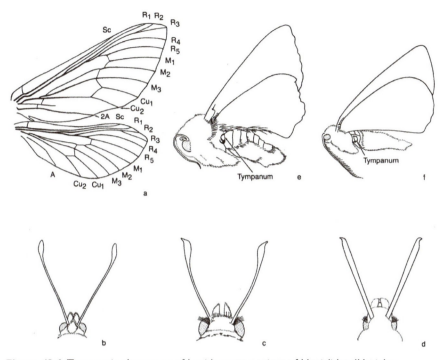

Figure 43.6 Taxonomic characters of Lepidoptera: a, wings of Hepialidae (*Hepialus sequoiolus*); b, antenna of Papilionidae (*Papilio*); c, antenna of Hesperiidae (*Hesperia*); d, antenna of Sphingidae (*Celerio*); e, tympanal organ of Noctuidae (*Orthosia pacifica*); f, tympanal organ of Geometridae (*Drepanulatrix*).

Forewings and hind wings conspicuously different in venation (Fig. 43.7 to 43.11); usually different in shape and size, or narrow, lanceolate (Fig. 43.8f); hind wings with 9 or fewer veins reaching margin   **4**

2(1). Large moths at least 10 mm long; maxillary palpi minute or vestigial
                    Monotrysia, in part
Small moths less than 6 mm long; maxillary palpi 5-segmented, conspicuous (Fig. 43.2b); iridescent, diurnal moths   **3**

3(2). Middle tibiae without spurs (apical tufts of hairs may be present); proboscis absent
                    Zeugloptera
Middle tibiae with single spur; proboscis short, noncoiled   **Dacnonypha**

4(1). Wings narrow, lanceolate (Fig. 43.8f); hind wings with fringe of setae broader than wing membrane; small moths less than 6 mm long   **5**

Wings broader, ovate or truncate; hind wings with setal fringe narrower than membrane; small to very large lepidopterans   **6**

5(4). Proboscis absent or vestigial; maxillary palpi 5-segmented, conspicuous; female with sclerotized, large external ovipositor
                    Monotrysia, in part
Proboscis present; maxillary palpi usually very small, 1 to 4-segmented; ovipositor membranous, inconspicuous or absent
                    Ditrysia, in part

6(4). Maxillary palpi large, folded, with 5 segments (as in Fig. 43.7c); if palpi inconspicuous, antennae much longer than body; ovipositor large, sclerotized, conspicuous; small moths 5 to 10 mm long
                    Exoporia and **Monotrysia**, in part
Maxillary palpi usually inconspicuous, with 1 to 4 segments (Figs. 43.7f, 43.8b); if

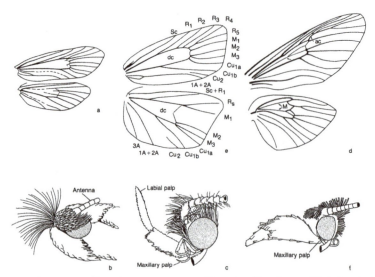

Figure 43.7 Taxonomic characters of Lepidoptera: a, wings of Tineidae (*Scardia gracilis*); b, head of Lyonetiidae (*Bucculatrix althaeae*); c, head of Tineidae (*Scardia gracilis*); d, wings of Cossidae (*Prionoxystus robiniae*; arrows indicate vein M in discal cells); e, wings of Torticidae (*Sparganothis*); f, head of Tortricidae (*Sparganothis senecionana*). Left labial palp (near side) removed on heads. ac, accessory cell; dc, discal cell.

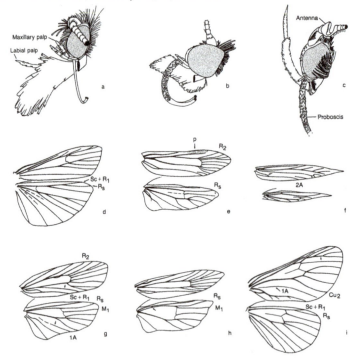

Figure 43.8 Taxonomic characters of Lepidoptera: a, head of Yponomeutidae (Abebaea cervella); b, head of Pyralidae (Nomophila nearctica); c, head of Blastobasidae (Holcocera gigantella); d, wings of Pyralidae (Dioryctria); e, wings of Blastobasidae (Holocera); f, wings of Coleophoridae (Coleophora; arrow indicates basal fork of 2A); g, wings of Oecophoridae (Agonopteryx fusciterminella; arrows indicate veins 1A in fore and hind wings); h, wings of Gelechiidae (Filatima demissae; arrow indicates short R_1); i, wings of Megalopygidae (Norape tener; arrow indicates branched radial vein). Left labial palp (near side) removed on heads. p, pterostigma.

Figure 43.9 Representative Lepidoptera: a, Hepialidae (*Hepialus sequoiolus*); b, Incurvariidae (*Adela trigrapha*); c, Cossidae (*Prionoxystus robiniae*); d, Tineidae (*Tinea pellionella*); e, Tortricidae (*Archips argyrospilus*); f, Yponomeutidae (*Atteva punctella*); g, Gelechiidae (*Phthorimaea operculella*); h, Sesiidae (*Ramosia resplendens*); i, Pyralidae (*Udea profundalis*); j, Pyralidae (*Desmia funeralis*); k, Pterophoridae (*Platyptilia carduidactyla*); l, Ethmiidae (*Ethmia arctostaphylella*).

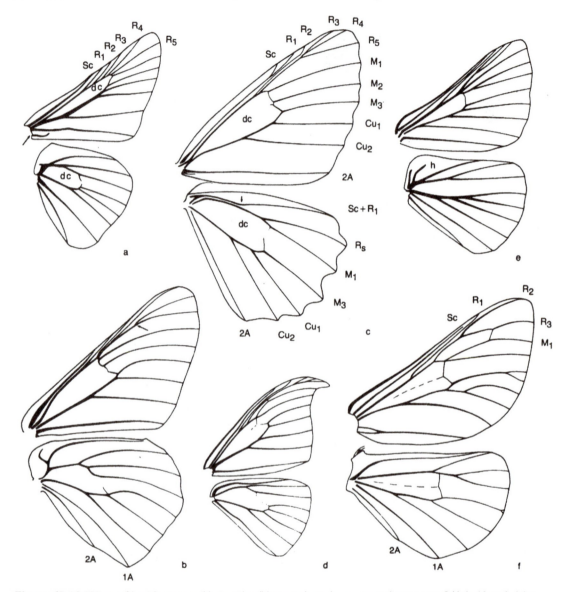

Figure 43.10 Wings of Lepidoptera: a, *Hesperiidae (Hesperia harpalus;* arrow indicates vein 3A); b, *Nymphalidae (Euphydryas chalcedona;* arrow indicates branched vein); c, *Geometridae (Neoalcis californica;* arrow indicates Sc+R₁ recurved toward Rs); d, *Drepanidae (Drepana);* e, *Lasiocampidae (Malacosoma californica);* f, *Saturniidae (Hemileuca nevadensis).* dc, discal cell; h, humeral veins.

palpi 5-segmented, then antennae shorter than body and ovipositor membranous, inconspicuous or absent; small to very large lepidopterans 3 to 50 mm long

Ditrysia, in part

Suborder Zeugloptera

Small moths without proboscis but with functional mandibles in pupa and adult (Fig. 43.2a); adult with ocelli, 5-segmented maxillary palp, 4-segmented labial palp; subcosta branching near middle, jugum

Figure 43.11 Representative Lepidoptera, Papilionoidea and Hesperioidea: a, Hesperiidae (*Hesperia harpalus*); b, Hesperiidae (*Pyrgus scriptura*); c, Papilionidae (*Papilio rutulus*); d, Nymphalidae (*Cynthia virginiensis*); e, Nymphalidae (*Euphydryas editha*); f, Pieridae (*Colias philodice*); g, Lycaenidae (*Strymon melinus*); h, Lycaenidae (*Apodemia mormo*); i, Satyridae (*Cercyonis peglerae*).

with humeral vein. Larva sluglike with prolegs on first 8 abdominal segments, large suckers on ninth and tenth segments. Pupa exarate.

The family Micropterygidae contains about 100 species in 9 genera of diurnal, frequently iridescent moths that occur sporadically throughout the world. Adults feed on pollen of various flowers; larvae feed on moss, liverworts, and possibly detritus and angiosperms. Pupation is in a tough, parchmentlike cocoon that is cut open with the pupal mandibles. The North

American fauna includes three or four species of *Epimartyria*, which are mostly encountered in damp woodlands and bogs and have a 2-year life cycle. *Agathiphaga* and *Heterobathmia*, each with a single genus in Australia and South America, respectively, have functional mandibles as adults but differ as larvae and are placed in the suborders Aglossata and Heterobathmiina in some classifications. *Agathiphaga* mines the seeds of *Agathis* (Araucariaceae); *Heterobathmia* mines the leaves of *Nothophagus* (southern beech). Both hosts have typical Gondwanan distributions, emphasizing the early divergence of these primitive moths.

Suborder Dacnonypha

Small moths with or without proboscis, mandibles vestigial; maxillary palpi 5-segmented, labial palpi 3- to 4-segmented; forewing with fingerlike jugum; larva apodous; pupa exarate with large functional mandibles.

Dacnonypha are small or very small, mostly diurnal moths, often with iridescent vestiture. The suborder contains an assemblage of primitive moths with many generalized features. *Eriocraniella* and *Dyseriocrania* (Eriocraniidae), with species throughout North America, have vestigial adult mandibles and maxillae lacking the laciniae. The galea is modified as a short proboscis. Larvae initiate a linear mine in leaves of oak, chestnut, or other woody plants, later expanding the mine into a large blotch, then leaving the mine to pupate in the soil. Adults appear in late winter or early spring. *Acanthopteroctetes* (Acanthopteroctetidae), with four species in western North America, forms blotch mines on *Ceanothus* (Rhamnaceae). The two families of Dacnonypha may be distinguished by the presence (Eriocraniidae) or absence (Acanthopteroctetidae) of ocelli. Dacnonypha are restricted to the Northern Hemisphere.

Suborder Exoporia

Exoporia are very largely restricted to the southern continents. The superfamily Hepialoidea, with the single family Hepialidae, is represented by about 20 species in North America. Hepialids are moderate or large moths with hairy, elongate bodies and long narrow wings (Fig. 43.9a). The proboscis is vestigial or absent, and adults do not feed. Venation is very similar in the forewings and hind wings (Fig. 43.6a); wing coupling is with a jugum (Fig. 43.3b). The family is cosmopolitan but especially well represented in the Indo-Australian region, where larval feeding habits include external foliage feeding as well as boring in stems or trunks of woody plants and some adults are showy, diurnal moths. North American species are borers in the stems or root crowns of various shrubs and trees or feed externally on their roots. Pupation is in a cocoon in the larval tunnel, which is also lined with silk. Adults are somber colored, and most are crepuscular. Hepialids are unusual in broadcasting the numerous eggs in the general vicinity of the larval food rather than ovipositing on the host plant.

Suborder Monotrysia

Very small to large moths with mandibles and laciniae absent, galeae as a short proboscis or absent; maxillary palpi 1- to 5-segmented, frequently minute; labial palpi with 2 to 3 segments. Larva caterpillarlike (Fig. 43.5a) or flattened, apodous (leafminers: Fig. 43.5c,d); pupa adecticous, obtect.

The Monotrysia includes extremely diverse taxa and is probably polyphyletic.

Nepticulidae and Opostegidae are small to exceedingly small moths with linear wings with reduced venation, characteristics common to various unrelated small lepidopterous leafminers. Adults have rudimentary mouthparts, and many probably do not feed. The ovipositor is reduced and the eggs deposited on the surface of the feeding substrate. Larvae of a few species cause twig or petiole galls or mine in bark, but the great majority excavate serpentine mines in the leaves of a broad array of woody plants. Larvae leave the mine to pupate in a tough cocoon in soil or leaf litter.

Larvae of Tischeriidae mine the leaves of shrubs of several families of angiosperms and also occur on annual Compositae. Eggs are deposited on the leaf surface, the newly emerged larvae boring into the leaf, eventually forming trumpet-shaped mines. The larval mine is lined with silk and may include a silken retreat used when the larva is threatened. Pupation is in the mine.

Females of the superfamily Incurvarioidea have sclerotized ovipositors, which are used to insert the eggs into plant tissues. Heliozelidae and some Incurvariidae are leaf miners as early instar larvae, later leaving the mine and feeding externally, usually cutting out oval shelters from the mine. Larvae of *Adela* (Incurvariidae) (Fig. 43.9b) first feed in developing ovules, later inhabiting cases constructed from leaf or flower parts, and feeding externally, often on detritus. Other Incurvariidae (*Tegeticula* and *Prodoxus*) bore the seed pods or stems of *Yucca*, the adult moths ovipositing in the immature flower ovaries or stalks. *Tegeticula* females have specialized, elongate maxillary palpi used to collect the glutinous pollen of *Yucca*, which is pollinated only by these moths. The suborder Monotrysia includes about 1000 species worldwide.

Suborder Ditrysia

This is an exceedingly variable assemblage characterized by the ditrysian organization of the female reproductive organs (Fig. 43.4). Ditrysia have the galea produced as a proboscis (occasionally atrophied). The wings are scaled (rarely with a few aculeae), and wing coupling is either frenulate or amplexiform (never jugate). Wing venation is usually strikingly different in the fore and hind wings (heteroneurous), in contrast to the similar, homoneurous venation of the more primitive suborders.

The Ditrysia comprise the great majority of Lepidoptera, including nearly all the familiar or economically important species. Venation, an important differentiating feature for many families, is frequently invisible without descaling the wings. The following key, based as far as possible on characteristics other than venation, is modified in part from Common (1970).

KEY TO THE NORTH AMERICAN FAMILIES OF THE SUBORDERS MONOTRYSIA AND EXOPORIA

1. Mouthparts atrophied or absent; wings with fingerlike jugum (Fig. 43.3b); head small, attached anteroventrally

   Hepialidae

   Mouthparts with haustellum and maxillary and labial palps; wings with frenulum; head large, attached anteriorly.   **2**

2(1). Antenna with scape expanded into broad eye cap (Fig. 43.7b); forewing lacking discal cell (superfamily **Nepticuloidea**)   **3**

   Antennae with scape not expanded; forewing with discal cell (Fig. 43.7e) (superfamilies **Incurvarioidea** and **Tischerioidea**) **4**

3(2). Forewing without branched veins

   Opostegidae

   Forewing with some veins branched

   Nepticulidae

4(2). Cranium with erect, irregular hairs

   Incurvariidae and **Tischeriidae**

   Cranium with smooth, flattened scales

   Heliozelidae

KEY TO THE COMMON NORTH AMERICAN SUPERFAMILIES OF THE DITRYSIA SUBORDER

1. Wings absent or vestigial   **24**

   Wings large, suitable for flying   **2**

2(1). Antennae threadlike, with abruptly swollen knob at tip, or with tip abruptly hooked (Fig. 43.6b,c); frenulum and ocelli absent **3**

   Antennae filiform, feathery, or gradually enlarged, very rarely knobbed; if thickened apically, frenulum present; ocelli present or absent   **4**

3(2). Distance between bases of antennae greater than diameter of eye (Fig. 43.6c); hind tibia usually with intermediate spur; all veins arising individually from discal cell in both wings (Fig. 43.10a)   **Hesperioidea**

   Distance between bases of antennae less than diameter of eye; hind tibiae without intermediate spur; forewings with some peripheral veins branched (Fig. 43.10b, at arrow)   **Papilionoidea**

4(2). Antennae spindle-shaped, thickest near middle or in apical third (Fig. 43.6d); body

long, stout, pointed at both ends; ocelli absent; heavy-bodied moths more than 20 mm long   **Sphingoidea**
Antennae threadlike, featherlike, or comb-like, never spindle-shaped; body rarely spindle-shaped; ocelli present or absent   **5**

5(4). Wings deeply cleft (Fig. 43.9k) with hind wing divided into 3 or more lobes   **6**
Wings not deeply cleft; hind wings consisting of single undivided lobe   **7**

6(5). Hind wing divided into 3 lobes (Fig. 43.9k)
                    Pterophoroidea
Hind wing divided into 6 to 7 lobes
          Copromorphoidea (Alucitidae)

7(5). Hind wing narrow or linear (Fig. 43.8f), with marginal fringe of hairs or scales subequal to or broader than wing membrane; venation frequently reduced; small to minute moths with wingspan almost always less than 15 mm   **8**
Hind wing oval or triangular, with marginal fringe of scales much shorter than wing breadth; venation usually complete; wingspan usually greater than 10 mm   **12**

8(7). Hind wing with vein $Sc+R_1$ fused with R_s beyond discal cell, then diverging (Fig. 43.8d); abdomen with paired tympana on first sternite (Fig. 43.6f)   **Pyraloidea**
Hind wing with veins $Sc + R_1$ and R_s separate beyond discal cell (Fig. 43.8e,g) or Sc occasionally joined to discal cell by short R_1 (Fig. 42.5h, at arrow); abdomen without tympana   **9**

9(8). Proboscis not covered by scales; maxillary palpi large or small   **10**
Proboscis densely covered by overlapping scales, at least basally (Fig. 43.8b,c); maxillary palpi short, 4-segmented, folded about base of proboscis (Fig. 43.8c)
                    Gelechioidea, in part

10(9). Antenna with scape enlarged, cup-shaped on underside, forming cover over eye (Fig. 43.7b)   **Tineoidea**, in part
Antennae with scape not expanded, not cup-shaped   **11**

11(10). Maxillary palpi usually long, 5-segmented, folded in repose (Fig. 43.7c); hind tibiae with

long, stiff bristles only on dorsal surface, or smooth-scaled **Tineoidea**, in part
Maxillary palpi with 3 to 4 segments, not folded in repose (Fig. 43.8a); hind tibiae frequently with whorls of long stiff bristles
Yponomeutoidea and **Copromorphoidea**, in part

12(7). Tympanal organs present in thorax or abdomen (Fig. 43.6e,f)   **13**
Tympanal organs absent   **15**

13(12). Tympanal organs in thorax (Fig. 43.6e)
                    Noctuoidea
Tympanal organs in abdomen (Fig. 43.6f)
                    14

14(13). Maxillary palpi minute or vestigial; proboscis bare, not covered with scales
          Geometroidea and **Drepanoidea**
Maxillary palpi with 3 or 4 segments, usually about as long as head and held horizontally in front of head; proboscis densely covered with scales at base (Fig. 43.8b)
                    Pyraloidea

15(12). Vein M present in discal cell of one or both wings (Fig. 43.7d, at arrows)   **16**
Vein M absent from discal cell of both wings (Fig. 43.7e)   **20**

16(15). Vein M branched in discal cell of both wings; R with 5 branches in forewing (Fig. 43.7d, at arrows); proboscis vestigial or absent; large, heavy-bodied moths at least 20 mm long   **Cossoidea**
Vein M not branched in discal cell of both wings; R usually with 4 branches in forewing (Fig. 43.7a); mostly small, frail moths less than 15 mm long   **17**

17(16). Hind wing with marginal hairs or scales much longer near wing base than at apex   **18**
Hind wing with marginal hairs or scales near wing base subequal in length to those at apex   **19**

18(17). Labial palpi held horizontally in front of head (Fig. 43.7f); frequently longer than head   **Tortricoidea**, in part
Labial palpi drooping or upright, not held horizontally (Fig. 43.7c); usually short
                    Tineoidea, in part

19(17). Forewing with vein Cu_2 intersecting 1A before margin   **Tineoidea**, in part
Forewing with vein Cu_2 not intersecting 1A, separate to margin (as in Fig. 43.8g–i, vein 1A at arrow)   **Zygaenoidea**

20(15). Proboscis bare or with few, sparse scales near base (Fig. 43.8a); if scaled, ocelli are large and prominent and maxillary palpi are not folded about base of proboscis (Fig. 43.8b)   **21**
Proboscis densely covered by overlapping scales, at least basally (Fig. 43.8b,c); ocelli always small or absent; maxillary palpi short, folded around base of proboscis (Fig. 43.8c)   **Gelechioidea**

21(20). Hind wing with marginal hairs or scales much longer near wing base than at apex   **22**
Hind wing with marginal hairs or scales near wing base subequal in length to those at apex   **23**

22(21). Labial palpi held horizontally in front of head, frequently as long as head or longer (Fig. 43.7f)   **Tortricoidea**[1]
Labial palpi drooping or upright, not held horizontally (Fig. 43.8a), usually shorter than head   **Yponomeutoidea and Copromorphoidea**, in part

23(21). Forewing at least 4 times as long as broad; wings held together by recurved bristles along margins and by frenulum (as in Fig. 43.3a)   **Sesioidea** (Sesiidae)
Forewing no more than twice as long as broad; wings lacking both marginal bristles and frenulum, or with vestigial frenulum   **Bombycoidea**

24(1). Legs greatly reduced, nonfunctional; moth developing in case constructed by larva   **Tineoidea**
Legs well developed, used in walking; moth not developing in larval case   **25**

25(24). Proboscis vestigial or absent   **Noctuoidea**

---

[1] Carposinidae (Copromorphoidea) with seven North American species will also key here.

Proboscis present, longer than head   **Geometroidea**

*Superfamily **Cossoidea**.* Small to large moths, North American species moderate or large (Fig. 43.9c). Body stout, long, usually exceeding hind wings; head with erect hairs or scales, ocelli absent, proboscis very short or absent, antennae bipectinate in male; forewing with large accessory cell, discal cell divided by forked M vein (Fig. 43.7d, at arrows). Larva stout, cylindrical, with prognathous mandibles; thorax with large, sclerotized dorsal shield.

The family Cossidae occurs in North America. These moths show many primitive features, particularly the extensive wing venation with poorly defined discal cell and the chaetotaxy of the larva. At the same time they are specialized in the reduction of the mouthparts and loss of ocelli. Adult cossids are short-lived. They produce large numbers of eggs (up to 18,000 have been recorded from a single individual). Larvae bore in heartwood or beneath the bark of various trees or shrubs or invade the crowns of woody herbs, requiring up to several years to mature. Frass and debris are pushed outside the tunnel, and pupation is within the larval gallery. The family is particularly diversified in Australia. About 40 species are known in North America.

*Superfamily **Tortricoidea**.* Small moths, mostly with broad, truncate, gray, brown, or tan mottled wings (Fig. 43.7e). Head with rough scales on vertex, ocelli and chaetosemata usually present, proboscis short or vestigial (Fig. 43.7f); maxillary palpi minute, labial palpi 3-segmented, porrect. Forewings with accessory cell present (as in Fig. 43.7d), discal cell sometimes divided by unbranched M.

The single family, Tortricidae, comprises a large cosmopolitan group of considerable economic importance. The Tortricinae and Olethreutinae (sometimes recognized as families) contain many species whose larvae bore into fruit, nuts, or seeds, including the codling moth, *Cydia pomonella*, and the oriental fruit moth, *Grapholita molesta*, among the more destructive forms. The movements of Mexican jumping beans are caused by the sudden jerks of the larvae of *Cydia deshaisiana* that live in the seeds. Many other tortricid larvae feed externally, usually sheltering in rolled or tied leaves, and some species feed internally in early instars then externally in later instars. The external feeders also include many economically important species, such as the spruce budworm, *Choristoneura fumiferana*, and the fruit tree leaf roller, *Archips argyrospilus*. Pupation usually occurs within the larval food or shelter. In the Phaloniinae the forewing is usually buckled and bent downward at the end of the discal cell (wings usually flat in Tortricinae and Olethreutinae). The larvae hollow seeds, root crowns, or foliage terminals of a variety of herbs, trees, and shrubs.

*Superfamily **Tineoidea.*** Very small to medium-sized moths, sually somber gray or brown with elongate rounded or pointed wings (Fig. 43.9d). Head rough-scaled (most families, Fig. 43.7c) or smooth-scaled (Gracilariidae); ocelli present or absent, chaetosemata absent; proboscis usually present, without covering of scales; maxillary palpi large, usually 5-segmented, labial palpi short, drooping or ascending (not porrect) (Fig. 43.7c). Forewings with extensive venation (Fig. 43.7a), frequently including branched or unbranched M vein in discal cell; accessory cell absent.

Both generalized and highly specialized taxa are included in this superfamily. As in most other primitive groups of Lepidoptera, the larvae tend to occur in protected situations. For example, Tineidae includes species that tunnel in fungi or fungus-infested wood, others that scavenge beneath loose bark, in bat caves and animal burrows, or in human habitations; several species are major pests of stored grains. Many members of this family are capable of digesting keratin and consume material of animal origin, especially fur and feathers, including articles of apparel; hence their common name, "clothes moths." Many tineid larvae construct tubes over their food or inhabit a portable protective case, in which pupation later occurs. In the Psychidae, which are plant feeders, this case-making behavior

is universal. The cases, sometimes of bizarre configuration, may be constructed entirely of silk or may incorporate fragments of plants or substrate material. A wide variety of angiosperms serve as hosts, and a few species feed on gymnosperms, lichens, and mosses. Females of all North American species are wingless, remaining in the case, where the eggs are deposited. However, in primitive forms, including many Australian species, the females are winged. Psychidae do not feed as adults, and the proboscis is vestigial or absent.

Larvae of Lyonetiidae and Gracilariidae are phytophagous. Most species mine leaves; they may be highly modified with legs reduced or absent. Some species form serpentine mines, while others later enlarge the mine into a blotch. *Bucculatrix*, probably the most familiar genus, initiates a serpentine mine, then abandons the mine to tie adjacent leaves together and skeletonize them. Adult lyonetiids, like many other very small moths, have narrow, lanceolate wings with reduced venation. The wing tip is suddenly attenuate in many species. Many gracilariid larvae show striking hypermetamorphosis. The early leaf-mining instars, which feed on sap, often lack both prolegs and thoracic legs, and have reduced head capsule and mouthparts. The full-grown larva has normal head and mouthparts, thoracic legs with claws, and prolegs on abdominal segments 3 to 5 and 10.

Ochsenheimeriidae (not included in following key) is represented in eastern North America by a single introduced species, *Ochsenheimeria vacculella*, whose larvae mine the stems of a variety of grasses. Other members of this small family are Palearctic.

KEY TO THE NORTH AMERICAN FAMILIES OF TINEOIDEA

1. Wings about as long as body or longer   **2**
   Wings absent or shorter than body
                              Psychidae (females only)
2(1). Antenna with scape enlarged, ventrally concave, forming eyecap (Fig. 43.7b); very small moths (<4 mm long) with lanceolate wings   **Lyonetiidae**

Antenna with scape slender, cylindrical or slightly flattened, not forming cover for eye; size variable **3**

3(2). Head with erect, bushy scales covering vertex and face (Fig. 43.7c) **4**

Head with smooth, appressed scales at least on face (as in Fig. 43.7b); vertex smooth-scaled or with posterior fringe of erect scales; very small moths (<4 mm long) with lanceolate wings **Gracilariidae**

4(3). Proboscis and maxillary palps vestigial; antennal scape without pecten **Psychidae** (males)

Proboscis and maxillary palps almost always large, functional (Fig. 43.7c); if vestigial, pecten present on antennal scape **Tineidae**

Superfamilies **Yponomeutoidea** *and* **Copromorphoidea** *(Fig. 43.9f,h).* Small or occasionally moderate in size; head with smooth scales; ocelli and chaetosemata present or absent; proboscis moderate, without covering of scales; labial palpi short or prominent, either drooping or ascending (Fig. 43.8a). Forewing usually elongate with rounded or angulate apex; discal cell rarely divided by M vein.

Yponomeutoidea and Copromorphoidea include moths that are superficially similar, and several families have been included in both superfamilies in different classifications. The arrangement followed here is conservative.

The moths of these superfamilies are mostly uncommon or obscure, with some notable exceptions mentioned below. Many species, including most Glyphipterigidae and Heliodinidae, are brightly colored, diurnal insects. In contrast to the morphologically similar Tineoidea, which contains many scavengers, nearly all Yponomeutoidea and Copromorphoidea are phytophagous.

Douglasiidae (Yponomeutoidea) mine leaves or stems as larvae. Heliodinidae and Epermeniidae include leaf miners, borers in fruit terminals, and leaf skeletonizers, and Yponomeutidae show a comparable diversity in larval habits, including external foliage feeding. The external feeders usually spin a more or less extensive web over the food, sometimes tying leaves together. Although many are widely distributed, such as the cosmopolitan diamondback moth (*Plutella xylostella*) (Yponomeutidae), which feeds on most Cruciferae, few are of economic significance.

Copromorphoidea are represented by a few small families of rarely encountered moths in North America. Alucitidae are readily distinguished from all other Lepidoptera in having the hind wing divided into at least six plumes. Larval habits are largely unknown. Carposinidae and Copromorphidae (not included in following key) are superficially similar to Tortricidae but lack vein M in the hind wing. Known larvae of both families are internal feeders in fruits, buds, galls, or necrotic bark. Fewer than 10 species are known from North America. Larvae of Glyphipterigidae are mostly seed and stem borers, especially in grasses and sedges.

KEY TO THE MORE COMMON NORTH AMERICAN FAMILIES OF YPONOMEUTOIDEA AND COPROMORPHOIDEA

1. Hind wings entire **2**

Hind wings divided into at least six plumes **Alucitidae** (Copromorphoidea)

2(1). Tibia and tarsus of hind legs bearing stiff, erect bristles

Heliodinidae and **Epermeniidae**

Tibia and tarsus of hind legs with smooth scales **3**

3(2). Ocelli large, conspicuous; maxillary palpi minute (as in Fig. 43.7f) **4**

Ocelli absent; if ocelli present, maxillary palpi large (Fig. 43.8a) **Yponomeutidae** (including Argyresthiidae, Acrolepidae, Plutellidae)

4(3). Body length less than 3 mm; hind wings narrow, lanceolate, with fringe of setae broader than membrane **Douglasiidae**

Body length greater than 3 mm; hind wings with membrane broader than fringe of setae

Glyphipterigidae (incl. Choreutidae)

Superfamily **Sesioidea** *(Fig. 43.9h).* Small to moderate-sized moths; ocelli usually present, chaetosemata present or absent; proboscis naked or scaled at base; maxillary palpi small, labial palpi recurved, ascending; forewing with discal cell divided by M vein or not.

Sesiidae (fig. 43.9h) includes several species destructive to agricultural crops. The moths of this family are mostly diurnal, fast-flying mimics of stinging Hymenoptera and usually have the wing membrane free of scales except at the margins. Most North American species occur in the arid west. Larvae bore in trunks, bark, stems, or roots of herbs, shrubs, or trees, causing significant damage to peach, currant, gooseberry, blackberry, and to squash and other cucurbits. Pupation is in the larval gallery. Adults usually visit flowers.

Superfamily **Gelechioidea** *(Fig. 43.9g,l).* Small or very small moths; ocelli present or absent, chaetosemata absent. Proboscis clothed with scales, at least basally, and embraced by small, 3- to 4-segmented maxillary palpi; labial palpi recurved over head, usually extending to vertex. (Fig. 43.8c). Forewing with discal cell undivided by M vein.

The Gelechioidea represents the largest superfamily of "microlepidoptera," the family Gelechiidae alone containing at least 4000 species. North American forms are predominantly small, somber-colored, rather obscure moths, but some Old World species reach 75 mm in wingspan, with bright, conspicuous markings. In terms of larval feeding, the Gelechioidea are perhaps more diverse than any other superfamily. Included are leaf miners, tiers, and rollers; borers in stems, seeds, nuts, tubers, and fruits; gall makers, scavengers, and predators of scale insects. In addition many species feed exposed or protected by a silken web or case. Many of the larvae wriggle vigorously when exposed, often causing them to bounce erratically. Most gelechioid families, especially those with large numbers of species, are variable both in feeding mode and host-plant preferences. Several of the smaller, more specialized families are more restricted. For example, Coleophoridae are leaf miners in the first instar, then construct a case from silk, frass, and plant material in later instars. Elachistidae are predominantly leaf miners, but some bore into stems. The more primitive species feed in dicotyledonous plants, but most form blotch mines on the leaves of grasses, sedges, and rushes. Pupation occurs in the larval shelter in many species, or exposed in others, sometimes with a silk girdle attaching the pupa to the substrate. Blastobasidae include a few species that feed in flowers of fruits of *Yucca* and *Agave*, but most appear to be opportunistic scavengers that inhabit diverse situations such as the webs of foliage feeding caterpillars, galleries of other insects, or even feeding on scale insects or other small prey. Most Scythridae spin frail webs from which they feed on buds and leaves of a variety of plants. *Areniscythris brachypteris* of central coastal California is notable for the greatly abbreviated wings of both sexes. The larvae inhabit silken tubes, which are attached to sand dune vegetation and covered by shifting sand.

Several families, notably Gelechiidae and, to a lesser extent, Oecophoridae and Cosmopterygidae show a marked tendency to attack the reproductive structures of their hosts (buds, flowers, tubers, etc.), but other species are leaf miners, leaf tiers or rollers, stem borers, or gall formers. Some species feed communally beneath webs on foliage, and some are scavengers. Most economically important species feed in reproductive structures and include the pink bollworm, *Pectinophora gossypiella* (Gelechiidae), possibly the most important insect pest on cotton. Gelechiids of lesser importance include the potato tuber worm, *Phthorimaea operculella* (Fig. 43.9g), which mines leaves as well as boring into tubers, and the Angoumois grain moth, *Sitotroga cerealella*, which infests various whole grains, either fresh or dried. A few Oecophoridae and Cosmopterygidae occasionally cause minor damage on a few crops or ornamentals. *Euclemensia* and *Antequera* (Cosmopterygidae) are unusual in being internal parasitoids of *Kermes* (Coccidae) on oaks.

KEY TO THE COMMON NORTH AMERICAN FAMILIES OF GELECHIOIDEA

1. Hind wings narrow, linear (Fig. 43.8e,f), with posterior fringe of setae much wider than membrane   **2**

   Hind wing broader, oval, with setal fringe narrower than membrane   **7**

2(1). Forewing with thickening (pterostigma) on costal margin near apex (Fig. 43.8e); vein R_2 arising from near apex of discal cell; antennae with scape usually expanded, sometimes concave beneath (Fig. 43.8a)   **Blastobasidae**

   Forewing without pterostigma; antenna with scape slender, never concave beneath   **3**

3(2). Maxillary palpi with two segments, exceedingly small, apparently absent   **4**

   Maxillary palp with 3 to 4 segments; folded around base of proboscis (as in Fig. 43.8a)   **5**

4(3). Forewings with vein 2A forked at base (Fig. 43.8f, at arrow)   **Coleophoridae**

   Forewings with vein 2A undivided at base   **Elachistidae**

5(3). Forewing with vein R_2 arising well before apex of discal cell (Fig. 43.8g)   **6**

   Forewing with vein R_2 arising very near apex of discal cell (as in Fig. 43.8e); mostly small, dark moths; hind wing with closed discal cell   **Scythridae**

6(5). Hind wing with veins R_s and M_1 arising separately on discal cell (Fig. 43.8g); abdomen often with dorsal spines on segments 1 to 7   **Oecophoridae**

   Hind wing with veins R_s and M_1 arising conjointly from discal cell (as in Fig. 43.8h); abdomen without dorsal spines **Cosmopterygidae** (incl. Momphidae, Walshiidae)

7(1). Both wings with vein 1A present, at least in distal half of wing; hind wing not terminating as a narrow appendage (Fig. 43.8g, at arrows) **Oecophoridae** (incl. Ethmiidae, Stenomidae)

Both wings with vein 1A absent; hind wing often narrowed, apically appendiculate (Fig. 43.8h)   **Gelechiidae**

Superfamily Zygaenoidea. Small to medium-sized moths with bipectinate antennae, at least in males; ocelli, chaetosemata present or absent; proboscis large (Zygaenidae) or absent; maxillary palpi small or absent, labial palpi small or minute. Forewings and hind wings with discal cells divided by M vein (Fig. 43.8i).

Adults of this small, predominantly tropical superfamily are mostly densely hairy, heavy-bodied moths, frequently brownish or tan in color, sometimes with pectinate antennae. Zygaenidae, however, include brightly colored, diurnal species that visit flowers for nectar. Larvae are highly modified, often sluglike in appearance, with semiconcealed head and sticky cuticle or tufts or bands of hairs, which may produce severe skin irritation on contact. Most zygaenid caterpillars are leaf skeletonizers, especially on plants of the grape family. Caterpillars of *Harrisina brillians*, the grape leaf skeletonizer, which often causes economic damage in North America, feed communally, lining up side by side and moving backward as a leaf is consumed. The larvae of Epipyropidae are ectoparasites of Fulgoroidea, primarily in the Old World tropics. Large numbers of eggs (up to 3000) are laid on foliage frequented by the hosts, which are located by the active first instar. Later instars are grublike, the caterpillar residing beneath one wing of the host and lapping up hemolymph oozing through punctures made by the styliform mandibles. A single species, *Fulgoraecia exigua*, occurs in North America. The larvae of Megalopygidae (puss caterpillars) are covered with tufts of dense, silky setae that often form a posterior tail-like extension. Beneath the silky tufts are strongly urticating setae. Puss caterpillars feed exposed on a variety of trees and shrubs. Limacodidae and Dalceridae (slug caterpillars) feed exposed on woody plants or occasionally on herbs. Limacodidae include naked, sluglike species and others with spiny lobes, which are sometimes dehiscent and urticating. Dalcerid caterpillars (one species in extreme southern Arizona, not included in following key) are covered with soft, detachable tubercles of jellylike consistency, presumable defensive. About 100 species of Zygaenoidea occur in North America, mostly in the families Zygaenidae and Limacodidae.

KEY TO THE NORTH AMERICAN FAMILIES OF ZYGAENOIDEA

1. Proboscis and maxillary palpi vestigial or absent; vertex without chaetosemata **2**
   Proboscis and maxillary palpi large, functional; chaetosemata large **Zygaenidae**
2(1). Hind wing with veins Sc + R_1 and R_s fused along most of the discal cell (Fig. 43.8i) **Megalopygidae**
   Hind wing with veins Sc + R_1 and R_s separate or fused for a very short distance near base (as in Fig. 43.8g) **3**
3(2). Tibial spurs absent; all mouthparts absent or vestigial; forewing with all veins arising separately from discal cell **Epipyropidae**
   Tibial spurs almost always present on middle and hind legs; maxillary and labial palpi usually apparent; forewing with at least one radial vein branched after leaving discal cell (as in Fig. 43.8i, at arrow)**Limacodidae**

Superfamily Pyraloidea. Small to large moths with ocelli and chaetosemata present or absent; maxillary palpi usually moderate, 4-segmented; labial palpi usually beaklike, held porrect in front of head (Fig. 43.8b); legs usually long, slender; wings without M vein in discal cell (Fig. 43.8d).

Three families occur in North America. Thyrididae, including small, dusky moths, usually with transparent patches in the wings, are distinguished from Pyralidae in lacking vein CuP in the hind wing. Fewer than 10 species occur in North America. Thyridid larvae bore into stems or roll leaves of various plants. Hyblaeidae, with a single introduced species in southeastern North America, are superficially similar to Noctuidae but have two anal veins in the forewings. In contrast, the Pyralidae constitute an enormous grouping, with over 1400 species recorded north of Mexico. Adults are mostly small, dull-colored insects with elongate, triangular wings (Fig. 43.9i,j) but large, bright-colored or metallic forms occur, especially in the tropics. Pyralidae are unique among "microlepidoptera" in possessing tympanal auditory organs on the abdomen. Analogous structures occur on the metathorax of Geometroidea and Noctuoidea.

Larval biology is highly variable, including aquatic as well as terrestrial forms, external feeders on webbed foliage, borers in stems, seeds, or fruit, gall-inhabiting forms, scavengers, and a few predators. Many species fold leaves or construct silk shelters or tubes, usually incorporating frass or food particles. Most Pyralidae are solitary, but dense populations of larvae are common in forms that infest concentrated food sources such as stored products. A few foliage-feeding species construct communal webs. Pupation is in a silk cocoon, either on the host or in or on surrounding litter or soil. Aquatic species occupy habitats ranging from lakes to hot springs to swiftly flowing streams, mostly feeding on submerged vegetation, which they web or tunnel, but females of *Petrophila* crawl down into the water to oviposit beneath stones, where the larvae spin silk webs and feed on algae and diatoms. Pupation is in a silk cocoon. The newly emerged adult swims to the surface with the posterior pairs of legs.

Numerous economically important species attack stored products, in addition to a wide array of agricultural crops and ornamental plants. More destructive species in North America include *Ostrinia nubilalis*, the European corn borer, and *Desmia funeralis*, the grape leaf folder, as well as several cosmopolitan granary pests (*Pyralis farinalis*, *Ephestia kuehniella*, and *Plodia interpunctella*). *Galleria mellonella* and *Achroia grisella* infest beehives and are capable of digesting wax with the aid of intestinal bacteria. A few species are beneficial, including *Cactoblastis cactorum*, which has been utilized to control *Opuntia* cactus in Australia.

Superfamily Pterophoroidea. Small to moderate moths, ocelli and chaetosemata absent; maxillary palpi minute, labial palpi long, held horizontally in front of head; forewings and hind wings each divided into 2 to 4 lobes.

The only family, Pterophoridae, includes slender-bodied moths with characteristic, deeply cleft wings, which are held at right angles to the body at rest (Fig. 43.9k). The wings of pterophorids are split into two or three plumes, while those of alucitids (Copromorphoidea) are divided into at least six plumes. Pterophorid larvae are sometimes leaf miners initially, later folding or rolling leaves, boring in stems or occasionally feeding exposed. In North America fewer than 100 species utilize a variety of hosts in about 15 families of plants.

*Superfamily **Hesperioidea** (Skippers).* Medium-sized, with ocelli absent, chaetosemata present; antennae scaled, gradually clubbed, usually with hooked, bare tip (Fig. 43.6c); maxillary palpi absent, labial palpi large, erect; forewing with 5 radial veins arising separately from discal cell (Fig. 43.10a).

A single family, Hesperiidae (Fig. 43.11a,b). About 3500 species of these stout-bodied butterflies are known, with about 200 in the United States. North American species are diurnal, but some tropical species fly at dusk or night. The wingbeat is rapid and flight direct, rather than fluttering, as in most butterflies (Papilionoidea). Adults sip nectar from flowers; the larvae mostly feed from webbed foliage. The more primitive species utilize a variety of dicotyledonous plants, but the more specialized ones, comprising about 60 percent of the superfamily, eat monocotyledonous plants, especially grasses. Larvae of the subfamily Megathyminae (sometimes treated as a family) bore in stalks and roots of *Yucca* and *Agave*. Pupation is in a cocoon of webbed foliage or in the larval gallery.

*Superfamily **Papilionoidea** (Butterflies) (Fig. 43.11c–i).* Medium to large, occasionally small lepidopterans; ocelli absent, chaetosemata present; antennae gradually clubbed (Fig. 43.6b), rarely with hooked tip; maxillary palpi small, 1-segmented or absent; labial palpi short to much longer than head; forewing with 2 or more radial veins arising conjointly from discal cell (Fig.43.10b).

The butterflies, because of the aesthetic appeal of their brilliant colors or striking patterns, have received more attention than any other group of insects. Accurate species lists exist for most temperate regions, and comprehensive biological information is available for many species. Recent treatments of the North American species include Ferris and Brown (1981), Howe (1975), Opler and Krizek (1984), Opler and Malikul (1992), and Pyle (1981). The Papilionoidea are among the most specialized Lepidoptera in terms of behavior and ecology. Perhaps more than any other insects, butterflies display aposematic or warning coloration and engage in mimicry relationships. In North America the most familiar examples involve (1) the monarch (*Danaus plexippus*), queen (*Danaus gilippus*), and the viceroy (*Limenitis archippus*) (Nymphalidae), and (2) *Battus philenor* (Papilionidae) and *Limenitis astyanax* (Nymphalidae). Mimicry is especially common in tropical regions. At single localities in Central and South America, up to 20 mimetic species of Ithomiinae and Heliconiinae (Nymphalidae) are commonly encountered, along with occasional members of other families, including moths of families such as Ctenuchidae. Such mimetic complexes include both Müllerian relationships, in which several distasteful species are mutually protected by displaying a common, easily recognized pattern, and Batesian relationships, in which edible species resemble distasteful ones. Batesian mimics may be polymorphic, with the different color forms resembling distasteful species with strikingly different patterns, as in the African *Papilio dardanus* (Papilionidae), in which males have the appearance of typical swallowtails while females lack tails on the wings and mimic several species of distasteful Nymphalidae. Seasonal polymorphism, regulated by external factors such as temperature and photoperiod, and usually comprising paler and darker generations also occurs in a number of butterflies, especially Pieridae.

Adults of several species are migratory. Most notable is the monarch, which has spread over the globe, following introductions of its host plants. In North America two-way migrations between Mexico and southern Canada annually cover at least 2000 km. In North America *Cynthia*, *Libytheana*, and *Colias* migrate over shorter distances. Local movements include aggregation on hilltops and patrolling behavior in males, both involved with mate location. Males of many species have a home range in which they occupy centralized perches from which exploratory flights are made to locate females.

Larvae of butterflies are predominantly external foliage feeders. Most are cryptically colored, but species feeding on distasteful plants may be brightly ringed or banded. Papilionidae evert an odoriferous osmeterium when disturbed, while larvae of many Nymphalidae are protected by hairs or spines. Certain lycaenid larvae are myrmecophilous, producing special glandular secretions and sometimes stridulatory signals to attract ants, which protect them from predators and parasites. Most of these myrmecophilous caterpillars are foliage feeders, but some of them spend daylight hours in the host galleries, migrating nightly to feed. A few species reside permanently in the host nest, feeding on ant larvae or pupae. A few other lycaenids are predatory on aphids or coccids. Naked pupae or chrysalises, attached to the substrate by the cremaster, are typical of Papilionoidea. Some chrysalises are also supported by a silken girdle about the midregion.

The Papilionidae, or swallowtails, are among the most familiar butterflies, but constitute a relatively small family, with only about 30 species north of Mexico. Pieridae, the whites and sulphurs, number only about 60 North American species, but include the economically important *Pieris rapae*, the cabbage butterfly, whose caterpillars feed widely on garden crucifers. Libytheidae, the snout butterflies are primarily tropical, with a single North American species, *Libytheana bachmanii*. The largest families of butterflies are the Lycaenidae and Nymphalidae, each with over 200 species in the United States.

The butterflies were formerly classified as the suborder Rhopalocera, a special recognition that is contradicted by their overall similarity to certain families of moths, notably Hedylidae, a small neotropical family superficially similar to Geometridae (Scoble, 1986). Recent molecular phylogenetic work supports a relationship between Hedylidae and butterflies and some microlepidopterans, rather than other macrolepidopterans. Some authorities recognize considerably more families than treated here, but recent comparative studies of skeletal

anatomy show that most of the obvious variation involves superficial characteristics.

KEY TO THE FAMILIES OF PAPILIONOIDEA

1. Hind wing with 2 anal veins (Fig. 43.10b); fore tibiae without epiphysis (articulated spur)   **2**

   Hind wing with 1 anal vein; fore tibiae with epiphysis   **Papilionidae**

2(1). Tarsal claws simple, not divided   **3**

   Tarsal claws bifid   **Pieridae**

3(2). Labial palpi shorter than thorax; usually erect   **4**

   Labial palpi longer than thorax, held horizontally in front of head   **Libytheidae**

4(3). Antennal base and eye contiguous; eye usually notched around antennal base

   Lycaenidae (including Riodinidae)
   Antennal base and eye not contiguous; eye not notched around antenna **(Nymphalidae)**   **5**

5(4). Hind wing with humeral vein recurved toward wing base

   Heliconiinae
   Hind wing with humeral vein straight or curved toward wing apex   **6**

6(5). Forewing with vein 3A absent; antennae with scales dorsally (Fig. 44.10b)   **7**

   Forewing with vein 3A appearing as a short basal fork of A$_2$ (as in Figs. 43.10a,c, at arrow in a)   **Danainae**

7(6). Forewing with vein Sc strongly swollen at base (Fig. 43.11i)   **Satyrinae**

   Forewing with veins at most slightly swollen at base   **Nymphalinae**

*Superfamilies **Geometroidea** and **Drepanoidea**.* Mostly small to medium-sized moths with broad wings and slender body; ocelli usually absent; chaetosemata present in Geometroidea, absent in Drepanoidea, except in Drepanidae; maxillary palpi minute or absent; labial palpi usually short, ascending; abdomen with tympanal auditory organs located ventrolaterally in second segment (Fig. 43.6f).

In North America these moths are represented predominantly by Geometridae (Fig. 43.13d,e), one of the largest families of moths, with about 12,000 species, 1400 recorded north of Mexico. Accumulations of moths at lights normally contain a high proportion of Geometridae. Many species assume characteristic postures with the wings pressed against the substrate or held at various angles to the long axis of the body and spend the daylight hours resting on bark, lichen, twigs, or other surfaces where posturing apparently disrupts the body configuration, helping to camouflage the moths. The abdominal tympanal organs serve the same function as those of the Noctuidae—avoidance of bats by detection of the high-pitched sounds the mammals produce to echolocate their prey. Females of several species are brachypterous, living only long enough to mate and deposit their eggs near the pupation site.

Geometrid larvae are commonly known as inchworms or measuring worms, in reference to their looping gait, which is necessitated by the lack of abdominal prolegs in the middle of the body (Fig. 42.5b). Many bear a strong resemblance to twigs, which is enhanced by the rigid postures adopted at rest. Caterpillars of *Eupithecia* in the Hawaiian Islands have utilized this cryptic behavior to become predators of Drosophilidae and other small flies, which are seized with sudden strikes when they alight on the caterpillar's hind quarters. Several species, such as the omnivorous looper, cause minor feeding damage on ornamentals; the cankerworms *Paleacrita vernata* and *Alsophila pometaria* become serious pests, but most Geometridae do not cause appreciable economic loss. Pupation is usually in a flimsy cocoon.

Whereas most members of the superfamily have cryptically patterned brown, gray, or green wings and bodies, the tropical, diurnal Uraniinae, frequently with brilliantly metallic wings, often resemble papilionid butterflies. Larvae of many species feed on Euphorbiaceae, which contain alkaloids toxic to vertebrates. Presumably these toxins are retained by the adults. Larvae of Thyatiridae, Drepanidae, and Epiplemidae differ from those of Geometridae in having four pairs of abdominal prolegs (one or two pairs or one or two pairs plus

two or three reduced pairs in Geometridae). They feed on foliage of various shrubs and trees.

KEY TO THE COMMON NORTH AMERICAN FAMILIES OF GEOMETROIDEA AND DREPANOIDEA[1]

1. Wings about as long as body   2
    Wings less than half as long as body, not functional in flying   **Geometridae**, in part
2(1). Hind wings with veins $Sc + R_1$ and R_s diverging or parallel throughout length; frenulum always present   **Epiplemidae**
    Hind wings with veins $Sc + R_1$ recurved toward vein R_s at or beyond discal cell (Fig. 43.10c,d, at arrow in c); frenulum present or absent   3
3(2). Hind wings with veins $Sc + R_1$ and R_s parallel or fusing beyond middle of discal cell (Fig. 43.10d); proboscis usually absent; forewings usually with hooked apex
                                        Drepanidae
    Hind wings with veins $Sc + R_1$ and R_s usually contiguous or fused before or near the middle of discal cell (Fig. 43.10c, at arrow); proboscis almost always present; forewings not usually hooked
    Geometridae, in part (including Uraniidae)

Superfamily Bombycoidea. Medium-sized to large, stout-bodied moths with broad wings and without ocelli or chaetosemata (Fig. 43.13g,h); proboscis and maxillary palpi rudimentary or absent; antennae bipectinate, featherlike; frenulum reduced or absent; tympanal organs absent.

Adult Bombycoidea are specialized for reproduction, with atrophied mouthparts. The frenulum is almost always absent, wing coupling being accomplished by overlap, as in butterflies. Nevertheless,

---

[1] Sematuridae, lacking abdominal tympanal organs, are classified as Geometroidea. A single species occurs in southern Arizona. Thyatiridae, represented by about 15 uncommon species in North America, superficially resemble noctuids but have the cubitus three-branched in the forewings and four-branched in the hind wings.

many of these moths are strong fliers, the males traversing distances of several miles to reach females. Males fly upwind and locate nearby females by following gradients of pheromones released by the moths. Despite their inability to feed, the moths may remain active for many days, subsisting on stored fats. Females emerge with the eggs at an advanced stage of development and usually remain near the pupation site until mating occurs.

Larvae feed predominantly on foliage of shrubs and trees. Most species are solitary, but some Saturniidae feed gregariously, and Lasiocampidae, the tent caterpillars, construct large, communal webs or tents enveloping entire branches and frequently cause severe defoliation. They are the only significantly destructive members of the superfamily. In North America several species of *Malacosoma* (Fig. 43.13h) are important tent caterpillars.

The pupal period is passed in a tough cocoon, and *Bombyx mori*, the silkworm (Bombycidae), is propagated commercially for its silk.

In North America Saturniidae is the most common family. Many of the species are elegant moths, subtly shaded in muted colors with contrasting eye spots in one or both wings (Fig. 43.13g). Mimallonidae and Apatelodidae are primarily tropical families with only a few genera and species in North America. Mimallonidae are notable for the portable cases, open at both ends, constructed by their larvae. When resting, the larva blocks the anterior opening with its head capsule and the posterior opening with the circular, thickened posterior abdominal segment. Adults of both families are similar to small Saturniidae.

KEY TO THE NORTH AMERICAN FAMILIES OF BOMBYCOIDEA

1. Hind wing with costal margin greatly expanded basally and supported by 1 to 3 short, stout humeral veins (Fig. 43.10e, marked "h") **Lasiocampidae**
   Hind wing with costal margin not broadened, without humeral veins (Fig. 43.10f) **2**

2(1). Forewing with vein R with 3 branches (Fig. 43.10f); hind wing usually with 1 anal vein **Saturniidae** (incl. Citheroniidae)
   Forewing with vein R with 4 to 5 branches; hind wing with 2 anal veins **3**

3(2). Forewing with vein R_{2+3} and R_{4+5} with separate stalks; hind wing without crossvein between veins Sc and R_s **Mimallonidae**
   Forewing with veins R_2, R_3, R_4, and R_5 arising from a common stalk (as in Fig. 43.12c) **Bombycidae[1]** and **Apatelodidae**

Superfamily **Sphingoidea.** Medium-sized to large, heavy-bodied moths with long, narrow forewings; ocelli and chaetosemata absent; proboscis usually very long; antennae usually thickened in middle (Fig. 43.6d); maxillary palpi 1-segmented, labial palpi stout, 2- or 3-segmented.

The single family Sphingidae is most diverse in tropical regions, but the large, fast-flying moths are familiar insects throughout the world (Fig. 43.13f). The larger species exceed small hummingbirds in size and share the avian habit of extracting nectar while hovering above flowers. Certain plants, including some Onagraceae in temperate regions, are adapted for pollination by sphingid moths, with deep-throated, usually pale-colored flowers that open at dusk as the moths become active. Other sphingids are diurnal, some closely resembling bumblebees or other stinging Hymenoptera.

The stout-bodied larvae often bear a tapering caudal projection; hence the common name "horn worms." They consume prodigious quantities of foliage, and *Manduca quinquemaculata* and *M. sexta* are pests on solanaceous plants such as tobacco, tomato, and potato. Pupation occurs free in a cell in soil or in a wispy cocoon.

Superfamily **Noctuoidea.** Small to large, mostly stout-bodied moths without chaetosemata, ocelli absent; proboscis long or reduced or absent, maxillary palpi minute or absent; tympanal organs present in metathorax (Fig. 43.6e).

---

[1] *Bombyx mori*, the silkworm (Bombycidae) is sporadically reared in North America.

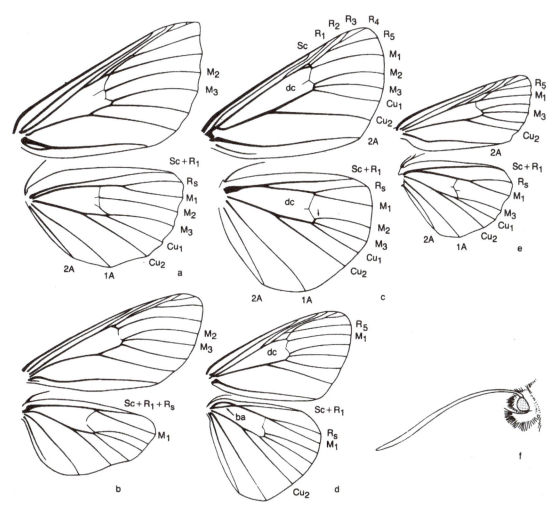

Figure 43.12 Taxonomic characters of Lepidoptera: a to e, wing venation; a, Notodontidae (*Nadata gibbosa*); b, Ctenuchidae (*Ctenucha rubroscapus*); c, Arctiidae (*Apantesis ornata*; arrow indicates basal arch of M2); d, Lymantriidae (*Orgyia vetusta*); e, Noctuidae (*Lacinipolia quadrilineata*); f, antenna of Agaristidae (*Alypia ridingsi*). ba, basal areole; dc, discal cell.

The Noctuoidea constitute the largest group of Lepidoptera, with more than 20,000 species recognized. Included are many extremely familiar insects, such as the gypsy moth, *Lymantria dispar* (Lymantriidae, Fig. 43.13i), and various species of cutworms (Noctuidae, Fig. 43.13a,b). In terms of economic damage to crops, the Noctuoidea are easily the most important group of Lepidoptera in North America.

Adults are typically strong-flying, nocturnal insects that almost always constitute a major proportion of the insects attracted to lights. Certain species of Noctuidae annually migrate distances of many hundreds of miles north from mild winter regions in the southeastern United States. Similar migrations have been documented in other regions. Tympanal organs (Fig. 44.6e) are present in all but a few exceptional species. In Noctuidae they have been shown to function in detecting the high-pitched sounds emitted by bats to locate potential prey. The moths respond by evasive flight patterns, including sudden diving or erratic

Figure 43.13 Representative Lepidoptera: a, Noctuidae (*Trichoplusia brassicae*); b, Noctuidae (*Catocala andromache*); c, Arctiidae (*Apantesis ornata*); d, Geometridae (*Sabulodes caberata*); e, Geometridae (*Dichorda iridaria*); f, Sphingidae (*Celerio lineata*); g, Saturniidae (*Hemileuca electra*); h, Lasiocampidae (*Malacosoma californica*); i, Lymantriidae (*Lymantria dispar*).

looping. Aposematic Arctiidae produce ultrasonic signals to warn bats of their inedibility. Females of some species, including many Lymantriidae (tussock moths), are brachypterous or have such heavy bodies that they fly short distances at most. Early-instar larvae of such species are dispersed by wind.

Most Noctuoidea are mottled brown or black, very often with prominent tufts or mounds of hair or scales on the thorax or wings. As in other moths that rest on irregular surfaces, such patterning is important in camouflaging the body and disrupting its outline and contours. Many Arctiidae

(Fig. 43.13c) and most Ctenuchidae are aposematically patterned in white, yellow, or red and black. Many species, especially of Ctenuchidae, are diurnal, sometimes with clear wings, providing a close resemblance to stinging Hymenoptera. Some arctiid adults feign death when disturbed, distending the abdomen to reveal bright-colored patches of cuticle. The chemical basis for such apparently aposematic behavior in these families is largely unknown.

Larvae of Noctuoidea are herbivores on a great variety of plants, including herbaceous and woody angiosperms and gymnosperms. Many species are highly polyphagous, greatly augmenting their destructive potential. For example, the noctuid *Helicoverpa zea* is known as the corn earworm, the tomato fruitworm, and the cotton bollworm in different parts of the country or by different growers. Several species of Lymantriidae, most notably the introduced gypsy moth, are important defoliators of shade trees in eastern North America. Noctuid, notodontid, and agaristid larvae are typically smooth-skinned, whereas arctiid, ctenuchid, and lymantriid caterpillars are closely covered with bristly or silky hairs, which are brightly colored and urticating in some species. Noctuids frequently hide under debris or litter, emerging at night to feed on or near the surface, where they often chew through stems—hence the name "cutworms." Species of several families, including Noctuidae, Nolidae, and Arctiidae, construct feeding webs on foliage; some Arctiidae feed communally in webs.

Pupation occurs naked in a cell in the ground or bark (Agaristidae, some Noctuidae) or in a cocoon. Larval hair is usually incorporated into the cocoon, which may then contain little or no silk. The urticating hairs of some Lymantriidae are transferred by the adult female from the cocoon to her freshly deposited egg mass.

KEY TO COMMON NORTH AMERICAN FAMILIES OF NOCTUOIDEA

1. Wings absent or much less than half length of body   **Lymantriidae** (females only)

Wings about as long as body or longer   2
2(1). Forewing with veins M_2 and M_3 parallel (Fig. 43.12a); tympanal organ directed ventrally   **Notodontidae**[1]
Forewing with veins M_2 and M_3 converging near discal cell, frequently adjacent (Fig. 43.12b–e); tympanal organ directed posteriorly   3
3(2). Hind wing with veins $Sc + R_1$ and R_s confluent (Fig. 43.12b); brightly colored, frequently metallic, or wasplike diurnal species
    Ctenuchidae (often treated as subfamily of Arctiidae)
Hind wing with veins $Sc + R_1$ and R_s separate beyond base (Fig. 43.12c–e); usually gray or brown, rarely metallic or wasplike   4
4(3). Hind wing with vein M_2 arising closer to M_3 than M_1; M_2 usually arched basally (Fig. 43.12c–e, arch at arrow in c)   5
Hind wing with vein M_2 arising midway between M_1 and M_3, or closer to M_1; M_2 not usually arched (as in Fig. 43.12a)
    Arctiidae, in part
5(4). Antennae gradually enlarged apically (Fig. 43.12f); veins $Sc + R_1$ and R_s confluent in basal fourth of discal cell in hind wing; black moths with large white or yellow wing spots   **Agaristidae**
Antennae threadlike or feathery, not gradually enlarged; veins $Sc + R_1$ and R_s variable   6
6(5). Ocelli present   7
Ocelli absent   9
7(6). Hind wing with vein $Sc + R_1$ usually swollen, bulbous at base (Fig. 43.12c), usually fused with R_s to about middle of discal cell   **Arctiidae**, in part
Hind wing with vein $Sc + R_1$ not noticeably swollen at base; separating before middle of discal cell   8

---

[1] Dioptidae key here. The single North American species, *Phryganidia californica*, occurs in California. The slender-bodied, diurnal adults are common about the larval food plants—various species of oaks.

8(7). Black moths with several large white dots on wings; tympanal hoods very large, prominent, dorsolateral on first abdominal segment **Pericopidae**

Dark brown or gray moths with mottled color pattern or contrastingly banded or patterned, but not with large, white dots; tympanal hoods lateral **Noctuidae**

9(6). Forewings with smooth scaling **10**

Forewings with tufts or lines of raised scales **Nolidae**

10(9). Proboscis vestigial or absent; hind wings with Sc + R_1 and R_s forming basal areole (Fig. 43.12d) **Lymantriidae**

Proboscis large, functional, coiled; veins Sc + R_1 and R_s not separated at base (Fig. 43.12e) **Noctuidae**

CHAPTER 44

Order Trichoptera (Caddisflies)

ORDER TRICHOPTERA

Adult. Small to moderate-sized Endopterygota with 2 pairs of wings. Head free, mobile, with large compound eyes, 2 to 3 ocelli present or absent, antennae filiform, elongate, with numerous segments; mandibles vestigial; maxillae with single small lobe or mala, maxillary palpi 3- to 5-segmented; labium with large mentum and 3-segmented palps. Thorax with first segment reduced, collarlike; mesothorax largest, metathorax large to small; wings with macrotrichia or hairs densely covering both veins and membrane; venation mostly of longitudinal veins with very few crossveins; legs long, slender, with long coxae; tibiae with both apical and preapical spurs; tarsi with 5 segments; abdomen with 9 segments in males, 10 segments in females.

Larvae. Eruciform or campodeiform, frequently occurring in protective case or covering; head capsule sclerotized, with strongly developed mandibles, maxillae, and labium, very short antennae and lateral ocelli. Three distinct, short thoracic segments of variable sclerotization; legs long, slender, usually with some segments subdivided; abdomen soft, weakly sclerotized, with hook-shaped prolegs on terminal segment.

Pupa. Decticous, exarate, occurring in larval case or in specially constructed pupal case or cocoon.

With an estimated 13,000 species in 45 families worldwide, over 1200 in North America, the Trichoptera are one of the smaller orders of Endopterygota. The adults are slender, mostly somber-colored insects that somewhat resemble moths but never have the maxillae modified to form a proboscis, and have hair rather than scales on the body and wings. The basal portions of the radial and median veins are not obliterated to form a discal cell, as in most Lepidoptera, and in Trichoptera veins 2A and 3A in the forewing are usually looped forward to vein 1A (Fig. 44.1b). A useful diagnostic feature is the length of the antennae, which exceeds that of the forewing in most Trichoptera, being shorter in most Lepidoptera.

By day, adult caddisflies rest concealed on vegetation, under bridges or in culverts and are seldom seen without searching. In contrast, the larvae, which often inhabit portable cases, are conspicuous in many aquatic habitats, where they are of considerable importance as food for fish and other predators. The great majority of species inhabit fresh water, but smaller numbers occur in brackish coastal waters, and *Philanisus* lives in intertidal rock pools in Australia and New Zealand.

Adult Trichoptera are relatively homogeneous in external morphological features, seemingly adapted for a relatively short life. Although the mouthparts appear to be adapted for lapping up liquids, few

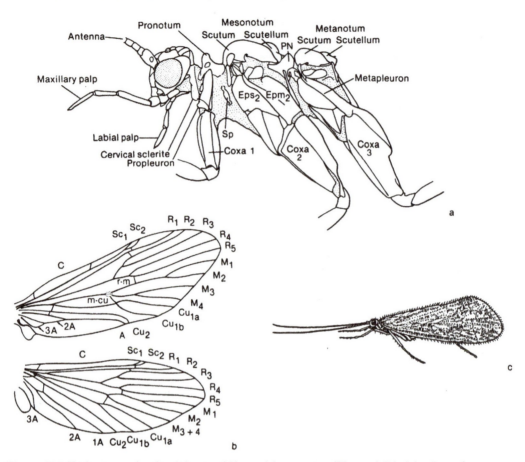

Figure 44.1 Trichoptera: a, head and thorax of *Rhyacophila torrentium* (Rhyacophilidae); b, wings of *Rhyacophila torrentium*; c, *Frenesia missa* (Limnephilidae). Eps, episternum; Epm, epimeron; PN, postnotum; Sp, spiracle.

Source: a, b, redrawn from Schmid, 1970; c, redrawn from H.H. Ross, 1944.

Trichoptera have been observed to feed, and most species probably reproduce without taking any sustenance. Most caddisflies are weak, fluttery fliers, but others may fly great distances. In general, the weaker fliers have relatively broad forewings, broadly overlapping the hind wings or with the jugal lobe enlarged, as in primitive Lepidoptera. In stronger-flying species the macrotrichia of the hind-wing base are developed to form a frenulum, which interlocks with the jugal lobe of the forewing. In several families a row of enlarged, upcurved macrotrichia along the costal margin of the hind wing hooks over a ventral, longitudinal ridge on the anal region of the forewing. In Leptoceridae, which

include many active, strong-flying diurnal species, and in Helicopsychidae, the modified hind-wing macrotrichia are developed as hamuli (Fig. 44.6c), much as in Hymenoptera, and the antennae are as much as three times the length of the body.

Larvae are of two general types. Free-living and net-spinning larvae are campodeiform, while those that inhabit cases tend to be eruciform. However, many intermediate forms are known. In Hydropsychidae, which appear to be primitive in a number of features, all of the thoracic terga, as well as the prosternum, are sclerotized. The legs are robust and widely spaced, suited for walking, and the anal prolegs are stout with simple claws

(Fig. 44.2a). Many of these non–case-making larvae are predatory and have slender elongate mandibles, superficially similar to those of Carabidae and other predatory Coleoptera. In some Rhyacophilidae the fore femora and tibiae are strongly raptorial, for grasping prey (Table 44.1).

Several modifications are apparent in larvae that construct cases. The terga of the meso- and metathorax and the prosternum are membranous. The thoracic segments are short, so that the legs are attached close together near the anterior end of the body and can be easily protruded from the case (Fig. 43.2b). The middle and especially the hind legs are relatively long, further facilitating movement from within the case. In most caddisfly larvae the forelegs are relatively short and are used more for manipulation of food than for walking.

In case-making forms the abdomen is always extremely soft and vulnerable. Gills are more frequently present than in free-living species, and in some case-makers a lateral fringe of abdominal hairs is used to circulate water through the case. Three tubercles, which position the larva within the case, are frequently present on the first abdominal segment (Fig. 44.2b). The terminal prolegs are fused basally, with the appendagelike distal portions ending in complexly pointed claws adapted for holding the larva inside the case.

Case construction and configuration are quite variable and apparently have arisen independently several times. Hydrobiosidae and Rhyacophilidae are free-living, while the closely related Glossosomatidae construct saddle-shaped cases of silk and sand grains, open at both ends (Fig. 44.3b), the

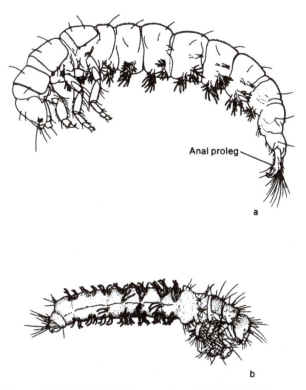

Figure 44.2 Trichoptera larvae: a, Rhyacophilidae (*Smicridea fusciatella*); b, Limnephilidae (*Philarctus quaeris*).
Source: a, redrawn from Flint, 1974; b, redrawn, with permission, from Larvae of the North American Caddisfly Genera (Trichoptera) *by Glenn B. Wiggins, University of Toronto Press, 2nd ed., 1996..*

Table 44.1 Characteristics of Larval Trichoptera of North America

Family	Case structure	Case material	Current	Temperature	Substrate
Philopotamidae	Long tunnels or nets	Silk, no extraneous materials	Rapids, riffles	Cold or cool	Stony
Psychomyiidae	Tunnels or nets	Silk	Streams, ponds, lakes	Variable	Sandy bottoms, stones, aquatic vegetation
Hydropsychidae	Nets	Silk	Rivers, streams, moderate current	Variable	Stones, logs, debris
Rhyacophilidae	Free-living	No case	Riffles, moving water	Cold	Irregular bottoms, stones, debris
Glossomatidae	Tortoise shell-shaped; open both ends	Small stones, pebbles	Moving water	Cold	Stony, irregular bottoms, stones, debris
Hydrobiosidae	Free-living	No case	Moving water	Variable	Irregular bottoms, stones, debris
Hydroptilidae	Barrel-shaped or purse-shaped, open both ends	Variable	Variable	Variable	Variable
Phryganeidae	Stout, tubular case	Grass stem segments, spirally arranged	Marshes, stream eddies	Cold	Plant debris, silt
Brachycentridae	Cylindrical or square cases	Mostly silk (cylindrical cases) or rocks or woody material (square cases)	Rapids, riffles in streams, rivers	Cold	Variable
Limnephilidae	Tubular; variable in proportions	Extremely variable	Variable	Variable	Variable
Lepidostomatidae	Four-sided cases or slender tubes, or irregular cases	Sticks, twigs, sand, or varied materials	Springs, streams, rivers, lake shores	Cold	Variable
Calamoceratidae	Tubular, often triangular in cross section	Sticks, leaf fragments	Springs, slow streams	Cool or cold	Variable
Helicopsychidae	Helical spiral, snail shell-shaped cases	Stone	Springs, rapid streams, lake margins	Variable	Variable
Odontoceridae	Elongate tubular cases	Minute stones, compactly webbed together	Rapid streams	Cold	Sand, gravel, stones
Sericostomatidae	Stout, tubular cases	Minute stones	Rapid streams, springs	Cold	Sand, gravel, stones
Leptoceridae	Variable	Variable	Lakes, streams	Variable, often warm	Variable
Molannidae	Flattened, oblong cases	Minute stones, sand	Lakes, streams	Cool to cold	Stony
Beraeidae	Curved, tapering cases	Sand grains	Springs, seeps, small streams	Cool	Plant debris

protruding abdominal apex being used in locomotion. Hydroptilidae are free-living in early instars and morphologically similar to free Rhyacophilidae. In later instars they spin-flattened cases from silk, sometimes incorporating fine sand or other material, and the abdomen enlarges. Hydroptilid cases are purse-shaped in general configuration, with the head and legs projecting from a slit in the anterior end and the abdominal prolegs projecting from a slit in the posterior margin (Fig. 44.4a).

Hydropsychidae, Psychomyiidae, and Philopotamidae construct fixed shelters or retreats and feed by filtering small particulate matter from the water. These larvae are similar to the free-living Rhyacophilidae. Details of shelter construction

vary enormously, from masonry tubes or burrows strengthened with silk to a variety of silken webs, funnels, or tubes (Figs. 44.3c–e). The constructions are arranged so that water moves through a net, either placed at right angles to the stream flow or spun across a burrow.

Cases of other families are generally in the form of elongate, portable tubes completely enclosing their inhabitants (Fig. 44.4c–e), Construction

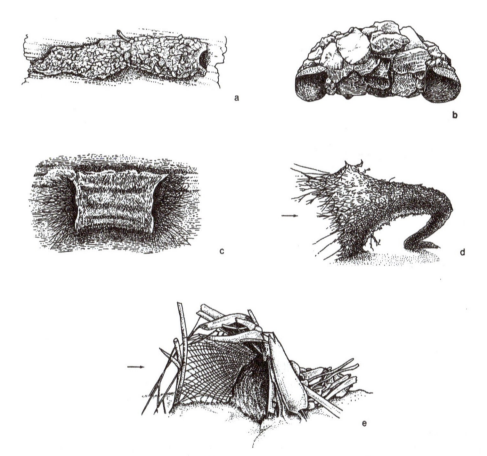

Figure 44.3 Larval cases and retreats of Trichoptera: a, retreat of *Lype* sp. (Psychomyiidae) is on a piece of wood, open at both ends, with the roof of silk and bits of detritus; b, case of *Culoptila* sp. (Glossosomatidae) is made of small stones; c, retreat of *Nyctiophlax* sp. (Polycentropinae, Psychomyiidae) has a silk roof, open at both ends, over a depression in a rock or piece of wood, from which the larva attacks prey that touch the threads at the entrances; d, retreat of *Neureclipsis* sp. (Polycentropinae, Psychomyiidae) is a funnel-shaped silk tube which serves to filter prey from water entering the large opening; e, capture net or retreat of *Hydropsyche* sp. (Hydropsychidae) is made in flowing water and catches small prey.
Source: Redrawn, with permission, from Larvae of the North American Caddisfly Genera (Trichoptera) *by Glenn B. Wiggins, University of Toronto Press, 2nd ed., 1996.*

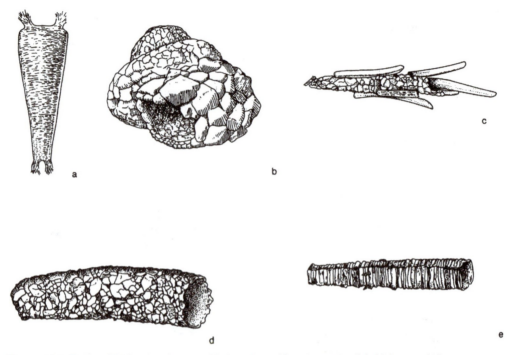

Figure 44.4 Cases of Trichoptera larvae: a, Hydroptilidae (*Oxyethira serrata*); b, Helicopsychidae (*Helicopsyche borealis*); c, Leptoceridae (*Mystacides sepulchralis*); d, Limnephilidae (*Pseudostenophylax edwardsi*); e, Brachycentridae (*Brachycentrus*).
Source: a, b, redrawn from Ross, 1944; c, redrawn from Yamamoto and Wiggins, 1964; d, redrawn from Wiggins and Anderson, 1968; e, redrawn from Wiggins, 1965.

always involves binding particles from the substrate with silk produced from labial glands and extruded through a spinneret. Larvae of these typical case-makers have soft, membranous abdomens and hypognathous heads as well as a number of other specialized features mentioned above, and apparently represent a monophyletic lineage. Configuration of the cases varies from stout, straight tubes (many Limnephilidae) to long, slender, curved tubes (Leptoceridae). Cases are usually circular in cross section but may be polygonal [Brachycentridae, Leptoceridae (Fig. 44.4e)]. Members of the Helicopsychidae use sand grains or small pebbles to build spiral tubes (Fig. 44.4b), which resemble snail shells, and Calamoceratidae build flat shelters from two pieces of leaf. In general, species inhabiting riffles and rapids build the most solidly constructed cases, and those occupying lakes or ponds the loosest ones. Choice of building

materials depends in part upon local availability, so that significant variation exists within species. Yet most Trichoptera choose particles within a certain size range and with a certain composition, and cases are often somewhat characteristic at the species level.

It is unclear whether the free-living mode of life or the case-making habit is ancestral in Trichoptera, and the evolutionary relationships among the various types of case makers are disputed.

Caddisfly adults are encountered on vegetation and on stone overhangs and similar places near water. Most are nocturnal or fly at dusk or dawn, and many are attracted to lights. Copulation takes place in the air or on substrates such as low vegetation, and swarming behavior occurs in some families, including Leptoceridae. A few forms, such as *Dolophilodes* (Philopotamidae), are brachypterous or apterous. About 100 to over 1000 eggs, depending on the

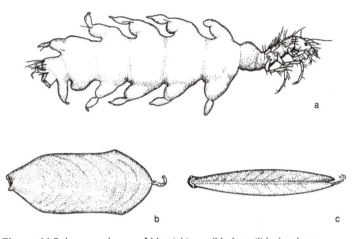

Figure 44.5 Larva and case of *Ithytrichia* sp. (Hydroptilidae): a, larva, lateral view; b, case, ventral view; c, case, lateral view.
Source: Redrawn, with permission, from Larvae of the North American Caddisfly Genera (Trichoptera) *by Glenn B.Wiggins, University of Toronto Press, 2nd ed., 1996.*

species, are laid in compact masses or long strings, directly into the water by most species, and females may oviposit while submerged. The eggs are enclosed in a polysaccharide matrix, which swells to form a protective covering in the water. Limnephilidae may deposit egg masses on objects above the water, even on twigs high in trees. Apparently rain dissolves the matrix holding the eggs, which then drop into the water below. The Australian genus *Philanisus*, whose larvae inhabit intertidal rock pools, oviposits into the coelomic cavity of starfish. The larvae are free-living, feeding on coralline algae.

Larvae pass through five to seven instars. Non–case-making forms may shelter beneath submerged stones or in debris (Rhyacophilidae) or may spin nets used for filtering food from moving water, as described above (Psychomyiidae, Hydropsychidae). In numbers of individuals, these filter feeders are the dominant Trichoptera in larger streams in North America. The larvae or their exuviae may become so abundant as to clog intakes in generating plants or municipal water works. Case-bearing species may occur exposed on stony, sandy, or silty bottoms, shelter beneath stones or logs, or clamber over vegetation, depending on the species. Some Leptoceridae are capable of active swimming by using the fringed hind legs. Free-living and net-spinning forms tend

to be predatory, while most case bearers are probably scavengers. Algae and diatoms are commonly consumed by many species, as well as detritus of plant and animal origin. Larval instars are similar in most species, but in the Hydroptilidae, the first four instars are free-living with a short abdomen and long, slender abdominal prolegs. The fifth and sixth instars, which have a bloated, soft abdomen with short prolegs, inhabit cases (Fig. 44.5). Early instars of some Lepidostomatidae build cases of sand grains, then switch to cases of leaf fragments as they grow.

Distribution of larval caddisflies is strongly influenced by the qualities of the substrate as well as by current, presence and type of vegetation, and water temperature. Different portions of one stream or closely adjacent bodies of water that differ in one or more of these characteristics may support extremely different faunas. Because of this specificity, caddisflies are important indicators of water quality.

Pupation occurs in water within a shelter—either the larval case or retreat, which may be shortened and closed, or a special pupal case. The saddle case makers (Rhyacophilidae, Glossosomatinae) remove the bottom of the case, attach the top to a rock, and spin a cocoon inside. Trichoptera use the pupal mandibles to open the case just before adult emergence. The pharate adult swims to the surface, using setal fringes on the middle legs, and immediately emerges and takes flight, or first climbs onto emergent vegetation or other objects.

Identification of most Trichoptera depends upon structures that shrivel in dried specimens; both larvae and adults should be collected into liquid preservatives. Pupae can frequently be identified by the features of the pharate adult. They are a valuable means of associating adults and larvae, since the cast skin of the last larval instar frequently remains in the case. The following key is based on those of Ross (1944, 1959), Wiggins (1977, 1984, 1987), and Morse and Holzenthal (1996).

Among living orders, Lepidoptera are clearly the closest relatives of Trichoptera, and it has even been previously proposed that the two orders be merged despite the difference in larval biology. The order has received some excellent taxonomic and phylogenetic attention in recent years; the reader is referred to the comprehensive reviews by Morse (1997) and Holzenthal et al. (2007).

KEY TO THE COMMON NORTH AMERICAN FAMILIES OF TRICHOPTERA

Adults[1]

1. Wings without clubbed hairs; antennae about as long as wings or much longer; body length 5 to 40 mm   2
   Wings with numerous erect clubbed hairs; antennae shorter than
   wings; densely hairy caddisflies less than 6 mm long   **Hydroptilidae**
2(1). Ocelli present   3
   Ocelli absent   9
3(2). Maxillary palpi with 3 to 4 segments   4
   Maxillary palpi with 5 segments   5
4(3). Maxillary palpi with 3 segments
   Limnephilidae (males; including Goeridae)
   Maxillary palpi with 4 segments
       Phryganeidae (males)

---

[1] The families Hydrobiosidae, Uenoidae, and Xiphocentridae, each with a few species in North America, are omitted.

5(3). Maxillary palpi with fifth (apical) segment subequal to fourth segment   6
   Maxillary palpi with fifth segment 2 to 3 times longer than fourth segment
       Philopotamidae
6(5). Maxillary palpi with first and second segments short, subquadrate, approximately equal in length   7
   Maxillary palpi with second segment about twice the length of first segment, distinctly longer than wide   8
7(6). Fore tibia with preapical spur   21
   Fore tibia without preapical spur
       Glossosomatidae
8(6). Fore tibiae with at least 2 spurs; middle tibiae with 4 spurs   **Phryganeidae** (females)
   Fore tibiae with 1 spur or without spurs; middle tibiae with 2 to 3 spurs
   Limnephilidae (females; including Goeridae)
9(2). Maxillary palpi with at least 5 segments   10
   Maxillary palpi with 3 to 4 segments   12
10(9). Maxillary palpi with fifth (apical) segment at least twice as long as fourth segment; fifth segment annulate, appearing multisegmented   11
   Maxillary palpi with fifth segment subequal to fourth segment, not annulate   12
11(10). Hind wings at least as broad as forewings, with anal region occupying one-third to one-half of wing; mesoscutum without raised, wartlike areas
       Hydropsychidae
   Hind wings narrower than or occasionally subequal to forewings, with anal region usually occupying less than one-fourth of wing area; mesoscutum with a pair of raised, wartlike tubercles (as in Fig. 44.6b)
   Psychomyiidae (incl. **Polycentropodidae**)
12(9,10). Middle tibiae without preapical spurs   13
   Middle tibiae with preapical spurs   16
13(12). Antennae 1.5 to 3 times length of forewing; mesoscutum with irregular raised spots bearing setae (Fig. 44.6a)
       Leptoceridae
   Antennae about as long as forewing; mesoscutum with small paired tubercles (Fig. 44.6b) or without tubercles   14

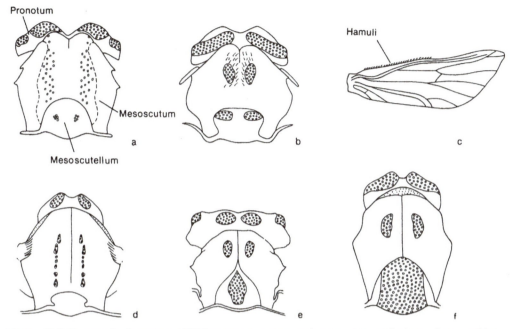

Figure 44.6 Taxonomic characters of Trichoptera, pronotum and mesonotum, and wings: a, Leptoceridae (*Athripsodes tarsipunctatus*); b, Helicopsychidae (*Helicopsyche borealis*); c, Helicopsychidae (*Helicopsyche*); d, Calamoceratidae (*Ganonema americanum*); e, Limnephilidae (*Goera calcarata*); f, Odontoceridae (*Psilotreta frontalis*).
Source: Redrawn from Ross, 1944.

14(13). Hind wings with anterior margin straight or evenly curved   **15**
Hind wings with anterior margin straight basally, with row of stout curved setae, then abruptly curved (Fig. 44.6c)
Helicopsychidae

15(14). Middle tibiae with apical spurs about one half length of basal tarsomere; distal segments of middle and hind tarsi with spines in apical crown on each segment   **Beraeidae**
Middle tibiae with apical spurs about one fourth length of basal tarsomere; distal segments of middle and hind tarsi with scattered spines
Sericostomatidae and **Brachycentridae**

16(12). Middle femora with a row of 6 to 10 dark spines on apex of anterior face
Molannidae
Middle femora with 2 or fewer dark spines on anterior face   **17**

17(16). Mesoscutellum small, arcuate (Fig. 44.6d); mesoscutum with a pair of longitudinal rows of setate spots   **Calamoceratidae**
Mesoscutellum large, triangular or acutely rounded, pointed apically; mesoscutum with tubercles round or oval (Fig. 44.6e,f)   **18**

18(17). Mesoscutellum with a single large, raised area medially (Fig. 44.6e,f)   **19**
Mesoscutellum with paired, lateral raised areas   **20**

19(18). Tibial spurs hairy; mesoscutellum triangular, with medial raised area (Fig. 44.6e) some   **Limnephilidae**
Tibial spurs without hairs; mesoscutellum acutely rounded with raised wart occupying most of area (Fig. 44.6f)
Odontoceridae

20(18). Middle tibiae with irregular row of spines; middle tarsi with double row of spines
Brachycentridae

Middle tibiae without spines; middle tarsi with sparse, scattered spines

Lepidostomatidae

21(7). Maxillary palpi with second segment rounded and globose   **Rhyacophilidae**
Maxillary palpi with second segment cylindrical, similar to first

Hydrobiosidae

Larvae

1. Metathorax with single tergal plate   2
Metathorax entirely membranous or with dorsum divided into at least 2 sclerites   3
2(1). Abdomen without gills; abdominal leg with 2 to 3 bristles at base of claw

Hydroptilidae

Abdomen with numerous branched gills ventrally; abdominal leg with broad fan of setae at base of claw   **Hydropsychidae**
3(1). Abdomen with sclerotized plate on ninth segment   5
Abdomen entirely membranous   4
4(3). Labrum membranous, narrowed basally

Philopotamidae

Labrum sclerotized, broadest basally

Psychomyiidae (incl. **Polycentropodidae**)

5(3). Abdominal prolegs fused basally, appearing as tenth abdominal segment; claws directed laterally, much shorter than basal portion of proleg   17
Abdominal prolegs free; claws directed ventrally, about as long as basal portion of proleg   6
6(5). Claws on all legs similar in structure   7

Claws of hind legs much smaller than those of front and middle legs, modified as short, setose clubs   **Molannidae**

7(6). Antennae 1 to 4 times as long as wide   8
Antennae at least 8 times as long as wide

Leptoceridae

8(7). Mesonotum with conspicuous sclerotized plates   9
Mesonotum membranous or with minute sclerites   **Phryganeidae**

9(8). Mesonotum with a single pair of narrow, arcuate sclerites oriented longitudinally near midline (Fig. 44.7a)   **Leptoceridae**
Mesonotum with 1 to several subquadrate, transverse, or oval sclerites   10

10(9). Labrum with about 16 long, stout setae in a closely set, transverse row across middle (Fig. 44.7b)   **Calamoceratidae**
Labrum with 6 to 8 large setae arranged in a loose, transverse arc or irregularly scattered (Fig. 44.7c)   11

11(10). Larva inhabiting a spirally twisted case (Fig. 44.4b); claws of abdominal legs with a comb of subequal teeth

Heliocopsychidae

Larval case not spirally twisted; claws of abdominal legs with teeth not arranged as a comb   12

12(11). Metanotum with broad anterior sclerite, longitudinal lateral sclerites, and narrow, transverse posterior sclerite

Odontoceridae

Metanotum usually with 1 to several small, rounded poorly defined sclerites   13

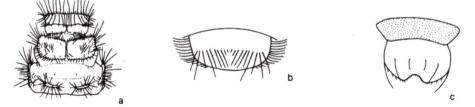

Figure 44.7 Taxonomic characters of Trichoptera larvae: a, thorax of Leptoceridae (*Athripsodes*); b, labrum of Calamoceratidae (*Ganonema*); c, labrum of Limnephilidae (*Limnephilus*). *Source: Redrawn from H.H. Ross, 1944.*

13(12). Pronotal sclerite bearing strong, transverse furrow   **14**

Pronotal sclerite without transverse furrow
                                                    15

14(13). Hind leg with tarsal claw longer than tibia; claw without basal tooth   **Beraeidae**

Hind leg with tarsal claw much shorter, usually about one-half length of tibia; claw with prominent basal tooth

Brachycentridae

15(13). Antennae inserted adjacent to anterior margin of eye   **Lepidostomatidae**

Antennae inserted midway between eyes and mandibles or closer to mandibles   **16**

16(15). Antennae inserted midway between eyes and mandibles; prosternum with distinct, medial hornlike projection

Limnephilidae (including **Goeridae**)

Antennae inserted very close to base of mandibles; prosternum without hornlike projection   **Sericostomatidae**

17(5). Anal prolegs broadly joined with abdominal segment 9; anal claw with one or more dorsal accessory hooks

Glossosomatidae

Anal prolegs free from segment 9; anal claw without dorsal accessory hooks   **18**

18(17). Free part of abdominal prolegs much longer than fused portion; claw without small dorsal accessory hook

Rhyacophilidae

Free part of abdominal proleg shorter than fused portion; claw with small dorsal accessory hook   **Hydrobiosidae**

CHAPTER 45

Collecting and Preservation

Because of the extraordinarily large number of species, the study of insects usually entails some collecting of specimens. Collecting is an integral part of systematics but is also often necessary in ecological studies, including those of an applied nature. Species identification in many groups is extremely difficult, even for specialists, and it is not uncommon for specimens to be misidentified or for classifications to change due to the discovery of new species. Even in physiological or morphological work carried out entirely in the laboratory, it is important to preserve voucher specimens to verify the identity of the species at a later date. Collecting is perhaps the most effective way to associate taxonomic names with the insects themselves and to gain a knowledge of insect natural history. Living animals are always more interesting than preserved specimens, and characteristic behaviors such as the clicking of elaterid beetles are usually much easier to recall than morphological details found in keys. Finally, the pleasure and satisfaction that come from pursuits centered on natural history are considerable. Insect collecting or photography is a lifelong avocation for many people who are not professional entomologists and has the advantage that it can be enjoyed almost anywhere, so ubiquitous are the subjects.

It should be kept in mind, however, that insects are living animals, and their lives should only be taken when there is some real scientific or practical benefit to be gained. In addition, while it is not known that collecting by itself has endangered the survival of any insect species (with the possible exception of a few spectacular prize butterflies and beetles), a number of species are having enough trouble on their own in response to habitat loss and other environmental changes so that unnecessary taking of specimens may in the long run be detrimental. Fortunately, given recent technological advances, many of the specimens already existing in collections may be useful for a wide spectrum of scientific inquiries, including studies of their DNA, when once it was necessary to collect new fresh specimens for such purposes.

COLLECTING EQUIPMENT

Basic Equipment

The most basic item for most collectors is the aerial net, which is used to snare flying insects and also to capture specimens that are perched on flowers, branch tips, or on the ground. Nets are conveniently purchased from supply houses or easily constructed from locally available materials (Fig. 45.1a). The exact dimensions are not critical, but handles about 1 meter long and rims about 35 centimeters in diameter are preferred by most collectors. The bags of aerial nets are sewn from an open mesh netting that allows the captured insects to be seen. When an insect is caught the net should be swung vigorously to force the specimen to the bottom. The net is then turned so the bag folds across the rim, preventing the insect from flying out the opening. The tip of the net and the enclosed insect may then be placed into a killing jar, or one may slip the jar into the net and maneuver the insect into it.

The basic net may be modified in various ways. Longer handles may be used to advantage to catch insects that often perch out of reach, such as dragonflies, or those that visit the flowers of trees., such as butterflies or bees. Nets with bags of muslin or very light canvas are used to sweep insects from coarse vegetation that would damage the fabric of an aerial net. Such sweep nets often have a larger-diameter rim and shorter handle than a standard aerial net. Sweeping usually accumulates vegetation and flowers as well as insects; normally the entire contents are placed into a large killing jar until the insects are stunned and can be sorted from the debris. Aerial nets may also be used in water, but a separate aquatic net is convenient. Aquatic nets may be constructed in the manner of aerial nets but using a heavier netting material. Alternatively, the aquatic net bag may be a cylinder of muslin with a flat bottom of netting (Fig. 45.1b).

Many insects hide in leaf axils, flower clusters, or otherwise concealed spots on plants, and many drop to the ground when the plant is disturbed, where they are difficult to find without extensive searching. The sweep net described above is used on low, soft vegetation, but stiff shrubs, brambles, or the lower branches of trees are most conveniently sampled with the beating sheet, which is a rectangle of muslin or light canvas suspended from crossed sticks (Fig. 45.2). In use the beating sheet is held under the canopy of a shrub or beneath a branch, which is then tapped with a stick or the net handle. The insects fall onto the sheet, where they are readily seen. Beating sheets work especially well during cool, overcast weather or early in the morning when active, flying insects are sluggish. Beating sheets may also serve to separate insects from tangles of vines, vegetation overhanging streams, or road cuts and provide a useful surface on which to break apart rotten wood, fungus, or other materials containing hidden insects.

An aspirator (Fig. 45.3) or pooter is an extremely useful device for removing small insects from a beating sheet. The aspirator is also useful for gathering insects from light sheets or from irregular surfaces such as bark or fungus or from cracks and crevices in rotten wood. Sold by supply houses, aspirators are also easily constructed from plastic tubing, corks or rubber stoppers, and pill bottles (Fig. 45.3). The tube should be long enough to allow one's arm to be extended, and the corks should fit tightly against the bottle so as not to leave a crack in which small insects will wedge themselves and be crushed when the aspirator is opened. A number of insects may be aspirated into the collecting vial before it is emptied into a killing jar.

The collector of moths or other nocturnal insects will probably wish to employ a light sheet. This is simply a muslin bedsheet or other large cloth hung vertically with one or more lights suspended at or near the top. The most effective

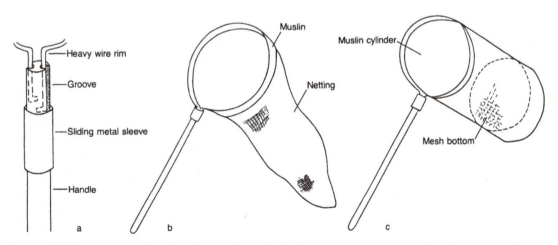

Figure 45.1 Insect nets. a, b Aerial net. Inset a shows detail of attachment of rim to handle. c, Acquatic net. The muslin cylinder extends below the net bottom.

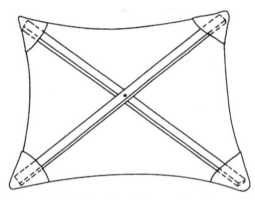

Figure 45.2 Beating sheet in inverted position to show arrangement of crossed sticks held in reinforced pockets of sheet.

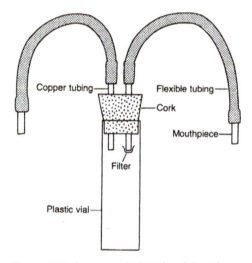

Figure 45.3 Aspirator: the lengths of the tubes may be customized to fit individual preferences. The neck of the plastic receiving vial should be reinforced with tape.

lights are mercury vapor and ultraviolet. Mercury vapor lights require high current, necessitating a generator if an electrical outlet is unavailable. Inexpensive devices to convert direct current to alternating allow ultraviolet lights to be run from a car battery. Most insects usually alight on the sheet near the light, but many beetles sit on the ground or shrubbery several meters away and must be located by flashlight.

A stout trowel, small pry bar, or similar tool will be very useful for removing the bark from dead trees, breaking apart rotten wood, doing light digging, or extracting insects from many of the other places in which they hide.

Traps

Many insects are most effectively captured with special trapping devices. Even experienced entomologists are often surprised by the number and diversity of the catches of various traps, and some taxa are seldom encountered without trapping. While these devices are more elaborate than the simple equipment described above, most may be constructed from commonly available materials.

The Malaise trap, consisting of vertical panels of fine netting intersecting at right angles with a catcher at the top (Fig. 45.4), is extremely useful for capturing a variety of flying insects. In use the trap is pegged to the ground and insects fly against the panels, then move upward, eventually entering the catcher, which contains a killing agent. An I-shaped modification of the Malaise trap designed by Henry Townes is now the preferred model.

Malaise traps are highly effective for Diptera, Hymenoptera, some Lepidoptera, and other aerial insects. For Coleoptera and other insects that drop to the ground when their flight is interrupted, the window pane trap is more useful. Window pane traps consist of vertical transparent panels (usually plastic) with a trough at the bottom to catch the falling insects (Fig. 45.5). The trough may be filled with water containing a little soap solution (to relieve tension at the surface film) if emptied daily. Window pane traps may be left for long periods if the trough is filled with a nonvolatile preservative such as ethylene glycol (antifreeze). If rain is possible, the trap should be shielded to prevent the trough from filling with water. Small yellow pans with salt water are also effective for ground-inhabiting insects.

Insects that inhabit leaf litter or soil are most conveniently extracted with Berlese funnels, also known as separators. This device consists of a cone or cylinder, usually made of sheet metal, with a coarse mesh wire platform for holding samples of litter or soil (Fig. 45.6). A light bulb or heating

Figure 45.4 Two hymenopterists set up a Townes-style Malaise trap in a clearing in Ecuadorean montane forest.

element placed above the sample drives the insects and other arthropods downward, where they are collected in a catcher containing alcohol. Berlese funnels or similar devices provide practically the only means of collecting Protura, Diplura, and many families of beetles and other insects that inhabit litter. Rodent or bird nests, some types of fungus, and other porous substrates may also be processed using Berlese funnels.

Drift nets are fine mesh traps left in streams to capture insects carried by currents. Old nylon stockings are effective for this purpose. Pitfalls are simply containers that are buried to the rim and left to trap ground-dwelling arthropods. Pitfalls may be baited with dung, rotten meat, or other substances to lure specific types of insects, but often they are used without attractants. The containers may be shielded with covers supported by small stones or blocks to deter insectivorous mammals and to deflect rain. Unless they are to be tended frequently, pitfalls should be partially filled with

a nonvolatile preservative such as ethylene glycol, otherwise many of the trapped insects will be damaged. Pitfalls are especially effective in catching beetles and work especially well in sand dunes or very sandy soils.

Light traps enhance the catch of nocturnal insects. The typical light trap entails a mercury vapor or ultraviolet light suspended above a funnel with a catcher at the lower end (Fig. 45.7). The trap is covered to keep out rain and may have a pair of vertical vanes to obstruct flying insects and direct them downward. A hardware cloth screen placed in the funnel will prevent large beetles from entering the catcher and damaging more delicate insects.

The Lundgren trap (Fig. 45.8) consists of a series of conical sleeves, open at the bottom and arranged in a column with spacers between each cone. At the bottom of the column is a catcher containing killing agent. The entire apparatus is suspended by cords or wire and hung from a convenient support such

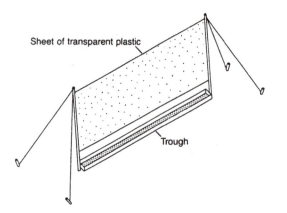

Figure 45.5 Diagram of a window pane trap: the vertical dimension is usually about 2 to 3 feet. If left for more than a few days or depending on weather, the trap should be covered with a rain deflector.

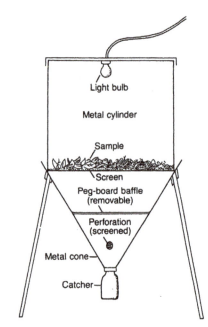

Figure 45.6 Diagram of a Berlese funnel: The peg board baffle prevents debris from entering the catcher. The metal cone should have several perforations to ensure good air circulation through the sample material, greatly speeding operation.

as a tree limb. The Lundgren trap is especially proficient at catching forest Coleoptera, which apparently mistake the columnar shape for a tree trunk and attempt to alight, falling through the series of cones to the bottom container. Lundgren traps may

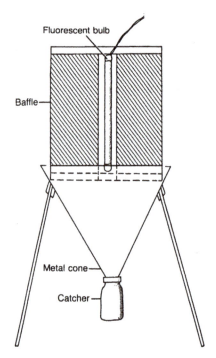

Figure 45.7 Diagram of a light trap: there are four baffles, radiating at right angles from the fluorescent tube. A rain deflector may be added.

be left for long periods if a nonvolatile preservative such as ethylene glycol is used in the catcher.

KILLING METHODS

For purposes of general collecting, all harder-bodied insects are best killed dry in killing jars. Several different chemicals may be used to quickly kill captured insects. Typically the toxic agent is held in the bottom of the jar by a layer of cardboard or plaster (Fig. 45.9). The insects are placed in the space above the plaster. The most effective killing substance is potassium or calcium cyanide, which has the disadvantage of extreme toxicity to humans. Killing jars containing cyanide must have the bottoms securely covered with tape to prevent breakage and must have the cyanide held beneath plaster. Cyanide jars will last for weeks or months before needing replacement and do not sweat like jars charged with ethyl acetate or other liquids. Most insects may be left in cyanide jars for several hours without physical deterioration, but the yellow pigment in many Hymenoptera irreversibly

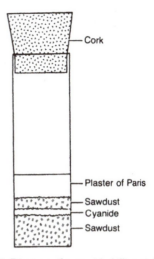

Figure 45.9 Diagram of a cyanide killing vial: the function of the sawdust is top prevent direct contact of the cyanide with the wet plaster. The portion of the vial containing cyanide should be securely taped to prevent breakage and loss of the contents.

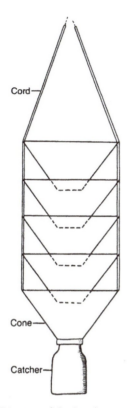

Figure 45.8 Diagram of the Lundgren trap: The number of cones may be varied, but at least six are normally used.

turns red with prolonged exposure to cyanide. Cyanide also causes most specimens to become rigid, so that relaxing them before further processing may be desirable.

Ethyl acetate kills insects more slowly than cyanide and causes the jars to sweat but is relatively nontoxic. Specimens killed in ethyl acetate are relaxed and may be immediately processed. The disadvantage is that the specimens may decompose more extensively after death and are often far less suitable for any subsequent DNA studies.

Several killing jars are needed to accommodate a variety of types and sizes of insects. Small delicate flies or wasps should be kept separate from beetles, grasshoppers, or other large, robust insects. Butterflies and moths should be placed individually in small killing vials, then transferred to a larger holding jar once they are stunned. All killing jars should contain a loose, shock-absorbing wad of tissue paper

to keep specimens apart and take up excess moisture. Jars should never be overloaded with specimens or the more delicate ones will surely be damaged.

Larvae and soft-bodied adults (Table 45.1) should be killed in 95 percent ethyl alcohol or in special solutions such as Kahle's solution or Bouin's fluid. Large specimens need to be slit open or punctured to allow penetration of the preservative and prevent deterioration due to enzymatic action. Excellent preservation of color and body configuration is usually obtained if larger specimens are dipped momentarily into boiling water, then transferred to 80 percent alcohol. After a day or so the old alcohol should be replaced with fresh to avoid dilution by water contained in the specimens.

PRESERVING INSECTS

All insects may be preserved in liquids such as alcohol, which must be used for specimens in which the tissues are to be examined or soft part anatomy observed. For general use 80 percent ethyl alcohol is excellent, but for cytological purposes or DNA extraction mixtures such as Kahle's solution, Bouin's fluid, etc., are preferable. These fixatives, which contain formalin and acetic acid, result in better tissue preservation but cause excessive

Table 45.1 Killing and Preserving of Insects for Study

Order	Recommended method
Protura, Diplura, Collembola Archeognatha, Thysanura, Ephemeroptera	Kill in 95% ethanol and store in 80% ethanol
Odonata	Larvae as above; use killing jar for adults and pin (option: spread wings) or fold wings and store in envelopes
Blattodea	Killing jar and pin (option: eviscerate large specimens); or kill in 95% ethanol and store in 80% ethanol
Mantodea	Killing jar and pin (option: eviscerate large specimens)
Isoptera	Kill in 95% ethanol and store in 80% ethanol
Grylloblattodea	As above
Dermaptera	Killing jar and pin; or collect in 95% ethanol and store in 80% ethanol
Plecoptera	Kill in 95% ethanol and store in 80% ethanol
Embioptera	As above
Orthoptera	Killing jar and pin (options: eviscerate large specimens or spread left wings)
Phasmatodea	As above
Zoraptera	Kill in 95% ethanol and store in 80% ethanol
Psocoptera	As above
Phthiraptera	As above
Hemiptera, Heteroptera terrestrial species: aquatic species: parasites of vertebrates:	Killing jar and pin; small species mounted on paper points Killing jar and pin; or collect in 95% ethanol and store in 80% ethanol Kill in 95% ethanol and store in 80% ethanol
Hemiptera, Homoptera larger, hard-bodied species: aphids: scales, whiteflies, mealy bugs: winged males of scales, etc.:	Killing jar and pin; small species mounted on paper points (option: collect on piece of food plant) Kill in 95% ethanol and store in 80% ethanol (option: collect on piece of food plant) place in dry box or as above Place in dry box or as above (option: collect on piece of food plant) Kill in 95% ethanol and store in 80% ethanol
Thysanoptera	Kill in 95% ethanol and store in 80% ethanol
Megaloptera, Raphidioptera Neuroptera	Larvae as above; adults in killing jar and pin or store large specimens in envelopes
Coleoptera	Larvae in 95% ethanol and store in 80% ethanol; adults in killing jar and pin; small species glued to paper points or put in 80% ethanol
Strepsiptera	Kill in 95% ethanol and store in 80% ethanol
Mecoptera, Diptera	Larvae in 95% ethanol and store in 80% ethanol; adults in killing jar and pin; small species mounted on points or minuten pins
Siphonaptera	Kill in 95% ethanol and store in 80% ethanol
Lepidoptera	Larvae in 95% ethanol and store in 80% ethanol; adults in killing jar and pin (option: spread both wings) or fold wings and store in envelopes; small species mounted on minuten pins
Trichoptera	Kill in 95% ethanol and store in 80% ethanol
Hymenoptera	Larvae in 95% ethanol and store in 80% ethanol; adults in killing jar and pin; small species mounted on paper points or minuten pins

hardening of muscles and other tissues over time. Cytology manuals provide additional information. Larvae and soft-bodied adults (Table 45.1) are permanently preserved in liquids, usually 80 percent alcohol. To ensure the integrity of alcohol-preserved specimens, it is only necessary to maintain the level of alcohol in the containers.

Specimens from which later samples for DNA work may be taken are best preserved in 95 to 100 percent ethanol; generally the lower amount is adequate and the specimens do not become quite as brittle over time. It is also possible to first fix the specimens briefly in 95 to 100 percent ethanol, then transfer to a 70 to 80 percent solution. For any studies in which preservation of colors or DNA is desired, specimens preserved in ethanol should be kept cold (in a refrigerator) and preferably out of the light.

The great majority of adult insects are mounted on pins (Fig. 45.10). Dried, pinned insects retain much of their natural appearance and may be conveniently arranged in boxes or trays. The pin provides a handle so that the specimen does not need to be touched when it is moved or examined. Nonrusting pins are specially manufactured in various sizes for mounting insects and are longer than sewing pins, which are unsuitable. Insect pin sizes 2 and 3 are used for practically all routine pinning. Specimens should be mounted so that about 1 cm of the pin extends above the dorsal surface. A convenient surface for mounting is a block of high-density foam, but many persons prefer to hold the specimen between the thumb and fingers while inserting the pin with the other hand. Most specimens are pinned in horizontal position with the pin just to the right of the midline of the mesothorax. Grasshoppers, roaches, and beetles are usually pinned through the right wing or elytra, and Heteroptera through the right side of the scutellum. Dragonflies are best pinned through the sides with the wings folded over the back. Lepidoptera are traditionally pinned with the wings spread horizontally, exposing the venation and color pattern. Special spreading boards (Fig. 45.11) are used

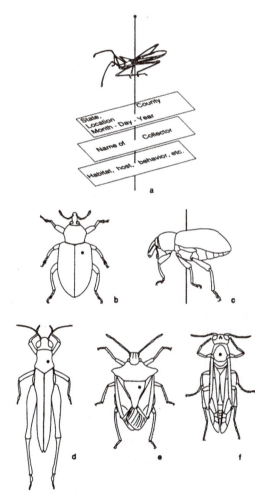

Figure 45.10 Position where various insects should be pinned.

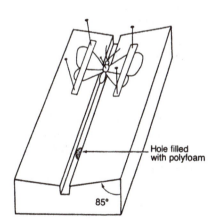

Figure 45.11 Diagram of a spreading block for Lepidoptera: several blocks of different dimension are needed to spread specimens of different size.

to keep the wings at a slight dihedral angle, and paper strips held in place with pins retain them in a symmetrical position with the trailing edges of the forewings at approximately right angles to the body. Spread specimens should be dried for about 3 to 7 days, depending on size, before removal from the boards. Insects other than Lepidoptera and Odonata are not usually arranged in any particular manner, but for purposes of handling and storage the appendages and wings should be drawn close to the body and the abdomen should not sag. Appendages may be retained with pins until the specimen dries and strips of stiff paper pinned beneath the specimen will hold the abdomen in horizontal position.

Standard pinning techniques work well with most insects larger than about 5 mm, but damage or destroy many smaller specimens, which are best prepared with double mounts. Double mounts are mostly of two types. Paper points, small triangles cut or punched from light card stock, are used with small Coleoptera, Hemiptera, Hymenoptera, Diptera, and optionally for most other orders. The point is pinned through the broad end and the specimen glued to the pointed end, which is bent with fine forceps to conform to the side of the specimen (Fig. 45.12b). Fingernail polish is a convenient source of glue, or special glues that remain flexible may be purchased from supply houses. Standard water-soluble Elmer's glue is suitable in some cases but can soften and loosen over time in more humid conditions. Some small Diptera and all small Lepidoptera are usually mounted through the venter or the sides on tiny pins called minuten, which are first thrust through a small block of pith or polyporus mounted on a standard insect pin (Fig. 45.12a,c).

Insects should be pinned as soon as possible after death. If pinning is not possible within about 12 hours, specimens should be held in a relaxing chamber or placed in a deep freeze. Relaxing chambers may be made of any tight-fitting ceramic or plastic container in which the specimens may be held apart from the water that saturates the atmosphere. If water alone is present, specimens may be held for only about a day, so that a saturated solution of phenol is normally used and will prevent

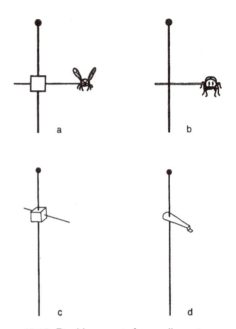

Figure 45.12 Double mounts for small specimens. a,c, Minuten and pith block, used especially for small Diptera and Lepidoptera. b,d, Paper point, used for small Coleoptera, Hymenoptera, and Hemiptera. The tip of the point is bent to roughly fit the specimen.

decay for about a week. Frozen specimens may be held indefinitely but will gradually dry and require relaxing.

LABELING AND STORAGE

All pinned specimens and all alcohol vials should receive an individual label bearing the collection locality and date. The locality should include the country, state or province, county or district (if appropriate), and a specific locality with orientation to a place name on standard maps. Now that global postioning systems make obtaining actual grid references for field sites easy to obtain, it has become standard to include the geographical coordinates on the label as well. These coordinates then make it straightforward to later locate the collecting site both on a map and in the field, even if place names change over time. Dates should include year, month, and day. This locality label is usually placed on the pin just below the specimen. The collector's name often appears on a smaller separate label but may be included on the locality label if space permits. Additional information, such as a host plant name, time of day, other specific ecological information, or voucher numbers for any DNA work, appears on one or more additional labels. For large lots of insects, labels are typically batch printed, while individual specimens or very small lots can be labeled by hand. In either case it is important to keep the size of the label as small as possible (usually no more than about 3 cm long or 1.5 cm broad). Labels should be printed on acid-free, light cardstock using indelible black ink for long-term preservation.

Practically any tight-fitting, lidded box may be used to house a small insect collection. Bottoms suitable for pinning may be fashioned from polyethylene foam, balsa wood, or other materials. A difficulty with most such containers is that they do not exclude dermestids and other museum pests, which will destroy unprotected insects. High-quality boxes or trays purchased from supply houses give better protection but are expensive. Regardless of the type of container, periodic inspection is necessary to detect and eliminate dermestids and other pests. The safest way of killing such invaders is to place each container in a freezer overnight. Alternatively containers may be fumigated with ethylene dichloride, carbon disulfide, or vapona-type insecticide strips.

HOW TO FIND INSECTS

Despite the frequently written advice that insects are everywhere, much experience is required to efficiently sample an area. The continual discovery of new species, even in relatively well-studied regions such as North America, testifies to the difficulty of finding many insects.

Probably the greatest diversity of insects of all types is encountered on plants, especially those in flower or in fruit. Some orders, such Hemiptera, occur predominantly on plants, and many are quite host-specific so that as many plants as possible should be sampled, either by searching or by using the sweep net or beating sheet. Flowering plants are a productive place to search for Hymenoptera, especially Apoidea, which, once again, are often highly host-specific. Other insects will be found

in or on other plant parts, such as bark, leaf axils, rolled or tied leaves, in mines in the leaves or stems, or on the roots.

Two other specific terrestrial habitats that are especially productive are the subcortical region of dying or dead trees and leaf litter and the upper layers of soil. The subcortical habitat is most effectively sampled by hand-searching. Rotting wood of all ages should be investigated, since a succession of insects appears as decay proceeds. Insects inhabiting the subcortical region are predominantly Coleoptera but include many adult and immature Apterygota, Orthoptera, Dermaptera, and Hemiptera, as well as larvae of many other orders. Litter and soil are best sampled with Berlese funnels (described above) and produce a multitude of insects and other arthropods. Especially predominant are Myriapoda, Parainsecta, Entognatha, Apterygota, larval Diptera, and Coleoptera, but some members of practically every order are occasionally encountered. Even a few species of normally aquatic groups such as Trichoptera and Odonata (immatures) inhabit saturated litter in high-rainfall areas such as Hawaii. Some orders, such as Diplura and Protura, are found almost exclusively in these habitats.

Several orders and some members of most other orders are aquatic and are usually specific to certain types of habitats, such as vegetation in ponds, sandy or stony bottoms in either still or running water, on or under submerged stones, or in the interstices of gravelly bottoms. Tree holes, vernal pools, and the marine intertidal also support a variety of aquatic or subaquatic insects. A habitat that should not be overlooked is the littoral zone along streams, ponds, and sea coasts, where numerous insects will be found on the soil surface as well as burrowing into sandy or muddy edges.

Rotting or decaying plant or animal material is inhabited by many scavenging insects as well as their predators and parasites. A backyard compost bin or carcass beside a highway will often produce many insects rarely encountered anywhere else. Deteriorating mushrooms support a host of insects, especially Diptera, and the bracket fungi are host to numerous others, especially Coleoptera.

Many insects are uncommonly encountered as adults but are readily reared from larvae that are easily found. Perhaps the best example is gall formers, which are usually small, are often relatively short-lived as adults, and may be active in winter. The galls, however, are relatively conspicuous and if gall-bearing twigs are placed in water, they will often produce adults of the gall makers as well as parasitoids and predators. Parasitoid forms in general are often easier to obtain by rearing from hosts than by general collecting, and this method provides valuable biological information as well. Lepidopterous caterpillars are especially productive of parasitoids, but practically every insect that has been intensively studied has proven to have at least one parasitoid.

Perhaps the most rewarding method of collecting is searching. All sorts of hiding places should be investigated, including animal burrows or nests, undersides of logs, stones, boards, cardboard, old buildings, animal dung, old wasp's or bee's nests, ant's nests, inside crevices in rocks, and in plant parts such as cones, galls, or hollow stems or twigs, in pads of moss or lichens, around the bases of plants, in the middens left by rodents, or in practically any other place offering food or shelter. Insects even occur in the briny evaporating ponds used to produce salt and in pools of raw petroleum.

LEGAL REQUIREMENTS

Until recently insects could be freely collected in most parts of the world and moved from one nation to another as long as they were killed and properly prepared. Shrinking natural habitats and consequent declines in populations of many organisms, including some insects, have prompted passage of guidelines or rules regulating collecting and transportation of practically all natural history objects, including preserved insects. The intent of these laws is to protect endangered or threatened species or populations. While protection may seem unnecessary for abundant organisms such as insects, many occur only in localized, narrow habitats. Many insects have been listed as threatened or endangered by the U.S. Fish and Wildlife Service, based on detailed studies of distribution and life history, and at least one species has become extinct.

Laws vary greatly from one country to another, and regulations may vary among states or districts in the United States. Furthermore, many of the regulations, particularly governing transportation of specimens, are under review and subject to change. In general, however, a few practices will satisfy the legal requirements.

- No collecting should be undertaken on publicly owned land without written consent of the agency responsible. Parks (national, state, and local), wildlife refuges and reserves, and national forests are all public lands. National forests are typically quite open to collecting as a part of their multiuse mandate.
- Collecting on private land is unregulated except for endangered species. Permission is needed from the landowner.
- Collecting in nearly all foreign nations requires permission from the responsible government agency and usually entails a fee, sometimes considerable. Collecting without the necessary permits risks prosecution by local wardens or police, as well as by United States federal officials if the specimens are brought into this country.

Abe, Y., D. E. Bignell, and T. Higashi, eds. 2000. *Termites: Evolution, Sociality, Symbioses, Ecology.* Kluwer Academic Publishers, Dordrecht, Netherlands.

Achtelig, M., and N. P. Kristensen. 1973. A reexamination of the relationships of the Raphidioptera (Insecta). Z. Zool. Syst. Evol. Forsch. 11:268–274.

Ackery, P. R. 1984. *Milkweed Butterflies.* Br. Mus. Nat. Hist., London.

Adams, J. (ed.) 1992. *Insect Potpourri: Adventures in Entomology.* Sandhill Crane Press, Gainesville, FL.

Adams, M. E. 2009. Hormonal control of development. In *Encyclopedia of Insects,* V. Resh and R. Cardé (eds.). Academic Press, San Diego, CA. pp. 261–266.

Afzelius, B. A., and R. Dallai. 1988. Spermatozoa of Megaloptera and Raphidioptera (Insecta, Neuropteroidea). J. Ultrastruct. Mol. Struct. Res. 101:185–191.

Afzelius, B. A., and R. Dallai. 1994. Characteristics of the flagellar axoneme in Neuroptera, Coleoptera, and Strepsiptera. J. Morphol. 219:15–20.

Agrawal, A. A. 2006. Macroevolution of plant defense strategies. Trends in Ecology and Evolution 22:103–109.

Alba-Tercedor, J., and A. Sanchez-Ortega. 1991. *Overview and Strategies of Ephemeroptera and Plecoptera.* Sandhill Crane Press, Gainesville, FL.

Albrecht, F. O. 1953. *The Anatomy of the Migratory Locust.* Athlone Press, London.

Alcock, J. 1994. Postinsemination associations between males and females in insects: the mate-guarding hypothesis. Annu. Rev. Entomol. 39:1–21.

Aldrich, J. R. 1988. Chemical ecology of the Heteroptera. Annu. Rev. Entomol. 33:211–238.

Alexander, R. D. 1957. Sound production and associated behavior in insects. Ohio J. Sci. 57:101–113.

Alexander, R. D. 1962. Evolutionary change in cricket acoustical communication. Evolution 16:443–467.

Alexander, R. D., and T. E. Moore. 1962. The evolutionary relationships of 17-year and 13-year cicadas, and three new species (Homoptera, Cicadidae, *Magicicada*). Misc. Publ. Mus. Zool. Univ. Michigan 121:1–59.

Alexander, R. D., T. E. Moore, and R. E. Woodruff. 1963. The evolutionary differentiation of stridulatory signals in beetles (Insecta: Coleoptera). Anim. Behav. 11:111–115.

Alford, D. V. 1975. *Bumblebees.* Davis-Poynter, London.

Allan, J. D. 1995. *Stream Ecology. Structure and Function of Running Waters.* Chapman & Hall, London.

Allen, C. T. 2008. Boll weevil eradication: an areawide pest management effort. In *Areawide Pest Management,* O. Koul, G. W. Cuperus, and N. Elliot (eds.). CABI, Oxfordshire, UK. pp. 467–559.

Alonso-Mejia, A., and L. P. Brower. 1994. From model to mimic: age-dependent unpalatability in monarch butterflies. Experientia (Basel) 50:176–181.

Amos, W. H. 1967. *The Life of the Pond.* McGraw-Hill, New York.

Ananthakrishnan, T. N. 1984. *Bioecology of Thrips.* Indira Publishing House, Oak Park, MI.

Ananthakrishnan, T. N. 1990. *Reproductive Biology of Thrips.* Indira Publishing House, Oak Park, MI.

Ananthakrishnan, T. N. 1993. Bionomics of thrips. Annu. Rev. Entomol. 38:71–92.

Ananthakrishnan, T. N., and R. Gopichandran. 1993. *Chemical Ecology in Thrips—Host Plant Interactions.* International Science Publishers, New York.

Ananthakrishnan, T. N., and A. Raman. 1989. *Thrips and Gall Dynamics.* E. J. Brill, Leiden.

Ananthakrishnan, T. N., and A. Raman. 1993. *Chemical Ecology of Phytophagous Insects.* International Science Publishers, New York.

Andersen, N. M., and T. A. Weir. 1994. *Austrobates rivularis,* gen. et sp. nov., a freshwater relative of *Halobates* Eschscholtz (Hemiptera: Gerridae), with a new perspective on the evolution of sea skaters. Invertebr. Taxon. 8:1–15.

Anderson, D. T. 1973. *Embryology and Phylogeny in Annelids and Arthropods*. Pergamon Press, Oxford.

Anderson, N. H., and J. R. Sedell. 1979. Detritus processing by macroinvertebrates in stream ecosystems. Annu. Rev. Entomol. **24**:351–377.

Andersson, M. 1984. The evolution of eusociality. Annu. Rev. Entomol. **15**:165–189.

Ando, H. 1962. *The Comparative Embryology of Odonata, with Special Reference to a Relic Dragonfly, Epiphlebia superstes Selys*. Society for the Promotion of Science, Tokyo, Japan.

Ando, H. 1982. *Biology of the Notoptera*. Kashiyo Insatsu, Nagano, Japan.

Angelini, D. R., and Kaufman, T. C. 2005. Comparative developmental genetics and the evolution of arthropod body plans. Annu. Rev. Genet. **39**:95–119.

Anstey M. L., S. M. Rogers, S. R. Ott, M. Burrows, and S. J. Simpson. 2009. Serotonin mediates behavioral gregarization underlying swarm formation in desert locusts. Science **323**:627–630.

Arnett, R. H., Jr. 1960. *The Beetles of the United States (A Manual for Identification)*. Catholic University Press, Washington, DC. (Reprinted, 1968, American Entomological Institute, Ann Arbor, MI.)

Arnett, R. H., Jr. 1985. *American Insects: A Handbook of the Insects of America North of Mexico*. Van Nostrand Reinhold, New York.

Arnol'di, L. V., and N. J. Vandenberg. 1992. *Mesozoic Coleoptera*. Smithsonian Institution Libraries, Washington, DC.

Askew, R. R. 1971. *Parasitic Insects*. American Elsevier, New York.

Askew, R. R. 1984. The biology of gall wasps. In *Biology of Gall Insects*, T. N. Ananthakrishnan (ed.). Edward Arnold, London. pp. 223–271.

Askew, R. R., and M. R. Shaw. 1986. Parasitoid communities: their size, structure and development. In *Insect Parasitoids*, J. Waage and D. Greathead (eds.). Academic Press, London. pp. 225–264.

Aspock, H. 1986. The Raphidioptera of the world: a review of present knowledge. In *Recent Research in Neuropterology. Proc. 2nd Int. Symp. Neuropterology, Hamburg*, J. Gepp (ed.). Privately published, Graz. pp. 15–29.

Aspock, H., and U. Aspock. 1975. The present state of knowledge on the Raphidioptera of America (Insecta: Neuropteroidea). Pol. Pismo Entomol. **45**:537–546.

Atkins, M. D. 1978. *Insects in Perspective*. Macmillan, New York.

Atkins, M. D. 1980. *Introduction to Insect Behavior*. Macmillan, New York.

Atkinson, T. H., P. G. Koehler, and R. S. Patterson. 1991. *Catalog and Atlas of the Cockroaches (Dictyoptera) of North America North of Mexico*. Misc. Publ. Entomol. Soc. Am. no. 78. Entomological Society of America, Lanham, MD.

Baccetti, B. 1987. *Evolutionary Biology of Orthopteroid Insects*. Ellis Horwood series in entomology and acarology. Halsted Press, Chichester, UK.

Bailey, S. F. 1957. The thrips of California, Part 1: Suborder Terebrantia. Bull, California Insect Survey **4**:143–220.

Bailey, W. J., and D. C. Rentz. 1990. *The Tettigoniidae: Biology, Systematics, and Evolution*. Crawford House Press, Bathurst, UK.

Balduf, S. 2003. Phylogeny for the faint of heart: a tutorial. Trends Genet. **19**:345–351.

Balduf, W. V. 1935. *The Bionomics of Entomophagous Coleoptera*. John S. Swift Co., St. Louis, MO.

Balduf, W. V. 1939. *The Bionomics of Entomophagous Insects, part II*. John S. Swift Co., St. Louis, MO.

Ball, E. D., E. R. Tinkham, R. Flock, and C. T. Vorheis. 1942. The grasshoppers and other Orthoptera of Arizona. Arizona Agric. Exp. Stn. Tech. Bull. **93**:257–373.

Banks, N. 1927. Revision of nearctic Myrmeleontidae. Bull. Mus. Comp. Zool., Harvard Univ. **68**:1–84.

Bao, N., and W. R. Robinson. 1990. Morphology and mating configuration of genitalia of the oriental cockroach *Blatta orientalis* (Blattodea; Blattidae). Proc. Entomol. Soc. Washington **92**:416–421.

Barbosa, P., and D. K. Letourneau. 1988. *Novel Aspects of Insect-Plant Interactions*. John Wiley & Sons, New York.

Barker, S. C. 1994. Phylogeny and classification, origins, and evolution of host associations of lice. Int. J. Parasitol. **24**:1285–1291.

Barker, S. C., M. Whiting, K. P. Johnson, and A. Murrell. 2003. Phylogeny of the lice (Insecta, Phthiraptera) inferred from small subunit rRNA. Zoologica Scripta **32**:407–414.

Barlet, J. 1989. Study of the musculature and nervous system of the japygid (Insecta, Apterygotes, Diplura). Bull. Inst. R. Sci. Nat. Belg. Entomol. **59**:23–42.

Barnhart, C. S. 1961. The internal anatomy of the silverfish

Ctenolepisma campbelli and *Lepisma saccharinum* (Thysanura: Lepismatidae). Ann. Entomol. Soc. Am. **54**:177–196.

Baron, S. 1972. *The Desert Locust*. Eyre Methuen, London.

Barr, T. C., Jr. 1967. Observations on the ecology of caves. Am. Nat. **101**:475–491.

Barr, T. C., Jr. 1974. Cave insects. Insect World Digest (September/October):11–21.

Barth, F. G. 1991. *Insects and Flowers*. Princeton University Press, Princeton, NJ.

Barth, R. 1954. Untersuchungen an den Tarsaldrusen von *Embolyntha batesi* MacLachlan,1877(Embioidea). Zool. Jahrb. **74**:172–188.

Bateman, M. A. 1972. The ecology of fruit flies. Annu. Rev. Entomol. **17**:493–518.

Bates, H. W. 1862. Contributions to an insect fauna of the Amazon Valley. Lepidoptera: Heliconidae. Trans. Linn. Soc. Zool. **23**:495–566.

Baumann, P., L. Baumann, C.-Y. Lai, D. Rouhbakhsh, N. A. Moran, and M. A. Clark. 1995. Genetics, physiology, and evolutionary relationships of the genus *Buchnera*: intracellular symbionts of aphids. Annu. Rev. Microbiol. **49**:55–94.

Baumann, R. W. 1987. Order Plecoptera. In *Immature Insects*, vol. 1, F. W. Stehr (ed.). Kendall/Hunt, Dubuque, IA. pp. 186–195.

Baust, J. G. 1972a. Influence of low temperature acclimation in cold hardiness in *Pterostichus brevicornis*. J. Insect Physiol. **18**:1935–1947.

Baust, J. G. 1972b. Mechanisms of insect freezing protection: *Pterostichus brevicornis*. Nature **236**:219–221.

Baust, J. G. 1976. Temperature buffering in an Arctic microhabi-

tat. Annu. Entomol. Soc. Am. **69**:117–119.

Baust, J. G., and L. K. Miller. 1970. Seasonal variations in glycerol content and its influence in cold hardiness in the Alaskan carabid beetle, *Pterostichus brevicornis*. J. Insect Physiol. **16**:979–990.

Bay, E. C. 1974. Predator-prey relationships among aquatic insects. Annu. Rev. Entomol. **19**:441–454.

Baz, A. 1991. Observations on the biology and ecology of Psocoptera found in different kinds of leaf litter in East-Central Spain. Pedobiologia **35**:89–100.

Baz, A. 1992. Phenology and crypsis as possible determinants of habitat selection in populations of *Hemineura bigoti* from central Spain (Psocoptera: Elipsocidae). Entomol. Gen. **17**:293–298.

Beardsley, J. W., Jr., and R. H. Gonzalez. 1975. The biology and ecology of armored scales. Annu. Rev. Entomol. **20**:47–73.

Beattie, A. J. 1985. *The Evolutionary Ecology of Ant-Plant Mutualisms*. Cambridge Studies in Ecology. Cambridge University Press, Cambridge.

Beck, S. D. 1965. Resistance of plants to insects. Annu. Rev. Entomol. **10**:207–232.

Beckage,N.E.,S.N.Thompson,and B. A. Federici. 1993. *Parasites and Pathogens of Insects, vol. 2. Pathogens*. Academic Press, New York.

Bedford,G.O.1978.Biologyandecology of the Phasmatodea. Annu. Rev. Entomol. **23**:125–130.

Begon, M., M. Mortimer, and D. J. Thompson. 1996. *Population Ecology: A Unified Study of Animals and Plants*. 3rd ed. Blackwell Scientific, Oxford.

Beier, M. 1955. Ordnung: Saltatoptera m. (Saltatoria Latrielle 1817). Bronn's

Kl. Ordn. Tierreichs (5) (3)**6**:34–304.

Bell,W.J.,L.M.Roth,andC.A.Nalepa. 2007. *Cockroaches: Ecology, Behavior and Natural History*. Johns Hopkins University Press, Baltimore, MD.

Bellows, T. S., and D. H. Headrick. 1999. Arthropods and vertebrates in biological control of plants. In *Handbook of Biological Control: Principles and Applications of Biological Control*, T. S. Bellows and T. W. Fisher (eds.). Academic Press, San Diego, CA. pp. 505–516.

Ben-Dov, Y. 1993. *A Systematic Catalogue of the Soft Scale Insects of the World (Homoptera:Coccoidea:Coccidae): With Data on Geographical Distribution, Host Plants, Biology, and Economic Importance*. Flora & Fauna Handbook no. 9. Sandhill Crane Press, Gainesville, FL.

Benedetti,R.1973.Notesonthebiology of *Neomachilis halophila* on a California sandy beach (Thysanura: Machilidae). Pan-Pac. Entomol. **49**:246–249.

Berenbaum, M. 1995. Putting on airs. Am. Entomol. **41**:199–200.

Berenbaum,M.2008.Insect conservation and the Entomological Society of America. Am. Entomologist **54**:117–120.

Bernard, E. C., and S. L. Tuxen. 1987. Class and Order Protura. In *Immature Insects*, vol. 1, F. W. Stehr (ed.). Kendall/Hunt, Dubuque, IA. pp. 47–54.

Bernays, E. A. (ed.). 1989. *Insect-Plant Interactions*. CRC Press, Boca Raton, FL. vols. 1 and 2.

Bernays,E.A.2009.Phytophagous insects. In *Encyclopedia of Insects*, V. Resh and R. Cardé (eds.). Academic Press, San Diego, CA. pp. 787–800.

Bernays, E. A., and M. Graham. 1988. On the evolution of host specificity in phytophagous insects. *Ecology* **69**:886–892.

Bernays, E. A., and R. F. Chapman. 1994. *Host-Plant Selection by Phytophagous Insects.* Contemporary Topics in Entomology, vol. 2. Chapman & Hall, New York.

Berner, L. 1950. *The Mayflies of Florida.* University of Florida Press, Gainesville.

Berner, L. 1959. A tabular summary of the biology of North American mayfly nymphs (Ephemeroptera). Bull. Florida State Mus. Biol. Ser. **4**:1–58.

Berner, L., and M. L. Pescador. 1988. *The Mayflies of Florida.* University Presses of Florida, Gainesville.

Berry, R. J. 1990. Industrial melanism and peppered moths *Biston betularia* L. Biol. J. Linn. Soc. **39**:301–322.

Beutel, R. G., and F. Haas. 2000. Phylogenetic relationships of the suborders of Coleoptera (Insecta). Cladistics **16**:103–141.

Billen, J. 1992. *Biology and Evolution of Social Insects.* Leuven University Press, Leuven, Belgium.

Billingsley, P. F. 1990. The midgut ultrastructure of hematophagous insects. Annu. Rev. Entomol. **35**:219–248.

Binnington, K., and A. Retnakaran. 1991. *Physiology of the Insect Epidermis.* Commonwealth Scientific and Industrial Research Organization, Canberra, Australia.

Birch, C., G. M. Poppy, and T. C. Baker. 1990. Scents and eversible scent structure of male moths. Annu. Rev. Entomol. **35**:25–58.

Birch, M. C., and A. Hefetz. 1987. Extrusible organs in male moths and their role in court-ship behavior. Bull. Entomol. Soc. Am. **33**:222–229.

Birch, M. C., and D. L. Wood. 1975. Mutual inhibition of the attractant pheromone response by two species of *Ips* (Coleoptera: Scolytidae). J. Chem. Ecol. **1**:101–113.

Birket-Smith, S. J. P. 1974. On the abdominal morphology of Thysanura (Archeognatha and Thysanura s. str.). Entomol. Scand. Suppl. **6**:1–67.

Bishop, J. A., and L. M. Cook. 1975. Moths, melanism and clean air. Sci. Am. **232**(1):90–99.

Bitsch, J., C. Bitsch, T. Bourgoin, and C. D'Haese. 2004. The phylogenetic position of early hexapod lineages: morphological data contradict molecular data. Syst. Entomol. **29**:433–440.

Blackman, R. 1974. *Aphids.* Ginn and Company, London.

Blackman, R. L., and V. F. Eastop. 1994. *Aphids on the World's Trees: An Identification and Information Guide.* CAB International, Wallingford, UK.

Blatchley, W. D. 1920. *Orthoptera of Northeastern America.* Nature Publishing Company, Indianapolis, IN.

Blatchley, W. D. 1926. *Heteroptera or True Bugs of Eastern North America, with Special Refrence to the Faunas of Indiana and Florida.* Nature Publishing Company, Indianapolis, IN.

Blest, A. D. 1957. The function of eye-spot patterns in the Lepidoptera. Behaviour **11**:209–256.

Blum, M. S. 1987. Biosynthesis of arthropod exocrine compounds. Annu. Rev. Entomol. **32**:381–413.

Bohart, R. M. 1941. A revision of the Strepsiptera with special reference to the species of North America. Univ. California Publ. Entomol. **7**:91–160.

Bohart, R. M., and A. S. Menke. 1976. *Sphecid Wasps of the World.* University of California Press, Berkeley.

Bolton, B. 1995. *Identification Guide to the Ant Genera of the World.* Harvard University Press, Cambridge, MA.

Bonhag, P. F. 1949. The thoracic mechanism of the adult horse-fly. Cornell Univ. Agric. Exp. Stn. Mem. **285**:1–39.

Borden, J. H. 1982. Aggregation pheromones. In *Bark Beetles in North American Conifers. A System for the Study of Evolutionary Biology*, J. B. Mitton and K. B. Sturgeon (eds.). University of Texas Press, Austin. pp. 74–139.

Borden, J. H. 1984. Semiochemical-mediated aggregation and dispersion in the Coleoptera. In *Insect Communication*, T. Lewis (ed.). Academic Press, London. pp. 123–149.

Borror, D. J., C. A. Triplehorn, and N. F. Johnson. 1989. *An Introduction to the Study of Insects.* 6th ed. Saunders College Publishing, Philadephia.

Boucek, Z. 1988. An overview of the higher classification of the Chalcidoidea (parasitic Hymenoptera). In *Advances in Parasitic Hymenoptera Research*, V.-K. Gupta (ed.). E. J. Brill, Leiden. pp. 11–23.

Boucias, D. G., and J. C. Pendland. 1998. *Principles of Insect Pathology.* Kluwer Academic, Boston.

Böving, A. G., and F. C. Craighead. 1930. An illustrated synopsis of the principal larval forms of the order Coleoptera. Entomol. Am. **11**:1–351.

Bowles, D. E., and R. T. Allen. 1992. Life histories of six species of caddisflies (Trichoptera) in an Ozark stream, USA. J. Kansas Entomol. Soc. **65**:174–184.

Bradley, T. J. 1987. Physiology of osmoregulation in mosquitoes. Annu. Rev. Entomol. **32**:439–462.

Brady, S. G., T. R. Schultz, B. L. Fisher, and P. S. Ward. 2006. Evaluating alternative hypotheses for the evolution and diversification of ants. Proc. Natl. Acad. Sci. U.S.A. **103**:18172–18177.

Brakefield, P. M. 1987. Industrial melanism: do we have the answers? Trends Ecol. and Evol. **2**:117–122.

Brakefield, P. M. 1990. A decline of melanism in the peppered moth *Biston betularia* in The Netherlands. Biol. J. Linn. Soc. **39**:327–334.

Brakefield, P. M., Gates, J., Keys, D., et al. 1996. Development, plasticity and evolution of butterfly eyespot patterns. Nature 384:236–242.

Brakefield, P. M., and D. R. Lees. 1987. Melanism in *Adalia* ladybirds and declining air pollution in Birmingham, England [UK]. Heredity **59**:273–278.

Breed, M. D., and R. E. Page. 1989. *The Genetics of Social Evolution*. Westview Studies in Insect Biology. Westview Press, Boulder, CO.

Breed, M. D., C. D. Michener, and H. E. Evans (eds.). 1982. *The Biology of Social Insects*. Westview Press, Boulder, CO.

Breznak, J. A., and A. Brune. 1994. Role of microorganisms in the digestion of lignocellulose by termites. Annu. Rev. Entomol. **39**:453–487.

Briceno, R. D., and W. G. Eberhard. 1994. The functional morphology of male cerci and associated characters in 13 species of tropical earwigs (Dermaptera: Forficulidae, Labiidae, Carcinophoridae, Pygidicranidae). Smithsonian Contrib. Zool. **555**:1–63.

Bright, D. E. 1993. *The Weevils of Canada and Alaska*. The Insects and Arachnids of Canada, part 21. Research Branch, Agriculture Canada, Ottawa.

Brinck, P. 1957. Reproductive systems and mating in Ephemeroptera. Opusc. Entomol. **22**:1–37.

Brindle, A. 1987. Dermaptera. In *Immature Insects*, vol. 1, F. W. Stehr (ed.). Kendall/Hunt, Dubuque, IA. pp. 171–178.

Britt, N. W. 1962. Biology of two species of Lake Erie mayflies, *Ephoron album* (Say) and *Ephemera simulans* Walker. Bull. Ohio Biol. Survey **1**:1–70.

Brittain, J. E. 1982. Biology of mayflies. Annu. Rev. Entomol. **27**:119–148.

Britton, E. B. 1970. Coleoptera (beetles). In *The Insects of Australia*, Commonwealth Scientific and Industrial Research Organization (ed.). Melbourne University Press, Carlton. pp. 495–621.

Britton, E. B. 1974. Coleoptera (beetles). In *The Insects of Australia, Supplement 1974*, CSIRO (ed.). Melbourne University Press, Carlton. pp. 62–89.

Britton, W. E., et al. 1923. The Hemiptera or sucking insects of Connecticut. Guide to the insects of Connecticut. Connecticut State Geol. Nat. Hist. Survey **4**:1–807.

Brodsky, A. K. 1994. *The Evolution of Insect Flight*. Oxford University Press, Oxford.

Brooks, A. R. 1958. Acridoidea of southern Alberta, Saskatchewan, and Manitoba (Orthoptera). Can. Entomol. Suppl. **9**:1–92.

Brooks, A. R., and L. A. Kelton. 1967. Aquatic and semiaquatic Heteroptera of Alberta, Saskatchewan, and Manitoba (Hemiptera). Mem. Entomol. Soc. Can. **51**:1–92.

Brothers, D. J. 1975. Phylogeny and classification of the aculeate Hymenoptera with special reference to Mutillidae. Univ. Kansas Sci. Bull. **50**:483–648.

Brower, L. P., and J. V. Z. Brower. 1964. Birds, butterflies, and plant poisons: a study in ecological chemistry. Zoologica **49**:137–159.

Brower, L. P., W. N. Ryerson, L. L. Coinger, and S. C. Glazier. 1968. Ecological chemistry and the palatability spectrum. Science **161**:1349–1351.

Brower, L. P., J. Alcock, and J. V. Z. Brower. 1971. Avian feeding behavior and the selective advantage of incipient mimicry. In *Ecological Genetics and Evolution*, R. Creed (ed.). Appleton-Century-Crofts. New York. pp. 261–274.

Brown, H. P. 1987. Biology of riffle beetles. Annu. Rev. Entomol. **32**:253–273.

Brown, K. S., Jr. 1981. The biology of *Heliconius* and related genera. Annu. Rev. Entomol. **26**:427–456.

Brown, W. L., Jr., T. Eisner, and R. H. Whittaker. 1970. Allomones and kairomones: transspecific chemical messengers. BioScience **20**:21.

Brues, C. T., A. L. Melander, and F. M. Carpenter. 1954. Classification of insects. Bull. Mus. Comp. Zool. Harvard Univ. **73**:1–917.

Brushwein, J. R. 1987. Bionomics of *Lomamayia hamata* (Neuroptera: Berothidae). Ann. Entomol. Soc Am. **80**:671–679.

Buchmann, S. L. 1987. The ecology of oil flowers and their bees. Annu. Rev. Ecol. Syst. **18**:343–369.

Buchmann, S. L., and G. P. Nabhan. 1996. *The Forgotten Pollinators*. Island Press, Washington, DC.

Buchner, P. 1965. *Endosymbiosis of Animals with Plant Microorganisms*, rev. ed. Interscience Publishers, New York.

Buckley, R. C. 1987. Interactions involving plants, Homoptera, and ants. Annu. Rev. Ecol. Syst. **18**:111–136.

Burks, B. D. 1953. The mayflies, or Ephemeroptera, of Illinois. Illinois Nat. Hist. Survey Bull. **26**:1–216.

Burn, A. J., T. H. Coaker, and P. C. Jepson (eds.). 1987. *Integrated Pest Management*. Academic Press, London.

Busvine, J. R. 1948. The "head" and "body" races of *Pediculus humanus* L. Parasitology **39**:1–16.

Butcher, J. W., R. Snider, and R. J. Snider. 1971. Bioecology of edaphic Collembola and Acarina. Annu. Rev. Entomol. **16**:249–288.

Buxton, P. A. 1947. *The Louse. An Account of the Lice Which Infest Man, Their Medical Importance and Control*. 2nd ed. Edward Arnold, London.

Byers, G. W. 1963. The life history of *Panorpa nuptialis* (Mecoptera: Panorpidae). Ann. Entomol. Soc. Am. **56**:142–149.

Byers, G. W. 1965. Families and genera of Mecoptera. Proc. 12th Int. Congr. Entomol. **1964**:123.

Byers, G. W. 1971. Ecological distribution and structural adaptation in the classification of Mecoptera. Proc. 13th Int. Congr. Entomol. 1968, **1**:486.

Byers, G. W. 1989. Homologies in wing venation of primitive Diptera and Mecoptera. Proc. Entomol. Soc. Washington **91**:497–501.

Byers, G. W., and R. Thornhill. 1983. Biology of the Mecoptera. Annu. Rev. Entomol. **28**:203–228.

Byers, J. A. 1991. Pheromones and chemical ecology of locusts. Biol. Rev. Cambridge Philos. Soc. **66**:347–378.

Byers, J. R., and C. F. Hinks. 1973. The surface sculpturing of the integument of lepidopterous larvae and its adaptive significance. Can. J. Zool. **51**:1171–1179.

Byrne, D. N., and T. S. Bellows. 1991. Whitefly biology. Annu. Rev. Entomol. **36**:431–457.

Caltagirone, L. E. 1981. Landmark examples in classical biological control. Annu. Rev. Entomol. **26**:213–232.

Caltagirone, L. E., and R. L. Doutt. 1989. The history of the vedalia beetle importation to California and its impact on the development of biological control. Annu. Rev. Entomol. **34**:1–16.

Cameron, E. 1961. *The Cockroach (Periplaneta americana, L.), An Introduction to Entomology for Students of Science and Medicine.* William Heinemann, London.

Cameron, S. A. 2004. Phylogeny and biology of neotropical orchid bees (Euglossini). Annu. Rev. Entomol. **49**:377–404.

Cameron, S. A., H. M. Hines, and P. H. Williams. 2007. A comprehensive phylogeny of the bumble bees (*Bombus*). Biol. Linn. Soc. **91**:161–188.

Cameron, S. A., and P. Mardulyn. 2001. Multiple molecular data sets suggest independent origins of highly eusocial behavior in bees (Hymenoptera: Apinae). Syst. Biol. **50**:194–214.

Cameron, S. L., S. C. Barker, and M. F. Whiting. 2006. Mitochondrial genomics and the new insect order Mantophasmatodea. Mol. Phylogenet. Evol. **38**:274–279.

Campbell, B. C. 1989. On the role of microbial symbiotes in herbivorous insects. In *Insect-Plant Interactions*, vol. 1, E. A. Bernays (ed.). CRC Press, Boca Raton, FL. pp. 1–44.

Campbell, B. C., J. D. Steffen-Campbell, and R. J. Gill. 1994. Evolutionary origin of whiteflies (Hemiptera: Sternorrhyncha: Aleyrodidae) inferred from 18s rDNA sequences. Insect Mol. Biol. **3**:73–88.

Campbell, B. C., J. D. Steffen-Campbell, J. T. Sorensen, and R. J. Gill. 1995. Paraphyly of Homoptera and Auchenorrhyncha inferred from 18s rDNA nucleotide sequences. Syst. Entomol. **20**:175–194.

Campos-Ortega, J. A., and V. Hartenstein. 1985. *The Embryonic Development of Drosophila melanogaster*. Springer-Verlag, Berlin.

Cappuccino, N., and P. W. Price (eds.). 1995. *Population Dynamics: New Approaches and Synthesis*. Academic Press, New York.

Carayon, J. 1966. Traumatic insemination and the paragenital system. In *Monograph of Cimicidae*. R. L. Usinger (ed.). Entomological Society of America, Baltimore. pp. 81–166.

Carey, J. R. 1993. *Applied Demography for Biologists*. Oxford University Press, Oxford.

Carlberg, U. 1987. Chemical defense in Phasmida vs. Mantodea (Insecta). Zool. Anz. **218**:369–373.

Carpenter, F. M. 1931. Revision of nearctic Mecoptera. Bull. Mus. Comp. Zool. Harvard Univ. **72**:205–277.

Carpenter, F. M. 1936. Revision of the nearctic Raphidioidea (recent and fossil). Proc. Am. Acad. Arts Sci. **71**:89–157.

Carpenter, F. M. 1940. Revision of the nearctic Hemerobiidae, Berothidae, Sisyridae, Polystoechotidae, and Dilaridae (Neuroptera). Proc. Am. Acad. Arts Sci. **74**:193–280.

Carpenter, F. M. 1953. The evolution of insects. Am. Sci. **41**:256–270.

Carpenter, F. M. 1970. Adaptations among paleozoic insects. Proc. North Am. Paleontol. Conv., 1969 **2**:1236–1251.

Carpenter, F. M. 1977. Geological history and evolution of the insects. Proc. 15th Int. Congr. Entomol. Washington (1976):63–70.

Carpenter, F. M. 1992. Part R. Arthropoda 4. Vols. 3 and 4: Superclass Hexapoda. In *Treatise on Invertebrate Paleontology*, R. L. Kaesler (ed.). Geological Society of America, Boulder, CO. and Lawrence. KS. pp. 1–655.

Carroll, C. R., and D. H. Janzen. 1973. Ecology of foraging by ants. Annu. Rev. Ecol. Syst. **4**:231–257.

Carson, R. 1962. *Silent Spring*. Houghton Mifflin, Boston.

Carter, W. 1973. *Insects in Relation to Plant Disease*, 2nd ed. John Wiley & Sons, New York.

Carver, M., G. F. Ross, and T. E. Woodward. 1991. Hemiptera. In *The Insects of Australia, vol. I*, Commonwealth Scientific and Industrial Research Organization (ed.). Melbourne University Press, Carlton, pp. 429–509.

Cassagnau, P. 1990. Four hundred million years old Hexapoda, the Collembola: 2. Biogeography and ecology. Annee Biol. **29**:39–70.

Catts, E. P., and M. L. Goff. 1992. Forensic entomology in criminal investigations. Annu. Rev. Entomol. **37**:253–272.

Catts, E. P., and G. R. In Mullen. 2002. Myiasis (Muscoidea, Oestroidea). In *Medical and Veterinary Entomology*, G. Mullen and L. Durden (eds.). Academic Press, San Diego, CA. pp. 317–348.

Caudell, A. N. 1920. Zoraptera not an apterous order. Proc. Entomol. Soc. Washington **22**:84–97.

Caveney, S. 1986. The phylogenetic significance of ommatidium structure in the compound eyes of polyphagan beetles. Can. J. Zool. **64**:1787–1819.

Chandler, H. P. 1956. Megaloptera. In *Aquatic Insects of California*, R. L. Usinger (ed.). University of California Press, Berkeley. pp. 229–233.

Chandler S. M, T. L. Wilkinson, and A. E. Douglas. 2008. Impact of plant nutrients on the relationship between a herbivorous insect and its symbiotic bacteria. Proc. R. Soc. Lond. B. **275**:565–570.

Chapco, W., R. A. Kelln, and D. A. Mcfadyen. 1994. Mitochondrial DNA variation in North American melanopline grasshoppers. Heredity **72**:1–9.

Chapela, I. H., S. A. Rehner, T. R. Schultz, and U. G. Mueller. 1994. Evolutionary history of the symbiosis between fungus-growing ants and their fungi. Science **266**:691–694.

Chapman, A. D. 2006. *Numbers of Living Species in Australia and the World*. Australian Biological Resources Study. ISBN 978-0-642-56850-2.

Chapman, R. F. 1991. General anatomy and function. In *The Insects of Australia*, vol. 1, Commonwealth Scientific and Industrial Research Organization (ed.). Melbourne University Press, Melbourne. pp. 33–67.

Chapman, R. F. 1998. *The Insects: Structure and Function*. 4th ed. Cambridge University Press, Cambridge, UK.

Chapman, R. F., and G. de Boer (eds.) 1995. *Regulatory Mechanisms in Insect Feeding*. Chapman & Hall, New York.

Chapman, R. F., and A. Joern. 1990. *Biology of Grasshoppers*. John Wiley & Sons, New York.

Cheetham, T. B. 1988. *Male Genitalia and Phylogeny of the Pulicoidea (Siphonaptera)*. Theses zoologicae, vol. 8. Koeltz Scientific Books, Koenigstein, Germany.

Chen, D., C. B. Montllor, and A. H. Purcell. 2000. Fitness effects of two facultative endosymbiontic bacteria on the pea aphid, *Acyrthosiphon pisum*, and the blue alfalfa aphid. *A. kondoi*. Entomol. Exp. Appl. **95**:315–323.

Cheng, L. (ed.). 1976. *Marine Insects*. North-Holland Publishing Company, Amsterdam.

Cheng, L. 1985. Biology of *Halobates* (Heteroptera:Gerridae). Annu. Rev. Entomol. **30**:111–135.

Cherry, R. H. 1987. History of sericulture. Bull. Entomol. Soc. Am. **33**:83–84.

Child, C.M. 1894. Bieträgzur Kenntnis der antennalen Sinnesorgane der Insekten. Zeischr. Wiss. Zool., 58: 475–528.

Choe, J. C. 1989. *Zorotypus gurneyi*, new species, from Panama and redescription of *Z. barberi* Gurney (Zoraptera: Zorotypidae). Ann. Entomol. Soc. Am. **82**:149–155.

Choe, J. C. 1994a. Sexual selection and mating system in *Zorotypus gurneyi* Choe (Insecta: Zoraptera): I. Dominance hierarchy and mating success. Behav. Ecol. Sociobiol. **34**:87–93.

Choe, J. C. 1994b. Sexual selection and mating system in *Zorotypus gurneyi* Choe (Insecta: Zoraptera): II. Determinants and dynamics of dominance. Behav. Ecol. Sociobiol. **34**:233–237.

Choe, J. C. 1994c. Communal nesting and subsociality in a webspinner, *Anisembia texana* (Insecta: Embiidina: Anisembiidae). Anim. Behav. **47**:971–973.

Chopard, L. 1938. La biologie des Orthopteres. Encycl. Entomol. (A) **20**:1–541.

Chopard, L. 1949a. Ordre des Dictyopteres Leach, 1818

(= Blattaeformia Werner, 1906; = Oothecaria Karny, 1915). Traite Zool. **9**:355–407.

Chopard, L. 1949b. Ordre de Cheleutopteres. Traite Zool. **9**:594–616.

Chopard, L. 1949c. Ordre de Orthopteres. Traite Zool, **9**:617–722.

Chopard, L. 1949d. Ordre de Dermapteres. Traite Zool. **9**:745–770.

Christiansen, K. 1964, Bionomics of Collembola. Annu. Rev. Entomol. **9**:147–178.

Christiansen, K. 1992. Springtails. Kansas School Nat. **39**:1–16.

Christiansen, K., and P. Bellinger. 1988. Marine littoral Collembola of North and Central America. Bull. Mar. Sci. **42**:215–245.

Christiansen, K. A., and P. F. Bellinger. 1992. *Collembola*. Insects of Hawaii, vol. 15. University of Hawaii Press, Honolulu.

Claassen, P. W. 1931. Plecoptera nymphs of America (North of Mexico). Thomas Say Found. Publ. **3**:1–199.

Claridge, M. F., and J. Den Hollander, 1982. Virulence to rice cultivars and selection for virulence in populations of the brown planthopper, *Nilaparvata lugens*. Entomol. Exp. Appl. **32**:213–221.

Clark, E. J. 1948. Studies in the ecology of British grasshoppers. Trans. R. Entomol. Soc. London **99**:173–222.

Clarke, C. A., F. M. M. Clarke, and H. C. Dawkins. 1990. *Biston betularia* (the peppered moth) in West Kirby, Wirral, 1959–1989: updating the decline in f. *carbonaria*. Biol. J. Linn. Soc. **39**:323–326.

Clausen, C. P. 1940. *Entomophagous Insects*. McGraw-Hill, New York.

Clausen, C. P. 1976. Phoresy among entomophagous insects. Annu. Rev. Entomol. **21**:343–368.

Clay, T. 1970. The Amblycera (Phthiraptera: Insecta). Bull. Br. Mus. Nat. Hist. Entomol. **25**:73–98.

Cleveland, L. R., S. R. Hall, E. P. Saunders, and J. Collier. 1934. The wood-feeding roach, *Cryptocercus*, its protozoa, and the symbiosis between protozoa and roach. Mem. Am. Acad. Sci. **17**:185–342.

Cloudsley-Thompson, J. L. 1975. Adaptations of arthropoda to arid environments. Annu. Rev. Entomol. **20**:261–283.

Coates, P. 2006. *American Perceptions of Immigrant and Invasive Species*, University of California Press, Berkeley.

Cobben, R. H. 1968. *Evolutionary Trends in Heteroptera*. Part I: *Eggs, Architecture of the Shell, Gross Embryology, and Eclosion*. Centre for Agricultural Publishing and Documentation, Wageningen, Netherlands.

Cobben, R. H. 1978. Evolutionary trends in Heteroptera. Part II. Mouthpart structures and feeding strategies. Meded. Landb.-Hoogesch. Wageningen **78**:1–407.

Cochran, D. G. 1985. Nitrogen excretion in cockroaches. Annu. Rev. Entomol. **30**:29–49.

Cohen, A. C. 1995. Extra-oral digestion in predaceous terrestrial Arthropoda. Annu. Rev. Entomol. **40**:85–103.

Cole, F. R. 1969. *The Flies of Western North America*. University of California Press, Berkeley.

Colless, D. H., and D. K. McAlpine. 1991. Diptera (flies). In *The Insects of Australia*, vol. 2. Commonwealth Scientific and Industrial Research Organization (ed.). Melbourne University Press, Carlton. pp. 717–786.

Common, I. F. B. 1970. Lepidoptera (moths and butterflies). In *The Insects of Australia*, Commonwealth Scientific and Industrial Research Organization (ed.). Melbourne University Press, Carlton. pp. 765–866.

Common, I. F. B. 1975. Evolution and classification of the Lepidoptera. Annu. Rev. Entomol. **20**:183–203.

Comstock, J. H. 1918. *The Wings of Insects*. Comstock Publishing, Ithaca, NY.

Cook, B. J., and G. M. Holman. 1985. Peptides and kinins. In *Comprehensive Insect Physiology, Biochemistry, and Pharmacology*, vol. 11. Pergamon Press Ltd, Oxford, UK. pp. 531–559.

Cook, J. M. 1993. Sex determination in the Hymenoptera: a review of models and evidence. Heredity **71**:421–435.

Cook, M. A., and M. J. Scoble. 1992. Tympanal organs of geometrid moths: a review of their morphology, function, and systematic importance. Syst. Entomol. **17**:219–232.

Coope, G. R. 1979. Late Cenozoic fossil Coleoptera: evolution, biogeography, and ecology. Annu. Rev. Ecol. Syst. **10**:247–268.

Cooper, K. W. 1972. A southern California *Boreus*, *B. notoperates* n. sp. I. Comparative morphology and systematics (Mecoptera: Boreidae). Psyche **79**:269–283.

Cooper, K. W. 1974. Sexual biology, chromosomes, development, life histories and parasites of *Boreus*, especially *B. notoperates*, a southern California *Boreus*. II. (Mecoptera: Boreidae). Psyche **81**:84–120.

Cope, O. B. 1940. The morphology of *Psocus confraternus* Banks. Microentomology **5**:91–115.

Corbet, P. S. 1962. *A Biology of Dragonflies*. Quadrangle Books, Chicago.

Corbet, P. S. 1980. Biology of Odonata. Annu. Rev. Entomol. **25**:189–217.

Corbet, P. S., C. Longfield, and N. W. Moore. 1960. *Dragonflies*. Collins, London.

Corbet, P. S., C. Longfield, and N. W. Moore. 1985. *Dragonflies*. Collins, London.

Cordero, A., and P. L. Miller. 1992. Sperm transfer, displacement and precedence in *Ischnura graellsii* (Odonata: Coenagrionidae). Behav. Ecol. Sociobiol. **30**:261–267.

Cornwell, P. B. 1968. *The Cockroach*. Vol. I. *A Laboratory Insect and an Industrial Pest*. Hutchinson Publishing Group, London.

Costa, C., S. A. Vanin, and S. A. Casari-Chen. 1988. *Larvas de Coleoptera do Brasil*. Universidad de Sao Paulo, Sao Paulo.

Cox, G. W. 2004. *Alien Species and Evolution*. Island Press, Washington, DC.

Cox, J. M. 1987. *Pseudococcidae (Insecta: Hemiptera)*. Fauna of New Zealand, no. 11. Science Information Publishing Centre, DSIR, Wellington, N.Z.

Craighead, F. C. 1950. Insect Enemies of Eastern Forests. U.S. Dept. Agric. Misc. Publ. **657**:1–679.

Crampton, G. C., et al. 1942. Guide to the insects of Connecticut. Part VI. The Diptera or true flies of Connecticut. First fascicle. Bull. Connecticut Geol. Nat. Hist. Survey **64**:1–509.

Crane, E. 1990. *Bees and Beekeeping. Science, Practice, and World Resources*. Comstock Publishing Associates, Ithaca, NY.

Crane, J. 1952. A comparative study of innate defensive behavior in Trinidad mantids (Orthoptera, Mantodea). Zoologica **37**:239–293.

Cranshaw, W. 2004. *Garden Insects of North America*. Princeton University Press, Princeton, NJ.

Creighton, W. S. 1950. The ants of North America. Bull. Mus. Comp. Zool., Harvard Univ. **104**:1–585.

Crespi, B. 1987. Fighting behavior in male tubuliferan thrips. In *Population Structure, Genetics and Taxonomy of Aphids and Thysanoptera; International Symposia, Smolenice, Czechoslovakia, September 9–14, 1985*. J. Holman (ed.). Spb Academic Publishing. The Hague. pp. 424–425.

Crespi, B. J. 1992a. Behavioural ecology of Australian gall thrips (Insecta, Thysanoptera). J. Nat. Hist. **26**:769–809.

Crespi, B. 1992b. Eusociality in Australian gall thrips. *Nature* **359**:724.

Croft, B. A. 1990. *Arthropod Biological Control Agents and Pesticides. Environmental Science and Technology*. John Wiley & Sons, New York.

Cromartie, R. I. T. 1959. Insect pigments. Annu. Rev. Entomol. **4**:59–76.

Crowson, R. A. 1955. *The Natural Classification of the Families of Coleoptera*. Nathaniel Lloyd, London.

Crowson, R. A. 1960. The phylogeny of the Coleoptera. Annu. Rev. Entomol. **5**:111–134.

Crowson, R. A. 1970. *Classification and Biology*. Heinemann Educational Books, London.

Crowson, R. A. 1981. *The Biology of the Coleoptera*. Academic Press, London.

Cruz, Y. P. 1981. A sterile defender morph in a polyembryonic hymentopterous parasite. Nature **294**:446–447.

Commonwealth Scientific and Industrial Research Organization (CSIRO). 1991. *The Insects of Australia*, 2 vols. 2nd ed. Melbourne University Press, Carlton.

Cummins, K. W. 1973. Trophic relations of aquatic insects. Annu. Rev. Entomol. **18**:183–206.

Curran, C. H. 1934. *The Families and Genera of North American Diptera*. Privately published, New York.

Dadant and Sons (eds.). 1975. *The Hive and the Honey Bee*. Dadant and Sons, Hamilton, IL.

Dafni, A. 1984. Mimicry and deception in pollination. Annu. Rev. Ecol. Syst. **15**:259–278.

Daily, G. C., ed. 1997. *Nature's Services: Societal Dependence on Natural Ecosystems*. Island Press, Washington, DC.

Daly, H. V. 1963. Close-packed and fibrillar muscles of the Hymenoptera. Ann. Entomol. Soc. Am. **56**:295–306.

Daly, H. V. 1964. Skeleto-muscular morphogenesis of the thorax and wings of the honey bee, *Apis mellifera* (Hymenoptera: Apidae). Univ. Calif. Publ. Entomol. **39**:1–77.

Damen, W. G. M., Saridaki, T., and Averof, M. 2002. Diverse adaptations of an ancestral gill: a common origin for wings, breathing organs, and spinnerets. Current Biology **12**:1711–1716.

Danforth, B. D. 1989. The evolution of hymenopteran wings: the importance of size. J. Zool. (London) **218**:247–276.

David, D. T., and M. C. Birch. 1989. Pheromones and insect behavior. In *Insect Pheromones in Plant Protection*, A. R. Jutsum and R. F. S. Gordon (eds.). John Wiley & Sons Ltd., London. pp. 17–35.

Davis, C. 1940. Family classification of the order Embioptera. Ann. Entomol. Soc. Am. **33**:677–682.

Davis, D. R. 1967. A revision of the moths of the subfamily Prodoxinae (Lepidoptera: Incurvariidae). U.S. Natl. Mus., Bull. **255**:1–170.

Davis, D. R. 1986. A new family of monotrysian moths from austral South America (Lepidoptera: Palaephatidae), with a phylogenetic review of the monotrysia. Smithsonian Contrib. Zool. **282**:1–39.

Davis, M. A. 2009. *Invasion Biology*. Oxford University Press, Oxford, UK.

Day, W. C. 1956. Ephemeroptera. In *Aquatic Insects of California*, R. L. Usinger (ed.). University of California Press, Berkeley. pp. 79–105.

DeBach, P. (ed.). 1964. *Biological Control of Insect Pests and Weeds*. Chapman & Hall, London.

DeBach, P. 1974. *Biological Control by Natural Enemies*. Cambridge University Press, New York.

DeBach, P., and D. Rosen. 1991. *Biological Control by Natural Enemies*. 2nd ed. Cambridge University Press, Cambridge.

DeBach, P., and R. A. Sundby. 1963. Competitive displacement between ecological homologues. Hilgardia **34**:105–166.

DeCoursey, R. M. 1971. Keys to the families and subfamilies of the nymphs of North American Hemiptera-Heteroptera. Proc. Entomol. Soc. Washington **73**:413–428.

Decu, V., M. Gruia, S. L. Keffer, and S. M. Sarbu. 1994. Stygobiotic waterscorpion, *Nepa anophthalma*, n. sp. (Heteroptera: Nepidae), from a sulfurous cave in Romania. Ann. Entomol. Soc. Am. **87**:755–761.

DeFoliart, G. R. 1989. The human use of insects as food and as animal feed. Bull. Entomol. Soc. Am. **35**(1):22–35.

Delamare-Deboutteville, C. 1948. Observations sur l'écologie et l'éthologie des Zoraptères. La question de leur vie social et de leurs prétendus rapports avec les termites. Rev. Entomol. **19**:347–352.

Delany, M. J. 1957. Life histories in the Thysanura. Acta Zool. Cracov. **2**:61–90.

Delany, M. J. 1959. The life histories and ecology of two species of *Petrobius* Leach, *P. brevistylis* and *P. maritimus*. Trans. R. Soc. Edinburgh **63**:501–533.

Delcomyn, F. 1985. Factors regulating insect walking. Annu. Rev. Entomol. **30**:239–256.

Delcomyn, F. 2004. Insect walking and robotics. Annu. Rev. Entomol. **49**:51–70.

DeLong, D. M. 1971. The bionomics of leafhoppers. Annu. Rev. Entomol. **16**:179–210.

Demerec, M. (ed.). 1950. *Biology of Drosophila*. John Wiley & Sons, New York.

De Moraes, C. M., and M. C. Mescher. 2004. Biochemical crypsis in the avoidance of natural enemies by an insect herbivore. Proc. Natl. Acad. Sci. **101**:8993–8997.

Denholm, I., and M. W. Rowland. 1992. Tactics for managing pesticide resistance in arthropods: theory and practice. Annu. Rev. Entomol. **37**:91–112.

Denlinger, D. 2009. Cold and heat tolerance. In *Encyclopedia of Insects*, V. Resh and R. Cardé (eds.). Elsevier Academic Press, San Diego, CA. pp. 179–183.

Denning, D. G. 1956. Trichoptera. In *Aquatic Insects of California*. R. L. Usinger (ed.). University of California Press, Berkeley. pp. 237–270.

Dennis, R. L. H. (ed.). 1992. *The Ecology of Butterflies in Britain*. Oxford University Press. Oxford.

Dent, D. 1991. *Insect Pest Management*. CAB International, Oxon, UK.

Desalle, R., and D. A. Grimaldi. 1991. Morphological and molecular systematics of the Drosophilidae Annu. Rev. Ecol. Syst. **22**:447–476.

Desalle, R., J. Gatesy. W. Wheeler, and D. Grimaldi. 1992. DNA sequences from a fossil termite in Oligo-Miocene amber and their phylogenetic implications. Science **257**:1933–1936.

Desender, K. 1994. *Carabid Beetles: Ecology and Evolution*. Series Entomologica 51. Kluwer Academic Publishers, Dordrecht, Netherlands.

Dethier, V. G. 1947. The response of hymenopterous parasites to chemical stimulation of the ovipositor. J. Exp. Zool. **105**:199–208.

Dethier, V. G. 1954. Evolution of feeding preferences in phytophagous insects. Evolution **8**:33–54.

Dethier, V. G. 1992. *Crickets and Katydids, Concerts and Solos*. Harvard University Press, Cambridge. MA.

Dethier, V. G., L. Barton-Browne, and C. N. Smith 1960. The designation of chemicals in terms of the responses they elicit from insects. J. Econ. Entomol **53**:134–136.

Dettner, K. 1987. Chemosystematics and evolution of beetle chemical defenses. Annu. Rev. Entomol **32**:17–48.

Dettner, K., and C. Liepert. 1994. Chemical mimicry and camouflage. Annu. Rev. Entomol. **39**:129–154.

Devonshire, A. L., and L. M. Field. 1991. Gene amplification and insecticidal resistance. Annu. Rev. Entomol. **36**:1–23.

DeVries, P. J. 1990. Enhancement of symbioses between butterfly caterpillars and ants by vibrational communication. Science **248**:1104–1106.

DeVries, P. J. 1992. Singing caterpillars, ants and symbioses. Sci. Am. **249**:76–82.

DeWalt, R. E., C. Favret, and D. W. Webb. 2005. Just how imperiled are aquatic insects? A case study of stoneflies (Plecoptera) in Illinois. Ann. Entomol. Soc. Am. **98**:941–950.

DeWalt, R. E., U. Neu-Becker, and G. Steuber. 2009. *Plecoptera Species File Online*. Version 1.1/3.5. [11.09.2009]. <http://Plecoptera.SpeciesFile.org>.

Dewalt, R. E., and K. W. Stewart. 1995. Life histories of stoneflies (Plecoptera) in the Rio Conejos of southern Colorado. Great Basin Nat. **55**:1–18.

Dickinson, M. 2001. Solving the mystery of insect flight. Sci. Am. **June**:34–41.

Dickinson, M., and R. Dudley. 2009. Flight. In *Encyclopedia of Insects*, V. Resh and R. Cardé (eds.). Academic Press, San Diego, CA. pp. 364–372.

Dickson, R. 1976. *A Lepidopterist's Handbook*. The Amateur Entomologists' Society, Hanworth, Middlesex, UK.

Dillon, E. S., and L. S. Dillon. 1961. *A Manual of Common Beetles of Eastern North America*. Row, Peterson & Company, Evanston, IL.

Dirsh, V. M. 1975. *Classification of the Acridomorphoid Insects*. E. W. Classey, Faringdon, Oxon., UK.

Dixon, A. F. G. 1973. *Biology of Aphids*. Edward Arnold Publishers, London.

Dixon, A. F. G. 1985. *Aphid Ecology*. Blackie, Glasgow.

Dolling, W. R. 1991. *The Hemiptera*. Natural History Museum Publications, Oxford University Press, London.

Douglas, A. E. 1989. Mycetocyte symbiosis in insects. Biol. Rev. Cambridge Philos. Soc. **64**:409–434.

Douglas, A. E. 1994. *Symbiotic Interactions*. Oxford University Press, New York.

Dow, J. A. T., and S. A. Davies. 2001. The *Drosophila melanogaster* Malpighian tubule. Adv. Insect Physiol. **28**:1–83.

Downes, B. J., and J. Jordan. 1993. Effects of stone topography on abundance of net-building caddisfly larvae and arthropod diversity in an upland stream. Hydrobiologia **252**:163–174.

Downes, J. A. 1969. The swarming and mating flight of Diptera. Annu. Rev. Entomol. **14**:271–298.

Downes, J. A. 1971. The ecology of blood-sucking Diptera: An evolutionary perspective. In *Ecology and Physiology of Parasites*, A. M. Fallis (ed.). University of Toronto Press, Toronto. pp. 232–258.

Doyen, J. T. 1966. The skeletal anatomy of *Tenebrio molitor* (Coleoptera: Tenebrionidae). Misc. Publ. Entomol. Soc. Am. **5**:103–150.

Dressler, R. L. 1982. Biology of orchid bees. Annu. Rev. Ecol. Syst. **13**:373–394.

Drosopoulos, S. and M. F. Claridge (eds.) 2006. *Insect Sounds and Communication: Physiology, Behaviour, Ecology and Evolution*. Taylor and Francis, Boca Raton, FL.

Dudley, R. 2000. *The Biomechanics of Insect Flight: Form, Function, Evolution*. Princeton University Press, Princeton, NJ.

Dugdale, J. S. 1974. Female genital configuration in the classification of Lepidoptera. N. Z. J. Zool. **1**:127–146.

Dukas, R. 2008. Evolutionary biology of insect learning Annu. Rev. Entomol. **53**: 145–160.

Dunkel, S. W. 2000. *Dragonflies Through Binoculars: A Field Guide to Dragonflies of North America*. Oxford University Press, Oxford, UK.

Dunn, R. R. 2005. Modern insect extinctions, the neglected majority. Conserv. Biol. **19**:1030–1036.

Duporte, E. M. 1960. Evolution of cranial structures in adult Coleoptera. Can. J. Zool. **38**:655–675.

Durden, L. A., and G. G. Musser. 1994. The sucking lice (Insecta, Anoplura) of the world: a taxonomic checklist with records of mammalian hosts and geographical distributions. Bull. Am. Mus. Nat. Hist. **218**:1–90.

Eaton, J. L. 1988. *Lepidopteran anatomy*. John Wiley & Sons, New York.

Ebeling, W. 1968. Termites: Identification, biology, and control of termites attacking buildings. Calif. Agric. Exp. Stn. Ext. Serv. Man. **38**:1–68.

Eberhard, M. J. W. 1975. The evolution of social behavior by kin selection. Q. Rev. Biol. **50**:1–34.

Eberhard, W. G. 1979. The function of horns in *Podischnus agenor* (Dynastinae) and other beetles. In *Sexual Selection and Reproductive Competition in Insects*. M. S. Blum and N. A. Blum (eds.). Academic Press. New York. pp. 231–258.

Eberhard, W. G. 1982. Beetle horn dimorphism: making the best of a bad lot. Am. Nat. **119**:420–426.

Eberle, M. W., and D. L. McLean. 1983. Observation of symbiote migration in human body lice with scanning and transmission microscopy. Can. J. Microbiol. **29**:755–762.

Edgerly, J. S. 1988. Maternal behavior of a webspinner (order Embiidina): mother-nymph associations. Ecol. Entomol. **13**:263–272.

Edmunds, G. F., Jr. 1959. Ephemeroptera. In *Freshwater Biology*, W. T. Edmondson (ed.). John Wiley & Sons, New York. pp. 908–916.

Edmunds, G. F. 1972. Biogeography and evolution of Ephemeroptera. Annu. Rev. Entomol. **17**:21–42.

Edmunds, G. F., Jr., and R. K. Allen. 1987. Order Ephemeroptera. In *Immature Insects*, vol. 1, F. W. Stehr (ed.). Kendall/Hunt, Dubuque, IA. pp. 75–94.

Edmunds, G. F., Jr., and W. P. McCafferty, 1988. The mayfly subimago. Annu. Rev. Entomol. **33**:509–529.

Edmunds, G. F., Jr., and J. R. Traver. 1954. The flight mechanics and evolution of the wings of Ephemeroptera, with notes on the archetype insect wing. J. Washington Acad. Sci. **44**:390–400.

Edmunds, G. F., Jr., K. Allen, and W. L. Peters. 1963. An annotated key to the nymphs of the families and subfamilies of mayflies (Ephemeroptera). Univ. Utah Biol. Ser. **13**:1–55.

Edmunds, G. F., Jr., S. L. Jensen, and L. Berner. 1976. *The Mayflies of North and Central America*. University of Minnesota Press, Minneapolis.

Edmunds, M., and D. Dudgeon. 1991. Cryptic behavior in the oriental leaf mantis *Sinomantis denticulata* Beier (Dictyoptera, Mantodea). Entomol. Mon. Mag. **127**:45–48.

Edney, E. B. 1974. Desert Arthropods. In *Desert Biology, vol. 2*, G. W. Brown, Jr. (ed.). Academic Press, New York. pp. 311–384.

Edwards, J. S. 1987. Arthropods of alpine aeolian eco-systems. Annu. Rev. Entomol. **32**:163–179.

Edwards, J. S. 1992. Adhesive function of coxal vesicles during ecdysis in *Petrobius brevistylis* Carpenter (Archaeognatha, Machilidae). Int. J. Insect Morphol. Embryol. **21**:369–371.

Ehrlich, P. R. 1958. The comparative morphology, phylogeny and higher classification of the butterflies (Lepidoptera: Papilionoidea). Univ. Kansas Sci. Bull. **39**:305–370.

Ehrlich, P. R. 1961. Intrinsic barriers to dispersal in checkerspot butterfly. Science **134**:108–109.

Ehrlich, P. R. 1970. Coevolution and the biology of communities. In *Biochemical Coevolution*, K. L. Chambers (ed.). Oregon State University Press, Corvallis. pp. 1–11.

Ehrlich, P. R., and A. H. Ehrlich. 1961. *How to Know the Butterflies*. Wm. C. Brown, Dubuque, IA.

Ehrlich, P. R., and P. Raven. 1967. Butterflies and plants. Sci. Am. **216**:104–113.

Ehrlich, P. R., R. R. White, M. C. Singer, S. W. McKechnie, and L. E. Gilbert. 1975. Checkerspot butterflies: a historical perspective. Science **188**:221–228.

Eickwort, G. C., and H. J. Ginsberg. 1980. Foraging and mating behavior in Apoidea. Annu. Rev. Entomol. **25**:421–446.

Eisner, T. 1960. Defense mechanisms of arthropods. II. The chemical and mechanical weapons of an earwig. Psyche **67**:62–70.

Ekblom, T. 1926. Morphological and biological studies of the Swedish families of Hemiptera Heteroptera. Parts I and II. Zool. Bidr. **10**:31–180, **12**:113–150.

Eldridge, B. F., and J. D. Edman (eds.). 2004. *Medical Entomology*. Kluwer Academic, Norwell, MA.

Elias, S. A. 1994. *Quaternary Insects and Their Environments*. Smithsonian Institution Press, Washington. DC.

Elliott, J. M. 1968. The daily activity patterns of mayfly nymphs (Ephemeroptera). J. Zool. **155**:201–221.

Elzinga, R. J. 1987. *Fundamentals of Entomology*. 3rd ed. Prentice-Hall, Englewood Cliffs, NJ.

Emerson, A. E. 1952. The biogeography of termites. Bull. Am. Mus. Nat. Hist. **99**:217–225.

Emerson, A. E. 1955. Geographical origins and dispersions of termite genera. Fieldiana, Zool. **37**:465–521.

Emerson, A. E. 1967. Cretaceous insects from Labrador. 3. A new genus and species of termite (Isoptera: Hodotermitidae). Psyche **74**:276–289.

Emerson, K. C. 1956. Mallophaga (chewing lice) occurring on the domestic chicken. J. Kansas Entomol. Soc. **35**:63–79.

Emmel, T. C. 1997. *Butterfly Gardening*. Friedman/Fairfax, New York.

Engel, M. S., and D. A. Grimaldi. 2002. The first Mesozoic Zoraptera (Insecta). *American Museum Novitates* **3362**:1–20.

Engel, M., and D. Grimaldi. 2006. The earliest webspinners (Insecta: Embiodea). Am. Mus. Novit. **3514**:1–15.

Engels, W., and A. Buschinger. 1990. *Social Insects: An Evolutionary Approach to Castes and Reproduction*. Springer-Verlag, Berlin.

Ennos, A. R., and R. J. Wootton. 1989. Functional wing morphology and aerodynamics of *Panorpa germanica* (Insecta: Mecoptera). J. Exp. Biol. **143**:267–284.

Eriksen, C. H. 1966. Ecological significance of respiration and substrate for burrowing Ephemeroptera. Can. J. Zool. **46**:93–103.

Erwin, T. L. 1983. Tropical forest canopies: the last biotic frontier. Bull. Entomol. Soc. Am. **29**:14–19.

Erwin, T. L. 1997. Biodiversity at its utmost: tropical forest beetles. In *Biodiversity II*, M. L. Reaka-Kudla, D. E. Wilson, and E. O. Wilson (eds.). Joseph Henry Press, Washington, DC. pp. 27–40.

Essig, E. O. 1930. *A History of Entomology*. Macmillan, New York.

Evans, C. J., S. A. Sinenko, L. Mandal, J. A. Martinez-

Agosto, V. Hartenstein, and U. Banerjee. 2002. Genetic dissection of hematopoiesis using *Drosophila* as a model system. Adv. Develop. Biol. **18**:259–299.

Evans, D. L., and J. O. Schmidt (eds.) 1990. *Insect Defenses. Adaptive Mechanisms and Strategies of Prey and Predators*. State University of New York Press, Albany.

Evans, H. E. 1984. *Insect Biology: A Textbook of Entomology*. Addison-Wesley, Reading, MA.

Evans, H. E. 1985. *The Pleasures of Entomology*. Smithsonian Institution Press, Washington, DC.

Evans, H. E., and M. J. W. Eberhard. 1970. *The Wasps*. University of Michigan Press, Ann Arbor.

Evans, J. W. 1963. The phylogeny of the Homoptera Annu. Rev. Entomol. **8**:77–94.

Evans, J. W. 1982. A review of the present knowledge of the family Peloridiidae and new genera and species from New Zealand and New Caledonia (Hemiptera: Insecta). Rec. Aust. Mus. **34**:381–406.

Evans, M. E. G. 1961. On the muscular and reproductive systems of *Atomaria ruficornis* (Marshall) (Coleoptera, Cryptophagidae). Trans. R. Soc. Edinburgh **64**:297–399.

Evans, M. E. G. 1975. *The Life of Beetles*. George Allen & Unwin, Ltd., London.

Ewing, H. E. 1940. The Protura of North America Ann. Entomol. Soc. Am. **33**:495–551.

Ewing, H. E., and I. Fox. 1943. The fleas of North America. U. S. Dept. Agric. Misc. Publ. **500**:1–128.

Farrow, R. A. 1990. Flight and migration in acridoids. In *Biology of Grasshoppers*, R. F. Chapman and A Joern (eds.).

John Wiley & Sons, Somerset, NJ. pp 227–314.

Felsenstein, J. 2004. *Inferring Phylogenies*. Sinauer Associates, Sunderland, MA.

Felt, E. P. 1940. *Plant Galls and Gall Makers*. Comstock Publishing Associates, Ithaca, NY.

Fenton, M. B., and K. D. Roeder. 1974. The microtymbals of some Arctiidae. J. Lepid. Soc. **28**:205–211.

Ferrar, P. 1988. A guide to the breeding habits and immature stages of the Diptera Cyclorrhapha. Entomograph **8**:1–907.

Ferris, C. D., and F. M. Brown. 1981. *Butterflies of the Rocky Mountain States*. University of Oklahoma Press, Norman.

Ferris, G. F. 1931. The louse of elephants, *Haematomyzus elephantis*. Parasitology **23**:112–127.

Ferris, G. F. 1937–1955. *Atlas of the Scale Insects of North America*. Stanford University Press, Palo Alto, CA.

Ferris, G. F. 1940. The morphology of *Plega signata* (Hagen) (Neuroptera, Mantispidae). Microentomology **5**:33–56.

Ferris, G. F. 1951. The sucking lice. Mem. Pac. Coast Entomol. Soc. **1**:1–320.

Ferris, G. F., and P. Pennebaker. 1939. The morphology of *Agulla adnixa* (Hagen) (Neuroptera, Raphidiidae). Microentomology **5**:33–56.

Field, S. A., and A. D. Austin. 1994. Anatomy and mechanics of the telescopic ovipositor system of *Scelio* Latreille (Hymenoptera: Scelionidae) and related genera. Int. J. Insect Morphol. Embryol. **23**:135–158.

Fishelson, L. 1985. *Orthoptera, Acridoidea*. Fauna Palaestina. Insecta, 3. Israel Academy of Sciences and Humanities, Jerusalem.

Fisher, B. L., and S. P. Cover. 2007. *Ants of North America: A*

Guide to the Genera. University of California Press, Berkeley, CA.

Fisk, F. W. 1987. Order Blattodea. In *Immature Insects*, vol. 1, F. W. Stehr (ed.). Kendall/Hunt, Dubuque, IA. pp. 120–131.

Fleck, G., B. Ullrich, M. Brenk, C. Wallnisch, M. Orland, S. Bleidissel, and B. Misof. 2008. A phylogeny of anisopterous dragonflies (Insecta: Odonata) using mtRNA genes and mixed nucleotide/doublet models. J. Zool. Syst. Evol. Res. **46**:310–322.

Fletcher, D. J. C., and K. G. Ross. 1985. Regulation of reproduction in eusocial Hymenoptera. Annu. Rev. Entomol. **30**:319–344.

Flook, P. K., S. Klee, and C. H. F. Rowell. 1999. Combined molecular phylogenetic analysis of the Orthoptera (Arthropoda, Insecta) and its implications for their higher systematics. Sys. Biol. **48**:233–253.

Flower, J. W. 1964. On the origin of flight in insects. J. Insect Physiol. **10**:81–88.

Fochetti and Tierno de Figueroa. 2008. Global diversity of stoneflies (Plecoptera; Insecta) in freshwater. *Hydrobiologia* **595**:365–377.

Foote, R. H., F. L. Blanc, and A. L. Norrbom. 1993. *Handbook of the Fruit Flies (Diptera: Tephritidae) of America North of Mexico*. Comstock Publishing Associates, Ithaca, NY.

Forbes, W. T. M. 1923–1960. Lepidoptera of New York and neighboring states. Mem. Cornell Univ. Agric. Exp. Stn. **68**:1–729; **274**:1–263; **329**:1–433; **371**:1–188.

Ford, E. B. 1971. *Ecological Genetics*. 3rd ed. Chapman & Hall, London.

Ford, N. 1926. On the behavior of *Grylloblatta*. Can. Entomol. **58**:66–70.

Fowler, H. G., and J. C. A. Pinto. 1989. Life history of the "Northern" mole cricket, *Neocurtilla hexadactyla* (Perty) (Orthoptera, Gryllotalpidae), in southeastern Brazil. Rev. Bras. Entomol. **33**:143–148.

Fox, H. M., and G. Vevers. 1960. *The Nature of Animal Colours*. Macmillan, New York.

Fracker, S. B. 1930. The classification of lepidopterous larvae. Contrib. Entomol. Lab., Univ. Illinois. **43**:1–161.

François, J., R. Dallai, and W. Y. Yin. 1992. Cephalic anatomy of *Sinentomon erythranum* Yin (Protura: Sinentomidae). Int. J. Insect Morphol. Embryol. **21**:199–213.

Frankie, G., and R. Thorp. 2009. Pollination and pollinators. In *Encyclopedia of Insects*, V. Resh and R. Cardé (eds.). Academic Press, San Diego, CA. pp. 813–819.

Fraser, F. C. 1957. *A Reclassification of the Order Odonata*. Royal Zoological Society of New South Wales, Sydney.

Free, J. B. 1987. *Pheromones of Social Bees*. Comstock Publishing Associates, Ithaca, NY.

Free, J. B. 1993. *Insect Pollination of Crops*. 2nd ed. Academic Press, London.

Freilich, J. E. 1991. Movement patterns and ecology of *Pteronarcys* nymphs (Plecoptera): Observations of marked individuals in a Rocky Mountain stream. Freshwater Biol. **25**:379–394.

Fremling, C. R. 1960. Biology of a large mayfly, *Hexagenia bilineata* (Say), of the upper Mississippi River. Iowa State Univ. Agric. Home Econ. Exp. Stn. Res. Bull. **482**:841–852.

Frisch, K. von. 1967. *The Dance Language and Orientation of Bees*. Harvard University Press, Cambridge, MA.

Frisch, K. von. 1971. *Bees, Their Vision, Chemical Senses, and Language*. Cornell University Press, Ithaca, NY.

Frisch, K. von. 1974. Decoding the language of the bee. Science **185**:663–668.

Frison, T. H. 1935. The stoneflies, or Plecoptera, of Illinois. Illinois Nat. Hist. Survey Bull. **20**:281–471.

Frison, T. H. 1942. Studies on North American Plecoptera with special reference to the fauna of Illinois. Illinois Nat. Hist. Survey Bull. **22**:235–355.

Froeschner, R. C. 1954. The grasshoppers and other Orthoptera of Iowa. Iowa State Coll. J. Sci. **29**:163–354.

Froeschner, R. C. 1960. Cydnidae of the Western Hemisphere. Proc. U.S. Natl. Mus. **111**:337–680.

Fulton, B. B. 1924. Some habits of earwigs. Ann. Entomol. Soc. Am. **17**:357–367.

Furman, D. P., and E. P. Catts. 1961. *Manual of Medical Entomology*. National Press Books, Palo Alto, CA.

Fuzeau-Braesch, S. 1972. Pigments and color changes. Annu. Rev. Entomol. **17**:403–424.

Gadakgar, R. 1993. And now…eusocial thrips! Curr. Sci. **64**:215–216.

Gagne, R. J. 1989. *The Plant-Feeding Gall Midges of North America*. Comstock Publishing Associates, Ithaca, NY.

Galil, J., and G. Neeman. 1977. Pollen transfer and pollination in the common fig (*Ficus carica* L.). New Phytol. **79**:163–171.

Garman, P. 1927. The Odonata or dragonflies of Connecticut. Guide to the Insects of Connecticut. Connecticut Geol. Nat. Hist. Survey **5**:1–331.

Gaston, K. J. (ed.) 1996. *Biodiversity: A Biology of Numbers and Difference*. Blackwell Science, Oxford, UK.

Gatehouse, A. M. R., and V. A. Hilder. 1994. Genetic manipulation of crops for insect resistance. In *Molecular Biology in Crop Protection*, G. Marshall and D. Walters (eds.). Chapman & Hall, London. pp. 177–201.

Gaufin, A. R., A. V. Nebeker, and J. Sessions. 1966. The stoneflies (Plecoptera) of Utah. Univ. Utah Biol. Ser. **14**:1–93.

Gaugler, R., and H. K. Kaya (eds.) 1990. *Entomopathogenic Nematodes in Biological Control*. CRC Press, Boca Raton, FL.

Gauld, I., and B. Bolton. 1988. *The Hymenoptera*. Oxford University Press, Oxford.

Gaunt M. W., and M. A. Miles. 2002. An insect molecular clock dates the origin of the insects and accords with palaeontological and biogeographic landmarks. Mol. Biol. Evol. **19**:748–761.

Georghiou, G. P., and A. Lagunes-Tejeda. 1991. *The Occurrence of Resistance to Pesticides in Arthropods*. Food and Agriculture Organization of the United Nations, Rome.

Giberson, D. J., and R. J. Mackay. 1991. Life history and distribution of mayflies (Ephemeroptera) in some acid streams in south central Ontario, Canada. Can. J. Zool. **69**:899–910.

Gibson, G. A. P., J. T. Huber, and J. B. Woolley, eds. 1997. *Annotated Keys to the Genera of Nearctic Chalcidoidea*. NRC Research Press, Ottawa, ON.

Gilbert, C. 1994. Form and function of stemmata in larvae of holometabolous insects. Annu. Rev. Entomol. **39**:323–349.

Gilbert, L. E., and M. C. Singer. 1975. Butterfly ecology. Annu. Rev. Ecol. Syst. **6**:365–397.

Giles, E. T. 1963. The comparative external morphology and affinities of the Dermaptera. Trans. R. Entomol. Soc. London **115**:95–164.

Giles, E. T. 1974. The relationship between Hemimerina and the other Dermaptera: a case for reinstating Hemimerina within the Dermaptera, based on numerical procedure. Trans. R. Entomol. Soc. London **126**:189–206.

Gillespie, J. P., and M. R. Kanost. 1997. Biological mediators of insect immunity. Annu. Rev. Entomol. **42**:611–643.

Gillett, J. D. 1971. *Mosquitoes*. Weidenfeld and Nicolson, London.

Gillett, J. D. 1972. *The Mosquito: Its Life, Activities and Impact on Human Affairs*. Doubleday, Garden City, NY.

Gillott, C. 1995. *Entomology*. 2nd ed. Plenum Publishing, New York.

Glassberg, J. 1999. *Butterflies Through Binoculars: A Field Guide to Butterflies of Eastern North America*. Oxford University Press, New York.

Glick, P. A. 1939. The distribution of insects, spiders, and mites in the air. U.S. Dept. Agric., Tech. Bull. **673**:1–150.

Gloyd, L. K., and M. Wright. 1959. Odonata. In *Freshwater Biology*, W. T. Edmondson (ed.). John Wiley & Sons, New York. pp. 917–940.

Godfray, H. C. J. 1994. *Parasitoids, Behavioral and Evolutionary Ecology*. Monographs in Behavior and Ecology. Princeton University Press, Princeton, NJ.

Godfray, H. J. C. 2002. Challenges for taxonomy. Nature **417**:17–19.

Goldsmith, M. R., and A. S. Wilkins (eds.). 1995. *Molecular Model Systems in the Lepidoptera*. Cambridge University Press, Cambridge.

Goodman, L. J., and R. C. Fisher. 1991. *The Behaviour and Physiology of Bees*. CAB International. Wallingford, Oxon, UK.

Gordh, G. and D.H. Headrick 2001. A dictionary of Entomology. CABI Publishing, New York, NY. 1032 pp.

Gordon, H. T. 1961. Nutritional factors in insect resistance to chemicals. Annu. Rev. Entomol. **6**:27–54.

Gorham, J. R. (ed.). 1991. *Insect and Mite Pests in Food. An Illustrated Key*, U. S. Departments of Agriculture and Health and Human Services, Agriculture Handbook No. 655, 2 vols. Washington, DC. pp. 1–310, 311–767.

Gould, J. L. 1975. Honey bee recruitment: the dance-language controversy. Science **189**:685–693.

Goulet, H., and J. T. Huber. 1993. *Hymenoptera of the World: An Identification Guide to Families*. Research Branch, Agriculture Canada, Publication 1894/E.

Graham, S. A., and F. B. Knight. 1965. *Principles of Forest Entomology*, 4th ed. McGraw-Hill, New York.

Grandcolas, P., and P. Deleport. 1992. La position systématique de *Cryptocercus* Scudder au sein de Blattes et ses implications évolutives. C. R. Acad. Sc., Paris **315**:317–322.

Grbic, M., P. J. Ode, and M. R. Strand. 1992. Sibling rivalry and brood sex ratios in polyembryonic wasps. Nature (London) **360** (6401):254–256.

Greathead, D. J., and A. H. Greathead. 1992. Biological control of insect pests by insect parasitoids and predators: the BIOCAT database. Biocontrol News Inf. **13** (4):61N–68N.

Greathead, D. J., and J. K. Waage. 1986. *Insect Parasitoids: 13th Symposium of the Royal Entomological Society of London, 18–19 September 1985 at the Department of Physics Lecture Theatre, Imperial College, London*. Academic Press, London.

Greenslade, P. J. 1991. Collembola. In *The Insects of Australia*, vol. 1, Commonwealth Scientific and Industrial Research Organization (ed.). Melbourne University Press, Carlton. pp. 252–264.

Greenwood, S. R. 1988. Habitat stability and wing length in two species of arboreal Psocoptera. Oikos **52**:235–238.

Griffiths, G. C. D. 1972. The phylogenetic classification of the Diptera Cyclorrhapha with special reference to the structure of the male postabdomen. Ser. Entomol. Hague **8**:1–340.

Griffiths, G. C. D. 1980–. *Flies of the Nearctic Region*. E. Schweizerbartsche Verlag, Stuttgart.

Griffiths, G. C. D. 1994. Relationships among the major subgroups of Brachycera (Diptera): a critical review. Can. Entomol. **126**:861–880.

Grimaldi, D., and M. S. Engel 2005. *Evolution of the Insects*. Cambridge University Press, Cambridge, UK.

Grissell, E. 2001. *Insects and Gardens*. Timber Press, Portland, OR.

Grissell, E. 2010. *Bees, Wasps and Ants: The Indispensible Role of Hymenoptera in Gardens*. Timber Press, Portland, OR.

Grissell, E. E., and M. E. Schauff. 1990. *A Handbook of the Familes of Nearctic Chalcidoidea (Hymenoptera)*. Entomological Society of Washington, Washington, DC. Handbook No. 1.

Grodnitsky, D. L. 1999. *Form and Function of Insect Wings: The Evolution of Biological Structures*. Johns Hopkins University Press, Baltimore, 1999.

Grote, A. R. 1971. *Noctuidae of North America*. E. W. Classey, London.

Guenther, K. K., and K. Herter. 1974. Ordnung Dermaptera (Ohrwurmer). Handb. Zool. **4** (2):1–158.

Gullen, P. J., and P. S. Cranston. 1994. *The Insects: An Outline of Entomology*. Chapman & Hall, London.

Guppy, R. 1950. Biology of *Anisolabis maritima* (Gene) the seaside earwig, on Vancouver Island (Dermaptera, Labiidae). Proc. Entomol. Soc. Br. C. **46**:14–18.

Gupta, A. P. 1985. Cellular elements in the hemolymph. In *Comprehensive Insect Physiology, Biochemistry and Pharmacology*, vol. 3, G. A. Kerkut and L. I. Gilbert (eds.). Pergamon Press, New York. pp. 401–451.

Gupta, A. P. (ed.) 1990. *Morphogenetic Hormones of Arthropods*. Rutgers University Press, New Brunswick, NJ.

Gurney, A. B. 1938. A synopsis of the order Zoraptera, with notes on the biology of *Zorotypus hubbardi* Caudell. Proc. Entomol. Soc. Washington **40**:57–87.

Gurney, A. B. 1948a. Praying mantids of the United States. Smithsonian Inst. Rep. **1950**:339–362.

Gurney, A. B. 1948b. The taxonomy and distribution of the Grylloblattidae. Proc. Entomol. Soc. Washington **50**:86–102.

Gurney, A. B. 1950. Corrodentia. In *Pest Control Technology*. Entomological Section. National Pest Control Association, New York. pp. 129–163.

Gurney, A. B., and F. W. Fisk. 1986. Chap. 2. Cockroaches. In *Insects and Mites in Foods.*, J. R. Gorham (ed.). U.S. Department of Agriculture, Washington, DC.

Guthrie, D. M., and A. R. Tindall. 1968. *The Biology of the Cockroach*. St. Martins Press, New York.

Gwynne, D. T., and G. K. Morris (eds.). 1983. *Orthoptera Mating Systems. Sexual Competition in a Diverse Group of Insects*. Westview Press, Boulder, CO.

Haas, F., D. Waloszek and R. Hartenberger. 2003. *Devonohexapodus bocksbergensis,* a new marine hexapod from the Lower Devonian Husrück slates, and the origin of Atelocerata and Hexapoda organisms, Div. Evol. 3:39–54.

Hadley, N. F. 1994. *Water Relations of Terrestrial Arthropods.* Academic Press, San Diego, CA.

Hagan, H. R. 1951. *Embryology in the Viviparous Insects*. Ronald Press, New York.

Hagen, K. 1987. Nutritional ecology of terrestrial insect predators. In *Nutritional Ecology of Insects, Mites, Spiders, and Related Invertebrates*, F. J. Slansky and J. G. Rodriguez (eds.). John Wiley & Sons, New York. pp. 533–577.

Hagen, K. S., T. L. Tassan, and E. F. Sawall, Jr. 1970. Some ecophysiological relationships between certain *Chrysopa*, honeydews and yeasts. Boll. Lab. Entomol. Agric., Portici **28**:113–134.

Hagen, K. S., S. Bombosch, and J. A. McMurtry. 1976. The biology and impact of predators. In *Theory and Practice of Biological Control*, C. B. Huffaker and P. S. Messenger (eds.). Academic Press, New York. pp. 93–142.

Hajek, A. E. 2005. *Natural Enemies: An Introduction to Biological Control.* Cambridge University Press, Cambridge, UK.

Haine, E. R., Y. Moret, M. T. Siva-Jothy, and J. Rolff. 2008. Antimicrobial defense and persistent infection in insects. Science **322**:1257–1259.

Hale, W. G. 1965. Observations on the breeding biology of Collembola. Pedobiologia **5**:146–152.

Hall, D. G. 1948. The blow flies of North America. Thomas Say Found. Publ. **4**:1–477.

Halstead, S. B. 2008. Dengue virus–mosquito interactions. Annu. Rev. Entomol. **53**:273–291.

Hamilton, K. G. A. 1971. The insect wing. I. Origin and development of wings from notal lobes. J. Kansas Entomol. Soc. **44**:421–433.

Hamilton, K. G. A. 1972. The insect wing. Part IV. Venational trends and the phylogeny of the winged orders. J. Kansas Entomol. Soc. **45**:295–308.

Hamilton, K. G. A. 1981. Morphology and evolution of the rhynchotan head (Insecta: Hemiptera, Homoptera). Can. Entomol. **113**:953–974.

Hamilton, K. G. A. 1983. Classification, morphology and phylogeny of the family Cicadellidae (Rhynchota: Homoptera). In *Proceedings of the 1st International Workshop on Biotaxonomy, Classification and Biology of Leafhoppers and Planthoppers (Auchenorrhyncha) of Economic Importance*, W. J. Knight (ed.). Commonwealth Institute of Entomology, London. pp. 15–37.

Handlirsch, A. 1908. *Die fossilen Insekten und die Phylogenie der rezenten Formen*. W. Englemann, Leipzig.

Handlirsch, A. 1926. Vierter Unterstamm des Stammes der Arthropoda. Insecta = Insekten. Handbuch Zool. **4**:403–592.

Hanover, J. W. 1975. Physiology of tree resistance to insects. Annu. Rev. Entomol. **20**:75–95.

Hansen, M. 1991. The hydrophiloid beetles: Phylogeny, classification and a revision of the genera (Coleoptera, Hydrophiloidea).

K. Dan. Vidensk. Selsk. Biol. Skr. **40**:1–367.

Hanski, I., and Y. Cambefort. 1991. *Dung Beetle Ecology*, Princeton University Press, Princeton, NJ.

Happ, G. M. 1992. Maturation of the male reproductive system and its endocrine regulation. Annu. Rev. Entomol. **37**:303–320.

Harden, P. H., and C. E. Mickel. 1952. The stoneflies of Minnesota (Plecoptera). Univ. Minn. Agric. Exp. Stn. Tech. Bull. **201**:1–84.

Harland, W. B., R. L. Armstrong, A. V. Cox, L. E. Craig, A. G. Smith, and D. G. Smith. 1990. *A Geologic Time Scale 1989*. Cambridge University Press, Cambridge.

Harris, K. F. (ed.). 1986–. *Advances in Vector Research*. Springer-Verlag, New York.

Harris, W. V. 1961. *Termites. Their Recognition and Control*. Longmans Group, London.

Hartley, S. E., and S. E. Gange. 2009. Impacts of plant symbiotic fungi on insect herbivores: mutualism in a multitrophic context. Annu. Rev. Entomol. **54**:323–342.

Hassell, K. A. 1990. *The Biochemistry and Uses of Insecticides*. 2nd ed. VCH, Weinheim.

Hatch, M. H. 1957–1973. *The Beetles of the Pacific Northwest*. 5 vols. University of Washington Publications Biology and University of Washington Press, Seattle.

Haub, F. 1980. Concerning "phylogenetic" relationships of parasitic Psocodea and taxonomic position of the Anoplura by K. C. Kim and H. W. Ludwig. Ann. Entomol. Soc. Am. **73**:3–6.

Haukioja, E., V. Ossipov, and K. Lempa. 2002. Interactive effects of leaf maturation and phenolics on consumption and growth of a geometrid

moth. Entomol. Exp. Appl. **104**:1125–1136.

Haunerland, N. H., and P. D. Shirk. 1995. Regional and functional differentiation in the insect fat body. Annu. Rev. Entomol. **40**:121–145.

Hayashi, F. 1988. Prey selection by the dobsonfly larva. *Protohermes grandis* (Megaloptera: Corydalidae). Freshwater Biol. **20**:19–30.

Hayashi, F. 1989. Microhabitat selection by the fishfly larva, *Parachauliodes japonicus*, in relation to its mode of respiration. Freshwater Biol. **21**:489–496.

Hayashi, F. 1992. Large spermatophore production and consumption in dobsonflies *Protohermes* (Megaloptera, Corydalidae). Jpn. J. Entomol. **60**:59–66.

Hayashi, F., and M. Nakane. 1989. Radio tracking and activity monitoring of the dobsonfly larva, *Protohermes grandis* (Megaloptera: Corydalidae). Oecologia (Berlin) **78**:468–472.

Hebard, M. 1934. The Dermaptera and Orthoptera of Illinois. Illinois Nat. Hist. Survey Bull. **20**:125–279.

Heinrich, B. 1973. The energetics of the bumblebee. Sci. Am. **228**(4):96–102.

Heinrich, B. 1974. Thermoregulation in endothermic insects. Science **185**:747–756.

Heinrich, B. 1993. *The Hot-Blooded Insects: Strategies and Mechanisms of Thermoregulation*. Harvard University Press, Cambridge, MA.

Heinrich, B. 2009. Thermoregulation. Pp. 993–999 in: *Encyclopedia of Insects*, V. Resh and R. Cardé, eds. Academic Press, San Diego, CA.

Heinrich, B., and G. A. Bartholomew. 972. Temperature control in flying moths. Sci. Am. **226**(6):69–77.

Heise, B. A., and J. F. Flannagan. 1987. Life histories of *Hexagenia limbata* and *Ephemera simulans* (Ephemeroptera) in Dauphin Lake, Manitoba [Canada]. J. North Am. Benthol. Soc. **6**:230–240.

Helfer, J. R. 1987. *How to Know the Grasshoppers, Crickets, Cockroaches, and Their Allies*. Dover Publications, New York.

Heliövaara, K., and R. Vaisanen. 1993. *Insects and Pollution*. CRC Press, Boca Raton, FL.

Hellenthal, R. A., and R. D. Price. 1991. Biosystematics of the chewing lice of pocket gophers. Annu. Rev. Entomol. **36**:185–203.

Hendry, L. B., B. Piatek, L. E. Browne, D. L. Wood, J. A. Byers, R. H. Fish, and R. A. Hicks. 1980. *In vivo* conversion of a labelled host plant chemical to pheromones of the bark beetle *Ips paraconfusus*. Nature **284**:485.

Hengstenberg, R. 1998. Controlling the fly's gyroscopes. *Nature* **392**:757–758.

Hennig, W. 1948–1952. *Die Larvenformen der Dipteren*. Part 1. Akademie-Verlag, Berlin.

Hennig, W. 1953. Kritische Bemerkungen zum phylogenetischen System der Insekten. Beitr. Entomol. **3**:1–85.

Henry, C. S. 1989. The unique purring song of *Chrysoperla comanche* (Banks), a green lacewing of the Rufilabris spp. group (Neuroptera: Chrysopidae). Proc. Entomol. Soc. Washington **91**:133–142.

Henry, T. J. 2009. Biodiversity of the Heteroptera. In *Insect Biodiversity: Science and Society*, R. G. Foottit and P. H. Adler (eds.). Wiley-Blackwell, Oxford, UK. pp. 223–263.

Henry, T. J., and R. C. Froeschner (eds.). 1988. *Catalog of the*

Heteroptera, or True Bugs, of Canada and the Continental United States. St. Lucie Press, Delray Beach, FL.

Hepburn, H. R. 1969. The skeletomuscular system of Mecoptera: The head. Kansas Univ. Sci. Bull. **48**:721–765.

Hering, M. 1951. *Biology of the Leaf Miners.* The Hague, Netherlands.

Hermann, H. R. (ed.). 1979–82. *Social Insects,* vols. 1–4. Academic Press, New York.

Herring, J. L., and P. D. Ashlock. 1971. A key to the nymphs of the families of Hemiptera (Heteroptera) of America north of Mexico. Florida Entomol. **54**:207–213.

Hershberger, W., and L. Elliott. 2006. *The Songs of Insects.* Houghton Mifflin, New York (with sound CD of calls).

Hespenheide, H. H. 1991. Bionomics of leaf-mining insects. Annu. Rev. Entomol. **36**:535–560.

Hildrew, A. G., and R. Wagner. 1992. The briefly colonial life of hatchlings of the net-spinning caddisfly *Plectrocnemia conspersa.* J. North Am. Benthol. Soc. **11**:60–68.

Hille Ris Lambers, D. 1966. Polymorphism in Aphididae. Annu. Rev. Entomol. **11**:47–78.

Hines, H. M., J. H. Hunt, T. K. O'Connor, J. J. Gillespie, and S. A. Cameron. 2007. Multigene phylogeny reveals eusociality evolved twice in vespid wasps. Proc. Natl. Acad. Sci. U.S.A. **104**:3295–3299.

Hinton, H. E. 1945. *A Monograph of the Beetles Associated with Stored Products.* British Museum of Natural History, London.

Hinton, H. E. 1946. On the homology and nomenclature of the setae of the lepidopterous larvae, with some notes on the phylogeny of the Lepidoptera.

Trans. R. Entomol. Soc. London **97**:1–37.

Hinton, H. E. 1955a. On the structure and distribution of the prolegs of the Panorpoidea with a criticism of the Berlese-Imms theory. Trans. R. Entomol. Soc. London **106**:455–540.

Hinton, H. E. 1955b. On the respiratory adaptions, biology and taxonomy of the Psephenidae, with notes on some related families (Coleoptera). Proc. Zool. Soc. London **125**:543–568.

Hinton, H. E. 1958. The phylogeny of the panorpoid orders. Annu. Rev. Entomol. **3**:181–206.

Hinton, H. E. 1963a. The origin and function of the pupal stage. Proc. R. Entomol. Soc. London (A) **38**:77–85.

Hinton, H. E. 1963b. The origin of flight in insects. Proc. R. Entomol. Soc. London (C) **28**:23–32.

Hinton, H. E. 1969a. Plastron respiration in adult beetles of the suborder Myxophaga. J. Zool. Soc. London **159**:131–137.

Hinton, H. E. 1969b. Respiratory systems of insect egg shells. Annu. Rev. Entomol. **14**:343–368.

Hinton, H. E. 1971. Some neglected phases in metamorphosis. Proc. R. Entomol. Soc. London (C) **35**:55–64.

Hinton, H. E. 1977. Enabling mechanisms. XV Int. Congr. Entomol., 71–83.

Hinton, H. E. 1981. *Biology of Insect Eggs.* Pergamon Press, Oxford. 3 vols.

Hodek, I. 1973. *Biology of Coccinellidae.* Academia, Prague, and Junk's, The Hague.

Hodges, R. W., et al. 1983. *Checklist of the Lepidoptera of America North of Mexico.* Wedge Entomological Research Foundation, National Museum of

Natural History, Washington, DC.

Hodgson, C. J. 1994. *The Scale Insect Family Coccidae: An Identification Manual to Genera.* CAB International, Wallingford, Oxon, UK.

Hodkinson, T. R., and J. A. N. Parnell (eds.). 2007. *Reconstructing the Tree of Life: Taxonomy and Systematics of Species Rich Taxa.* Systematics Association Special Volume 72. CRC Press, Boca Raton, FL.

Holdsworth, R. P. 1941. The life history and growth of *Pteronarcys proteus* Newman (Pteronarcyidae: Plecoptera). Ann. Entomol. Soc. Am. **34**:394–502.

Holland, G. P. 1949. The Siphonaptera of Canada. Can. Dept. Agric. Publ. 817, Tech. Bull. **70**:1–306.

Holland, G. P. 1964. Evolution, classification and host relationships of Siphonaptera. Annu. Rev. Entomol. **9**:123–146.

Holland, W. J. 1968. *The Moth Book.* Dover Publications, New York.

Hölldobler, B., and E. O. Wilson. 1990. *The Ants.* Belknap Press, Cambridge, MA.

Holman, G. M., R. J. Nachman, and M. S. Wright. 1990. Insect neuropeptides. Annu. Rev. Entomol. **35**:201–217.

Holsinger, J. R. 1988. Troglobites: The evolution of cave-dwelling organisms. Am. Sci. **76**:147–153.

Holzenthal, R. W. 2009. Trichoptera (caddisflies). In *Encyclopedia of Inland Waters,* vol. 2, G. E. Likens (ed.). Elsevier, Oxford, UK. pp. 456–467.

Holzenthal, R. W., R. J. Blahnik, A. L. Prather, and K. M. Kjer. 2007. Order Trichoptera Kirby, 1813 (Insecta), caddisflies. *Zootaxa* **1668**:639–698.

Homberg, U., T. A. Christensen, and J. G. Hildebrand. 1989. Structure and function of the deutocerebrum in insects. Annu. Rev. Entomol. **34**:477–501.

Hopkins, G. H. E. 1949. The host associations of the lice of mammals. Proc. Zool. Soc. London **119**:387–604.

Horsfall, W. R., Jr., H. W. Fowler, L. J. Moretti, and J. R. Larsen. 1973. *Bionomics and Embryology of the Inland Floodwater Mosquito Aedes vexans*. University of Illinois Press, Urbana.

Houk, E. J., and G. W. Griffiths. 1980. Intracellular symbiotes of the Homoptera. Annu. Rev. Entomol. **25**:161–187.

Howarth, F. G. 1983. Ecology of cave arthropods. Annu. Rev. Entomol. **28**:365–389.

Howarth, F. G. 1991. Environmental impacts of classical biological control. Annu. Rev. Entomol. **36**:485–509.

Howe, W. H. 1975. *The Butterflies of North America*. Doubleday, Garden City, NY.

Howse, P. E. 1970. *Termites: A Study in Social Behavior*. Hutchinson University Library, London.

Hoy, M. A. 1994. *Insect Molecular Genetics: An Introduction to Principles and Applications*. Academic Press, San Diego, CA.

Hoy, M. A., and Herzog, D. C. (eds.). 1985. *Biological Control in Agricultural IPM Systems*. Academic Press, New York.

Hsu, C.-Y., and C.-W. Li. 1994. Magnetoreception in honeybees. Science **265**:95–97.

Hubbard, C. A. 1947. *Fleas of Western North America*. Iowa State College Press, Ames.

Hubbard, M. D. 1990. *Mayflies of the World: A Catalog of the Family and Genus Group Taxa (Insecta: Ephemeroptera)*. Flora & Fauna Handbook No. 8. Sandhill Crane Press, Gainesville, FL.

Hubbell, S. 1993. *Broadsides from the Other Orders. A Book of Bugs*. Random House, New York.

Huber, I., E. P. Masler, and B. R. Rao. 1990a. *Cockroaches as Models for Neurobiology: Applications in Biomedical Research*. CRC Press., Boca Raton, FL. 2 vols.

Huber, F., T. E. Moore, and W. Loher. 1990b. *Cricket Behavior and Neurobiology*. Cornell University Press, Ithaca, NY.

Huber, J. T. 1998. The importance of voucher specimens, with practical guidelines for preserving specimens of the major invertebrate phyla for identification. *J. Nat. History* **32**:367–385.

Huelsenbeck, J. P. 1998. Systematic bias in phylogenetic analysis: is the Strepsiptera problem solved? Syst. Biol. **47**:519–537.

Huffaker, C. B. (ed.). 1971. *Biological Control*. Plenum Publishing, New York.

Huffaker, C. B., and P. S. Messenger (eds.). 1976. *Theory and Practice of Biological Control*. Academic Press, New York.

Huisken J., J. Swoger, F. Del Bene, J. Wittbrodt, and E. H. K. Stelzer. 2004. Optical sectioning deep inside live embryos by selective plane illumination microscopy. Science **305**:1007–1009.

Hughes, J. M., P. B. Mather, A. L. Sheldon, and F. W. Allendorf. 1999. Genetic structure of the stonefly, *Yoraperla brevis*, populations: the extent of gene flow among adjacent montane streams. Freshwater Biol. **41**:63–72.

Hull, F. M. 1962. Robberflies of the world: the genera of the family Asilidae. U.S. Natl. Mus. Bull. **224**:1–907 (2 vols.).

Hungerford, H. B. 1959. Hemiptera. In *Freshwater Biology*, W. T. Edmondson (ed.). John Wiley & Sons, New York. pp. 958–972.

Hunt, B. P. 1950. The life history and economic importance of the burrowing mayfly, *Hexagenia limbata*, in southern Michigan lakes. Inst. Fisheries Res. Bull. Michigan Dept. Conserv. **4**:1–151.

Hunt, J. H., and C. A. Nalepa (eds.) 1994. *Nourishment and Evolution in Insect Societies. Studies in Insect Biology*. Westview Press, Boulder, CO.

Hunt, T., J. Bergsten, Z. Levkanicova, A. Papadopoulou, O. St. John, R. Wild, P. M. Hammond, A. Ahrends, M. Balke, M. S. Caterino, J. Gomez-Zurita, I. Ribera, T. G. Barraclough, M. Bocakova, L. Bocak, and A. P. Vogler. 2007. A comprehensive phylogeny of beetles reveals the evolutionary origins of a super-radiation. *Science* **318**:1913–1916.

Huxley, C. R., and D. F. Cutler. 1991. *Ant-Plant Interactions*. Oxford University Press, Oxford.

Hynes, H. B. N. 1970. The ecology of stream insects. Annu. Rev. Entomol., **15**:25–42.

Hynes, H. B. N. 1976. Biology of Plecoptera. Annu. Rev. Entomol. **21**:135–154.

Ide, F. P. 1935. The effect of temperature on the distribution of the mayfly fauna of a stream. Univ. Toronto Studies, Biol. Ser. No. 39, Publ. Ontario Fisheries Res. Lab. **50**:3–76.

Illies, J. 1965. Phylogeny and zoogeography of the Plecoptera. Annu. Rev. Entomol. **10**:117–140.

Imms, A. D. 1936. The ancestry of insects. Trans. Soc. Br. Entomol. **3**:1–32.

International Commission on Zoological Nomenclature. 2000. *International Code of Zoological Nomenclature*. 4th ed. International Trust for

Zoological Nomenclature, London.

Inward, D., G. Beccaloni, and P. Eggleton. 2007. Death of an order: a comprehensive phylogenetic study confirms that termites are eusocial cockroaches. Biol. Lett. **3**:331–335.

Iseley, F. B. 1944. Correlation between mandibular morphology and food specificity in grasshoppers. Ann. Entomol. Soc. Am. **37**:47–67.

Istock, C. A. 1966. The evolution of complex life cycle phenomena: an ecological perspective. Evolution **21**:592–605.

Ito, Y. 1993. *Behaviour and Social Evolution of Wasps: The Communal Aggregation Hypothesis*. Oxford Series in Ecology and Evolution. Oxford University Press, Oxford.

Iwata, K. 1976. *Evolution of Instinct: Comparative Ethology of Hymenoptera*. Amerind Publishing, New Delhi.

Jacobi, D. I., and A. C. Benke. 1991. Life histories and abundance patterns of snag-dwelling mayflies in a blackwater Coastal Plain river. J. North Am. Benthol. Soc. **10**:372–387.

James, M. T. 1947. The flies that cause myiasis in man. U.S. Dept. Agric. Misc. Publ. **631**:1–175.

James, M. T., and R. F. Harwood. 1969. *Herm's Medical Entomology*. 6th ed. Macmillan, London.

Janzen, D. H. 1966. Coevolution of mutualism between ants and acacias in Central America. Evolution **20**:249–275.

Janzen, D. H. 1967a. Fire, vegetation structure, and the ant-acacia interaction in Central America. Ecology **48**:26–35.

Janzen, D. H. 1967b. Interaction of the bull's-horn acacia (*Acacia cornigera* L.) with an ant inhabitant (*Pseudomyrmex ferruginea* F. Smith) in eastern

Mexico. Kansas Univ. Sci. Bull. **47** (6):15–558.

Jarvis, K. J., F. Haas, and M. F. Whiting. 2005. Phylogeny of earwigs (Insecta: Dermaptera) based on molecular and morphological evidence: reconsidering the classification of Dermaptera. Syst. Entomol. **30**:442–453.

Jarvis, K. J., and M. F. Whiting. 2006. Phylogeny and biogeography of ice crawlers (Insecta: Grylloblattodea) based on six molecular loci: designating conservation status for Grylloblattodea species. Mol. Phylogenet. Evol. **41**:222–237.

Jeanne, R. L. 1980. Evolution of social behavior in the Vespidae. Annu. Rev. Entomol. **25**:371–396.

Jewett, S. G. 1956. Plecoptera. In *Aquatic Insects of California*. R. L. Usinger (ed.). University of California Press, Berkeley. pp. 155–181.

Jewett, S. G. 1959. The stoneflies (Plecoptera) of the Pacific Northwest. Oregon State Mongr. Stud. Entomol. **3**:1–95.

Jewett, S. G., Jr. 1963. A stonefly aquatic in the adult stage. Science **139**:484–485.

Johannsen, O. A., and F. H. Butt. 1941. *Embryology of Insects and Myriapods*. McGraw-Hill, New York.

Johnson, J. B., and K. S. Hagen. 1981. A neuropteran larva uses an allomone to attack termites. Nature **289**:506–507.

Johnson, K. P., K. Yoshizawa, and V. S. Smith. 2004. Multiple origins of parasitism in lice. Proc. R. Soc. Lond. Series B **217**:1771–1776.

Jolivet, P. 1996. *Biologie de Coleoptères Chrysomélides*. Boubée, Paris.

Jones, T. 1954. The external morphology of *Chirothrips hematus* (Trybom) (Thysanoptera).

Trans. R. Entomol. Soc. London **105**:163–187.

Joosse, E. N. G. 1976. Littoral apterygotes (Collembola and Thysanura). In *Marine Insects*, L. Cheng (ed.). North-Holland Publishing Company, Amsterdam. pp. 151–186.

Joosse, E. N. G. 1983. New developments in the ecology of Apterygota. Pedobiologia **25**:217–234.

Joosse, E. N. G., and H. A. Verhoef. 1987. Developments in ecophysiological research on soil invertebrates. Adv. Ecol. Res. **16**:175–248.

Jost, M. K., and K. L. Shaw. 2006. Phylogeny of Ensifera (Hexapoda: Orthoptera) using three ribosomal loci, with implications for the evolution of acoustic communication. Mol. Phylogenet. Evol. **38**:510–530.

Judd, W. W. 1948. A comparative study of the proventriculus of orthopteroid insects with reference to its use in taxonomy. Can. J. Res. **26**:93–161.

Juniper, B., and R. Southwood (eds.). 1986. *Insects and the Plant Surface*. Edward Arnold, London.

Kaiser, W., and J. Steiner-Kaiser. 1987. Sleep research on honeybees: neurophysiology and behavior. In *Neurobiology and Behavior of Honeybees*, R. Menzel and A. Mercer (eds.). Springer-Verlag, Berlin. pp. 112–120.

Kamp, J. W. 1963. Descriptions of two new species of Grylloblattidae and of the adult of *Grylloblatta barberi*, with an interpretation of their geographic distribution. Ann. Entomol. Soc. Am. **56**:53–68.

Kamp, J. W. 1973. Numerical classification of the orthopteroids, with special reference to the Grylloblattodea. Can. Entomol. **105**:1235–1249.

Karban, R., and I. T. Baldwin. 1997. *Induced Responses to Herbivory*. University of Chicago Press, Chicago.

Kathirithamby, J. 1989. Review of the order Strepsiptera. Syst. Entomol. **14**:41–92.

Kathirithamby, J., and M. Carcupino. 1992. Comparative spermology of four species of Strepsiptera and comparison with a species of primitive Coleopters Rhipiphoridae. Int. J. Insect Morph. Embryol. **22**:459–470.

Kauffman, W. C., and J. E. Nechols (eds.). 1992. *Selection Criteria and Ecological Consequences of Importing Natural Enemies*. Proceedings, Thomas Say Publications in Entomology, Entomological Society of America, Lanham, MD.

Kaufmann, T. 1971. Hibernation in the arctic beetle. *Pterostichus brevicornis*, in Alaska. J. Kansas Entomol. Soc. **44**:81–92.

Keister, A. R., and E. Strates. 1984. Social behavior in a thrips from Panama. J. Nat. Hist. **18**:304–314.

Keler, S. von. 1969. 17. Ordnung Mallophaga (Federlinge und Haarlinge). Handb. Zool. **4**(2):1–72.

Kelsey, L. P. 1954. The skeleto-motor mechanism of the Dobson fly, *Corydalus cornutus*. Part I. Head and prothorax. Mem. Agric. Exp. Stn. Cornell Univ **334**:1–51.

Kelsey, L. P. 1957. The skeleto-motor mechanism of the Dobson fly, *Corydalus cornutus*. Part II. Pterothorax. Mem. Agric. Exp. Stn. Cornell Univ. **346**:1–31.

Kennedy, J. S., and H. L. G. Stroyan. 1959. Biology of aphids. Annu. Rev. Entomol. **4**:139–160.

Kent, D. K., and J. A. Simpson. 1992. Eusociality in the beetle *Austroplatypus incompertus* (Coleoptera. Curculionidae). Naturwissenschaften **79**:86–87.

Kerkut, G. A., and L. I. Gilbert (eds.). 1985. *Comprehensive Insect Physiology, Biochemistry, and Pharmacology*. Pergamon Press, Oxford.

Kettlewell, H. B. D. 1959. Darwin's missing evidence. Sci. Am. **200**:48–53.

Kettlewell, H. B. D. 1961. The phenomenon of industrial melanism in Lepidoptera. Annu. Rev. Entomol **6**:245–262.

Kettlewell, H. B. D. 1973. *The Evolution of Melanism* Clarendon Press, Oxford.

Kevan, D. K. McE. 1962. *Soil Animals*. Philosophical Library, New York.

Kevan, D. K. McE. 1977. The higher classification of the orthopteroid insects. Mem. Lyman Entomol. Mus Res. Lab. **4**:1–52.

Kevan, D. K. McE. 1982. Orthoptera (pp. 352–379), Phasmatoptera (pp. 379–383). In *Synopsis and Classification of Living Organisms*, S. P. Parker (ed.). McGraw-Hill, New York.

Kevan, P. G., and H. G. Baker. 1983. Insects as flower visitors and pollinators. Annu. Rev. Entomol. **28**:407–453.

Kim, K. C., and H. W. Ludwig. 1982. Parallel evolution, cladistics, and classification of parasitic Psocodea. Ann. Entomol. Soc. Am. **75**:537–548.

Kim, K. C., H. D. Pratt, and C. J. Stojanovich. 1986. *The Sucking Lice of North America: An Illustrated Manual for Identification*. Pennsylvania State University Press, University Park.

King, P. E., and K. S. Ahmed. 1989. Sperm structure in the Psocoptera. Acta Zool. (Stockholm) **70**:57–62.

Kingsolver, J. G., and M. A. R. Koehl. 1985. Aerodynamics, thermoregulation, and the evolution of insect wings: differential scaling and evolutionary change. Evolution. **39**:488–504.

Kingsolver, J. G., and M. A. R. Koehl. 1994. Selective factors in the evolution of insect wings. Annu. Rev. Entomol. **39**:425–451.

Kinzelbach, R. 1990. The systematic position of Strepsiptera (Insecta). Am. Entomol. **36**:292–303.

Kinzelbach, R., and H. Pohl. 1994. The fossil Strepsiptera (Insecta: Strepsiptera). Ann. Entomol. Soc. Am. **87**:59–70.

Kinzelbach, R. K. 1971a. Strepsiptera (Facherflugler). Handb. Zool. **42**(24):1–68.

Kinzelbach, R. K. 1971b. Morphologische Befunde an Facherfluglern und ihre phylogenetische Bedeutung (Insecta: Strepsiptera). Zoologica **41**(119):1–256.

Kinzelbach, R. K. 1978. Insecta: Facherflugler (Strepsiptera). Tierwelt Dtschl. **65**:1–166.

Kirchner, W. H. 1993. Acoustical communication in honeybees. Apidologie **24**:297–307.

Kirchner, W. H., and W. F. Towne. 1994. The sensory basis of the honeybee's dance language. Sci. Am. **270**:74–80.

Kirchner, W. H., M. Lindauer, and A. Michelsen. 1988. Honeybee dance communication. Acoustical indication of direction in round dances. Naturwissenschaften **75**:629–630.

Kiriakoff, S. G. 1963. The tympanic structures of Lepidoptera and the taxonomy of the order. J. Lepid. Soc. **17**:1–6.

Kistner, D. H. 1979. Social and evolutionary significance of social insect symbionts. In *Social Insects*, vol. 1, H. R. Herman (ed.). Academic Press, New York. pp. 339–415.

Kistner, D. H. 1982. The social insects bestiary. In *Social Insects*, vol. 2, H. R. Herman (ed.). Academic Press, New York. pp. 1–244.

Kitching, I. J., P. L. Forey, C. J. Humphries, and D. M. Williams. 1998. *Cladistics: The Theory and Practice of Parsimony Analysis.* Systematics Association Publication No. 11. Oxford University Press, Oxford, UK.

Kjer, K. M., R. J. Blahnik, and R. W. Holzenthal. 2001. Phylogeny of Trichoptera (caddisflies): characterization of signal within multiple datasets. Syst. Biol. **50**:781–816.

Kjer, K. M., F. L. N. Carle, J. Litman, and J. Ware 2006. A molecular phylogeny of Hexapoda. Arthr. Syst. Phylog. **64**:35–44.

Klass, K.-D., O. Zompro, N. P. Kristensen, and J. Adis. 2002. Mantophasmatodea: a new insect order with extant members in the Afrotropics. *Science* **296**:1456–1459.

Klimaszewski, J., and D. K. McE. Kevan. 1985. The brown lacewing flies of Canada and Alaska (Neuroptera: Hemerobiidae). Memoir, Lyman Entomological Museum and Research Laboratory, no. 15.

Klots, A. B. 1951. *A Field Guiide to the Butterflies.* Houghton Mifflin, Boston.

Klowden, M. J. 2007. *Physiological Systems in Insects.* Academic Press, San Diego, CA.

Knight, K. L. (ed.). 1989. Entomology. Serving Society 1889–1989. Bull. Entomol. Soc. Am. **35** (3):1–215.

Knoll, A. H. and S. B. Carroll. 1999. Early animal evolution: emerging views from comparative biology and geology. Science **284**:2129–2137.

Kofoid, C. A. (ed.). 1934. *Termites and Termite Control.* University of California Press, Berkeley.

Kormondy, E. J. 1961. Territoriality and dispersal in dragonflies (Odonata). J. New York. Entomol. Soc. **69**:42–52.

Koss, R. W. 1968. Morphology and taxonomic use of Ephemeroptera eggs. Ann. Entomol. Soc. Am. **61**:696–721.

Kosztarab, M. 1987. Everything unique or unusual about scale insects (Homoptera: Coccoidea). Bull. Entomol. Soc. Am. (winter):215–220.

Kosztarab, M. 1996. *Scale Insects of Northeastern North America: Identification, Biology, and Distribution.* Virginia Museum of Natural History, Martinsville.

Kosztarab, M., and C. W. Schaefer (eds.). 1990. *Systematics of the North American Insects and Arachnids: Status and Needs.* Virginia Agric. Exp. Stn. Inf. Ser. 90–1. Blacksburg.

Kosztarab, M., and M. P. Kosztarab. 1988. *A Selected Bibliography of the Coccoidea (Homoptera): Third Supplement (1970–1985).* Studies on the Morphology and Systematics of Scale Insects no. 14. Virginia Polytechnic Institute and State University, Blacksburg.

Kosztarab, M., Y. Ben-Dov, M. P. Kosztarab, and H. Morrison, 1986. *An Annotated List of Generic Names of the Scale Insects (Homoptera, Coccoidea): Third Supplement.* Studies on the Morphology and Systematics of Scale Insects no. 13. Virginia Polytechnic Institute and State University, Blacksburg.

Kramer, L. D., L. M. Styer, and G. D. Ebel. 2008. A global perspective on the epidemiology of West Nile virus. Annu. Rev. Entomol. **53**:61–81.

Kramer, S. 1950. Morphology and phylogeny of the auchenorrhynchous Homoptera (Insecta). Illinois Biol. Monogr. **20**:1–78.

Krishna, K. 1990. Isoptera. Bull. Am. Mus. Nat. Hist. **195**:76–81.

Krishna, K., and F. M. Weesner. 1969a. *Biology of Termites*, vol. 1. Academic Press. New York.

Krishna, K., and F. M. Weesner. 1969b. *Biology of Termites*, vol. 2. Academic Press, New York.

Kristensen, N. P. 1968. The morphological and functional evolution of the mouthparts in adult Lepidoptera. Opusc. Entomol. **33**:69–72.

Kristensen, N. P. 1971. The systematic position of the Zeugloptera in the light of recent anatomical investigations. Proc. 13th Int. Congr. Entomol. (1968) **1**:261.

Kristensen, N. P. 1975. The phylogeny of hexapod "orders": a critical review of recent accounts. Z. Zool. Syst. Evol. Forsch. **13**:1–44.

Kristensen, N. P. 1981. Phylogeny of insect orders. Annu. Rev. Entomol. **26**:135–158.

Kristensen, N. P. 1984. Studies on the morphology and systematics of primitive Lepidoptera (Insecta). Steenstrupia **10**:141–191.

Kristensen, N. P. 1991. Phylogeny of extant hexapods. In *The Insects of Australia*, vol. 1, Commonwealth Scientific and Industrial Research Organization (ed.). Melbourne University Press, Carlton. pp. 125–140.

Kristensen, N. P., M. J. Scoble, and O. Karsholt. 2007. Lepidoptera phylogeny and systematics: the state of inventorying moth and butterfly diversity. *Zootaxa* **1118**:699–747.

Krombein, K. V. 1967. *Trap-Nesting Wasps and Bees. Life Histories, Nests, and Associates.* Smithsonian Institution, Washington, DC.

Kuhnelt, W. 1961. *Soil Biology, with Special Reference to the Animal Kingdom.* Faber & Faber, London.

Kukalová-Peck, J. 1978. Origin and evolution of insect wings and their relation to metamorphosis, as documented by fossil record. J. Morphol. **156**:53–126.

Kukalová-Peck, J. 1983. Origin of the insect wing and wing articulation from the arthropodan leg. Can. J. Zool. **61**:1618–1669.

Kukalová-Peck, J. 1985. Ephemeroid wing venation based on new gigantic Carboniferous mayflies and basic morphology, phylogeny, and metamorphosis of pterygote insects (Insecta, Ephemerida), Can. J. Zool. **63**:933–955.

Kukalová-Peck, J. 1987. New Carboniferous Diplura, Monura and Thysanura, the hexapod ground plan, and the role of thoracic side lobes in the origin of wings (Insecta). Can. J. Zool. **65**:2327–2345.

Kukalová-Peck, J. 1991. Fossil history and the evolution of hexapod structures. In *The Insects of Australia*, vol. 1, Commonwealth Scientific and Industrial Research Organization (ed.). Melbourne University Press, Carlton. pp. 141–179.

Kukalová-Peck, J. 1992. The "Uniramia" do not exist: the ground plan of the Pterygota as revealed by Permian Diaphanopterodea from Russia (Insecta: Paleodictyopteroidea). Can. J. Zool. **70**:236–255.

Kukalová-Peck, J., and C. Brauckmann. 1992. Most Paleozoic Protorthoptera are ancestral hemipteroids: major wing braces as clues to a new phylogeny of Neoptera (Insecta). Can. J. Zool. **70**:2452–2473.

Kukalová-Peck, J., and J. F. Lawrence. 1993. Evolution of the hind wing in Coleoptera. Can. Entomol. **125**:181–258.

Kukalová-Peck, J., and S. Peck. 1993. Zoraptera wing structure: evidence for new genera and relationship with the blattoid orders (Insecta: Blattoneoptera). Syst. Entomol. **18**:333–350.

Kuris, A. M. 2003. Did biological control cause extinction of the coconut moth, *Levuana iridescens*, in Fiji? Biol. Invasions B:133–141.

Labandeira, C. C., and J. J. Sepkoski. 1993. Insect diversity in the fossil record. Science **261**:310–315.

Labandeira, C., B. Beall, and P. Hueber. 1988. Early insect diversification: evidence from a lower Devonian bristletail from Quebec. Science **242**:913–916.

Labandeira, C. C., D. L. Dilcher, D. R. Davis, and D. L. Wagner. 1994. Ninety-seven million years of angiosperm-insect association: paleobiological insights into the meaning of coevolution. Proc. Natl. Acad. Sci. U.S.A. **91**:12278–12282.

Lakshminarayana, K. V. 1985. Glossary of taxonomic characters for the study of chewing-lice (Phthiraptera: Insecta). Rec. Zool. Survey India., Misc. Publ., Occas. Pa. **82**:1–59.

Land, M. F., and D.-E. Nilsson. 2002. *Animal Eyes*. Oxford University Press, London.

Landa, V., and T. Soldan. 1985. *Phylogeny and Higher Classification of the Order Ephemeroptera: A Discussion from the Comparative Anatomical Point of View*. Academia, Prague.

Langston, R. L., and J. A. Powell. 1975. The earwigs of California (order Dermaptera). Bull. California Insect Survey **20**:1–25.

Larsen, O. 1966. On the morphology and function of the locomotor organs of the Gyrinidae and other Coleoptera. Opusc. Entomol. Suppl. **30**:1–242.

Lavialle, M., and B. Dumortier. 1975. Mise en evidence de deux caractères inhabituels dans les rhythmes circadiens: existence d'un photorécepteur extraoculaire et particularités du libre cours dans le rhythme de ponte du phasme: *Carausius morosus*. C. R. Acad. Soc. Paris **281**:1489–1492.

Lavine, M. D., and N. E Beckage. 1995. Polydnaviruses: potent mediators of host immune dysfunction. Parasitol. Today **11**:368–378.

Lawrence, J. F. 1993a. *Beetle Larvae of the World*. Commonwealth Scientific and Industrial Research Organization, Division of Entomology, Canberra, Australia.

Lawrence, J. F. 1993b. *Beetle Larvae of the World: Interactive Identification and Information Retrieval for Families and Sub-families*. Commonwealth Scientific and Industrial Research Organization, Division of Entomology, Canberra, Australia.

Lawrence, J. F. 1982. Coleoptera. In *Synopsis and Classification of Living Organisms*, vol. 2, S. P. Parker (ed.). McGraw-Hill, New York. pp. 482–553.

Lawrence, J. F. 1989. Mycophagy in the Coleoptera: feeding strategies and morphological adaptations. In *Insect-Fungus Interactions*, N. Wilding (ed.). Royal Entomological Society of London no. 14. pp. 1–23.

Lawrence, J. F., and E. B. Britton. 1991. Coleoptera (beetles). In *The Insects of Australia*, vol. 2, Commonwealth Scientific and Industrial Research Organization (ed.). Melbourne University Press, Carlton. pp. 543–683.

Lawrence, J. F., and E. B. Britton. 1994. *Australian Beetles*. Melbourne University Press, Carlton.

Lawrence, J. F., and A. F. Newton, Jr. 1982. Evolution and

classification of beetles. Annu. Rev. Ecol. Syst. **13**:261–290.

Lawrence, P. A. 1992. *The Making of a Fly: The Genetics of Animal Design*. Blackwell Scientific, Oxford.

Lawrence, R. F. 1953. *The Biology of the Cryptic Fauna of Forests*. A. A. Balkema, Cape Town, South Africa.

Layne, J. N. 1971. Fleas (Siphonaptera) of Florida. Florida Entomol. **54**:35–51.

Leather, S. R., Y. Bassettt, and B. A. Hawkins. Insect conservation: finding the way forward. Insect Conserv. Diver. **1**:67–69.

Le Ceheric, F., M-T. Guillam, F. Beuron, A. Cavalier, D. Thomas, J. Gouranton, and J. F. Hubert. 1997. Aquaporin-related proteins in the filter chamber of homopteran insects. Cell Tissue Res. **290**:143–151.

Lee, K. E., and T. G. Wood. 1971. *Termites and Soils*. Academic Press, London.

Leech, H. B., and H. P. Chandler. 1956. Aquatic Coleoptera. In *Aquatic Insects of California*, R. L. Usinger (ed.). University of California Press, Berkeley. pp. 293–371.

Leech, H. B., and M. W. Sanderson, 1959. Coleoptera. In *Freshwater Biology*, W. T. Edmondson (ed.). John Wiley & Sons, New York. pp. 981–1023.

Lehane, M. J. 1991. *Biology of Blood-Sucking Insects*. HarperCollins Academic, London.

Leonard, D. E. 1974. Recent developments in ecology and control of the gypsy moth. Annu. Rev. Entomol. **19**:197–230.

Leonard, J. W., and F. A. Leonard. 1962. *Mayflies of Michigan Trout Streams*. Cranbrook Institute of Science, Bloomfield Hills.

Levin, M. D. 1983. Value of bee pollination to U.S. agriculture. Bull. Entomol. Soc. Am. **29**(4):50–51.

Lewis, R. E., J. H. Lewis, and C. Maser. 1988. *The Fleas of the Pacific Northwest*. Oregon State University Press, Corvallis, OR.

Lewis, T. 1973. *Thrips, Their Biology, Ecology and Economic Importance*. Academic Press, London.

Lewis, T. (ed.). 1984. *Insect Communication*. Academic Press, London.

Libersat, F., A. Delago, and D. Gal. 2009. Manipulation of host behavior by parasitic insects and insect parasites. Annu. Rev. Entomol. **54**:189–207.

Liebhold, A., V. Mastro, and P. W. Schaefer. 1989. Learning from the legacy of Leopold Trouvelot. Bull. Entomol. Soc. Am. **35**(2):20–22.

Linit, M. J. 1988 Nematode-vector relationships in the pine wilt disease system. J. Nematol. **20**:227–235.

Linsley, E. G. 1959a. Mimetic form and coloration in the Cerambycidae (Coleoptera). Ann. Entomol. Soc. Am. **52**:125–131.

Linsley, E. G. 1959b. Ecology of Cerambycidae. Annu. Rev. Entomol. **4**:99–138.

Linsley, E. G., and J. W. MacSwain. 1957. Observations on the habits of *Stylops pacifica* Bohart. Univ. Calif. Publ. Entomol. **11**:395–430.

Littig, K. S. 1942. External anatomy of the Florida walking stick *Anisomorpha buprestoides* Stoll. Florida Entomol. **25**:33–41.

Lloyd, J. E. 1971. Bioluminescent communication in insects. Annu. Rev. Entomol. **16**:97–122.

Lloyd, J. E. 1983. Bioluminescence and communication in insects. Annu. Rev. Entomol. **28**:131–160.

Lloyd, J. E. 1984. Occurrence of aggressive mimicry in fireflies. Florida Entomol. **67**:368–376.

Lloyd, J. L. 1921. The biology of North American caddis fly larvae. Lloyd Libr. Bot., Pharm. Mater. Med. Bull. **21**:1–124.

Lo, N., G. Tokuda, H. Watanabe, H. Rose, M. Slaytor, K. Maekawa, C. Banda, and H. Noda. 2000. Evidence from multiple gene sequences indicates that termites evolved from wood-feeding cockroaches. Curr. Biol. **10**:801–804.

Lockwood, J. A., L. D. Debrey, C. D. Thompson, C. M. Love, R. A. Nunamaker, S. R. Shaw, S. P. Schell, and C. R. Bomar. 1994. Preserved insect fauna of glaciers of Fremont County in Wyoming: insights into the ecology of the extinct rocky mountain locust. Environ. Entomol. **23**:220–235.

Lockwood, J. A. 2004. *Locust*. Basic Books, New York.

Lockwood, J. L., M. F. Hoopes, and M. P. Marchetti. 2007. *Invasion Ecology*. Blackwell Publishing, Malden, MA.

Loher, W. 1960. The chemical acceleration of the maturation process and its hormonal control in the male of the desert locust. Proc. R. Entomol. Soc. London, Ser. B **153**:380–397.

Loher, W. 1972. Circadian control of stridulation in the cricket *Teleogryllus commodus* Walker. J. Comp. Physiol. **79**:173–190.

Loher, W., I. Ganjian, I. Kubo, D. Stanley-Samuelson, and S. S. Tobe. 1981. Prostaglandins: their role in egg-laying of the cricket *Teleogryllus commodus*. Proc. Natl. Acad. Sci. U.S.A. **78**:7835–7838.

Lomer, C. J., and C. Prior. 1992. *Biological Control of Locusts and Grasshoppers: Proceedings of a Workshop Held at the International Institute of Tropical Agriculture, Cotonou, Republic of Benin, 29 April–1 May 1991*. CAB. International, Wallingford, Oxon, UK.

Louw, G. N., and M. K. Seely. 1982. *Ecology of Desert Organisms.* Longmans Groups, London.

Lozier, J. C., and S. A. Cameron. 2009. Comparative genetic analyses of historical and comtemporary collections highlight contrasting demographic histories for the bumble bees *Bombus pensylvanicus* and *B. impatiens* in Illinois. Mol. Ecol. **18**:1875–1886.

Luan, Y.-X., J. M. Mallatt, R.-D. Xie, Y.-M. Tang, and W.-Y. Yin. 2005. The phylogenetic positions of three basal-hexapod groups (Protura, Diplura and Collembola) based on ribosomal RNA gene sequences. Mol. Biol. Evol. **22**:1579–1592.

Lundquist, John E., and B. J. Bentz, 2009. Bark beetles in a changing climate. *U.S. Forest Service Pacific Northwest Research Station General Technical Report PNW-GTR Issue:* 39–49.

Luscher, M. 1961. Social control of polymorphism in termites. Symp. R. Entomol. Soc. London **1**:57–67.

Lyal, C. H. C. 1985. Phylogeny and classification of the Psocodea with particular reference to the lice (Psocodea: Phthiraptera). Syst. Entomol. **10**:145–165.

Lyman, F. E. 1955. Seasonal distribution and life cycles of Ephemeroptera. Ann. Entomol. Soc. Am. **48**:380–391.

Mackay, R. J., and G. B. Wiggins. 1979. Ecological diversity in Trichoptera. Annu. Rev. Entomol. **24**:185–208.

MacKerras, I. M. 1970. Evolution and classification of the insects. In *The Insects of Australia,* Commonwealth Scientific and Industrial Research Organization (ed.). Melbourne University Press, Carlton. pp. 152–167.

MacLeod, E. G., and P. A. Adams. 1967. A review of the taxonomy and morphology of the Berothidae, with a description of a new subfamily from Chile (Neuroptera). Psyche **74**:237–265.

Magurran, A. E., 2004. *Measuring Biological Diversity.* Blackwell Publishing, Malden, MA. p. 256.

Mahy, B. W. J. 2004. Vector-borne diseases. In Microbe-vector Interactions in Vector-borne Diseases, S. H. Gillespie, G. L. Smith, and A. Osbourne (eds.). Cambridge University Press, West Nyack, NY. pp. 1–18.

Malcolm, S., and P. Zalucki. 1993. *Biology and Conservation of the Monarch Butterfly.* Natural History Museum of Los Angeles, Los Angeles.

Mallis, A. 1971. *American Entomolo gists.* Rutgers University Press, New Brunswick, NJ.

Malyshev, S. I. 1968. *Genesis of t he Hymenoptera and Phases of Their Evolution.* Methuen, London.

Mangan, B. P. 1994. Pupation ecology of the dobsonfly *Corydalus cornutus* (Corydalidae: Megaloptera) along a large river. J. Freshwater Ecol. **9**:57–62.

Mani, M. S. 1962. *Introduction to High Altitude Entomology.* Methuen, London.

Mani, M. S. 1968. *Ecology and Biogeography of High Altitude Insects.* Series Entomologica. Junk, The Hague.

Mani, M. S., and L. E. Giddings. 1980. *Ecology of Highlands.* Monographiae Biologicae. Junk, The Hague.

Manton, S. M. 1964. Mandibular mechanisms and the evolution of arthropods. Philos. Trans. R. Soc. London, Ser. B **247**:1–183.

Manton, S. M. 1972. The evolution of arthropod locomotory mechanisms. Part 10. Locomotory habits, morphology and evolution of the hexapod classes. Zool. J. Linn. Soc. **51**:203–400.

Manton, S. M. 1973. Arthropod phylogeny—a modern synthesis. J. Zool. **171**:111–130.

Manuel, K. L., and R. M. Bohart. 1993. First report of a twisted-wing insect (Strepsiptera) larva in a caddisfly (Trichoptera). Entomol. News **104**:139.

Marden, J. H., and M. G. Kramer. 1994. Surface-skimming stoneflies: a possible intermediate stage in insect flight evolution. Science **266**:427–430.

Margulis, L. and R. Fester (eds.). 1991. *Symbiosis as a Source of Evolutionary Innovation: Speciation and Morphogenesis.* MIT Press, Cambridge, MA.

Markin, G. P. 1970. The seasonal life cycle of the Argentine ant, *Iridomyrmex humilis* (Hymenoptera: Formicidae). in southern California. Ann. Entomol. Soc. Am. **63**:1238–1242.

Marquardt, W. C (ed.). 2005. *Biology of Disease Vectors.* 2nd ed. Elsevier Academic Press, San Diego, CA.

Marshall, A. G. 1981. *The Ecology of Ectoparasitic Insects.* Academic Press, London.

Marshall, A. G. 1987. Nutritional ecology of ectoparasitic insects. In *Nutritional Ecology of Insects, Mites. Spiders, and Related Invertebrates,* F. Slansky. Jr. and J. G. Rodriguez (eds.). John Wiley & Sons, New York. pp. 721–739.

Marshall, A. T. 1966. Histochemical studies on a muco-complex in the Malpighian tubules of cercopid larvae. J. Insect Physiol. **12**:925–932.

Martin, J. E. H. 1977. *Collecting, Preparing and Preserving Insects, Mites and Spiders.* Insects and Arachnids of Canada Series, Agriculture Canada, Ottawa, ON.

Martin, M. 1934. Life history and habits of the pigeon

louse [*Columbicola columbae* (Linn.)]. Can. Entomol. **66**:1–16.

Martin, M. M. 1987. *Invertebrate-Microbial Interactions.* Explorations in Chemical Ecology. Cornell University Press, Ithaca, NY.

Martynov, A. V. 1925. Über zwei Grundtypen des Flugel bei den Insekten und ihre Evolution. Zh. Morphol. Oekol. Tiere **4**:465–501.

Martynova, O. 1961. Palaeoentomology. Annu. Rev. Entomol. **6**:285–294.

Matsuda, R. 1965. Morphology and evolution of the insect head. Mem. Am. Entomol. Inst. **1**:1–334.

Matsuda, R. 1970. Morphology and evolution of the insect thorax. Mem. Entomol. Soc. Can. **76**:1–431.

Matthews, G. A., and J. P. Tunstall (eds.). 1994. *Insect Pests of Cotton.* CAB International, Oxon, UK.

Matthews, R. W. 1974. Biology of Braconidae. Annu. Rev. Entomol. **19**:15–32.

Matthews, R. W., and J. R. Matthews. 1978. *Insect Behavior.* John Wiley & Sons, New York.

Matthysse, J. G. 1946. Cattle lice, their biology and control. Cornell Univ. Agric. Exp. Stn. Bull. **823**:1–67.

May, R. M. 1988. How many species are there on earth? *Science* **241**:441–1449.

Mayhew, P. J. 2007. Why are there so many insect species? Perspectives from fossils and phylogenies. Biol. Rev. **82**:425–454.

Maynard, E. A. 1951. *A Monograph of the Collembola, or Springtails, of New York State.* Comstock Publishing Associates, Ithaca, NY.

Mazzini, M., and V. Scali. 1987. *Stick Insects: Phylogeny and Reproduction.* University of Siena, Bologna.

McAlpine, J. F., B. V. Peterson, G. E. Shewell. H. J. Teskey, J. R. Vockeroth, and D. M. Wood (eds.). 1981–89. *Manual of Nearctic Diptera.* Canadian Govt. Publications, Agriculture Canada, Quebec. Monographs 27, 28, 32; 3 vols.

McCafferty, W. P. 1998. *Aquatic Entomology: The Fishermen's and Ecologists' Illustrated Guide to Insects and Their Relatives.* Jones and Bartlett, Sudbury, MA.

McGavin, G., and K. Preston-Mafham. 1993. *Bugs of the World.* Facts on File, New York.

McIver, J. D., and G. Stonedahl. 1993. Myrmecomorphy: morphological and behavioral mimicry of ants. Annu. Rev. Entomol. **38**:351–379.

McKenzie, H. L. 1967. *The Mealybugs of California.* University of California Press, Berkeley.

McKeown, K. C., and V. H. Mincham. 1948. The biology of an Australian mantispid (*Mantisspa vittata* Guerin). Aust. Zool. **11**:207–224.

McKittrick, F. A. 1964. Evolutionary studies of cockroaches. Cornell Univ. Agric. Exp. Stn., Mem. **389**:1–197.

Mead-Briggs, A. R. 1977. The European rabbit, the European rabbit flea and myxomatosis. Appl. Biol. **2**:183–261.

Meinander, M. 1972. A revision of the family Coniopterygidae (Planipennia). Acta Zool. Fenn. **136**:1–357.

Melzer, R. R., H. F. Paulus, and N. P. Kristensen. 1994. The larval eye of nannochoristid scorpionflies (Insecta, Mecoptera). Acta Zool. (Copenhagen) **75**:201–208.

Membiela, P. 1990. The mating calls of *Perla madritensis* Rambur, 1842 (Plecoptera, Perlidae). Aquat. Insects **12**:223–226.

Menzel, R., and A. Mercer (eds.). 1987. *Neurobiology and Behavior of Honeybees.* Springer-Verlag, Berlin.

Merritt, R. W., and K. W. Cummins. 2006. *An Introduction to the Aquatic Insects of North America.* Kendall/Hunt, Dubuque, IA.

Metcalf, R. L., and W. H. Luckmann (eds.). 1982. *Introduction to Insect Pest Management.* 2nd ed. John Wiley & Sons, New York.

Metcalf, R. L., and R. A. Metcalf. 1993. *Destructive and Useful Insects: Their Habits and Control.* 5th ed. McGraw-Hill, New York.

Michener, C. D. 1944. Comparative external morphology, phylogeny, and a classification of the bees (Hymenoptera). Bull. Am. Mus. Nat. Hist. **82**:151–326.

Michener, C. D. 1953. Comparative morphological and systematic studies of bee larvae with a key to the families of hymenopterous larvae. Univ. Kansas Sci. Bull. **35**:987–1102.

Michener, C. D. 1969. Comparative social behavior of bees. Annu. Rev. Entomol. **14**:299–342.

Michener, C. D. 1974. *The Social Behavior of the Bees.* Harvard University Press, Cambridge, MA.

Michener, C. D. 1979. Biogeography of the bees. Ann. Missouri Bot. Garden **66**:277–347.

Michener, C. D. 2007. *Bees of the World.* 2nd ed. Johns Hopkins University Press, Baltimore, MD.

Michener, C. D., and D. A. Grimaldi. 1988. The oldest fossil bee: apoid history, evolutionary stasis, and antiquity of social behavior. Proc. Natl. Acad. Sci. U.S.A. **85**:6424–6426.

Michener, C. D., and S. F. Sakagami. 1990. Classification of the Apidae (Hymenoptera). Univ. Kansas Sci. Bull. **54**(4):76–164.

Michener, C. D., R. J. McGinley, and B. N. Danforth. 1994. *The Bee Genera of North and Central America (Hymenoptera: Apoidea).* Smithsonian Institution Press, Washington, DC.

Mickoleit, G. 1973. Über den Ovipositor des Neuropteriodea und Coleoptera und seine phylogenetische Bedeutung (Insecta, Holometabola). Z. Morphol. Tiere **74**:37–64.

Miller, D. R., and M. Kosztarab. 1979. Recent advances in the study of scale insects. Annu. Rev. Entomol. **24**:1–27.

Miller, L. K. 1969. Freezing tolerance in an adult insect. Science **166**:105–106.

Miller, N. C. E. 1956. *Biology of the Heteroptera.* Leonard Hill, London.

Mills, H. B. 1932. The life history and thoracic development of *Oligotoma texana* (Mel.). Ann. Entomol. Soc. Am. **35**:648–652.

Mills, H. B. 1934. *A Monograph of the Collembola of Iowa.* Iowa State College Press, Ames.

Minks, A., and P. Harrewijn. 1987. *Aphids: Their Biology, Natural Enemies and Control.* Elsevier Science Publishers, Amsterdam.

Miskimen, G. W., N. L. Rodriguez, and M. L. Nazario. 1983. Reproductive morphology and sperm transport facilitation and regulation in the female sugarcane borer, *Diatrea saccharalis* (F.) (Lepidoptera: Crambidae). Ann. Entomol. Soc. Am. **76**:248–252.

Mitchell, B. K. 2003. Chemoreception. In *Encyclopedia of Insects,* V. Resh and R. Cardé (eds.). Academic Press, San Diego, CA. pp. 169–174.

Mitchell, R. W. 1969. A comparison of temperate and tropical cave communities. Southwest. Nat. **14**:73–88.

Mitchell, T. B. 1960–1962. Bees of eastern United States. 2 vols. North Carolina. Agric. Exp. Stn. Tech. Bull. **141**:1–538.

Mitton, J. B., and K. B. Sturgeon (eds.). 1982. *Bark Beetles in North American Conifers. A System for the Study of Evolutionary Biology.* University of Texas Press, Austin.

Mizuta, K. 1988a. Adult ecology of *Ceriagrion melanurum* Selys and *Ceriagrion nipponicum* Asahina (Zygoptera: Coenagrionidae): 1. Diurnal variations in reproductive behavior. Odonatologica (Utrecht) **17**:195–204.

Mizuta, K. 1988b. Adult ecology of *Ceriagrion melanurum* Selys and *Ceriagrion nipponicum* Asahina (Zygoptera: Coenagrionidae): 2. Movements and distribution. Odonatologica (Utrecht) **17**:357–364.

Mockford, E. L. 1951. The Psocoptera of Indiana. Proc. Indiana Acad. Sci. **60**:192–204.

Mockford, E. L. 1957. Life history studies on some Florida insects of the genus *Archipsocus* (Psocoptera). Bull. Florida State Mus. **1**:253–274.

Mockford, E. L. 1993. *North American Psocoptera (Insecta).* Flora & Fauna Handbook no. 10. Sandhill Crane Press. Gainesville, FL.

Mockford, E. L., and A. B. Gurney. 1956. A review of the psocids or book-lice and barklice of Texas (Psocoptera). J. Washington Acad. Sci. **46**:353–368.

Monteiro, A. 2008. Alternative models for the evolution of eyespots and of serial homology on lepidopteran wings. Bioessays **30**:358–366.

Montgomery, S. L. 1983. Carnivorous caterpillars: the behavior, biogeography and conservation of *Eupithecia* in

the Hawaiian Islands. Geojour. **7.6**:549–556.

Moore, J. C., and D. E. Walter. 1988. Arthropod regulation of micro- and mesobiota in below-ground detrital food webs. Annu. Rev. Entomol. **33**:419–439.

Moore, J. M., and M. D. Picker. 1991. Heuweltjies (earth mounds) in the Clanwilliam district, Cape Province, South Africa: 4000-year-old termite nests. Oecologia (Heidelburg) **86**:424–432.

Moore, T. E., W. Loher, and F. Huber. 1989. *Cricket Behavior and Neurobiology.* Comstock Publishing Associates, Ithaca, NY.

Moran, N. A. 1992. The evolution of aphid life cycles. Annu. Rev. Entomol. **37**:321–348.

Moran, N. A., and P. H. Degnan. 2006. Functional genomics of *Buchnera* and the ecology of aphid hosts. Mol. Ecol. **15**:1251–1261.

Morgan, F. D. 1968. Bionomics of Siricidae. Annu. Rev. Entomol. **13**:239–256.

Moritz, R. F. A., and E. E. Southwick. 1992. *Bees as Superorganisms: An Evolutionary Reality.* Springer-Verlag, Berlin.

Morse, J. C. (ed.). 1984. *Proceedings, IV International Symposium on Trichoptera, Clemson University, South Carolina.* Junk, The Hague.

Morse, J. C. 1997. Phylogeny of Trichoptera. Annu. Rev. Entomol. **42**:427–450.

Morse, J. C., and R. W. Holzenthal. 1996. Trichoptera genera. In *An Introduction to the Aquatic Insects of North America,* R. W. Merritt and K. W. Cummins (eds.). Kendall/Hunt, Dubuque, IA. pp. 350–386.

Morton, D. B., and M. L. Hudson 2002. Neurotransmitter transporters in the insect nervous system. Adv. Insect Physiol. **31**:55–150.

Mosher, E. 1916. A classification of the Lepidoptera based on characters of the pupa. Bull. Illinois Nat. Hist. Survey **12**:15–159.

Mound, L. A. 2009. Thysanoptera. In *Encyclopedia of Insects*, V. Resh and R. Cardé (eds.). Academic Press, San Diego, CA. pp. 999–1003.

Mound, L. A., B. S. Heming, and J. M. Palmer. 1980. Phylogenetic relationships between the families of recent Thysanoptera (Insecta). Zool. J. Linn. Soc. **69**:111–141.

Mukerji, S. 1927. On the morphology and bionomics of *Embia minor* sp. n., with special reference to its spinning organs. Rec. Indian Mus., Calcutta **29**:253–282.

Mukhopadhyay, A. 1993. Chemical ecology of some seed-feeding Hemiptera. In *Chemical Ecology of Phytophagous Insect.*, T. N. Ananthakrishnan and A. A. Raman (eds.). International Science Publisher, New York. pp. 179–195.

Mulkern, G. B. 1967. Food selection by grasshoppers. Annu. Rev. Entomol. **12**:59–78.

Mullen, G., and L. Durden (eds.). 2009. *Medical and Veterinary Entomology*. 2nd ed. Academic Press, New York.

Müller, F. 1879. *Ituna* and *Thyridia*; a remarkable case of mimicry in butterflies. Proc. Entomol. Soc. London:xx–xxix.

Murray, D. R. 2003. *Seeds of Concern: The Genetic Manipulation of Plants*. CABI Publishing, Wallington, UK.

Murray, M. D. 1960. The ecology of lice on sheep. I-II. Aust. J. Zool. **5**:13–29, 173–187.

Murray, M. D. 1987. Arthropods—the pelage of mammals as an environment. Int. J. Parasitol. **17**:191–195.

Murrell, A., and S. C. Barker. 2005. Multiple origins of parasitism in lice: phylogenetic analysis of SSU rDNA indicates that the Phthiraptera and Psocoptera are not monophyletic. Parasitol. Res. **97**:274–280.

Mutanen, M., N. Wahlberg, and L. Kaila. 2010. Comprehensive gene and taxon coverage elucidates radiation patterns in moths and butterflies. Proc. R. Soc. Series B **277**:2839–2848.

Myers, J. G. 1929. *Insect Singers. A Natural History of the Cicadas*. George Routledge and Sons, London.

Nagy, L., and M. Grbic. 2003. Embryogenesis. In *Encyclopedia of Insects*, V. Resh and R. Cardé (eds.). Academic Press, San Diego, CA. pp. 359–364.

Narang, S. K., A. C. Bartlett, and R. M. Faust (eds.). 1993. *Applications of Genetics to Arthropods of Biological Control Significance*. CRC Press, Boca Raton, FL.

Nardi, J. B. 2007. *Life in the Soil: A Guide for Naturalists and Gardeners*. University of Chicago Press, Chicago, IL.

Naskrecki, P. 2005. *The Smaller Majority*. Belknap Press, Cambridge, MA.

Nation, J. L. 2008. *Insect Physiology and Biochemistry*. CRC Press, Boca Raton, FL.

National Research Council. 2000. *Genetically Modified Pest-Protected Plants*. National Academy Press, Washington, DC.

Nault, L. R., and J. G. Rodriguez (eds.). 1985. *The Leafhoppers and Planthoppers*. Wiley, New York.

Needham, G. R., R. E. Page, Jr., M. Delfinado-Baker, and C. E. Bowman (eds.). 1988. *Africanized Honey Bees and Bee Mites*. Ellis Harwood Ltd., Chichester, UK.

Needham, J. G., and P. W. Claassen. 1925. A monograph of the Plecoptera or stoneflies of America north of Mexico. Thomas Say Found. Publ. **2**:1–397.

Needham, J. G., and M. J. Westfall, Jr. 1955. *A Manual of Dragonflies of North America (Anisoptera)*. University of California Press, Berkeley.

Needham, J. G., J. R. Traver, and Y.-C. Hsu. 1935. *The Biology of Mayflies*. Comstock Publishing Associates, Ithaca, NY.

Nelson, C. H. 2008. Hierarchical relationships of North American states and provinces: an area cladistic analysis based on the distribution of stoneflies (Insecta: Plecoptera). *Illiesia* **4**(18):176–204.

Neumann, D. 1976. Adaptations of chironomids to intertidal environments. Annu. Rev. Entomol. **21**:387–414.

New, T. R. 1975. The biology of Chrysopidae and Hemerobiidae (Neuroptera), with reference to their usage as biocontrol agents: a review. Trans. R. Entomol. Soc. London **127**:115–140.

New, T. R. 1987. Biology of the Psocoptera. Orient. Insects **21**:1–109.

New, T. R. 1991a. *Butterfly Conservation*. Oxford University Press, Oxford.

New, T. R. 1991b. *Insects as predators*. Australian Studies in Biology Series 5. New South Wales University Press, Kensington, Australia.

New, T. R., and G. Theischinger. 1993. *Megaloptera (Alderflies, Dobsonflies)*. Handbuch der Zoologie. Band IV, Arthropoda: Insecta. T.33. de Gruyter, Berlin.

Newcomer, E. G. 1918. Some stoneflies injurious to vegetation. J. Agric. Res. **13**:37–41.

Nichols, S. W. 1989. *The Torre-Bueno Glossary of Entomology*. The New York Entomological Society in cooperation with the American Museum of Natural History, New York.

Nicolas, G., and D. Sillans. 1989. Immediate and latent

effects of carbon dioxide on insects. Annu. Rev. Entomol. **34**:97–116.

Nieh, J. C. 1993. The stop signal of honey bees: reconsidering its message. Behav. Ecol. Sociobiol. **33**:51–56.

Nielsen, E. S., and I. F. B. Common. 1991. Lepidoptera (moths and butterflies). In *The Insects of Australia.* vol. 2, Commonwealth Scientific and Industrial Research Organization (ed.). Melbourne University Press, Carlton. pp. 817–915.

Niering, W. A. 1966. *The Life of the Marsh.* McGraw-Hill, New York.

Nijhout, H. F. 1991. *The Development and Evolution of Butterfly Wing Patterns.* Smithsonian Institution Press, Washington, DC.

Nijhout, H. F. 1994. *Insect Hormones.* Princeton University Press, Princeton, NJ.

Nilsson, D.-E. 1989. Optics and evolution of the compound eye. In *Facets of Vision*, G. V. Stavenga and R. C. Hardie (eds.). Springer, Berlin. pp. 30–73.

Noble-Nesbitt, J. 1989. Spiracular closing mechanisms in the firebrat, *Thermobia domestica* (Packard) (Thysanura). Tissue Cell **21**:93–100.

Norris, D. M. 1986. Anti-feeding compounds. In *Chemistry of Plant Protection*, vol. 1. *Sterol Biosynthesis Inhibitors and Anti-feeding Compounds.* G. Haug and H. Hoffmann (eds.). Springer-Verlag, Berlin. pp. 97–146.

Norris, K. R. 1965. The bionomics of blow flies. Annu. Rev. Entomol. **10**:47–68.

Norris, R. F., E. P. Caswell-Chen, and M. Kogan. 2003. *Concepts in Integrated Pest Management.* Prentice Hall, Upper Saddle River, NJ.

Nur, U. 1980. Evolution of unusual chromosome systems in scale insects (Coccoidea: Homoptera). In *Insect Cytogenetics*, R. L. Blackman (ed.). Blackwell, Oxford. pp. 97–117.

O'Donnell, M. 2008. Insect excretory mechanisms. Adv. Insect Physiol. **35**:1–122.

Ogden, T. H., and M. F. Whiting. 2005. Phylogeny of Ephemeroptera (mayflies) based on molecular evidence. Mol. Phylogene. Evol. **37**:625–643.

Oliver, K. M., J. A. Russell, N. A. Moran, and M. S. Hunter. 2003. Facultative bacterial symbionts in aphids confer resistance to parasitic wasps. Proc. Natl. Acad. Sci. USA **100**:1803–1807.

O'Neill, K. M. 2001. *Solitary Wasps: Behavior and Natural History.* Comstock Publishing, Ithaca, NY.

O'Neill, S. L., H. A. Rose, and D. Rugg. 1987. Social behavior and its relationship to field distribution in *Panesthia cribrata* Saussure (Blattodea: Blaberidae). J. Aust. Entomol. Soc. **26**:313–321.

O'Toole, C., and A. Raw. 1991. *Bees of the World.* Blandford. London.

Ohnishi, E., and H. Ishizaki (eds.). 1990. *Molting and Metamorphosis.* Springer-Verlag, Berlin.

Oken, L. 1831. *Lehrbuch der Naturphilosophie.* 2nd ed. Jena, Germany.

Oldroyd, H. 1964. *The Natural History of Flies.* Weidenfeld and Nicolson, London.

Oliver, D. R. 1971. Life history of the Chironomidae. Annu. Rev. Entomol. **16**:211–230.

Opler, P. A., and G. O. Krizek. 1984. *Butterflies East of the Great Plains.* Johns Hopkins University Press, Baltimore.

Opler, P. A., and V. Malikul. 1992. *Eastern Butterflies.* Peterson Field Guides, Houghton Mifflin, New York.

Oseto, C. Y., and T. J. Helms. 1976. Anatomy of the adults of *Loxagrotis albicosta.* Univ. Nebraska Stud. **52**:1–127.

Ossiannilsson, F. 1949. Insect drummers: a study of the morphology and function of the sound-producing organs of Swedish Homoptera, Auchenorrhyncha. Opuse. Entomol. Suppl. **10**:1–146.

Oswald, J. D. 1993a. Revision and cladistic analysis of the world genera of the family Hemerobiidae (Insecta: Neuroptera). J. New York Entomol. Soc. **101**:143–299.

Oswald, J. D. 1993b. Phylogeny, taxonomy, and biogeography of extant silky lacewings (Insecta: Neuroptera: Psychopsidae). Mem. Am. Entomol. Soc. **40**:1–65.

Otte, D. 1981a. *The North American Grasshoppers.* vol. 1. *Acrididae: Gomphocerinae and Acridinae.* Harvard University Press, Cambridge, MA.

Otte, D. 1981b. *The North American Grasshoppers.* vol. 2. *Acrididae: Oedipodinae.* Harvard University Press. Cambridge, MA.

Pages, J. 1989. Abdominal sclerites and appendages of Diplura (Insecta, Apterygota). Arch. Sci. (Geneva) **42**:509–552.

Painter, R. H. 1951. *Insect Resistance to Crop Plants.* University of Kansas Press. Lawrence.

Painter, R. H. 1958. Resistance of plants to insects. Annu. Rev. Entomol. **3**:267–290.

Pakaluk, J., and S. A. Slipinski. 1995. *Biology, Phylogeny, and Classification of Coleoptera: Papers Celebrating the 80th Birthday of Roy A. Crowson.* Muzeum i Instytut Zoologii PAN, Warsaw. 2 vols.

Palmer, J. M., L. A. Mound, G. J. Du Heaume, and C. R. Betts. 1989. *Thysanoptera*. CIE Guides to Insects of Importance to Man, vol. 2. CAB International, Wallingford, UK.

Pannabecker, T. 1995. Physiology of the Malpighian tubule. Annu. Rev. Entomol. **40**:493–510.

Papaj, D. R., and A. C. Lewis (eds.). 1993. *Insect Learning. Ecological and Evolutionary Perspectives.* Chapman & Hall, New York.

Parfin, S. 1952. The Megaloptera and Neuroptera of Minnesota. Am. Midl. Nat. **47**:421–434.

Pasteels, J. M., J.-C. Gregoire, and M. Rowell-Rahier. 1983. The chemical ecology of defense in arthropods. Annu. Rev. Entomol. **28**:263–289.

Pasteur, G. 1982. A classificatory review of mimicry systems. Annu. Rev. Entomol. **13**:169–199.

Paulson, D. R. 1974. Reproductive isolation in damselflies. Syst. Zool. **23**:40–49.

Payne, T. L., M. C., Birch, and C. E. J. Kennedy. 1989. *Mechanisms of Insect Olfaction.* Clarendon Press, Oxford.

Pearman, J. V. 1928. On sound-production in the Psocoptera and on a presumed stridulatory organ. Entomol. Mon. Mag. **64**:179–186.

Pearson, D. L. 1988. Biology of tiger beetles. Annu. Rev. Entomol. **33**:123–147.

Pearson, D. L., and F. Cassola. 1992. World-wide species richness patterns of tiger beetles (Coleoptera: Cicindelidae): indicator taxon for biodiversity and conservation studies. Conserv. Biol. **6**:376–391.

Peck, O., Z. Boucek, and A. Hoffer. 1964. Keys to the Chalcidoidea of Czechoslovakia. Mem. Entomol. Soc. Can. **34**:1–120.

Pedigo, L. P. 1989. *Entomology and Pest Management.* Macmillan. Inc., New York.

Pellmyr, O., J. Leebens-Mack, and C. J. Huth. 1996. Non-mutualistic yucca moths and their evolutionary consequences. Nature **380**:155–156.

Penny, N. D. 1975. Evolution of the extant Mecoptera. J. Kansas Entomol. Soc. **48**:331–350.

Peters, W. L., and P. G. Peters (eds.). 1973. *Proceedings of the First International Conference on Ephemeroptera.* E. J. Brill, Leiden.

Petersen, H. 1994. A review of collembolan ecology in ecosystem context. Acta Zool. Fenn. **195**:111–118.

Peterson, A. 1948. *Larvae of Insects.* Part I. *Lepidoptera and Plant Infesting Hymenoptera.* Edwards Brothers, Ann Arbor, MI.

Peterson, A. 1951. *Larvae of Insects.* Part II. *Coleoptera, Diptera, Neuroptera, Siphonaptera, Mecoptera, Trichoptera.* Edwards Brothers, Ann Arbor, MI.

Pfadt, R. E. (ed.). 1985. *Fundamentals of Applied Entomology.* Macmillan, New York.

Picker, M., C. Griffiths, and A. Weaving. 2004. *Field Guide to Insects of South Africa.* Struik, Cape Town, South Africa.

Pickett, J. A., L. J. Wadhams, C. M. Woodcock, and J. Hardie. 1992. The chemical ecology of aphids. Annu. Rev. Entomol. **37**:67–90.

Pinder, L. C. V. 1986. Biology of freshwater Chironomidae. Annu. Rev. Entomol. **31**:1–23.

Platt, A. P. 1983. Evolution of North American admiral butterflies. Bull. Entomol. Soc. Am. **29**(3):11–22.

Plumstead, E. P. 1963. The influence of plants and environment on the developing animal life of Karroo times. S. Afr. J. Sci. **59**:147–152.

Pohl, H. 2009. The oldest fossil strepsipteran larva (Insecta:Strepsiptera) from the Geisel Valley, Germany (Eocene). Insect Syst. Evol. **40**:333–347.

Poinar, G. O., Jr. 1979. *Nematodes for Biological Control of Insects.* CRC Press, Boca Raton, FL.

Poinar, G. O., Jr. 1983. *The Natural History of Nematodes.* Prentice-Hall, Englewood Cliffs, NJ.

Poinar, G. O. Jr. 1988. *Zorotypus palaeus*, new species, a fossil Zoraptera (Insecta) in Dominican amber. J. New York Entomol. Soc. **96**:253–259.

Poinar, G. O., Jr. 1992. *Life in Amber.* Stanford University Press, Palo Alto, CA.

Poinar, G. O., Jr. 1993. Insects in amber. Annu. Rev. Entomol. **39**:145–159.

Poisson, R., and P. Pesson. 1951. Super-ordre des Hemipteroides (Hemiptera Linné, 1758, Rhynchota Bermeister, 1835). Traite Zool. **10**:1385–1803.

Polcyn, D. M. 1994. Thermoregulation during summer activity in Mojave desert dragonflies (Odonata: Anisoptera). Funct. Ecol. **8**:441–449.

Polhemus, J. T. 1996. Aquatic and semiaquatic Hemiptera. In *An Introduction to the Aquatic Insects of North America*, R. W. Merritt and K. W. Cummins (eds.). Kendall/Hunt, Dubuque, IA. pp. 267–297.

Popham, E. J. 1965. A key to the Dermaptera subfamilies. Entomologist **98**:126–136.

Poulson, D. F. 1950. Histogenesis, organogenesis and differentiation in the embryo of *Drosophila melanogaster* (Meigen). In *Biology of Drosophila*, M. Demerec (ed.). John Wiley & Sons, New York. pp. 168–274.

Powell, J. A. 1964. Biological and taxonomic studies on tortricine moths, with reference to the species in California. Univ.

California Publ. Entomol. **32**:1–317.

Powell, J. A. 1980. Evolution of larval food preferences in microlepidoptera. Annu. Rev. Entomol. **25**:133–160.

Powell, J. A. 1984. Biological inter-relationships of moths and *Yucca schottii*. Univ. California Publ. Entomol. **100**:1–93.

Powell, J. A. 1992. Interrelationships of yuccas and yucca moths. Trends Ecol. Evol. **7**:10–15.

Powell, J. A., and R. A. Mackie. 1966. Biological interrelation-ships of moths and *Yucca whipplei*. Univ. California Publ. Entomol. **42**:1–59.

Preston-Mafham, K., and R. Preston-Mafham. 1990. *Grasshoppers and Mantids of the World*. Facts on File, New York.

Prestwich, G. D. 1983. Chemical systematics of termite exo-crine secretions. Annu. Rev. Ecol. Syst. **14**:287–312.

Prestwich, G. D. 1984. Defense mechanisms of termites. Annu. Rev. Entomol. **29**:201–232.

Prezler, R. W., and P. W. Price. 1988. Host quality and sawfly populations: a new approach to life table analysis. Ecology **69**:2012–2020.

Price, P. W. (ed.). 1974. *Evolutionary Strategies of Parasitic Insects and Mites*. Plenum Press, New York.

Price, P. W. 1984. *Insect Ecology*. 2nd ed. John Wiley & Sons, New York.

Price, P. W., W. J. Mattson, and Y. N. Baranchikov 1994. *The Ecology and Evolution of Gall-Forming Insects*. General technical report NC, 174. U.S. Dept of Agriculture, Forest Service, North Central Forest Experiment Station, St. Paul, MN.

Pringle, J. A. 1938. A contri-bution to the knowledge of *Micromalthus debiliis*

Lec. (Coleoptera). Trans. R. Entomol. Soc. London. **87**:271–286.

Pritchard, G. 1983. Biology of Tipulidae. Annu. Rev. Entomol. **28**:1–22.

Proctor, M., and P. Yeo. 1973. *The Pollination of Flowers*. William Collins Sons, London.

Prokopy, R. J. 1991. Epideictic pher-omones that influence spac-ing patterns of phytophagous insects. In *Semiochemicals: Their Role in Pest Control*, D. A. Nordlund, R. L. Jones, and W. J. Lewis (eds.). John Wiley & Sons, New York. pp. 181–197.

Purcell, A. H. 1989. Homopteran transmission of xyleminhab-iting bacteria. In *Advances in Disease Vector Research*, vol. 6, K. F. Harris (ed.). Springer-Verlag, New York. pp. 243–266.

Purcell, A. H. 2009. Plant disease and insects. In *Encyclopedia of Insects*, V. Resh and R. Cardé (eds.). Elsevier Academic Press, San Diego, CA. pp 302–302.

Pyle, R. M. 1981. *The Audubon Society Field Guide to North American Butterflies*. Alfred A. Knopf, New York.

Quicke, D. L. J. 1997. *Parasitic Wasps*. Chapman and Hall, London.

Quicke, D. L. J. 1999. Preservation of hymenopteran specimens for subsequent molecular and morphological study. Zool. Scr. **28**:261–267.

Quicke, D. L. J., S. N. Ingram, H. S. Baillie, and P. V. Gaitens. 1992. Sperm structure and ultra-structure in the Hymenoptera (Insecta). Zool. Scripta **21**:381–402.

Raffa, K. F., T. W. Phillips, and S. M. Salom. 1993. Strategies and mechanisms of host colonization by bark beetles. In *Beetle-Pathogen Interactions in Conifer*

Forests. T. D. Schowalter and G. M. Filip (eds.). Academic Press, London. pp. 103–128.

Raikhel, A. S., and T. S. Dhadialla. 1992. Accumulation of yolk proteins in insect oocytes. Annu. Rev. Entomol. **37**:217–251.

Rainey, R. C. 1965. The origin of insect flight: some implica-tions of recent findings from palaeoclimatology and locust migration. Proc. 12th Int. Congr. Entomol. **1964**:134.

Ramsay, G. W. 1990. *Mantodea (Insecta): With a Review of Aspects of Functional Morphology and Biology*. Fauna of New Zealand no. 19. DSIR Pub., Wellington, N.Z.

Rasnitsyn, A. P., and D. L. J. Quicke. 2002. *History of Insects*. Kluwer Scientific, Dordrecht, Netherlands.

Rathman, R. J., R. D. Akre, and J. F. Brunner. 1988. External morphology of a species of *Metajapyx* (Diplura: Japygidae) from Washington [USA]. Pan-Pac, Entomol. **64**:185–192.

Redak, R. A., A. H. Purcell, J. R. S. Lopes, M. J. Blua, R. F. Mizell III, and P. C. Andersen. 2003. The biology of xylem fluid-feeding insect vectors and their relation to disease epide-miology. Annu. Rev. Entomol. **49**:243–270.

Redborg, Kurt E., and Ellis G. MacLeod. 1985. The developmental ecology of *Mantispa uhleri* Banks (Neuroptera:Mantispidae). Illinois Biolo. Monog. **53**:1–130.

Regier, J. C., J. W. Schultz, and R. E. Kambic. 2004. Phylogeny of basal hexapod lineages and estimates of divergence times. Ann. Entomol. Soc. Am. **97**:411–419.

Regier, J. C., A. Zwick, M. P. Cummings, A. Y. Kawahara, S. Cho, S. Weller, A. Roe, J.

Baixeras, J. W. Brown, C. Parr, D. R. Davis, M. Epstein, W. Hallwachs, A. Hausmann, D. H. Janzen, I. J. Kitching, M. A. Solis, S.-H. Yen, A. L. Bazinet, and C. Mitter. 2009. Towards reconstructing the evolution of advanced moths and butterflies (Lepidoptera: Ditrysia): an initial molecular study. BMC Evol. Biol. **9**:208.

Rehn, A. C. 2003. Phylogenetic analysis of high-level relationships of Odonata. Syst. Entomol. **28**:181–239.

Rehn, J. A. G., and H. J. Grant, Jr. 1961. A monograph of the Orthoptera of North America (North of Mexico), vol. I. Monogr. Acad. Natl. Sci., Philadelphia **12**:1–255.

Rehn, J. W. H. 1939. Studies in North American Mantispidae (Neuroptera). Trans. Am. Entomol. Soc. **65**:237–263.

Rehn, J. W. H. 1950. A key to the genera of North American *Blattaria*, including established adventives. Entomol. News **61**:64–67.

Rehn, J. W. H. 1951. Classification of the *Blattaria* as indicated by their wings (Orthoptera). Mem. Am. Entomol. Soc. **14**:1–134.

Reitze, M., and W. Nentwig. 1991. Comparative investigations into the feeding ecology of six Mantodea species. Oecologia (Heidelburg) **86**:568–574.

Rembold, H. 1991. Roles of morphogenetic hormones in caste polymorphism in stingless bees. In *Morphogenetic Hormones of Arthropods*, vol. 3, A. P. Gupta (ed.). Rutgers University Press, New Brunswick. NJ. pp. 325–345.

Remington, C. L. 1954. The suprageneric classification of the order Thysanura (Insecta). Ann. Entomol. Soc. Am. **47**:277–286.

Remington, C. L. 1956. The "Apterygota." In *A Century of Progress in the Natural Sciences, 1853–1953*, E. L. Kessel (ed.). California Academy of Science, San Francisco. pp. 495–505.

Rentz, D. C. F., and D. K. Kevan. 1991. Dermaptera. In *The Insects of Australia: A Textbook for Students and Researchers*. Cornell University Press, Ithaca, NY. pp. 360–368.

Resh, V. R., and D. M. Rosenberg (eds.). 1984. *The Ecology of Aquatic Insects*. Praeger Publishers, New York.

Retnakaran, A., and K. Binnington. 1991. *Physiology of the Insect Epidermis*. Commonwealth Scientific and Industrial Research Organization, East Melbourne. Victoria.

Ribeiro, J. M. C. 1987. Role of saliva in blood-feeding by arthropods. Annu. Rev. Entomol. **32**:463–478.

Richards, O. W. 1956. *Hymenoptera. Introduction and Key to Families*. Royal Entomological Society of London, Handbook for Identifying British Insects, no. 6, London.

Richards, O. W., and R. G. Davies. 1977. *Imms's General Textbook of Entomology*. 10th ed. Chapman & Hall, London.

Richards, W. R. 1968. Generic classification, evolution, and biogeography of the Sminthuridae of the world (Collembola). Mem. Entomol. Soc. Can. **53**:1–54.

Richardson, J. W., and A. R. Gaufin. 1971. Food habits of some western stonefly nymphs. Trans. Am. Entomol. Soc. **97**:91–121.

Richman, D. B. 1993. *A Manual of the Grasshoppers of New Mexico (Orthoptera: Acrididae and Romaleidae)*. New Mexico State University Cooperative Extension Service Handbook. New Mexico State University, Las Cruces.

Ricker, W. E. 1952. Systematic studies in Plecoptera. Indiana Univ. Stud. Sci. Ser. **18**:1–200.

Ricker, W. E. 1959. Plecoptera. In *Freshwater Biology*, W. T. Edmondson (ed.). John Wiley & Sons., New York. pp. 941–957.

Riddiford, L. M. 1995. Hormonal regulation of gene expression during lepidopteran development. In *Molecular Model Systems in the Lepidoptera*, M. R. Goldsmith and A. S. Wilkins (eds.). Cambridge University Press, Cambridge. pp. 293–322.

Riegel, G. T. 1963. The distribution of *Zorotypus hubbardi* (Zoraptera). Ann. Entomol. Soc. Am. **56**:744–747.

Riek, E. F., and J. Kukalová-Peck. 1984. A new interpretation of dragonfly wing venation, based upon early Upper Carboniferous fossils from Argentina (Insecta: Odonatoidea) and basic character states in pterygote wings. Can. J. Zool. **62**:1150–1166.

Rienks, J. H. 1985. Phenotypic response to photoperiod and temperature in a tropical pierid butterfly. Aust. J. Zool. **33**:837–847.

Riley, C. V. 1892. The yucca moth and yucca pollination. Rept. Missouri Bot. Gdn. **3**:99–158.

Ritcher, P. O. 1958. Biology of Scarabaeidae. Annu. Rev. Entomol. **3**:311–334.

Ritchie, M., and D. Pedgley. 1989. Desert locusts cross the Atlantic. Antenna **13**:10–12.

Ritland, D. B. 1994. Variation in palatability of queen butterflies *(Danaus gilippus)* and implications regarding mimicry. Ecology **75**:732–746.

Ritland, D. B. 1995. Comparative unpalatability of mimetic viceroy butterflies *(Limenitis archippus)* from four southeastern United States populations. Oecologia **103**:327–336.

Ritland, D. B., and L. P. Brower. 1991. The viceroy butterfly is not a Batesian mimic. Nature **350**:497.

Robert, A. 1962. *Les Libellules du Québec*. Ministère du Tourisme, de la Chasse et de la Pêche, Province de Québec.

Roberts, D. B. 1986. *Drosophila: A Practical Approach*. The Practical Approach Series. IRL Press, Oxford [Oxfordshire].

Roberts, M. J. 1971. The structure of the mouthparts of some calyptrate dipteran larvae in relation to their feeding habits. Acta Zool. **52**:171–188.

Robinson, G. E. 1992. Regulation of division of labor in insect societies. Annu. Rev. Entomol. **37**:637–665.

Robinson, G. E. 1996. Chemical communication in honeybees. Science **271**:1824–1825.

Roeder, K. D. 1935. An experimental analysis of the sexual behavior of the praying mantis *(Mantis religiosa,* L.). Biol. Bull. **69**:203–220.

Roeder, K. D. 1965. Moths and ultrasound. Sci. Am. **212**:94–102.

Roeder, K. D. 1966. Auditory system of noctuid moths. Science **154**:1515–1521.

Rohdendorf, B. B. 1944. A new family of Coleoptera from the Permian of the Urals. C. R. Acad. Sci. URSS **44**:252–262.

Rohdendorf, B. B. 1969. Phylogenie. Handb. Zool. **4**:1/4.

Romeis, J., R. G. Van Driesche, B. I. P. Barratt, and F. Bigler. 2008. Insect-resistant transgenic crops and biological control. In *Integration of Insect-Resistant Genetically Modified Crops within IPM Programs*, J. Romeis, A. M. Shelton, and G. G. Kennedy (eds.). Springer, New York, pp. 87–118.

Romoser, W. S. and J. G. Stoffolano, Jr. 1998. *The Science of Entomology*, 4th ed. WGB McGraw-Hill, Boston, MA.

Room, P. M. 1990. Ecology of a simple pant-herbivore system: biological control of *Salvinia*. Trends Ecol. Evol. **5**:74–79.

Roonwal, M. L. 1962. Recent developments in termite systematics (1949–60). In *Proceedings of the New Delhi Symposium, 1960, Termites in the Humid Tropics*. UNESCO, Paris. pp. 31–50.

Rosen, D. 1990. *Armored Scale Insects: Their Biology, Natural Enemies, and Control*. World Crop Pests, 4. Elsevier, Amsterdam. 2 vols.

Rosenthal, G. A. 1986. The chemical defenses of higher plants. Sci. Am. **254**(1):94–99.

Ross, E. S. 1940. A revision of the Embioptera of North America. Ann. Entomol. Soc. Am. **33**:629–676.

Ross, E. S. 1944. A revision of the Embioptera, or webspinners, of the new world. Proc. U.S. Natl. Mus. **94**:401–504.

Ross, E. S. 1957. The Embioptera of California. Bull. California Insect Survey **6**:51–57.

Ross, E. S. 1970. Biosystematics of the Embioptera. Annu. Rev. Entomol. **15**:157–172.

Ross, E. S. 1984a. A synopsis of the Embiidina of the United States. Proc. Entomol. Soc. Washington **86**:82–93.

Ross, E. S. 1984b. A classification of the Embiidina of Mexico with descriptions of new taxa. Occas. Pap. California Acad. Sci. **140**:1–54.

Ross, E. S. 1991. Embioptera, Embiidina (Embiids, webspinners, foot-spinners). In *The Insects of Australia*, vol. 1. Commonwealth Scientific and Industrial Research Organization (ed.). Melbourne University Press, Carlton. pp. 405–409.

Ross, H. H. 1937a. A generic classification of Nearctic sawflies (Hymenoptera, Symphyta). Illinois Biol. Monogr. **15**:1–173.

Ross, H. H. 1937b. Studies of Nearctic aquatic insects. I. Nearctic alder flies of the genus *Sialis* (Megaloptera, Sialidae). Illinois Nat. Hist. Survey Bull. **21**:57–78.

Ross, H. H. 1944. The caddisflies, or Trichoptera, of Illinois. Bull. Illinois Nat. Hist. Survey **23**:1–326.

Ross, H. H. 1956. *Evolution and Classification of the Mountain Caddisflies*. University of Illinois Press, Urbana.

Ross, H. H. 1959. Trichoptera. In *Freshwater Biology*, W. T. Edmondson (ed.). John Wiley & Sons, New York. pp. 1024–1049.

Ross, H. H. 1964. Evolution of caddisworm cases and nets. Am. Zool. **4**:209–220.

Ross, H. H. 1967. The evolution and past dispersal of Trichoptera. Annu. Rev. Entomol. **12**:169–206.

Ross, H. H., and W. E. Ricker. 1971. The classification, evolution and dispersal of the winter stonefly genus Allocapnia. Ill. Biol. Monogr. No. **43**.

Ross, K. G., and R. W. Matthews. 1991. *The Social Biology of Wasps*. Comstock Publishing Associates, Ithaca, NY.

Roth, L. M. 1970. Evolution and taxonomic significance of reproduction in *Blattaria*. Annu. Rev. Entomol. **15**:75–96.

Roth, L. M., and H. B. Hartman. 1967. Sound production and its evolutionary significance in the *Blattaria*. Ann. Entomol. Soc. Am. **60**:740–752.

Roth, L. M., and E. K. Willis. 1960. The biotic associations of cockroaches. Smithsonian Misc. Collect. **141**:1–470.

Rothschild, M. 1965. The rabbit flea and hormones. Endeavor **24**:162–168.

Rothschild, M. 1973. Secondary plant substances and warning colouration in insects. Symp. R. Entomol. Soc. London **6**:59–83.

Rothschild, M. 1975. Recent advances in our knowledge of the order Siphonaptera. Annu. Rev. Entomol. **20**:241–259.

Rothschild, M. 1985. Mimicry, butterflies and plants. Symb. Bot. Uppsala **22**:82–99.

Rothschild, M., and T. Clay. 1952. *Fleas, Flukes and Cuckoos: A Study of Bird Parasites.* William Collins Sons, London.

Roubik, D. W. 1988. *Ecology and Natural History of Tropical Bees.* Cambridge University Press, Cambridge.

Rousch, D. K., and J. A. McKenzie. 1987. Ecological genetics of insecticide and acaricide resistance. Annu. Rev. Entomol. **32**:361–380.

Rowell, C. H. F. 1971. The variable coloration of the acridoid grasshoppers. Adv. Insect Physiol. **8**:146–199.

Rudd, R. L. 1964. *Pesticides in the Living Landscape.* University of Wisconsin Press, Madison.

Rudinsky, J. A. 1962. Ecology of the Scolytidae. Annu. Rev. Entomol. **7**:327–348.

Rudolph, D. 1982. Occurrence, properties and biological implications of the active uptake of water vapor from the atmosphere in Psocoptera. J. Insect Physiol. **28**:111–121.

Rudolph, D. 1983. The water uptake system of the Phthiraptera. J. Insect Physiol. **29**:15–25.

Rugg, D., and H. A. Rose. 1991. Biology of *Macropanesthia rhinoceros* Saussure (Dictyoptera: Blaberidae). Ann. Entomol. Soc. Am. **84**:575–582.

Russell, L. K. 1982. The life history of *Caurinus dectes* Russell, with a description of the immature stages (Mecoptera: Boreidae). Entomol. Scand. **13**:225–235.

Sacca, G. 1964. Comparative bionomics in the genus *Musca.* Annu. Rev. Entomol. **9**:341–358.

Sailer, R. I. 1950. A thermophobic insect. Science **112**:743.

Sakai, S. 1982. A new proposed classification of the Dermaptera with special reference to the check list of the Dermaptera of the world. Bull. Daito Bunka University **20**:1–108.

Salmon, J. T. 1951. Keys and bibliography to the Collembola. Victoria Univ. College Publ. Zool., Wellington, NZ No. **8**:1–82.

Salmon, J. T. 1956. Keys and bibliography to the Collembola. First supplement. Victoria Univ. College. Publ. Zool., Wellington, NZ. No. **20**:1–35.

Salt, G. 1970. *The Cellular Defense Reactions of Insects.* Cambridge University Press, New York.

Samways, M. J. 2005. *Insect Diversity Conservation.* Cambridge University Press, Cambridge, UK.

Saux, C., C. Simon, and G. S. Spicer. 2003. Phylogeny of the dragonfly and damselfly order Odonata as inferred by mitochondrial 12S ribosomal RNA sequences. Ann. Entomol. Soc. Am. **96**:693–699.

Schaefer, C. W. (ed.) 1996. *Studies on Hemipteran Phylogeny.* Thomas Say Found. Publ. Proc., Entomol. Soc. Amer., Lanham, MD.

Schaefer, C. W. 2009. Prossorhyncha. In *Encyclopedia of Insects*, V. Resh and R. Cardé (eds.). Academic Press, San Diego, CA. pp. 839–855.

Schal, C., J.-Y. Gautier, and W. J. Bell. 1984. Behavioral ecology of cockroaches. Biol. Rev. **59**:209–254.

Schaller, F. 1971. Indirect sperm transfer by soil arthropods. Annu. Rev. Entomol. **16**:407–446.

Scharrer, B. 1951. The woodroach. Sci. Am. **185**:58–62.

Schneider, D. 1992. 100 years of pheromone research: An essay on Lepidoptera. Naturwissenschaften **79**:241–250.

Schoonhoven, M., J. J. A. van Loon, and M. Dicke. 2005. *Insect-Plant Biology.* Oxford University Press, Oxford.

Schowalter, T. D., and G. M. Filip (eds.). 1993. *Beetle-Pathogen Interactions in Conifer Forests.* Academic Press, London.

Schuh, R. T., and J. H. Slater. 1995. *True Bugs of the World (Hemiptera: Heteroptera): Classification and Natural History.* Cornell University Press, Ithaca, NY.

Schultheis, A. S., A. C. Hendricks, and L. A. Weigt. 2002. Genetic evidence for 'leaky' cohorts in the semivoltine stonefly *Peltoperla tarteri* (Plecoptera: Peltoperlidae). Freshwater Biol. **47**:367–376.

Schwartz, M. P., M. H. Richards, and B. N. Danforth. 2007. Changing paradigms in insect social evolution: insights from halictine and allodapine bees. Annu. Rev. Entomol. **52**:127–150.

Scoble, M. J. 1986. The structure and affinities of the Hedyloidea: A new concept of the butterflies. Bull. Br. Mus. Nat. Hist. (Entomol.) **53**:251–286.

Scoble, M. J. 1992. *The Lepidoptera: Form, Function, and Diversity.* Natural History Museum Publications. Oxford University Press, Oxford.

Scott, H. G. 1961. Collembola: Pictorial keys to the Nearctic genera. Ann. Entomol. Soc. Am. **54**:104–113.

Scriber, J. M., and F. Slansky, Jr. 1981. The nutritional ecology of immature insects. Annu. Rev. Entomol. **26**:183–211.

Scudder, G. G. E. 1971. Comparative morphology of insect

genitalia. Annu. Rev. Entomol. **16**:379–406.

Seastedt, T. R. 1984. The role of microarthropods in decomposition and mineralization processes. Annu. Rev. Entomol. **29**:25–46.

Seeley, T. D. 1992. The tremble dance of the honey bee: message and meanings. Behav. Ecol. Sociobiol. **31**:375–383.

Seguy, E. 1950. La biologie de Diptères. Encycl. Entomol. **26**:1–609.

Seguy, E. 1951. Ordre des Dipteres (Diptera Linne, 1758). Traite Zool. **10**:449–744.

Service, M. 2008. *Medical Entomology for Students.* Cambridge University Press, New York. 289 pp.

Setty, L. R. 1940. Biology and morphology of some North American Bittacidae. Am. Midl. Nat. **23**:257–353.

Severin, H. H. P. 1911. The life-history of the walkingstick, *Diaphemeromera femorata* Say. J. Econ. Entomol. **4**:307–320.

Seybold, S. J., D. R. Quilici, J. A. Tillman, D. Vanderwel, D. L. Wood, and G. J. Blomquist. 1995. *De novo* biosynthesis of the aggregation pheromone components ipsenol and ipsdienol by the pine bark beetles *Ips paraconfusus* Lanier and *Ips pini* (Say) (Coleoptera: Scolytidae). Proc. Natl. Acad. Sci. U.S.A. **92**:8393–8397.

Shapiro, J. P., J. H. Law, and M. A. Wells. 1988. Lipid transport in insects. Annu. Rev. Entomol. **33**:297–318.

Sharov, A. G. 1966. *Basic Arthropodan Stock with Special Reference to Insects.* Pergamon Press, Oxford.

Sharov, A. G. 1971. *Phylogeny of the Orthopteroidea.* Israel Program for Scientific Translations, Jerusalem.

Sharov, A. G. 1975. The phylogenetic relations of the order

Thysanoptera. Entomol. Rev. **51**:506–508.

Shattuck, S. O. 1992. Review of the dolichoderine ant genus *Iridomyrmex* Mayr with descriptions of three new genera (Hymenoptera: Formicidae). J. Aust. Entomol. Soc. **31**:13–18.

Shear, W. A., and Kukalová-Peck. 1990. The ecology of paleozoic terrestrial arthropods: the fossil evidence. Can. J. Zool. **68**:1807–1834.

Shear, W. A., P. M. Bonamo, J. D. Grierson, W. D. I. Rolfe, E. L. Smith, and R. H. Norton. 1984. Early land animals in North America: evidence from Devonian age arthropods from Gilboa, New York. Science **224**:492–494.

Sheppard, P. M. 1962. Some aspects of the geography, genetics and taxonomy of a butterfly. In *Taxonomy and Geography*, D. Nichols (ed.). Systematics Association, London. Publ. 4. pp. 135–152.

Shetlar, D. J. 1978. Biological observations on *Zorotypus hubbardi* Caudell (Zoraptera). Entomol. News **89**:217–223.

Siemann, E., D. Tilman, and J. Haarstad. 1996. Insect species diversity, abundance and body size relationships. *Nature* **380**:704–706.

Silberglied, R. E., and T. Eisner. 1968. Mimicry of Hymenoptera by beetles with unconventional flight. Science **163**:486–488.

Simberloff, D. 2009. Introduced insects. In *Encyclopedia of Insects*, V. Resh and R. Cardé (eds.). Academic Press, San Diego, California. pp. 529–533.

Skaife, S. H. 1961. *Dwellers in Darkness.* Doubleday, Garden City, NY.

Skarlato, Orest Aleksandrovich. 1985. *Systematics of Diptera (Insecta): Ecological and*

Morphological Principles. Oxonian Press, New Delhi.

Slabaugh, R. E. 1940. A new thysanuran, and a key to the domestic species of Lepismatidae (Thysanura) found in the United States. Entomol. News **51**:95–98.

Slater, J. A., and R. M. Baranowski. 1978. *How to Know the True Bugs (Hemiptera-Heteroptera).* Wm. C. Brown, Dubuque, IA.

Smart, J., and N. F. Hughes. 1973. The insect and the plant: progressive palaeoecological integration. In *Insect/Plant Relationships*, H. F. van Emden (ed.). John Wiley & Sons, New York. pp. 143–155.

Smart, J., R. J. Wootton, and R. A. Crowson. 1967. Insecta. In *The Fossil Record. A Symposium with Documentation*, W. B. Hailand (ed.). Geological Society, London. pp. 508–534.

Smith, C. F., R. W. Eckel, and E. Lampert. 1992. A key to many of the common alate aphids of North Carolina (Aphididae: Homoptera). North Carolina Agric. Res. Serv. Tech. Bull. **299**:1–92.

Smith, C. M. 1989. *Plant Resistance to Insects: A Fundamental Approach.* John Wiley & Sons, New York.

Smith, E. L. 1969. Evolutionary morphology of external insect genitalia. I. Origin and relationships to other appendages. Ann. Entomol. Soc. Am. **62**:1051–1079.

Smith, E. L. 1970. Biology and structure of some California bristletails and silverfish (Apterygota: Microcoryphia, Thysanura). Pan-Pac. Entomol. **46**:212–225.

Smith, K. G. V. 1986. *A Manual of Forensic Entomology.* British Museum (Natural History). London.

Smith, R. C. 1922. The biology of the Chrysopidae. Cornell

Univ. Agric. Exp. Stn. Mem. **58**:1291–1372.

Smith, R. F., T. E. Mittler, and C. N. Smith (eds.). 1973. *History of Entomology*, Annual Reviews, Palo Alto, CA.

Smith, R. I., and J. T. Carlton. 1975. *Light's Manual: Intertidal Invertebrates of the Central California Coast.* 3rd ed. University of California Press, Berkeley.

Smithers, C. N. 1990. Keys to the families and genera of Psocoptera (Arthropoda, Insecta). Tech. Rep. Aust. Mus. **2**:1–82.

Smithers, C. N., and C. Lienhard. 1992. A revised bibliography of the Psocoptera (Arthropoda: Insecta). Tech. Rep. Austr. Mus. **6**:1–86.

Snodgrass, R. E. 1935. *Principles of Insect Morphology*, McGraw-Hill, New York.

Snodgrass, R. E. 1944. The feeding apparatus of biting and sucking insects affecting man and animals. Smithsonian Misc. Collect. **104**:1–113.

Snodgrass, R. E. 1946. The skeletal anatomy of fleas (Siphonaptera). Smithsonian Misc. Collect. **104**:1–89.

Snodgrass, R. E. 1950. Comparative studies on the jaws of mandibulate arthropods. Smithsonian Misc. Collect. **116**:1–185.

Snodgrass, R. E. 1952. *A Textbook of Arthropod Anatomy.* Comstock Publishing Associates, Ithaca, NY.

Snodgrass, R. E. 1954. The dragonfly larva. Smithsonian Misc. Collect. **123**:1–38.

Snodgrass, R. E. 1956. *Anatomy of the Honey Bee.* Cornell University Press, Ithaca, NY.

Snodgrass, R. E. 1958. Evolution of arthropod mechanisms. Smithsonian Misc. Collect. **138**:1–77.

Snodgrass, R. E. 1963. A contribution toward an encyclopedia of insect anatomy. Smithsonian Misc. Collect. **146**:1–48.

Snyder, T. E. 1954. *Order Isoptera— The Termites of the United States and Canada.* National Pest Control Association, New York.

Snyder, T. E. 1956. Annotated, subject-heading bibliography of termites 1350 B.C. to A.D. 1954. Smithsonian Misc. Collect. **130**:1–305.

Snyder, T. E. 1968. Second supplement to the annotated, subject-heading bibliography of termites 1961–1965. Smithsonian Misc. Collect. **152**:1–188.

Solbrig, O. T., and D. J. Solbrig. 1979. *Population Biology and Evolution.* Addison-Wesley Publishing, Reading, MA.

Sommerman, K. M. 1943. Bionomics of *Ectopsocus pumilis* (Banks) (Corrodentia, Caeciliidae). Psyche **50**:53–63.

Sorensen, J. T. 1980. An integumental anatomy for the butterfly *Glaucopsyche lygdamus* (Lepidopteras Lycaenidae): a morphological terminology and homology. Zool. J. Linn. Soc. **70**:55–101.

Sorensen, J. T., B. C. Campbell, R. J. Gill, and J. D. Steffen-Campbell. 1995. Non-monophyly of Auchenorrhyncha ("Homoptera"), based upon 18S rDNA phylogeny: eco-evolutionary and cladistic implications within pre-Heteropterodea Hemiptera (S. L.) and a proposal for new monophyletic suborders. Pan-Pac. Entomol. **71**:31–60.

Sosa, O., Jr. 1989. Carlos J. Finlay and yellow fever: a discovery. Bull. Entomol. Soc. Am. **35**(2):23–25.

Southwood, T. R. E., and P. A. Henderson. 2000. *Ecological Methods.* 3rd ed. Blackwell Science, Oxford, UK.

Southwood, T. R. E., and D. Leston. 1959. *Land and Water Bugs of the British Isles.* Frederick Warne and Co., London.

Spangler, H. G. 1988. Moth hearing, defense, and communication. Annu. Rev. Entomol. **33**:59–82.

Spence, J. R., and N. M. Andersen. 1994. Biology of water striders: interactions between systematics and ecology. Annu. Rev. Entomol. **39**:101–128.

Spencer, K. A. 1990. *Host Specialization in the world Agromyzidae (Diptera).* Series Entomologica, vol. 45. Kluwer Academic Publishers, Dordrecht, Netherlands.

Spencer, K. A., and G. C. Steyskal. 1986. *Manual of the Agromyzidae (Diptera) of the United States.* Agriculture Handbook no. 638. U.S. Department of Agriculture, Agricultural Research Service, Washington, DC.

Spivak, M., D. J. C. Fletcher, and M. D. Breed (eds.). 1991. The *"African" Honey Bee.* Westview Press, Boulder, CO.

Spooner, C. S. 1938. The phylogeny of the Hemiptera based on a study of the head capsule. Illinois Biol. Monogr. **16**:1–102.

Spradberry, J. P. 1973. *Wasps. An Account of the Biology and Natural History of Social and Solitary Wasps.* University of Washington Press, Seattle.

Srygley, R.B., and A. L. R. Thomas. 2002. Unconventional lift-generating mechanisms in free-flying butterflies. Nature **420**:660–664.

St Clair, R. M. 1993. Life histories of six species of Leptoceridae (Insecta: Trichoptera) in Victoria. Aust. J. Mar. Freshwater Res. **44**:363–379.

Staddon, B. W. 1986. Biology of scent glands in the Hemiptera-Heteroptera. Ann. Soc. Entomol. France **22**:183–190.

Stamp, N. E., and T. M. Casey (eds.) 1993. *Caterpillars. Ecology and*

Evolutionary Constraints on Foraging. Chapman & Hall, NY.

Stannard, L. J., Jr. 1956. The relationship of the hemipteroid insects. Syst. Zool. **5**:94–95.

Stannard, L. J., Jr. 1957. The phylogeny and classification of the North American genera of the suborder Tubulifera (Thysanoptera). Illinois Biol. Monogr. **25**:1–200.

Stannard, L. J., Jr. 1968. The thrips, or Thysanoptera, of Illinois. Illinois Nat. Hist. Surv. Bull. **29**:215–552.

Starks, K. J., R. Muniappan, and R. D. Eikenbary. 1972. Interaction between plant resistance and parasitism against greenbug on barley and sorghum. Ann. Entomol. Soc. Am. **65**:650–655.

Steffan, W. A. 1966. A generic revision of the family Sciaridae (Diptera) of America north of Mexico. Univ. California Publ. Entomol. **44**:1–77.

Stehr, F. W. (ed.). 1987. *Immature Insects*. vol. 1. Kendall/Hunt, Dubuque, IA.

Stehr, F. W. (ed.). 1991. *Immature Insects*. vol. 2. Kendall/Hunt, Dubuque, IA.

Steinhaus, E. A. 1946. *Insect Microbiology*. Comstock Publishing Associates, Ithaca, NY.

Steinhaus, E. A. 1949. *Principles of Insect Pathology*. McGraw-Hill, New York.

Stephen, F. M., C. W. Berisford, D. L. Dahlsten, P. Fenn, and J. C. Moser. 1993. Invertebrate and microbial associates. In *Beetle-Pathogen Interactions in Conifer Forests*, T. D. Schowalter and G. M. Filip (eds.). Academic Press, London. pp. 129–153.

Stephen, W. P., G. E. Bohart, and P. F. Torchio. 1969. *The Biology and External Morphology of Bees with a Synopsis of the Genera of Northwestern America*. Oregon State University, Agricultural Experiment Station, Corvallis.

Stern, D. L., and W. A. Foster. 1996. The evolution of soldiers in aphids. Biol. Rev. Cambridge Philos. Soc. **71**:27–79.

Stewart, A. J. A., T. R. New, and O. T. Lewis. 2007. *Insect Conservation Biology*. CABI, Wallingford, UK.

Stewart, K. W. 2001. Vibrational communication (drumming) and mate-searching behavior of stoneflies (Plecoptera); evolutionary considerations. pp. 217–226. In *Trends in Research in Ephemeroptera and Plecoptera*, E. Dominguez (ed.). Kluwer Academic/Plenum Publishers, New York. pp. 217–226.

Stewart, K. W., and B. P. Stark. 2002. *Nymphs of North American Stonefly Genera (Plecoptera)*. 2nd ed. The Caddis Press, Columbus, OH.

Stewart, K. W., and B. P. Stark. 2007. Plecoptera. In *An Introduction to the Aquatic Insects of North America*. 4th ed., R. W. Merritt, K. W. Cummings, and M. B. Berg (eds.). Kendall/Hunt Publishing Co., Dubuque, Iowa. pp. 311–384.

Stewart, K. W., B. P. Stark, and J. A. Stanger. 1988. *Nymphs of North American Stonefly Genera (Plecoptera)*. Thomas Say Foundation (Series), vol. 12. Entomological Society of America, College Park, MD. (also reprinted, 1993, University of North Texas Press).

Stokes, D. W. 1983. *A Guide to Observing Insect Lives*. Little, Brown, Boston.

Stoltz, D. B., and J. B. Whitfield. 2009. Making nice with viruses. *Science* **323**: 884–885.

Stout, M. J. 2007. Types and mechanisms of rapidly induced plant resistance to herbivorous arthropods. In *Induced Resistance for Plant Defense*, D. Walters, A. Newton, and G. Lyon (eds.). Blackwell, Oxford, UK. pp. 89–107.

Stouthamer, R., R. F. Luck, and W. D. Hamilton. 1990. Antibiotics cause parthenogenetic *Trichogramma* to revert to sex. Proc. Natl. Acad. Sci. U.S.A. **87**:2424–2427.

Stowe, M. K., T. C. J. Turlings, J. H. Lougrin, W. J. Lewis, and J. H. Tumlinson. 1995. The chemistry of eavesdropping, alarm, and deceit. Proc. Nat. Acad. Sci. U.S.A. **92**:23–28.

Strohecker, H. F., W. W. Middlekauff, and D. C. Rentz. 1968. The grasshoppers of California. Bull. California Insect Survey **10**:1–171.

Strong, D. R., J. H. Lawton, and R. Southwood. 1984. *Insects on Plants. Community Patterns and Mechanisms*. Harvard University Press, Cambridge, MA.

Sturm, H. 1987. Das Paarungsverhalten von *Thermobia domestica* (Packard) (Lepismatidae, Zygentoma, Insecta). Braunschweig. Naturk. Schr. **2**:673–711.

Sturm, H. 1992. Mating behavior and sexual dimorphism in *Promesomachilis hispanica* Silvestri, 1923 (Machilidae, Archeognatha, Insecta). Zool. Anz. **228**:60–73.

Sturm, H., and C. Bach De Roca. 1993. On the systematics of the Archaeognatha (Insecta). Entomol. Gen. **18**:55–90.

Suarez, A. V., and N. D. Tsutsui. 2004. The value of museum collections for research and society. *Bioscience* **54**:66–74.

Sudd, J. H., and N. R. Franks. 1987. *The Behavioural Ecology of Ants. Tertiary Level Biology*. Blackie, Glasgow.

Sun, L., A. Sabo, M. D. Meyer, R. P. Randolph, L. M. Jacobus, W. P. McCafferty, and V. R. Ferris. 2006. Tests of current hypotheses of mayfly (Ephemeroptera) phylogeny using molecular (18s rDNA) data. Ann. Entomol. Soc. Am. **99**:241–252.

Svenson, G. J., and M. F. Whiting. 2004. Phylogeny of Mantodea based on molecular data: evolution of a charismatic predator. Syst. Entomol. **29**: 359–370.

Sweeney, B. W., D. H. Funk, and L. J. Standley. 1993. Use of the stream mayfly *Cloeon triangulifer* as a bioassay organism: life history response and body burden following exposure to technical chlordane. Environ. Toxicol. Chem. **12**:115–125.

Symmons, S. 1952. Comparative anatomy of the mallophagan head. Trans. Zool. Soc. London **27**:349–436.

Szumik, C., J. S. Edgerly, and C. Y. Hayashi. 2008. Phylogeny of embiopterans (Insecta). *Cladistics* **24**:993–1005.

Tabashnik, B. E. 1994. Evolution of resistance to *Bacillus thuringiensis*. Annu. Rev. Entomol. **39**:47–79.

Tallamy, D. W., and T. K. Wood. 1986. Convergence patterns in subsocial insects. Annu. Rev. Entomol. **31**:369–390.

Tanada, Y., and H. K. Kaya. 1993. *Insect Pathology.* Academic Press, New York.

Telfer, W. H., and J. G. Kunkel. 1991. The function and evolution of insect storage hexamers. Annu. Rev. Entomol. **36**:205–228.

Telford, M. J. and R. H. Thomas. 1998. Expression of homeobox genes shows chelicerate arthropods retain their deutocerebral segment. Proceedings of the National Academy of Sciences of the USA **95**:10671-10675.

Terra, P. S. 1992. Maternal care in *Cardioptera brachyptera* (Mantodea, Vatidae, Photininae). Rev. Bras. Entomol. **36**:493–503.

Terra, W. R. 1990. Evolution of digestive systems of insects. Annu. Rev. Entomol. **35**:181–200.

Terry, M. D., and M. F. Whiting. 2005. Mantophasmatodea and phylogeny of the lower neopterous insects. *Cladistics* **21**:240–257.

Thayer, M. K. 1992. Discovery of sexual wing dimorphism in Staphylinidae (Coleoptera): *"Omalium" flavidum*, and a discussion of wing dimorphism in insects. J. New York Entomol. Soc. **100**:540–573.

Thomas, P. A., and P. M. Room. 1986. Successful control of the floating weed *Salvinia molesta* in Papua New Guinea—a useful biological invasion neutralizes a disastrous one. Environ. Conserv. **13**:342–348.

Thompson, S. N., and S. J. Simpson. 2003. Nutrition. In *Encyclopedia of Insects*, V. Resh and R. Cardé (eds.). Academic Press, San Diego, CA. pp. 807–813.

Thorne, B. L. 1990. A case for ancestral transfer of symbionts between cockroaches and termites. Proc. R. Soc. London Ser. B. Biol. Sci. **241**:37–41.

Thorne, B. L. 1991. Ancestral transfer of symbionts between cockroaches and termites: an alternative hypothesis. Proc. R. Soc. London Ser. B Biol. Sci. **246**:191–196.

Thorne, B. L., and J. M. Carpenter. 1992. Phylogeny of the dictyoptera. Syst. Entomol. **17**:253–268.

Thornhill, R., and K. P. Sauer. 1991. The notal organ of the scorpionfly *(Panorpa vulgaris)*: an adaptation to coerce mating duration. Behav. Ecol. **2**:156–164.

Thornhill, R., and P. Sauer. 1992. Genetic sire effects on the fighting ability of sons and daughters and mating success of sons in a scorpionfly. Anim. Behav. **43**:255–264.

Thornton, I. W. B. 1985. The geographical and ecological distribution of arboreal

Psocoptera. Annu. Rev. Entomol. **30**:175–196.

Tichy, H. 1988. A kinematic study of front legs' movement in walking Protura (Insecta). J. Exp. Zool. **245**:130–136.

Tiegs, O. W., and S. M. Manton. 1958. The evolution of the Arthropoda. Biol. Rev. **33**:255–337.

Tietz, H. M. 1973. *An Index to the Described Life Histories, Early Stages, and Hosts of the Macrolepidoptera of the Continental United States and Canada.* E. W. Classey, Hampton, England.

Tilmon, K. (ed.). 2008. *Specialization, Speciation, and Radiation: The Evolutionary Biology of Herbivorous Insects.* University of California Press, Berkeley.

Timmermans, M. J. T. N., D. Roelofs, J. Marien, and N. M. van Straalen. 2008. Revealing pancrustacean relationships: phylogenetic analysis of ribosomal protein genes places Collembola (springtail) in a monophyletic Hexapoda and reinforces the discrepancy between mitochondrial and nuclear DNA markers. BMC Evol. Biol. **8**(83):1–10.

Tol, J. van, and M. J. Verdonk. 1988. *The Protection of Dragonflies (Odonata) and Their Biotopes.* Nature and Environment Series, no. 38. European Committee for the Conservation of Nature and Natural Resources, Strasbourg.

Townes, H. 1969. The genera of Ichneumonidae. Mem. Am. Entomol. Inst. **11**:1–300, **12**:10–537, **13**:1–307.

Traniello, J. F. A. 1989. Foraging strategies of ants. Annu. Rev. Entomol. **34**:191–210.

Traub, R., and H. Starcke (eds.). 1980. *Proceedings of the International Conference on Fleas, Ashton Wold, Peterborough, UK. 21–25 June 1977.* Balkema, Rotterdam.

Trivedi, M. C., B. S. Rawat, and A. K. Saxena. 1991. The distribution of lice (Phthiraptera) on poultry *(Gallus domesticus).* Int. J. Parasitol. **21**:247–250.

Troester, G. 1990. The head of *Hybophthirus notophallus* (Neumann) (Phthiraptera: Anoplura): an analysis of functional morphology and phylogeny. Stuttgart Beitr. Naturkd. Ser A (Biol.) **442**:1–89.

Trueman, J. W. H., B. E. Pfeil, S. A. Kelchner, and D. K. Yeates. 2004. Did stick insects really regain their wings? System. Entomol. **29**:138–139.

Truman, J. W. 1990. Neuroendocrine control of ecdysis. In *Molting and Metamorphosis*, E. Ohnishi and H. Ishizaki (eds.). Springer-Verlag, Berlin. pp. 67–82.

Truman, J. W. 1995. Lepidoptera as model systems for studies of hormone action on the central nervous system. In *Molecular Model Systems in the Lepidoptera*, M. R. Goldsmith and A. S. Wilkins (eds.). Cambridge University Press. Cambridge. pp. 323–339.

Truman, J. W. 1996. Ecdysis control sheds another layer. Science **271**:40–41.

Tschinkel, W. R. 1975a. A comparative study of the chemical defensive system of the tenebrionid beetles. I. Chemistry of the secretions. J. Insect Physiol. **21**:753–783.

Tschinkel, W. R. 1975b. A comparative study of the chemical defensive system of tenebrionid beetles. II. Defensive behavior and ancillary structures. Ann. Entomol. Soc. Am. **68**:439–453.

Turner, J. R. G. 1981. Adaptation and evolution in *Heliconius*: a defense of NeoDarwinism. Annu. Rev. Ecol. Syst. **12**:99–121.

Tuxen, S. L. 1959. The phylogenetic significance of entognathy in entognathous apterygotes. Smithsonian Misc. Collect. **137**:379–416.

Tuxen, S. L. 1964. *The Protura. A Revision of the Species of the World with Keys for Determination.* Hermann Press, Paris.

Tuxen, S. L. (ed.). 1970a. *Taxonomist's Glossary of Genitalia in Insects*, 2nd ed. rev. Ejnar Munksgaard, Copenhagen.

Tuxen, S. L. 1970b. The systematic position of entognathous Apterygotes. An. Esc. Nat. Cienc. Biol., Mex. **17**:65–69.

Tuxen, S. L. 1986. *Protura (Insecta).* Fauna of New Zealand, no. 9. Science Information Publishing Centre, DSIR, Wellington, N. Z.

Ulrich, W. 1966. Evolution and classification of the Strepsiptera. Proc. 1st Int. Congr. Parasitol. **1**:609–611.

Urbani, C. B. 1989. Phylogeny and behavioral evolution in ants with a discussion of the role of behavior in evolutionary processes. Ethol. Ecol. Evol. **1**:137–168.

Urbani, C. B. 1993. The diversity and evolution of recruitment behaviour in ants, with a discussion of the usefulness of parsimony criteria in the reconstruction of evolutionary histories. Insectes Soc. **40**:233–260.

Urquhart, F. A. 1987. *The Monarch Butterfly: International Traveler.* Nelson Hall, Chicago.

Usinger, R. L. (ed.). 1956a. *Aquatic Insects of California.* University of California Press, Berkeley.

Usinger, R. L. 1956b. Aquatic Hemiptera. In *Aquatic Insects of California*, R. L. Usinger (ed.). University of California Press, Berkeley. pp. 182–228.

Usinger, R. L. 1967. *The Life of Rivers and Streams.* McGraw-Hill, New York.

Uvarov, B. P. 1966. *Grasshoppers and Locusts. A Handbook of General Acridology.* vol 1. Anatomy, Physiology and Development, Phase Polymorphism, Introduction to Taxonomy. Cambridge University Press, Cambridge, MA.

van der Plank, J. E. 1963. *Plant Disease Epidemics and Control.* Academic Press, New York.

van Dreische, R. G., and T. S. Bellows (eds.). 1996. *Biological Control.* Chapman & Hall, New York.

van Driesche, R. G., and T. S. Bellows. 1993. *Steps in Classical Arthropod Biological Control.* Thomas Say Publications in Entomology. Proceedings. Entomological Society of America, Lanham, MD.

van Emden, H. F. (ed.). 1972. *Aphid Technology.* Academic Press, Inc., New York.

Vandel, A. 1965. *Biospeleology, the Biology of Cavernicolous Animals.* Pergamon Press. Oxford.

Vickery, V. R., D. K. McE. Kevan, and C. D. Dondale. 1985. *The Grasshoppers, Crickets, and Related Insects of Canada and Adjacent Regions: Ulonata, Dermaptera, Cheleutoptera, Notoptera, Dictuoptera, Grylloptera, and Orthoptera.* The Insects and Arachnids of Canada, pt. 14. Biosystematics Research Institute, Research Branch, Agriculture Canada, Ottawa.

Vigoreaux, J. O. 2006. *Nature's Versatile Engine: Insect Flight Muscle Inside and Out.* Springer Science, New York.

Villani, M. G., L. L. Allee, A. Díaz, and P. S. Robbins. 1999. Adaptive strategies of edaphic arthropods. Annu. Rev. Entomol. **44**:233–256.

Waage, J. K. 1979. Dual function of the damselfly penis: sperm removal and transfer. Science **203**:916–918.

Waage, J. K. 1984. Sperm competition and the evolution of odonate mating systems. In *Sperm Competition and the Evolution of Animal Mating Systems*, R. L. Smith (ed.). Academic Press, New York. pp. 251–290.

Wagner, M., and K. F. Raffa. 1993. *Sawfly Life History Adaptations to Woody Plants*. Academic Press, San Diego, CA.

Wajnberg, E., and S. A. Hassan. 1994. *Biological Control with Egg Parasitoids*. CAB International, Wallingford, Oxon, UK.

Waldbauer, G. 2003. *What Good Are Bugs? Insects in the Web of Life*. Harvard University Press, Cambridge, MA.

Waldorf, E. S. 1974. Sex pheromone in the springtail, *Sinella curviseta*. Environ. Entomol. **3**:916–918.

Walker, E. M. 1949. On the anatomy of *Grylloblatta campodeiformis* Walker. V. The organs of digestion. Can. J. Res. **27**:309–344.

Walker, E. M. 1953. *The Odonata of Canada and Alaska*, vol. 1, *General, the Zygoptera-Damselflies*. University of Toronto Press, Toronto, Canada.

Walker, E. M. 1958. *The Odonata of Canada and Alaska*, vol. 2, part III: *The Anisoptera—Four Families*. University of Toronto Press, Toronto, Canada.

Walker, E. M., and P. S. Corbet. 1975. *The Odonata of Canada and Alaska*, vol. 3, part III: *The Anisoptera—Three Families*. University of Toronto Press, Toronto, Canada.

Walker, T. J. 1963. The taxonomy and calling songs of the United States tree crickets (Orthoptera: Gryllidae: Oecanthinae). II. The *nigricornis* group of the genus *Oecanthus*. Ann. Entomol. Soc. Am. **56**:772–789.

Wallace, J. B., and R. W. Merritt. 1980. Filter-feeding ecology of aquatic insects. Annu. Rev. Entomol. **25**:103–132.

Wallwork, J. A. 1970. *Ecology of Soil Animals*. McGraw-Hill, New York.

Ward, J. V. 1992. *Aquatic Insect Ecology*. vol. 1. *Biology and Habitat*. John Wiley & Sons, New York.

Ward, J. V., and J. A. Stanford. 1982. Thermal responses in the evolutionary ecology of aquatic insects. Annu. Rev. Entomol. **27**:97–117.

Ware, G. W. 1989. *The Pesticide Book*. Thomson Publications, Fresno, CA.

Washburn, J. O., M. E. Gross, D. R Mercer, and J. R. Anderson. 1988. Predator-induced trophic shift of a free-living ciliate: parasitism of mosquito larvae by their prey. Science **240**:1193–1195.

Watson, J. A. L. 1965. The endocrine system of the lepismatid Thysanura and its phylogenetic implications. Proc. 12th Int. Congr. Entomol. **1964**:144.

Way, M. J. 1963. Mutualism between ants and honeydew-producing Homoptera. Annu. Rev. Entomol. **8**:307–344.

Way, M. J. 1968. Intra-specific mechanisms with special reference to aphid populations. In *Insect Abundance*, T. R. E. Southwood (ed.). Blackwell, Oxford. pp. 18–36.

Weaver, J. S., and J. C. Morse. 1986. Evolution of feeding and case-making behavior in Trichoptera. J. North Am. Benthol. Soc. **5**:150–158.

Weber, H. 1933. *Lehrbruch der Engomologie*. Gustav Fisher, Jena. 726 pp.

Wehner, R. 1984. Astronavigation in insects. Annu. Rev. Entomol. **29**:277–298.

Wenner, A. M. 1971. *The Bee Language Controversy, An Experience in Science*. Education Programs Improvement Corporation, Boulder, CO.

Wenner, A. M., and P. H. Wells. 1990. *Anatomy of a Controversy: The Question of a "Language" Among Bees*. Columbia University Press, New York.

Wenzel, J. W. 1990. A social wasp's nest from the Cretaceous period, Utah, USA, and its biogeographical significance. Psyche **97**:21–30.

Werren J. H. 1997. Biology of *Wolbachia*. Annu. Rev. Entomol. **42**:587–609.

West, L. S. 1951. *The Housefly*. Comstock Publishing Associates, Ithaca, NY.

Westbrook, A. L., and W. E. Bollenbacher. 1990. The prothoracicotropic hormone neuroendocrine axis in *Manduca sexta*: development and function. In *Molting and Metamorphosis*, E. Ohnishi and H. Ishizaki (eds.). Springer-Verlag, Berlin. pp. 3–16.

Westneat, M. W., O. Betz, R. W. Blob, K. Fezzaa, W. J. Cooper, and W-K. Lee. 2003. Tracheal respiration in insects visualized with synchrotron X-ray imaging. Science **299**:558–560.

Wheeler, W. C., R. T. Schuh, and R. Bang. 1993. Cladistic relationships among higher groups of Heteroptera: congruence between morphological and molecular data sets. Entomol. Scand. **24**:121–137.

Wheeler, W. C., M. F. Whiting, J. C. Carpenter, and Q. D. Wheeler. 2001. The phylogeny of the insect orders. Cladistics **17**:113–169.

Wheeler, W. M. 1930. *Demons of the Dust*. W. W. Norton, New York.

Whitcomb, R. F. 1981. The biology of spiroplasmas. Annu. Rev. Microbiol. **26**:397–425.

White, R. E. 1983. *A Field Guide to the Beetles*. Peterson Field Guide Series, Houghton Mifflin, Boston.

Whitfield, J. B. 1990. Parasitoids, polydnaviruses and endosymbiosis. Parasitology Today **6**:381–384.

Whitfield, J. B. 1998. Phylogeny and evolution of the host/parasitoid relationship in the Hymenoptera. Annu. Rev. Entomol. **43**:129–151.

Whitfield, J. B. 1999. Destructive sampling and information management in molecular systematic research: an entomological perspective. In *Managing the Modern Herbarium: An Interdisciplinary Approach*. Byers, S. and D. Metsger (eds.). Society for Preservation of Natural History Collections and Royal Ontario Museum. pp. 301–314.

Whitfield, J. B. 2003. Phylogenetic insights into the evolution of parasitism in Hymenoptera. In *The Evolution of Parasitism—A Phylogenetic Approach* (T. J. Littlewood, ed.). Adv. Parasitol. **54**:69–100.

Whiting, M. F. 1991. A distributional study of *Sialis* (Megaloptera: Sialidae) in North America. Entomol. News **102**:50–56.

Whiting, M. F. 1994. Cladistic analysis of the alderflies of America North of Mexico (Megaloptera: Sialidae). Syst. Entomol. **19**:77–91.

Whiting, M. F. 2002. Mecoptera is paraphyletic: multiple genes and phylogeny of Mecoptera and Siphonaptera. Zool. Scr. **31**:93–104.

Whiting, M. F., S. Bradler, and T. Maxwell. 2003. Loss and recovery of wings in stick insects. *Nature* **421**:264–267.

Whiting, M. F., and J. Kathirithamby. 1995. Strepsiptera do not share hind wing venation synapomorphies with Coleoptera: a reply to Kukalová-Peck and Lawrence. J. New York Entomol. Soc. **103**:1–14.

Whiting, M. F., and W. C. Wheeler. 1994. Insect homeotic transformation. *Nature* **368**:696.

Whiting, M. F., A. S. Whiting, M. W. Hastriter, and K. Dittmar. 2008. A molecular phylogeny of fleas (Insecta: Siphonaptera): origins and host associations. *Cladistics* **24**:1–31.

Whittaker, R. H., and P. P. Feeny. 1971. Allelochemics: chemical interactions between species. Science **171**:757–770.

Whitten, W. M., A. M. Young, and D. L. Stern. 1993. Nonfloral sources of chemicals that attract male euglossine bees (Apidae: Euglossini). J. Chem. Ecol. **19**:3017–3027.

Whitten, W. M., A. M. Young, and N. H. Williams. 1989. Function of glandular secretions in fragrance collection by male euglossine bees (Apidae: Euglossini). J. Chem. Ecol. **15**:1285–1296.

Whitman, D. W., and T. N. Ananthakrishnan (eds.). 2009. *Phenotypic Plasticity of Insects: Mechanisms and Consequences*. Science Publishers, Enfield, NH.

Wiebes, J. T. 1979. Co-evolution of figs and their insect pollinators. Annu. Rev. Ecol. Syst. **10**:1–12.

Wiegmann, B. M., M. D. Trautwein, J.-W. Kim, B. K. Cassel, M. A. Bertone, S. L. Winterton, and D. K. Yeates. 2009. Single-copy nuclear genes resolve the phylogeny of the holometabolous insects. BMC Biol. **7**:34.

Wiegmann, B. M., D. K. Yeates, J. L. Thorne, and H. Kishino. 2003. Time flies, a new molecular timescale for brachyceran fly evolution without a clock. Syst. Biol. **52**:745–756.

Wiens, J. J. (ed.). 2000. *Phylogenetic Analysis of Morphological Data*. Smithsonian Institution Press, Washington, DC.

Wiggins, G. B. 1996. *Larvae of the North American Caddisfly Genera (Trichoptera)*. University of Toronto Press, Toronto.

Wiggins, G. B. 1987. Order Trichoptera. In *Immature Insects*, vol. 1, F. Stehr (ed.). Kendall/Hunt, Dubuque, IA. pp. 253–287.

Wiggins, G. B., and W. Wichard. 1989. Phylogeny of pupation in Trichoptera, with proposals on the origin and higher classification of the order. J. North Am. Benthol. Soc. **8**:260–276.

Wigglesworth, V. B. 1972. *The Principles of Insect Physiology*, 7th ed. Chapman & Hall, London.

Wigglesworth, V. B. 1976. The evolution of insect flight. In *Insect Flight*, R. C. Rainey (ed.). John Wiley & Sons, New York. pp. 255–269.

Wilcox, J., N. Papavero, and T. Pimentel. 1989. *Studies of Mydidae (Diptera). IVb, Mydas and allies in the Americas (Mydinae, Mydini)*. Colecao Emilie Snethlage. SCT/PR, CNPq, Museu Paraense Emilio Goeldi, Belem-Para, Brazil.

Wild, A. L., and D. R. Maddison. 2008. Evaluating nuclear protein-coding genes for phylogenetic utility in beetles. Mol. Phylogenet. Evol. **48**:877–891.

Wilding, N., N. M Collins, P. M. Hammond, and J. F. Webber (eds.). 1989. *Insect-Fungus Interactions*. Academic Press, London.

Wilkins, A. S. 1993. *Genetic Analysis of Animal Development*. 2nd ed. Wiley-Liss, New York.

Will, K. W. 1995. Plecopteran surface-skimming and insect flight evolution. Science. **270**:1684–1685.

Wille, A. 1960. The phylogeny and relationships between the insect orders. Rev. Biol. Trop. **8**:93–123.

Willemstein, S. C. 1987. *An Evolutionary Basis for Pollination Ecology.* Leiden Botanical Series, vol. 10. E.J. Brill, Leiden.

Williams, D. D. 1987. *The Ecology of Temporary Waters.* Croom Helm Ltd., London.

Williams, D. D., and B. W. Feltmate. 1992. *Aquatic Insects.* CAB Int. Wallingford, Oxon, UK.

Williams, K. S., and C. Simon. 1995. The ecology, behavior, and evolution of periodical cicadas. Annu. Rev. Entomol. **40**:269–295.

Williams, M. A. J. 1994. *Plant Galls: Organisms, Interactions, Populations.* Oxford Science Publications. Oxford University Press, Oxford.

Williams, M. L., and M. Kosztarab. 1972. Morphology and systematics of the Coccidae of Virginia, with notes on their biology (Homoptera: Coccoidea). Virginia Polytech. Inst. State Univ. Res. Div. Bull. **74**:1–215.

William, R. 1987. The phylogenetic system of the Mecoptera. Syst. Entomol. **12**:519–524.

Wilson, E. O. 1971. *The Insect Societies.* Harvard University Press, Cambridge, MA.

Wilson, E. O. 1975. Slavery in ants. Sci. Am. **232**:232–236.

Wilson, E. O. 1985. The sociogenesis of insect colonies. Science **228**:1489–1495.

Wilson, E. O. 1987. The little things that run the world: the importance of conservation of invertebrates. Conserv. Biol. **1**:344–346.

Wilson, E. O. 1990. *Success and Dominance in Ecosystems: The Case of the Social Insects.* Excellence in Ecology, 2. Ecology Institute, Oldendorf/Luhe, Germany.

Wilson, E. O., F. M. Carpenter, and W. L. Brown. 1967. The first Mesozoic ants. Science **157**:1038–1040.

Wingstrand, K. G. 1973. The spermatozoa of the Thysanuran insects *Petrobius brevistylis* Carp. and *Lepisma saccharina* L. Acta Zool. **54**:31–52.

Winston, J. E. 1999. *Describing Species: Practical Taxonomic Procedure for Biologists.* Columbia University Press, New York.

Winston, M. L. 1987. *The Biology of the Honey Bee.* Harvard University Press, Cambridge, MA.

Winterton, S., and S. de Freitas. 2006. Molecular phylogeny of the green lacewings (Neuroptera: Chrysopidae). Aust. J. Entomol. **45**:235–243.

Winterton, S. L., N. B. Hardy, and B. M. Wiegmann. 2010. On wings of lace: phylogeny and Bayesian divergence time estimates of Neuropterida (Insecta) based on morphological and molecular data. Syst. Entomol. **35**:349–378.

Winterton, S. L., and V. N. Makarkin. 2010. Phylogeny of moth lacewings and giant lacewings (Neuroptera: Ithonidae, Polystoechotidae) using DNA sequence data, morphology, and fossils. Ann. Entomol. Soc. Am. **103**:511–522.

Wirth, W. W., and W. L. Grogan. 1988. *The Predaceous Midges of the World.* Flora & Fauna Handbook no. 4. E. J. Brill, Leiden.

Wirth, W. W., and A. Stone. 1956. Aquatic Diptera. In *Aquatic Insects of California*, R. L. Usinger (ed.). University of California Press, Berkeley. pp. 372–482.

Wissinger, S. A. 1989a. Comparative population ecology of the dragonflies *Libellula lydia* and *Libellula luctuosa* (Odonata: Libellulidae). Can. J. Zool. **67**:931–936.

Wissinger, S. A. 1989b. Seasonal variation in the intensity of competition and predation among dragonfly larvae. Ecology **70**:1017–1027.

Withycombe, C. L. 1925. Some aspects of the biology and morphology of the Neuroptera with special reference to the immature stages and their possible phylogenetic significance. Trans. Entomol. Soc. London **1924**:303–411.

Witteveen, J., H. A. Verhoef, and T. E. A. M. Huipen. 1988. Life history strategy and egg diapause in the intertidal collembolan *Anurida maritima*. Ecol. Entomol. **13**:443–452.

Woglum, R. S., and E. A. McGregor. 1958. Observations on the life history and morphology of *Agulla bractea* Carpenter. Ann. Entomol. Soc. Am. **51**:129–141.

Woglum, R. S., and E. A. McGregor. 1959. Observations on the life history and morphology of *Agulla astuta* (Banks) (Neuroptera: Raphidioidea: Raphidiidae). Ann. Entomol. Soc. Am. **52**:489–502.

Wood, D. L. 1982. The role of pheromones, kairomones, and allomones in the host selection behavior of bark beetles. Annu. Rev. Entomol. **27**:411–446.

Wood, D. M. and A. Borkent. 1989. Phylogeny and classification of the Nematocera. In *Manual of Nearctic Diptera*, J. F. McAlpine et al. (eds.). Canadian Govt. Publications, Quebec, vol. 3. pp. 1133–1170.

Woodard, S. H., B. J. Fischman, A. Venkat, M. E. Hudson, K. Verala, S. A. Cameron, A. G. Clark and G. E. Robinson. 2011. Genes involved in convergent evolution of eusociality in bees. *Proceedings of the National Academy of Science of the USA.* **108**:7472–7477.

Wootton, R. J. 1976. The fossil record and insect flight. In *Insect Flight*, R. C. Rainey (ed.). John Wiley & Sons, New York. pp. 235–254.

Wootton, R. J. 1981. Paleozoic insects. Annu. Rev. Entomol. **26**:135–158.

Wootton, R. J. 1992. Functional morphology of insect wings. Annu. Rev. Entomol. **37**:113–140.

Wootton, R. J., J. Kukalova-Peck, D. J. S. Newman, and J. Muzon. 1998. Smart engineering in the mid-Carboniferous: How well could Paleozoic dragonflies fly? *Science* **282**:749–751.

Wright, M., and A. Petersen. 1944. A key to the genera of anisopterous dragonfly nymphs of the United States and Canada (Odonata, suborder Anisoptera). Ohio J. Sci. **44**:151–166.

Wu, C. F. 1923. Morphology, anatomy and ethology of *Nemoura*. Bull. Lloyd Library, Cincinnati **23**:1–81.

Wygodzinsky, P. 1961. On a surviving representative of the Lepidotrichidae (Thysanura). Ann. Entomol. Soc. Am. **54**:621–627.

Wygodzinsky, P. 1972. A revision of the silverfish (Lepismatidae, Thysanura) of the United States and the Caribbean area. Am. Mus. Novitates **2481**:10–26.

Yager, D. D. 1990. Sexual dimorphism of auditory function and structure in praying mantises (Mantodea: Dictyoptera). J. Zool. (London) **221**:517–538.

Yang, Z. 2006. *Computational Molecular Evolution*. Oxford Series in Ecology and Evolution. Oxford University Press, Oxford, UK.

Yeates, D. K., and B. M. Wiegmann (eds.). 2005. *The Evolutionary Biology of Flies*. Columbia University Press, New York.

Yeates, D. K., B. M. Wiegmann, G. W. Courtney, R. Meier, C. Lambkin, and T. Pape. 2007. Phylogeny and systematic of Diptera: two decades of progress and prospects. *Zootaxa* **1668**:565–590.

Yoshizawa, K. 2007. The Zoraptera problem: evidence for Zoraptera + Embiodea from the wing base. Syst. Entomol. **32**:197–204.

Yoshizawa, K., and K. P. Johnson. 2003. Phylogenetic position of Phthiraptera (Insecta: Paraneoptera) and elevated rate of evolution in mitochondrial 12S and 16S rDNA. Mol. Phylogenet. Evol. **29**:102–114.

Yoshizawa, K., and K. P. Johnson. 2005. Aligned 18S for Zoraptera (Insecta): phylogenetic position and molecular evolution. Mol. Phylogenet. Evol. **37**:572–580.

Zablotny, J. E. 2003. Sociality. In *Encyclopedia of Insects*, V. H. Resh and R. T. Cardé (eds.). Academic Press, New York. pp. 1044–1053.

Zacharuk, R. Y., and V. D. Shields. 1991. Sensilla of immature insects. Annu. Rev. Entomol. **36**:331–354.

Zimmerman, E. C. 1991–4. *Australian Weevils, v. 1–3, 5–6.* Commonwealth Scientific and Industrial Research Organization, Melbourne.

Zinsser, H. 1935. *Rats, Lice and History: Being a Study in Biography, Which, after Twelve Preliminary Chapters Indispensible for the Preparation of the Lay Reader, Deals with the Life History of Typhus Fever.* Little, Brown, New York.

Zitnan, D., T. G. Kingan, J. L. Hermesman, and M. E. Adams. 1996. Identification of ecdysis-triggering hormone from an epitracheal endocrine system. Science **271**:88–91.

Zompro, O., J. Adis, and W. Weitschat. 2002. A review of the order Mantophasmatodea (Insecta). Zool. Anz. **241**:269–279.

Zwick, P. 1990. Emergence, maturation and upstream oviposition flights of Plecoptera from the Breitenbach [Hesse, Germany], with notes on the adult phase as a possible control of stream insect populations. Hydrobiologia **194**:207–224.

Zwick, P. 2000. Phylogenetic system and zoogeography of the Plecoptera. Ann. Rev. Entomol. **45**:709–746.

The Taxonomic Index includes all scientific names of organisms in the text. Entries in *italics* indicate genera and species; entries in **bold** indicate living orders and families. Page numbers in **bold** indicate main discussions. Page numbers in *italics* indicate illustrations/photos/figures

SUBJECT INDEX

For scientific names, see Taxonomic Index. Page numbers in **bold** indicate main discussions. Page numbers in *italics* indicate illustrations/photos/figures.

Abdomen, **39–43**, *40*
  cerci, 18, 39, *43*
  modifications, 6–7, 41–43
  ovipositor, 7, 39–41, *42*
  pregenital segments, 6, 39
  sternum, 19, 39
  tergum, 19, 39
Absorption
  in alimentary canal, 93
  in excretory systems, 105
Accessory glands, 101
Acorn weevil. *See Curculio uniformis*
    in Taxonomic index
Acrocerid fly. *See Lasia kletti* in
    Taxonomic index
Acrosternite, 19
Acrotergite, 19
Action potential, 120
Adecticous pupae, 68
Adenotrophic viviparity, 74
Aedeagus, 7
Aerial nets, 558–59
Aeropyles, 57
African dung beetle. *See* Scarabaeidae
    in Taxonomic index
African honey bees, 161–62
Age polytheism, 163
Aggregating scents, 140–41
Agricultural crops, destruction of,
    12–13
Air sacs, 96
Alarm pheromones, 139
Alderflies. *See* Megaloptera in
    Taxonomic index
Alfalfa weevils. *See Hyperica postica* in
    Taxonomic index
Alimentary canal, *88–89*
  absorption, 93
  anatomy, 6, 87–96
  digestion, 91–92
  Malpighian tubules, 6, 87, *89*
  modifications of, 90
  peritrophic membrane, 91
  salivary glands, 90–91
Alinotum, 32
Alkaloids, 205

Allochemicals, 139
Allodapine bees, 159–60
Allomones, 139–40
Ambulatory legs, 35
American burying beetle. *See*
    Nicrophorus americanus in
    Taxonomic index
American grasshopper. *See Tropidacris*
    latreillei in Taxonomic index
Ametabolous metamorphosis, 61, *62*
Amino acids, 86
Amnion, 59
Amphientometae, 361
Amplexiform coupling, of butterflies,
    38, 39
Anal fold, 39
Anal lobe, 39
Analogy, 15
Anamorphic development, 60
Anamorphosis, 264
Anatomy, **5–7**
  cellular respiration needs, 6
  elements, *7*
  exoskeleton, 6
  scaling effect, 6
  terms for anatomical position,
    15–16, *16*
Anemophily, 192–93
Antarctica, insects in, 2
Anteclypeus, 22
Antecosta, 19
Antecostal suture, 19
Antennae, 6, **24–25**
  modifications, 23
  types, *24*
Antennal socket, 24
Antennifer, 24
Anterior notal wing processes, 32
Anterior tentorial pits, 21
Antibiosis, 204, 251
Antifeeding compounds, 205
Antixenosis, 204, 251
Antlions. *See* Neuroptera in
    Taxonomic index
Ants. *See also* Hymenoptera in
    Taxonomic index

biology of, **156–59**
  castes among, 157–58
  colony establishment, 157–58
  ergonomic stage, 157
  foundling stage, 156
  life history of, 158–59
  males, 157
  queens, 157, 475
  reproductive stage, 157
  slavery by, 210
  soldiers, 457
  workers, 157–58, 457
Aorta, 101
Aphids. *See also* Aphididae in
    Taxonomic index
  seasonal forms of, 79–80
Apical cell, 69
Apiculture, 160
Apneustic respiratory system, 97
Apodemes, 17
Apod larvae, 69
Apodous, 35
Apolysis, 51
Apomictic parthenogenesis, 75
Apophyseal pits, 34
Apophyses, 17
Aposematic behavior, in butterflies,
    190, 523
Apparency, 206
Apple-grain aphid. *See Rhopalosiphum*
    padi in Taxonomic index
Apposition image, 128
Apposition type, of compound eye,
    127–28
Apterous wings, 37, 39
Aquatic beetle, *See Tropisternus* sp. in
    Taxonomic index
Aquatic insects, 5, **174–79**, *175. See also*
    Lakes
  fluid balance in, 175
  gas exchange in, 97–99, *100*
  hyporheic zone, 178
  in lentic water, 178–79
  in lotic water, 176–78
  in marine environments, 175–76
Aquatic nets, 559